T0189305

Lecture Notes in Artificial Intelligence 10841

Subseries of Lecture Notes in Computer Science

LNAI Series Editors

Randy Goebel
University of Alberta, Edmonton, Canada
Yuzuru Tanaka
Hokkaido University, Sapporo, Japan
Wolfgang Wahlster
DFKI and Saarland University, Saarbrücken, Germany

LNAI Founding Series Editor

Joerg Siekmann
DFKI and Saarland University, Saarbrücken, Germany

More information about this series at http://www.springer.com/series/1244

Leszek Rutkowski · Rafał Scherer
Marcin Korytkowski · Witold Pedrycz
Ryszard Tadeusiewicz · Jacek M. Zurada (Eds.)

Artificial Intelligence and Soft Computing

17th International Conference, ICAISC 2018
Zakopane, Poland, June 3–7, 2018
Proceedings, Part I

 Springer

Editors
Leszek Rutkowski
Częstochowa University of Technology
Częstochowa
Poland

and

University of Social Sciences
Lodz
Poland

Rafał Scherer
Częstochowa University of Technology
Częstochowa
Poland

Marcin Korytkowski
Częstochowa University of Technology
Częstochowa
Poland

Witold Pedrycz
University of Alberta
Edmonton, AB
Canada

Ryszard Tadeusiewicz
AGH University of Science and Technology
Kraków
Poland

Jacek M. Zurada
University of Louisville
Louisville, KY
USA

ISSN 0302-9743 ISSN 1611-3349 (electronic)
Lecture Notes in Artificial Intelligence
ISBN 978-3-319-91252-3 ISBN 978-3-319-91253-0 (eBook)
https://doi.org/10.1007/978-3-319-91253-0

Library of Congress Control Number: 2018942345

LNCS Sublibrary: SL7 – Artificial Intelligence

Printed on acid-free paper

This Springer imprint is published by the registered company Springer Nature Switzerland AG
The registered company address is: Gewerbestrasse 11, 6330 Cham, Switzerland

Preface

This volume constitutes the proceedings of 17th International Conference on Artificial Intelligence and Soft Computing ICAISC 2018, held in Zakopane, Poland, during June 3–7, 2018. The conference was organized by the Polish Neural Network Society in cooperation with the University of Social Sciences in Łódź, the Institute of Computational Intelligence at the Częstochowa University of Technology, and the IEEE Computational Intelligence Society, Poland Chapter. Previous conferences took place in Kule (1994), Szczyrk (1996), Kule (1997) and Zakopane (1999, 2000, 2002, 2004, 2006, 2008, 2010, 2012, 2013, 2014, 2015, 2016, and 2017) and attracted a large number of papers and internationally recognized speakers: Lotfi A. Zadeh, Hojjat Adeli, Rafal Angryk, Igor Aizenberg, Cesare Alippi, Shun-ichi Amari, Daniel Amit, Albert Bifet, Piero P. Bonissone, Jim Bezdek, Zdzisław Bubnicki, Andrzej Cichocki, Swagatam Das, Ewa Dudek-Dyduch, Włodzisław Duch, Pablo A. Estévez, João Gama, Erol Gelenbe, Jerzy Grzymala-Busse, Martin Hagan, Yoichi Hayashi, Akira Hirose, Kaoru Hirota, Adrian Horzyk, Eyke Hüllermeier, Hisao Ishibuchi, Er Meng Joo, Janusz Kacprzyk, Jim Keller, Laszlo T. Koczy, Tomasz Kopacz, Zdzislaw Kowalczuk, Adam Krzyzak, Rudolf Kruse, James Tin-Yau Kwok, Soo-Young Lee, Derong Liu, Robert Marks, Evangelia Micheli-Tzanakou, Kaisa Miettinen, Krystian Mikołajczyk, Henning Müller, Ngoc Thanh Nguyen, Andrzej Obuchowicz, Erkki Oja, Witold Pedrycz, Marios M. Polycarpou, José C. Príncipe, Jagath C. Rajapakse, Šarunas Raudys, Enrique Ruspini, Jörg Siekmann, Roman Słowiński, Igor Spiridonov, Boris Stilman, Ponnuthurai Nagaratnam Suganthan, Ryszard Tadeusiewicz, Ah-Hwee Tan, Shiro Usui, Thomas Villmann, Fei-Yue Wang, Jun Wang, Bogdan M. Wilamowski, Ronald Y. Yager, Xin Yao, Syozo Yasui, Gary Yen, Ivan Zelinka, and Jacek Zurada. The aim of this conference is to build a bridge between traditional artificial intelligence techniques and so-called soft computing techniques. It was pointed out by Lotfi A. Zadeh that "soft computing (SC) is a coalition of methodologies which are oriented toward the conception and design of information/intelligent systems. The principal members of the coalition are: fuzzy logic (FL), neurocomputing (NC), evolutionary computing (EC), probabilistic computing (PC), chaotic computing (CC), and machine learning (ML). The constituent methodologies of SC are, for the most part, complementary and synergistic rather than competitive." These proceedings present both traditional artificial intelligence methods and soft computing techniques. Our goal is to bring together scientists representing both areas of research. This volume is divided into three parts:

- Neural Networks and Their Applications
- Evolutionary Algorithms and Their Applications
- Pattern Classification

The conference attracted a total of 242 submissions from 48 countries and after the review process, 140 papers were accepted for publication.

I would like to thank our participants, invited speakers, and reviewers of the papers for their scientific and personal contribution to the conference. The Program Committee and additional reviewers were very helpful in reviewing the papers.

Finally, I thank my co-workers Łukasz Bartczuk, Piotr Dziwiński, Marcin Gabryel, Marcin Korytkowski and the conference secretary, Rafał Scherer, for their enormous efforts to make the conference a very successful event. Moreover, I appreciate the work of Marcin Korytkowski, who was responsible for the Internet submission system.

June 2018   Leszek Rutkowski

Organization

ICAISC 2018 was organized by the Polish Neural Network Society in cooperation with the University of Social Sciences in Łódź and the Institute of Computational Intelligence at Częstochowa University of Technology.

ICAISC Chairs

Honorary Chairmen

Hojjat Adeli	Ohio State University, USA
Witold Pedrycz	University of Alberta, Edmonton, Canada
Jacek Żurada	University of Louisville, USA

General Chairman

Leszek Rutkowski	Częstochowa University of Technology, Poland and University of Social Sciences, Łodz, Poland

Co-chairmen

Wlodzislaw Duch	Nicolaus Copernicus University, Torun, Poland
Janusz Kacprzyk	Systems Research Institute, Polish Academy of Sciences, Poland
Józef Korbicz	University of Zielona Góra, Poland
Ryszard Tadeusiewicz	AGH University of Science and Technology, Poland

ICAISC Program Committee

Rafał Adamczak, Poland
Cesare Alippi, Italy
Shun-ichi Amari, Japan
Rafal A. Angryk, USA
Jarosław Arabas, Poland
Robert Babuska, The Netherlands
Ildar Z. Batyrshin, Russia
James C. Bezdek, Australia
Marco Block-Berlitz, Germany
Leon Bobrowski, Poland
Piero P. Bonissone, USA
Bernadette Bouchon-Meunier, France
Tadeusz Burczynski, Poland
Andrzej Cader, Poland
Juan Luis Castro, Spain

Yen-Wei Chen, Japan
Wojciech Cholewa, Poland
Kazimierz Choroś, Poland
Fahmida N. Chowdhury, USA
Andrzej Cichocki, Japan
Paweł Cichosz, Poland
Krzysztof Cios, USA
Ian Cloete, Germany
Oscar Cordón, Spain
Bernard De Baets, Belgium
Nabil Derbel, Tunisia
Ewa Dudek-Dyduch, Poland
Ludmiła Dymowa, Poland
Andrzej Dzieliński, Poland
David Elizondo, UK

Meng Joo Er, Singapore
Pablo Estevez, Chile
David B. Fogel, USA
Roman Galar, Poland
Adam Gaweda, USA
Joydeep Ghosh, USA
Juan Jose Gonzalez de la Rosa, Spain
Marian Bolesław Gorzałczany, Poland
Krzysztof Grąbczewski, Poland
Garrison Greenwood, USA
Jerzy W. Grzymala-Busse, USA
Hani Hagras, UK
Saman Halgamuge, Australia
Rainer Hampel, Germany
Zygmunt Hasiewicz, Poland
Yoichi Hayashi, Japan
Tim Hendtlass, Australia
Francisco Herrera, Spain
Kaoru Hirota, Japan
Adrian Horzyk, Poland
Tingwen Huang, USA
Hisao Ishibuchi, Japan
Mo Jamshidi, USA
Andrzej Janczak, Poland
Norbert Jankowski, Poland
Robert John, UK
Jerzy Józefczyk, Poland
Tadeusz Kaczorek, Poland
Władysław Kamiński, Poland
Nikola Kasabov, New Zealand
Okyay Kaynak, Turkey
Vojislav Kecman, New Zealand
James M. Keller, USA
Etienne Kerre, Belgium
Frank Klawonn, Germany
Jacek Kluska, Poland
Przemysław Korohoda, Poland
Jacek Koronacki, Poland
Jan M. Kościelny, Poland
Zdzisław Kowalczuk, Poland
Robert Kozma, USA
László Kóczy, Hungary
Dariusz Król, Poland
Rudolf Kruse, Germany
Boris V. Kryzhanovsky, Russia
Adam Krzyzak, Canada

Juliusz Kulikowski, Poland
Věra Kůrková, Czech Republic
Marek Kurzyński, Poland
Halina Kwaśnicka, Poland
Soo-Young Lee, South Korea
Antoni Ligęza, Poland
Simon M. Lucas, UK
Jacek Łęski, Poland
Bohdan Macukow, Poland
Kurosh Madani, France
Luis Magdalena, Spain
Witold Malina, Poland
Jacek Mańdziuk, Poland
Urszula Markowska-Kaczmar, Poland
Antonino Marvuglia, Luxembourg
Andrzej Materka, Poland
Jacek Mazurkiewicz, Poland
Jaroslaw Meller, Poland
Jerry M. Mendel, USA
Radko Mesiar, Slovakia
Zbigniew Michalewicz, Australia
Zbigniew Mikrut, Poland
Wojciech Moczulski, Poland
Javier Montero, Spain
Eduard Montseny, Spain
Kazumi Nakamatsu, Japan
Detlef D. Nauck, Germany
Antoine Naud, Poland
Ngoc Thanh Nguyen, Poland
Robert Nowicki, Poland
Andrzej Obuchowicz, Poland
Marek Ogiela, Poland
Erkki Oja, Finland
Stanisław Osowski, Poland
Nikhil R. Pal, India
Maciej Patan, Poland
Leonid Perlovsky, USA
Andrzej Pieczyński, Poland
Andrzej Piegat, Poland
Vincenzo Piuri, Italy
Lech Polkowski, Poland
Marios M. Polycarpou, Cyprus
Danil Prokhorov, USA
Anna Radzikowska, Poland
Ewaryst Rafajłowicz, Poland
Sarunas Raudys, Lithuania

Olga Rebrova, Russia
Vladimir Red'ko, Russia
Raúl Rojas, Germany
Imre J. Rudas, Hungary
Enrique H. Ruspini, USA
Khalid Saeed, Poland
Dominik Sankowski, Poland
Norihide Sano, Japan
Robert Schaefer, Poland
Rudy Setiono, Singapore
Paweł Sewastianow, Poland
Jennie Si, USA
Peter Sincak, Slovakia
Andrzej Skowron, Poland
Ewa Skubalska-Rafajłowicz, Poland
Roman Słowiński, Poland
Tomasz G. Smolinski, USA
Czesław Smutnicki, Poland
Pilar Sobrevilla, Spain
Janusz Starzyk, USA
Jerzy Stefanowski, Poland
Vitomir Štruc, Slovenia
Pawel Strumillo, Poland
Ron Sun, USA
Johan Suykens, Belgium
Piotr Szczepaniak, Poland
Eulalia J. Szmidt, Poland

Przemysław Śliwiński, Poland
Adam Słowik, Poland
Jerzy Świątek, Poland
Hideyuki Takagi, Japan
Yury Tiumentsev, Russia
Vicenç Torra, Spain
Burhan Turksen, Canada
Shiro Usui, Japan
Michael Wagenknecht, Germany
Tomasz Walkowiak, Poland
Deliang Wang, USA
Jun Wang, Hong Kong, SAR China
Lipo Wang, Singapore
Paul Werbos, USA
Slawo Wesolkowski, Canada
Sławomir Wiak, Poland
Bernard Widrow, USA
Kay C. Wiese, Canada
Bogdan M. Wilamowski, USA
Donald C. Wunsch, USA
Maciej Wygralak, Poland
Roman Wyrzykowski, Poland
Ronald R. Yager, USA
Xin-She Yang, UK
Gary Yen, USA
Sławomir Zadrożny, Poland
Ali M. S. Zalzala, United Arab Emirates

ICAISC Organizing Committee

Rafał Scherer, Secretary
Łukasz Bartczuk
Piotr Dziwiński
Marcin Gabryel, Finance Chair
Rafał Grycuk
Marcin Korytkowski, Databases and Internet Submissions
Patryk Najgebauer

Additional Reviewers

J. Arabas
T. Babczyński
M. Baczyński
Ł. Bartczuk
P. Boguś
B. Boskovic
J. Botzheim
J. Brest
T. Burczyński
R. Burduk
L. Chmielewski
W. Cholewa
K. Choros
P. Cichosz
P. Ciskowski
B. Cyganek
J. Cytowski
I. Czarnowski
K. Dembczynski
J. Dembski
N. Derbel
L. Diosan
G. Dobrowolski
A. Dockhorn
A. Dzieliński
P. Dziwiński
B. Filipic
M. Gabryel
E. Gelenbe
M. Giergiel
P. Głomb
F. Gomide
Z. Gomółka
M. Gorzałczany
D. Grabowski
M. Grzenda
J. Grzymala-Busse
L. Guo
H. Haberdar
C. Han
Y. Hayashi
T. Hendtlass
Z. Hendzel

F. Hermann
H. Hikawa
K. Hirota
A. Horzyk
E. Hrynkiewicz
J. Ishikawa
D. Jakóbczak
E. Jamro
A. Janczak
W. Kamiński
E. Kerre
J. Kluska
L. Koczy
Z. Kokosinski
A. Kołakowska
J. Konopacki
J. Korbicz
P. Korohoda
J. Koronacki
M. Korytkowski
M. Korzeń
J. Kościelny
L. Kotulski
Z. Kowalczuk
J. Kozlak
M. Kretowska
D. Krol
R. Kruse
B. Kryzhanovsky
A. Kubiak
E. Kucharska
P. Kudová
J. Kulikowski
O. Kurasova
V. Kurkova
M. Kurzyński
J. Kusiak
H. Lenz
Y. Li
A. Ligęza
J. Łęski
B. Macukow
W. Malina

J. Mańdziuk
M. Marques
F. Masulli
A. Materka
R. Matuk Herrera
J. Mazurkiewicz
V. Medvedev
M. Mernik
J. Michalkiewicz
Z. Mikrut
S. Misina
W. Mitkowski
W. Moczulski
F. Mokom
W. Mokrzycki
O. Mosalov
W. Muszyński
H. Nakamoto
G. Nalepa
M. Nashed
S. Nemati
F. Neri
M. Nieniewski
R. Nowicki
A. Obuchowicz
S. Osowski
E. Ozcan
M. Pacholczyk
W. Palacz
G. Paragliola
A. Paszyńska
K. Patan
A. Pieczyński
A. Piegat
Z. Pietrzykowski
P. Prokopowicz
A. Przybył
R. Ptak
E. Rafajłowicz
E. Rakus-Andersson
A. Rataj
Ł. Rauch
L. Rolka

Contents – Part I

Neural Networks and Their Applications

Three-Dimensional Model of Signal Processing in the Presynaptic
Bouton of the Neuron 3
    Andrzej Bielecki, Maciej Gierdziewicz, and Piotr Kalita

The Parallel Modification to the Levenberg-Marquardt Algorithm 15
    Jarosław Bilski, Bartosz Kowalczyk, and Konrad Grzanek

On the Global Convergence of the Parzen-Based Generalized
Regression Neural Networks Applied to Streaming Data 25
    Jinde Cao and Leszek Rutkowski

Modelling Speaker Variability Using Covariance Learning 35
    Moses Ekpenyong and Imeh Umoren

A Neural Network Model with Bidirectional Whitening 47
    Yuki Fujimoto and Toru Ohira

Block Matching Based Obstacle Avoidance for Unmanned Aerial Vehicle . . . 58
    Adomas Ivanovas, Armantas Ostreika, Rytis Maskeliūnas,
    Robertas Damaševičius, Dawid Połap, and Marcin Woźniak

Prototype-Based Kernels for Extreme Learning Machines and Radial
Basis Function Networks 70
    Norbert Jankowski

Supervised Neural Network Learning with an Environment Adapted
Supervision Based on Motivation Learning Factors 76
    Maciej Janowski and Adrian Horzyk

Autoassociative Signature Authentication Based on Recurrent
Neural Network 88
    Jun Rokui

American Sign Language Fingerspelling Recognition Using Wide
Residual Networks 97
    Kacper Kania and Urszula Markowska-Kaczmar

Neural Networks Saturation Reduction 108
    Janusz Kolbusz, Pawel Rozycki, Oleksandr Lysenko,
    and Bogdan M. Wilamowski

Learning and Convergence of the Normalized Radial Basis
Functions Networks.   118
   Adam Krzyżak and Marian Partyka

Porous Silica-Based Optoelectronic Elements as Interconnection Weights
in Molecular Neural Networks   130
   Magdalena Laskowska, Łukasz Laskowski, Jerzy Jelonkiewicz,
   Henryk Piech, and Zbigniew Filutowicz

Data Dependent Adaptive Prediction and Classification
of Video Sequences.   136
   Amrutha Machireddy and Shayan Srinivasa Garani

Multi-step Time Series Forecasting of Electric Load Using Machine
Learning Models.   148
   Shamsul Masum, Ying Liu, and John Chiverton

Deep Q-Network Using Reward Distribution   160
   Yuta Nakaya and Yuko Osana

Motivated Reinforcement Learning Using Self-Developed Knowledge
in Autonomous Cognitive Agent.   170
   Piotr Papiez and Adrian Horzyk

Company Bankruptcy Prediction with Neural Networks.   183
   Jolanta Pozorska and Magdalena Scherer

Soft Patterns Reduction for RBF Network Performance Improvement   190
   Pawel Rozycki, Janusz Kolbusz, Oleksandr Lysenko,
   and Bogdan M. Wilamowski

An Embedded Classifier for Mobile Robot Localization Using Support
Vector Machines and Gray-Level Co-occurrence Matrix.   201
   Fausto Sampaio, Elias T. Silva Jr, Lucas C. da Silva,
   and Pedro P. Rebouças Filho

A New Method for Learning RBF Networks by Utilizing Singular Regions . . .   214
   Seiya Satoh and Ryohei Nakano

Cyclic Reservoir Computing with FPGA Devices for Efficient
Channel Equalization.   226
   Erik S. Skibinsky-Gitlin, Miquel L. Alomar, Christiam F. Frasser,
   Vincent Canals, Eugeni Isern, Miquel Roca, and Josep L. Rosselló

Discrete Cosine Transform Spectral Pooling Layers for Convolutional
Neural Networks.   235
   James S. Smith and Bogdan M. Wilamowski

Extreme Value Model for Volatility Measure in Machine
Learning Ensemble   247
   Ryszard Szupiluk and Paweł Rubach

Deep Networks with RBF Layers to Prevent Adversarial Examples   257
   Petra Vidnerová and Roman Neruda

Application of Reinforcement Learning to Stacked Autoencoder
Deep Network Architecture Optimization   267
   Roman Zajdel and Maciej Kusy

Evolutionary Algorithms and Their Applications

An Optimization Algorithm Based on Multi-Dynamic Schema
of Chromosomes   279
   Radhwan Al-Jawadi and Marcin Studniarski

Eight Bio-inspired Algorithms Evaluated for Solving
Optimization Problems   290
   Carlos Eduardo M. Barbosa and Germano C. Vasconcelos

Robotic Flow Shop Scheduling with Parallel Machines and No-Wait
Constraints in an Aluminium Anodising Plant with the CMAES Algorithm . . .   302
   Carina M. Behr and Jacomine Grobler

Migration Model of Adaptive Differential Evolution Applied
to Real-World Problems   313
   Petr Bujok

Comparative Analysis Between Particle Swarm Optimization Algorithms
Applied to Price-Based Demand Response   323
   Diego L. Cavalca, Guilherme Spavieri, and Ricardo A. S. Fernandes

Visualizing the Optimization Process for Multi-objective
Optimization Problems   333
   Bayanda Chakuma and Mardé Helbig

Comparison of Constraint Handling Approaches
in Multi-objective Optimization   345
   Rohan Hemansu Chhipa and Mardé Helbig

Genetic Programming for the Classification of Levels
of Mammographic Density   363
   Daniel Fajardo-Delgado, María Guadalupe Sánchez,
   Raquel Ochoa-Ornelas, Ismael Edrein Espinosa-Curiel,
   and Vicente Vidal

Feature Selection Using Differential Evolution for Unsupervised
Image Clustering.   376
    Matheus Gutoski, Manassés Ribeiro, Nelson Marcelo Romero Aquino,
    Leandro Takeshi Hattori, André Eugênio Lazzaretti,
    and Heitor Silvério Lopes

A Study on Solving Single Stage Batch Process Scheduling Problems
with an Evolutionary Algorithm Featuring Bacterial Mutations   386
    Máté Hegyháti, Olivér Ősz, and Miklós Hatwágner

Observation of Unbounded Novelty in Evolutionary Algorithms
is Unknowable   395
    Eric Holloway and Robert Marks

Multi-swarm Optimization Algorithm Based on Firefly and Particle
Swarm Optimization Techniques.   405
    Tomas Kadavy, Michal Pluhacek, Adam Viktorin, and Roman Senkerik

New Running Technique for the Bison Algorithm.   417
    Anezka Kazikova, Michal Pluhacek, Adam Viktorin, and Roman Senkerik

Evolutionary Design and Training of Artificial Neural Networks.   427
    Lumír Kojecký and Ivan Zelinka

Obtaining Pareto Front in Instance Selection with Ensembles
and Populations   438
    Mirosław Kordos, Marcin Wydrzyński, and Krystian Łapa

Negative Space-Based Population Initialization Algorithm (NSPIA).   449
    Krystian Łapa, Krzysztof Cpałka, Andrzej Przybył, and Konrad Grzanek

Deriving Functions for Pareto Optimal Fronts Using Genetic Programming . . .   462
    Armand Maree, Marius Riekert, and Mardé Helbig

Identifying an Emotional State from Body Movements
Using Genetic-Based Algorithms.   474
    Yann Maret, Daniel Oberson, and Marina Gavrilova

Particle Swarm Optimization with Single Particle Repulsivity
for Multi-modal Optimization   486
    Michal Pluhacek, Roman Senkerik, Adam Viktorin, and Tomas Kadavy

Hybrid Evolutionary System to Solve Optimization Problems.   495
    Krzysztof Pytel

Horizontal Gene Transfer as a Method of Increasing Variability
in Genetic Algorithms   505
    Wojciech Rafajłowicz

Evolutionary Induction of Classification Trees on Spark   514
    Daniel Reska, Krzysztof Jurczuk, and Marek Kretowski

How Unconventional Chaotic Pseudo-Random Generators Influence
Population Diversity in Differential Evolution.   524
    Roman Senkerik, Adam Viktorin, Michal Pluhacek, Tomas Kadavy,
    and Ivan Zelinka

An Adaptive Individual Inertia Weight Based on Best, Worst and Individual
Particle Performances for the PSO Algorithm   536
    G. Spavieri, D. L. Cavalca, R. A. S. Fernandes, and G. G. Lage

A Mathematical Model and a Firefly Algorithm for an Extended Flexible
Job Shop Problem with Availability Constraints   548
    Willian Tessaro Lunardi, Luiz Henrique Cherri, and Holger Voos

On the Prolonged Exploration of Distance Based Parameter Adaptation
in SHADE   561
    Adam Viktorin, Roman Senkerik, Michal Pluhacek, and Tomas Kadavy

Investigating the Impact of Road Roughness on Routing Performance:
An Evolutionary Algorithm Approach   572
    Hulda Viljoen and Jacomine Grobler

Pattern Classification

Integration Base Classifiers in Geometry Space by Harmonic Mean   585
    Robert Burduk

Similarity of Mobile Users Based on Sparse Location History   593
    Pasi Fränti, Radu Mariescu-Istodor, and Karol Waga

Medoid-Shift for Noise Removal to Improve Clustering.   604
    Pasi Fränti and Jiawei Yang

Application of the Bag-of-Words Algorithm in Classification the Quality
of Sales Leads   615
    Marcin Gabryel, Robertas Damaševičius, and Krzysztof Przybyszewski

Probabilistic Feature Selection in Machine Learning   623
    Indrajit Ghosh

Boost Multi-class sLDA Model for Text Classification   633
    Maciej Jankowski

Multi-level Aggregation in Face Recognition   645
    Adam Kiersztyn, Paweł Karczmarek, and Witold Pedrycz

Direct Incorporation of L_1-Regularization into Generalized Matrix Learning
Vector Quantization.   657
  Falko Lischke, Thomas Neumann, Sven Hellbach, Thomas Villmann,
  and Hans-Joachim Böhme

Classifiers for Matrix Normal Images: Derivation and Testing   668
  Ewaryst Rafajłowicz

Random Projection for k-means Clustering.   680
  Sami Sieranoja and Pasi Fränti

Modified Relational Mountain Clustering Method   690
  Kristina P. Sinaga, June-Nan Hsieh, Josephine B. M. Benjamin,
  and Miin-Shen Yang

Relative Stability of Random Projection-Based Image Classification   702
  Ewa Skubalska-Rafajłowicz

Cost Reduction in Mutation Testing with Bytecode-Level
Mutants Classification   714
  Joanna Strug and Barbara Strug

Probabilistic Learning Vector Quantization with Cross-Entropy
for Probabilistic Class Assignments in Classification Learning   724
  Andrea Villmann, Marika Kaden, Sascha Saralajew,
  and Thomas Villmann

Multi-class and Cluster Evaluation Measures Based on Rényi
and Tsallis Entropies and Mutual Information.   736
  Thomas Villmann and Tina Geweniger

Verification of Results in the Acquiring Knowledge Process Based
on IBL Methodology.   750
  Lukasz Was, Piotr Milczarski, Zofia Stawska, Slawomir Wiak,
  Pawel Maslanka, and Marek Kot

A Fuzzy Measure for Recognition of Handwritten Letter Strokes   761
  Michał Wróbel, Katarzyna Nieszporek, Janusz T. Starczewski,
  and Andrzej Cader

Author Index   771

Contents – Part II

Computer Vision, Image and Speech Analysis

Moving Object Detection and Tracking Based on Three-Frame Difference
and Background Subtraction with Laplace Filter 3
  Beibei Cui and Jean-Charles Créput

Robust Lane Extraction Using Two-Dimension Declivity 14
  Mohamed Fakhfakh, Nizar Fakhfakh, and Lotfi Chaari

Segmentation of the Proximal Femur by the Analysis of X-ray Imaging
Using Statistical Models of Shape and Appearance 25
  Joel Oswaldo Gallegos Guillen, Laura Jovani Estacio Cerquin,
  Javier Delgado Obando, and Eveling Castro-Gutierrez

Architecture of Database Index for Content-Based Image
Retrieval Systems 36
  Rafał Grycuk, Patryk Najgebauer, Rafał Scherer,
  and Agnieszka Siwocha

Symmetry of Hue Distribution in the Images 48
  Piotr Milczarski

Image Completion with Smooth Nonnegative Matrix Factorization 62
  Tomasz Sadowski and Rafał Zdunek

A Fuzzy SOM for Understanding Incomplete 3D Faces 73
  Janusz T. Starczewski, Katarzyna Nieszporek, Michał Wróbel,
  and Konrad Grzanek

Feature Selection for 'Orange Skin' Type Surface Defect
in Furniture Elements 81
  Bartosz Świderski, Michał Kruk, Grzegorz Wieczorek, Jarosław Kurek,
  Katarzyna Śmietańska, Leszek J. Chmielewski, Jarosław Górski,
  and Arkadiusz Orłowski

Image Retrieval by Use of Linguistic Description in Databases 92
  Krzysztof Wiaderek, Danuta Rutkowska, and Elisabeth Rakus-Andersson

Bioinformatics, Biometrics and Medical Applications

On the Use of Principal Component Analysis and Particle Swarm
Optimization in Protein Tertiary Structure Prediction   107
  Óscar Álvarez, Juan Luis Fernández-Martínez, Celia Fernández-Brillet,
  Ana Cernea, Zulima Fernández-Muñiz, and Andrzej Kloczkowski

The Shape Language Application to Evaluation of the Vertebra
Syndesmophytes Development Progress   117
  Marzena Bielecka, Rafał Obuchowicz, and Mariusz Korkosz

Analytical Realization of the EM Algorithm for Emission
Positron Tomography   127
  Robert Cierniak, Piotr Dobosz, Piotr Pluta, and Zbigniew Filutowicz

An Application of Graphic Tools and Analytic Hierarchy Process
to the Description of Biometric Features   137
  Paweł Karczmarek, Adam Kiersztyn, and Witold Pedrycz

On Some Aspects of an Aggregation Mechanism in Face
Recognition Problems   148
  Paweł Karczmarek, Adam Kiersztyn, and Witold Pedrycz

Nuclei Detection in Cytological Images Using Convolutional
Neural Network and Ellipse Fitting Algorithm   157
  Marek Kowal, Michał Żejmo, and Józef Korbicz

Towards the Development of Sensor Platform for Processing Physiological
Data from Wearable Sensors.   168
  Krzysztof Kutt, Wojciech Binek, Piotr Misiak, Grzegorz J. Nalepa,
  and Szymon Bobek

Severity of Cellulite Classification Based on Tissue Thermal Imagining.   179
  Jacek Mazurkiewicz, Joanna Bauer, Michal Mosion,
  Agnieszka Migasiewicz, and Halina Podbielska

Features Selection for the Most Accurate SVM Gender Classifier
Based on Geometrical Features   191
  Piotr Milczarski, Zofia Stawska, and Shane Dowdall

Parallel Cache Efficient Algorithm and Implementation of
Needleman-Wunsch Global Sequence Alignment   207
  Marek Pałkowski, Krzysztof Siedlecki, and Włodzimierz Bielecki

Using Fuzzy Numbers for Modeling Series of Medical Measurements
in a Diagnosis Support Based on the Dempster-Shafer Theory   217
  Sebastian Porebski and Ewa Straszecka

Averaged Hidden Markov Models in Kinect-Based Rehabilitation System . . .   229
  Aleksandra Postawka and Przemysław Śliwiński

Genome Compression: An Image-Based Approach   240
  Kelvin Vieira Kredens, Juliano Vieira Martins, Osmar Betazzi Dordal,
  Edson Emilio Scalabrin, Roberto Hiroshi Herai,
  and Bráulio Coelho Ávila

Stability of Features Describing the Dynamic Signature
Biometric Attribute   250
  Marcin Zalasiński, Krzysztof Cpałka, and Konrad Grzanek

Data Mining

Text Categorization Improvement via User Interaction   265
  Jakub Atroszko, Julian Szymański, David Gil, and Higinio Mora

Uncertain Decision Tree Classifier for Mobile Context-Aware Computing . . .   276
  Szymon Bobek and Piotr Misiak

An Efficient Prototype Selection Algorithm Based
on Dense Spatial Partitions   288
  Joel Luís Carbonera and Mara Abel

Complexity of Rule Sets Induced by Characteristic Sets and Generalized
Maximal Consistent Blocks   301
  Patrick G. Clark, Cheng Gao, Jerzy W. Grzymala-Busse,
  Teresa Mroczek, and Rafal Niemiec

On Ensemble Components Selection in Data Streams Scenario
with Gradual Concept-Drift   311
  Piotr Duda

An Empirical Study of Strategies Boosts Performance of Mutual
Information Similarity   321
  Ole Kristian Ekseth and Svein-Olav Hvasshovd

Distributed Nonnegative Matrix Factorization with HALS Algorithm
on Apache Spark   333
  Krzysztof Fonał and Rafał Zdunek

Dimensionally Distributed Density Estimation   343
  Pasi Fränti and Sami Sieranoja

Outliers Detection in Regressions by Nonparametric Parzen
Kernel Estimation   354
  Tomasz Galkowski and Andrzej Cader

Application of Perspective-Based Observational Tunnels Method
to Visualization of Multidimensional Fractals   364
    Dariusz Jamroz

Estimation of Probability Density Function, Differential Entropy
and Other Relative Quantities for Data Streams with Concept Drift   376
    Maciej Jaworski, Patryk Najgebauer, and Piotr Goetzen

System for Building and Analyzing Preference Models Based on Social
Networking Data and SAT Solvers   387
    Radosław Klimek

On Asymmetric Problems of Objects' Comparison   398
    Maciej Krawczak and Grażyna Szkatuła

A Recommendation Algorithm Considering User Trust and Interest.   408
    Chuanmin Mi, Peng Peng, and Rafał Mierzwiak

Automating Feature Extraction and Feature Selection in Big
Data Security Analytics   423
    Dimitrios Sisiaridis and Olivier Markowitch

Improvement of the Simplified Silhouette Validity Index   433
    Artur Starczewski and Krzysztof Przybyszewski

Feature Extraction in Subject Classification of Text Documents in Polish. . . .   445
    Tomasz Walkowiak, Szymon Datko, and Henryk Maciejewski

Efficiency of Random Decision Forest Technique in Polish Companies'
Bankruptcy Prediction   453
    Joanna Wyrobek and Krzysztof Kluza

TUP-RS: Temporal User Profile Based Recommender System   463
    Wanling Zeng, Yang Du, Dingqian Zhang, Zhili Ye, and Zhumei Dou

Feature Extraction of Surround Sound Recordings for Acoustic
Scene Classification.   475
    Sławomir K. Zieliński

Artificial Intelligence in Modeling, Simulation and Control

Cascading Probability Distributions in Agent-Based Models:
An Application to Behavioural Energy Wastage   489
    Fatima Abdallah, Shadi Basurra, and Mohamed Medhat Gaber

Symbolic Regression with the AMSTA+GP in a Non-linear Modelling
of Dynamic Objects.   504
    Łukasz Bartczuk, Piotr Dziwiński, and Andrzej Cader

A Population Based Algorithm and Fuzzy Decision Trees
for Nonlinear Modeling   516
    Piotr Dziwiński, Łukasz Bartczuk, and Krzysztof Przybyszewski

The Hybrid Plan Controller Construction for Trajectories
in Sobolev Space   532
    Krystian Jobczyk and Antoni Ligęza

Temporal Traveling Salesman Problem – in a Logic-
and Graph Theory-Based Depiction   544
    Krystian Jobczyk, Piotr Wiśniewski, and Antoni Ligęza

Modelling the Affective Power of Locutions in a Persuasive
Dialogue Game.   557
    Magdalena Kacprzak, Anna Sawicka, and Andrzej Zbrzezny

Determination of a Matrix of the Dependencies Between Features
Based on the Expert Knowledge   570
    *Adam Kiersztyn, Paweł Karczmarek, Khrystyna Zhadkovska,
    and Witold Pedrycz*

Dynamic Trust Scoring of Railway Sensor Information   579
    *Marcin Lenart, Andrzej Bielecki, Marie-Jeanne Lesot, Teodora Petrisor,
    and Adrien Revault d'Allonnes*

Linear Parameter-Varying Two Rotor Aero-Dynamical System Modelling
with State-Space Neural Network   592
    Marcel Luzar and Józef Korbicz

Evolutionary Quick Artificial Bee Colony for Constrained Engineering
Design Problems.   603
    *Otavio Noura Teixeira, Mario Tasso Ribeiro Serra Neto,
    Demison Rolins de Souza Alves, Marco Antonio Florenzano Mollinetti,
    Fabio dos Santos Ferreira, Daniel Leal Souza,
    and Rodrigo Lisboa Pereira*

Various Problems of Artificial Intelligence

Patterns in Video Games Analysis – Application of Eye-Tracker
and Electrodermal Activity (EDA) Sensor   619
    Iwona Grabska-Gradzińska and Jan K. Argasiński

Improved Behavioral Analysis of Fuzzy Cognitive Map Models   630
    *Miklós F. Hatwagner, Gyula Vastag, Vesa A. Niskanen,
    and László T. Kóczy*

On Fuzzy Sheffer Stroke Operation   642
   Piotr Helbin, Wanda Niemyska, Pedro Berruezo, Sebastia Massanet,
   Daniel Ruiz-Aguilera, and Michał Baczyński

Building Knowledge Extraction from BIM/IFC Data for Analysis
in Graph Databases   652
   Ali Ismail, Barbara Strug, and Grażyna Ślusarczyk

A Multi-Agent Problem in a New Depiction   665
   Krystian Jobczyk and Antoni Ligęza

Proposal of a Smart Gun System Supporting Police Interventions   677
   Radosław Klimek, Zuzanna Drwiła, and Patrycja Dzienisik

Knowledge Representation in Model Driven Approach in Terms
of the Zachman Framework   689
   Krzysztof Kluza, Piotr Wiśniewski, Antoni Ligęza, Anna Suchenia,
   and Joanna Wyrobek

Rendezvous Consensus Algorithm Applied to the Location of Possible
Victims in Disaster Zones   700
   José León, Gustavo A. Cardona, Luis G. Jaimes, Juan M. Calderón,
   and Pablo Ospina Rodriguez

Exploiting OSC Models by Using Neural Networks with an Innovative
Pruning Algorithm   711
   Grazia Lo Sciuto, Giacomo Capizzi, Christian Napoli, Rafi Shikler,
   Dawid Połap, and Marcin Woźniak

Critical Analysis of Conversational Agent Technology for Intelligent
Customer Support and Proposition of a New Solution   723
   Mateusz Modrzejewski and Przemysław Rokita

Random Forests for Profiling Computer Network Users   734
   Jakub Nowak, Marcin Korytkowski, Robert Nowicki, Rafał Scherer,
   and Agnieszka Siwocha

Leader-Follower Formation for UAV Robot Swarm Based on
Fuzzy Logic Theory   740
   Wilson O. Quesada, Jonathan I. Rodriguez, Juan C. Murillo,
   Gustavo A. Cardona, David Yanguas-Rojas, Luis G. Jaimes,
   and Juan M. Calderón

Towards Interpretability of the Movie Recommender Based
on a Neuro-Fuzzy Approach   752
   Tomasz Rutkowski, Jakub Romanowski, Piotr Woldan,
   Paweł Staszewski, and Radosław Nielek

Dual-Heuristic Dynamic Programming in the Three-Wheeled Mobile
Transport Robot Control   763
    Marcin Szuster

Stylometry Analysis of Literary Texts in Polish   777
    Tomasz Walkowiak and Maciej Piasecki

Constraint-Based Identification of Complex Gateway Structures
in Business Process Models   788
    Piotr Wiśniewski and Antoni Ligęza

Developing a Fuzzy Knowledge Base and Filling It with Knowledge
Extracted from Various Documents   799
    *Nadezhda Yarushkina, Vadim Moshkin, Aleksey Filippov,
    and Gleb Guskov*

Correction to: Analytical Realization of the EM Algorithm for Emission
Positron Tomography C1
    Robert Cierniak, Piotr Dobosz, Piotr Pluta, and Zbigniew Filutowicz

Author Index   811

Neural Networks and Their Applications

Three-Dimensional Model of Signal Processing in the Presynaptic Bouton of the Neuron

Andrzej Bielecki[1], Maciej Gierdziewicz[1(✉)], and Piotr Kalita[2]

[1] Chair of Applied Computer Science, Faculty of Automation, Electrical Engineering,
Computer Science and Biomedical Engineering,
AGH University of Science and Technology,
Al. Mickiewicza 30, 30-059 Kraków, Poland
{bielecki,gierdzma}@agh.edu.pl
[2] Chair of Computer Mathematics, Institute of Computer Science and
Computational Mathematics, Faculty of Mathematics and Computer Science,
Jagiellonian University, Łojasiewicza 6, 30-348 Kraków, Poland
kalita@ii.uj.edu.pl

Abstract. In this paper the model of a signal transmission in a synapse of the neuron is studied. The model is based on partial differential equations. The three-dimensional simulations based on the model are presented and discussed in details. The simulations enabled to estimate the value of the coefficient of diffusion transmission of neurotransmitters in the presynaptic bouton.

Keywords: Presynaptic bouton · Neurotransmitters
Differential diffusive model · Numerical three-dimensional simulations

1 Introduction

Modelling of transport processes in the neuron is one of the key topics in computational biology. The models are based either on differential equations or on stochastic approach. In differential models ordinary differential equations are commonly used as the basis of the model [8]. Such approach is sufficient if spatial aspects of the phenomenon can be neglected. They have to be taken into consideration, however, if phenomena that concern spatial aspects of the transport inside the neuron are studied. Such aspects, in turn, are necessary to be studied in order to construct fully functional artificial neuron that reflects all properties of the biological original. In such a case a model based on partial differential equations is needed. Such models were introduced in [4] for the neurotransmitter transport in the presynaptic bouton and in [7] for the neuropeptide transport. Neural simulations for two dimensions for the neurotransmitter flow model were presented in [6] and three-dimensional simulations for the same model were described in [11].

© Springer International Publishing AG, part of Springer Nature 2018
L. Rutkowski et al. (Eds.): ICAISC 2018, LNAI 10841, pp. 3–14, 2018.
https://doi.org/10.1007/978-3-319-91253-0_1

In this paper 3D simulations of neurotransmitter signaling in the presynaptic bouton is presented. The paper is organized in the following way. In the next section the motivations are discussed. The model is recollected briefly in Sect. 3 whereas the results are presented in Sect. 5.

2 Motivations

Neural structures and processes are performed for a few reasons. First of all, they sometimes allow researchers to understand deeply the natural phenomena [13]. If the model is adequate, then the behavior of the modeled structure and, as a consequence, the dynamics of the studied process, can be tested in dependence on, for instance, the model parameters [5,6]. Such approach enables, among others, estimation of the values of the biological parameters which cannot be measured - for instance in the paper [6] the diffusion coefficient for neurotransmitter transport in the presynaptic bouton was estimated. Secondly, such studies are the basis for creating artificial neural structures both on software and hardware level. The last one can be a starting point for neural structures prosthetic as well as for electronic equivalents of neural structures for robots. The second possibility is important in the context of the studies that concern artificial autonomous agents; the more so because that the mechanisms which are patterned on reflexes and automatized reactions are investigated in the context of autonomous robots and results are promising [2,3]. Such systems are presently implemented as a software but founding them on electronic artificial neural systems would make them far faster.

3 The Model

The model is based on partial differential equations that describe diffusion-type processes inside the presynaptic bouton of the neuron. These processes are the basis of the signal processing and transmission in biological reality.

The presynaptic bouton $\Omega \subset \mathbb{R}^3$ is a bounded open polyhedral domain; $\partial\Omega$ is a finite sum of flat polygons. The vesicles are produced in a polyhedral domain $\Omega_1 \subset \Omega$ and released from the site $\Gamma_d \subset \partial\Omega$ on the bouton boundary, where Γ_d is a finite sum of flat polygons. The function $f : \Omega \to \mathbb{R}$ of probability of synthesizing vesicles filled with neurotransmitter in Ω is defined as $f(x) = \beta > 0$ on Ω_1 and $f(x) = 0$ on $\Omega \backslash \Omega_1$. The synthesis threshold $\bar{\varrho}$ is the maximum value of ϱ, above which new vesicles are not synthesized. The α coefficient is the exocytosis rate. Isotropy and homogeneity of the diffusion are assumed for simplicity, with $a > 0$ denoting the diffusion coefficient. If the presence of cytoskeleton and organelles are taken into account, the 3×3 symmetric diffusion tensor, possibly depending on $x \in \Omega$, will be introduced. The assumed simplification. However, improves the clarity of the model. For modeling vesicle concentration the following assumptions have been made:

1. The continuous unknown variable of the model

$$\varrho : \Omega \times (0,T) \to \mathbb{R}$$

denotes the vesicle concentration.
2. The number of vesicles is proportional to the amount of neurotransmitter.
3. The number of vesicles is large, so the fact that it is a natural variable, whereas the model has a continuous character, does not create any problems.

The formula of the model is:

$$\varrho_t(x,t) = a\Delta\varrho(x,t) + f(x)(\bar{\varrho} - \varrho(x,t))^+ \qquad (1)$$
$$\text{in} \quad \Omega \times (0,T).$$

with the following initial and boundary conditions:

$$\varrho(x,0) = \varrho_0(x) \quad \text{in} \quad \Omega,$$

$$-\frac{\partial\varrho(x,t)}{\partial\nu} = 0 \quad \text{on} \quad (\partial\Omega\backslash\Gamma_d) \times (0,T),$$

$$-\frac{\partial\varrho(x,t)}{\partial\nu} = \eta(t)a\varrho(x,t) \quad \text{on} \quad \Gamma_d \times (0,T),$$

where $\partial/\partial\nu$ is a directional derivative in the outer normal direction ν. The function $\eta(t) = 1$ if the time t belongs to one of the time intervals $[t_n, t_n + \tau]$ where t_n are the action potential arrival moments and τ is the release time, and $\eta(t) = 0$ otherwise. The weak form of the problem is derived in [4] and in [7].

Naturally, to solve the derivative equations describing a biological model, the initial and boundary conditions have to be set to enable calculations with the results which reflect reality. However, it should be emphasized that while the boundary conditions, especially time-dependent, govern all the process and may have strong influence on it, the initial conditions are often not so important since the modeled objects sometimes tend to "forget" the initial values of the variables. This should also be true for signal processing in neurons which is essentially continuous over a long period of time, and setting the initial condition means merely choosing the time of beginning the observation.

At the presented stage of research the bouton geometry was simplified to create a reference point for further simulations. This allowed us to examine the influence of changes in bouton shape on the number of released neurotransmitter vesicles and, consequently, on the synaptic depression. The geometry was based on 3D ball. The bouton skeleton consisted of two spheres located concentrically. The outer sphere (the bouton membrane) had the radius R and the inner one (the boundary of the synthesis zone of the neurotransmitter) had the radius r. We assumed that $R = 1.6\,\mu\text{m}$, following the results described in [14] where the neuromuscular junction of *Drosophila melanogaster* was presented in the fluorescence microscopy photograph. For the inner sphere the radius value $r = 0.64\,\mu\text{m}$ which falls in the range of the parameters of the numerical experiment described in [6]. It was also assumed that the neurotransmitter is secreted from the zone situated around the "South Pole" up to the (45°S) latitude on the outer sphere.

The models of the surfaces of both spheres was built of flat triangles and trapezoids. First, the points located at cross-sections of the circles of latitude and the meridians were chosen, both in 15° intervals, except two single points at the poles. In such a way the total of 266 points at each sphere were chosen. The meridians and the circles of latitude made up a mesh. Vertexes of its cells were connected by line segments. The cells with one of the poles as a node were approximated by triangles, and the remaining ones - by trapezoids. As it has been aforementioned, the generation of a good quality three dimensional mesh is not a trivial task. The quality of the constructed mesh must adhere to the defined standards. The problem has already been discussed in [6] so it is not presented here in detail.

The ratio of the longest edge to the radius of the inscribed sphere $\tilde{R}$ for each tetrahedron is the parameter commonly used to estimate the mesh quality for the purpose of finite element computations. It should be as close as possible to the ideal value of 4.899 for a regular tetrahedron. Hence, the mesh should be constructed in such a way that the tetrahedra differ from regular ones as little as possible. The mesh described in this paper have sufficiently good quality for computations - see Table 2. The fact that the mean and median of the quality parameter of the elements is not much higher than the ideal, confirms that the mesh is sufficiently good. All parameters of the surface meshes are summarized in Table 1 and those of the generated three dimensional mesh - in Table 2.

Table 1. Geometric parameters of the wireframe structure of the bouton

Parameter name	Symbol (formula)	Parameter value
The radius of the outer sphere	R	$1.6\,\mu m$
The radius of the inner sphere	r	$0.64\,\mu m$
The angle between meridians	α_{mer}	$15° = \pi/12$
The angle between circles of latitude	α_{lat}	$15° = \pi/12$
The latitude of "North Pole"	θ_{NP}	$90° = \pi/2$
The latitude of "South Pole"	θ_{SP}	$-90° = -\pi/2$
The number of meridians	$N_{mer} = 2\pi/\alpha_{mer}$	24
The number of circles of latitude*	$N_{lat} = (\theta_{NP}-\theta_{SP})/\alpha_{lat}-1$	11
The minimum latitude of the release zone	θ_{SP}	$-\pi/2$
The maximum latitude of the release zone	$\theta_{SP}/2$	$-\pi/4$

* - excluding poles

4   Simulations

Piecewise linear C^0 tetrahedra and exact formulas for integrals of products of shape functions and their gradients were used in the simulation equations.

The parameters chosen for the 3D simulation correspond to those of the 2D experiment in [7]. The initial density distribution $\varrho_0(x)$, where $x = (x_1, x_2, x_3)$, was obtained by using the bimodal distribution being a sum of two terms:

Table 2. Geometric characteristics of the tetrahedral mesh (E/R - the length of the longest tetrahedron edge divided by the radius of the inscribed circle)

Parameter name	Symbol	Parameter value
The mesh presented in Fig. 1	Ω	–
The volume of the mesh	$m(\Omega)$	$16.673\,\mu m^3$
The mesh boundary	$\partial\Omega$	–
The boundary area	$m(\partial\Omega)$	$32.159\,\mu m^2$
The release zone	Γ_d	–
The area of the release zone	$m(\Gamma_d)$	$4.628\,\mu m^2$
The synthesis domain	Ω_1	–
The volume of the synthesis domain	$m(\Omega_1)$	$1.067\,\mu m^3$
Mesh quality measure (optimal)	E/R	4,899
Number of nodes (vertices)	N_n	7523
Number of faces (triangles)	N_f	4090
Number of elements (tetrahedra)	N_e	42801
Mesh quality measure (minimum)	E/R	4.988
Mesh quality measure (maximum)	E/R	44.790
Mesh quality measure (average)	E/R	7.939
Mesh quality measure (median)	E/R	7.536

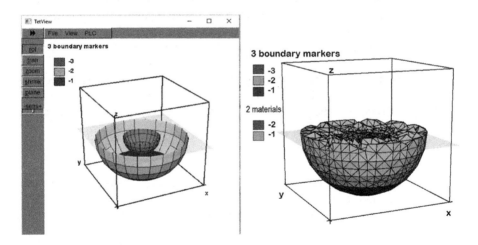

Fig. 1. Geometric configuration of the model of the presynaptic bouton used in simulations: the cross-section of the wireframe model of the bouton (left) and the cross-section of its tetrahedralization (right). The radius of the outer sphere is $1.6\,\mu m$ whereas that of the inner sphere is $0.64\,\mu m$.

ϱ_{01} and ϱ_{02}, each with Gaussian distribution centered at $x_1 = (x_{11}, x_{12}, x_{13})$ and $x_2 = (x_{21}, x_{22}, x_{23})$, respectively. It appeared a reasonable assumption to assume higher concentration values in the synthesis domain. With that choice of parameters, the total number of vesicles in the bouton was about 78500 which is in agreement with the data from the neuromuscular junction in *Drosophila melanogaster* [12]. The number of vesicles in a bouton synapse ranges approximately from 100 to 500000 and in this paper a relatively high value was chosen because the boutons with large number of vesicles are better described, presumably, with the diffusive model. The assumption that about 70000 to 75000 vesicles are uniformly distributed within the reserve pool [12] resulted in estimating the expected production threshold. Noteworthy is that the model presented in this paper introduces the production pool and the intermediate pool. These two vesicle pools taken together correspond to the reserve pool described in [12].

In the preliminary analysis we have studied the sensitivity of the model to $\bar{\varrho}$ and other parameters. The exocytosis rate α was chosen such that the number of vesicles emptying their content during one action potential was about 200, in agreement with [4,12]. The diffusion coefficient of free acetylcholine is known to be equal to $300\,\mu\text{m}^2\text{s}^{-1}$. To account for the fact that neurotransmitter vesicles are slower than acetylcholine particles, the coefficient of diffusion was set a few orders of magnitude smaller.

The synthesis efficiency was assumed three times smaller than the quantity needed to counterbalance the release under constant stimulation. The release time of the vesicles was 0.5 ms and the stimulation paradigm was the same as in [1,6]. in the first half of each second the stimulation frequency was 40 Hz while in the second half it was chosen as 20 Hz. The maximal simulation time was 5 s.

The numerical tests we have run studied the sensitivity of the model to the choice of the diffusion coefficient: we assumed the values $10^{-2} \cdot 300\,\mu\text{m}^2\text{s}^{-1}$, $10^{-3} \cdot 300\,\mu\text{m}^2\text{s}^{-1}$ and $10^{-5} \cdot 300\,\mu\text{m}^2\text{s}^{-1}$. For The lowest assumed value was improper because the diffusion was too slow in comparison to other processes. For the highest value diffusion turned out to be almost instantaneous, dominating all other processes; that contradicts the inertia of the process of the pools replenishment [9,12]. Therefore the value:

$$a = 10^{-3} \cdot 300\,\mu\text{m}^2\text{s}^{-1}$$

was chosen to be the reference one for all next simulations, as well as the values of α, β and $\bar{\varrho}$, equal to:

$$\alpha = 41.6\,\mu\text{m}\,\text{s}^{-1},$$

$$\beta = 0.115\,\text{s}^{-1},$$

and

$$\bar{\varrho} = 5000\,[vesicles]\mu\text{m}^{-3},$$

respectively. The results of the simulation have been presented in Figs. 2 and 3. The simulation parameters are summarized in Table 3.

The calculations presented in this work were performed in Krakow, Poland at the Academic Computer Center ACK CYFRONET AGH. The hardware was

Table 3. Simulation parameters

Parameter name	Symbol	Value
Diffusion coefficient [$\mu m^2 s^{-1}$]	a	$10^{-2} \cdot 300$
Time step [s]	τ	$5 \cdot 10^{-4}$
Initial NT density parameters:		
1. Central point [μm]	x_1, x_2	$(-0.5, 0, 0), (0.4, 0, 0.4)$
2. Amplitude [*vesicles*]μm^{-2}	A_0	7500
3. Decay μm^{-2}	b_0	0.28
4. Noise parameter	c_0	0.1
Initial $\varrho(x, t)$ value	$\varrho_0(x) = \varrho(x, 0)$	$A_0 e^{-b_0|x - x_1|^2 + c_0 U(-1,1)} +$ $A_0 e^{-b_0|x - x_2|^2 + c_0 U(-1,1)}$
Initial total NT amount [*vesicles*]	$\int_\Omega \varrho_0(x)\, dx$	78400–78600*
Synthesis rate s^{-1}	β	0.115
Release rate $\mu m\, s^{-1}$	α	41.6
Production threshold [*vesicles*]	$\bar{\varrho}$	5000

* affected by noise

"Zeus" HP computer cluster with BL2x220c configuration, consisting of 234 two-processor nodes. With 6 threads per processor this gives 12 threads per node. Each Intel Xeon E5645 processor had 2.4 GHz clock and 24 GB memory. The operating system was Scientific Linux 6. The software used in this work was the SAS statistical package and the Python programming language. Each experiment used around 6 h CPU per one second of simulation, 340 MB of memory, and 994 MB of virtual memory. The program was single-threaded, however the parallel version is under development.

5   Results

First, a preliminary set of experiments has been performed to choose the values of the parameters of the model. The parameters taken into consideration were: the diffusion coefficient, the synthesis rate, the release rate and the synthesis threshold. These parameters are, in general, difficult to estimate. One of the goals of our research was to assess at the beginning at least their order of magnitude. The set of parameters which was eventually used for the analysis which we treated as the reference point for next investigations included: the diffusion coefficient, the synthesis rate and threshold, and the release (exocytosis) rate.

The simulated time was assumed to be 5 s. The results of the simulation are presented in Figs. 2 and 3.

The spatial distribution of the neurotransmitter density, for the reference values of the parameters a, α, β and $\bar{\varrho}$, as a function of time is presented in Fig. 2.

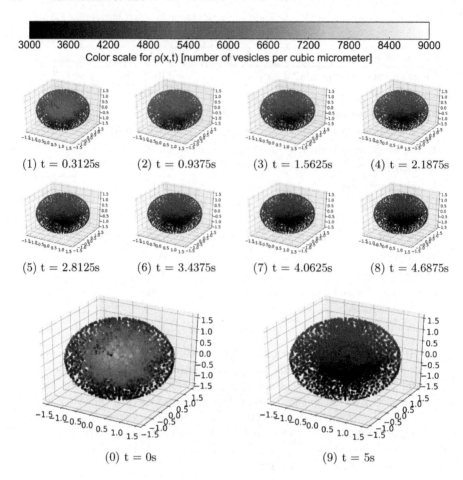

Fig. 2. Spatial distribution of neurotransmitter density $\varrho(x,t)$ changing in time, assuming diffusion coefficient $10^{-3} \cdot 300\,\mu\mathrm{m}^2\mathrm{s}^{-1}$. The two larger bottom graphs depict the density at the beginning ($\mathrm{t}=0\,\mathrm{s}$) and at the end ($\mathrm{t}=5\,\mathrm{s}$) of the simulation, the rest - in intermediate, equally distributed, time intervals. Panels (0) and (9) represent the very beginning and the very end of the simulation whereas the remaining panels illustrate time sequence between these two states. The scale of all axes, both in large panels (0 and 9) and the small ones is exactly the same - from $-1.5\,\mu\mathrm{m}$. to $+1.5\,\mu\mathrm{m}$.

During the process the vesicles diffuse towards the walls and the concentration is such distributed that it is lower near the release site. The distribution changes with the periodical exocytosis at the bottom part.

In Fig. 3 the radial distribution of vesicles concentration during the simulation is presented. Random effects in the initial distribution are visible but after less than $0.3125\,\mathrm{s}$ the randomness in the initial data appears to have no influence on the density distribution. The release effect is clearly visible in Fig. 3, timeshots (1), (4) and (7). The release effect can be directly deduced by analyzing the

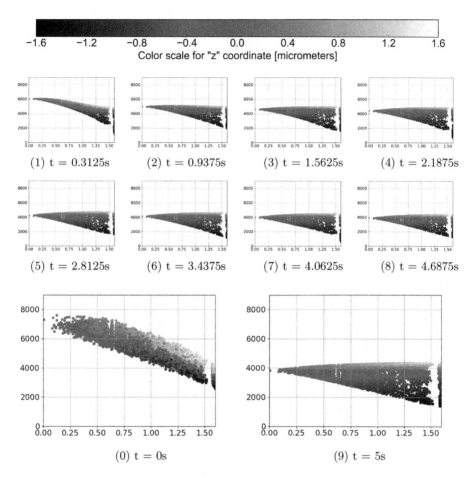

Fig. 3. Spatial distribution of neurotransmitter density $\varrho(x, t)$ changing in time, assuming diffusion coefficient $10^{-3} \cdot 300\,\mu\mathrm{m}^2\mathrm{s}^{-1}$. The two larger bottom graphs depict the density at the beginning ($\mathrm{t} = 0\,\mathrm{s}$) and at the end ($\mathrm{t} = 5\,\mathrm{s}$) of the simulation, the rest - in intermediate, equally distributed, time intervals. Panels (0) and (9) represent the very beginning and the very end of the simulation whereas the remaining panels illustrate time sequence between these two states. The horizontal axis in every plot denotes the radius in $\mu\mathrm{m}$, whereas the vertical one - the vesicle concentration in $[vesicles]\mu\mathrm{m}^{-3}$. The scales of corresponding axes, both in large panels (0 and 9) and the small ones are exactly the same - from $0.0\,\mu\mathrm{m}$. to $+1.5\,\mu\mathrm{m}$. and from 0 to 8000 vesicles.

mentioned timeshots. Namely, it can be easily observed that for $r > 1.5$ $\mu\mathrm{m}$ significant inhomogeneity in the graph exists. It manifests itself by the fact that the neurotransmitter density significantly drops near the boundary. This is visible as the rightmost end of plots is not only locally wider but also dark in its bottom part which means that the significant amount of neurotransmitter has just been released.

Table 4. Coordinates of the center and the radius of synthesis zone and simulation results for the standard parameter set: number of vesicles released in the last impulse of 40 Hz− and 20 Hz− stimulation, number of vesicles in the docked pool, synthesis pool and reservoir pool at the end ($t = 1$ s) of simulation

Parameters				40 Hz	20 Hz	Docked	Synthesis	Reservoir
x_r	y_r	z_r	r					
0.0	0.0	0.0	0.64	149	170	2109	5069	66213

During the 40 Hz stimulation period the number of released vesicles decreases in time. During 20 Hz stimulation periods, however, the rate of change of the number of released vesicles is small. Since at 20 Hz synthesis and diffusion is able to counterbalance the release with given coefficients of diffusion and production, the number of released vesicles can sometimes increase (see Table 4). For the same coefficients, however, at 40 Hz the release process is clearly dominating. This is consistent with the results of [1,6]. It is also visible that the area far from the release site, not necessarily within the synthesis zone, becomes the reservoir of the vesicles. It can be seen in Figs. 2 and 3 that the concentration is higher in the top part of the bouton; brighter points, depicting locations far from the release site, are located on top of other points, and their y coordinates, meaning vesicle concentration, grow with their distance from the middle of the bouton.

The total number of vesicles in the presynaptic bouton significantly decreased which is consistent with the phenomenon of synaptic depression as it was noticed in [15]. Furthermore, the dynamics of the decrease is faster during the 40 Hz periods. In general, the number of vesicles dropped from the initial value to about three fourths of that level within 5 s.

The research concerning the 3D model of processes in the presynaptic bouton based on diffusive equations has already been investigated in [11]. It should be mentioned, however, that the model in this paper and the one presented in [11] differ significantly. First, in the numerical calculations in [11] the source term which may represent the recycling or synthesis of vesicles is assumed to have the value of zero, so the recovery of the bouton after the neurotransmitter release does not take place. Conversely, the source term in our model is a nonlinear function which accounting for the saturation of vesicles synthesis and the recovery dynamics after the release can be modeled. In addition, in [11] the release term is probabilistic, without spatial aspects, which leads to homogeneous boundary conditions with no flux through the boundary of the bouton. The aforementioned approach was chosen in order to reflect the probabilistic aspects of the release. In the model presented in this paper precise boundary conditions are used and the release is represented in a robust way. All in all, the two papers mentioned above are, in a way, complementary.

6   Concluding Remarks

Furthermore, assuming isotropic and homogeneous diffusion has been undoubtedly a simplification because of the presence of cytoskeleton and organelles. We also assumed that the transport of vesicles is only diffusive. However, in [10] it was observed that the diffusive approximation can be a reasonable first step of the process modeling. On the basis of our findings it can be observed that the value of the diffusion coefficient which is approximately $10^{-3} \cdot 300 \, \mu m^2 s^{-1}$ is biologically realistic. This is justified both by the biological experiments and by properties of relaxation and vesicular transportation. Numerical experiments, founded on a good mathematical model should supply us with the reliable value of this parameter. They also provide knowledge about how to control synaptic conductivity effectively which, in turn, seems to be important in examining conductivity-dependent neural disorders.

References

1. Aristizabal, F., Glavinovic, M.I.: Simulation and parameter estimation of dynamics of synaptic depression. Biol. Cybern. **90**, 3–18 (2004)
2. Bielecki, A.: A model of human activity automatization as a basis of artificial intelligence systems. IEEE Trans. Auton. Ment. Dev. **6**, 169–182 (2014)
3. Bielecki, A., Bielecka, M., Bielecki, P.: Conditioned anxiety mechanism as a basis for a procedure of control module of an autonomous robot. Lect. Notes Artif. Intell. **10246**, 390–398 (2017)
4. Bielecki, A., Kalita, P.: Model of neurotransmitter fast transport in axon terminal of presynaptic neuron. J. Math. Biol. **56**, 559–576 (2008)
5. Bielecki, A., Kalita, P.: Dynamical properties of the reaction-diffusion type model of fast synaptic transport. J. Math. Anal. Appl. **393**, 329–340 (2012)
6. Bielecki, A., Kalita, P., Lewandowski, M., Siwek, B.: Numerical simulation for a neurotransmitter transport model in the axon terminal of a presynaptic neuron. Biol. Cybern. **102**, 489–502 (2010)
7. Bielecki, A., Kalita, P., Lewandowski, M., Skomorowski, M.: Compartment model of neuropeptide synaptic transport with impulse control. Biol. Cybern. **99**, 443–458 (2008)
8. Bui, L., Glavinovic, M.: Temperature dependence of vesicular dynamics at excitatory synapses of rat hippocampus. Cogn. Neurodyn. **8**, 277–286 (2014)
9. Denker, A., Rizzoli, S.O.: Synaptic vesicle pools: an update. Front. Synaptic Neurosci. **2**, 135 (2010)
10. Joensuu, M., Padmanabhan, P., Durisic, N., Adekunle, T.D., Bademosi Cooper-Williams, E., Morrow, I.C., Harper, C.B., Jung, W., Parton, R.G., Goodhill, G.J., Papadopulos, A., Meunier, F.A.: Subdiffractional tracking of internalized molecules reveals heterogeneous motion states of synaptic vesicles. J. Cell Biol. **215**, 277–292 (2016)
11. Knödel, M.M., Geiger, R., Ge, L., Bucher, D., Grillo, A., Wittum, G., Schuster, C., Queisser, G.: Synaptic bouton properties are tuned to best fit the prevailing firing pattern. Front. Comput. Neurosci. **8**, 101 (2014)
12. Rizzoli, S.O., Betz, W.J.: Synaptic vesicle pools. Nat. Rev. Neurosci. **6**, 57–60 (2005)

13. Tadeusiewicz, R.: New trends in neurocybernetics. Comput. Methods Mater. Sci. **10**, 1–7 (2010)
14. von Gersdorff, H., Matthews, G.: Inhibition of endocytosis by elevated internal calcium in a synaptic terminal. Nature **370**, 652–655 (1994)
15. Wang, Y., Manis, P.B.: Short-term synaptic depression and recovery at the mature mammalian endbulb of held synapse in mice. J. Neurophysiol. **100**(3), 1255–1264 (2008)

The Parallel Modification
to the Levenberg-Marquardt Algorithm

Jarosław Bilski[1]($\boxtimes$), Bartosz Kowalczyk[1], and Konrad Grzanek[2,3]

[1] Institute of Computational Intelligence, Częstochowa University of Technology,
Częstochowa, Poland
{Jaroslaw.Bilski,Bartosz.Kowalczyk}@iisi.pcz.pl
[2] Information Technology Institute, University of Social Sciences,
Łódź, Poland
[3] Clark University, Worcester, MA 01610, USA
kongra@gmail.com

Abstract. The paper presents a parallel approach to the Levenberg-Marquardt algorithm (also called LM or LMA). The first section contains the mathematical basics of the classic LMA. Then the parallel modification to LMA is introduced. The classic Levenberg-Marquardt algorithm is sufficient for a training of small neural networks. For bigger networks the algorithm complexity becomes too big for the effective teaching. The main scope of this paper is to propose more complexity efficient approach to LMA by parallel computation. The proposed modification to LMA has been tested on a few function approximation problems and has been compared to the classic LMA. The paper concludes with the resolution that the parallel modification to LMA could significantly improve algorithm performance for bigger networks. Summary also contains a several proposals for the possible future work directions in the considered area.

Keywords: Feed-forward neural network
Parallel neural network training algorithm · Optimization problem
Levenberg-Marquardt algorithm · QR decomposition · Givens rotation

1 Introduction

The methods of Artificial Intelligence are broadly used in the modern world. Genetic algorithms, neural networks and their applications are the subject of many researches [11,12,14,15]. The most common use of the modern AI can be found in the areas of optimization, classification and recognition of the given patterns [1,2,13,16,17,21]. The Levenberg-Marquardt algorithm (also called LM or LMA) is a highly efficient and a popular teaching method for feedforward neural networks [9]. It is classified as a quasi-Newton method which finds a wide usage in areas of function minimization problems such as neural networks teaching [8]. One of the biggest downfalls of the LMA is a huge complexity contrary to

© Springer International Publishing AG, part of Springer Nature 2018
L. Rutkowski et al. (Eds.): ICAISC 2018, LNAI 10841, pp. 15–24, 2018.
https://doi.org/10.1007/978-3-319-91253-0_2

classical teaching methods as e.g. the well known Back Propagation algorithm with all its variants [3,7,10]. That limitation makes LMA impractical as a teaching algorithm for bigger networks. The following paper describes the parallel modification for LMA which can be applied to any feedforward multi-layered neural network.

2   The Classic Levenbert-Marquardt Algorithm

The Levenberg-Marquard method is an iterational algorithm mostly used for non-linear function optimization. It combines the advantages of a steepest descent and the Gauss-Newton methods. In order to teach a neural network, LMA is used to minimize the following error function

$$E\left(\mathbf{w}\left(n\right)\right) = \frac{1}{2}\sum_{t=1}^{Q}\sum_{r=1}^{N_L}\varepsilon_r^{(L)^2}\left(t\right) = \frac{1}{2}\sum_{t=1}^{Q}\sum_{r=1}^{N_L}\left(y_r^{(L)}\left(t\right) - d_r^{(L)}\left(t\right)\right)^2 \tag{1}$$

where $\varepsilon_i^{(L)}$ stands for non-linear neuron error and can be depicted as

$$\varepsilon_r^{(L)}(t) = \varepsilon_r^{(Lr)}(t) = y_r^{(L)}(t) - d_r^{(L)}(t) \tag{2}$$

while $d_r^{(L)}(t)$ is the r-th expected network response to the t-th teaching sample. In the Levenberg-Marquardt algorithm weights delta is given by

$$\Delta\left(\mathbf{w}(n)\right) = -\left[\nabla^2\mathbf{E}\left(\mathbf{w}(n)\right)\right]^{-1}\nabla\mathbf{E}\left(\mathbf{w}(n)\right) \tag{3}$$

where $\nabla\mathbf{E}\left(\mathbf{w}(n)\right)$ stands for the gradient vector

$$\nabla\mathbf{E}\left(\mathbf{w}(n)\right) = \mathbf{J}^T\left(\mathbf{w}(n)\right)\varepsilon\left(\mathbf{w}(n)\right) \tag{4}$$

and $\nabla^2\mathbf{E}\left(\mathbf{w}(n)\right)$ stands for the Hessian matrix

$$\nabla^2\mathbf{E}\left(\mathbf{w}(n)\right) = \mathbf{J}^T\left(\mathbf{w}(n)\right)\mathbf{J}\left(\mathbf{w}(n)\right) + \mathbf{S}\left(\mathbf{w}(n)\right) \tag{5}$$

while $\mathbf{J}\left(\mathbf{w}(n)\right)$ is the Jacobian matrix

$$\mathbf{J}(\mathbf{w}\left(n\right)) = \begin{bmatrix} \frac{\partial\varepsilon_1^{(L)}(1)}{\partial w_{10}^{(1)}} & \frac{\partial\varepsilon_1^{(L)}(1)}{\partial w_{11}^{(1)}} & \cdots & \frac{\partial\varepsilon_1^{(L)}(1)}{\partial w_{ij}^{(k)}} & \cdots & \frac{\partial\varepsilon_1^{(L)}(1)}{\partial w_{N_L N_{L-1}}^{(L)}} \\ \vdots & \vdots & \vdots & \vdots & \vdots & \vdots \\ \frac{\partial\varepsilon_{N_L}^{(L)}(1)}{\partial w_{10}^{(1)}} & \frac{\partial\varepsilon_{N_L}^{(L)}(1)}{\partial w_{11}^{(1)}} & \cdots & \frac{\partial\varepsilon_{N_L}^{(L)}(1)}{\partial w_{ij}^{(k)}} & \cdots & \frac{\partial\varepsilon_{N_L}^{(L)}(1)}{\partial w_{N_L N_{L-1}}^{(L)}} \\ \vdots & \vdots & \vdots & \vdots & \vdots & \vdots \\ \frac{\partial\varepsilon_{N_L}^{(L)}(Q)}{\partial w_{10}^{(1)}} & \frac{\partial\varepsilon_{N_L}^{(L)}(Q)}{\partial w_{10}^{(1)}} & \cdots & \frac{\partial\varepsilon_{N_L}^{(L)}(Q)}{\partial w_{ij}^{(k)}} & \cdots & \frac{\partial\varepsilon_{N_L}^{(L)}(Q)}{\partial w_{N_L N_{L-1}}^{(L)}} \end{bmatrix}. \tag{6}$$

The non-linear errors $\varepsilon_i^{(lr)}$ of the neurons in hidden layers are calculated as follows

$$\varepsilon_i^{(lr)}(t) \triangleq \sum_{m=1}^{N_{l+1}} \delta_i^{(l+1,r)}(t)\, w_{mi}^{(l+1)}, \tag{7}$$

$$\delta_i^{(lr)}(t) = \varepsilon_i^{(lr)}(t)\, f'\left(s_i^{(lr)}(t)\right). \tag{8}$$

That creates a possibility to calculate contents of the Jacobian matrix across all weights in the network

$$\frac{\partial \varepsilon_r^{(L)}(t)}{w_{ij}^{(l)}} = \delta_i^{(lr)}(t)\, x_j^{(l)}(t). \tag{9}$$

In the classic LMA all weights of a neural network are stored as a single vector. The $\mathbf{S}\left(\mathbf{w}(n)\right)$ factor from Eq. (5) is depicted as

$$\mathbf{S}\left(\mathbf{w}(n)\right) = \sum_{t=1}^{Q}\sum_{r=1}^{N_L} \varepsilon_r^{(L)}(t)\nabla^2 \varepsilon_r^{(L)}(t). \tag{10}$$

In the Gauss-Newton method it is assumed that $\mathbf{S}\left(\mathbf{w}(n)\right) \approx 0$, so the Eq. (3) can be simplified

$$\varDelta\left(\mathbf{w}(n)\right) = -\left[\mathbf{J}^T\left(\mathbf{w}(n)\right)\mathbf{J}\left(\mathbf{w}(n)\right)\right]^{-1}\mathbf{J}^T\left(\mathbf{w}(n)\right)\varepsilon\left(\mathbf{w}(n)\right). \tag{11}$$

For the Levenberg-Marquardt algorithm needs it is assumed that $\mathbf{S}\left(\mathbf{w}(n)\right) = \mu\mathbf{I}$ so the Eq. (3) takes the form

$$\varDelta\left(\mathbf{w}(n)\right) = -\left[\mathbf{J}^T\left(\mathbf{w}(n)\right)\mathbf{J}\left(\mathbf{w}(n)\right) + \mu\mathbf{I}\right]^{-1}\mathbf{J}^T\left(\mathbf{w}(n)\right)\varepsilon\left(\mathbf{w}(n)\right). \tag{12}$$

Let

$$\mathbf{A}(n) = -\left[\mathbf{J}^T\left(\mathbf{w}(n)\right)\mathbf{J}\left(\mathbf{w}(n)\right) + \mu\mathbf{I}\right]$$
$$\mathbf{h}(n) = \mathbf{J}^T\left(\mathbf{w}(n)\right)\varepsilon\left(\mathbf{w}(n)\right) \tag{13}$$

so the Eq. (12) can be depicted as

$$\varDelta\left(\mathbf{w}(n)\right) = \mathbf{A}(n)^{-1}\mathbf{h}(n). \tag{14}$$

At this stage the QR decomposition can be used to solve Eq. (14). This results with obtaining a desired weight update vector $\varDelta\left(\mathbf{w}(n)\right)$.

$$\mathbf{Q}^T(n)\,\mathbf{A}(n)\,\varDelta\left(\mathbf{w}(n)\right) = \mathbf{Q}^T(n)\,\mathbf{h}(n), \tag{15}$$

$$\mathbf{R}(n)\,\varDelta\left(\mathbf{w}(n)\right) = \mathbf{Q}^T(n)\,\mathbf{h}(n). \tag{16}$$

For the results presented in Sect. 4 the QR decomposition based on Givens rotations has been used. The summary of the Levenberg-Marquardt teaching algorithm can be presented in 5 steps:

1. Calculate network outputs and errors across all teaching samples and solve the goal criterion.
2. Run backpropagation method and calculate the jacobian matrix.
3. Perform QR decomposition in order to obtain $\Delta(\mathbf{w}(n))$.
4. Apply $\Delta(\mathbf{w}(n))$ for a neural network and calculate the goal criterion once again. If the new error is smaller than the original one commit the weights, divide μ by β and proceed to step 1. If the new error is bigger than the original one, multiply μ by β and go back to step 3.
5. The algorithm is deemed to be finished once the gradient is reduced below the accepted threshold or the network error satisfies the predefined error goal.

3   Parallel Modification

As shown in the previous section (especially in Eq. (6)), the classic LMA requires computing a jacobian, which is a $[no \cdot np] \times [nw]$ matrix of error derivatives. Where no stands for a number of network outputs, np is a number of teaching samples and nw is a total number of weights in considered network. Then the Eq. (14) needs to be solved what involves the inversion of $[nw]$ x $[nw]$ size matrix. The clue of a presented modification is to create a set of jacobians, unique for each neuron of the network. Then the computation can be done in parallel for all neurons. In such case the algorithm complexity is significantly reduced to the biggest weight vector for the respective neuron.

   The neural network shown in Fig. 1 is taken into further considerations. The network contains 2 inputs and 3 layers. The two hidden layers consist of 100 neurons while the output layer contains a single neuron. To simplify the calculations teaching vector is assumed to contain 100 samples. In the classic Levenberg-Marquardt algorithm (from now on also called LMC) the jacobian size is the following

$$[no \cdot np] \times \left[\sum_{l=1}^{L} N_l \cdot (N_{l-1} + 1) \right]. \tag{17}$$

By filling the above equation with considered network's values the jacobian size calculates as follows

$$[1 \cdot 100] \times [100 \cdot 3 + 100 \cdot 101 + 1 \cdot 101] = [100] \times [10501]. \tag{18}$$

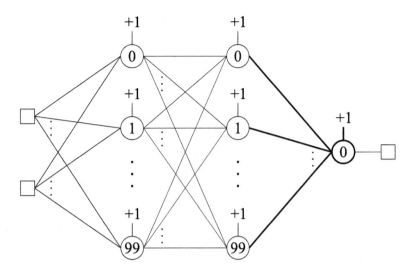

Fig. 1. The considered neural network. One of the neurons with the biggest weight vector and its connections are highlighted.

In a parallel approach to the Levenberg-Marquardt algorithm (from now on also called LMP) the separate jacobian matrix is created for each neuron of the network so it's size can be depicted as

$$[np] \times [\max{(N_{l-1} + 1)}], \tag{19}$$

which in considered network translates into the biggest jacobian of size

$$[100] \times [101]. \tag{20}$$

It is easy to see that the smaller jacobian matrix is, the faster QR decomposition is completed. Table 1 shows the comparison of jacobians sizes across all layers in the considered network for both LMC and LMP. Table 2 presents the average times of solving Eq. 14 for matrices of size $[10501] \times [10502]$ and $[101] \times [102]$. Intel core i7 CPU with the frequency set to 4.40 GHz has been used for computation.

Table 1. Sizes of jacobians in LMC and LMP comparison (LMP is divided into layers).

LMC	LMP		
—	1	2	3
$[100] \times [10501]$	$[100] \times [3]$	$[100] \times [101]$	$[100] \times [101]$

Table 2. Average times for solving Eq. (14) for matrices of size [10501] × [10502] and [101] × [102] in milliseconds.

LMC	LMP
628791,2	0,7942649

4   Results

4.1   Approximating $y = 4x(1 - x)$ Function

The first teaching problem is a logistic curve approximation given by the formula $y = 4x(1 - x)$. A fully connected 1-5-1 network has been used. Teaching set consists of 11 samples to cover the argument range in $x \in [0, 1]$. As a teaching goal the maximum error of value 0.001 has been set. The best teaching results LMP algorithm achieved with the narrow initial weight values $w_{init} \in [-0.5, 0.5]$ and the small value of $\beta \in [1.2, 4]$ factor. Figure 2 shows the networks outputs trained by LMC, LMP and the expected curve. Figure 3 shows the error for all teaching samples after a neural network training with LMP algorithm.

4.2   Approximating $y = \sin x \cdot \log x$ Function

The second teaching problem is a curve approximation given by the formula $y = \sin x \cdot \log x$. A fully connected 1-15-1 network has been used. Teaching set consists of 40 samples to cover the argument range in $x \in [0.1, 4]$. As a teaching

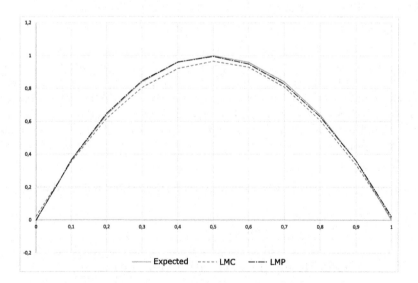

Fig. 2. Comparison of network outputs for $y = 4x(1 - x)$ function approximation problem. Teaching target is set as max epoch error < 0.001.

goal the maximum error of value 0.001 has been set. The best teaching results LMP algorithm achieved with the narrow initial weight values $w_{init} \in [-0.5, 0.5]$ and the small value of $\beta \in [1.2, 4]$ factor. Figure 4 shows the networks outputs trained by LMC, LMP and the expected curve. Figure 5 shows the error for all teaching samples after a neural network training with LMP algorithm.

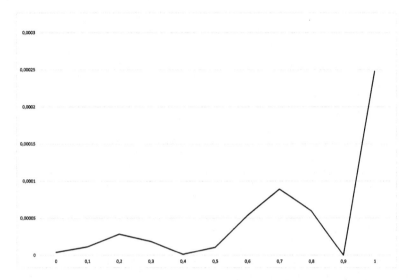

Fig. 3. Network error for $y = 4x(1 - x)$ function approximation problem trained by LMP. Teaching target is set as max epoch error < 0.001.

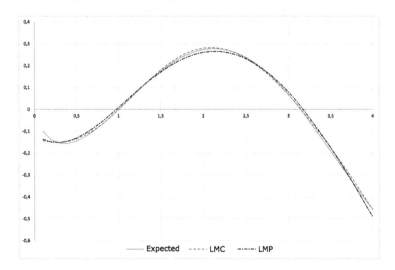

Fig. 4. Comparison of network outputs for $f(x) = \sin x \cdot \log x$ function approximation problem. Teaching target is set as max epoch error < 0.001.

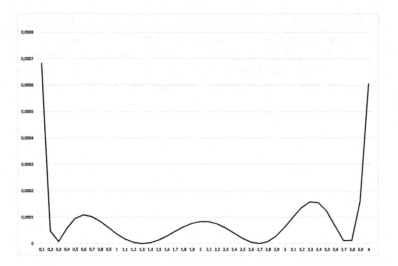

Fig. 5. Network error for $f(x) = \sin x \cdot \log x$ function approximation problem trained by LMP. Teaching target is set as max epoch error < 0.001.

5 Conclusion

Parallelization of the teaching algorithms seems to be a good direction for neural networks training optimization. This topic has been approached multiple times by many researchers e.g. [5,18–20]. The parallel approach to the popular Levenberg-Marquardt neural network teaching algorithm has been presented in this paper. In the classic form the LMA is characterized by a very high computational complexity which goes to the square of a network size. The presented parallel modification to the Levenberg-Marquardt algorithm seems to overcome this limitation maintaining the reasonable accuracy and performance of the archetype. As shown in Sect. 4 in both approximation problems, networks trained by LMP achieved the assumed requirements. The LMP method seems to be a good direction for Levenberg-Marquardt algorithm optimization.

In the near future additional consideration in this matter can be taken. First, the complete parallel implementation of the LMP algorithm can be performed e.g. using CUDA platform and GPU capabilities. Then, more complex approximation problems can be trained e.g. higher dimension functions. Also in deep analysis of influence of parameter β and initial network weights can be performed. Finally LMP implementation can be compared to similar neural network training algorithm parallel optimizations e.g. [4,6].

References

1. Starczewski, A.: A new validity index for crisp clusters. Pattern Anal. Appl. **20**(3), 687–700 (2015)
2. Starczewski, A., Krzyżak, A.: Improvement of the validity index for determination of an appropriate data partitioning. In: Rutkowski, L., Korytkowski, M., Scherer, R., Tadeusiewicz, R., Zadeh, L., Zurada, J. (eds.) ICAISC 2017. LNCS (LNAI), vol. 10246, pp. 159–170. Springer, Cham (2017). https://doi.org/10.1007/978-3-319-59060-8_16
3. Bilski, J., Wilamowski, B.M.: Parallel Levenberg-Marquardt algorithm without error backpropagation. In: Rutkowski, L., Korytkowski, M., Scherer, R., Tadeusiewicz, R., Zadeh, L., Zurada, J. (eds.) ICAISC 2017. LNCS (LNAI), vol. 10245, pp. 25–39. Springer, Cham (2017). https://doi.org/10.1007/978-3-319-59063-9_3
4. Bilski, J., Kowalczyk, B., Żurada, J.M.: Parallel implementation of the givens rotations in the neural network learning algorithm. In: Rutkowski, L., Korytkowski, M., Scherer, R., Tadeusiewicz, R., Zadeh, L., Zurada, J. (eds.) ICAISC 2017. LNCS (LNAI), vol. 10245, pp. 14–24. Springer, Cham (2017). https://doi.org/10.1007/978-3-319-59063-9_2
5. Bilski, J., Smoląg, J.: Parallel realisation of the recurrent RTRN neural network learning. In: Rutkowski, L., Tadeusiewicz, R., Zadeh, L., Zurada, J. (eds.) ICAISC 2008. LNCS (LNAI), vol. 5097, pp. 11–16. Springer, Heidelberg (2008). https://doi.org/10.1007/978-3-540-69731-2_2
6. Bilski, J., Smoląg, J.: Parallel architectures for learning the rtrn and elman dynamic neural network. IEEE Trans. Parallel Distrib. Syst. **26**(9), 2561–2570 (2015)
7. Bilski, J., Smoląg, J., Żurada, J.M.: Parallel approach to the Levenberg-Marquardt learning algorithm for feedforward neural networks. In: Rutkowski, L., Korytkowski, M., Scherer, R., Tadeusiewicz, R., Zadeh, L., Zurada, J. (eds.) ICAISC 2015. LNCS (LNAI), vol. 9119, pp. 3–14. Springer, Cham (2015). https://doi.org/10.1007/978-3-319-19324-3_1
8. Marqardt, D.: An algorithm for last-sqares estimation of nonlinear paeameters. J. Soc. Ind. Appl. Math. **11**(2), 431–441 (1963)
9. Hagan, M.T., Menhaj, M.B.: Training feedforward networks with the Marquardt algorithm. IEEE Trans. Neural Netw. **5**(6), 989–993 (1994)
10. Werbos, J.: Beyond Regression: New Tools for Prediction and Analysis in the Behavioral Sciences. Harvard University, Cambridge (1974)
11. Cpałka, K., Łapa, K., Przybył, A.: A new approach to design of control systems using genetic programming. Inf. Technol. Control **44**(4), 433–442 (2015)
12. Łapa, K., Cpałka, K.: On the application of a hybrid genetic-firework algorithm for controllers structure and parameters selection. In: Borzemski, L., Grzech, A., Świątek, J., Wilimowska, Z. (eds.) Information Systems Architecture and Technology: Proceedings of 36th International Conference on Information Systems Architecture and Technology – ISAT 2015 – Part I. AISC, vol. 429, pp. 111–123. Springer, Cham (2016). https://doi.org/10.1007/978-3-319-28555-9_10
13. Łapa, K., Cpałka, K., Galushkin, A.I.: A new interpretability criteria for neurofuzzy systems for nonlinear classification. In: Rutkowski, L., Korytkowski, M., Scherer, R., Tadeusiewicz, R., Zadeh, L., Zurada, J. (eds.) ICAISC 2015. LNCS (LNAI), vol. 9119, pp. 448–468. Springer, Cham (2015). https://doi.org/10.1007/978-3-319-19324-3_41

14. Khan, N.A., Shaikh, A.: A smart amalgamation of spectral neural algorithm for nonlinear Lane-Emden equations with simulated annealing. J. Artif. Intell. Soft Comput. Res. **7**(3), 215–224 (2017)
15. Liu, H., Gegov, A., Cocea, M.: Rule based networks: an efficient and interpretable representation of computational models. J. Artif. Intell. Soft Comput. Res. **7**(2), 111–123 (2017)
16. Notomista, G., Botsch, M.: A machine learning approach for the segmentation of driving Maneuvers and its application in autonomous parking. J. Artif. Intell. Soft Comput. Res. **7**(4), 243–255 (2017)
17. Rotar, C., Lantovics, L.B.: Directed evolution - a new Metaheuristc for optimization. J. Artif. Intell. Soft Comput. Res. **7**(3), 183–200 (2017)
18. Rutkowska, D., Nowicki, R., Hayashi, Y.: Parallel processing by implication-based neuro-fuzzy systems. In: Wyrzykowski, R., Dongarra, J., Paprzycki, M., Waśniewski, J. (eds.) PPAM 2001. LNCS, vol. 2328, pp. 599–607. Springer, Heidelberg (2002). https://doi.org/10.1007/3-540-48086-2_66
19. Smoląg, J., Bilski, J.: A systolic array for fast learning of neural networks. In: V NNSC, pp. 754–758 (2000)
20. Smoląg, J., Bilski, J., Rutkowski, L.: Systolic array for neural networks. In: IV KSNiIZ, pp. 487–497 (1999)
21. Villmann, T., Bohnsack, A., Kaden, M.: Can learning vector quantization be an alternative to SVM and deep learning? Recent trends and advanced variants of learning vector quantization for classification learning. J. Artif. Intell. Soft Comput. Res. **7**(1), 65–81 (2017)

On the Global Convergence of the Parzen-Based Generalized Regression Neural Networks Applied to Streaming Data

Jinde Cao[1] and Leszek Rutkowski[2,3]([⊠])

[1] School of Mathematics, Southeast University, Nanjing 211189, China
[2] Institute of Computational Intelligence, Czestochowa University of Technology,
Al. Armii Krajowej 36, 42-200 Czestochowa, Poland
leszek.rutkowski@iisi.pcz.pl
[3] Information Technology Institute, Academy of Social Sciences, 90-113 Łódź, Poland

Abstract. In the paper we study global (integral) properties of the Parzen-type recursive algorithm dealing with streaming data in the presence of the time-varying noise. The mean integrated squared error of the regression estimate is shown to converge under several conditions. Simulations results illustrate asymptotic properties of the algorithm and its convergence for a wide spectrum of a time-varying noise.

Keywords: Stream data mining · Parzen-type estimator
Global convergence · Generalized regression neural network

1 Introduction

Data stream is a potentially infinite sequence of data. In many cases, the characteristics of data evolve over time, what in literature is known under the name concept drift. Typical examples of changing environments include economical data analysis, aging effects in sensors or biological models characterized by time-varying delays, see [2,3,21,22], arising naturally by taking into account fluctuations of the environment. Various classical methods of soft computing, see e.g. [5,26,34], can be adopted to deal with stream data. A comprehensive survey of data stream mining algorithms, dealing with classification and regression problems, can be found in [9,20]. For classification problems the existing data stream mining algorithms can be divided into three groups: (i) Ensemble algorithms. In this group of algorithms, data stream is partitioned into data chunks. For each chunk, a single classifier is built and the results obtained for subsequent chunks are aggregated in an appropriate manner, see e.g. [25,33]. (ii) Online algorithms. In these algorithms each data element is processed at most once. The most commonly known is the VFDT algorithm [10] based on the Hoeffding inequality. In our works [11,19,30,31] we showed that the approach developed in [10] is wrong

© Springer International Publishing AG, part of Springer Nature 2018
L. Rutkowski et al. (Eds.): ICAISC 2018, LNAI 10841, pp. 25–34, 2018.
https://doi.org/10.1007/978-3-319-91253-0_3

and we proposed to use the McDiarmid's inequality instead. We also obtained splitting criteria using the Gaussian approximation, see [28,29]. The idea used in the VFDT algorithm was an inspiration for many other algorithms. (iii) Algorithms with sliding windows. In this kind of algorithms only recent data, stored in a window, are used to learn. The most popular algorithm in this field is the CVFDT algorithm [17]. It is a modification of the VFDT algorithm. The application of sliding window makes the online decision tree more effective in dealing with concept drift.

In the case of regression problems only few attempts have been made to deal with streaming data, including decision trees [8] and nonparametric techniques [12–14,18,24]. In [13] the authors have proven pointwise convergence of the Parzen-type recursive algorithm dealing with streaming data in the presence of the time-varying noise. The goal of this paper is to extend that result to the global (integral) convergence. In the paper, the appropriate theorem, along with an illustrative example and numerical simulations, will be presented.

2   Algorithm

Let us consider a stationary system with time-varying noise in the form

$$Y_n = \phi(X_n) + Z_n, \qquad n = 1, 2, \ldots, \tag{1}$$

where $\phi(\cdot)$ is an unknown function, $X_1, X_2, \ldots$ is a sequence of independent and identically distributed random variables in $\mathbb{R}^p$ with density function $f(x)$, $Z_1, Z_2, \ldots$, are independent random variables with zero expected value and the variance of Z_n is equal to d_n:

$$\mathbb{E}Z_n = 0, \qquad Var(Z_n) = d_n, \qquad n = 1, 2, \ldots. \tag{2}$$

It should be emphasized that the variance of Z_n changes over time. For various examples of systems working in the presence of noise with time-varying variances the reader is referred to [15,23,35]. The problem is to estimate the unknown function $\phi(x)$. We will solve the problem representing function $\phi(x)$ in the form

$$\phi(x) = \frac{\phi(x)f(x)}{f(x)} = \frac{R(x)}{f(x)} \tag{3}$$

at each point x at which $f(x) \neq 0$.

To estimate the regression function $\phi(x)$ we use the following formula

$$\hat{\phi}_n(x) = \frac{\hat{R}_n(x)}{f(x)} \tag{4}$$

assuming that $f(x) \neq 0$ and $\hat{R}_n(x)$ is the estimator of function $R(x) = \phi(x)f(x)$. In the sequel we assume that the density f is known. If it is unknown it can be estimated in a recursive manner, see e.g. [6,7,16].

To estimate the regression function $\phi(x)$ we apply the Parzen kernels in the form

$$K_n(x, u) = h_n^{-p} K\left(\frac{x - u}{h_n}\right), \tag{5}$$

where h_n is a certain sequences of numbers (smoothing parameters called the bandwidths) and K is an appropriately selected function satisfying the following conditions

$$||K||_\infty < \infty, \tag{6}$$

$$\int_{\mathbb{R}^p} |K(x)| dx < \infty, \tag{7}$$

$$\int_{\mathbb{R}^p} K(x) dx = 1. \tag{8}$$

To solve the considered problem we propose to use the recursive version of the Parzen-type generalized regression neural network [32]. Here the estimator $\hat{R}_n(x)$ of function $R(x)$ is in the form

$$\hat{R}_n(x) = \hat{R}_{n-1}(x) + \gamma_n \left[Y_n K_n(x, X_n) - \hat{R}_{n-1}(x) \right], \tag{9}$$

where γ_n is a sequence of positive numbers.

We assume that function K in kernel (5) meets conditions (6)–(8), and moreover it is of the form $\prod_{i=1}^{p} G(x_i)$ where

$$||G||_\infty < \infty, \tag{10}$$

$$\int_{\mathbb{R}} G(v) dv = 1, \tag{11}$$

$$\int_{\mathbb{R}} G(v) v^j dv = 0, \quad j = 1, \ldots, r - 1, \tag{12}$$

$$\int_{\mathbb{R}} |G(v) v^k| dv < \infty, \quad k = 1, \ldots, r. \tag{13}$$

$$D_n^{\mathbf{i}} = \int \left[\frac{\partial^r}{\partial x_{i_1} \ldots \partial x_{i_r}} R(x) \right]^2 dx < \infty, \tag{14}$$

where $\mathbf{i} = (i_1, \ldots, i_r)$, $i_k = 1, \ldots, p$, $k = 1, \ldots, r$, and r is the smoothness parameter associated with function R $(n = 1, 2, \ldots)$.

Example 1. It can be easily checked that conditions (10)–(13) are satisfied

(a) for $r = 2$ by the Gaussian kernel,
(b) for $r = 4$ one can take 'Mexican hat' given by

$$G(v) = \frac{1}{2\sqrt{2\pi}} \left(3 - v^2\right) \exp\left(-\frac{v^2}{2}\right), \tag{15}$$

For other examples of kernel functions satisfying conditions (10)–(13) the reader is referred to [16].

3 Global Convergence of Algorithm (4)

Let us define

$$r_n(x) = \mathbb{E}[Y_n K_n(x, X_n)] \tag{16}$$

Theorem 1. *Suppose that conditions (10)–(13) are satisfied. If*

$$\int \phi^2(x) f(x) dx < \infty \tag{17}$$

$$\gamma_n h_n^{-p}(d_n + 1) \to 0 \tag{18}$$

$$\gamma_n^{-2} h_n^{2r} \to 0 \tag{19}$$

then

$$\mathbb{E}\left[\int (\hat{\phi}_n(x) - \phi(x))^2 f^2(x) dx\right] \xrightarrow{n} 0 \tag{20}$$

Proof. Observe that

$$\int var\left[Y_n K_n(x, X_n)\right] dx \le \int \mathbb{E}\left[Y_n K_n(x, X_n)\right]^2 dx \tag{21}$$

$$= \int \mathbb{E}\left[(\phi(X_n) + Z_n) K_n(x, X_n)\right]^2 dx \tag{22}$$

$$= \int \mathbb{E}\left[\phi(X_n) K_n(x, X_n)\right]^2 dx + 2 \int \mathbb{E}\left[\phi(X_n) Z_n K_n^2(x, X_n)\right] dx \tag{23}$$

$$+ \int \mathbb{E}\left[Z_n^2 K_n^2(x, X_n)\right] dx$$

Since X_n and Z_n are independent random variables, using (2) one has

$$\int var\left[Y_n K_n(x, X_n)\right] dx \le \tag{24}$$

$$h_n^{-2p}\left[\int\int \phi(u) K^2(\frac{x-u}{h_n}) f(u) du dx + d_n \int\int K^2(\frac{x-u}{h_n}) f(u) du dx\right]$$

After some simple algebra one gets

$$\int var\left[Y_n K_n(x, X_n)\right] dx \le h_n^{-p}\left[\int \phi^2(u) f(u) du + d_n\right] \int K^2(x) dx \tag{25}$$

Moreover, in view of assumptions (10)–(13), using the Taylor's multidimensional expansion, one has

$$\int (r_n(x) - R(x))^2 dx = O(h_n^{2r}). \tag{26}$$

Now the convergence (20) is a consequence of Theorem 4.3 in [27].

Example 2. Let $d_n = n^t$, $t > 0$, $h_n = Dn^{-H}$, H and D are positive constants and $\gamma_n = kn^{-\gamma}$, $k > 0$ and $0 < \gamma < 1$. Then conditions (18) and (19) are satisfied if

$$Hp + t - \gamma < 0 \tag{27}$$

and

$$\gamma - Hr < 0 \tag{28}$$

It is easily seen that if $t < 1$ then we can choose the values of parameters H and γ such that conditions (27) and (28) are satisfied and convergence (20) holds.

4   Experimental Results

In this section, we will demonstrate the performance of the proposed algorithm. The considered dataset has 1000000 data elements. The dependent variables were generated from the mixture of the Gaussian distributions. In particular

$$X \sim 0.2 \cdot \mathcal{N}(2,0) + 0.15 \cdot \mathcal{N}(1,3) + 0.25 \cdot \mathcal{N}(-2.5, 0.5) + 0.4 \cdot \mathcal{N}(6,2). \tag{29}$$

The probability density function in depicted in Fig. 1. We investigate the function $\phi(x)$ of the following form

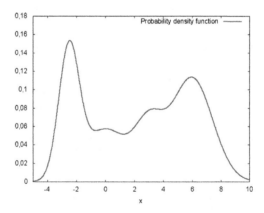

Fig. 1. Considered probability density function

$$\phi(x) = (x^4 - 3x^3 - 6x^2 + 2x - 1)\exp(-x^2/7) \tag{30}$$

and the noise was generated from the normal distribution with expected value equal to 0 and $d_n = n^{0.4}$.

The parameters of estimator (9), according to Example 2, were set as follows: $D = 10$, $H = 0,35$, $k = 1$ and $\gamma = 0,95$. The Parzen kernel in the form of

Mexican-hat was used, see (15). This choice of parameters ensures fulfillment of assumptions of Theorem 1.

In Fig. 2 the result of the estimation, along with training data, is presented. One can see that despite the fact that the data are largely noisy, the estimator seams to adjust to changes of data.

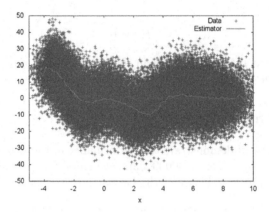

Fig. 2. Generated data and obtained estimator

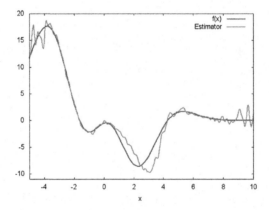

Fig. 3. Function $\phi(x)$ and obtained estimator

The comparison of function $\phi(x)$ and obtained estimator can be seen in Fig. 3. The estimator mimics function $\phi(x)$ in a satisfactory way. The process of the estimation of the function value in one point ($x = -2.5$) is depicted in Fig. 4. One can see that estimation has provided satisfactory result even at the beginning of the process and was stabilized after 30000 data elements.

To compute the mean squared error, the results of estimation were recorded in 450 points. The MSE values, for n running from 1 to 100000, are depicted in Fig. 5.

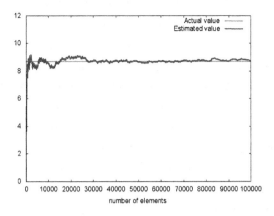

Fig. 4. The actual and the estimated values of function $\phi(x)$ for $x = -2.5$.

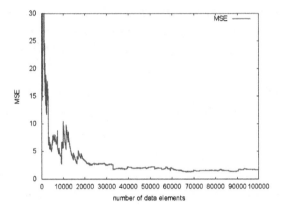

Fig. 5. The dependence between the number of data elements n and the value of MSE.

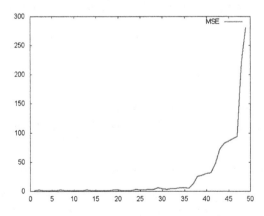

Fig. 6. The dependency between d_n and MSE

These results confirm that with the growth of a number of data elements, the accuracy increases.

The last experiment demonstrates the dependency between the value of d_n and obtained MSEs. For this purpose, we have assumed that $d_n = n^t$. The value of parameter t was changed from 0 to 1 with the step 0.02. For every single value of parameter t, the new training set was generated. The parameters of estimator (9) are the same as in the previous experiments. To demonstrate the results, the experiment was conducted 20 times. Figure 6 presents the averages obtained for every value of parameter t.

5   Conclusions

We have proven that the Parzen-type regression estimator is globally convergent despite the presence of a time-varying noise in the regression model (1). In our future works we will attempt to investigate the speed of the convergence of algorithm (9) and extend the results on regression models of a type $Y_n = \phi_n(X_n) + Z_n$, $n = 1, 2, \ldots$. Moreover, we will also attempt to apply various deep learning techniques, see e.g. [1,4], to deal with concept-drift.

Acknowledgments. This work was supported by the Polish National Science Center under Grant No. 2014/15/B/ST7/05264.

References

1. Bologna, G., Hayashi, Y.: Characterization of symbolic rules embedded in deep DIMLP networks: a challenge to transparency of deep learning. J. Artif. Intell. Soft Comput. Res. **7**(4), 265–286 (2017)
2. Cao, J., Wang, J.: Global asymptotic stability of a general class of recurrent neural networks with time-varying delays. IEEE Trans. Circ. Syst. I Fundam. Theory Appl. **50**(1), 34–44 (2003)
3. Cao, J., Wang, J.: Global asymptotic and robust stability of recurrent neural networks with time delays. IEEE Trans. Circ. Syst. I Regul. Pap. **52**(2), 417–426 (2005)
4. Chang, O., Constante, P., Gordon, A., Singana, M.: A novel deep neural network that uses space-time features for tracking and recognizing a moving object. J. Artif. Intell. Soft Comput. Res. **7**(2), 125–136 (2017)
5. Devi, V.S., Meena, L.: Parallel MCNN (pMCNN) with application to prototype selection on large and streaming data. J. Artif. Intell. Soft Comput. Res. **7**(3), 155–169 (2017)
6. Devroye, L., Krzyżak, A.: On the hilbert kernel density estimate. Stat. Probab. Lett. **44**(3), 299–308 (1999)
7. Devroye, L., Krzyżak, A.: New multivariate product density estimators. J. Multivar. Anal. **82**(1), 88–110 (2002)
8. Diam, A., Last, M., Kandel, A.: Knowledge discovery in data streams with regression tree methods. WIREs Data Min. Knowl. Discov. **2**, 69–78 (2012). https://doi.org/10.1002/widm.51

9. Ditzler, G., Roveri, M., Alippi, C., Polikar, R.: Learning in nonstationary environments: a survey. IEEE Comput. Intell. Mag. **10**(4), 12–25 (2015)
10. Domingos, P., Hulten, G.: Mining high-speed data streams. In: Proceedings of the 6th ACM SIGKDD International Conference on Knowledge Discovery and Data Mining, pp. 71–80 (2000)
11. Duda, P., Jaworski, M., Pietruczuk, L., Rutkowski, L.: A novel application of Hoeffding's inequality to decision trees construction for data streams. In: 2014 International Joint Conference on Neural Networks (IJCNN), pp. 3324–3330. IEEE (2014)
12. Duda, P., Jaworski, M., Rutkowski, L.: Knowledge discovery in data streams with the orthogonal series-based generalized regression neural networks. Inf. Sci. (2017). https://doi.org/10.1016/j.ins.2017.07.013
13. Duda, P., Jaworski, M., Rutkowski, L.: Convergent time-varying regression models for data streams: tracking concept drift by the recursive parzen-based generalized regression neural networks. Int. J. Neural Syst. **28**(02), 1750048 (2018)
14. Duda, P., Pietruczuk, L., Jaworski, M., Krzyzak, A.: On the Cesàro-means-based orthogonal series approach to learning time-varying regression functions. In: Rutkowski, L., Korytkowski, M., Scherer, R., Tadeusiewicz, R., Zadeh, L.A., Zurada, J.M. (eds.) ICAISC 2016. LNCS (LNAI), vol. 9693, pp. 37–48. Springer, Cham (2016). https://doi.org/10.1007/978-3-319-39384-1_4
15. Ellis, P.: The time-dependent mean and variance of the non-stationary Markovian infinite server system. J. Math. Stat. **6**, 68–71 (2010)
16. Greblicki, W., Pawlak, M.: Nonparametric System Identification. Cambridge University Press, Cambridge (2008)
17. Hulten, G., Spencer, L., Domingos, P.: Mining time-changing data streams. In: Proceedings of the 7th ACM SIGKDD International Conference on Knowledge Discovery and Data Mining, pp. 97–106 (2001)
18. Jaworski, M., Duda, P., Rutkowski, L., Najgebauer, P., Pawlak, M.: Heuristic regression function estimation methods for data streams with concept drift. In: Rutkowski, L., Korytkowski, M., Scherer, R., Tadeusiewicz, R., Zadeh, L.A., Zurada, J.M. (eds.) ICAISC 2017. LNCS (LNAI), vol. 10246, pp. 726–737. Springer, Cham (2017). https://doi.org/10.1007/978-3-319-59060-8_65
19. Jaworski, M., Duda, P., Rutkowski, L.: New splitting criteria for decision trees in stationary data streams. IEEE Trans. Neural Netw. Learn. Syst. **PP**(99), 1–14 (2017). https://doi.org/10.1109/TNNLS.2017.2698204
20. Krawczyk, B., Minku, L.L., Gama, J., Stefanowski, J., Wozniak, M.: Ensemble learning for data stream analysis: a survey. Inf. Fusion **37**, 132–156 (2017)
21. Li, R., Cao, J., Alsaedi, A., Alsaadi, F.: Exponential and fixed-time synchronization of cohen-grossberg neural networks with time-varying delays and reaction-diffusion terms. Appl. Math. Comput. **313**, 37–51 (2017)
22. Manivannan, R., Samidurai, R., Cao, J., Alsaedi, A., Alsaadi, F.E.: Delay-dependent stability criteria for neutral-type neural networks with interval time-varying delay signals under the effects of leakage delay. Adv. Differ. Equ. **2018**(1), 53 (2018)
23. Phillips, P.C.: Impulse response and forecast error variance asymptotics in nonstationary VARs. J. Econom. **83**(1), 21–56 (1998)
24. Pietruczuk, L., Rutkowski, L., Jaworski, M., Duda, P.: The Parzen kernel approach to learning in non-stationary environment. In: 2014 International Joint Conference on Neural Networks (IJCNN), pp. 3319–3323. IEEE (2014)
25. Pietruczuk, L., Rutkowski, L., Jaworski, M., Duda, P.: How to adjust an ensemble size in stream data mining? Inf. Sci. **381**, 46–54 (2017)

26. Riid, A., Preden, J.S.: Design of fuzzy rule-based classifiers through granulation and consolidation. J. Artif. Intell. Soft Comput. Res. **7**(2), 137–147 (2017)
27. Rutkowski, L.: New Soft Computing Techniques for System Modeling, Pattern Classication and Image Processing. Springer, Heidelberg (2004). https://doi.org/10.1007/978-3-540-40046-2
28. Rutkowski, L., Jaworski, M., Pietruczuk, L., Duda, P.: The CART decision tree for mining data streams. Inf. Sci. **266**, 1–15 (2014)
29. Rutkowski, L., Jaworski, M., Pietruczuk, L., Duda, P.: Decision trees for mining data streams based on the Gaussian approximation. IEEE Trans. Knowl. Data Eng. **26**(1), 108–119 (2014)
30. Rutkowski, L., Jaworski, M., Pietruczuk, L., Duda, P.: A new method for data stream mining based on the misclassification error. IEEE Trans. Neural Netw. Learn. Syst, **26**(5), 1048–1059 (2015)
31. Rutkowski, L., Pietruczuk, L., Duda, P., Jaworski, M.: Decision trees for mining data streams based on the McDiarmid's bound. IEEE Trans. Knowl. Data Eng. **25**(6), 1272–1279 (2013)
32. Specht, D.F.: A general regression neural network. IEEE Trans. Neural Netw. **2**(6), 568–576 (1991)
33. Street, W.N., Kim, Y.: A streaming ensemble algorithm (SEA) for large-scale classification. In: Proceedings of the Seventh ACM SIGKDD International Conference on Knowledge Discovery and Data Mining, pp. 377–382. ACM (2001)
34. Villmann, T., Bohnsack, A., Kaden, M.: Can learning vector quantization be an alternative to SVM and deep learning? - recent trends and advanced variants of learning vector quantization for classification learning. J. Artif. Intell. Soft Comput. Res. **7**(1), 65–81 (2017). https://doi.org/10.1515/jaiscr-2017-0005
35. Wong, K.F.K., Galka, A., Yamashita, O., Ozaki, T.: Modelling non-stationary variance in EEG time series by state space garch model. Comput. Biol. Med. **36**(12), 1327–1335 (2006)

Modelling Speaker Variability Using Covariance Learning

Moses Ekpenyong[1]([✉]) [ID] and Imeh Umoren[2]

[1] University of Uyo, Nwaniba Campus, Uyo, Nigeria
`mosesekpenyong@uniuyo.edu.ng`
[2] Akwa Ibom State University, Ikot Akpaden Campus, Mkpat-Enin, Nigeria
`imehumoren@aksu.edu.ng`

Abstract. In this contribution, we investigate the relationship between speakers and speech utterance, and propose a speaker normalization/adaptation model that incorporates correlation amongst the utterance classes produced by male and female speakers of varying age categories (children: 0–15; youths: 16–30; adults: 31–50; seniors: >50). Using Principal Component Analysis (PCA), a speaker space was constructed, and based on the speaker covariance matrix obtained directly from the speech data signals, a visualisation of the first three principal components (PCs) was achieved. For effective covariance learning, a component-wise normalisation of each vector weights of the covariance matrix was performed, and a machine learning algorithm (the SOM: self organising map) implemented to model selected speaker features (F0, intensity, pulse) variability. Results obtained reveal that, for the features selected, F0 gave the most variance, as both genders exhibited high variability. For male speakers, PC1 captured the most variance of 87%, while PC2 and PC3 captured the least variances of 7% and 3%, respectively. For female speakers, PC1 captured the most variance of 97%, while PC2 and PC3 captured the least variances of 2% and 1%, respectively. Further, intensity and pulse features show close similarity patterns between the speech features, and are not most relevant for speaker variability modelling. Component planes visualisation of the respective speech patterns learned from the features covariance revealed consistent patterns, and hence, useful in speaker recognition systems.

Keywords: Component visualisation · Machine learning · PCA
SOM · Speech variability

1 Introduction

Speech features can be characterised according to their compactness, correlation, distribution or spread behaviour, and relevance to the intended task. Compactness relates to the number of feature sets derived from the speech corpus. Correlation measures the relationship between relevant speech features. Distribution behaviour is important for efficient feature set abstraction, and necessary for

© Springer International Publishing AG, part of Springer Nature 2018
L. Rutkowski et al. (Eds.): ICAISC 2018, LNAI 10841, pp. 35–46, 2018.
https://doi.org/10.1007/978-3-319-91253-0_4

modelling - to ensure few training samples and parameters. Relevance refers to the requirement that abstracted features (must) contain significant information for the task performed. Four types of speech variability can be identified [1]. They include linguistic variability, speaker variability, channel variability, and residual variability. Linguistic variability occurs due to variation across the language's phonemes. Speaker variability arises due to variations in diverse speaker characteristics. Channel variability is attributed to the variations in communication channels (e.g., recording device). Residual variability reveals the effect of all the unaccounted sources such as phonetic context and co-articulation. Apart from source variabilities, source dependencies can also be studied, and these dependencies are collectively classified as interaction variabilities. Speaker variability (gender, accent, age, speech rate, and phones realizations) constitutes one of the greatest challenges in speech processing. Chen et al. [2] introduced the adapted Gaussian mixture model (GMM)-based speaker representation for speaker variability analysis. They used the principal component analysis (PCA) and independent component analysis (ICA) to extract dominant sources of speaker variability, and applied analysis of variance (ANOVA) to evaluate the factor dominance in certain principal/independent components. They observed that variability due to gender and accent appeared most dominant across speakers, but variability due to speaking rate was not so evident. Kajarekar [1] decomposed the total (speech) variability space using a multivariate analysis of variance (MANOVA). Three databases were used for the analysis (HTIMIT, OGI stories and OGI numbers), and the variability in commonly used features was measured within the spectral and temporal domains. Their results showed consistency across different databases and datasets, and aligned with previous studies.

This paper proposes the use of two unsupervised techniques: PCA and self organising map (SOM) for efficient modelling of speaker variability patterns. First, we exploit the power of PCA to extract dominant sources of speaker variability across varying age categories of male and female speakers, and use the SOM to learn patterns from the resulting features covariance. The rest of the paper is organised as follows: Sect. 2 formulates the speaker normalisation/adaptation model. Section 3 describes the PCA and SOM implementation. Section 4 introduces covariance learning. Section 5 presents the results obtained from experimental data. Section 6 concludes on the research and points to future directions.

2   Model Formulation

Modelling and analysing speaker variability is vital for in-depth understanding of inter-speaker variances and enhancing speech/speaker recognition systems [2,3]. Let S represent a set of s speakers ($S = 1, 2, 3, \ldots, s$), each speaker producing an utterance/phrase of w words ($P = 1, 2, 3, \ldots, w$). To construct a space describing the speaker variability, we begin by representing the speaker class using a vector concatenation construct of the mean features of word frames extracted from the speakers speech signal, thus:

$$\mu_1^s \mu_2^s \ldots \mu_i^s \ldots \mu_w^s \tag{1}$$

where μ_i^s is the mean feature vector of class $i \in P$ for speaker $s \in S$. The cumulative vector (μ^s) derived from the various speakers now forms the required vector space. If there exists w classes, and each feature vector in a class has D dimensions, then speaker s has a $1 \times w^D$ vector dimension, and the covariance matrix of all speakers (Σ_0) (an ordered set of eigen vectors (ϕ_i) represented in decreasing direction) constitutes the variances among speakers contributing most to the between speakers variability. To demonstrate the foregoing, we consider a typical example. Assuming that there are two phrase (or utterance) classes, p_1 and p_2 (i.e., $P = \{p_1, p_2\}$), where $\{p_1, p_2\}$ represents the following word sets $\{(w_{11}, w_{12}, \ldots, w_{1n}), (w_{21}, w_{22}, \ldots, w_{2n})\}$, and our feature of interest is the average F0 value. The speaker space in this case is a two dimensional vector space which axes represent the F0 value for the utterance classes p_1 and p_2. Each speaker within the speaker set can then be expressed as a sum of the group mean (μ_0) and the speaker deviation from the group mean (d^s):

$$\widehat{\mu}^s = \mu_0 + d^s \tag{2}$$

where $\widehat{\mu}^s$ is an estimate of the true speaker mean μ^s. The model can also be viewed as the sum of the speaker independent part (μ_0) and speaker dependent part (d^s). Now, taking μ_0 as the speaker independent model, the formulation in Eq. (2) describes a speaker adaptation process, where $\widehat{\mu}^s$ represents the adapted speaker specific model. Hence we rewrite Eq. (2) as,

$$\widehat{\mu}_0 = \mu^s - d^s \tag{3}$$

Equation (3) describes the speaker normalization process, where the speaker dependent vector d^s is exploited to normalise the speaker's data in the feature space $(\widehat{\mu}_0)$. The deviation vector d^s can then be expressed as a projection onto the orthogonal principal components as,

$$d^s = \phi\alpha^s = \sum_{i=1}^{qD} \phi_i.\alpha_i^s \tag{4}$$

where ϕ is the eigen-matrix of Σ_0, ϕ_i represents the ith ordered eigen vector, α_i^s defines the projection of the speaker difference from the group mean $(d^s - \mu_0)$ onto the eigen vectors. Consequently, the adaptation expression in Eq. (2) becomes,

$$\widehat{\mu}^s = \mu_0 + \sum_{i=1}^{qD} \phi_i.\alpha_i^s \tag{5}$$

When Σ_0 reaches its full rank, the projections onto the eigenvectors are computed as follows,

$$\alpha^s = \phi^T(\mu^s - \mu_0) \tag{6}$$

Substituting Eq. (6) into Eq. (5), we obtain,

$$\widehat{\mu}^s = \mu_0 + \phi^T(\mu^s - \mu_0) \tag{7}$$

$$= \mu^s \tag{8}$$

The estimation in the above scenario is thus confirmed to be perfect, *ceteris peribus*, and the underlying assumption proposed in this model is that features which represent classes of same speakers are well correlated, and all speaker variances can be described by ϕ.

3   Model Implementation

3.1   PCA Implementation

Theoretically, perfect normalization is achievable if the complete data for all classes used to construct the speaker space are available for a specific speaker, and all the eigen components of the covariance matrix are explored. This implies that the speaker vector can be successfully mapped onto the group mean, but the principal components typically constitute majority of the total variance of the speaker space, which shows that not all eigen components are necessary to sufficiently describe the variances among the speakers. PCA is an optimal variance compressor, and components corresponding to the largest eigen values, and comprise the subspace with the most required variance. Using PCA, Eq. (5) can be approximated as,

$$\widehat{\mu}^s = \mu_0 + \sum_{i=1}^{n} \phi_i.\alpha_i^s \tag{9}$$

The required number of eigen vectors (n) is then determined by the contributions of the dominant eigen vectors to the variance between the speakers. Selecting the first few principal components has further advantage in that less data are required to model the adaptation, but the projection coefficients α_i^s may not be perfect for excluded directions. During speaker recognition, a test speaker provides speech for adaptation, be it supervised or unsupervised. In supervised adaptation, the speech content is known, whereas in unsupervised adaptation, the recognition output is used to estimate what was spoken. Models are then adapted – using the supplied data to recognize the test speaker's data. With limited adaptation data, the projections (α_i^s) can be estimated using,

$$\widehat{\mu}^{s'} = \mu'_0 + \sum_{i=1}^{n} \phi_i.\alpha_i^s \quad n \ll wD \tag{10}$$

3.2   SOM Implementation

SOM [4] is among the most commonly used connectionist models for data clustering and visualization. SOM models comprise an important class of competitive neural models. The main difference between SOM and standard competitive networks is that the output neurons are arranged in specific geometrical forms [5]. Each neuron c in the SOM is associated with a weight vector,

$wc = [w_{c_1}, w_{c_2}, \ldots, w_{c_n}]^T \in \mathbf{R}^n$, and of same dimension as the input vector, $I = [I_1, I_2, \ldots, I_n]^T \in \mathbf{R}^n$. Using an unsupervised learning process, the output neurons are tuned and organized after several presentations of the data. The learning algorithm that produces the self organised map can be summarized in two steps:

– A winning or best-matching unit (denoted as $\delta(I)$) is found by using a similarity measure such as the Euclidian distance between the input and the weight vectors, hence:

$$\delta(I) = \frac{\arg\min}{c \in A} \ dist(I, w_c(t)) \tag{11}$$

where A is the set of neurons and $dist$ is the Euclidian distance function;
– The winner and its neighbours in the map have their weights $w_r(t)$ updated toward the current input I:

$$w_c(t+1) = w_c(t) + \nu(t) h_{ce} [I - w_c], \quad c \in A \tag{12}$$

where $\nu(t)$ is the learning rate and h_{ce} is the neighbourhood function.

4   Covariance Learning

Majority of neural models are concerned with the learning of static mappings to outputs that converge to a stable point given the required input patterns, which then act as memories for a set of static patterns. Real world applications however present dynamic data such that a particular pattern cannot be assumed to be independent of its antecedents. Processing dynamic patterns differs fundamentally from processing static data because the temporal ordering and correlation of the observed patterns are important considerations during learning. Neural networks also suffer from top-down learning, i.e., their 'loss-function' or 'cost-function' (ability of the network to predict correctly) represents the error found at the output layer. Covariance on the other hand eliminates top-down learning by finding errors at any layer (not just the top-most output layer). Whereas the loss-function is only defined at the output layer, covariance is defined at every neuron in the network. With covariance, learning happens wherever it's required, irrespective of the layer and negative covariance – being an error-detector, points specifically to the neurons 'responsible' for the errors. The benefits of using covariance learning include: (i) significant reduction in training and retraining time; (ii) suspension of feature detection to the lowest layer and delayed overfitting; and (iii) fast convergence. In Le Cun et al. [6], eigenvalues of covariance matrices have been applied to neural network learning. The form of eigenvalue distribution derived suggests new technique for accelerating the learning process. Park et al. [7] applied convolution neural network to image feature covariance matrix and variation of energy in each sub-band – to resolve the issue of common sounds. CNN was applied with several techniques to reduce training time and resolve problems of initialization and local optimization. In Zehraoui and Bennani [5], approaches to sequence clustering and classification are presented using the self organizing map (SOM). Inputs to the map were modelled to consider correlation of patterns contained in the sequences. The first set of approaches considered inputs as vectors or covariance matrix to account for correlations between the components, but do not take into account the temporal order in the sequences (the dynamics). The second set of approaches introduced dynamics in the covariance matrix. Results of their experiments produced good results.

In this paper, input to our SOM is the weighted covariance matrix obtained from PCA. Note that only half of the covariance matrix is required because of its symmetric nature. Hence, for each input sequence, $X = x(1),\ x(2),\ x(3),\ldots,\ x(pX)$, we compute the weighted covariance matrix as follows:

$$COV_X = \frac{1}{pX}\sum_{i=1}^{pX}\lambda^{pX-i}(x(i)-\overline{x})(x(i)-\overline{x})^T \qquad (13)$$

where $\overline{x}\frac{1}{pX}\sum_{i=1}^{pX}x(i)$ represents the mean vector, and $0 < \lambda \leq 1$. The SOM algorithm is then applied to model the speakers variability patterns.

5   Results

5.1   Overall Average Analysis

Using Praat scripting, selected acoustic features (F0, intensity and pulse) were extracted for analysis. In this subsection, we compare the average features for various age categories, for both gender, to see what can be deduced from these information. Tables 1 and 2 present the pulled averages from each speech signal, represented as (m1–m10) for male speakers, and (f1–f10) for female speakers, repectively. The overall mean is at the far right column of the two tables.

Table 1. Average speech features for male speakers.

Class	m1	m2	m3	m4	m5	m6	m7	m8	m9	m10	Mean
F0(0–15)	292.63	131.55	141.48	302.73	294.52	346.58	292.63	148.46	141.48	155.47	224.75
F0(16–30)	127.95	145.83	153.18	131.32	117.66	118.77	134.25	130.09	220.91	129.56	140.95
F0(31–50)	103.73	240.73	167.26	137.31	144.56	153.08	275.25	184.98	141.46	128.98	167.73
F0(>50)	139.03	159.52	119.66	305.45	103.96	135.92	166.41	163.24	123.39	156.05	157.26
Int(0–15)	86.48	80.63	81.80	78.71	78.72	73.66	73.91	83.57	81.86	87.10	80.64
Int(16–30)	64.75	72.03	70.81	71.66	70.38	69.46	82.77	75.33	85.36	69.50	73.20
Int(31–50)	69.83	71.87	76.45	73.04	73.33	80.85	86.57	73.08	66.19	70.53	74.17
Int(>50)	80.16	83.65	80.27	88.35	78.79	81.48	77.33	75.44	71.43	75.30	79.22
Pulse(0–15)	2.23	2.16	2.02	1.93	2.86	2.94	2.09	1.98	2.02	1.42	2.16
Pulse(16–30)	1.72	1.42	1.63	1.59	2.66	1.85	1.71	1.36	1.21	2.24	1.74
Pulse(31–50)	2.78	3.37	2.94	2.90	2.33	2.27	1.74	2.50	2.41	2.31	2.55
Pulse(>50)	1.19	1.66	0.70	0.90	1.93	1.09	1.08	1.06	1.22	2.00	1.28

Table 2. Average speech features for female speakers.

Class	f1	f2	f3	f4	f5	f6	f7	f8	f9	f10	Mean
F0(0–15)	294.52	250.35	385.80	241.67	263.36	335.68	311.05	214.22	254.76	262.27	281.37
F0(16–30)	230.85	215.92	229.11	233.48	240.99	249.51	249.51	235.09	214.49	221.37	232.03
F0(31–50)	203.49	196.89	194.80	212.78	156.55	205.52	308.63	199.26	196.45	185.47	205.98
F0(>50)	146.45	213.59	213.59	157.37	145.96	165.69	98.10	183.39	193.26	146.96	166.44
Int(0–15)	84.30	79.39	83.09	75.63	73.81	74.47	75.69	80.61	79.46	74.17	78.06
Int(16–30)	68.92	68.72	73.98	87.14	78.23	80.27	87.78	84.68	74.15	77.47	78.13
Int(31–50)	70.48	74.94	64.87	82.06	67.98	66.32	90.97	76.27	73.35	73.52	74.07
Int(>50)	67.66	77.84	83.92	74.91	63.40	75.55	78.30	67.14	79.96	64.40	73.31
Pulse(0–15)	0.90	1.07	2.17	1.69	2.09	1.94	2.32	1.61	1.03	1.40	1.62
Pulse(16–30)	2.12	3.03	1.45	1.05	1.23	1.13	1.19	2.29	2.50	1.99	1.80
Pulse(31–50)	1.70	2.01	2.88	1.62	1.47	2.31	1.74	1.14	0.97	1.25	1.71
Pulse(>50)	1.95	1.67	1.42	1.43	1.81	1.41	1.45	1.83	1.57	1.21	1.58

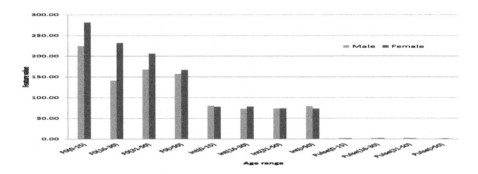

Fig. 1. Overall average plot

5.2   Principal Component Visualisation

In Table 3, an extraction of the first three principal components of selected speech features (F0, intensity and pulse), obtained for different age ranges (children: 0–15; youth: 16–30; adult: 31–50; senior: >50) is analysed. We observed that for both gender, the three principal components (PC1–PC3) showed no significance for intensity and pulse features, for the age ranges studied. Hence, only F0 exhibited major significant differences among the principal components (i.e., have eigenvalues of at least 1).

Visualisation plots in Fig. 2 reveal that feature patterns of intensity and pulse features for male and female speakers (Fig. 2(a) and (b)), tend to cluster together, meaning that speech features in this class maintained similar speech profiles. Hence, F0 feature in male speakers showed high variability with PC1 (87%) capturing the most variance, while PC2 (7%) and PC3 (3%) captured the least variances, while in female speakers: F0 showed high variability, with PC1 (97%) capturing the most variance, while PC2 (2%) and PC3 (1%) captured the least variances. As such, F0 feature exhibits high variability in speech profiles, and is useful for investigating speech feature variability in tone languages.

Table 3. First three principal component extraction for selected features

	Principal Component					
	Male speaker			Female speaker		
Speech feature	1	2	3	1	2	3
F0(0-15)	1.9676	-2.3392	-0.7569	1.8646	-0.5147	-2.4839
F0(16-30)	0.6697	1.2531	-0.0686	1.3281	0.2831	1.0194
F0(31-50)	1.0683	1.7902	-1.4543	1.0788	2.1586	0.7926
F0(>50)	0.9828	0.3086	2.8645	0.6621	-2.3933	1.4561
Int(<15)	-0.0704	0.189	0.0026	-0.2223	-0.1553	0.1779
Int(16-30)	-0.1571	0.2205	-0.1573	-0.2222	0.2538	0.3063
Int(31-50)	-0.1354	0.0365	-0.2569	-0.2638	0.4092	0.441
Int(>50)	-0.0737	0.0505	0.0303	-0.2665	-0.0988	-0.0128
Pulse(<15)	-1.0596	-0.3829	-0.0629	-0.9901	0.0308	-0.4413
Pulse(16-30)	-1.0653	-0.3847	-0.054	-0.9889	0.0004	-0.3857
Pulse(31-50)	-1.0555	-0.3642	-0.035	-0.989	0.009	-0.4558
Pulse(>50)	-1.0715	-0.3774	-0.0515	-0.9909	0.0174	-0.4138

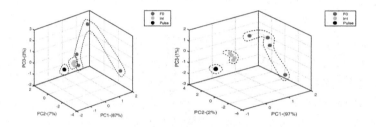

(a) Male speakers   (b) Female speakers

Fig. 2. Component visualisation of selected speech features

5.3   SOM Covariance Visualisation

Tables 4 and 5 are covariance matrices representing the symmetry of male and female speakers according to their age ranges and selected features. The covariance is used to determine the direction of linear relationship between two variables as follows: (i) if both variables tend to increase or decrease together, the coefficient is positive; (ii) if one variable tends to increase as the other decreases, the coefficient is negative.

Before covariance learning, the vector weights for each column were first normalised (component-wise normalisation) using the *zscore* function of MATLAB 2015a. The normalisation process indeed allows the SOM network to learn the data in a more productive way, else, the network will process points from different parts of the input space with different bias (i.e., larger values will have

Table 4. Covariance matrix for male speakers

Class	F0 (0–15)	F0 (16–30)	F0 (31–50)	F0 (>50)	Int (0–15)	Int (16–30)	Int (31–50)	Int (>50)	Pulse (0–15)	Pulse (16–30)	Pulse (31–50)	Pulse (>50)
F0(0–15)	7575.60	−1436.52	−669.69	1109.97	−229.28	−145.23	248.94	177.91	23.81	15.47	−15.43	−4.79
F0(16–30)	−1436.52	906.12	−25.21	−304.99	21.47	129.50	−79.22	−70.03	−3.81	−7.64	1.24	−2.29
F0(31–50)	−669.69	−25.21	2794.49	45.89	−125.00	168.06	192.20	5.52	−0.39	−6.20	−3.53	−1.98
F0(>50)	1109.97	−304.99	45.89	3141.73	−30.86	−7.20	11.68	167.70	−8.92	−6.10	5.53	−7.12
Int(0–15)	−229.28	21.47	−125.00	−30.86	20.97	−9.43	−20.61	−6.57	−1.24	−0.09	0.79	0.54
Int(16–30)	−145.23	129.50	168.06	−7.20	−9.43	40.04	5.09	−15.93	−0.53	−1.26	−1.27	−0.51
Int(31–50)	248.94	−79.22	192.20	11.68	−20.61	5.09	34.21	5.60	0.82	0.35	−1.34	−0.84
Int(>50)	177.91	−70.03	5.52	167.70	−6.57	−15.93	5.60	22.68	0.45	0.07	1.19	−0.48
Pulse(0–15)	23.81	−3.81	−0.39	−8.92	−1.24	−0.53	0.82	0.45	0.20	0.06	−0.03	0.00
Pulse(16–30)	15.47	−7.64	−6.20	−6.10	−0.09	−1.26	0.35	0.07	0.06	0.19	−0.07	0.12
Pulse(31–50)	−15.43	1.24	−3.53	5.53	0.79	−1.27	−1.34	1.19	−0.03	−0.07	0.21	−0.02
Pulse(>50)	−4.79	−2.29	−1.98	−7.12	0.54	−0.51	−0.84	−0.48	0.00	0.12	−0.02	0.19

Table 5. Covariance matrix for female speakers

Class	F0 (0–15)	F0 (16–30)	F0 (31–50)	F0 (>50)	Int (0–15)	Int (16–30)	Int (31–50)	Int (>50)	Pulse (0–15)	Pulse (16–30)	Pulse (31–50)	Pulse (>50)
F0(0–15)	2594.93	210.33	424.42	42.15	37.75	−60.15	−142.18	187.28	12.44	−15.62	24.51	−3.98
F0(16–30)	210.33	154.83	211.13	−276.50	−18.29	54.59	17.71	−10.87	4.76	−6.73	1.84	0.00
F0(31–50)	424.42	211.13	1533.10	−771.09	−9.21	134.22	244.59	114.16	6.57	−6.76	2.08	−2.17
F0(>50)	42.15	−276.50	−771.09	1267.16	67.35	−131.40	−154.37	102.18	−6.18	13.22	5.81	0.71
Int(0–15)	37.75	−18.29	−9.21	67.35	14.68	−14.45	−7.41	7.87	−0.82	1.27	0.50	0.45
Int(16–30)	−60.15	54.59	134.22	−131.40	−14.45	47.62	33.18	−3.25	2.22	−2.98	−0.81	−0.51
Int(31–50)	−142.18	17.71	244.59	−154.37	−7.41	33.18	61.19	5.93	0.51	−0.44	−1.59	−0.27
Int(>50)	187.28	−10.87	114.16	102.18	7.87	−3.25	5.93	50.82	0.56	−0.01	2.25	−0.58
Pulse(0–15)	12.44	4.76	6.57	−6.18	−0.82	2.22	0.51	0.56	0.26	−0.28	0.12	−0.04
Pulse(16–30)	−15.62	−6.73	−6.76	13.22	1.27	−2.98	−0.44	−0.01	−0.28	0.47	−0.13	0.06
Pulse(31–50)	24.51	1.84	2.08	5.81	0.50	−0.81	−1.59	2.25	0.12	−0.13	0.33	−0.03
Pulse(>50)	−3.98	0.00	−2.17	0.71	0.45	−0.51	−0.27	−0.58	−0.04	0.06	−0.03	0.05

larger effect). Tables 6 and 7 show the normalised covariance matrices for male and female speakers, respectively.

Using the batch unsupervised weight/bias algorithm (*trainbu*) – where weights and biases are only updated after all the inputs are fed into the network, a training of each model, holding a Kohonen 10×10 SOM was performed. The batch version of the SOM algorithm is computationally more efficient [8]. During each training step, all the input data vectors are simultaneously used to update the weight vectors. On obtaining the SOM results, the map was then transformed into $m = 10$ clusters, using minimum Euclidian distance, and each set or cluster used to update the corresponding weight vector. The updated weight vectors are computed as follows:

$$m_i(t+1) = \frac{\sum_{j=1}^{m} n_j h_{ij}(t)\overline{x}_j}{\sum_{j=1}^{m} n_j h_{ij}(t)} \tag{14}$$

where $\overline{x}_j$ is the mean of the n data vectors in group j, $h_{ij}(t)$ denotes the value of the neighborhood function at cluster unit j when the neighbourhood function is centered on cluster unit i. A constant learning rate was adopted in the

Table 6. Normalised covariance matrix for male speakers

Class	FO (0–15)	FO (16–30)	FO (31–50)	FO (>50)	Int (0–15)	Int (16–30)	Int (31–50)	Int (>50)	Pulse (0–15)	Pulse (16–30)	Pulse (31–50)	Pulse (>50)
F0(0–15)	3.07	−2.64	−1.03	0.81	−2.71	−2.05	2.34	2.11	2.97	2.73	−2.88	−1.44
F0(16–30)	−0.87	1.89	−0.26	−0.69	0.73	1.50	−1.22	−1.29	−0.61	−1.24	0.46	−0.37
F0(31–50)	−0.53	0.09	3.07	−0.31	−1.28	2.00	1.72	−0.25	−0.16	−0.99	−0.49	−0.24
F0(>50)	0.24	−0.45	−0.18	2.96	0.01	−0.27	−0.23	1.97	−1.27	−0.97	1.32	−2.44
Int(0–15)	−0.34	0.18	−0.38	−0.40	0.72	−0.30	−0.58	−0.42	−0.27	0.06	0.37	0.85
Int(16–30)	−0.31	0.39	−0.04	−0.37	0.30	0.34	−0.30	−0.55	−0.18	−0.14	−0.04	0.40
Int(31–50)	−0.13	−0.01	−0.01	−0.35	0.15	−0.11	0.01	−0.25	−0.01	0.13	−0.06	0.25
Int(>50)	−0.16	0.00	−0.23	−0.19	0.34	−0.38	−0.30	−0.02	−0.05	0.08	0.45	0.41
Pulse(0–15)	−0.23	0.13	−0.23	−0.37	0.42	−0.18	−0.35	−0.32	−0.09	0.08	0.21	0.61
Pulse(16–30)	−0.24	0.13	−0.24	−0.37	0.43	−0.19	−0.35	−0.33	−0.10	0.11	0.20	0.66
Pulse(31–50)	−0.25	0.14	−0.24	−0.36	0.44	−0.19	−0.37	−0.31	−0.12	0.06	0.25	0.61
Pulse(>50)	−0.24	0.14	−0.24	−0.37	0.44	−0.18	−0.37	−0.34	−0.11	0.09	0.21	0.69

Table 7. Normalised covariance matrix for female speakers

Class	FO (0–15)	FO (16–30)	FO (31–50)	FO (>50)	Int (0–15)	Int (16–30)	Int (31–50)	Int (>50)	Pulse (0–15)	Pulse (16–30)	Pulse (31–50)	Pulse (>50)
F0(0–15)	3.11	1.41	0.52	0.06	1.29	−1.01	−1.48	2.36	2.35	−2.11	3.06	−2.66
F0(16–30)	−0.09	0.98	0.10	−0.63	−1.03	0.78	0.13	−0.77	0.67	−0.78	−0.15	0.40
F0(31–50)	0.20	1.41	2.65	−1.70	−0.66	2.02	2.41	1.20	1.07	−0.79	−0.12	−1.26
F0(>50)	−0.31	−2.36	−1.79	2.72	2.52	−2.13	−1.60	1.01	−1.71	2.20	0.41	0.95
Int(0–15)	−0.32	−0.36	−0.32	0.12	0.33	−0.30	−0.12	−0.48	−0.54	0.41	−0.34	0.75
Int(16–30)	−0.45	0.20	−0.04	−0.31	−0.87	0.67	0.29	−0.65	0.12	−0.22	−0.53	0.01
Int(31–50)	−0.56	−0.08	0.17	−0.36	−0.58	0.44	0.57	−0.51	−0.26	0.16	−0.64	0.20
Int(>50)	−0.12	−0.31	−0.08	0.19	0.05	−0.13	0.01	0.20	−0.24	0.22	−0.09	−0.04
Pulse(0–15)	−0.35	−0.18	−0.29	−0.04	−0.31	−0.04	−0.04	−0.59	−0.31	0.18	−0.39	0.37
Pulse(16–30)	−0.39	−0.27	−0.31	0.00	−0.22	−0.12	−0.05	−0.60	−0.43	0.29	−0.43	0.45
Pulse(31–50)	−0.34	−0.21	−0.30	−0.02	−0.25	−0.09	−0.06	−0.57	−0.34	0.20	−0.36	0.38
Pulse(>50)	−0.38	−0.22	−0.31	−0.03	−0.26	−0.08	−0.05	−0.61	−0.37	0.23	−0.42	0.45

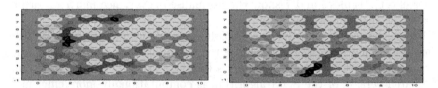

Fig. 3. SOM neighbour weight distances (Color figure online)

experiment, and the performance metric used to evaluate the SOM was the Mean Squared Error (MSE).

In Fig. 3, the SOM neighbour weight distances for male and female speakers are presented. The neurons are presented as gray-blue patches and their direct neighbour relations with red lines. The neighbour patches are coloured from black to yellow to show how close each neuron's weight vector is to its neighbours. We found that both maps clustered the input data into clusters of varying degree of distances.

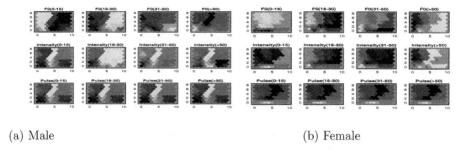

(a) Male   (b) Female

Fig. 4. SOM component planes visualisation

To effectively visualise the resulting patterns of each feature, the component plane maps for the various categories of speakers are presented in Fig. 4, for male and female speakers (see Fig. 4(a) and (b)). We observed that on the average, the intensity and pulse planes showed more similar patterns, compared to F0 features, hence confirming our PCA results. A closer look at each pattern reveals that for male speakers, the intensity pattern for youths (16–30) speakers deviated from other age ranges indicating some defects in the data, probably due to suboptimal conditions such as background noise, poor recording, and more. Similar observation goes for female speakers, which intensity patterns of children (0–15) and seniors (>50) differed from youths (31–50) and adults (31–50). Investigation into speaker defects shall be handled in a future paper.

6   Conclusion

Speaker variability modelling is one of the greatest challenges in speech processing, and is vital in the design and implementation of speaker recognition systems. This paper proposed the modelling of speaker variability using covariance learning. Two unsupervised techniques: PCA and SOM, were adopted – PCA was used to select the first three principal component features – necessary for the study of variance between the selected speech features (F0, intensity and pulse), while SOM was used to visualise the feature component planes – useful for observing inherent patterns in the speech features. Results obtained reveal interesting discovery for African tone language research. A future research direction is a study of speaker variability compromise between native and non-native speakers.

Acknowledgments. This research is funded by the Tertiary Education Trust Fund (TETFund), Nigeria. We also appreciate the students, staff and other participants who accepted to offer their voices for this experiment.

References

1. Kajarekar, S.S.: Analysis of variability in speech with applications to speech and speaker recognition. Ph.D. thesis, Oregon Health and Science University, Oregon (2002)
2. Chen, T., Huang, C., Chang, E., Wang, J.: On the use of Gaussian mixture model for speaker variability analysis. In: 17th International Conference on Spoken Language Processing, Denver, Colorado, USA, pp. 1–4 (2002)
3. Huang, C., Chen, T., Li, S., Chang, E., Zhou, J.: Analysis of speaker variability. In: 7th European Conference on Speech Communication and Technology, Scandinavia, pp. 1–4 (2001)
4. Kohonen, T.: MATLAB Implementations and Applications of the Self-organizing Map. Unigrafia Oy, Helsinki (2014)
5. Zehraoui, F., Bennani, Y.: M-SOM: matricial self organizing map for sequence clustering and classification. In: Proceedings of IEEE International Joint Conference on Neural Networks, Budapest, Hungary, vol. 1, pp. 763–768 (2004)
6. Le Cun, Y., Kanter, I., Solla, S.A.: Eigenvalues of covariance matrices: application to neural-network learning. Phys. Rev. Lett. **66**(18), 2396 (1991)
7. Park, S., Mun, S., Lee, Y., Ko, H.: Acoustic scene classification based on convolution neural network using double image features. In: Proceedings of Detection and Classification of Acoustic Scenes and Events Workshop, Munich, Germany, pp. 1–5 (2017)
8. Vesanto, J., Himberg, J., Alhoniemi, E., Parhankangas, J.: Self-organizing map in MATLAB: the SOM Toolbox. In: Proceedings of MATLAB DSP Conference, Espoo, Finland (1999)

A Neural Network Model
with Bidirectional Whitening

Yuki Fujimoto$^{(\boxtimes)}$ and Toru Ohira

Graduate School of Mathematics, Nagoya University, Nagoya, Japan
{m15042x,ohira}@math.nagoya-u.ac.jp

Abstract. We present here a new model and algorithm which performs an efficient Natural gradient descent for multilayer perceptrons. Natural gradient descent was originally proposed from a point of view of information geometry, and it performs the steepest descent updates on manifolds in a Riemannian space. In particular, we extend an approach taken by the "Whitened Neural Networks" model. We make the whitening process not only in the feed-forward direction as in the original model, but also in the back-propagation phase. Its efficacy is shown by an application of this "Bidirectional Whitened Neural Networks" model to a handwritten character recognition data (MNIST data).

1 Introduction

Interests for developing and efficient learning algorithm for multilayer neural networks have grown rapidly due to recent upheaval of the deep learning and other machine learnings. Natural gradient descent (NGD) is considered as one of the strong methods. It was proposed from a point of view of information geometry [1], where neural networks are considered as manifolds in a Riemannian space with a measure given by the Fisher information matrix (FIM). Then, the learning process can be interpreted as an optimization problem of a function in a Riemannian space. The idea of applying the NGD to multilayer neural networks was initiated by Amari. Recently, it has regained interests from machine learning researchers [6,9].

However, difficulty exists for using the NGD: the computational costs of estimating the FIM and obtaining its inverse is high. Much attention and research efforts have gone into solving this difficulty [4,5,7,8,10].

In this paper, we will focus on one of such approaches, and extend the work of [4]. In their approach "Whitened Neural Networks" model was proposed. There, a neural network architecture, whose FIM is closer to the identity matrix with less computational demands, is explored. Extra neurons and connections are added to achieve this whitening approximation. In particular, they have used this scheme for the forward direction of inputs to neurons and achieved lower computational costs.

Our main proposal in this paper is to further push the approximation of the FIM being closer to the identity by implementing the whitening process also in

© Springer International Publishing AG, part of Springer Nature 2018
L. Rutkowski et al. (Eds.): ICAISC 2018, LNAI 10841, pp. 47–57, 2018.
https://doi.org/10.1007/978-3-319-91253-0_5

the back-propagation phase. This model, which we term as the "Bidirectional Whitened Neural Networks" model, will be described in the following. Its efficacy is also shown through its application to a handwritten character recognition data (MNIST data).

2   Multilayer Perceptron and Natural Gradient Descent

We present here a brief review of the Multilayer Perceptron and the Natural Gradient Descent, which we focus on this paper. The approximation for the FIM is also discussed.

2.1   Multilayer Perceptron

Multilayer Perceptron is a model of neural networks which has feed-forward structure with no recurrent loops. Let us consider a N layer Perceptron, and set the values of the input as $z^{(0)} = x$, the hidden layer values as $z^{(i)} = h^{(i)}$, $(1 \leq i \leq N - 1)$, and the output of the entire network as $z^{(N)} = f(x; w)$.

This $f(x; w)$ can be viewed as a function of x by fixing the parameters w, and thus called as a "multilayer Perceptron function". The rules of computing the value of the i layer from the $i - 1$ layer in the network is given as follows $(1 \leq i \leq N)$.

$$a^{(i)} = W^{(i)} z^{(i-1)} + b^{(i)} = \bar{W}^{(i)} \bar{z}^{(i-1)} \tag{1}$$

$$z^{(i)} = \phi^{(i)}(a^{(i)}) \tag{2}$$

Here, $\phi^{(i)}(\cdot)$ is an activation function applied to each element of a. Also, the right hand side of (1) is a shortened notation by setting $\bar{W}^{(i)} \equiv (b^{(i)}, W^{(i)})$, $\bar{z}^{(i)} \equiv (1, z^{(i)T})^T$.

Hence, the multilayer Perceptron function (MPF) is defined by setting $\{(W^{(i)}, b^{(i)})\}$. It is often convenient to denote these parameters by w, defined by

$$w \equiv (\text{vec}(\bar{W}^{(1)})^T, \ldots, \text{vec}(\bar{W}^{(N)})^T)^T \tag{3}$$

where $\text{vec}(A)$ means a compound vector of column vectors of a matrix A.

The learning process of multilayer perceptrons is the following optimization problem.

$$w^* \equiv \arg \min_{w \in \Theta} \sum_{k=1}^{K} -\log p(x_k, y_k; w) \equiv \arg \min_{w \in \Theta} M(w) \tag{4}$$

where, $\{(x_k, y_k)\}_{k=1}^{K}$ is the training data, $p(x, y; w)$ is a joint probability density function associated with MPF, and $\Theta \subset \mathbb{R}^M$ is a set of parameters. Also, we have set the target function to minimize as $M(w)$. Research on efficient algorithms for this optimization problem is the central issue in the following.

2.2   Natural Gradient Method

Natural Gradient Method is a steepest descent method in a Riemannian space. Let us start by defining the Fisher information matrix and the Natural Gradient Descent [2].

Definition: Fisher Information Matrix

We set $l(\boldsymbol{x}; \boldsymbol{w}) \equiv \log p(\boldsymbol{x}; \boldsymbol{w})$. For $\boldsymbol{w} \in \Theta$, a square matrix $G(\boldsymbol{w}) = (g_{ij}(\boldsymbol{w}))$ is defined as follows.

$$G(\boldsymbol{w}) \equiv E\left[\nabla l(X; \boldsymbol{w})\nabla l(X; \boldsymbol{w})^T\right] \tag{5}$$

(5) can be expressed by each elements as,

$$g_{ij}(\boldsymbol{w}) = E\left[\frac{\partial l}{\partial w_i}(X; \boldsymbol{w})\frac{\partial l}{\partial w_j}(X; \boldsymbol{w})\right] \tag{6}$$

$$= \int \frac{\partial l}{\partial w_i}(\boldsymbol{x}; \boldsymbol{w})\frac{\partial l}{\partial w_j}(\boldsymbol{x}; \boldsymbol{w})p(\boldsymbol{x}; \boldsymbol{w})d\boldsymbol{x} \tag{7}$$

We call this matrix G the Fisher information matrix (FIM).

Definition: Natural Gradient Descent

We call the following gradient method as the Natural Gradient Descent (NGD).

$$\boldsymbol{w}(t+1) = \boldsymbol{w}(t) - \eta(t)G^{-1}(\boldsymbol{w}(t))\nabla M(\boldsymbol{w}(t)) \tag{8}$$

Here $\eta(t)$ is a rate of the learning.

Then, $-G^{-1}(\boldsymbol{w}(t))\nabla M(\boldsymbol{w}(t))$ is the direction of the maximal decrease of the target function M given a fixed step size. We note that this NGD reduces to the ordinary gradient descent, when G is the identity matrix.

2.3   Approximation of the Fisher Information Matrix

As discussed in the previous section, the FIM and its inverse play important roles in the calculation in the NGD. We, thus, present a preliminary approximation of the FIM in order to lessen the computational burdens [7].

Let us first compute the FIM for the multilayer perceptrons. If we set $\delta_j^{(i)} = \frac{\partial l}{\partial a_j^{(i)}}$, then the FIM for the MLP is given as follows.

$$G(\boldsymbol{w}) = \begin{pmatrix} G_{1,1} & G_{1,2} & \cdots & G_{1,N} \\ G_{2,1} & G_{2,2} & \cdots & G_{2,N} \\ \vdots & \vdots & \ddots & \vdots \\ G_{N,1} & G_{N,2} & \cdots & G_{N,N} \end{pmatrix} \tag{9}$$

$$G_{i,j} \equiv E\left[\bar{\boldsymbol{z}}^{(i-1)}\bar{\boldsymbol{z}}^{(j-1)T} \otimes \boldsymbol{\delta}^{(i)}\boldsymbol{\delta}^{(j)T}\right] \tag{10}$$

Here, $\otimes$ is the Kronecker product. Hence, the FIM for the MLP is composed of the block matrices $G_{i,j}$.

For the efficient computation, it is essential to approximate this FIM. Firstly, we define $\tilde{G}_{i,j}$ as an approximation of $G_{i,j}$ as follows.

$$\tilde{G}_{i,j} \equiv E\left[\bar{z}^{(i-1)}\bar{z}^{(j-1)^T}\right] \otimes E\left[\delta^{(i)}\delta^{(j)^T}\right] \tag{11}$$

This approximation means that we are inter-changing the expectation of the Kronecker products with the Kronecker products of the expectations. Then we define $\check{G}$ as an approximation of G as follows.

$$\check{G} \equiv \mathrm{diag}(\tilde{G}_{1,1}, \tilde{G}_{2,2}, \ldots, \tilde{G}_{N,N}) \tag{12}$$

Here, $\mathrm{diag}(\cdots)$ denotes a block diagonal matrix, whose non-zero diagonals are given by the elements. This approximated FIM is the same as used in [4,7].

3   Whitened Neural Networks

In this section, we present algorithms which aim to perform Natural Gradient Descent efficiently with the approximated FIM, $\check{G}$.

3.1   Natural Gradient Descent by Whitening

Let us first describe Whitened Neural Networks [4]. The main idea of this method is to perform the NGD by reconfiguring the network and parameters, so that the FIM becomes closer to the identity matrix. When the FIM is the identity matrix, the NGD is the same as the ordinary gradient descent, thus can be implemented simply with less computational costs.

Whitened Neural Networks. The architecture of the Whitened Neural Networks (WNN) is obtained by changing (1), (2) into the following form.

$$z^{\dagger(i-1)} = U^{(i-1)}(z^{(i-1)} - c^{(i-1)}) \tag{13}$$

$$a^{(i)} = W^{\dagger(i)}z^{\dagger(i-1)} + b^{\dagger(i)} \tag{14}$$

$$z^{(i)} = \phi^{(i)}\left(a^{(i)}\right) \tag{15}$$

Here $\{(U^{(i-1)}, c^{(i-1)})\}$ are the new parameters introduced as "Whitening" parameters. $\{(W^{\dagger(i)}, b^{\dagger(i)})\}$ are the new model parameters associated with this new architecture. These are the ones which we want to estimate and update using gradient descent methods as in the normal multilayer perceptrons.

We present in the Fig. 1 the new architecture defined by (13), (14), (15). It shows the $i-1$th layer to the ith layer. We note the gray layer in the Fig. 1 is the new inserted layer for the purpose of "Whitening". This change of network configuration is the essence of WNN.

By (11), then, the approximated FIM $\tilde{G}_{i,i}$ in the WNN, is expressed as in the following.

$$\tilde{G}_{i,i} = E\left[\bar{z}^{\dagger(i-1)}\bar{z}^{\dagger(i-1)^T}\right] \otimes E\left[\boldsymbol{\delta}^{(i)}\boldsymbol{\delta}^{(i)^T}\right] \tag{16}$$

The essential idea of the whitening is to make $\check{G}$ closer to the identity by defining the whitening parameters $\{(U^{(i-1)}, \boldsymbol{c}^{(i-1)})\}$ as

$$E\left[\bar{z}^{\dagger(i-1)}\bar{z}^{\dagger(i-1)^T}\right] = I \tag{17}$$

for each i and performs the gradient descent. (Our idea, which will be described later in Sect. 3.2, is to further extend the whitening to the latter factor $E\left[\boldsymbol{\delta}^{(i)}\boldsymbol{\delta}^{(i)^T}\right]$ in (16)).

Updating of the Whitening Parameters. We calculate here explicitly $\{(U^{(i-1)}, \boldsymbol{c}^{(i-1)})\}$, which satisfies the condition (17).

As $\bar{z}^{\dagger(i-1)} = (1, z^{\dagger(i-1)^T})^T$, (17) can be decomposed into

$$\begin{pmatrix} 1 & E\left[z^{\dagger(i-1)^T}\right] \\ E\left[z^{\dagger(i-1)}\right] & E\left[z^{\dagger(i-1)}z^{\dagger(i-1)^T}\right] \end{pmatrix} = I \tag{18}$$

Thus,

$$E\left[z^{\dagger(i-1)}\right] = \boldsymbol{0} \tag{19}$$

$$E\left[z^{\dagger(i-1)}z^{\dagger(i-1)^T}\right] = U^{(i-1)}\check{Z}_{i-1,i-1}U^{(i-1)^T} = I \tag{20}$$

are required to satisfy this condition. Here, we set the matrix $\check{Z}_{i-1,i-1}$ by the following

$$\check{Z}_{i-1,i-1} \equiv E\left[(z^{(i-1)} - \boldsymbol{c}^{(i-1)})(z^{(i-1)} - \boldsymbol{c}^{(i-1)})^T\right] \tag{21}$$

Hence, by updating $\{(U^{(i-1)}, \boldsymbol{c}^{(i-1)})\}$ as follows, (19), (20) is approximately satisfied.

$$\boldsymbol{c}_{new}^{(i-1)} = E\left[z^{(i-1)}\right] \tag{22}$$

$$U_{new}^{(i-1)} = (\Lambda + \varepsilon I)^{-\frac{1}{2}} \cdot P^T \tag{23}$$

Here, Λ, P are the diagonalized and the orthogonal matrices associated with $\check{Z}_{i-1,i-1}$, and ε is the small positive parameter to avoid a division by zero.

By this process, called the whitening process, according to (22) and (23), we update the whitening parameters satisfying (17). We note that, in this updating, the calculation of $z^{(i-1)}$ in feed-forward phase is essential.

Updating of the Model Parameters. We now turn our attention to the updating of the model parameters $\{(W^{\dagger^{(i)}}, b^{\dagger^{(i)}})\}$. We need to pay attention so that the inclusion of the whitening process and the associated layer does not change the value of the multilayer Perceptron function (MPF) itself. In concrete, we need to do the following. Let us assume the whitening parameters $\{(U^{(i-1)}, c^{(i-1)})\}$ are updated to $\{(U_{new}^{(i-1)}, c_{new}^{(i-1)})\}$. We want to keep the value of (14) unchanged by this updating. This places constrains in the way we update the model parameters $\{(W^{\dagger^{(i)}}, b^{\dagger^{(i)}})\}$ as follows.

$$W_{new}^{\dagger^{(i)}} = W^{\dagger^{(i)}} U^{(i-1)} U_{new}^{(i-1)^{-1}} \tag{24}$$

$$b_{new}^{\dagger^{(i)}} = b^{\dagger^{(i)}} - W^{\dagger^{(i)}} U^{(i-1)} c^{(i-1)} + W_{new}^{\dagger^{(i)}} U_{new}^{(i-1)} c_{new}^{(i-1)} \tag{25}$$

Thus, we can keep MPF the same by updating whitening parameters first as in (22) and (23) and then update model parameters with (24) and (25).

As we change model parameters, the values of $E[z^{(i-1)}], \check{Z}_{i-1,i-1}$ changes, which in turn requires the update of the whitening parameters to keep the FIM close to the identity matrix. However, it is computationally expensive to update both set of parameters at every iterations. Thus, in actual implementations, the update of the whitening parameters are performed at certain fixed time intervals [4], though this makes a gradual digression from the NGD for that time interval between the successive updating of the whitening parameters.

The method and algorithm described above is called "Projected Natural Gradient Descent"(PRONG) [4], which is outlined in Algorithm 1.

Algorithm 1. Projected Natural Gradient Descent (PRONG) [4].

Input: training set D, initial parameter $w(0)$
Hyper parameters: updating period of whitening parameters τ
- $U^{(i)} \leftarrow I; c^{(i)} \leftarrow 0; t \leftarrow 0$

while ending condition not satisfied **do**
    if $\mod(t, \tau) = 0$ **then**
        for all layers i **do**
            - Estimations of $E[z^{(i-1)}], \check{Z}_{i-1,i-1}$.
            - Updating of the Whitening parameters $\{(U^{(i-1)}, c^{(i-1)})\}$.
            - Updating of the model parameters $\{(W^{\dagger^{(i)}}, b^{\dagger^{(i)}})\}$.
        end for
    end if
    - Updating of $\{(W^{\dagger^{(i)}}, b^{\dagger^{(i)}})\}$ by the ordinary gradient descent.
    - $t \leftarrow t + 1$
end while

3.2 Extension of Whitening

Here, we describe our proposal of the new extended whitening algorithms based on Sect. 3.1.

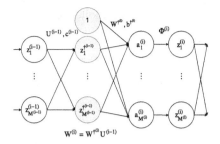

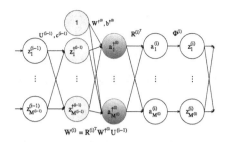

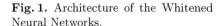

Fig. 1. Architecture of the Whitened Neural Networks.

Fig. 2. Architecture of the Bidirectional Whitened Neural Networks.

In the whitening method described above, in order to keep the approximated FIM, $\check{G}$, closer to the identity matrix, updating of the whitening parameters $\{(U^{(i)}, c^{(i)})\}$ are performed. This makes the first factor $E\left[\bar{z}^{\dagger(i-1)} \bar{z}^{\dagger(i-1)^T}\right]$ in

$$\tilde{G}_{i,i} = E\left[\bar{z}^{\dagger(i-1)} \bar{z}^{\dagger(i-1)^T}\right] \otimes E\left[\delta^{(i)} \delta^{(i)^T}\right] \tag{26}$$

closer to the identity matrix.

The main idea of our method is to make the second factor $E\left[\delta^{(i)} \delta^{(i)^T}\right]$ toward the identity as well, so that $\tilde{G}_{i,i}$ is even better approximated by the identity matrix. This turns out that we implement whitening process not only in the feed-forward phase but also in the back-propagating phase.

Bidirectional Whitened Neural Networks. In order to perform the back-whitening, we modify the forward-whitening process described by (13), (14) and (15) into the following.

$$z^{\dagger(i-1)} = U^{(i-1)}(z^{(i-1)} - c^{(i-1)}) \tag{27}$$

$$a^{\dagger(i)} = W^{\dagger(i)} z^{\dagger(i-1)} + b^{\dagger(i)} \tag{28}$$

$$a^{(i)} = R^{(i)^T} a^{\dagger(i)} \tag{29}$$

$$z^{(i)} = \phi^{(i)}\left(a^{(i)}\right) \tag{30}$$

Here, $\{R^{(i)^T}\}$ is a newly introduced parameter, called the back-whitening parameter.

We show the architecture of this extended method defined by (27), (28), (29), (30) in the Fig. 2. The dark gray part in the Fig. 2 is the newly introduced layer to accommodate the back-whitening parameter $\{R^{(i)^T}\}$.

As mentioned above, this proposed method performs whitening process both in feed-forward and back-propagating phase. Thus, we call this new architecture as the Bidirectional Whitened Neural Networks (BWNN).

We introduce a new parameter $\boldsymbol{\delta^{\dagger}}^{(i)}$ in place of $\boldsymbol{\delta}^{(i)}$ as in the following.

$$\boldsymbol{\delta^{\dagger}}^{(i)} \equiv \frac{\partial l}{\partial \boldsymbol{a^{\dagger}}^{(i)}} \tag{31}$$

Then, the approximation of $\tilde{G}_{i,i}$ is then expressed as

$$\tilde{G}_{i,i} = E\left[\bar{z}^{\dagger(i-1)} \bar{z}^{\dagger(i-1)^T}\right] \otimes E\left[\boldsymbol{\delta^{\dagger}}^{(i)} \boldsymbol{\delta^{\dagger}}^{(i)^T}\right] \tag{32}$$

In analogy with Sect. 3.1, we will fix the back-whitening parameter $\{R^{(i)^T}\}$ so that

$$E\left[\boldsymbol{\delta^{\dagger}}^{(i)} \boldsymbol{\delta^{\dagger}}^{(i)^T}\right] = I \tag{33}$$

Updating of the Back-Whitening Parameter. Let us explicitly find $\{R^{(i)^T}\}$ to satisfy (33). From (31), we have

$$\boldsymbol{\delta^{\dagger}}^{(i)} = R^{(i)} \boldsymbol{\delta}^{(i)} \tag{34}$$

Thus, $\boldsymbol{\delta^{\dagger}}^{(i)}$ is a linear transformation of $\boldsymbol{\delta}^{(i)}$.

If we set $D_{i,i} \equiv E\left[\boldsymbol{\delta}^{(i)} \boldsymbol{\delta}^{(i)^T}\right]$ and insert (34) into (33), we obtain

$$E\left[\boldsymbol{\delta^{\dagger}}^{(i)} \boldsymbol{\delta^{\dagger}}^{(i)^T}\right] = R^{(i)} D_{i,i} R^{(i)^T} = I \tag{35}$$

Hence, in analogy with (23), $R^{(i)}$ which satisfies (33) is given by the following

$$R_{new}^{(i)} = (\Lambda + \varepsilon I)^{-\frac{1}{2}} \cdot P^T \tag{36}$$

Here, Λ, P are the diagonalized and the orthogonal matrices associated with $D_{i,i}$, and ε is the small positive parameter to avoid a division by zero.

Altogether, as in the case of the forward-whitening parameters, (33) is satisfied by updating of the back-whitening parameters according to (36), which, in turn, depends on the calculation of $\boldsymbol{\delta}^{(i)}$ in the back-propagating phase.

Updating of the Model Parameters. As in the feed-forward phase, we update the model parameters $\{(W^{\dagger(i)}, \boldsymbol{b}^{\dagger(i)})\}$ so that the values of the multilayer Perceptron function are kept the same when the back-whitening parameters are updated.

In order to achieve this, the model parameters $\{(W^{\dagger(i)}, \boldsymbol{b}^{\dagger(i)})\}$ need to be updated as follows, given the back-whitening parameters are updated from $R^{(i)}$ to $R_{new}^{(i)}$.

$$W_{new}^{\dagger(i)} = (R_{new}^{(i)^T})^{-1} R^{(i)^T} W^{\dagger(i)} \tag{37}$$
$$\boldsymbol{b}_{new}^{\dagger(i)} = (R_{new}^{(i)^T})^{-1} R^{(i)^T} \boldsymbol{b}^{\dagger(i)} \tag{38}$$

Algorithm 2. Bidirectional Projected Natural Gradient Descent(BPRONG).

Input: training set D, initial parameter $\boldsymbol{w}(0)$
Hyper parameters: parameters for forward-whitening τ_1, c_1,parameters for back-whitening τ_2, c_2
- $U^{(i)} \leftarrow I; \boldsymbol{c}^{(i)} \leftarrow \boldsymbol{0}; R^{(i)^T} \leftarrow I; t \leftarrow 0$
while ending condition not satisfied **do**
  if $\mathrm{mod}(t, \tau_1) = c_1$ **then**
    - forward-whitening (cf. Algorithm 1).
  end if
  if $\mathrm{mod}(t, \tau_2) = c_2$ **then**
    for all layers i **do**
      - Estimation of $D_{i,i}$.
      - Computation of the back-whitening parameters $\{R_{new}^{(i)^T}\}$.
      - Updating the model parameters $\{(W^{\dagger^{(i)}}, \boldsymbol{b}^{\dagger^{(i)}})\}$.
      - Updating the back-whitening parameters $\{R^{(i)^T}\}$.
    end for
  end if
  - Updating of $\{(W^{\dagger^{(i)}}, \boldsymbol{b}^{\dagger^{(i)}})\}$ by the ordinary gradient descent.
  - $t \leftarrow t + 1$
end while

We will call the above algorithm as "Bidirectional Projected Natural Gradient Descent"(BPRONG). Its outline is shown in Algorithm 2. Also, as in the forward-whitening, we can perform the back-whitening update in a fixed intervals. They can both be done at the same time, or independently. In the following section, we will employ the latter method for a numerical application.

4 Numerical Experiment

In order to see the efficacy of our proposed method BPRONG in Sect. 3.2, we have applied it to a problem of hand-written character (digits) recognition using the MNIST data set (http://yann.lecun.com/exdb/mnist/) and compared against three other methods: ordinary Stochastic Gradient Descent(SGD), Batch Normalization(BN)[5], and PRONG. More specifically, we confirm that (a) better approximating Natural Gradient Descent, thereby decreasing the objective function value further at each step and, (b) the computation time and generalization capabilities are reasonable.

The network architecture is common to all the compared methods with 5 layers of 784-100-100-100-10 neurons from input to output. Also, common learning rate of 0.01 is taken and the mini-batch size is 100. The training data contains 60000 sets and the test data has 10000. We call updates of 600 as 1 epoch, and plot, at each epoch, the training loss with the training set, and the validation loss with the test data sets.

We observe the advantage of BPRONG with respect to the iteration numbers both in the training and the validation losses as shown in Figs. 3 and 4. With respect to the actual computation times, BPRONG is faster than PRONG, and about the same speed as the BN (Figs. 5 and 6). This is due to the fact that eigen-value decomposition associated with the whitening is computationally costly to offset the advantage over BN with respect to iteration numbers.

Altogether, our proposed method, BPRONG, has shown its potential. If we can find methods to speed up the whitening process, BPRONG can gain its effectiveness further.

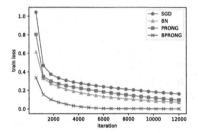

Fig. 3. Training loss as a function of the iteration numbers.

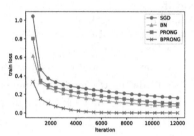

Fig. 4. Validation loss as a function of the iteration numbers.

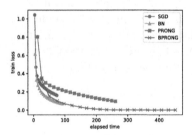

Fig. 5. Training loss as a function of the computational time.

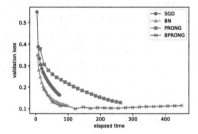

Fig. 6. Validation loss as a function of the computational time.

5   Discussion

We presented here an extended model of the previously proposed Whitened Neural Networks [4] as a method to realize the Natural Gradient Descent. Our extension, which we call Bidirectional Whitened Neural Networks, aims to make the Fisher Information Matrix closer to the identity matrix. It has shown its potential as an efficient method thorough a numerical application to a hand-written digits recognition problem.

We note two points as topics to be investigated further. First, the proposed model should be tested for larger and deeper network architectures for a check of its efficacy and stability. It may require further modifications for improvements on these aspects, particularly by exploring matrix decomposition methods. Secondly, we want to find more dynamical way for the whitening process. In other words, we would like to keep the Fisher Information Matrix constantly closer to the identity by continuous whitenings. Though it is computationally more expensive, we may build on some previous studies, such as adaptive calculations of transforming matrices [3].

Acknowledgement. This work was supported by Grant-in-Aid for Scientific Research from Japan Society for the Promotion of Science KAKENHI No. 16H03360 and No. 16H01175.

References

1. Amari, S.-I.: Natural gradient works efficiently in learning. Neural Comput. **10**(2), 251–276 (1998)
2. Amari, S.-I., Nagaoka, H.: Methods of Information Geometry (Translations of Mathematical Monographs). American Mathematical Society, Providence (2007)
3. Cardoso, J.-F., Laheld, B.H.: Equivariant adaptive source separation. IEEE Trans. Signal Process. **44**(12), 3017–3030 (1996)
4. Desjardins, G., Simonyan, K., Pascanu, R., Kavukcuoglu, K.: Natural neural networks. In: Cortes, C., Lawrence, N.D., Lee, D.D., Sugiyama, M., Garnett, R. (eds.) Advances in Neural Information Processing Systems, vol. 28, pp. 2071–2079. Curran Associates Inc. (2015)
5. Ioffe, S., Szegedy, C.: Batch normalization: accelerating deep network training by reducing internal covariate shift. In: Proceedings of the 32nd International Conference on Machine Learning (ICML-15), pp. 448–456 (2015)
6. Martens, J.: New insights and perspectives on the natural gradient method. arXiv preprint arXiv:1412.1193 (2014)
7. Martens, J., Grosse, R.: Optimizing neural networks with kronecker-factored approximate curvature. arXiv preprint arXiv:1503.05671 (2015)
8. Park, H., Amari, S.-I., Fukumizu, K.: Adaptive natural gradient learning algorithms for various stochastic models. Neural Netw. **13**(7), 755–764 (2000)
9. Pascanu, R., Bengio, Y.: Revisiting natural gradient for deep networks. arXiv preprint arXiv:1301.3584 (2013)
10. Salimans, T., Kingma, D.P.: Weight normalization: a simple reparameterization to accelerate training of deep neural networks. In: Lee, D.D., Sugiyama, M., Luxburg, U.V., Guyon, I., Garnett, R. (eds.) Advances in Neural Information Processing Systems, vol. 29, pp. 901–909. Curran Associates Inc. (2016)

Block Matching Based Obstacle Avoidance
for Unmanned Aerial Vehicle

Adomas Ivanovas[1], Armantas Ostreika[1], Rytis Maskeliūnas[1], Robertas Damaševičius[1],
Dawid Połap[2], and Marcin Woźniak[2(✉)]

[1] Department of Multimedia Engineering, Kaunas University of Technology,
51368 Kaunas, Lithuania
{adomas.ivanovas,armantas.ostreika,rytis.maskeliunas,
robertas.damasevicius}@ktu.lt
[2] Institute of Mathematics, Faculty of Applied Mathematics,
Silesian University of Technology, Gliwice, Poland
{dawid.polap,marcin.wozniak}@polsl.pl

Abstract. Unmanned aerial vehicles (UAVs) are becoming very popular now. They have a variety of applications: search and rescue missions, crop inspection, 3D mapping, surveillance and military applications. However, many of the lower-end UAV do not have obstacle avoidance systems installed, which can lead to broken equipment or people may get injured. In this paper, we describe the design of low-cost UAV with computer vision based obstacle avoidance system. We used Block Match (BM) and Semi Global Block Match (SGBM) algorithms for detection of obstacles in stereo images. We constructed custom UAV platform, and demonstrated the effectiveness of UAV with an obstacle avoidance system in real-world field testing conditions.

Keywords: Stereo vision · Obstacle avoidance · Unmanned aerial vehicle
Drone

1 Introduction

The development and deployment of small unmanned aerial vehicles (UAVs) (aka drones) have become increasingly popular in the last few years. UAV is a semi-autonomous aircraft that can be controlled and operated remotely by using a computer with a radio link [1]. Radio-controlled aircraft are a fast-paced platform. Currently, UAVs already have many different applications: aerial photography, crop photography, intelligence and search missions, 3D mapping and amateur flying. Applications of UAVs include Search and Rescue Operations [2], remote sensing for agricultural applications [3], wildlife research and management [4], and military applications [5]. The UAVs will be an integral part of both military and civilian operations in the future. In these missions, the UAV has to fly at low altitude near the surface of the earth at risk of collision with obstacles such as trees, buildings, and other UAVs. The awareness of the UAV is based on the perception of the immediate environment to estimate collision and to plan its path appropriately. There are various methods to model actions for multi-agent systems [23].

© Springer International Publishing AG, part of Springer Nature 2018
L. Rutkowski et al. (Eds.): ICAISC 2018, LNAI 10841, pp. 58–69, 2018.
https://doi.org/10.1007/978-3-319-91253-0_6

Ensuring a hazard-free, safe flight is a prime concern for UAV operators. It refers to the ability of UAVs to detect and avoid collisions with static and moving obstacles in the environment [6]. Obstacle avoidance is especially relevant for small UASs, which are typically flown at much lower altitudes than standard aircraft. In general we use a model of leader-follower communication system for end-to-end robot communications [20]. Obstacle avoiding is important for detecting and avoiding Air-to-Air Counter Unmanned Aerial Systems (CUAS) [7], which may be deployed by an adversary. To solve this problem, it is necessary to create obstacle avoidance systems. On the other hand diversity in communication systems influence performance [19]. Such systems can use various locating methods and can be implemented using a variety of sensors. A sensor is a device that senses a temperature, pressure, movement, light, magnetic field, or some other effect, and informs about it by an electrical signal. Video, radio waves, ultrasound and light sensors are used to detect and avoid obstacles.

Computer vision has already been proven as an effective method for the prevention of collisions of UAVs with obstacles. Detection of UAVs using computer vision falls into two main categories. Object detection refers to detection of an object without regard to specific features but comparing the shape with other known shapes, while feature detection is concerned with finding specific features (i.e. edges, corners, points or lines) [8]. There are various interesting methods for classification of object from images [24, 25] and detection of important features by locating key points [21, 22].

The solutions to obstacle avoidance by UAVs include the potential field principle [9], random trees [10], Global Position System (GPS) [11], millimeter wave radar [12], the Scanning Laser Range Finder (SLRF) [13], Structure from Motion (SfM) [14], and Deep Convolutional Neural Network (CNN) to predict depth map from a single image [15]. These solutions have both advantages and disadvantages. For example, in SfM-based obstacle avoidance, the drone is limited to short flight hops rather than long-range continuous flying. Time of flight sensors like Light Detection and Ranging (LIDAR) and structured light sensors can be have large weight and power requirements. Sonar and IR are only suitable for very short range (within a meter) depth sensing, and their field of sensing is narrow. Computer visions methods, which use stereo images provide high quality depth, provides there is large enough separation between cameras on the drone surface.

In this paper, we describe the design of UAV and apply computer vision algorithms to UAVs (Unmanned Aerial Vehicles) for path planning and obstacle avoidance.

2   Method

2.1   Video Camera Calibration

In order to generate an undistorted spatial image, it is necessary to perform camera calibratio. Modern pinhole cameras distort the image due to the shape of the lens. When making a photo, the straight lines appear curved, and the image shows the radiant distortion of the lines (Fig. 1).

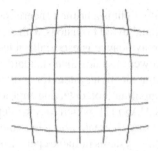

Fig. 1. Camera image distortion

This distortion can eliminated by applying the formulas:

$$x_c = x\left(1 + k_1 r^2 + k_2 r^4 + k_3 r^6\right)$$
$$y_c = y\left(1 + k_1 r^2 + k_2 r^4 + k_3 r^6\right)$$

(1)

here x_c and y_c are the coordinate without distortion, x and y are the coordinates of the distorted image, and k_1, k_2, k_3 are the distortion coefficients.

Tangential distortion that occurs when the lens is not aligned parallel to the imaging plane. This problem is solved using the formulas:

$$x_c = x + \left(2p_1 xy + p_2\left(r^2 + 2x^2\right)\right)$$
$$y_c = y + \left(p_1\left(r^2 + 2y^2\right) + 2p_2 xy\right)$$

(2)

here p_1 and p_2 is the coefficients of tangential distortion.

Therefore, five distortion factors to eliminate distortion of the captured image:

$$K = \left(k_1\ k_2\ p_1\ p_2\ k_3\right)$$

(3)

For calibration, the external and internal camera parameters are required. The internal settings for each camera are different, and include focal length and optical center information, which make the camera matrix. The camera matrix is:

$$\begin{bmatrix} f_x & 0 & c_x \\ 0 & f_y & c_y \\ 0 & 0 & 1 \end{bmatrix}$$

(4)

here f_x, f_y is the focal length, and c_x, c_y is the optical center.

The chess board (Fig. 2) is used for obtaining camera parameters. Using a picture with a chess board can detect camera distortion. The edges of the board and the points of the merger of the squares are represented by coordinates that can be used to determine the distortion level.

Fig. 2. Calibration using chessboard

The chess board template with nine square columns and seven rows of squares is used for calibration. The size of the square is 11 mm, and the size of the board itself is 100×77 cm. For calibration of the camera it is necessary to obtain the 3D coordinates of the board points. This is done by fixing the board at various angles and rotations. Each vector from the center q to the point p, q within the search window is perpendicular to the gradient of the image:

$$M_i = DI_{pi}^T \cdot (q - p_i) \tag{5}$$

here DI_{pi} is the gradient of a picture in one of the dots located near the q point. The q value is found, so that the function of M_i can be minimized. The function system is represented by the formula:

$$\sum_i \left(DI_{pi} \cdot DI_{pi}^T \right) - \sum_i DI_{pi} \cdot DI_{pi}^T \cdot p_i \tag{6}$$

here the gradients are accumulated in the q point scan window. When calling the first gradient value G and the second one, b is obtained by the latter formula:

$$q = G^{-1} \cdot b \tag{7}$$

The algorithm sets the new scan center at q and runs until the center is reached at the specified threshold size.

According to the obtained coefficients, we can calibrate the stereo camera. This method determines the transformation between two cameras. The image of both cameras is deformed so that the epipolar lines of the cameras match each other (Fig. 3).

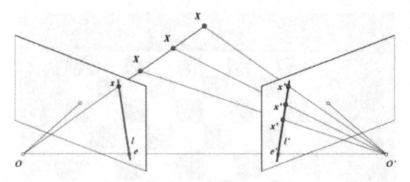

Fig. 3. Epipolar lines of stereo camera

2.2 Block Match Algorithm

Block matching algorithm is a way to find matching macroblocks in digital pictures. This way you can create a cloud of spatial images and points. For this algorithm, two spatial pairs are used that contain block comparisons and calculate the sum of the difference in blocks [16]. The block with the smallest amount indicates the corresponding macroblock of another photo. The size of the macroblock displacement specifies the depth of the spatial image in that macroblock. After all macroblocks are checked, a spatial image is obtained. In order to create a spatial image, two cameras are directed in one direction in parallel with each other. A spatial image is obtained according to the formula:

$$D = x - x' = \frac{Bf}{Z};\qquad(8)$$

here D is a spatial image; x, x' - difference between points in pictures; B - difference between two video cameras; Z - depth; f is the focal length.

The BM algorithm can be implemented by calculating the Macroblock Sums Difference Modulus (SAD) or the Macroblock Sums Difference in Square (SSD). The SAD method is expressed by the formula:

$$\sum_{(i,j)\in W} |I_1(i,j) - I_2(x+i, y+j)|\qquad(9)$$

here i is the x coordinate of the picture, j is the y coordinate of the picture, I_1 is the first picture, I_2 is the second picture, x is the x coordinate of the macroblock, y is the y coordinate of the macroblock. The SSD values are calculated using a similar method:

$$\sum_{(i,j)\in W} \left(I_1(i,j) - I_2(x+i, y+j)\right)^2\qquad(10)$$

The depth of the pixels depends on the distance between the same pixels in different camera shots. This method allows obtaining a spatial image. The generated spatial representation divides the scene into zones of different depths, in which light colors represent areas that are close, and dark areas are far away. A high-precision obstacle avoidance system can be created with the use of high-frequency spatial cameras and a high-speed computer through the block comparison method.

Local Stereo Matching [17] is a dynamic scene-based algorithm considers motion information in dynamic video sequences and adds motion flow for estimating disparity.

2.3 Semi-Global Block Match (SGBM) Algorithm

The SGBM method [18] combines global and local spatial methods. Algorithms use pairs of paintings with known internal and external object orientations for calculations. For calculation of algorithm the formula is used:

$$E(D) = \sum_p \left(C(p, D_p) + \sum_{q \in N_p} P_1 I \left[\left| D_p - D_q \right| = 1 \right] + \sum_{q \in N_p} P_2 I \left[\left| D_p - D_q \right| > 1 \right] \right) \tag{11}$$

here $E(D)$ - the spatial image energy; p and q are pixel indices of the image; N_p is the closest neighbor of the pixel; $C(p, D_p)$ is the value of the pixel identity with depth value D_p; P_1 is a coefficient indicating pixel changes between neighbors equal to one; P_2 is a coefficient indicating pixel changes between neighbors that are more than one; $I[.]$ is function, which returns one if the argument is true, or zero, if it is the opposite.

The minimized function generates a spatial image. The resulting spatial image is more accurate than using the BM method, but the algorithm's calculation time is longer.

3   Design of UAV with Obstacle Avoidance System

3.1   Hardware

For the design of the drone we used the following hardware components: LRP F-1400 Airplane, Odroid XU4 computer, Pixhawk autopilot, Playstation 4 camera, Voltage converter 5.5 V 7 A, Voltage distribution module, Radio receiver, Telemetry transmitter, Mobius camera, GPS receiver, LiPo battery of 2200 by 7.4 V (see Fig. 4).

Fig. 4. Hardware components: Pixhawk controller and Odroid XU4 computer, Playstation 4 eye camera

The Odroid XU4 computer uses two four-core ARM processors (A15 and A7), and has 2 GB DDR3 RAM memory. Odroid XU4 has been installed with Ubuntu server 14.04 operating system. The image processing uses the OpenCV library, which has a number of computer vision algorithms.

The computer and autopilot communicate by the Dronekit software. Dronekit provides the ability to create applications that combine flight control controls and flight computers. Software communicates with the MAVLink protocol that provides access to the telemetry data, parameters, and control of the aircraft. Dronekit can be applied to computer vision, 3D modeling, and flight planning applications.

Pixhawk is an advanced autopilot system designed as an open source project for the PX4 and manufactured by the 3D Robotics company. Autopilot features advanced sensors, ST Microelectronics® processor, and Nuttix real-time operating system. Several peripheral modules can be connected to the Pixhawk module, such as: digital speed sensor, color LED Indicator, external magnetometer. GPS, telemetry and radio are connected to the autopilot, which allows communication with external devices.

The Playstation 4 eye camera has three work modes (1280 × 800 pixels, 60 Hz, 640 × 400 pixels, 120 Hz, 320 × 192 pixels, 240 Hz). Camera uses the Playstation AUX connector, so to connect camera to the Odroid XU4, we had to change the AUX connection to a USB 3.0 port. A USB TTL signal converter is used to connect the computer to the autopilot, which allows the exchange of MAVlink data. The converter is connected to the flight controller's TELEM2 port.

The Flight Controller manages four servo motors and an electronic speed controller. The controller controls the freezer AC motor. The components are powered by a 2,200 mAh 7.4 V LiPo battery. Power is distributed using the energy distribution module. After the modification, a 7 A 5 V voltage converter is connected to the power distribution module. The power converter supplies power to the Odroid computer. The eight-channel Turnigy 9x radio transmitter is used for manual flight management.

Selected components have influenced the selection of the airplane model. The airplane model used is LRP F-1400. The airplane belongs to the "trainer" category because it flies steadily, allowing the airplane to fly for a long time. Also, the propeller is at the top, not the front, which, in case of control error after hitting the ground, will not be damaged. All equipment is fitted in a fuselage and is protected from impact. The video camera is installed at the front of the mountaineer, above the wing, to prevent shocks and damage to the airplane balance center. The original wings were replaced with longer and wider wings in order to generate a higher lifting force. The airplane wings are 180 cm long and the length of the wreath is 95 cm. The model prepared for the flight is presented in the Fig. 5. Total airplane weight with all equipment is 1304 g. The test flight found that the weight of the airplane does not interfere with the performance of the experiment is fast and maneuverable.

3.2 Software

The developed software included algorithms for Camera calibration; Spatial image generation with selected algorithms; Identification of obstacles from the spatial image; and Computer and autopilot communication. Python programming language was used

Fig. 5. UAV model

for implementation, since it allows a smooth interconnection between the camera, the computer and the autopilot. Computer vision algorithms were created using OpenCV 2.4.12.1 library.

Designing an obstacle avoidance system requires the selection of accurate spatial resolution parameters. First, the system was tested in a non-flight mode using an image with an object located ten meters away, and parameters of the system were calibrated. The objects in the spatial image are displayed at a certain intensity. Objects were marked as obstacles if the intensity of pixels ranged from 0.1 to 0.2 (see Fig. 6).

Fig. 6. Detection of obstacles: the red spots mark the detected obstacle on the road. (Color figure online)

4   Experimental Validation of Obstacle Avoidance System

4.1   Experimental Setting

The experiment is conducted in an open space with some obstacles on the flight path. The autopilot uses Mission Planner software, which provides the flight path that the aircraft will fly before the task is completed. An airplane has been planned an autonomous flight path (GPS: lat = 54.9369, lon = 23.8674). Flight path is selected so that the aircraft will fly to the height of five meters (Fig. 7). The path was completed three times.

Fig. 7. Flight plan (left) and actual flight (right)

To test the avoidance experiment, several parameters were set as follows: Window size - 319 × 192; Algorithm - "Block Match"; Parallel calculation - off.

The control and switching of obstacle avoidance algorithms is carried out by the Turnigy 9x radio transmitter using the eighth radio channel. The airplane sent back the result of obstacle detection (an image) as well as several camera and flight parameters such as frames per second; Delay, ms; Speed, m/s; Height, m; GPS coordinates.

Weather conditions during the experiment were: Air temperature - 20 °C; wind speed - 2 m/s; visibility - good; precipitation - 0%; sky - clear.

4.2   Results

The experiment was performed smoothly. The aircraft flight path was followed without major deviations. Actual flight height met the planning requirements. Using the Mobius video camera a live video has been recorder, while some shots are given in Fig. 8. Average speed of the drone during the flight was 11.8 m/s (min - 8 m/s, max - 30 m/s), while mean height during the flight was 5.19 m. This speed is sufficient for testing obstacle avoidance system. Average delay of the obstacle avoidance algorithm was 19.3 ms, which gives us a minimum obstacle distance of 0.22 m.

The image of the obstacles found during the obstacle avoidance is shown in the Fig. 9. In the image, there are two obstacles identified (shown in red color).

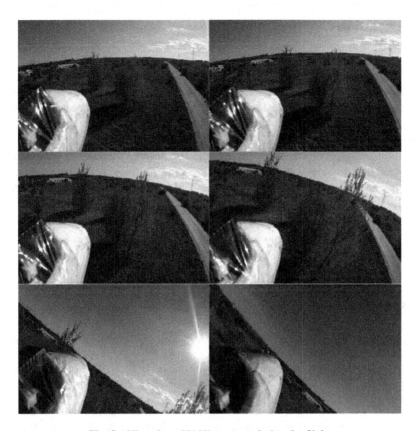

Fig. 8. View from UAV's camera during the flight

Fig. 9. Obstacles captured real-life UAV flight (Color figure online)

5   Conclusion

We have described the development of low-cost unmanned aerial vehicle (UAV) with the obstacle avoidance system. The system uses stereo vision based Block matching algorithm. The algorithm managed to identify and avoid all obstacles in the planned flight path of the UAV. Average drone speed during obstacle avoidance experiment was 11.8 m/s, average flight height was 5.19 m, frame interval 33.12 ms, and window size for image analysis was 319×192 pixels.

In future work, we plan to further explore the possibilities for the application of object detection methods to further develop and improve the system of obstacle avoidance. Linking ultrasound sensors to spatial obstacle avoidance techniques could increase the likelihood and accuracy of detecting obstacles.

Acknowledgements. The Authors would like to acknowledge contribution to this research from the "Diamond Grant 2016" No. 0080/DIA/2016/45 funded by the Polish Ministry of Science and Higher Education.

References

1. Austin, R.: Unmanned Aircraft Systems: UAVS Design, Development and Deployment. Wiley, Hoboken (2010)
2. Naidoo, Y., Stopforth, R., Bright, G.: Development of an UAV for search & rescue applications. In: AFRICON 2011, Livingstone, pp. 1–6 (2011). https://doi.org/10.1109/afrcon.2011.6072032
3. Zhang, C., Walters, D., Kovacs, J.M.: Applications of low altitude remote sensing in agriculture upon farmers' requests-a case study in northeastern Ontario, Canada. PLoS One 9(11), e112894 (2014). https://doi.org/10.1371/journal.pone.0112894
4. Jones, G.P., Pearlstine, L.G., Percival, H.F.: An assessment of small unmanned aerial vehicles for wildlife research. Wildl. Soc. Bull. **34**(3), 750–758 (2006)
5. Cermak, P., Martinu, J.: Component based design of mini UAV. In: International Conference on Military Technologies, ICMT 2015, pp. 1–5 (2015). https://doi.org/10.1109/miltechs.2015.7153714
6. Ashraf, A., Majd, A., Troubitsyna, E.: Towards a realtime, collision-free motion coordination and navigation system for a UAV fleet. In: Rysavy, O., Vranić, V., Papadopoulos, G.A. (eds.) Proceedings of the Fifth European Conference on the Engineering of Computer-Based Systems (ECBS 2017), Article no. 11, p. 9. ACM, New York (2017). https://doi.org/10.1145/3123779.3123805
7. Goppert, J.M., Wagoner, A.R., Schrader, D.K., Ghose, S., Kim, Y., Park, S., Gomez, M., Matson, E.T., Hopmeier, M.J.: Realization of an autonomous, air-to-air counter unmanned aerial system (CUAS). In: First IEEE International Conference on Robotic Computing (IRC), Taichung, pp. 235–240 (2017)
8. Wagoner, A.R., Schrader, D.K., Matson, E.T.: Survey on detection and tracking of UAVs using computer vision. In: First IEEE International Conference on Robotic Computing (IRC), Taichung, pp. 320–325 (2017). https://doi.org/10.1109/irc.2017.15

9. Budiyanto, A., Cahyadi, A., Adji, T.B., Wahyunggoro, O.: UAV obstacle avoidance using potential field under dynamic environment. In: International Conference on Control, Electronics, Renewable Energy and Communications, ICCEREC 2015, pp. 187–192 (2015). https://doi.org/10.1109/iccerec.2015.7337041

10. Borenstein, J., Everett, H.R., Feng, L., Wehe, D.: Mobile robot positioning: sensors and techniques. J. Robot. Syst. **14**(4), 231–249 (1997)

11. Jian, L., Xiao-min, L.: Vision-based navigation and obstacle detection for UAV. In: 2011 International Conference on Electronics, Communications and Control, pp. 1771–1774 (2011)

12. Kwag, Y.K., Choi, M.S., Jung, C.H., Hwang, K.Y.: Collision avoidance radar for UAV. In: 2006 CIE International Conference on Radar, pp. 1–4 (2006). https://doi.org/10.1109/icr.2006.343231

13. Luo, D., Wang, F., Wang, B., Chen, B.M.: Implementation of obstacle avoidance technique for indoor coaxial rotorcraft with scanning laser range finder. In: Proceedings of the 31st Chinese Control Conference, Hefei, pp. 5135–5140 (2012)

14. Mader, D., Blaskow, R., Westfeld, P., Maas, H.: UAV-based acquisition of 3D point cloud - a comparison of a low-cost laser scanner and SFM-tools. Int. Arch. Photogrammetry Remote Sens. Spatial Inf. Sci. - ISPRS Arch. **40**(3W3), 335–341 (2015). https://doi.org/10.5194/isprsarchives-xl-3-w3-335-2015

15. Chakravarty, P., Kelchtermans, K., Roussel, T., Wellens, S., Tuytelaars, T., Van Eycken, L.: CNN-based single image obstacle avoidance on a quadrotor. In: IEEE International Conference on Robotics and Automation, pp. 6369–6374 (2017). https://doi.org/10.1109/icra.2017.7989752

16. Je, C., Park, H.-M.: Optimized hierarchical block matching for fast and accurate image registration. Sign. Process.: Image Commun. **28**, 779–791 (2013)

17. Yang, J., Wang, H., Ding, Z., Lv, Z., Wei, W., Song, H.: Local stereo matching based on support weight with motion flow for dynamic scene. IEEE Access **4**, 4840–4847 (2016). https://doi.org/10.1109/ACCESS.2016.2601069

18. Hirschmüller, H.: Stereo processing by semi-global matching and mutual information. IEEE Trans. Pattern Anal. Mach. Intell. **30**(2), 328–341 (2008)

19. Esmaeili, A., et al.: The impact of diversity on performance of holonic multi-agent systems. Eng. Appl. Artif. Intell. **55**, 186–201 (2016)

20. Min, B.-C., et al.: A directional antenna based leader–follower relay system for end-to-end robot communications. Robot. Auton. Syst. **101**, 57–73 (2018)

21. Łągiewka, M., Korytkowski, M., Scherer, R.: Distributed image retrieval with color and keypoint features. In: 2017 IEEE International Conference on INnovations in Intelligent SysTems and Applications (INISTA). IEEE (2017)

22. Najgebauer, P., Rutkowski, L., Scherer, R.: Novel method for joining missing line fragments for medical image analysis. In: 2017 22nd International Conference on Methods and Models in Automation and Robotics (MMAR). IEEE (2017)

23. Esmaeili, A., et al.: A socially-based distributed self-organizing algorithm for holonic multi-agent systems: case study in a task environment. Cogn. Syst. Res. **43**, 21–44 (2017)

24. Gabryel, M., Damaševičius, R.: The image classification with different types of image features. In: Rutkowski, L., Korytkowski, M., Scherer, R., Tadeusiewicz, R., Zadeh, L.A., Zurada, J.M. (eds.) ICAISC 2017. LNCS (LNAI), vol. 10245, pp. 497–506. Springer, Cham (2017). https://doi.org/10.1007/978-3-319-59063-9_44

25. Grycuk, R., et al.: Content-based image retrieval optimization by differential evolution. In: 2016 IEEE Congress on Evolutionary Computation (CEC). IEEE (2016)

Prototype-Based Kernels for Extreme Learning Machines and Radial Basis Function Networks

Norbert Jankowski[✉]

Department of Informatics, Nicolaus Copernicus University, Toruń, Poland
norbert@is.umk.pl

Abstract. Extreme learning machines or radial basis function networks depends on kernel functions. If the kernel set is too small or not adequate (for the problem/learning data) the learning can be fruitless and generalization capabilities of classifiers do not become rewarding.

The article presents a method of automatic stochastic selection of kernels. Thanks to the proposed scheme of kernel function selection we obtain the proper number of kernels and proper placements of kernels. Evaluation results clearly show that this methodology works very well and is superior to standard extreme learning machine, support vector machine or k nearest neighbours method.

Keywords: Extreme learning machines · Kernel methods
Prototypes · Prototype selection · Machine learning
k nearest neighbours method

1 Introduction

For simplicity assume we consider classification problems which base on the learning data $\mathcal{D}$ which consists of the learning vectors $\mathbf{x}_i$ ($\mathbf{x}_i \in R^n, i \in [1, \ldots, m]$) with corresponding class labels y_i ($\mathbf{y} = [y_1, \ldots, y_m]$).

The Extreme learning machines (ELM) [1,2] and Radial Basis Function Networks (RBFN) [3] are general learning scheme:

$$F(\mathbf{x}; \mathbf{w}) = \sum_j w_j g_j(\mathbf{x}),\tag{1}$$

where w_j are weights and $g_j(\mathbf{x})$ are kernel functions ($j \in [1, l]$). l defines the number of kernels. Sigmoidal function was the original kernel in the ELM but it can be also the gaussian function (most typical for RBFN) or another kernel as well [2].

The learning of ELM (estimation of $\mathbf{w}$) can be defined by kernel selection and the minimization of the goal:

$$J(\mathbf{w}) = ||\mathbf{G}\mathbf{w} - \mathbf{y}||^2 = \sum_{i=1}^{m} \left(\sum_{j=1}^{l} w_j g_j(\mathbf{x}_i) - y_i \right)^2 \tag{2}$$

© Springer International Publishing AG, part of Springer Nature 2018
L. Rutkowski et al. (Eds.): ICAISC 2018, LNAI 10841, pp. 70–75, 2018.
https://doi.org/10.1007/978-3-319-91253-0_7

over the kernels. Let's define $G_{ij} = g_j(\mathbf{x}_i)$. After $\nabla J(\mathbf{w}) = 0$ we obtain

$$\mathbf{w} = (G^T G)^{-1} G^T \mathbf{y} = G^\dagger \mathbf{y} \tag{3}$$

Pseudo-inverse matrix $G^\dagger$ can be efficiently computed by singular value decomposition (SVD) algorithm in $O(ml^2)$ time complexity. The kernels can be constructed at random or as (especially in gaussian kernel case) a subset of vectors of $\mathcal{D}$. In the second case we obtain equivalence with the original radial basis function network [3]. To keep the learning time really efficient we should try to use as few kernels as possible, because complexity depends on square of kernel number and linearly with instance count. It can be noted that in nontrivial learning problems the gaussian kernel in EML can be slightly more efficient [4].

The second type of algorithms used in next sections are the algorithms of prototypes selection. By prototype selection we means selection of vectors from the learning data $\mathcal{D}$. The selected subset of $\mathcal{D}$ must still be enough to use with the k nearest neighbours algorithm for efficient classification. Such selection can be seen as selection of most important instances from $\mathcal{D}$. For an overview of prototype selection algorithm please see [5,6]. This time we will focus on the DROP2 algorithm of prototype selection proposed in [7]. The main idea of the DROP2 algorithm lies in the definition of a set $\mathcal{A}(\mathbf{x}, k)$ which consists of neighbours of $\mathbf{x}$ for which $\mathbf{x}$ is of k nearest neighbours. The scheme of DROP2 is presented below:

```
1  function DROP2(D,k)
2      do {
3          foreach xᵢ in D in dist−order
4              delete xᵢ if it will not change classification of instance
5                  from A(xᵢ, k)
6      } while(changes in D)
7      return D;
8  end
```

The main concept of DROP2 is to delete all vectors which do not change the classification of the rest of $\mathcal{D}$. This idea produce the definition of set $\mathcal{A}$ which simplifies the test of classification changes to few elements of $\mathcal{D}$ in contrary to testing of the whole $\mathcal{D}$. The 'dist-order' defines the descending order of instances (in $\mathcal{D}$) to their nearest enemy (to the nearest instance from opposite class). The outer loop usually iterates a few times. The inner loop iterates for each instance of $\mathcal{D}$. The default complexity is $O(m^2 n)$. The reduction of $\mathcal{D}$ is quite strong, for detail please see [7] or [6].

The next section presents the main idea of proposed algorithm and motivations. Further sections are devoted to the analysis of the new algorithm on several data benchmarks and comparison with best known classification algorithms.

2 Prototype-Based Kernels for Extreme Learning Machines

The main goal of the algorithm presented in this section base on few remarks:

- To construct efficient learning algorithm for ELM without the need of manual tuning of free parameters. Currently the complexity is $O(ml^2 + m^2 n)$, but it can be done in estimated complexity $O(ml^2 + nm \log m)$. Additionally we try to keep $O(l^2)$ to be smaller than $O(m)$ (in other case it is hard to use this algorithm for huge data sets).
- To construct learning algorithm which automatically selects kernels (the placements of gaussian functions) and their number.

In case of standard ELM or RBFN the numer of kernels must be estimated manually or (for example) in inner cross-validation. To eliminate this disadvantage we can first start the DROP2 algorithm to selects prototypes which will be next used to construct kernels for the ELM or RBFN. The scheme of such algorithm:

```
1  function ProtoELM(𝒟)
2      [x'₁,...,x'ₗ] = DROP2(𝒟);  // prototypes selected from D
3      Gᵢⱼ = gⱼ(xᵢ)
4      pinvG = svd_pseudo_inv(G);
5      w = pinvG y
6      return kernels g and weights w
7  end
```

where $g_j(\mathbf{x})$ is the (gaussian) kernel placed in $\mathbf{x}'_j$.

Such scheme of a learning algorithm guaranties that the construction of kernels is done in one shot and next we can simply focus on the learning phase of ELM/RBFN on selected kernels as in normal ELM/RBFN with random gaussian kernels.

3 Comparison of New Algorithm with Others

To present a trustful comparison of algorithm we have take around 40 data sets from the UCI machine learning repository [8] devoted on classification problems. Data sets differ in origin, goal, in the number of instances, features and classes, to presents objectively real behavior of proposed algorithm.

All tests were conducted on the base of 10 times repeater 10-fold cross-validation. For each test the data set is standardized. The new algorithm (DROP2-NN) was compared with linear discriminant (LDP) learned by pseudo-inverse, extreme learning machines with sigmoidal kernel, RBFN/ELM with gaussian kernel (ELMg), k Nearest Neighbours ($k = 5$, KNN), a Support Vector Machine with a linear kernel (L-SVM) and with a gaussian kernel (SVM).

Each learning algorithm was always used with the same learning parameters (no manual parameter tuning was done). The ELMs and RBFN were learned with 160 random kernels based on randomly selected vectors of given benchmark data set. The kNN was used with $k = 5$ and Euclidean metric. Each parameter of SVM was fixed in learning.

To visualize the performance of all algorithms we present average accuracy for each benchmark data set and for each learning machine. Additionally we present the *rank* for each machine for given data set. The rank is calculated as follows. First for a given data set $\mathcal{D}$ the averaged accuracies of all leaning machines are sorted in descending order. Then the first machine with highest accuracy is ranked 1. Additionally next machines whose results were not statistically different (paired t-test was used) are also ranked with 1. Then the first machine which result was statistically different starts the next group with rank equal to 2, and again the following machines whose results were not statistically different are also ranked with 2, and so on. Thanks to the concept of the rank we recognize not only winners and defeated, but more groups depending on really significant differences. It help to see how deeply given machine defeats another in the meaning of statistical differences and selected learning algorithms.

Notice that each cell of the main part of Table 1 is in form: $acc + std(rank)$, where acc is average accuracy (for given data set and given learning machine), std is its standard deviation and $rank$ is the rank describe just above. If a given cell of the table is in bold it means that this result is the best for given data set or not worse then the best one (rank 1 = winners).

The last three rows of Table 1 present cumulative results. The first of those rows presents averaged accuracies over all data sets and their standard deviations. The *mean rank* row presents most significant information about ranks of machines—for each machine its average rank over all data sets is presented with standard deviation. And the third row presents the numer of wins (how many times the given machine was the best or was not significantly worse) for the given machine and in brackets the number of unique wins. By the unique machine win we mean that all other machine are significantly different (worse). In simpler words if only winer machine achieved the best score (or statistically not different), it is a unique winner.

From the *mean rank* row we can find that best mean rank (1.98) is assigned to the new method (column DROP2-NN). Second best results (2.07) is assigned to RBFN and third to LDA (2.82). But in the *wins* row we see that the new method wins 19 times, RBFg wins 16 times and linear discriminant wins 15 times. On next places are L-SVM, ELM, SVM and kNN. Also the number of unique wins (6) is biggest in case of the new method, and 4 unique wins has SVM. Here we can see that in 3 cases the RBFg has significantly better results over 40 data sets. It means that at least in those 3 cases the prototype selection could be working better.

It is quite clear that the proposed algorithm outperforms the rest of the learning machines and there is still a gap to do something in a better way. The reader should notice that it was not obvious that connection of prototype selection with ELM/RBFN will be fruitful. In case of connection of prototype selection with the SVM the results were worse [6] than using a SVM without using prototype selection.

74   N. Jankowski

Table 1. Comparison of Prototype-based kernels for ELM/RBFN with LDA, ELM, RBFg, kNN, linear SVM and SVM with gaussian kernels

	Drop2-NN	LDA	ELM	RBFg	kNN	L-SVM	SVM
arrhythmia	32±8.9(4)	**53±20(1)**	26.1±17(5)	44.4±16(3)	**52.4±16(1)**	49.7±20(2)	0±0(6)
autos	72.2±12(2)	64.1±10(4)	67.2±9.6(3)	**81.7±11(1)**	62.6±12(4)	53±11(6)	57±12(5)
balance-scale	**90.8±1.9(1)**	86.6±2.7(4)	86.6±2.7(4)	**90.8±1.9(1)**	87.6±2.9(3)	84.5±3.1(5)	89.6±2(2)
blood-transfusion	79.1±3.6(2)	77.3±1.9(3)	77±2.1(3)	**79.5±3.7(1)**	76.3±4.2(4)	76.1±0.62(5)	76.8±2(4)
breast-cancer-diagnostic	96.6±2.4(2)	95.7±2.7(3)	94.9±2.9(4)	**97.4±2.2(1)**	**97±2.1(1)**	**97.3±2.2(1)**	96.2±2.5(3)
breast-cancer-original	96.8±2(2)	96±2(4)	96.3±2(3)	96.1±2.2(3)	96.7±1.9(2)	96.6±2(2)	**97±2.1(1)**
breast-cancer-prognostic	77.4±7(2)	**80±8.1(1)**	78.3±8.6(2)	72.7±8.9(4)	76.2±6.3(3)	**80.5±8.3(1)**	76.6±3.7(2)
breast-tissue	61.7±14(3)	66.2±13(2)	**68±12(1)**	54.6±16(4)	65.9±13(2)	43.6±8.5(5)	42.3±8.4(6)
car-evaluation	90.9±2.2(3)	84.2±2(5)	84.2±2(5)	92.4±1.7(2)	**93.1±1.4(1)**	82.2±3.5(6)	88±2(4)
cardiotocography-1	**84.5±2.2(1)**	66.4±2.8(6)	67.2±3.1(5)	80.9±2.4(2)	75.1±2.7(3)	58.2±2.7(7)	70.5±2.8(4)
cardiotocography-2	**92.7±1.6(1)**	86.5±1.8(7)	86.8±2(6)	91.4±1.9(2)	90.8±1.8(3)	87.4±1.9(5)	90.4±1.8(4)
chess-rook-vs-pawn	**98.9±0.64(1)**	94.1±1.4(6)	94±1.5(6)	95±1.2(4)	94.6±1.2(5)	96.8±0.98(3)	98.3±0.76(2)
cmc	49±4.3(3)	50.4±3.6(2)	50.4±3.9(2)	**53.4±4.1(1)**	46.8±4(4)	18.7±2.8(6)	30.6±3(5)
congressional-voting	96.5±3.7(2)	**97±3.6(1)**	**97±3.6(1)**	95.3±4.4(3)	92.1±5.1(4)	95.4±4.7(3)	96.3±3.9(2)
connectionist-bench-sonar	**84.5±6.6(1)**	75.2±9.7(4)	74.1±10(4)	**84.9±7.5(1)**	81.3±7.6(2)	74.6±9(4)	78.4±6.9(3)
connectionist-bench-vowel	**96.7±2.5(1)**	47.6±5.5(5)	47.7±5.5(5)	95.4±3.2(2)	93.4±3.4(3)	25.7±4.1(6)	60.9±4.9(4)
cylinder-bands	68.9±4.4(3)	**74.5±7.1(1)**	64.5±8.1(5)	70.3±5.9(2)	62±8(6)	**75.1±6.9(1)**	66.7±3(4)
dermatology	**95.7±3.1(1)**	95±3.4(2)	95±3.5(2)	**95.7±3(1)**	92.5±3.6(4)	93.4±3.9(3)	86.7±4.9(5)
ecoli	**86.5±5.4(1)**	84.8±5.1(2)	84.8±5.1(2)	**86.1±5.2(1)**	85.6±4.7(2)	76.1±6.2(4)	83.1±5.3(3)
glass	**66±9.8(1)**	60.8±9.6(3)	62.1±9.7(2)	**65±9.2(1)**	**65.8±8(1)**	36.4±7(5)	56.8±7.9(4)
habermans-survival	**74±5.5(1)**	**74.2±4.2(1)**	**74.2±4.2(1)**	73.6±5.7(1)	71.1±6.5(4)	72.6±2.5(3)	73.4±3.8(2)
hepatitis	84.4±12(3)	83.1±11(3)	83.1±11(3)	**89.9±10(1)**	87±11(2)	81.6±10(4)	**88±8.3(1)**
ionosphere	93.2±3.9(2)	86.4±4.1(4)	86.4±4.6(4)	93.2±3.8(2)	84.5±4.6(5)	88.4±4.6(3)	**94.7±3.6(1)**
iris	**96.9±4.2(1)**	83±8.2(3)	83±8.2(3)	95±6.5(2)	95±5.6(2)	78.1±8.6(4)	**96.2±5.3(1)**
libras-movement	**86±5.7(1)**	57.7±7.6(5)	63.1±7.3(4)	84.4±6(2)	75.8±5.8(3)	49.5±6.4(6)	46.7±7.6(7)
liver-disorders	68.4±7.2(2)	68.9±6.9(2)	69±7(2)	67.8±7(2)	61.8±8.1(3)	69.1±7.3(2)	**71±7.2(1)**
lymph	**84.3±8.9(1)**	**83.7±8.7(1)**	**83.8±8.8(1)**	82.1±9.7(2)	80.1±9.7(2)	80.4±9.3(2)	79±9.4(3)
monks-problems-1	99.9±0.69(2)	74.6±4.5(4)	74.6±4.5(4)	99.9±0.35(2)	99.6±0.95(3)	74.6±4.5(4)	**100±0(1)**
monks-problems-2	59.5±6.9(3)	63.1±2.9(2)	63.1±2.9(2)	62.2±7.3(2)	54.5±5.6(4)	**65.7±0.79(1)**	60.5±4.2(3)
monks-problems-3	98.8±1.6(2)	96.4±2.5(3)	96.4±2.5(3)	98.8±1.5(2)	**98.9±1.5(1)**	**98.9±1.5(1)**	**98.9±1.5(1)**
parkinsons	89.3±6.3(2)	88.6±6.9(2)	88.4±7(2)	**92.1±6.4(1)**	**91.3±6.3(1)**	86.9±7.5(3)	89±5.7(2)
pima-indians-diabetes	72.5±4.6(4)	**77.3±4.6(1)**	**77.3±4.5(1)**	73.2±4.6(3)	74±4.8(3)	**77±4.5(1)**	76.1±4.6(2)
sonar	**84.5±6.6(1)**	75.2±9.7(4)	74.1±10(4)	**84.9±7.5(1)**	81.3±7.6(2)	74.6±9(4)	78.4±6.9(3)
spambase	91.2±1.2(3)	88.7±1.4(7)	89.9±1.4(6)	90.6±1.2(5)	90.9±1.4(4)	**92.9±1.1(1)**	91.6±1.4(2)
spect-heart	**82.8±6.1(1)**	83.4±5.3(1)	83.2±5.5(1)	82.1±6.9(2)	81.8±6.6(2)	81.7±6.1(2)	82.4±6.1(2)
spectf-heart	79.9±6.8(1)	77.2±5.8(2)	76.9±7.3(2)	**79±7.7(1)**	72.8±6.5(3)	**79.1±7.5(1)**	78±4.4(2)
statlog-australian-credit	84.2±4(3)	**85.6±4.1(1)**	85.1±4.3(2)	84.8±4.7(2)	79.6±5.4(5)	84.8±4(2)	82.9±4.2(4)
statlog-german-credit	74.2±4.3(4)	**76.9±3.8(1)**	76.5±3.9(2)	75.2±4(3)	72.4±4(5)	**76.6±3.9(1)**	75.3±3.3(3)
statlog-heart	81±7.4(3)	**84.3±7(1)**	**84.3±7(1)**	76.4±9.1(4)	82.1±7.1(2)	**83.7±7.1(1)**	82.9±6.9(2)
statlog-vehicle	**83.4±3.8(1)**	75.6±4.2(3)	77.1±4.2(2)	**83.3±3.8(1)**	73±4.1(4)	68.2±4.5(5)	65.8±3.8(6)
teaching-assistant	51±11(3)	**59.5±12(1)**	**60.1±12(1)**	**61.3±13(1)**	42.4±12(4)	53.9±11(2)	39.7±11(5)
thyroid-disease	**96.2±0.51(1)**	93.3±0.29(6)	93.6±0.26(5)	95.5±0.56(2)	94.9±0.46(3)	93.7±0.35(4)	95.4±0.41(2)
vote	**96.8±3.2(1)**	**97±3.1(1)**	**97±3.1(1)**	94.3±4(2)	92.1±5.4(3)	**96.9±3.2(1)**	**96.9±3.1(1)**
wine	97.8±3.5(2)	**98±2.4(1)**	**98.9±2.4(1)**	93.7±8(4)	96.7±3.7(3)	96.4±3.8(3)	**98.3±2.8(1)**
zoo	61.8±12(4)	**94.8±5.7(1)**	92.8±7.3(2)	71.6±13(3)	40.3±9.4(5)	**93.8±5.9(1)**	35.2±12(6)
Mean Accuracy	82±5.2	79.2±5.6	78.5±5.7	82.4±5.9	79.2±5.6	75.6±5.5	76±4.6
Mean Rank	1.98±0.15	2.82±0.27	2.89±0.24	2.07±0.16	3.02±0.19	3.16±0.27	3.04±0.25
Wins[unique]	19[6]	15[1]	10[1]	16[3]	6[1]	12[2]	9[4]

4 Summary

The Proposed algorithm in place of random selection of kernels for ELM use the stochastic prototype selection algorithm (DROP2) and then the selected prototypes compose the kernels for ELM learned by SVD as well. Thanks to this concept the new algorithm automatically selects kernels and its number. There is no manual tuning of the number of kernels. What's more there is no manual tuning of any other parameter in the new algorithm—it is a big advantage (no need for inner cross-validation). It is good alternative for cost manual or automatic (for example via inner CV learning) estimation of number of kernels.

Additionally the current complexity of new algorithm is $O(ml^2 + m3)$, however we work on new version of the DROP2 algorithm whose complexity should be reduced from $O(m^2n)$ to an estimated complexity $O(nm \log_2 m)$, then this algorithm will be useful even for huge data sets.

References

1. Huang, G.B., Zhu, Q.Y., Siew, C.K.: Extreme learning machine: a new learning scheme of feedforward neural networks. In: International Joint Conference on Neural Networks, pp. 985–990. IEEE Press (2004)
2. Huang, G.B., Zhu, Q.Y., Siew, C.K.: Extreme learning machine: theory and applications. Neurocomputing **70**, 489–501 (2006)
3. Broomhead, D.S., Lowe, D.: Multivariable functional interpolation and adaptive networks. Complex Syst. **2**, 321–355 (1988)
4. Kasun, L.L.C., Zhou, H., Huang, G.B.: Representational learning with ELMS for big data. IEEE Intell. Syst. **28**(6), 31–34 (2013)
5. Jankowski, N., Grochowski, M.: Comparison of instances seletion algorithms I. Algorithms survey. In: Rutkowski, L., Siekmann, J.H., Tadeusiewicz, R., Zadeh, L.A. (eds.) ICAISC 2004. LNCS (LNAI), vol. 3070, pp. 598–603. Springer, Heidelberg (2004). https://doi.org/10.1007/978-3-540-24844-6_90
6. Grochowski, M., Jankowski, N.: Comparison of instance selection algorithms II. Results and comments. In: Rutkowski, L., Siekmann, J.H., Tadeusiewicz, R., Zadeh, L.A. (eds.) ICAISC 2004. LNCS (LNAI), vol. 3070, pp. 580–585. Springer, Heidelberg (2004). https://doi.org/10.1007/978-3-540-24844-6_87
7. Wilson, D.R., Martinez, T.R.: Reduction techniques for instance-based learning algorithms. Mach. Learn. **38**, 257–286 (2000)
8. Merz, C.J., Murphy, P.M.: UCI repository of machine learning databases (1998). http://www.ics.uci.edu/~mlearn/MLRepository.html

Supervised Neural Network Learning with an Environment Adapted Supervision Based on Motivation Learning Factors

Maciej Janowski and Adrian Horzyk[(✉)]

AGH University of Science and Technology,
Mickiewicza Av. 30, 30-059 Krakow, Poland
maciejanowski@gmail.com, horzyk@agh.edu.pl

Abstract. This paper introduces an innovative approach for supervised learning systems in cases when we do not have initially defined training data sets, but we need to develop them gradually during training process on the basis of the motivation factors that come from the given environment. We suppose to gradually develop and update knowledge about the environment and use it for supervision of training MLP. In the beginning, the gradually gained knowledge does not have to be correct, but it allows to adapt a neural network still better and more efficiently in time. It is illustrated on the problem of acquiring the ability to return to the starting position optimally by a virtual robot from anywhere in an initially unknown and gradually explored maze. The proposed approach focuses on the attempt to reflect human cognitive abilities and motivation factors in an introduced model using artificial neural networks. This article presents a new approach in which the decision-making method arises from the supervised learning process controlled by the knowledge gained during maze exploration. This paper presents a model of maze exploration and knowledge-based adaptation of the neural network. The experimental results of the classical supervised learning approach and the proposed modified approach will be compared to demonstrate significant improvements.

Keywords: Neural network · Motivated learning
Supervised learning · Environment adapted supervision
Knowledge-based learning · Knowledge development
Brain-inspired computations · Cognitive systems

1 Introduction

This article further develops the research implemented in the engineering work [1] of the coauthor of this paper in accordance with the algorithmic learning strategies and ideas of the second coauthor. It contains the gained results and the most important conclusions as well as its further development based on the

© Springer International Publishing AG, part of Springer Nature 2018
L. Rutkowski et al. (Eds.): ICAISC 2018, LNAI 10841, pp. 76–87, 2018.
https://doi.org/10.1007/978-3-319-91253-0_8

problems encountered and ways to solve them. In the engineering work mentioned above, the virtual robot was motivated to achieve the assumed goal, i.e. gain the ability to return from anywhere in the environment to the place of battery charging. Two needs of the robot were defined to motivate the robot to move smartly [2]: (1) the need to explore and learn its environment and (2) the need to maintain the battery level above the critical level. The introduced algorithm for this robot can be applied and adapted to many other problems - presented maze problem only illustrates the approach.

There are often issues connected to robot navigation caused by the lack of an external positioning signal (e.g. GPS). In this case, the robot must independently create an environment representation and learn to navigate in it. One of the approaches is the use of Wi-Fi Fingerprinting [3]. It focuses on the analysis of signals sent by access points and then on the basis of the calculation results determines the robot's position relative to known coordinates of signal senders. This approach, despite wide applications, is not applicable in rooms with a large number of obstacles, walls, windows or doors. What is more, it is not possible to use in places without any Wi-Fi signals.

This article focuses on creating a new approach based on supervised learning to solve the group of problems mentioned above, which potential applications can be found in the exploration of places that are not easily accessible to humans, including planetary, underwater or underground exploration.

This is an extension of the Richard Morris' water maze experiment analyzing the rats' ability of spatial learning, by motivating animals to reach the labyrinth end platform [4]. According to [5], there are three basic strategies to accomplish the task: (1) remembering the movements necessary to achieve the platform, (2) using the sense of sight to navigate and (3) using reference points to locate oneself.

In this paper, strategy (1) has been applied which is based on the building of implicit memory on the basis of episodic memory [6], i.e. personal experiences [7].

Memory is the ability of the living organisms to acquire or create information about the environment, store them, use them when needed, and remove them when unnecessary [8]. The most complex way of creating memory is learning through understanding that is characteristic of humans, chimpanzees, interestingly crows and many other species [9]. It involves defining the nature of the problem and creating a solution. For less complex organisms, the term "learning through imitation"? has been defined. As the name suggests, it is based on presenting correct behaviors to the individual, who will then try to repeat them [10,11]. This method of learning was adapted as supervised learning [12,13].

The need to explore the environment is related to the cognitive concept of curiosity [14], which has been described in support of earlier comparison to the behavior of rats in the maze [15]. As studies [16,17] show, rats, despite the lack of motivation in the form of thirst or hunger, display a desire to recognize unvisited sections of the maze. Curiosity has been described as a perceptible tension due to unknown stimuli or difficulty in recognizing [18]. The result of this tension is the desire to reduce it through exploration. Exploration is a desire to acquire new information - knowledge [19,20]. In the described problem, a robot placed in an unknown environment is endowed with the need to satisfy its curiosity. It achieves this by visiting all sections of the maze guided by the motivation to visit the fields visited the least number of times from its neighborhood.

In this research, the classic feedforward neural network MLP [12,21,22] with the logistic neurons [23] was used. The network input layer consists of two neurons that receive coordinates (x, y) of the robot in the maze. The output layer consists of four neurons, which represent directions of its movement of the von Neumann neighborhood (east, west, north, and south). Based on the results supplied by the genetic algorithm [24,25], the topology of the neural network was defined by two hidden layers with 70 and 40 neurons respectively.

2   Limitations of the Supervised Learning Algorithms

Today, supervised learning algorithms are widely used to many computational intelligence tasks [26]. The supervised approach is the most often used, but it requires the tutor's knowledge [19] which defines a correct transformation from input data to the output results [22]. In most cases, it is possible to define such data sets, however, such an approach is not convenient and smart. We still need human support to define such data sets to be able to train a chosen method and find a solution. It stands in contradiction with the assumptions of artificial intelligence which should be able to self-adapt on the basis of the data coming from its surroundings and some motivation factors coming from its needs.

The most important contribution of this paper is a dynamic construction and gradual improvement of a training set for the supervised learning algorithm stimulated by motivation factors coming from the defined robot's needs and the knowledge gained during exploration of the robot's surroundings. The standard application of the supervised learning algorithm approach would require the need to analyze the robot's environment and define the training data sets in the form of the tuples (a robot position, a direction of optimal movement). Such an approach would not develop any cognitive skills of a robot. When the robot builds a training data set independently on the basis of its experience, it acquires recognition skills close to human. What is more, it gives potential applications in the unknown environment thanks to its adaptive skills. In this paper, we try to develop new methods and trends in supervised learning [27].

3   Environment Based Motivated Learning Factors

An environment model was created in order to implement the algorithms of environment recognition and return from a given position to the initial position. The environment is defined as a grid of fields, on which the robot can move, and the fields which are unobtainable, i.e. containing obstacles. The tutor forms knowledge about each field in the explored environment. The knowledge is represented by the number of visits of each field and a value specifying the current lowest known number of necessary steps to return to the starting position, according to the current exploration experience and the tutor's state of knowledge. On this basis, a matrix representing the mentioned maze by the grid of fields is created. An exploration matrix was created containing information on the number of field visits in the previous expeditions and determining the direction of environment recognition. In turn, the return matrix represents the tutor's knowledge about the environment and contains information about the number of steps needed to return to the starting position and determines the return direction for training the neural network of the robot.

The following algorithm describes how the values of both matrices are updated and how the directions for one exploration and return route are chosen:

1. Determine the maximum number of steps during one exploration journey, create and reset the counter of steps.
2. Create matrices with dimensions equal to the given dimensions of the environment. In the real solution, the initial matrices size should be assumed. Next, during the exploration, the matrices size should be increased when necessary.
3. Reset the values responsible for the number of visits in each field to zero, set the values responsible for the number of steps to the maximum value.
4. For the exploration matrix, the number of visits in the fields occupied by obstacles should be set to the maximum value. In the real solution, the robot will recognize the obstacle thanks to its sensors and only then will write to the given matrix field the maximum value. In the virtual environment, the recognition of the obstacles is replaced by reading the maximum value from the field representing the position of the obstacles.
5. Place the robot in the starting position.
6. Recognize the number of visits in each of the neighboring fields.
7. Select the minimum value and in the case when several fields have the minimum values, select the field randomly.
8. Move the robot in the selected direction and increment the counter of steps.
9. Increment the number of visits in the exploration matrix. If the value of the visited field in the return matrix is greater than the value of the counter of steps, the value of the field responsible for the number of steps is overwritten by the value of the counter.
10. Check if the value of the battery level has reached its critical level. If it does not, go to step 6. Otherwise, go to the return procedure (step 11).
11. Recognize the number of steps in each of the neighboring fields.
12. Select the minimum value and in the case when several fields have the minimum values, select the field randomly.
13. Move the robot in the selected direction.
14. Check if the value of the number of steps in the currently occupied field is equal to 0 if not, return to step 12.
15. Terminate the return algorithm and reset the counter of steps.

4   Environment Adapted Supervision of Neural Network

Modification of the learning process was based on the analogy to the way people learn a new vocabulary, i.e.:

The first phase of learning consists of the gradual collection of training data sets, which are created by the teacher and teaching them the neural network. As described, the response to some of this data will be changed in the following steps. The neural network at this stage learns "easy" examples, and for some "difficult" ones (changed in subsequent iterations or for which the network response is nonstandard, such as bypassing the obstacle) is not sure of the answer. We use the analogy: new but easy vocabulary is remembered quickly, but with difficult ones, problems still arise.

The teacher's virtual memory is created when the initial learning phase is over, i.e. after which the network is assumed to have problems only with the "difficult" examples. The teacher writes in it, with which examples the student has a problem, that is, for which input data is made a mistake bigger than assumed one. In the next learning iterations, these examples are reminded. We use the analogy with focusing on learning difficult vocabulary.

The last and shortest stage consists of releasing the teacher's memory, that is teaching without recalling. We can also find out an analogy with a random repetition of examples of the entire trained vocabulary.

The curve of the error function for the proposed learning model has been presented in Fig. 3. The trend line shows the superiority of this approach. Despite the less effective environment recognition (large initial changes in the response given by the teacher), the final error is almost half as much. Both graphs shown in Figs. 2 and 3 have the same scale of the ordinates and the initial value of the abscissa for easier comparison of the data.

5   Experiments Applied to a Robot in a Certain Environment

In this section, an example of the exploration route (Fig. 1a) is described and used to train the robot using the created application. Yellow arrows on the exploration matrix show the next directions chosen at each step. The stop sign marks the place where the robot starts the return algorithm. The route designated by the return algorithm is marked on the return matrix representation with green arrows. The stop sign indicates the place where the algorithm ends when reaching the starting position. Black fields represent fields which were not yet visited. The designated route is based on the knowledge of the previous expeditions and is as short as possible.

Using the above-introduced symbols, the next exploration route is presented in Fig. 1b. The exploration algorithm forces the robot to choose the "down" direction due to the difference in the number of visits. This specific route was selected to illustrate the essence of the return algorithm and the way of gathering knowledge.

We can notice that the route chosen by the return algorithm is not optimal (Fig. 1b). Instead of starting the return by selecting the "up" direction and as a result reaching the starting position in five steps, it selects "down" direction and returns in seven steps. Well, this is because the algorithm considers the

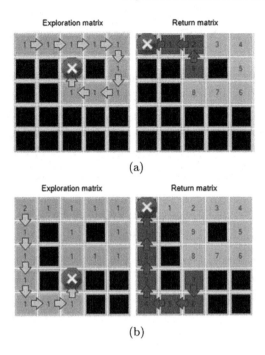

Fig. 1. Figures (a) and (b) show exemplary exploration and return routes and their representation matrices.

direction locally in its neighborhood - based only on the limited knowledge from the previous expeditions. It decides to move down because the field below the stop position (marked with the stop sign on the exploration matrix) has been reached in a smaller number of steps than one above.

The visualization of the algorithms that guide the robot's teacher during the selection of directions of movement reveals a significant problem. The initial return routes set by the teacher (training data) can be considered to be suboptimal. The teacher needs to perform many journeys to learn the neural network an optimal route based on the return matrix. Such a way of learning is very casual for people, especially children who learn from experience, not from a book of true samples.

The experiments were carried out for the same virtual environment. It was planned as a grid of 8×10 fields, where about 10% of the fields were occupied by obstacles of various shapes, which created problematic fields of a similar amount.

The learning result is shown in Fig. 2 based on the calculated total network error function (1). This error is determined after each expedition and return, in which the robot is taught, as the sum of errors of the network response and the expected response given by the teacher to each of the possible sets of input signals.

$$Q_{total} = \sum_{j=0}^{m-1} \sum_{i=0}^{n-1} \gamma(i,j) * \sum_{k=0}^{l-1} |z_k(i,j) - y_k(i,j)| \qquad (1)$$

$$\gamma(x,y) = \begin{cases} 1 & \text{if exploration matrix } (x,y) \text{ value is equal to zero} \\ 0 & \text{otherwise} \end{cases} \qquad (2)$$

where n and m are the dimensions of the matrix representation of the environment, l is the number of output neurons, $z_k(i,j)$ is the expected response of the neural network for k-th neuron, and $y_k(i,j)$ is the actual response of the network for input signals (i, j).

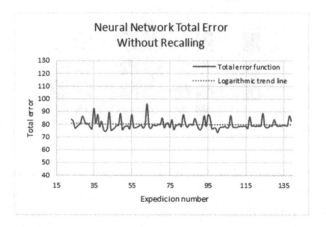

Fig. 2. The curve of the total error value based on (1) without recalling the problematic positions, i.e. changed in subsequent iterations or for which the network response is non-standard, such as bypassing the obstacle.

Error values are calculated after the discovery of all possible fields. It was performed in order to avoid the initial values obscuring the results, e.g.: after the first expedition the error is very small because few fields have been visited and the transfer of this knowledge to the network it is not burdened with a big error.

The characteristic spikes of the error function values (Figs. 2 and 3) only mean that during the return from a given expedition, the teacher recognized a better route and passed a new response to the observations. In analogy, the student during their studies needs to change his knowledge so far, so initially, he makes many mistakes using his previous knowledge.

6   Adaptive Knowledge-Based Learning

For the assumed experimental environment, after analysis of characteristics of the average total network error from many simulations, the parameters of the backpropagation algorithm were selected. In this experiment, we tried to implement a similar mechanism that works in human brain [28] on the basis of associations, associative neural systems [29] and the knowledge formed in an associative way [30] as in biological nervous systems [31]. In the experiment with a maze, the associations are taken from the proximity of the neighbor fields in a maze.

The learning rate, which determines the speed of changes in weights and thresholds in subsequent learning iterations, was chosen as 0.2. The selection of a larger parameter, which was determined based on the analysis of the error function for various parameters, leads to a longer learning time without significantly better results. During the entire learning process, the network error decreases very slowly, although there are large changes in successive iterations. Moreover, the correction of the wrongly chosen weights and thresholds along with subsequent iterations of the algorithm is becoming less and less effective. This is due to the setting of parameters with high values. Choosing a smaller parameter is the correct approach. The character of the average error chart has remained regular, and its value is gradually reduced. However, this approach requires a significant extension of the simulation time.

The momentum parameter determines the dependence on changes introduced in the previous iteration. The use of small parameters, which was also determined on the basis of the error function analysis for various parameters, does not lead to significant changes. Characteristics of functions present tendencies to frequent changes. For a parameter equal to 0.9, a significant improvement in the character of the function was observed after the estimated iteration, in which the teacher acquired the correct knowledge and did not introduce large changes in the student's knowledge. Therefore, this parameter value was selected.

It was also necessary, based on the conducted experiments, to estimate the number of exploratory expeditions necessary for the teacher to achieve knowledge that would ensure optimal decisions in most cases. To achieve this, the optimal total number of steps to return to the starting position from all allowed fields has been determined. Then, a series of simulations were carried out, for which the values of this sum were saved for random exploration runs. Finally, the average exploration number was determined, after which the sum of the number of steps written in the return matrix represents about 95% of the optimal sum of the number of steps. The estimated number of journeys was 40.

Another variable of the algorithm was the criterion of the classifying position as problematic. For each training sequence, weights and thresholds were adapted during the five learning iterations. After the last iteration, the average error (3) was calculated and compared to the assumed classification error value. If the error made by the network was greater than that value, the given learning sequence was saved in the teacher's memory. This classification error value was chosen based on several assumptions. In the output layer of the network, as mentioned, there are four neurons, each of which represents one of the movement directions. This direction is chosen for which the neuron value is the largest. In order for the choice to be considered certain, the neuron's response (belonging to the interval $(0, 1)$) must be at least 0.5. It was therefore assumed that the neuron response associated with the correct direction must also be at least equal to 0.5. The answers of the remaining neurons must be smaller. To obtain a significant difference between the incorrectly selected directions and the correct one maximum response value for the three remaining neurons was chosen as 0.35 (70% of minimum value for correct direction). Under such assumptions, the average network error is about 0.4, and this is an error dividing the positions for which the answer is well-defined and for which the answer is incorrect or uncertain.

$$Q_{average}(x, y) = \frac{\sum_{k=0}^{l-1} |z_k(x, y) - y_k(x, y)|}{l} \tag{3}$$

With such defined parameters, the total network error value, after which the network is able to determine the optimal return routes from each allowed environment position, was estimated. A total error value below 50 guarantees that this condition will be met. This error, in turn, is achieved on average in 70 exploratory expeditions. The course of the above-described learning process is shown in Fig. 3.

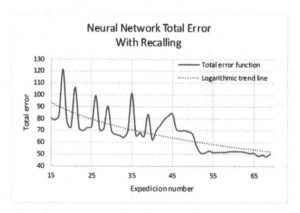

Fig. 3. The curve of the total error value based on (1) with recalling the problematic positions, i.e. changed in subsequent iterations or for which the network response is non-standard, such as bypassing the obstacle.

Charts in Fig. 4 present representations of the return matrices using the previously introduced symbols for a couple of experiments. Figure 4a shows the maze model, which was used to conduct initial experiments and on the basis of which all parameters of the learning algorithm and network structure were determined. Figure 4b shows the maze model introduced for validation test. Figure 4c and d show the applications in more complex mazes without parameters adjustment. All figures show correctly designated return routes from the chosen positions (green fields).

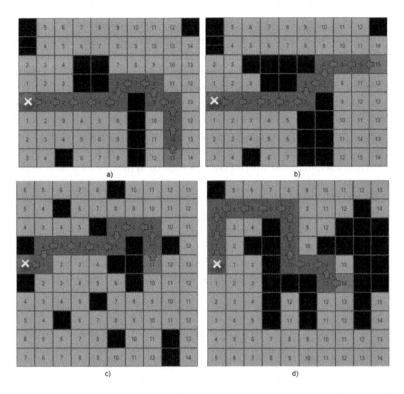

Fig. 4. Representation of the return matrices for four different experiments using the presented approach: (a) shows the basic maze used for the majority of tests, (b) shows the more complex maze introduced as the validation experiment, (c) and (d) show the most complex maze models. (Color figure online)

7   Conclusions and Remarks

This paper described new knowledge-based learning to determine the optimal return route to the initial position in a new environment. The adaptation algorithm was based on the classical supervised learning algorithm for MLP networks. The environment adapted supervision based on the motivation learning factors [32] was used to the dynamically developed and updated training data. Motivated factors cannot always be trustful [33], so must be selected with care or checked and updated as in the algorithm presented in this paper.

The proposed adaptation strategy and the modification of the supervised learning algorithm were based on the cognition of surroundings, the defined needs as motivation factors, and the properties of the memory mechanism: learn easy relationships quickly and repeat difficult ones to remember them correctly. All steps of this algorithm and the parameters necessary to determine or estimate were described in details. The results of the classic and proposed approaches were compared to identify new benefits coming from this new approach. In the case of a variable training data set, the classic approach does not solve the problem. The described modification not only guarantees success, but it also allows to define accuracy by changing the error value classifying the position as problematic.

The major achievement of the described approach makes it possible to exclude human influences from the process of learning completely to solve the problem by algorithmization of the parameters determination, i.e.: the number of journeys necessary to achieve the teacher's knowledge of the optimal routes, the maximum qualifying average error and the minimum total error of the network guaranteeing success. This paper showed that it was necessary to combine training with knowledge and motivation factors to achieve autonomic intelligent robot skills.

The described cognitive approach enables to automatically collect information about the maze using motivated learning, form the necessary knowledge and use it to control classic supervised training algorithms to learn the neural network the optimal solutions.

This work was supported by AGH 11.11.120.612 and the grant from the National Science Centre DEC-2016/21/B/ST7/02220.

References

1. Janowski, M.: Use of a neural network to move a robot back to the starting position in a maze using supervised learning. Engineering dissertation at the AGH University of Science and Technology in Krakow supervised and promoted by A. Horzyk (2018)
2. Tadeusiewicz, R., Horzyk, A.: Man-machine interaction improvement by means of automatic human personality identification. In: Saeed, K., Snášel, V. (eds.) CISIM 2014. LNCS, vol. 8838, pp. 278–289. Springer, Heidelberg (2014). https://doi.org/10.1007/978-3-662-45237-0_27
3. Jiang, P., Zhang, Y., Fu, W., Liu, H., Su, X.: Indoor mobile localization based on Wi-Fi fingerprint's important access point. Int. J. Distrib. Sens. Netw. **11**, 429104 (2015)
4. Nunez, M.J.: Water maze experiment. J. Vis. Exp. JoVE **19**, 897 (2008). https://doi.org/10.3791/897
5. Brandeis, R., Brandys, Y., Yehuda, S.: The use of the Morris water maze in the study of memory and learning. Int. J. Neurosci. **48**(1–2), 29–69 (1989)
6. Horzyk, A., Starzyk, J.A., Graham, J.: Integration of semantic and episodic memories. IEEE Trans. Neural Netw. Learn. Syst. **28**(12), 3084–3095 (2017). https://doi.org/10.1109/TNNLS.2017.2728203
7. Eichenbaum, H.: A cortical-hippocampal system for declarative memory. Nat. Rev. Neurosci. **1**, 41–50 (2000)
8. Kandel, E.R., Schwartz, J.H., Jessell, T.M., Siegelbaum, S.A., Hudspeth, A.J. (eds.): Principles of Neural Science. McGraw-Hill, New York (2013)
9. Heinrich, B., Bugnyar, T.: Just how smart are ravens? Sci. Am. **296**, 64–71 (2007)
10. Fiorito, G., Scotto, P.: Observational learning in Octopus vulgaris. Science **256**, 545–547 (1992)
11. Loukola, O.J., Perry, C.J., Coscos, L., Chittka, L.: Bumblebees show cognitive flexibility by improving on an observed complex behavior. Science **355**, 833 (2017)
12. Rutkowski, L.: Techniques and methods of artificial intelligence. PWN (2012)

13. Nielsen, M.A.: Neural Networks and Deep Learning. Determination Press (2015). http://neuralnetworksanddeeplearning.com/

14. Duch, W.: Brain-inspired conscious computing architecture. J. Mind Behav. **26**, 1–22 (2005)

15. Berlyne, D.E.: Novelty and curiosity as determinants of exploratory behavior. Br. J. Psychol. **41**, 68–80 (1950)

16. Dember, W.N.: Response by the rat to environmental change. J. Comp. Physiol. Psychol. **49**, 93 (1956)

17. Hughes, R.N.: Behaviour of male and female rats with free choice of two environments differing in novelty. Anim. Behav. **16**, 92–96 (1968)

18. Berlyne, D.E.: Conflict, Arousal, and Curiosity. McGraw Hill, New York (1960)

19. Larose, D.T.: Discovering knowledge from data. Introduction to Data Mining. PWN, Warsaw (2006)

20. Day, H.I.: Curiosity and the interested explorer. Nonprofit Manage. Leadersh. **21**, 19–22 (1982). https://doi.org/10.1002/pfi.4170210410

21. Hecht-Nielsen, R.: III.3 - Theory of the backpropagation neural network. In: Wechsler, H. (ed.) Neural Networks for Perception, pp. 65–93. Academic Press (1992)

22. Hertz, J., Krogh, A., Palmer, R.G.: Santa Fe Institute studies in the sciences of complexity: lecture notes, vol. 1 and computation and neural systems series. In: Introduction to the Theory of Neural Computation. Addison-Wesley/Addison Wesley Longman, Reading (1991)

23. Ke, Q., Oommen, B.J.: Logistic neural networks: their chaotic and pattern recognition properties. Neurocomputing **125**, 184–194 (2014)

24. Davis, L.D.: Handbook of Genetic Algorithms. Van Nostrand Reinhold, New York (1991)

25. Holland, J.: Adaptation in Natural and Artificial Systems, pp. 89–120. University of Michigan Press, Ann Arbor (1975)

26. Tadeusiewicz, R.: Introduction to Intelligent Systems, Fault Diagnosis. Models, Artificial Intelligence, Applications. CRC Press, Boca Raton (2011)

27. Tadeusiewicz, R.: New trends in neurocybernetics. Comput. Methods Mater. Sci. **10**(1), 17 (2010)

28. Horzyk, A.: How does generalization and creativity come into being in neural associative systems and how does it form human-like knowledge? Neurocomputing **144**, 238–257 (2014)

29. Horzyk, A.: Artificial associative systems and associative artificial intelligence, pp. 1–276. EXIT, Warsaw (2013)

30. Horzyk, A.: Human-like knowledge engineering, generalization, and creativity in artificial neural associative systems. In: Skulimowski, A.M.J., Kacprzyk, J. (eds.) Knowledge, Information and Creativity Support Systems: Recent Trends, Advances and Solutions. AISC, vol. 364, pp. 39–51. Springer, Cham (2016). https://doi.org/10.1007/978-3-319-19090-7_4

31. Kalat, J.W.: Biological Grounds of Psychology. PWN, Warsaw (2006)

32. Starzyk, J.A.: Motivated learning for computational intelligence. In: Computational Modeling and Simulation of Intellect: Current State and Future Perspectives. IGI Publishing, pp. 265–292 (2011). Red. B. Igelnik

33. Starzyk, J.A., Graham, J., Horzyk, A.: Trust in motivated learning agents. In: IEEE Symposium Series on Computational Intelligence, Greece, Athens, ISBN 978-1-5090-4239-5. IEEE Xplore (2016). https://doi.org/10.1109/SSCI.2016.7850027

Autoassociative Signature Authentication Based on Recurrent Neural Network

Jun Rokui[(⊠)]

Department of Mathematics and Computer Science Interdisciplinary,
Faculty of Science and Engineering, Shimane University,
Nishikawatsu chou 1060, Matsue, Shimane 690-8504, Japan
rokui@cis.shimane-u.ac.jp
http://www.cis.shimane-u.ac.jp/

Abstract. In online handwriting authentication, it is difficult to forge handwriting because stroke characteristics cannot be reproduced using only a handwriting trajectory. However, it is difficult to completely reproduce registered stroke characteristics, even when signers attempt to reproduce their own signatures. For this reason, the principal criteria for authentication must be lowered. In this study, we use a recurrent neural network to model the behavior of the musculoskeletal function for handwriting. The proposed model can represent the handwriting stroke process for a character visualized by the authenticator. This research is an anti-counterfeit effort to reduce the error between autoassociative stroke information and handwriting stroke information.

Keywords: Online handwriting authentication
Recurrent neural network · BPTT

1 Introduction

Online handwriting authentication is a personal authentication system that uses writing stroke traits as a biological feature [1,2]. In online handwriting authentication, it is not only difficult to forge signatures, but also to completely reproduce the registered stroke traits, even in cases in which a signer attempts to reproduce their own handwriting. Self-error, which is the fluctuation of a signer's stroke characteristics, makes it difficult to distinguish between genuine writing and false writing.

Improvements in handwriting skills, and the loss of those skills over time, typically cause these fluctuations. Therefore, to improve authentication performance, it is considered effective to reduce self-error. In the motion control process of neurophysiology, when a human moves, the body calculates a trajectory of purposeful behavior on a real space. After the human brain converts these trajectories into motion signals via musculoskeletal combinations, it propagates the motion signals in each musculoskeletal area [4]. Human musculoskeletal systems operate according to the target trajectory and kinetic instructions. The target trajectory is converted into kinetic instructions by giving teacher signals to

© Springer International Publishing AG, part of Springer Nature 2018
L. Rutkowski et al. (Eds.): ICAISC 2018, LNAI 10841, pp. 88–96, 2018.
https://doi.org/10.1007/978-3-319-91253-0_9

Purkinje cells in the cerebellar cortex from the olive nucleus under the medulla oblongata. One research effort studied the similarity between the cerebellum and perceptrons, based on the neural structure and behavior of the cerebellum [5]. Further, it has been reported that purposeful behavior empirically has an inverse dynamics model for learning the conversion systems of kinetic instructions and target trajectories [5]. Handwriting is a purposeful behavior using the forearm. Another research effort physiologically modeled the brain's motion control process associated with handwriting [6]. It is reasonable to assume that the conversion system of the target trajectory and kinetic instructions associated with handwriting has an optimized individuality determined by physical characteristics and experience. In fact, there is research that applies predictions about behavior patterns to personal authentication on smartphones, assuming that there is individuality in the surface EMG (electromyography) signal when the kinetic instruction from the musculoskeletal system occurs [3]. In this research, we model a system that converts kinetic instructions and the target trajectory for handwriting strokes by using observed stroke information, and propose an authentication method that uses handwriting trait data extracted by the model. Recurrent neural networks (RNN), along with long short-term memory cells, currently hold the best known results in unconstrained handwriting recognition [11]. In online handwriting methods used for recognizing words in the Mongolian language, RNNs achieved higher recognition performance than that provided by hidden Markov models (HMMs) [12]. In this paper, an RNN has been used as an autoassociative model to represent a time series of handwriting stroke operations. We propose a method to efficiently extract user handwriting traits from a handwriting stroke time series, using an RNN with an autoassociative structure.

2   Autoassociative Recurrent Neural Network Authentication (ARNN)

Online handwriting authentication is roughly divided into a learning phase and an authentication phase. In the learning phase, features are extracted from genuine handwriting after it has been normalized. A certifier model is learned from the features of the genuine handwriting, and is registered as a principal reference. In the authentication phase, features are extracted from stroke information after it has been normalized. The similarity index is calculated using the registered certifier model of these features. Authenticity is determined by comparing a predetermined threshold with the calculated similarity index.

2.1   Normalization and Feature Extraction

Handwriting stroke information is normalized to reduce fluctuations in the writing. Sampled stroke information is denoted by $sign$.

$$sign = \{x, y\} \tag{1}$$

$$x = x_1, \cdots, x_T \tag{2}$$

$$y = y_1, \cdots, y_T \tag{3}$$

T is the total number of samplings. x is the x-coordinate. y is the y-coordinate. x_t, y_t is sampling to t-th data. x'_t, y'_t after the area normalization process is as follows:

$$x'_t = \frac{x_t - \min_i x_i}{\max_i x_i - \min_i x_i}(i = 1, \cdots, T) \tag{4}$$

$$y'_t = \frac{y_t - \min_i y_i}{\max_i y_i - \min_i y_i}(i = 1, \cdots, T) \tag{5}$$

Even if the stroke information is from the same signer, the length of the time series data will vary according to differences in writing time. Time series normalization is a linear transformation of handwriting stroke information. Self-error is reduced by aligning the time series length.

$$x'_m = \frac{\sum_{i=mK}^{(m+1)K} x_{i+1}}{K}(m = 1, \cdots, M) \tag{6}$$

$$y'_m = \frac{\sum_{i=mK}^{(m+1)K} y_{i+1}}{K}(m = 1, \cdots, M) \tag{7}$$

$$K = \frac{T}{M} \tag{8}$$

M is the time series length after the time series normalization.

2.2   Reference Modeling

In this study, we have modeled the reference by using a recurrent neural network (RNN). An RNN is a neural network that interconnects units other than the input units to each other [7]. RNNs can process time series data by interconnecting the units. Figure 1 shows an overview of the ARNN proposed in this study.

In the proposed RNN, hidden units and output units are interconnected using self-connections. In this study, backpropagation through time (BPTT) was used as a learning algorithm [8]. The handwriting stroke information from input units is given as a supervised signal.

There are N input units in the input layer, and each unit is given the corresponding feature vector time series. Input value $u_i^{in}(t)$ of the input unit at time t is as follows:

$$u_i^{in}(t) = x_t \tag{9}$$

The output unit receives the input vector itself at the time the model is learned. The self-prediction vector applied to the input unit is output every time authentication is performed. There are N^{out} input units in the input layer. Each unit is given the corresponding feature vector time series. Back-propagated values $\sigma_i^{out}(t)$ in output value $u_i^{out}(t)$ of the output unit at time t are calculated as follows:

$$\sigma_i^{out}(t) = u_i^{out}(t) - x_t \tag{10}$$

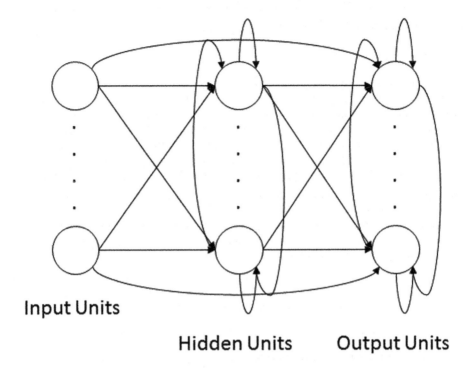

Fig. 1. Proposed ARNN Overview

In the past, feature vectors used for online handwriting authentication were extracted from handwriting stroke information that had been normalized. RNNs directly extracted feature vectors from the handwriting stroke time series. The proposed model extracts the elements in order from the beginning of the time series data, and outputs the prediction vector to associate the reference. The internal state of the model affects the next prediction through the forward propagation of elements. $x_i(t)(t = 1, \cdots , T)$ is the M-dimensional feature vector of stroke information. $y_i(t)(t = 1, \cdots , T)$ is the output of the model for $x_i(t)$. At this time, the prediction error *error* is calculated as follows:

$$error = \sum_t^T \sqrt{\sum_i^M (y_i(t) - x_i(t))^2} \tag{11}$$

Acceptance or non-acceptance is determined by comparing the threshold and *error* set in advance for each user.

3   Experimental Results

3.1   Experimental Conditions

For testing, this study used an experimental data set from SVC2004, a signature verification competition[1]. SVC2004 provided a database containing signatures from five users. For each user, there are 20 genuine handwriting samples and 20 "skilled forgeries." To account for the effects of forgetting over time, the signatures in the data set were collected at two different times. First, ten signatures were collected for each user; after a grace period of one week, the second set of ten was collected. A Wacom digital tablet was used for the signature collection process. X and Y-coordinate values, and the writing pressure at the time of pen-down, are recorded for each sampling time. In this study, we used the X and Y coordinates, the X-direction pen speed, and the Y-direction pen speed as the features. x_t, y_t are the elements of the feature vector x, y at time t. At this time, pen speeds vx_t, vy_t are calculated as follows:

$$vx_{t+1} = x_{t+1} - x_t \tag{12}$$

$$vy_{t+1} = y_{t+1} - y_t \tag{13}$$

3.2   Authentication Experiment

In this section, we describe the authentication accuracy verification results. A hidden Markov model (HMM) [9], dynamic time warping (DTW) [10], and a neural network (NN) [13, 14] were compared in this verification. Each model was learned by registering the genuine writing references into a data set. Five examples of genuine handwriting were selected as a training reference; the remainder of the genuine handwriting examples was used as test data. Because the data set was relatively small, the experiment was conducted using a cross-validation method. We prepared two types of models for the proposed autoassociative RNN. One model, NARNN, was given time-normalized stroke information. The other model, ARNN, was given stroke information that was not time-normalized. HMM and NN were given time-normalized stroke information. The HMM was

Table 1. Model internal parameters

	NARNN, ARNN	NN
Input unit	1	10
Hidden unit	100	100
Output unit	1	10
Learning rate	0.1	0.1
Epoch	5000	2000

---

[1] SVC2004, http://www.cs.ust.hk/svc2004/.

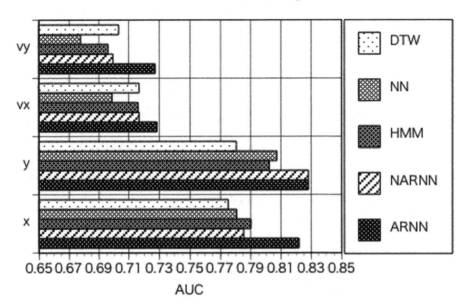

Fig. 2. AUC of ARNN, NARNN and HMM, NN and DTW

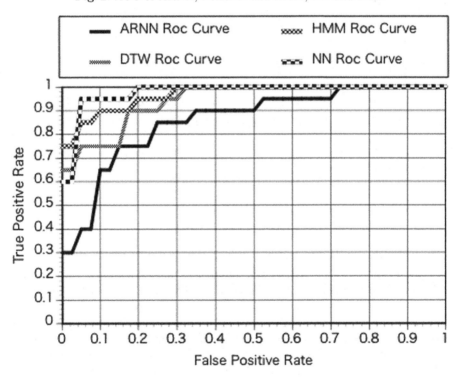

Fig. 3. X-Coordinate ROC curve for the features of the signer with the lowest AUC, according to ARNN

configured as a three-state, left-to-right type of HMM. The NN employed a three-layer configuration. Table 1 shows various internal parameters for the NARNN, ARNN, and NN models.

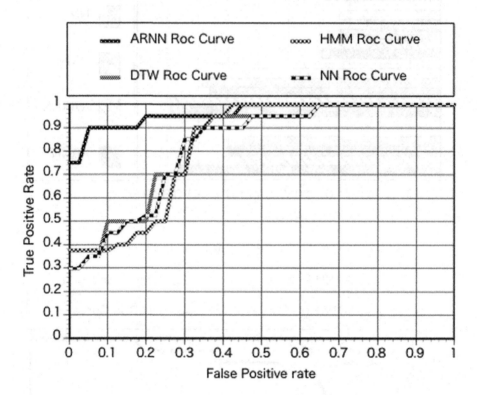

Fig. 4. X-Coordinate ROC curve for the features of the signer with the highest AUC, according to ARNN

Authentication accuracy was evaluated using the area under the curve (AUC) of a receiver operating characteristic (ROC) curve [15]. The AUC represents a complete classification at 1.0, and a random classification at 0.5.

Figure 2 shows the authentication accuracy verification results. The results of this experiment show that ARNN has the highest authentication accuracy of all the models. The authentication accuracy of NARNN was less than, or comparable to, that of HMM. It is highly probable that part of the time series information has been lost, owing to the influence of time normalization. Because NARNN and ARNN have common internal parameters, NARNN requires a parameter adjustment in consideration of the time normalization. ARNN provides high authentication accuracy without the need for such detailed parameter adjustments. Figure 3 shows the X-coordinate ROC curve for the features of the signer with the lowest AUC, as determined by the ARNN. Figure 4 shows the X-coordinate ROC curve for the features of the signer with the highest AUC, as determined by the ARNN. HMM and NN, which used time-normalized stroke information, describe similar trajectories. DTW, which used stroke information that had not been time normalized, has the lowest overall AUC. The ROC curve of the ARNN shows a trajectory that is different from those of the other models. In the past, time normalization was an effective preprocessing technique to accommodate signers with significantly different stroke operations than other signers. In other words, signers with extreme stroke operations can easily be discriminated by applying the general model. The autoassociative structure with ARNN can use time information to effectively extract personal signing traits for signers with traits that are difficult to extract.

4   Conclusion

In this paper, we modeled handwriting stroke operations based on cognitive neuroscience knowledge, and applied it to online handwriting authentication. Because movement instructions given to the musculoskeletal system constitute a time series signal, we proposed an autoassociative recurrent neural network (ARNN) based on a recurrent neural network suitable for time series signal utilization. Through experimental results using a public signature database, it was found that the ARNN provided high authentication accuracy for all types of feature vectors, compared with other well-known models used for comparison. Through verifications performed on ROC curve trajectories, it was found that each model extracts different features, even when using the same reference. The results of the experiments clearly show that the ARNN provides excellent feature extraction performance for signers with difficult-to-extract personal writing traits. The ARNN has the advantage of not requiring time normalization. However, ARNN cannot deliver superior performance for valid, time-normalized signatures by signers with significantly different stroke operations. Future research efforts will analyze methods of extracting a wider range of personal writing traits by combining two or more models.

Acknowledgments. In this research, I would like to thank S. Sugawara for receiving great advice on data collection and RNN construction.

References

1. Muramatsu, D., Kondo, M., Sasaki, M., Tachibana, S., Matsumoto, T.: A Markov chain Monte Carlo algorithm for bayesian dynamic signature verification. IEEE Trans. Inf. Forensic Secur. **1**(1), 22–34 (2006)
2. Koishi, K., Kinoshita, S., Muramatsu, D., Matsumoto, T.: Online signature verification based on user-generic fusion model with Markov Chain Monte Carlo, taking into account user individuality. J. Adv. Comput. Intell. Intell. Inf. **13**(4), 447–456 (2009)
3. Tsukamoto, M., Kaneoya, T., Mano, W.: Verification of humans using the Musculo-Skeletal model with electromyograms. J. IEICE J **J97–A**(11), 672–682 (2014)
4. Kosslyn, S.: On cognitive neuroscience. J. Cogn. Neurosci. **6**(3), 297–303 (1994)

5. Marr, D.: A theory of cerebellar cortex. J Physiol. (Lond) **202**, 437–470 (1969)
6. Taguchi, H., Fujii, K.: A functional description of brain mechanics in the writing movement control. J. IEICE **J70–D**(3), 640–649 (1987)
7. Fukuda, O., Tsuji, T., Bu, N., Kaneko, M.: Pattern discrimination of time series EEG signals using a recurrent neural network. In: Proceedings of Artificial Intelligence and Soft Computing, pp. 450–455 (2002)
8. Werbos, P.J.: Backpropagation through time: what it does and how to do it. Proc. IEEE **78**(10), 1550–1560 (1990)
9. Nakai, M., Akira, N., Shimodaira, H., Sagayama, S.: Substroke approach to HMM-based on-line Kanji handwriting recognition. In: Proceedings of Sixth International Conference on Document Analysis and Recognition, pp. 491–495 (2001)
10. Bahlmann, C., Burkhardt, H.: The writer independent online handwriting recognition system frog on hand and cluster generative statistical dynamic time warping. IEEE Trans. Pattern Anal. Mach. Intell. **26**(3), 299–310 (2004)
11. Pham, V., Bluche, T., Kermorvant, C., Louradour, J.: Dropout improves recurrent neural networks for handwriting recognition. In: Proceedings of 14th International Conference on Frontiers in Handwriting Recognition, pp. 285–290 (2014)
12. Wei, W., Guanglai, G.: Online handwriting Mongolia words recognition with recurrent neural networks. In: Proceedings of Fourth International Conference on Computer Sciences and Convergence Information Technology, pp. 165–167 (2009)
13. Goh, W.L., Mital, D.P., Babri, H.A.: An artificial neural network approach to handwriting recognition. In: Proceedings of Knowledge-Based Intelligent Electronic Systems vol.1, pp. 132–136 (1997)
14. Doetsch, P., Germany, A., Kozielski, M., Ney, H.: Fast and robust training of recurrent neural networks for offline handwriting recognition. In: Proceedings of 14th International Conference on Frontiers in Handwriting Recognition, pp. 279–284 (2014)
15. Fawcett, T.: An introduction to ROC analysis. Pattern Recogn. Lett. **27**, 861–874 (2006)

American Sign Language Fingerspelling Recognition Using Wide Residual Networks

Kacper Kania$^{(\boxtimes)}$ and Urszula Markowska-Kaczmar

Faculty of Computer Science and Management,
Wrocław University of Science and Technology, Wrocław, Poland
220873@student.pwr.edu.pl, urszula.markowska-kaczmar@pwr.edu.pl

Abstract. Despite existing solutions for accurate translation between written and spoken language, sign language is still not well-studied area. A reliable, robust and working in real-time translator of American Sign Language is a crucial bridge to facilitate communication between deaf and hearing people. In this paper we propose a method of sign language fingerspelling recognition using a modern architecture of convolutional neural network called Wide Residual Network trained with Snapshot Learning procedure. The model was trained on augmented datasets available at Surrey University and Massey University web pages using transfer learning. The final result is a robust classifier of all alphabet letters, which beats current state-of-the-art results. The outcomes encourage further research in this field for creating fully usable sign language translator.

1 Introduction

It is estimated that currently there are approximately ∼250,000–500,000 deaf people using American Sign Language (ASL) [1] and that ASL is a sixth most used language in the U.S. ASL fingerspelling contains 36 signs.

Still, there are very few solutions providing easy communication between deaf and hearing people. In order to overcome the problem, one uses different architectures of an artificial neural network. Lately, deep convolutional neural networks (CNN) are one of the most developing areas of machine learning. There are many applications in various domains which outperform humans in image recognition or natural language analysis tasks. There are many studies on ASL fingerspelling recognition using CNN [2–6] but results are not satisfactory.

Wide Residual Network [7] outperformed other architectures of CNN in many problems. Therefore, we decided to build ASL fingerspelling recognition system based on this relatively new CNN architecture. Thanks to the augmentation, we have also used much bigger training dataset in comparison to the previous works. The primary assumption in the system design was its low cost and high accuracy. The system uses video frames with gestures, processing flat RGB frames without time correlation. Decorrelation from image depth allows to use simple web camera to translate gestures. In the current system version, due to

© Springer International Publishing AG, part of Springer Nature 2018
L. Rutkowski et al. (Eds.): ICAISC 2018, LNAI 10841, pp. 97–107, 2018.
https://doi.org/10.1007/978-3-319-91253-0_10

the absence of a temporal feature, letters j and z, which require hand movement, are excluded from classifier output. This limitation is overcome with the help of a dictionary. Every part of the system was created to improve the accuracy of the final model and handle situations when signing hand is occluded, not visible or user is showing given gesture all the time without changing hand position.

The paper consists of seven sections. The next one presents the related papers. In Sect. 3 problem of recognizing American Sign Language fingerspelling is described. In the next one, applied deep learning techniques are shown. Section 5 describes the method, metrics and CNN architecture. In Sect. 6 description of experiments and achieved results are shown. The last section presents conclusions.

2   Related Work

Current research referring to the ASL recognition based on deep learning encompasses mainly two models: *Deep Belief Network* (DBN) [2] and CNN [3–5]. Authors of DBN fed extracted patches to this architecture and performed a classification task. The learned model got 77% of recall and 79% of precision on the test set. Better results were achieved with CNN, which preserves spatial relationship between pixels in the image and extracts more informative features. The VGG16 model in [4] is an example of CNN architecture that achieved the accuracy of 97.8% for recognizing letters in range of a–e and 72% for a–y. The paper [3] shows CNN architecture for classification of 20 Italian gestures. The author used a new end-to-end trainable neural network architecture integrating temporal convolutions with a bidirectional recurrent network. Authors achieved the accuracy of 91.7% using cross-validation procedure. Ameen and colleague in [6] used shallow CNN architecture with both depth and pixel intensity information as two types of inputs. Accuracy averaged on every user in the evaluated dataset from Massey University was equal to 80.34%.

Usually, CNN training needs a lot of data. In many computer vision applications, it is not possible to acquire such big dataset, for example in the medical domain. Therefore a typical practice is transfer learning, which improves generalization for the network. It relies on using a pretrained network and treating it as a starting point to learn a new task. Transfer learning for ASL recognition was applied by [4,8].

Using deep learning in ASL recognition needs a similar pipeline of processing modules as an application of traditional vision methods. First segmentation of a hand in a frame is performed. Next, classification process of a segmented area follows. Example solution using such basic workflow with DBN network can be found in work of Pigou et al. [3] but other research based on CNN uses a similar approach.

Gestures can be recorded using various devices. Authors of [3] used as input data images gathered by Microsoft Kinect device, capturing information about depth and color. The research described in [2] also used data recorded with Microsoft Kinect device. Based on image depth, patches dependent in-depth of the image were created. Patches were transformed to the binary color space. Many other researchers focused on these both sources of information [3,6,8]. In our case, the input is created by one information source only - a color of an image. This assumption enables using a simple web camera.

3   Problem Description

The objective of this work was to create a classifier of American Sign Language fingerspelling gestures, which performs in real time on a machine used to train the model. Input for a classifier is an RGB image of size $C \times H \times W$. Model outputs probability distribution over K classes. 24 signs are considered, due to exclusion of letters j and z, which require hand movement and temporal analysis. The problem of classification was narrowed down to fingerspelling. It was caused by two reasons:

1. Datasets containing full American Sign Language gestures are limited to very few words out of the full dictionary of ASL.
2. Fingerspelling is a primary step in learning speaking ASL, which is sufficient to describe unique proper nouns and simple thoughts.

Each frame captured by the web camera is processed by the classifier.

4   Applied Deep Learning Techniques

In this section, we present popular approaches to the CNN construction and training.

One of the most common problem in very deep CNNs is the vanishing gradient when more and more layers are added. This causes drop in model accuracy. To eliminate this problem, residual connections between layers were introduced in [9]. Residual connections enable building CNN even with 1000 layers without decreasing the accuracy. They pass the image signal without filtering, so values from the input of the network can be used in the latter layers. In downsampling layers, linear transform occurs as a residual connection.

Huge networks have a lot of parameters to train and huge training datasets are needed. For example, the most commonly used dataset ImageNet contains 14,197,122 images with 1000 different classes. Due to the size of the dataset, models learn general features in initial convolutional layers. Learned models are saved and often used in *transfer learning* technique, applied when available training dataset is small. During training on a smaller dataset, initial layers are frozen, so that weights do not change. Remaining layers are finetuned using the original small dataset. In case of discrepancy between dataset used for initial model training and target data, the *data augmentation* technique should be applied.

A beneficial technique applied in our method is *Snapshot Learning* introduced in [10]. The main idea is to use a cyclic learning rate schedule, starting at a relatively high value and then rapidly lowering it over time. Such learning rate schedule can be described by cosine annealing learning rate schedule presented in [10]. Every $\lceil T/M \rceil$ epochs learning rate resets to the initial value and continues with the same schedule, where T is the number of training epochs and M is the number of models to produce. Authors in [10] showed that even single model learned using this technique obtains lower error rate than the same architecture learned with standard methods.

5   Solution Description

As shown in Fig. 1, the fingerspelling recognition system is composed of a few modules creating a single pipeline. First, the image taken from a webcam is preprocessed. Then, the part of an image with the localized hand is sent to the deep classifier, where the deep convolutional network is applied. It produces a probability distribution over letters in the alphabet. This information, supported by the dictionary module is the base for the final sign recognition.

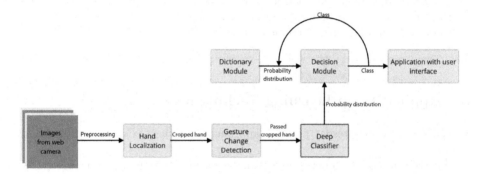

Fig. 1. Pipeline of the system

Below, there are detailed descriptions of the components.

Preprocessing. First, the frames captured by web camera are rescaled so that the next module can be applied correctly. Because translation must be performed in real time, therefore the size of images was limited to empirically defined values: the width is equal to 480 pixels and the height to such value, that aspect ratio is kept.

Hand Localization. The module is built upon Single Shot Multibox Detector [11] with MobileNet as an encoder [12]. The used implementation is available at tensorflow repository[1]. The model finds bounding boxes around hands in the image. A bounding box with the highest probability of being a hand is chosen. The frame is cropped to boundaries of the bounding box. Such crop is then passed to gesture classifier module.

---

[1] https://github.com/tensorflow/models/tree/master/research/object_detection.

Gesture Change Detection. This module works as a kind of attention system which aim is the detection of gesture change. It receives the crop and resizes it to 64×64 pixels. Pixel values are rescaled to range $[0, 1]$. Then, Siamese Network [13] with CNN as a feature extractor is used to detect whether the gesture was changed concerning the previous frames. If there are no previous frames, the crop is passed to the next module. Otherwise, the most recent frame is used as a first input and the current one as a second. It provides a binary result of gesture change detection. A positive class is returned by the network while the hand is moving on the web camera. If there is a transition between consecutive classification operations, such as the previous returned class was positive and the current is negative, then the hand is passed to the next module. Such transition means that user was changing the hand position and after that, he stabilized his hand to show an appropriate gesture.

Deep Classifier. The classifier inputs are scaled to 64×64 pixels, and input values are squashed to the range $[0, 1]$. The model returns probability distribution over letters due to use of the softmax activation function in the last layer.

CNN is built in reference to Wide Residual Network (WRN) specification [7]. The authors showed that widening convolutional layers perform better than very deep stacking. The widening layer means adding more filters. Moreover, such network performs fewer operations and is faster than standard ResNet with the same number of parameters. The example of WRN architecture is shown in Fig. 2. The Wide Residual Network is constructed with an initial convolutional layer and three blocks of convolutional layers. By default, in each block, there are $16 \times k$, $32 \times k$ and $64 \times k$ filters respectively for each convolutional layer in the block. The k variable is a widening factor. We follow the notation *WRN-l-k*, where l is the total number of layers.

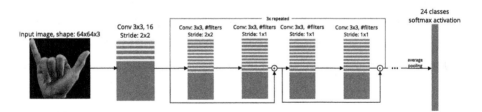

Fig. 2. Wide Residual Network in a configuration *WRN-16-8* for ASL fingerspelling classification task. Parameter *#filters* is equal to the number of filters for each repetition.

Dictionary Module. The module operates on a temporal context in the system which helps return the probability distribution over letters based on previous predictions. Input for the model is a tensor with dimensions $N \times 20 \times 27$, where N is a batch size, 20 is the size of the temporal context, and 27 is the total number of classes. The classes include letters *j*, *z* and a word delimiter.

Figure 3 shows the LSTM architecture used in the dictionary module. At the beginning, whole temporal context represented by matrix 20×27 is initialized with zeros. The first dimension in such matrix refers to time, and the second one - to the letters already predicted by the system encoded in one-hot manner represented by vectors $x_1, x_2, \ldots, x_n$. The matrix is fed to the model consisting of two bidirectional LSTM layers. The first one has 256 features. Values from both directions (forward and backward) are concatenated and passed as a sequence to the next BiLSTM layer which has 64 features. The last time step output concatenated with the results from the first layer is passed to a fully connected layer with softmax activation function. The calculated probability distribution is returned.

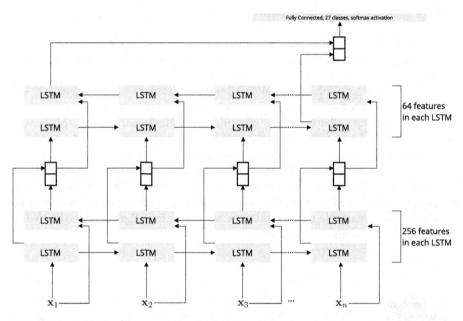

Fig. 3. The LSTM architecture applied in the dictionary module. x_1, x_2, ..., x_n are one-hot vectors of predicted by the system letters stored in the temporal context.

Decision Module. The decision module is the final prediction step. It receives probability distributions over letters from both dictionary and deep classifier modules. Since classifier outputs fewer classes than dictionary module, zeros are inserted in appropriate places of the vectors of CNN output, so that each index in both vectors corresponds to the same class. For first 5 time steps only classifier output is considered. After that, letter class is calculated using the weighted average from the classifier and the dictionary outputs letters with the Eq. 1. Weights for classifier w^{cls} and for classifier w^{dict} are presented in Table 1.

Predicted letter is then transformed to one-hot vector and stored in a temporal context of the decision module.

$$letter = \underset{c \in characters}{\arg\max} \ \frac{w_c^{dict} \cdot y_c^{dict} + w_c^{cls} \cdot y_c^{cls}}{w_c^{dict} + w_c^{cls}} \qquad (1)$$

Table 1. Weights for w_c^{dict} and w_c^{cls} for each character $c \in characters$.

Characters	{a, b, ..., x, y}/{j}	{j, z}	Word delimiter
w_c^{dict}	0.3	0.7	1.0
w_c^{cls}	0.7	0.3	0.0

6   Experiments

The aim of the experimental study was to evaluate the accuracy and the execution time for a single frame of the deep classifier of the proposed solution.

Datasets. Two datasets were used for models training and validation.

Exeter University Dataset [14] contains 65773 images of hands presenting 24 letters from ASL fingerspelling gestures. Images were captured in similar lighting conditions and similar background during 5 sessions. Depth channels were dropped from the dataset. Due to small variance in lighting and background, further processing and data augmentation were performed on this dataset.

Massey University Dataset [15] contains 2524 RGB images segmented from the background. Samples presenting digits were dropped so only 24 letters remained. Images were captured in different lighting conditions by 5 users. A green glow around inaccurately segmented hand images in this datasets was removed by the implemented algorithm using thresholding hand, an erosion of the background and a dilation to fill the holes. Then, the black background was substituted by a random image downloaded from Google Image search. The substitution was repeated 10 times with a different background image each time.

Both datasets were merged and split into a training set and validation set. The split was performed with respect to the ordinal number of user/session, i.e. 4 of them were used for training and remaining for validation. After splitting, there were 71274 samples in the training set and 13037 in the validation set. The same split procedure was by performed in [4].

During each training session, training data was heavily augmented by applying different methods: small random rotation, translation, horizontal flip, random scaling the V channel and random change in H channel in HSV color space, random gamma correction and adding shot noise. An empty space created by affine transforms was filled with values of the nearest pixels. Training dataset was augmented up to 100,000 samples per epoch. No augmentation was performed on the validation set. Each sample was uniquely generated without repetitions and values of pixels were normalized to the range $[0, 1]$.

Tested Models. In each experiment, different techniques and models were used. Baseline model was built to show that even a simple model with a massive data augmentation can outperform models shown in [4]. Next, Wide Residual Network is used, which evaluation results in the new state-of-the-art accuracy in American Sign Language fingerspelling recognition task. To improve results, a new technique of training ensembles called Snapshot Learning was implemented and used during experiments. Each model was trained with 24 classes of the alphabet letters excluding letters j and z.

Baseline Model was built with convolutional composites, each containing convolutional layer, batch normalization [16] and activation function ReLU. First composite has 7×7 kernels. Then, there are 10 layers split into 4 sequences of 2, 2, 3 and 3 composites with 32, 64, 128 and 64 filters for each composite in the corresponding sequence. Between each sequence max pooling occurs. After the last sequence, global average pooling is performed and averaged filters are fed to MLP classifier with 24 output neurons. Finally, softmax non-linearity is applied.

Pretrained Wide Residual Network in *WRN-16-8* configuration with 11.0M parameters was used. Dropout was applied between each pair of convolutional layers with the probability set to 0.5. A spatial modification of the dropout described in [17] was used, where instead of a single neuron, whole feature maps are dropped. Due to the number of parameters to optimize and small training dataset, *transfer learning* technique[2] with weights trained on the CIFAR-10 dataset was used. During training, the first 3 convolutional layers were frozen for the whole training procedure.

In the *Snapshot Learning* procedure the number M of produced models was set to 5 and the number of epochs $T = 100$. The *WRN-16-8* architecture was used. After training only model with the best performance from the whole produced ensemble was considered due to limitations of the performance of the classification module. The cosine annealing learning schedule introduced in [10] was applied. The base learning rate was set to 0.1.

Training the models was performed using Stochastic Gradient Descent optimizer with the learning rate set to 0.03 for first two models and 0.1 for the Snapshot Learning, momentum to 0.9 and Nesterov optimization [18]. The categorical cross entropy loss was chosen as a minimized metric. During training, loss on the validation set was measured. Early stopping procedure was applied, i.e., if there was no improvement for 6 epochs, learning was stopped, and the best model was saved. The rule was not applied to the Snapshot Learning.

Accuracy Evaluation. The results of accuracy evaluation are presented in Tables 2 and 3. In the first table, we have shown accuracies achieved by proposed models. In the second one, we compared the best model achievement with the results of the current state-of-the-art-models models, described in Sect. 2. The accuracy of our model is more than 10% higher about the models proposed so far.

---

[2] Weights were downloaded from https://github.com/titu1994/Wide-Residual-Networks/blob/master/weights/WRN-16-8%20Weights.h5.

Time Evaluation. We also evaluated models in reference to the time needed for prediction for a single frame. The test was performed on a machine with Intel(R) Core(TM) i7-6700 CPU @ 3.40 GHz and 32 GB RAM. GPU was disabled during testing. The first time measurement during the procedure was discarded since libraries with a static computational graph need initialization that gives high overhead to computation time and biases the results. The measurement was performed 10 times and then averaged. Results are shown in Table 4. As can be noticed, all models can be applied to the real-time application due to the short processing time (Fig. 4).

Table 2. The accuracy comparison of tested methods

Method	Accuracy [%]
Baseline model	88.5
Pretrained Wide Residual Network	91.1
Snapshot Learning	93.3

Table 3. The accuracy comparison of the Snapshot Learning model with the current state-of-the-art methods

Method	Accuracy [%]
Snapshot Learning	93.3
Bheda [5]	82.5
Garcia [4]	72.0

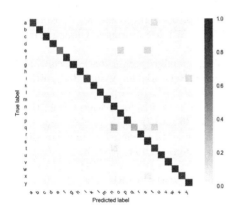

Table 4. Time evaluation of the tested methods for a single frame

Method	Time [ms]
Baseline model	0.46 ± 0.05
Pretrained WRN	4.48 ± 0.27
Snapshot Learning WRN	4.78 ± 0.46

Fig. 4. Confusion matrices for the Snapshot Learning model

7   Conclusions and Future Work

The work demonstrates a method of real time fingerspelling recognition. The main element of the system is deep classifier. We have evaluated three models in this role. All models achieved high results in terms of accuracy and the best one - Snapshot Learning model has about 10% higher accuracy than the best model proposed so far. We further demonstrated that the computation overhead caused by each model is small and a single model can work in the real-time application.

In the future work we would like to merge the hand localization module with deep classifier module. For this purpose, multiobjective loss function can be used in the Multi-task Learning manner [19], where both localization score and accuracy will be considered as a minimized metrics. Another idea is to merge a dictionary module with a classifier module. In this case, a 3D convolutional neural network can be considered, where 3rd dimension is time.

Acknowledgments. We thank Identt company for giving access to PC used to conduct experiments. Acknowledgments are directed also to dr Adam Gonczarek from the Wroclaw University of Science and Technology for leading the project of the recognition system. We thank Michał Kosturek and Piotr Grzybowski from scientific student assocation "medical.ml" at Wrocław University of Science and Technology, who implemented dictionary and hand localization modules respectively for the system.

References

1. Mitchell, R.E., Young, T.A., Bachleda, B., Karchmer, M.A.: How many people use ASL in the United States? Why estimates need updating. Sign Lang. Stud. **6**(3), 306–335 (2006)
2. Rioux-Maldague, L., Giguère, P.: Sign language fingerspelling classification from depth and color images using a deep belief network. CoRR, abs/1503.05830 (2015)
3. Pigou, L., Dieleman, S., Kindermans, P.-J., Schrauwen, B.: Sign language recognition using convolutional neural networks. In: Agapito, L., Bronstein, M.M., Rother, C. (eds.) ECCV 2014, Part I. LNCS, vol. 8925, pp. 572–578. Springer, Cham (2015). https://doi.org/10.1007/978-3-319-16178-5_40
4. Garcia, B., Viesca, S.: Real-time American sign language recognition with convolutional neural networks. In: Convolutional Neural Networks for Visual Recognition (2016)
5. Bheda, V., Radpour, D.: Using deep convolutional networks for gesture recognition in American sign language. CoRR, abs/1710.06836 (2017)
6. Ameen, S., Vadera, S.: A convolutional neural network to classify American sign language fingerspelling from depth and colour images. Expert Syst. **34**(3), e12197 (2017)
7. Zagoruyko, S., Komodakis, N.: Wide residual networks. CoRR, abs/1605.07146 (2016)
8. Kang, B., Tripathi, S., Nguyen, T.Q.: Real-time sign language fingerspelling recognition using convolutional neural networks from depth map. CoRR, abs/1509.03001 (2015)
9. He, K., Zhang, X., Ren, S., Sun, J.: Deep residual learning for image recognition. CoRR, abs/1512.03385 (2015)
10. Huang, G., Li, Y., Pleiss, G., Liu, Z., Hopcroft, J.E., Weinberger, K.Q.: Snapshot ensembles: train 1, get M for free. CoRR, abs/1704.00109 (2017)
11. Liu, W., Anguelov, D., Erhan, D., Szegedy, C., Reed, S.E., Fu, C.-Y., Berg, A.C.: SSD: single shot multibox detector. CoRR, abs/1512.02325 (2015)
12. Howard, A.G., Zhu, M., Chen, B., Kalenichenko, D., Wang, W., Weyand, T., Andreetto, M., Adam, H.: Mobilenets: efficient convolutional neural networks for mobile vision applications. CoRR, abs/1704.04861 (2017)

13. Chopra, S., Hadsell, R., LeCun, Y.: Learning a similarity metric discriminatively, with application to face verification. In: IEEE Computer Society Conference on Computer Vision and Pattern Recognition, CVPR 2005, vol. 1, pp. 539–546. IEEE (2005)
14. University of Exeter: ASL Finger Spelling Dataset, 2 November 2017
15. Barczak, A.L.C., Reyes, N.H., Abastillas, M., Piccio, A., Susnjak, T.: A new 2D static hand gesture colour image dataset for ASL gestures. Res. Lett. Inf. Math. Sci. **15**, 12–20 (2011)
16. Ioffe, S., Szegedy, C.: Batch normalization: accelerating deep network training by reducing internal covariate shift. CoRR, abs/1502.03167 (2015)
17. Tompson, J., Goroshin, R., Jain, A., LeCun, Y., Bregler, C.: Efficient object localization using convolutional networks. CoRR, abs/1411.4280 (2014)
18. Nesterov, Y.: Introductory Lectures on Convex Optimization. Springer US, New York (2004). https://doi.org/10.1007/978-1-4419-8853-9
19. Ruder, S.: An overview of multi-task learning in deep neural networks. CoRR, abs/1706.05098 (2017)

Neural Networks Saturation Reduction

Janusz Kolbusz[1], Pawel Rozycki[1(✉)], Oleksandr Lysenko[2],
and Bogdan M. Wilamowski[3]

[1] University of Information Technology and Management in Rzeszow,
Rzeszów, Poland
{jkolbusz,prozycki}@wsiz.rzeszow.pl
[2] National Technical University of Ukraine "Igor Sikorsky Kyiv Polytechnic
Institute", Kiev, Ukraine
allias@hotmail.com
[3] Auburn University, Auburn, AL 36849-5201, USA
wilambm@auburn.edu

Abstract. The saturation of particular neuron and a whole neural network is one of the reasons for problems with training effectiveness. The paper shows neural network saturation analysis, proposes a method for detection of saturated neurons and its reduction to achieve better training performance. The proposed approach has been confirmed by several experiments.

Keywords: Network training improvement · Saturation Reduction
Activation functions

1 Introduction

Artificial intelligence, including neural networks with deep architecture, are used to solve complex, non-linear problems. It has been shown that along with the increase in the depth of neural networks they can solve more and more complex non-linear problems and that the power of the neural network grows linearly with its width and exponentially with its depth [1–3]. Deep neural networks have achieved significant success in the last few years, giving good results in solving many difficult tasks [4–7], but there is still a lot of work to understand why deep architectures are able to learn and solve such non-linear problems. Neural networks with deep architecture are attractive because they provide much more computing power than shallow neural networks and are good candidates for modeling complex multidimensional non-linear systems, but the training process of these networks is very difficult [8–10]. It is practically impossible to train deep multilayer neural networks with traditional gradient methods. A standard teaching strategy based on randomly initiating network weights and applying a gradient using back propagation is well known, but it allows to obtain satisfactory results only when used for one or two hidden layers [8,11]. As the depth

This work was supported by the National Science Centre, Krakow, Poland, under-grant No.2015/17/B/ST6/01880.

of the network increases, the network has become less transparent in the case of training that hinders the convergence of deeper networks [12]. There is the so-called problem of the *vanishing gradient* [10], which is a well-known nuisance in multi-layer neural networks [13]. There are several approaches to reduce this effect in practice, for example by careful initiation [14] or surveillance of the hidden layer [15]. Recent research suggests solving this problem through standardized initialization [12,14] and batch normalization [16]. In order to solve the gradient vanishing problem, additional connections were introduced to improve information on several layers [17]. Research of this paper's authors confirms that the problem with the vanishing gradient can be largely eliminated by introducing additional connections between nonadjacent layers, using the so-called BMLP (Bridged MLP) architecture [18]. However, the vanishing gradient problem is not the only problem in training of deep neural networks. Research shows that the process of saturating neurons in the network architecture also has a significant impact on the success and training time [19].

The paper presents and describes the process of neuron saturation during learning of the deep neural network. The distribution of neurons saturation has been shown and some method of elimination of neuronal saturation has been proposed to improve the efficiency of deep neural network training.

2   Neural Network Saturation

In the case of a neural network, the phenomenon of neuron saturation concerns the condition in which the neuron generates values comparable to the asymptotic ends of the limited activation function. Saturation of the neuron reduces both information capacity and the ability to learn neural network [19]. The degree of saturation is an important feature of a neural network that can be used to understand the behavior of the network itself, as well as the learning algorithm used. The saturation of neurons during learning the neural network is an unfavorable phenomenon. For example, a saturated neuron in the output layer makes the whole network useless. Such a network always generates as a result 1 or 0 (in the case of unipolar sigmoid activation function) regardless of the input signal. Second, saturation is only a problem with nonlinear (e.g. sigmoidal or tanh), but not with linear activation function. Thus, activating functions such as linear (simple summator) or ReLU do not have this problem and usually learn faster in the initial phase of the training. However, these non-linear functions are used in networks that are able to solve strongly non-linear and complex problems, so replacing them with units with linear functions will not always be desirable.

The biggest problem with saturation is the gradient decrease to the value of 0. At this point, the learning process stops and the weights cease to update with iterations. In order for the weights to be updated, it must be a non-zero gradient. Algorithms that use the gradient method to update (and train) network weights encounter a problem, because in the case of zero gradients, the network does not learn.

Our proposition is to train a neural network with the function of sigmoidal activation with gradient methods, but with the detection of saturating neurons in a limited area of the activation function and move these neurons into different working space.

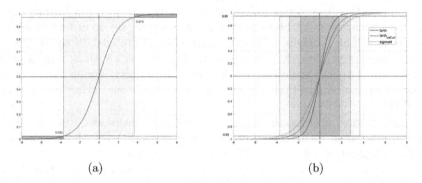

(a)   (b)

Fig. 1. Activation functions: (a)Sigmoid activation function with active space (blue) and saturated space (red), (b) Active ranges of different activation functions (Color figure online)

2.1   Neuron Saturation Detection

As mentioned in previous section the neuron seems to be in saturated state in the case when its output (or rather *net* input value) is not in active range of activation function – means, when outputs for the given data set are concentrated around the asymptotic ends. In the case of sigmoid activation function $\frac{1}{1+e^{-net}}$ that means output below given near 0 value and above given near 1 value. In Fig. 1a is shown active range (blue) and saturation ranges (red). Another activation functions: bipolar sigmoid function ($\frac{2}{1+e^{-net}}+1$), *tanh* function ($\frac{e^{net}-e^{-net}}{e^{net}+e^{-net}}$) and $tanh_{LeCun}$ that is *tanh* modified by LeCun in [20] function ($1.7159\tanh(\frac{2}{3}net)$), are shown in Fig. 1b. As can be observed active range is biggest for *sigmoid* function, and the smallest for *tanh* function.

In [14,19,21] can be found some analysis of neuron output values and output distribution in depends on used activation function and localization of given neuron in the network topology. Usually, given neuron output is different for different patterns, and usually part of these outputs is in the saturated space. Therefore, the saturation of neuron S_n can be calculated using Eq. 1

$$S_n = \frac{N_{SP}}{P} \tag{1}$$

where P is number of patterns and N_{SP} is number of patterns for which neuron's output value is saturated that means output value is over active output out_{sup} (Eq. 2) or below active output out_{sdown} (Eq. 3)

$$out_{sup} = OUT_{max} - SR(OUT_{max} - OUT_{min}) \tag{2}$$

$$out_{sdown} = OUT_{min} + SR(OUT_{max} - OUT_{min}) \qquad (3)$$

where OUT_{min} and OUT_{max} is minimal and maximal output values respectively and SR is saturation rate value that specify active and saturated spaces of neuron output (in Fig. 1a $SR = 0.025$ it means 2.5% of output range).

During training of given neural network some neurons are more saturated than others, and some of them are saturated for all training patterns while others are in active space for all cases. Thus, given neuron n can be treated as saturated if saturation is relatively high, that means, for most patterns the output is in the saturation space, so if $S_n > ST$ where ST (*Saturation Threshold*) is the percentage of saturated patterns.

2.2   Changing Working Space

Detection of saturated neurons is not enough to take any actions to get out of saturation. In most cases, saturation is a natural state during the training process and often the network is able to get out of saturation or solve the problem despite the high saturation of individual neurons. Therefore, apart from the high value of the ST parameter, actual training error and variability of weights on particular neurons should also be taken into account.

In the testing implementation of proposed approach that has been prepared for NBN algorithm [8, 22] some action for saturated neurons is done for $SR = 0.05$ and $ST = 0.80$ if training errors stay unchanged for a longer period of time (in practice relative change to some last iterations below 0.001). For such selected neurons and iteration the following action is made to get out selected neuron from saturation: (1) find weights that are unchanged for some last iterations, (2) change found weights to new randomly selected values.

3   Saturation Reduction – Example

As an example of Saturation Reduction (RS) mechanism has been prepared experiment with Parity-5 problem that should be resolved by MLP neural network in configuration with three hidden layers with two bipolar sigmoid activation function in each hidden layer (5-2-2-2-1) trained with NBN algorithm. Visual analysis of saturated neurons have been shown in Figs. 2, 3, 4. In each figure saturation of all 7 neurons (columns) in the key iterations of training are shown. First two columns contain saturation visualization of neurons in first layer, columns 3 and 4 show saturation of neurons in second layer and neurons in third layer are in column 5 and 6. Last 7 column show visualization of output neuron saturation. Each column contains visualization of the saturation of given neuron for each pattern. Yellow and blue color means that neuron is saturated for given pattern. On the bottom of each column is shown S_n value for given n neuron.

As can be seen already after 20 iterations, one of the neurons from the first layer is completely saturated and after 30 iterations both of them. However, the training error kept decreasing and with iteration 60 stabilized again and at

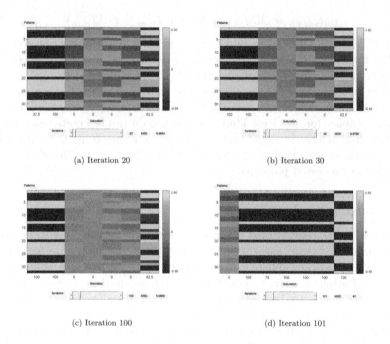

(a) Iteration 20   (b) Iteration 30

(c) Iteration 100   (d) Iteration 101

Fig. 2. Saturation analysis for Parity-5 problem – part 1.

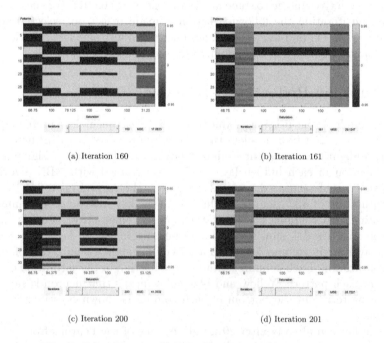

(a) Iteration 160   (b) Iteration 161

(c) Iteration 200   (d) Iteration 201

Fig. 3. Saturation analysis for Parity-5 problem – part 2.

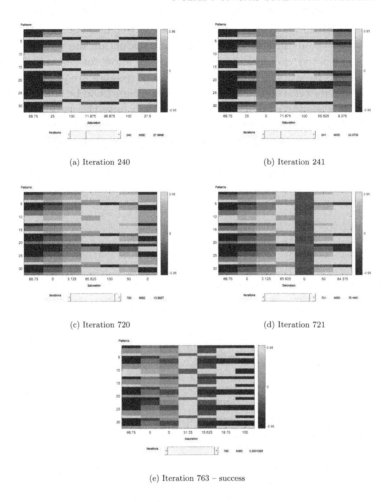

(a) Iteration 240

(b) Iteration 241

(c) Iteration 720

(d) Iteration 721

(e) Iteration 763 – success

Fig. 4. Saturation analysis for Parity-5 problem – part 3.

iteration 100 weights of the first saturated neuron stabilized enough that their weights change was made to excite the neuron. Excitation is performed only on one neuron closest to the input that meet conditions and consists in setting new random weights for those that have not changed for a long time (from the five iterations in this case). Already in the next iteration it can be seen that the excited neuron went out of saturation, however, the neurons being so far unsaturated entering into the state of saturation and training error increased from 6.667 to 40, but in subsequent iterates it began to decline and the network stabilized in iteration 160 enough to excite one of neurons – this time the another from the first layer. Repeated excitations in 200, 240 iterations (here the neuron in layer 2) and 720 (neuron in layer 3) that allows to achieve training success after 763 iterations, reaching an error below the assumed value of 0.002. Figure 5 shows training error for training without Saturation Reduction (a) and with this mechanism (b). Both cases started with the same weights. As can be observed the algorithm without implemented Saturation Reduction mechanism was not able to train with success, and modified algorithm allow for successful training reached after 763 iterations. Note, that changes of weights are good visible and as can be observed excite neurons by changing its weights can be efficient method to achieve better results of training. In this section only the example has been presented to show how the Saturation Reduction mechanism works. In the next section some experiments confirming proposed approach have been described.

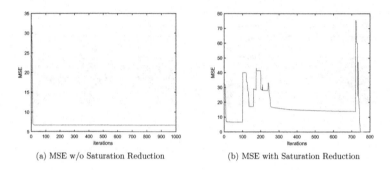

(a) MSE w/o Saturation Reduction   (b) MSE with Saturation Reduction

Fig. 5. Saturation analysis for Parity-5 problem – training errors (a) without and (b) with Saturation Reduction

4   Experimental Results

To confirm the effectiveness of the proposed method, several experiments were carried out on well-known benchmark data sets: Peaks function, Schwefel function and Two-Spirals shown in Fig. 6. The Matlab software with the NBN algorithm and the neural network in the MLP topology with 5 to 8 hidden layers with from 1 to 7 neurons in each layer has been used. In each

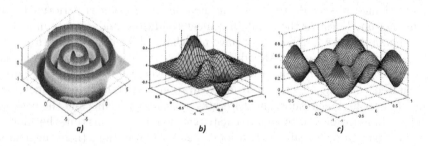

Fig. 6. Visualization of problems used in experiments: (a) Two-Spirals, (b) Peaks, (c) Schwefel function.

case, 100 trials were performed with and without Saturation Reduction mechanism. The measured value was the training efficiency $SuccessRate$ defined as $\frac{number\_of\_success\_trainings}{number\_of\_all\_tranings}$. Results are shown in Tables 1, 2, 3. In MLP columns are presented results achieved with NBN algorithm without Saturation Reduction mechanism, and in $MLPwS$ columns results for this mechanism are shown. As can be observed results achieved with proposed mechanism are a bit better in most cases. Improvement is maybe not significant, but seems to be stable. The network was also able occasionally to train with success also in some cases that ware untrainable without Saturation Reduction mechanism.

Table 1. Training success rates for peaks function

Layers	5		6		7		8	
Neurons	MLP	MLPwS	MLP	MLPwS	MLP	MLPwS	MLP	MLPwS
1	0	0	0	0	0	0	0	0
2	0	0	0	0.0	0	0.0	0	0.0
3	0	0.0	0.04	0.04	0.87	0.90	0.98	0.98
4	0.02	0.02	0.52	0.54	0.77	0.79	0.90	0.92
5	0.12	0.13	0.35	0.46	0.50	0.61	0.65	0.73
6	0.12	0.12	0.16	0.21	0.35	0.48	0.48	0.54
7	0.04	0.05	0.14	0.16	0.20	0.22	0.41	0.45
8	0.03	0.04	0.04	0.04	0.16	0.20	0.20	0.34

Table 2. Training success rates for two-spirals function

Layers	5		6		7		8	
Neurons	MLP	MLPwS	MLP	MLPwS	MLP	MLPwS	MLP	MLPwS
1	0	0	0	0	0	0	0	0
2	0	0	0	0	0	0.03	0.41	0.43
3	0.11	0.14	0.30	0.38	0.50	0.62	0.74	0.79
4	0.22	0.24	0.35	0.40	0.43	0.49	0.51	0.54
5	0.08	0.11	0.09	0.13	0.12	0.20	0.18	0.21
6	0	0.03	0	0.02	0	0.05	0	0.04
7	0	0.02	0	0.01	0	0.04	0	0.01
8	0	0.01	0	0	0	0	0	0

Table 3. Training success rates for Schwefel function

Layers	5		6		7		8	
Neurons	MLP	MLPwS	MLP	MLPwS	MLP	MLPwS	MLP	MLPwS
1	0	0	0	0	0	0	0	0
1	0	0	0	0	0	0	0	0
1	0	0.01	0.01	0.05	0	0	0	0
1	0.12	0.18	0.14	0.17	0	0.02	0	0.03
1	0.18	0.20	0.19	0.21	0.15	0.19	0	0.10
1	0.09	0.17	0.12	0.20	0.04	0.10	0	0.03
1	0.02	0.07	0.05	0.06	0	0.02	0	0.01
1	0	0.01	0	0.01	0	0.01	0	0

5   Conclusions

Saturation of individual neurons is a natural process during network training, which, however, significantly hinders and extends the training process. The paper presents a method for the analysis of the degree of saturation of individual neurons in the network and a method for its reduction and, consequently, the saturation reduction of entire neural network. Changes in the working space of the neuron by changing its weights causes the destabilization of the network, especially in deep-based networks, but it allows for better results, which was confirmed experimentally. Further work in this area will focus on the development of the proposed method of analysis and reduction of saturation in the hope that this will allow for better results.

References

1. Rozycki, P., Kolbusz, J., Wilamowski, B.M.: Dedicated deep neural network architectures and methods for their training. In: IEEE 19th International Conference on Intelligent Engineering Systems (INES 2015) Bratislava, 3–5 September 2015, pp. 73–78 (2015)
2. Hinton, G.E., Osindero, S., Teh, Y.W.: A fast learning algorithm for deep belief nets. Neural Comput. **18**, 1527–1554 (2006)
3. Mohamed, A., Dahl, G.E., Hinton, G.E.: Acoustic modeling using deep belief networks. IEEE Trans. Audio Speech Lang. Process. **20**, 14–22 (2012)
4. Krizhevsky, A., Sutskever, I., Hinton, G.E.: Imagenet classification with deep convolutional neural networks. In: Advances in Neural Information Processing Systems, pp. 1097–1105 (2012)
5. Simonyan K., Zisserman A.: Very deep convolutional networks for largescale image recognition. arXiv preprint arXiv:1409.1556 (2014)
6. Mnih, V., et al.: Human-level control through deep reinforcement learning. Nature **518**(7540), 529–533 (2015)
7. Silver, D., et al.: Mastering the game of go with deep neural networks and tree search. Nature **529**(7587), 484–489 (2016)

8. Wilamowski, B.M., Yu, H.: Neural network learning without backpropagation. IEEE Trans. Neural Netw. **21**(11), 1793–1803 (2010)

9. Hunter, D., Hao, Y., Pukish, M.S., Kolbusz, J., Wilamowski, B.M.: Selection of proper neural network sizes and architectures—A comparative study. IEEE Trans. Industr. Inf. **8**, 228–240 (2012)

10. Hochreiter, S.: The vanishing gradient problem during learning recurrent neural nets and problem solutions. Int. J. Uncertain. Fuzz. Knowl. Based Syst. **06**, 107 (1998)

11. Larochelle, H., et al.: Exploring strategies for training deep neural networks. J. Mach. Learn. Res. **10**(Jan), 1–40 (2009)

12. He, K., Zhang, X., Ren, S., Sun, J.: Delving deep into rectifiers: Surpassing human-level performance on imagenet classification. In: ICCV (2015)

13. Bengio, Y., Simard, P., Frasconi, P.: Learning long-term dependencies with gradient descent is dificult. IEEE Trans. Neural Netw. **5**(2), 157–166 (1994)

14. Glorot, X., Bengio, Y.: Understanding the difficulty of training deep feedforward neural networks. In: International Conference on Articial Intelligence and Statistics, pp. 249–256 (2010)

15. Lee, C.Y., Xie, S., Gallagher, P., Zhang, Z., Tu, Z.: Deeply-supervised nets. arXiv preprint arXiv:1409.5185 (2014)

16. Ioffe, S., Szegedy, C.: Batch normalization: Accelerating deep network training by reducing internal covariate shift. In: ICML (2015)

17. Srivastava, R.K., Greff, K., Schmidhuber, J.: Highway networks. arXiv preprint arXiv:1505.00387 (2015)

18. Kolbusz J., Różycki P., Wilamowski B.M.: The study of architecture MLP with linear neurons in order to eliminate the "vanishing gradient" problem. In: Artificial Intelligence and Soft Computing, ICAISC 2017, pp. 97–106 (2017)

19. Rakitianskaia, A., Engelbrecht, A.: Measuring saturation in neural networks. In: 2015 IEEE Symposium Series on Computational Intelligence, Cape Town, pp. 1423–1430 (2015)

20. LeCun, Y.A., Bottou, L., Orr, G.B., Müller, K.-R.: Efficient backprop. In: Montavon, G., Orr, G.B., Müller, K.-R. (eds.) Neural Networks: Tricks of the Trade. LNCS, vol. 7700, pp. 9–48. Springer, Heidelberg (2012). https://doi.org/10.1007/978-3-642-35289-8_3

21. Rakitianskaia, A., Engelbrecht, A.: Training high-dimensional neural networks with cooperative particle swarm optimiser. In: 2014 International Joint Conference on Neural Networks (IJCNN), Beijing, pp. 4011–4018 (2014)

22. Wilamowski, B.M., Yu, H.: Improved computation for levenberg Marquardt training. IEEE Trans. Neural Netw. **21**(6), 930–937 (2010)

Learning and Convergence
of the Normalized Radial Basis
Functions Networks

Adam Krzyżak[1,2]([✉]) and Marian Partyka[3]

[1] Department of Computer Science and Software Engineering, Concordia University,
1455 de Maisonneuve Blvd. West, Montreal H3G 1M8, Canada
krzyzak@cs.concordia.ca
[2] Department of Electrical Engineering, Westpomeranian University of Technology,
70-313 Szczecin, Poland
[3] Department of Knowledge Engineering,
Faculty of Production Engineering and Logistics,
Opole University of Technology, ul. Ozimska 75, 45-370 Opole, Poland
m.partyka@po.opole.pl

Abstract. In the paper we analyze convergence and rates of convergence
of the normalized radial basis function networks by relating their L_2
error to the L_2 error of the Wolverton-Wagner regression estimate. The
network parameters are learned by minimizing the empirical risk and are
applied in function learning and classification.

Keywords: Nonlinear regression · Classification
Wolverton-Wagner recursive radial basis function networks
MISE convergence · Strong convergence · Rates of convergence

1 Introduction

In artificial neural network literature several types of feed-forward neural networks are commonly considered. They include: multilayer perceptrons (MLP),
radial basis function (RBF) networks, normalized radial basis function (NRBF)
networks and deep networks. These neural network models have been applied
in different problems including interpolation, classification, data smoothing and
regression. Convergence analysis of MLP can be found among others in, e.g.,
Cybenko [8], White [47], Hornik et al. [25], Barron [3], Anthony and Bartlett
[1], Devroye et al. [11], Györfi et al. [22], Ripley [41], Haykin [24], Hastie et al.
[23] and Kohler and Krzyżak [26]. The latter paper is one of the first to analyze convergence of so called deep multilayer networks. RBF networks have been

A. Krzyżak—Research of the first author was supported by the Natural Sciences
and Engineering Research Council of Canada under Grant RGPIN-2015-06412.
He carried out this research at WUT during his sabbatical leave from Concordia
University.

© Springer International Publishing AG, part of Springer Nature 2018
L. Rutkowski et al. (Eds.): ICAISC 2018, LNAI 10841, pp. 118–129, 2018.
https://doi.org/10.1007/978-3-319-91253-0_12

considered in, e.g., Moody and Darken [38], Park and Sandberg [39,40], Girosi and Anzellotti [17], Xu et al. [49], Krzyżak et al. [29], Krzyżak and Linder [30], Krzyżak and Niemann [31], Györfi et al. [22], Krzyżak and Schäfer [35] and Krzyżak and Partyka [32].

In this paper we consider the normalized radial basis function (RBF) networks with one hidden layer of at most k nodes with a fixed kernel $\phi : \mathcal{R}_+ \to \mathcal{R}$:

$$f_k(x) = \frac{\sum_{i=1}^{k} w_i \phi_i \left(\|x - c_i\|_{A_i} \right)}{\sum_{i=1}^{k} \phi_i \left(\|x - c_i\|_{A_i} \right)} \tag{1}$$

where

$$\|x - c_i\|_{A_i}^2 = [x - c_i]^T A_i [x - c_i], \phi_i(x) = \frac{1}{h_i^d} \phi(x),$$

which are class of functions satisfying the following conditions:

(i) radial basis function condition: $\phi : \mathcal{R}_0^+ \to \mathcal{R}^+$ is a left-continuous, decreasing function, the so-called *kernel*.

(ii) centre condition: $c_1, \ldots, c_k \in \mathcal{R}^d$ are the so-called *centre vectors* with $\|c_i\| \leq R$ for all $i = 1, \ldots, k$.

(iii) receptive field condition: $A_1, \ldots, A_k$ are symmetric, positive definite, real $d \times d$-matrices each of which satisfies the eigenvalue inequalities $\ell \leq \lambda_{min}(A_i) \leq \lambda_{max}(A_i) \leq L$. Here, $\lambda_{min}(A_i)$ and $\lambda_{max}(A_i)$ are the minimal and the maximal eigenvalue of A_i, respectively. A_i specifies the *receptive field* about the centre c_i.

(iv) weight condition: $w_1, \ldots, w_k \in \mathcal{R}$ are the *weights* satisfying $|w_i| \leq B$ for all $i = 1, \ldots, k$.

(v) weight condition: $h_1, \ldots, h_k$ are the *bandwidth weights* satisfying the conditions listed in Sect. 4.

Throughout the paper we use the convention $0/0 = 0$. Common choices for the kernel satisfying (i) are:

– **Window type kernels.** These are kernels for which some $\delta > 0$ exists such that $\phi(t) \notin (0, \delta)$ for all $t \in \mathcal{R}_0^+$. The classical naive kernel $\phi(t) = \mathbf{1}_{[0,1]}(t)$ is a member of this class.

– **Non-window type kernels with bounded support.** These comprise all kernels with support of the form $[0, s]$ which are right-continuous in s. For example, for $\phi(t) = \max\{1 - t, 0\}$, $\phi(x^T x)$ is the Epanechnikov kernel.

– **Kernels with unbounded support,** i.e. $\phi(t) > 0$ for all $t \in \mathcal{R}_0^+$. The most famous example of this class is $\phi(t) = \exp(-t)$. Then $\phi(x^T x)$ is the classical Gaussian kernel.

Let us denote the parameter vector $(w_0, \ldots, w_k, c_1, \ldots, c_k, A_1, \ldots, A_k)$ by θ. It is assumed that the kernel is fixed, while network parameters $w_i, c_i, A_i, i = 1, \ldots, k$ are learned from the data. Normalized RBF networks are generalizations of standard RBF networks defined by

$$f_k(x) = \sum_{i=1}^{k} w_i \phi \left(\|x - c_i\|_{A_i} \right) + w_0. \tag{2}$$

The most popular choices of radial function ϕ are:

- $\phi(x) = e^{-x^2}$ (Gaussian kernel)
- $\phi(x) = e^{-x}$ (exponential kernel)
- $\phi(x) = (1 - x^2)_+$ (truncated parabolic or Epanechnikov kernel)
- $\phi(x) = \frac{1}{\sqrt{x^2+c^2}}$ (inverse multiquadratic)

All these kernels are nonincreasing. In the literature on approximation by means of radial basis functions the following monotonically increasing kernels were considered

- $\phi(x) = \sqrt{x^2 + c^2}$ (multiquadratic)
- $\phi(x) = x^{2n} \log x$ (thin plate spline)

They play important role in interpolation and approximation with radial functions [18], but are not considered in the present paper.

Standard RBF networks have been introduced by Broomhead and Lowe [7] and Moody and Darken [38]. Their approximation error was studied by Park and Sandberg [39,40]. These result have been generalized by Krzyżak et al. [29], who also showed weak and strong universal consistency of RBF networks for a large class of radial kernels in the least squares estimation problem and classification. The rate of approximation of RBF networks was investigated by Girosi and Anzellotti [17]. The rates of convergence of RBF networks trained by complexity regularization have been investigated in regression estimation problem by Krzyżak and Linder [30].

Normalized RBF networks (1) have been originally investigated by Moody and Darken [38] and Specht [44]. Further results were obtained by Shorten and Murray-Smith [43]. Some convergence results for the regression estimation problem have been discussed in [32,35].

Besides normalized RBF networks other classical nonparametric regression estimation techniques include Nadaraya-Watson kernel estimate and its recursive versions [21,22], nearest-neighbor estimate [12,14,22], partitioning estimate [4,22], orthogonal series estimate [20,22], tree estimate [6,23] and Breiman random forest [5,42].

This paper investigates the mean integrated square error (MISE) convergence and strong convergence as well as rates of convergence of the normalized recursive RBF network estimation in nonlinear function learning and classification by relating their MISE to MISE of the recursive Wolverton-Wagner type kernel regression estimate. The paper is organized as follows. In Sect. 2 the algorithm for nonlinear function learning is presented. In Sect. 3 the normalized recursive Wolverton-Wagner type RBF network classifier is discussed. In Sect. 4 convergence properties of the learning algorithms are investigated and Conclusions are given in Sect. 5.

2 Nonlinear Function Learning

Let $(X, Y), (X_1, Y_1), (X_2, Y_2), \ldots, (X_n, Y_n)$ be independent, identically distributed, $\mathcal{R}^d \times \mathcal{R}$–valued random variables with $\mathbf{E}Y^2 < \infty$, and let $R(x) =$

$\mathbf{E}(Y|X = x)$ be the corresponding nonlinear regression function. Let μ be the distribution of X. It is well-known that regression function R minimizes L_2 error:

$$\mathbf{E}|R(X) - Y|^2 = \min_{f:\mathcal{R}^d \to \mathcal{R}} \mathbf{E}|f(X) - Y|^2.$$

Our aim is to estimate R from the i.i.d. observations of random vector (X, Y)

$$D_n = \{(X_1, Y_1), \ldots, (X_n, Y_n)\}$$

using RBF network (1). We train the network by choosing its parameters that minimize the L_2 risk

$$\frac{1}{n} \sum_{j=1}^{n} |f(X_j) - Y_j|^2 \tag{3}$$

on the training data D_n, that is we choose RBF network m_n in the class

$$\mathcal{F}_n = \{f_k = f_\theta : \theta \in \Theta_n\} = \left\{ \frac{\sum_{i=1}^{k} w_i \phi_i (\|x - c_i\|_{A_i})}{\sum_{i=1}^{k} \phi_i (\|x - c_i\|_{A_i})} \right\}$$

where

$$\Theta_n = \{\theta = (w_1, \ldots, w_{k_n}, c_1, \ldots, c_{k_n}, A_1, \ldots, A_{k_n})\}.$$

so that

$$\frac{1}{n} \sum_{j=1}^{n} |m_n(X_j) - Y_j|^2 = \min_{f \in \mathcal{F}_n} \frac{1}{n} \sum_{j=1}^{n} |f_\theta(X_j) - Y_j|^2. \tag{4}$$

We measure the performance of the RBF network estimate by the MISE error

$$\mathbf{E}|m_n(X_1) - R(X_1)|^2 = \mathbf{E} \int |m_n(x) - m(x)|^2 \mu(dx).$$

Even though direct analysis of m_n has been carried out in [35] using Vapnik-Chervonenkis dimension and covering numbers [22, 45, 46] it is not fully satisfactory as it is pretty complex and learning of parameters by empirical risk minimization imposes heavy computational burden. In the reminder of the paper we will explore in the analysis of m_n its proximity to the recursive kernel regression function estimate

$$r_n(x) = \frac{\sum_{i=1}^{n} Y_i \frac{1}{h_i^d} K\left(\frac{x - X_i}{h_i}\right)}{\sum_{i=1}^{n} \frac{1}{h_i^d} K\left(\frac{x - X_i}{h_i}\right)} \tag{5}$$

where $K : \mathcal{R}^d \to \mathcal{R}$ is a kernel and $h_i, i = 1, 2, \ldots, n$ is a smoothing sequence (bandwidth) of positive real numbers. The estimate has been introduced by Wolverton and Wagner [48] and investigated by Greblicki [19] and Krzyżak and Pawlak [33]. It can be computed recursively as follows:

$$r_0(x) = g_0(x) = 0$$

$$g_n(x) = \left(\frac{h_n}{h_{n-1}}\right)^d g_{n-1}(x) + K_{h_n}(x - X_n)$$

and

$$r_n(x) = r_{n-1}(x) + g_n^{-1}(x)(Y_n - r_{n-1}(x))K_{h_n}(x - X_n)$$

where $K_h(x) = K(\frac{x}{h})$.

Estimator (5) need not be recomputed entirely when additional observation is combined with the previous ones. In addition, implementing a nonrecursive kernel regression estimate based on n observations requires storing the observations $X_1, \ldots, X_n$, whereas implementing m_n requires storing m_{n-1} and g_{n-1}.

The properties of the standard kernel regression estimate with applications to analysis of RBF networks have been discussed in detail in [31]. This approach leads to very simple and efficient training. Assume that K is spherically symmetric, i.e. $K(x) = K(||x||)$. Let parameters of (1) be trained as follows

$$k_n = n, A_i = \frac{1}{h_i^2}I, w_i = Y_i, \ c_i = X_i, \ i = 1, \cdots, n. \tag{6}$$

Thus we obtain the plug-in recursive RBF network

$$g_n(x) = \frac{\sum_{i=1}^n \phi_i(\frac{||x-X_i||}{h_i})Y_i}{\sum_{i=1}^n \phi_i(\frac{||x-X_i||}{h_i})} = \frac{\sum_{i=1}^n \frac{1}{h_i^d}K(\frac{x-X_i}{h_i})Y_i}{\sum_{i=1}^n \frac{1}{h_i^d}K(\frac{x-X_i}{h_i})}. \tag{7}$$

As a consequence of the simple bound

$$\mathbf{E}|m_n(X_1) - R(X_1)|^2 \le \mathbf{E}|g_n(X_1) - R(X_1)|^2 \tag{8}$$

plug-in recursive RBF network provides an upper bound on performance of m_n on D_n.

3   Recursive Classification Rules

Let (Y, X) be a pair of random variables taking values in the set $\{1, \ldots, M\}$, whose elements are called classes, and in R^d, respectively. The problem is to classify X, i.e. to decide on Y. Let us define *a posteriori* class probabilities

$$p_i(x) = P\{Y = i | X = x\}, i = 1, \cdots, M, x \in R^d.$$

The Bayes classification rule

$$\Psi^*(X) = i \ \text{ if } \ p_i(X) > p_j(X), j < i, \ \text{ and } p_i(X) > p_j(X), j > i$$

minimizes the probability of error. The Bayes risk L^* is defined by

$$P\{\Psi^*(X) \ne Y\} = \inf_{\Psi:R^d \to \{1,\ldots,M\}} P\{\Psi(X) \ne Y\}.$$

The local Bayes risk is equal to $P\{\Psi^*(X) \ne Y \mid X = x\}$. Observe that $p_i(x) = E\{I_{\{Y=i\}} \mid X = x\}$ may be viewed as a regression function of the indicator of the event $\{Y = i\}$. Given the learning sequence $V_n = \{(Y_1, X_1), \ldots, (Y_n, X_n)\}$ of independent observations of the pair (Y, X), we may learn $p_i(x)$ using recursive RBF nets mimicking (4), i.e.,

$$\frac{1}{n} \sum_{j=1}^{n} |\hat{p}_{in}(X_j) - I_{\{Y_j=i\}}|^2 = \min_{f \in \mathcal{F}_n} \frac{1}{n} \sum_{j=1}^{n} |f_Y(X_j) - Y_j|^2. \tag{9}$$

We construct an empirical recursive classification rule Ψ_n, which classifies every $x \in R^d$ to any class maximizing $\hat{p}_{in}$. In order to simplify learning process we will consider simple plug-in classification rules considered in Sect. 3. We propose plug-in recursive RBF classifier with parameters learned by (6) resulting in the recursive classification rule Ψ_n which classifies every $x \in R^d$ to any class maximizing

$$p_{in} = \frac{\sum_{j=1}^{n} I_{\{Y_j=i\}} \frac{1}{h_j^d} K\left(\frac{x-X_j}{h_j}\right)}{\sum_{j=1}^{n} \frac{1}{h_j^d} K\left(\frac{x-X_j}{h_j}\right)}. \tag{10}$$

The global performance of Ψ_n is measured by $L_n = P\{\Psi_n(X) \neq \theta \mid V_n\}$ and the local performance by $L_n(x) = P\{\Psi_n(x) \neq \theta \mid V_n\}$. A rule is said to be weakly, strongly, or completely Bayes risk consistent (BRC) if $L_n \to L^*$, in probability, almost surely, or completely, respectively, as $n \to \infty$. Thanks to relation (see [27])

$$|L_n(X) - L^*(X)| \leq \sum_{i=1}^{M} |p_{in}(X) - p_i(X)| \tag{11}$$

any convergence result obtained for regression estimate (7) is also valid for Ψ_n.

In the next section we will study convergence and rates of plug-in recursive RBF network regression estimate g_n and recursive RBF estimate m_n as well as classification rules induced by them.

4   Consistency and Rates of Convergence

In the first part of this section we present convergence results (consistency and the rates) for the recursive RBF learning function learning and classification algorithms. In the second part of the section we outline the proofs.

4.1   Convergence Results

We have the following convergence and rates of convergence results for the recursive plug-in RBF network g_n and classification rule Ψ_n. Inequality (8) enables us to apply convergence and rates results of g_n to the recursive RBF network m_n trained and evaluated on the sequence D_n. Likewise inequality (11) enables us to deduce convergence and rates of convergence for recursive RBF classification rules.

Theorem 1. *Let $EY^2 < \infty$,*

$$c_1 I_{S_{0,r}} \leq \phi(||x||) \leq c_2 I_{S_{0,R}}, \quad 0 < r < R < \infty, \quad c_1, c_2 > 0 \tag{12}$$

$$h_n \to 0, n^{-2} \sum_{i=1}^{n} h_i^{-d} \to 0 \quad as \ n \to \infty,$$

$$\limsup_n \frac{h_n^{-d}}{n^{-1} \sum_{i=1}^{n} h_i^{-d}} < \infty. \tag{13}$$

Then

$$E(g_n(X) - R(X))^2 \to 0 \quad as \ n \to \infty$$
$$E(m_n(X_1) - R(X_1))^2 \to 0 \quad as \ n \to \infty$$
and
$$E(L_n(X_1) - L^*(X_1))^2 \to 0 \quad as \ n \to \infty.$$

Theorem 1 provides MISE convergence of the recursive RBF plug-in estimates $g_n(x), m_n$ and L_n for all distributions of the data with the bounded second moment condition $EY^2 < \infty$. Radial functions satisfying condition (12) are functions with compact support separated away from zero at the origin. Such functions do not include Gaussian kernel. Using kernel trick introduced in [13] one can show that condition (12) can be relaxed to

$$\phi(||x||) \geq c I_{S_{0,r}}, \int \sup_{y \in S_{x,r}} \phi(||y||)dx < \infty, \quad c, r > 0. \tag{14}$$

However, enlarging the class of kernels results in necessity to impose stricter condition on the outputs, namely $|Y| \leq M < \infty$.

Condition (14) means that envelope of ϕ is bounded away from zero at the origin and is Riemann integrable and ϕ may have infinite support. It is satisfied by Gaussian and exponential kernels. Assumption (12) is satisfied for arbitrary finite ϕ with compact support and bounded away from zero at the origin.

Theorem 2 provides MISE convergence of the recursive RBF plug-in algorithms. Note that the rate of convergence is obtained for Lipschitz regression. As Devroye [9] points out there is no free-lunch, i. e., there are no distribution-free rates of convergence.

Theorem 2. *Let μ denote the probability measure of X with a compact support and let (12) hold. Let smoothing bandwidth h_i satisfy (13). Also let*

$$\sup_x E(Y^2 | X = x) \leq \sigma^2 < \infty \tag{15}$$
$$|R(x) - R(y)| \leq \beta ||x - y||^\alpha, \quad 0 < \alpha \leq 1, \quad \beta > 0.$$

Then

$$E(g_n(X) - R(X))^2 = O(n^{-\frac{2\alpha}{2\alpha+d}})$$
$$E(m_n(X_1) - R(X_1))^2 = O(n^{-\frac{2\alpha}{2\alpha+d}})$$
and
$$E(L_n(X_1) - L^*(X_1))^2 = O(n^{-\frac{2\alpha}{2\alpha+d}}). \tag{16}$$

The final result concerns an exponential bound from which almost sure convergence of the learning algorithms follows. For the sake of brevity we only present the result for g_n.

Theorem 3. *Let (12) and (13) hold. Then for every distribution of (X, Y) with $E|Y|^{2+\delta} < \infty$ with $\delta > 0$ and for every $\epsilon > 0$, there exist constants c and n_0 such that for all $n \geq n_0$,*

$$P\{\int |g_n(x) - g(x)|\mu(dx) > \epsilon\} \leq 2\exp(-c \cdot n). \tag{17}$$

4.2   Outlines of Proofs

The sketches of the proofs are given below. We only provide proofs for g_n as the proofs for remaining algorithms are similar.

Proof of Theorem 1. Let $K_i = \frac{1}{h_i^d}K((X_1 - X_i)/h_i), i = 1, \ldots, n$. We start by noticing that we have for any function g

$$E(R(X_1) - g_n(X_1))^2$$
$$\leq 4E(R(X_1) - g(X_1))^2 + 4E(\sum_{j=1}^{n}(Y_j - R(X_j))K_j/\sum_{j=1}^{n}K_j)^2$$
$$+4E\left(\frac{\sum_{j=1}^{n}(R(X_j)-g(X_j))K_j}{\sum_{j=1}^{n}K_j}\right)^2 + 4E\left(\frac{\sum_{j=1}^{n}g(X_j)K_j}{\sum_{j=1}^{n}K_j} - g(X_1)\right)^2$$
$$= 4(A + B + C + D).$$

For any $\epsilon > 0$ we can find continuous, compactly supported $g \in L_2(\mu)$ such that $\int (R(x) - g(x))^2 d\mu(x) < \epsilon/16$. Hence

$$A \leq \epsilon/16.$$

Using Jensen's inequality we can bound term C by $\epsilon/16$ and term D by

$$\sup_{x,y:||x-y||\leq\delta} |g(x) - g(y)|^2 < \epsilon/16.$$

Term B can be bounded by using truncation argument for Y, conditions (13).

Proof of Theorem 2. Let's bound MISE as follows

$$E(R(X_1) - g_n(X_1))^2$$
$$\leq 2E(\sum_{i=1}^{n}(Y_i - R(X_i))K_i/\sum_{i=1}^{n}K_i)^2 + 2E(\sum_{i=1}^{n}(R(X_i) - R(X_1))K_i/\sum_{i=1}^{n}K_i)^2$$
$$= 2(A + B).$$

Mimicking [33] one can show that compactness of μ implies

$$A = O\left(n^{-2}\sum_{i=1}^{n}h_i^{-d}\right). \tag{18}$$

Using Jensen's inequality and Lipschitz assumption for term B we have

$$B = O\left(\frac{\sum_{i=1}^n h_i^{-2d+2\alpha}}{\sum_{i=1}^n h_i^{-d}}\right). \tag{19}$$

The rate result (16) follows from (18) and (19).

Proof of Theorem 3. The exponential bound (17) follows from the exponential Hoeffding's type inequality for martingale difference sequences called Azuma inequality (see Azuma [2]) or in a more straightforward way from the fundamental McDiarmid's inequality [37], which can be stated as follows:

Let $X_1, \cdots, X_n$ be independent random variables and assume that

$$\sup_{x_i, x_i'} |f(x_1, \ldots, x_i, \ldots x_n) - f(x_1, \ldots, x_i', \ldots x_n)| \leq c_i, 1 \leq i \leq n. \tag{20}$$

Then

$$P\{|f(X_1, \cdots, X_n) - Ef(X_1, \cdots, X_n)| \geq \epsilon\} \leq 2\exp(-2\epsilon^2/\sum_{i=1}^n c_i^2). \tag{21}$$

Let $K((x - x_i)/h_i) = K_{h_i}(x - x_i)$ and take

$$f(X_1, Y_1, \cdots, X_n, Y_n) = \sum_{i=1}^n \int (Y_i - R(x))K_{h_i}(x - X_i)/\sum_{i=1}^n EK_{h_i}(x - X_i)\mu(dx).$$

Following [13]) the left-hand side of (20) is bounded by

$$\sup_{x_i, x_i', y_i, y_i'} \int \frac{|y_i K_{h_i}(x - x_i) - y_i' K_{h_i}(x - x_i')|}{\sum_{i=1}^n EK_{h_i}(x - X_i)}\mu(dx)$$
$$\leq 4M \sup_y \int \frac{K_{h_i}(x - y)}{\sum_{i=1}^n EK_{h_i}(x - X_i)}\mu(dx) \leq 4M\frac{\rho}{n}\left(\frac{h_i}{h}\right)^d.$$

Mimicking the proof of Theorem 1 in [28] (we omit the details) the result follows from (13) and (21).

5   Conclusions

We have analyzed MISE and strong convergence and the rates of convergence of the normalized radial basis function regression estimates and classification rules learned from data by the empirical risk minimization. The analysis has been simplified by taking advantage of the relationship between the empirical risk minimization and the recursive Wolverton-Wagner type kernel regression estimate.

References

1. Anthony, M., Bartlett, P.L.: Neural Network Learning: Theoretical Foundations. Cambridge University Press, Cambridge (1999)
2. Azuma, K.: Weighted sums of certain dependent random variables. Tohoku Math. J. **19**(3), 357–367 (1967)
3. Barron, A.R.: Universal approximation bounds for superpositions of a sigmoidal function. IEEE Trans. Inf. Theory **39**, 930–945 (1993)
4. Beirlant, J., Györfi, L.: On the asymptotic L_2-error in partitioning regression estimation. J. Stat. Plan. Inference **71**, 93–107 (1998)
5. Breiman, L.: Random forests. Mach. Learn. **45**, 5–32 (2001)
6. Breiman, L., Friedman, J.H., Olshen, R.A., Stone, C.J.: Classification and Regression Trees. Wadsworth Advanced Books and Software, Belmont, CA (1984)
7. Broomhead, D.S., Lowe, D.: Multivariable functional interpolation and adaptive networks. Complex Syst. **2**, 321–323 (1988)
8. Cybenko, G.: Approximations by superpositions of sigmoidal functions. Math. Control Sig. Syst. **2**, 303–314 (1989)
9. Devroye, L.: Any discrimination rule can have arbitrary bad probability of error for finite sample size. IEEE Trans. Pattern Anal. Mach. Intell. **PAMI-4**, 154–157 (1982)
10. Devroye, L.P., Wagner, T.J.: On the L1 convergence of the kernel estimators of regression functions with applications in discrimination. Zeitschrift für Wahrscheinlichkeitstheorie und verwandte Gebiete **51**(1), 15–25 (1980)
11. Devroye, L., Györfi, L., Lugosi, G.: Probabilistic Theory of Pattern Recognition. Springer, New York (1996). https://doi.org/10.1007/978-1-4612-0711-5
12. Devroye, L., Györfi, L., Krzyżak, A., Lugosi, G.: On the strong universal consistency of nearest neighbor regression function estimates. Ann. Stat. **22**, 1371–1385 (1994)
13. Devroye, L., Krzyżak, A.: An equivalence theorem for L_1 convergence of the kernel regression estimate. J. Stat. Plan. Inference **23**, 71–82 (1989)
14. Devroye, L., Biau, G.: Lectures on the Nearest Neighbor Method. Springer, New York (2015). https://doi.org/10.1007/978-3-319-25388-6
15. Duchon, J.: Sur l'erreur d'interpolation des fonctions de plusieurs variables par les D^m-splines. RAIRO-Analyse Numèrique **12**(4), 325–334 (1978)
16. Faragó, A., Lugosi, G.: Strong universal consistency of neural network classifiers. IEEE Trans. Inf. Theory **39**, 1146–1151 (1993)
17. Girosi, F., Anzellotti, G.: Rates of convergence for radial basis functions and neural networks. In: Mammone, R.J. (ed.) Artificial Neural Networks for Speech and Vision, pp. 97–113. Chapman and Hall, London (1993)
18. Girosi, F., Jones, M., Poggio, T.: Regularization theory and neural network architectures. Neural Comput. **7**, 219–267 (1995)
19. Greblicki, W.: Asymptotically Optimal Probabilistic Algorithms for Pattern Recognition and Identification. Monografie No. 3. Prace Naukowe Instytutu Cybernetyki Technicznej Politechniki Wroclawskiej, Nr. 18, Wroclaw, Poland (1974)
20. Greblicki, W., Pawlak, M.: Fourier and Hermite series estimates of regression functions. Ann. Inst. Stat. Math. **37**, 443–454 (1985)
21. Greblicki, W., Pawlak, M.: Necessary and sufficient conditions for Bayes risk consistency of a recursive kernel classification rule. IEEE Trans. Inf. Theory, **IT-33**, 408–412 (1987)

22. Györfi, L., Kohler, M., Krzyżak, A., Walk, H.: A Distribution-Free Theory of Nonparametric Regression. Springer, New York (2002). https://doi.org/10.1007/b97848

23. Hastie, T., Tibshirani, R., Friedman, J.: The Elements of Statistical Learning; Data Mining, Inference and Prediction, 2nd edn. Springer, New York (2009). https://doi.org/10.1007/978-0-387-84858-7

24. Haykin, S.O.: Neural Networks and Learning Machines, 3rd edn. Prentice-Hall, New York (2008)

25. Hornik, K., Stinchocombe, S., White, H.: Multilayer feed-forward networks are universal approximators. Neural Netw. **2**, 359–366 (1989)

26. Kohler, M., Krzyżak, A.: Nonparametric regression based on hierarchical interaction models. IEEE Trans. Inf. Theory **63**, 1620–1630 (2017)

27. Krzyżak, A.: The rates of convergence of kernel regression estimates and classification rules. IEEE Trans. Inf. Theory, **IT-32**, 668–679 (1986)

28. Krzyżak, A.: Global convergence of recursive kernel regression estimates with applications in classification and nonlinear system estimation. IEEE Trans. Inf. Theory **IT-38**, 1323–1338 (1992)

29. Krzyżak, A., Linder, T., Lugosi, G.: Nonparametric estimation and classification using radial basis function nets and empirical risk minimization. IEEE Trans. Neural Netw. **7**(2), 475–487 (1996)

30. Krzyżak, A., Linder, T.: Radial basis function networks and complexity regularization in function learning. IEEE Trans. Neural Netw. **9**(2), 247–256 (1998)

31. Krzyżak, A., Niemann, H.: Convergence and rates of convergence of radial basis functions networks in function learning. Nonlinear Anal. **47**, 281–292 (2001)

32. Krzyżak, A., Partyka, M.: Convergence and rates of convergence of recursive radial basis functions networks in function learning and classification. In: Rutkowski, L., Korytkowski, M., Scherer, R., Tadeusiewicz, R., Zadeh, L.A., Zurada, J.M. (eds.) ICAISC 2017. LNCS (LNAI), vol. 10245, pp. 107–117. Springer, Cham (2017). https://doi.org/10.1007/978-3-319-59063-9_10

33. Krzyżak, A., Pawlak, M.: Universal consistency results for the Wolverton-Wagner regression estimate with application in discrimination. Probl. Control Inf. Theory **12**, 33–42 (1983)

34. Krzyżak, A., Pawlak, M.: Distribution-free consistency of a nonparametric kernel regression estimate and classification. IEEE Trans. Inf. Theory **IT-30**, 78–81 (1984)

35. Krzyżak, A., Schäfer, D.: Nonparametric regression estimation by normalized radial basis function networks. IEEE Trans. Inf. Theory **51**, 1003–1010 (2005)

36. Lugosi, G., Zeger, K.: Nonparametric estimation via empirical risk minimization. IEEE Trans. Inf. Theory **41**, 677–687 (1995)

37. McDiarmid, C.: On the method of bounded differences. Surv. Comb. **141**, 148–188 (1989)

38. Moody, J., Darken, J.: Fast learning in networks of locally-tuned processing units. Neural Comput. **1**, 281–294 (1989)

39. Park, J., Sandberg, I.W.: Universal approximation using Radial-Basis-Function networks. Neural Comput. **3**, 246–257 (1991)

40. Park, J., Sandberg, I.W.: Approximation and Radial-Basis-Function networks. Neural Comput. **5**, 305–316 (1993)

41. Ripley, B.D.: Pattern Recognition and Neural Networks. Cambridge University Press, Cambridge (2008)

42. Scornet, E., Biau, G., Vert, J.-P.: Consistency of random forest. Ann. Stat. **43**(4), 1716–1741 (2015)

43. Shorten, R., Murray-Smith, R.: Side effects of normalising radial basis function networks. Int. J. Neural Syst. **7**, 167–179 (1996)
44. Specht, D.F.: Probabilistic neural networks. Neural Netw. **3**, 109–118 (1990)
45. Vapnik, V.N., Chervonenkis, A.Y.: On the uniform convergence of relative frequencies of events to their probabilities. Theory Probab. Appl. **16**, 264–280 (1971)
46. Vapnik, V.N.: Estimation of Dependences Based on Empirical Data. Springer, New York (1999). https://doi.org/10.1007/0-387-34239-7
47. White, H.: Connectionist nonparametric regression: multilayer feedforward networks that can learn arbitrary mappings. Neural Netw. **3**, 535–549 (1990)
48. Wolverton, C.T., Wagner, T.J.: Asymptotically optimal discriminant functions for pattern classification. IEEE Trans. Inf. Theory **IT-15**, 258–265 (1969)
49. Xu, L., Krzyżak, A., Yuille, A.L.: On radial basis function nets and kernel regression: approximation ability, convergence rate and receptive field size. Neural Netw. **7**, 609–628 (1994)

Porous Silica-Based Optoelectronic Elements as Interconnection Weights in Molecular Neural Networks

Magdalena Laskowska[1], Łukasz Laskowski[2(✉)], Jerzy Jelonkiewicz[2],
Henryk Piech[3], and Zbigniew Filutowicz[4,5]

[1] Institute of Nuclear Physics, Polish Academy of Sciences, 31-342 Krakow, Poland
[2] Department of Microelectronics and Nanotechnology,
Czestochowa University of Technology, Al. Armii Krajowej 36,
42-201 Czestochowa, Poland
[3] Institute of Computer Science, Czestochowa University of Technology,
Ul. Dabrowskiego 69, 42-201 Czestochowa, Poland
[4] Information Technology Institute, University of Social Sciences,
90-113 Łodz, Poland
`lukasz.laskowski@kik.pcz.pl`
[5] Clark University, Worcester, MA 01610, USA

Abstract. The paper describes a unique approach to optoelectronic elements application in artificial intelligence. Previously we considered molecular neural networks on the base of the functional porous silica thin films. But, for the successful molecular neural network design, we need efficient connections among them. Therefore we are presenting a material with tuneable non-linear optical (NLO) properties to be used for the optical signal transfer. The idea is briefly described and then followed by an experimental part to validate its feasibility. Promising results show that it is possible to design and synthesize the material with tuneable NLO properties.

Keywords: Artificial intelligence · Functional materials
Hopfield neural network · Spin-glass · Molecular magnet

1 Introduction

Optoelectronics is a branch of electronics that strongly overlaps with physics. Optoelectronic devices are electronic elements that operate on both light and electrical currents. Presently such elements become more and more popular for the reason of their usability in the modern IT devices. Particularly optoelectronic elements can be applied for construction of the intelligent systems [14], such as interconnection weights in Molecular Neural Networks [1–3,11,12].

Previously we reported how to create the molecular neural network on the base of the functional porous silica thin films [8,10]. Obtaining the bi-stable magnetic elements in a form of layout appears to be possible by means of the

© Springer International Publishing AG, part of Springer Nature 2018
L. Rutkowski et al. (Eds.): ICAISC 2018, LNAI 10841, pp. 130–135, 2018.
https://doi.org/10.1007/978-3-319-91253-0_13

bottom-up nanotechnology method. It will be described in details in our another article. Nevertheless, connections between neurons seem to be a challenging task. This problem can be solved by the application of optoelectronic elements, that can finely tune the optical signal. In the paper, we are presenting the material with tuneable non-linear optical properties. The starting point of the material's molecular structure design was self-assembly method that opens a promising way for the fabrication of materials with an extremely ordered structure on the molecular level. The properties of such materials are very sensitive to the supramolecular arrangement of the molecular units, that determines the molecular polarity and hence all molecular properties. The intermolecular interactions are responsible for cooperative phenomena and for the appearance of new features at the space charge transfer which is crucial for the NLO response [13]. Thus, the same molecule behaves substantially differently in other environments. This feature can be applied to the fabrication of materials with tuneable NLO properties, that can be practically used in optoelectronic devices. To obtain proper polar molecules and being able to precisely control their properties we propose to adjust the distance between them. We applied an appropriate surrounding matrix to control the intermolecular distances and so the interactions between active polar groups. The point, in this case, is a rigorous control of single functional molecules space distribution inside the matrix, in order to precisely adjust the distance between each other. The distance between the molecules determines supra-molecular interactions and therefore the final NLO response. Such a precisely assembled structure has not been found in the literature. Previously we proved that very accurate placement of functional units inside silica matrix is possible. In this way, we managed to avoid any agglomerations and obtain regularly distributed metal ions [6,7]. Having achieved this technology and after the modification of the silica matrix form, we are able to design a novel functional material in the form of thin films that shows the NLO response tuning.

According to the above-mentioned assumptions we designed the compound based on the porous silica matrix. This compound contains the polar functional units to be used as a material for desirable second and third order NLO. As a matrix we postulate mesoporous silica in the form of thin films, having 2D hexagonally distributed pores with the diameter of about 2 nm, aligned perpendicularly to the substrate. This material plays a role of the neutral (non-polar) matrix for the active polar units. One of a crucial feature of our matrix is its transparency in the visible spectral range which allows the functional groups, that are inside the material, to be excited efficiently.

For optically active centres we propose copper propyl phosphonate units that the play role of the polar - bonding between copper and oxygen atoms at the phosphonate units [9]. They have polarized covalent character. Such copper-containing functional groups are regularly distributed and anchored inside the silica matrix. The structure of the proposed material can be seen in Fig. 1.

By modification of the functional groups concentration inside the matrix, we are able to tune the degree of supramolecular interactions and also material's

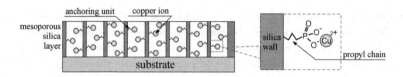

Fig. 1. Schematic representation of the ordered porous silica thin layers with vertically aligned pores, containing copper ions bounded via propyl-phosphonate groups.

molecular structure. Such a structure can be then used for tuning the NLO susceptibilities. Nevertheless, modification of functional groups content can result in systems reorganization what can lead to the quasi-phase transition. Therefore it is vital to investigate this phenomenon in details to take it into account during the NLO properties tuning.

We have shown that internal molecular structure of the species can define supramolecuar interactions, that can be fully controllable and strongly affects NLO properties of the samples.

2   Experimental Part

In order to check our assumptions, we prepared materials containing a various concentration of functional units according to the procedure described [4].

The structure of copper-containing mesoporous thin films was studied by the microscopic observation. We obtained correct geometry of silica film: 2D hexagonal well-ordered pores, creating irregular domains with diameter around 60–100 nm, containing pores lying perpendicularly to the substrate in the whole film's bulk. The distance between them was found to be close to 4 nm while their diameter was about 2.5 nm. Moreover, there were no impurities on the film surface, nor any agglomerations and crystals in the bulk.

The molecular structure of the samples was probed on the base of vibrational spectroscopy: Raman and infrared, supported by the numerical simulations (identification of vibrational modes). The procedure was applied earlier for similar systems with good efficiency [5,9]. The peak fitting analysis was applied to follow main silica spectral band position as well as full width and half maximum (FWHM). Hence, it was possible to trace precisely the increase in copper-containing functional group concentration and their impact on the silica matrix wall modification.

In order to explain vibrational spectroscopy results, we assumed presence of two different configurations of the copper containing functional groups:

- copper ion coordinated by single propyl-phosphonate unit – single-side-anchored-group – **SYSTEM SSA**,
- two copper ions coordinated by pair of propyl-phosphonate units from both sites – double-side-anchored-group – **SYSTEM DSA**.

To reveal, which system dominates in the samples, we prepared numerical models of both molecules. The optimized model structure of SYSTEM SSA and SYSTEM DSA can be seen in Fig. 2.

a)

b)

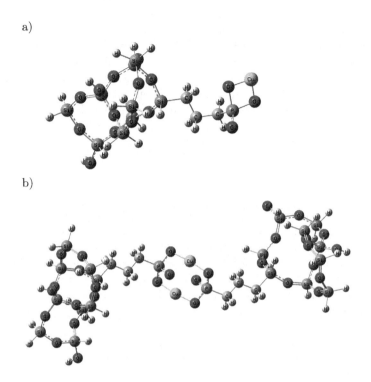

Fig. 2. Numerical model of functional groups attached to silica fragment for two configurations: SYSTEM SSA (a) and SYSTEM DSA (b)

Raman spectroscopy supported by DFT simulations provided information about the molecular structure of the specimens containing a different concentration of the functional groups. The bands' positions point out that for low functional groups concentration their position is close to those, calculated for SYSTEM SSA, while for the case of high functional groups concentration, their position is close to calculations results for SYSTEM DSA. Moreover, their positions change can be estimated by a Boltzman-like equation that describes phase-transition.

Taking into account preparation procedure, for low concentration of functional units, anchoring groups are well separated and placed relatively far from each other (anchoring points are further than 2.1 nm). Thus the copper ions cannot double the coordination by two propyl-phosphonate units. However, with increasing of the anchoring groups content, anchoring points become closer (denser distribution of anchoring groups). Taking under consideration length of SYSTEM DSA – double-side-anchored-groups – ~2.1 nm – double coordination of copper ions becomes possible. Of course, propyl chains can be bonded, so, with increasing of functional groups content, more conformations are possible. We can assume, that when the concentration is about 5% of molar we induce quasi-phase transition inside the system. It means that configuration of the internal active units changes from the single-side-anchored (dipolar) to the double-side-anchored (non-polar).

All these information can be used for fine tuning of nonlinear optical response. As we shown in our previous work [4], second and third harmonic signals increase with increasing of the functionalization degree, but only up to the limited degree of about 5% of the molar. Above this degree, the NL signals suddenly decrease. This phenomenon can be used for NLO response tuning by adjusting the functional groups concentration in the material.

3   Conclusion

In the paper, we clearly have shown that it is possible to design and synthesize the material with tunable NLO properties. Vibrational spectroscopy along with microscopic observations shown that we obtained the material with assumed microscopic and molecular structure. Moreover, we found that quasi-phase transition occurs in the material as a result of the functional groups concentration. Nevertheless, such a phenomenon can be taken into account during the NLO response tuning.

Aknowledgement. Financial support for this investigation has been provided by the National Centre of Science (Grant-No: 2015/17/N/ST5/03328).

References

1. Chang, O., Constante, P., Gordon, A., Singana, M.: A novel deep neural network that uses space-time features for tracking and recognizing a moving object. J. Artif. Intell. Soft Comput. Res. **7**(2), 125–136 (2017)
2. Ke, Y., Hagiwara, M.: An English neural network that learns texts, finds hidden knowledge, and answers questions. J. Artif. Intell. Soft Comput. Res. **7**(4), 229–242 (2017)
3. Khan, N.A., Shaikh, A.: A smart amalgamation of spectral neural algorithm for nonlinear Lane-Emden equations with simulated annealing. J. Artif. Intell. Soft Comput. Res. **7**(3), 215–224 (2017)
4. Laskowska, M., Kityk, I., Dulski, M., Jedryka, J., Wojciechowski, A., Jelonkiewicz, J., Wojtyniak, M., Laskowski, L.: Functionalized mesoporous silica thin films as a tunable nonlinear optical material. Nanoscale **9**(33), 12110–12123 (2017)
5. Laskowska, M., Laskowski, L., Jelonkiewicz, J.: SBA-15 mesoporous silica activated by metal ions - verification of molecular structure on the basis of raman spectroscopy supported by numerical simulations. J. Mol. Struct. **1100**, 21–26 (2015)

6. Laskowski, Ł., Laskowska, M.: Functionalization of SBA-15 mesoporous silica by Cu-phosphonate units: probing of synthesis route. J. Solid State Chem. **220**, 221–226 (2014)
7. Laskowski, Ł., Laskowska, M., Balanda, M., Fitta, M., Kwiatkowska, J., Dzilinski, K., Karczmarska, A.: Mesoporous silica SBA-15 functionalized by nickel-phosphonic units: Raman and magnetic analysis. Microporous Mesoporous Mater. **200**, 253–259 (2014)
8. Laskowski, Ł., Laskowska, M., Jelonkiewicz, J., Boullanger, A.: Molecular approach to hopfield neural network. In: Rutkowski, L., Korytkowski, M., Scherer, R., Tadeusiewicz, R., Zadeh, L.A., Zurada, J.M. (eds.) ICAISC 2015. LNCS (LNAI), vol. 9119, pp. 72–78. Springer, Cham (2015). https://doi.org/10.1007/978-3-319-19324-3_7
9. Laskowski, Ł., Laskowska, M., Jelonkiewicz, J., Dulski, M., Wojtyniak, M., Fitta, M., Balanda, M.: SBA-15 mesoporous silica free-standing thin films containing copper ions bounded via propyl phosphonate units-preparation and characterization. J. Solid State Chem. **241**, 143–151 (2016)
10. Laskowski, Ł., Laskowska, M., Jelonkiewicz, J., Piech, H., Galkowski, T., Boullanger, A.: The concept of molecular neurons. In: Rutkowski, L., Korytkowski, M., Scherer, R., Tadeusiewicz, R., Zadeh, L.A., Zurada, J.M. (eds.) ICAISC 2016. LNCS (LNAI), vol. 9693, pp. 494–501. Springer, Cham (2016). https://doi.org/10.1007/978-3-319-39384-1_43
11. Liu, H., Gegov, A., Cocea, M.: Rule based networks: an efficient and interpretable representation of computational models. J. Artif. Intell. Soft Comput. Res. **7**(2), 111–123 (2017)
12. Minemoto, T., Isokawa, T., Nishimura, H., Matsui, N.: Pseudo-orthogonalization of memory patterns for complex-valued and quaternionic associative memories. J. Artif. Intell. Soft Comput. Res. **7**(4), 257–264 (2017)
13. Terenziani, F., Painelli, A.: Supramolecular interactions in clusters of polar and polarizable molecules. Phys. Rev. B **68**(16), 165405 (2003)
14. Zalasiński, M., Cpałka, K., Er, M.J.: New method for dynamic signature verification using hybrid partitioning. In: Rutkowski, L., Korytkowski, M., Scherer, R., Tadeusiewicz, R., Zadeh, L.A., Zurada, J.M. (eds.) ICAISC 2014. LNCS (LNAI), vol. 8468, pp. 216–230. Springer, Cham (2014). https://doi.org/10.1007/978-3-319-07176-3_20

Data Dependent Adaptive Prediction and Classification of Video Sequences

Amrutha Machireddy$^{(\boxtimes)}$ and Shayan Srinivasa Garani

Department of Electronic Systems Engineering, Indian Institute of Science,
Bengaluru 560012, India
{amrutha,shayangs}@iisc.ac.in

Abstract. Convolutional neural networks (CNN) are popularly used for applications in natural language processing, video analysis and image recognition. However, the max-pooling layer used in CNNs discards most of the data, which is a drawback in applications, such as, prediction of video frames. With this in mind, we propose an adaptive prediction and classification network (APCN) based on a data-dependent pooling architecture. We formulate a combined cost function for minimizing prediction and classification errors. During testing, we identify a new class in an unsupervised fashion. Simulation results over a synthetic data set show that the APCN algorithm is able to learn the spatio-temporal information to predict and classify the video frames, as well as, identify a new class during testing.

Keywords: Data-dependent pooling · Adaptive network
Prediction and classification network

1 Introduction

With an increase in the number of videos available on the internet, algorithms to extract and analyze the content present in them are gaining increased prominence. Video feature extraction has been largely studied due to its applications in fields like video surveillance, scene identification [1], action recognition [2] etc. The high-level details in a video can be obtained by analyzing the interaction of features in a video sequence. These details define the class a particular video belongs to. There are two broad approaches for video classification, i.e., using supervised [1–3] and unsupervised [4] algorithms. The authors in [3] use volumetric features which are used to scan the videos to obtain spatio-temporal matching. The authors in [4] use spatio-temporal words to classify human action videos. CNNs [5] have shown promising results for applications such as scene labeling [6], image [7] and video classification [1] as they are capable of extracting spatial information. The above-mentioned architectures are capable of classifying a given video but fail at predicting the next video frame.

Inspired by language modeling, the authors in [8] proposed a recurrent convolutional neural network architecture for predicting video frames in the form of

© Springer International Publishing AG, part of Springer Nature 2018
L. Rutkowski et al. (Eds.): ICAISC 2018, LNAI 10841, pp. 136–147, 2018.
https://doi.org/10.1007/978-3-319-91253-0_14

patch clusters. However, this network is not capable of predicting video frames further ahead in time due to short-term memory in the network. To overcome this, Xingjian et al. [9] introduced convolutional long short-term memory (ConvLSTM) which is capable of learning both the spatial, as well as, the temporal correlations between video frames in a sequence. Using the ideas of ConvLSTM in an unsupervised architecture, authors in [10] consider prediction of consecutive video frames. However, none of the algorithms discussed till now are capable of both prediction, as well as, classification of video frames.

Consider an application, such as, fully automated traffic signal monitoring. Based on the video frames, the system would like to identify possible violations via classification and predict subsequent actions from an action space. In such cases, we would need a network that is capable of both predicting and classifying the video frames. The learning techniques thus far use max-pooling which is known to result in loss of feature details during training. Also, if the network is trained to identify P number of violations, it is possible that the network might see a new type of violation, resulting in misclassification. Motivated by such practical considerations, we propose a technique that considers the data present in the activation map to perform pooling. We also develop a technique to identify a new class during testing. We refer to our learning architecture as adaptive prediction and classification network (APCN).

This paper is organized as follows. In Sect. 2, we describe the APCN architecture. In Sect. 3, we set up the combined cost function for training and develop an algorithm to identify and learn new classes during testing. In Sect. 4, we perform experiments with a synthetic data set to validate our theory. Simulation results show that the APCN is capable of predicting and classifying the videos, as well as, adapt the network to accommodate new classes. We conclude the paper in Sect. 5.

2   Description of the APCN Architecture

Let us consider an input sequence of video frames $Y^{(1)}, \ldots, Y^{(t)}$ with each $Y^{(i)} \in \mathbb{R}^{m \times n}$. The network consists of two parts: feature extraction and feature consolidation. In the feature extraction part, we extract the spatio-temporal features from each video frame. To capture the spatial information, we learn a dictionary of features $\{D_j\}_{j=1}^q$ where each $D_j \in \mathbb{R}^{f \times f}$. For this purpose, the input frame $Y^{(t)}$ is convolved with kernels $\{D_j\}_{j=1}^q$ to obtain the corresponding coefficients $\{X_j\}_{j=1}^q$. The kernels $\{D_j\}_{j=1}^q$ can be considered as the low-level features. In the conventional CNN architecture, max-pooling is used for reducing the number of coefficients. This is a drawback during prediction. To overcome this, we formulate a new pooling mechanism which is data-dependent. We call this method as data-dependent pooling (DDP). In the case of classification, reducing the number of coefficients is helpful as the exact location of a feature is not necessary; on the other hand, loss of feature information makes it difficult for the algorithm to predict the next video frame. DDP strikes a balance between these two requirements to build a network capable of both prediction and classification. To capture the correlation between the consecutive frames of a video sequence, we use ConvLSTM. This structure of convolutional layer, DDP and ConvLSTM are repeated again to extract the higher-level features. The output from the 2nd ConvLSTM layer gives a spatio-temporal representation of the video frame. This forms the input to the feature consolidation part in which we extract the class information and predict the next video frame. The idea of our architecture can be visualized in Fig. 1. For classification, we use two fully connected layers and a softmax layer to identify the class label of each input video frame. For prediction of the next video frame $Y^{(t+1)}$, a set of deconvolutional layers and ConvLSTMs are used.

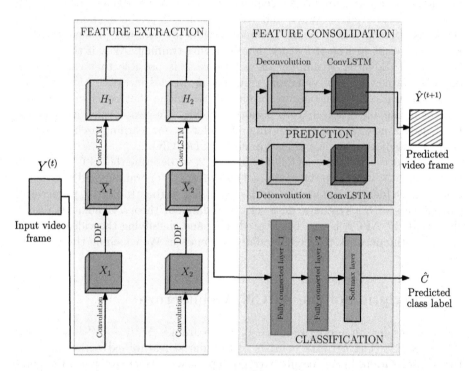

Fig. 1. Architecture of the APCN network: $Y^{(t)}$ is the input to the network. Feature extraction consists of the convolutional layers, data-dependent pooling layers and ConvLSTM layers. Feature consolidation for classification consists of fully connected layers concatenated with the softmax layer; and for prediction, it consists of deconvolutional layers concatenated with ConvLSTM layers. The outputs obtained from the network are $\hat{C}$, the class label of the video frame and $\hat{Y}^{(t+1)}$, the predicted video frame for the next time instance.

3   Algorithm

Consider N video sequences $\{Y_i\}_{i=1}^{N}$, where each Y_i consists of t video frames of size $m \times n$, i.e., $Y_i = \{Y_i^{(1)}, Y_i^{(2)}, \ldots, Y_i^{(t)}\}$. The video sequences $\{Y^{(i)}\}_{i=1}^{t}$

are given as input to the network. In the forward pass, for each video frame, we obtain the activation maps comprising of the coefficients $\{X_j\}_{j=1}^q$ corresponding to the convolution $(*)$ of the video frame $Y_i^{(t)}$ with the kernels $\{D_j\}_{j=1}^q$.

$$X_j = D_j * Y_i^{(t)} \; \forall \; j = 1, 2, \ldots, q. \tag{1}$$

3.1   Data-Dependent Pooling

We need to perform pooling over the activation maps which adaptively chooses the data that needs to be carried forward and the data to be removed. For this, we need a function that is capable of selectively keeping or discarding data. To achieve this, we use a Gaussian kernel of the form

$$G_{x,y,\sigma} = e^{\frac{-(x^2+y^2)}{2\sigma^2}}, \tag{2}$$

where x and y correspond to the positions of the kernel in $\mathbb{R}^2$ and σ^2 is the data-dependent variance. A lower σ^2 corresponds to a kernel which preserves the feature information locally. On the other hand, a higher σ^2 preserves the feature information globally. We need to compute the value of σ^2 to be chosen based on the values present in the activation map at the kernel location. To identify the amount of feature information present at the location of the kernel, we compute the sparsity parameter ρ given by

$$\rho = \frac{\text{Number of non-zero terms}}{\text{Total number of terms}}. \tag{3}$$

The variance of the kernel at a particular location is chosen such that $\sigma^2 = \gamma\rho$, where $\gamma \in (0, 10]$ is chosen heuristically based on the accuracy required. Higher γ corresponds to higher accuracy while lower γ corresponds to lower accuracy. DDP is performed after the convolutional layer by taking the Hadamard product $(\odot)$ at each location. The output from DDP, $\overline{X}$ is obtained as below:

$$\overline{X} = X \odot G_{x,y,\sigma}. \tag{4}$$

For example, suppose we have the activation map as given below

$$
\begin{array}{|c|c|c|}
\hline
1.0 & 0.0 & 0.0 \\\hline
0.0 & 0.0 & 0.0 \\\hline
0.0 & 0.0 & 7.0 \\\hline
\end{array}
\odot
\begin{array}{|c|c|c|}
\hline
0.2 & 0.4 & 0.2 \\\hline
0.4 & 1.0 & 0.4 \\\hline
0.2 & 0.4 & 0.2 \\\hline
\end{array}
\rightarrow
\begin{array}{|c|c|c|}
\hline
0.2 & 0.0 & 0.0 \\\hline
0.0 & 0.0 & 0.0 \\\hline
0.0 & 0.0 & 1.4 \\\hline
\end{array}
\tag{5}
$$

$$\text{Activation map, } X \qquad G_{x,y,\sigma} \qquad \text{Output from DDP, } \overline{X}$$

It can be seen from (5) that the amplitude of the coefficient corresponding to highest activation (depicted in red color) has become low (depicted in blue color) hence, we lose the feature information present therein. A higher value of the activation map coefficients corresponds to the presence of a particular feature in abundance. Keeping this in mind, the kernel center is placed at the location of the maximum coefficient (depicted in red color) as shown below

$$
\begin{array}{|c|c|c|}
\hline
1.0 & 0.0 & 0.0 \\
\hline
0.0 & 0.0 & 0.0 \\
\hline
0.0 & 0.0 & 7.0 \\
\hline
\end{array}
\odot
\begin{array}{|c|c|c|}
\hline
0.0 & 0.01 & 0.02 \\
\hline
0.01 & 0.16 & 0.4 \\
\hline
0.02 & 0.4 & 1.0 \\
\hline
\end{array}
\rightarrow
\begin{array}{|c|c|c|}
\hline
0.0 & 0.0 & 0.0 \\
\hline
0.0 & 0.0 & 0.0 \\
\hline
0.0 & 0.0 & 7.0 \\
\hline
\end{array}
\tag{6}
$$

Activation map, X $\qquad$ $G_{x,y,\sigma}$ $\qquad$ Output from DDP, $\overline{X}$

The drawback of this method is the loss of coefficients corresponding to certain features that are present only in this kernel location. To overcome this, a ranking matrix is created based on the entries of the activation map. Adjacent coefficient values are compared to decide the new rank. As the rank is increased, the σ^2 for the corresponding kernel is increased to preserve the feature information. After taking the Hadamard product with the kernel, the coefficients obtained are thresholded such that values below a threshold τ are made 0. The network was capable of extracting the spatial correlation information from the video frames. To extract the correlations between the consecutive frames the output from the DDP layer is then fed to the ConvLSTM layer. The main equations of ConvLSTM are given below:

$$
i^{(t)} = \sigma \left(\overline{X}^{(t)} * W_{xi} + H^{(t-1)} * W_{hi} + B_i \right) \tag{7}
$$

$$
f^{(t)} = \sigma \left(\overline{X}^{(t)} * W_{xf} + H^{(t-1)} * W_{hf} + B_f \right) \tag{8}
$$

$$
\tilde{C}^{(t)} = \tanh \left(\overline{X}^{(t)} * W_{x\tilde{c}} + H^{(t-1)} * W_{h\tilde{c}} + B_{\tilde{c}} \right) \tag{9}
$$

$$
C^{(t)} = \tilde{C}^{(t)} \odot i^{(t)} + C^{(t-1)} \odot f^{(t)} \tag{10}
$$

$$
o^{(t)} = \sigma \left(\overline{X}^{(t)} * W_{xo} + H^{(t-1)} * W_{ho} + B_o \right) \tag{11}
$$

$$
H^{(t)} = o^{(t)} \odot \tanh(C_t) \tag{12}
$$

The inputs to the ConvLSTM unit are the output from the DDP layer $\overline{X}^{(t)}$ and the hidden state from the previous time step $H^{(t-1)}$. It consists of a forget gate ($f^{(t)}$), an input gate ($i^{(t)}$) and an output gate ($o^{(t)}$) which enables the cell memory $C^{(t)}$ to selectively remember/forget the spatio-temporal correlations. These gates work like ON/OFF switches as they are computed using the sigmoid (σ) function whose output is in the range $[0, 1]$. The output from the unit is the hidden state $H^{(t)}$ which is a representation of the spatio-temporal correlations present in the video frames. Figure 2 shows the input-output relations of the ConvLSTM unit. The coefficients obtained from the hidden state of the ConvLSTM pass through the remaining layers as shown in Fig. 1.

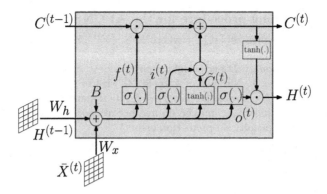

Fig. 2. Input-output relations in a ConvLSTM unit. The inputs to the unit are $\overline{X}^{(t)}$ and $H^{(t-1)}$ and the output is $H^{(t)}$. The spatial information is extracted through the convolutions at each step and the temporal information is extracted with the help of the cell memory $C^{(t)}$ which selectively remembers/forgets information based on the state of the forget gate ($f^{(t)}$), the input gate ($i^{(t)}$) and the output gate ($o^{(t)}$).

3.2   Cost Function

The cost function for classification E_c, is given below:

$$E_c = -\sum_i C_i \log(\hat{C}_i) , \qquad (13)$$

where, i is the number of neurons in the softmax layer, C_i is the ith coordinate of the true class label and $\hat{C}_i$ is the ith coordinate of output from the softmax layer for a given video frame. For prediction, the squared Euclidean distance is chosen as the cost function E_p and is given below:

$$E_p = \left\| \hat{Y}^{(t+1)} - Y^{(t+1)} \right\|_2^2 , \qquad (14)$$

where, $Y^{(t+1)}$ is the desired frame and $\hat{Y}^{(t+1)}$ is the predicted frame. We minimize E_c while keeping E_p below a threshold which is given by

$$\text{minimize } E_c$$
$$\text{subject to } E_p \leq \alpha, \qquad (15)$$

where α is a threshold on the prediction error. A regularized cost function is given by

$$E = E_c + \lambda(E_p - \alpha), \qquad (16)$$

where λ is the penalizing term which specifies the importance to be given to the prediction error. In the backward pass, the weights are updated similar to

---

Algorithm 1. Training procedure for the APCN network

Input: Video frame sequences $\{Y_i\}_{i=1}^{N}$ where $Y_i = \{Y_i^{(1)}, Y_i^{(2)}, \ldots, Y_i^{(T)}\}$ and corresponding labels $\{C_i\}_{i=1}^{N}$.

Output: Trained dictionary D and weight vectors.

    Initialization : Layer 1: Number of kernels N_1, size of kernels $f_1 \times f_1$, Layer 2: Number of kernels N_2, size of kernels $f_2 \times f_2$, dictionary D, initial learning rate η_{init}, final learning rate η_{final}, maximum number of epochs e_{max}, the threshold on prediction error α, penalty on prediction error λ.

1: **for** $e = 1$ to e_{max} **do**
2:    **for** $i = 1$ to N **do**
3:       **for** $t = 1$ to T **do**
4:          **for** $l = 1$ to 2 **do**
5:             Compute the coefficients using (1).

$$X_j = D_j * Y_i^{(t)} \ \forall \ j = 1, 2, \ldots, N_l.$$

6:             Perform data-dependent pooling to compute $\{\overline{X}_j\}_{j=1}^{N_l}$.
7:             Compute temporal coefficient $\{H_j\}_{j=1}^{N_l}$ using ConvLSTM.
8:          **end for**
9:          Compute the class label $\hat{C}_i$ and predicted video frame $\hat{Y}^{(t+1)}$.
10:         Compute classification error.

$$E_c = -\sum_i C_i \log(\hat{C}_i).$$

11:         Compute prediction error.

$$E_p = \left\| \hat{Y}^{(t+1)} - Y^{(t+1)} \right\|_2^2.$$

12:         Compute combined cost function.

$$E = E_c + \lambda(E_p - \alpha).$$

13:         Compute the gradient $\nabla_w E$ and update the weight vectors

$$w_{t+1} = w_t - \eta_e \nabla_w E.$$

14:      **end for**
15:    **end for**
16:    Update learning rate η

$$\eta_e = \eta_{init} \left(\frac{\eta_{final}}{\eta_{init}} \right)^{\frac{e}{e_{max}}}.$$

17: **end for**

---

backpropagation algorithm using the steepest gradient descent technique. The learning rate η is updated at the end of each epoch as follows:

$$\eta_e = \eta_{\text{init}} \left(\frac{\eta_{\text{final}}}{\eta_{\text{init}}} \right)^{\frac{e}{e_{\text{max}}}}, \tag{17}$$

where η_{init} and η_{final} are the initial and final learning rates respectively. e and e_{max} are the running epoch count and the maximum epochs respectively. Algorithm 1 describes the training procedure of the APCN.

The above algorithm is capable of predicting and classifying video frames of the classes it has learned during training. Since the network has been trained, we expect the classification error E_c to be less during testing. However, it is possible that some of the videos during testing belong to a class that the network has not seen during training. Such videos would get misclassified into one of the learned classes based on the final output at the softmax layer. We set an error threshold β on E_c of the network such that, when $E_c > \beta$ we identify the input to belong to a new class. We set ν to be the threshold on the number of misclassifications during testing. When ν inputs have been identified as belonging to a new class, the network is retrained to include a new class. Algorithm 2 describes the procedure of identifying and retraining the network for a new class.

Algorithm 2. Adaptive network for new class identification

Input: Video frame sequences $\{Y_i\}_{i=1}^{N_t}$ where $Y_i = \{Y_i^{(1)}, Y_i^{(2)}, \ldots, Y_i^{(T)}\}$ and corresponding labels $\{C_i\}_{i=1}^{N_t}$, $\mathcal{C}$ is the set of class labels, N_t is the number of test video sequences, β is the error threshold limit, k is misclassification counter and ν is the threshold for number of misclassifications.
    Initialization : $k = 0$, $C_0 = |\mathcal{C}|$.

1: **for** $r = 1$ to N_t **do**
2:     Test the input video Y_r with network trained using Algorithm 1.
3:     Compute classification error E_c.
4:     **if** $E_c \geq \beta$ **then**
5:         $k = k + 1$.
6:         Store the test video $S_k = Y_r$.
7:         **if** $k > \nu$ **then**
8:             $C_0 = C_0 + 1$.
9:             **retrain**(S_k, C_0)
10:       **end if**
11:     **end if**
12: **end for**

---

4   Simulation Setup and Results

In this section, we perform experiments with the synthetic data set and show the prediction and classification abilities of our algorithm along with the ability to learn a new class during testing.

4.1 Synthetic Data Set Generation

Due to computational constraints we validate our work using a synthetic dataset. We generate a video dataset corresponding to 10 classes, i.e., video sequences of digits from 0 to 9 moving inside a space of size 64×64. The generated dataset $\mathcal{S} = \{Y_i\}_{i=1}^{N}$ is such that each video sequence has 20 frames, i.e., $Y_i = \{Y_i^{(1)}, Y_i^{(2)}, \dots, Y_i^{(20)}\}$ and each $Y_i^{(t)} \in \mathbb{R}^{64 \times 64}$. For details on the images used for data generation, the reader is referred to [11,12]. The initial video frame is generated by placing the digit in the 64×64 lattice at random. The digit is then allowed to move in one of the four directions - northeast, northwest, southeast and southwest. The digit is allowed to move in that direction until it touches one of the edge of the lattice. The digit then moves in the next feasible direction chosen at random. In this way, a video consisting of 20 frames is generated. The above procedure is repeated for different classes to obtain the training and testing data.

4.2 Prediction Using Data Dependent Pooling

The goal of this experiment is to validate that the APCN is indeed capable of predicting the video frames using the ideas of data-dependent pooling. We use the synthetic data set generated above for this purpose. The APCN network was trained for 100 epochs. We choose the number of filters in each layer N_1 and N_2 to be 8 and 16 respectively. 3×3 filters were chosen in the first layer and filters of size 5×5 were chosen for the second layer. The initial and final learning rates are $\eta_i = 0.99$ and $\eta_f = 0.001$ respectively. The learning rate η is updated after each epoch. We train the network using $30,000$ video sequences. The video sequences are fed frame by frame to the network to enable it to learn the spatio-temporal correlations present between the frames. Figure 3 shows the convergence of the APCN algorithm. We observe that over the epochs, the error decreases. After completion of training, we test our algorithm with $N_t = 10,000$ video sequences.

Figure 4 shows the results for prediction of the video frames. From 4(d)–(f) and 4(j)–(l) we observe that the network is able to predict the location of the digit in the next video frame. It can be seen that the spatio-temporal information is being captured and the network is able to predict of the position of the digit in the next time step. Due to the lack of network architectures that perform both prediction and classification, we have not compared our work with any other algorithm.

4.3 Adaptability of the Network

The goal of this experiment is to test the adaptability of the network to identify and learn new classes during testing. We consider the synthetic dataset $\mathcal{S}$. To test the capability of identifying a new class of videos in the system, we train the network with only 9 classes, i.e., videos corresponding to digits 1 to 9. During testing, the training videos corresponding to the digit 0 are included along

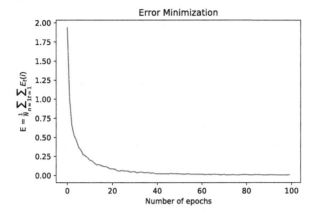

Fig. 3. Error versus epochs during training. It is observed that the error is decreasing monotonically towards the minimum.

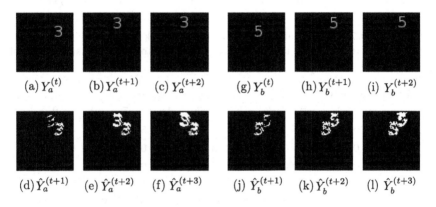

Fig. 4. Y_a and Y_b correspond to video sequences generated corresponding to digits 3 and 5 respectively. $Y^{(t)}$ and $\hat{Y}^{(t+1)}$ correspond to the input and output of the network at time t respectively. It is observed that the network in indeed able to predict the location of the digit in the next time frame. This shows that the spatio-temporal relation between video frames has been learned by the network.

with the test videos of digits from 1 to 9, i.e., the network is fed with videos corresponding to all the 10 classes. Since the network has been trained for 9 classes, irrespective of the input during testing, it will classify the test video into one of these 9 classes. The confusion matrix obtained during testing is shown in Table 1.

It is observed that videos corresponding to class 0 are misclassified into one of the trained class of the network. It can also be noted that some additional information about the new class can be obtained based on the confusion matrix, i.e., the new class 0 is similar to class 6 and class 8 spatially. The videos corresponding the new class are identified using Algorithm 2. The network is retrained using

Table 1. Confusion matrix during testing. It is observed that 0 is misclassified into one of the learned classes.

Actual	Predicted								
	1	2	3	4	5	6	7	8	9
0	17	431	14	17	29	1271	11	943	267
1	1000	0	0	0	0	0	0	0	0
2	0	1000	0	0	0	0	0	0	0
3	0	0	1000	0	0	0	0	0	0
4	0	0	0	1000	0	0	0	0	0
5	0	0	0	0	1000	0	0	0	0
6	0	0	0	0	0	1000	0	0	0
7	0	0	0	0	0	0	1000	0	0
8	0	0	0	0	0	0	0	1000	0
9	0	0	0	0	0	0	0	0	1000

Table 2. Confusion matrix after retraining the network using the videos identified as new class. It is observed that the network is able to learn the spatio-temporal information of the new class.

Actual	Predicted									
	0	1	2	3	4	5	6	7	8	9
0	989	0	1	0	0	0	5	0	4	1
1	0	1000	0	0	0	0	0	0	0	0
2	0	0	1000	0	0	0	0	0	0	0
3	0	0	0	1000	0	0	0	0	0	0
4	0	0	0	0	1000	0	0	0	0	0
5	0	0	0	0	0	1000	0	0	0	0
6	0	0	0	0	0	0	1000	0	0	0
7	0	0	0	0	0	0	0	1000	0	0
8	0	0	0	0	0	0	0	0	1000	0
9	0	0	0	0	0	0	0	0	0	1000

these videos and a new class is included in the output layer of the network. To test the performance after retraining, we use the test videos corresponding to the digit 0. The confusion matrix obtained is shown in Table 2. It is observed that the network is now able to classify class 0 as well as retain the information regarding the previously learned classes. This shows the adaptive spatio-temporal learning capability of the APCN network. The results show that our network is indeed capable of adapting itself to learn a new class during testing.

5   Conclusion

We formulated a cost function and derived an update rule for joint prediction and classification of video frames. The algorithm involves a data-dependent pooling layer which guides the network to preserve the spatio-temporal information content present in the video frames. The results obtained for prediction using the data-dependent pooling layer show that the network is indeed able to capture the spatio-temporal information. The second experiment shows that the network is capable of identifying and learning a new class on the fly. The future scope of this work would be to identify and learn multiple new classes during runtime.

Acknowledgments. S. S. Garani acknowledges IISc-start up funds for this project.

References

1. Karpathy, A., Toderici, G., Shetty, S., Leung, T., Sukthankar, R., Fei-Fei, L.: Large-scale video classification with convolutional neural networks. In: Proceedings of the IEEE Conference on Computer Vision and Pattern Recognition, pp. 1725–1732 (2014)
2. Ji, S., Xu, W., Yang, M., Yu, K.: 3D convolutional neural networks for human action recognition. IEEE Trans. Pattern Anal. Mach. Intell. **35**(1), 221–231 (2013)
3. Ke, Y., Sukthankar, R., Hebert, M.: Efficient visual event detection using volumetric features. In: 10th IEEE International Conference on Computer Vision, pp. 166–173 (2005)
4. Niebles, J.C., Wang, H., Fei-Fei, L.: Unsupervised learning of human action categories using spatial-temporal words. Int. J. Comput. Vision. **79**(3), 299–318 (2008)
5. LeCun, Y., Bottou, L., Bengio, Y., Haffner, P.: Gradient-based learning applied to document recognition. Proc. IEEE **86**(11), 2278–2324 (1998)
6. Farabet, C., Couprie, C., Najman, L., LeCun, Y.: Learning hierarchical features for scene labeling. IEEE Trans. Pattern Anal. Mach. Intell. **35**(8), 1915–1929 (2013)
7. Krizhevsky, A., Sutskever, I., Hinton, G.E.: Imagenet classification with deep convolutional neural networks. In: Advances in Neural Information Processing Systems, pp. 1097–1105 (2012)
8. Ranzato, M., Szlam, A., Bruna, J., Mathieu, M., Collobert, R., Chopra, S.: Video (language) modeling: a baseline for generative models of natural videos. arXiv preprint arXiv:1412.6604 (2014)
9. Xingjian, S., Chen, Z., Wang, H., Yeung, D., Wong, W., Woo, W.: Convolutional LSTM network: A machine learning approach for precipitation nowcasting. In: Advances in Neural Information Processing Systems, pp. 802–810 (2015)
10. Lotter, W., Kreiman, G., Cox, D.: Deep predictive coding networks for video prediction and unsupervised learning. arXiv preprint arXiv:1605.08104 (2016)
11. Bouncing digit dataset. https://github.com/ashwani-pandey/Bouncing-digit-dataset
12. The MNIST database. http://yann.lecun.com/exdb/mnist/

Multi-step Time Series Forecasting of Electric Load Using Machine Learning Models

Shamsul Masum[✉], Ying Liu, and John Chiverton

School of Engineering, University of Portsmouth,
Anglesea Building, Anglesea Road, Portsmouth PO1 3DJ, UK
{shamsul.masum,ying.liu,john.chiverton}@port.ac.uk

Abstract. Multi-step forecasting is very challenging and there are a lack of studies available that consist of machine learning algorithms and methodologies for multi-step forecasting. It has also been found that lack of collaborations between these different fields is creating a barrier to further developments. In this paper, multi-step time series forecasting are performed on three nonlinear electric load datasets extracted from Open-Power-System-Data.org using two machine learning models. Multi-step forecasting performance of Auto-Regressive Integrated Moving Average (ARIMA) and Long-Short-Term-Memory (LSTM) based Recurrent Neural Networks (RNN) models are compared. Comparative analysis of forecasting performance of the two models reveals that the LSTM model has superior performance in comparison to the ARIMA model for multi-step electric load forecasting.

Keywords: Time series analysis · Multi-step forecasting · ARIMA
LSTM

1 Introduction

Electric load forecasting plays a vital role in overall operation and planning of power systems. Accurate electric load forecasting helps to run the power system efficiently and effectively. Areas such as cause of power interruptions, coordination between supply and demand, operating costs, maintenance and infrastructure development can benefit from electric load forecasting. Electric load forecasting has therefore been a research subject of great interest for the past few decades [1]. However, electric load data is univariate which makes time series forecasting more challenging which can make it a difficult model to learn [2].

Duration of electric load forecasting can be mainly classified into two categories, short-term load forecasting and long-term load forecasting [3]. Single-step forecasting and multi-step forecasting are useful for short-term and long-term load forecasting respectively [4]. Several machine learning approaches have been considered for forecasting which can be divided into two broad categories, statistical techniques and soft computing techniques [5]. This paper will consider

© Springer International Publishing AG, part of Springer Nature 2018
L. Rutkowski et al. (Eds.): ICAISC 2018, LNAI 10841, pp. 148–159, 2018.
https://doi.org/10.1007/978-3-319-91253-0_15

Auto-Regressive Integrated Moving Average (ARIMA) as a statistical technique and Recurrent Neural Networks (RNN) as a soft computing technique applied here to forecasting electric load.

ARIMA is considered to be the most common approach for time series forecasting. The technique was introduced by Box and Jenkins. An ARIMA model considers the past values of the time series along with the errors in forecasting. It has been found to be effective for short-term forecasting [6,7]. However, it was found that ARIMA model performs better on linear time series data and stationary data compared to nonlinear and nonstationary data [8,9]. The ARIMA approach has been used by several researchers for short-term electric load forecasting [10–13].

RNN is a class of artificial neural network, which has been found to be an effective model for solving many forecasting problems in different applications. Researchers also found RNN as an effective model for electric load forecasting [14,15]. However, RNN seems to have difficulties in learning "long-term dependencies" as explored by Bengio et al. [16] in 1994. Long Short-Term Memory (LSTM) is a special kind of RNN introduced by Hochreiter and Schmidhuber in 1997, which solves the issue of learning long-term dependencies [17]. Applications of LSTM can be found in different areas. Sak et al. in 2014 used LSTM for speech recognition and concluded that LSTM outperforms a deep feed-forward neural network [18]. Marino et al. in 2016 compared standard LSTM and LSTM-based Sequence to Sequence (S2S) architecture for forecasting energy load and found LSTM-based Sequence to Sequence (S2S) performs better than the standard LSTM [19]. Wu et al. in 2016 found LSTM performs better than traditional deep neural network for forecasting wind power [20].

Comparison between ARIMA and RNN has been explored by many researchers in different fields. Ho et al. in 2002 compared ARIMA and RNN models for time series prediction and concluded that RNN performs better than ARIMA [21]. Fu et al. also found RNN model performed better than ARIMA model for predicting traffic flow [22] in 2016. Cao et al. explored RNN models and found that they outperform ARIMA models in forecasting wind speed [23]. Ma et al. compared LSTM with an ARIMA model in forecasting traffic speed prediction in 2015 and found LSTM outperforms the ARIMA model [24]. Tian et al. forecasted short-term traffic flow using LSTM and showed that LSTM performs better than most non-parameteric models [25].

It has been found that both ARIMA and RNN models have mostly been used for short-term electric load forecasting [10–15]. It has also been found that comparisons between ARIMA and RNN model have mostly been conducted for single-step ahead forecasting which is short-term [21–25]. In contrast to this here in this paper, the proposed ARIMA and LSTM forecasting models provide flexibility to forecast over both short-term and long-term. Long-term forecasting in particular, is considered to be a challenging task [26]. Comparison of the two models helps to identify the superiority between two models whilst also considering forecasting using multi-step.

2   Time Series Forecasting Strategies

Time series forecasting is considered to be an important area of machine learning with an ultimate goal of predicting the future. It is also called *"forecasting by exploring the pattern from past data"* [27]. Forecasting of future samples plays a vital role in guiding the decision making of selected areas. Data that is required for time series forecasting are classified into two types: one is time series data and another one is data with time points [28]. Time series data can be described as

$$X = (x_t; t = 1, \ldots, N) \tag{1}$$

where X is the time series, t is time over N observations during that time. Measurement at an individual time point is x_t for time point t. Shumway and Stoffer [29] defined time series as *"a collection of random variables indexed according to the order they are obtained in time"*.

2.1   Single-Step Forecasting

Selection of forecasting class and strategy depends on the requirement of the application field and frequency of collected data. Single-step forecasting is applicable where short-term forecasting is required. For example durations of several minutes, hours or days could all be considered short-term. For such a scenario, computing a one step ahead forecast is useful. One step forecast $(t+1)$ is achieved by passing the current and past observations $(t, t-1, \ldots, t-n)$ to a chosen model,

$$F(t + 1) = M(o(t), o(t - 1), o(t - 2), \ldots, o(t - n)) \tag{2}$$

where $F(t + 1)$ is the forecast for time $(t + 1)$, M is the model and $o(t)$ is an observation at time t.

2.2   Multi-step Forecasting

Multi-step forecasting is useful where the field of application requires long-term duration forecasting. For multiple steps ahead forecast computation Ben Taieb et al. [30] described five multi-step strategies, among them Direct H Step Strategy is considered in the work here.

Direct Strategy. Direct strategy develops N separate forecasting models to forecast N steps. For example, to forecast the next two points of any scenario using a Direct strategy then the first forecast point $F(t+1)$ needs to be calculated through a model and then a different model would be used to forecast the second point $F(t + 2)$. However, importantly, the second point is not dependent on the first point estimate. This example can be seen below.

$$F(t + 1) = M_1(o(t), o(t - 1), o(t - 2), \ldots, o(t - n)) \tag{3}$$
$$F(t + 2) = M_2(o(t), o(t - 1), o(t - 2), \ldots, o(t - n)). \tag{4}$$

The direct multi-step strategy can be expressed as

$$y_{t+h} = f_h(y_t, \ldots, y_{t-n+1}) \tag{5}$$

where h is the number of steps to forecast into the future, n is the autoregressive order of the model, f_h is any arbitrary learner.

The direct strategy does not consist of any accumulated errors because it does not use any forecast value as an input. However, it does not guarantee any statistical dependence between forecast points as every model is trained independently [31].

3   ARIMA Based Time Series Forecasting

The acronym of ARIMA is meaningful as it represents the key characteristics of the model which are [32]: AR(Autoregression), relying on a relationship between the current observation and past observations; I(Integrated): differencing of actual observations in order to make the time series stationary; MA(Moving Average): lags of the forecast errors of the moving average model.

These component are included in an ARIMA model as a set of parameters. The standard notation for the ARIMA model is usually given as ARIMA(p, d, q); where p is the number of lag observations, d is the degree of differencing and q is the size of the moving average window. In the ARMA model, the forecast value is a linear compound of past values and past errors, expressed as follows [32],

$$y_t = \theta_0 + \varphi_1 y_{t-1} + \ldots + \varphi_p y_{t-p} + \varepsilon_t - \theta_1 \varepsilon_{t-1} - \ldots - \theta_q \varepsilon_{t-q} \tag{6}$$

y_t is the current measured values at time t; ε_t is the random error at time t; φ_i and θ_j are known as coefficients; and p and q are the autoregressive and moving average specific parameters respectively. If the process of ARMA is dynamic and non-stationary, then a transformation of the series is introduced by Box and Jenkins to make it stationary resulting in the ARIMA model. This is achieved by replacing the measured values y_t with the results of a recursive differencing process $\nabla^d y_t$ where d is the number of times the differencing process has been applied. The first order differencing can be expressed as

$$\nabla^d y_t = \nabla^{d-1} y_t - \nabla^{d-1} y_{t-1}. \tag{7}$$

4   LSTM Based Time Series Forecasting

A traditional neural network considers all inputs and outputs as being independent of each other. For time series forecasting it would be unwise to use such a network. An alternative is to use a Recurrent Neural Network (RNN) which includes consideration of the dependencies for past observations and thus has been found to be more effective for time series forecasting [33]. The typical architecture of an RNN is shown in Fig. 1 which consists of some inputs which are fed into a neural network block A with an output O.

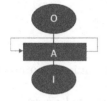

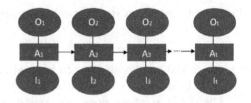

Fig. 1. Simple architecture of RNN   **Fig. 2.** Unrolled architecture of RNN

NN pass information through a loop. This looping process can be unrolled. The unrolled architecture of a full RNN network is illustrated in Fig. 2 for time steps 1, 2, 3 up to time t; where I_1, I_2, I_3, I_t are the inputs; A_1, A_2, A_3, A_t are the hidden states or the memory of the network; and O_1, O_2, O_3, O_t are the outputs. For example, A_2 is the hidden state at time step 2, where A_2 is calculated using a function (normally tanh) in which the function computes using a previously hidden state A_1 and the current input I_2 with,

$$A_2 = f(UI_2 + WA_1) \tag{8}$$

where U and W are parameters. It can therefore be concluded that RNN does capture information that has been calculated and uses it for long term forecasting. Based on the above description of an RNN it appears to passes some potentially useful properties for long-term forecasting and perhaps it is even capable of handling long-term dependencies. In practice however, these characteristics do not hold as shown by Bengio et al. [16,17]. LSTM introduced by Hochreiter and Schmidhuber [17] in 1997 is bale to learn long-term dependencies better than the RNN architecture.

The motivation behind developing LSTM was to remove the vanishing gradients issues that occur with RNN when processing long-term dependencies. The Standard RNN consists of a chain of repeating modules of the neural network, where each module consists of a structure. For example, in Fig. 3 module has a single tanh layer. Such module structures are found to be simple. Although LSTM has the same chain of repeating modules as an RNN except the LSTM module structure is relatively more complex. Each module consists of four layers rather a single layer as for an RNN module. Figure 4 consists of a simple LSTM network architecture. There modules or memory blocks consist of an input gate, a forget gate, an output gate and the cell state. All these layers interact in a particular way. Information that will be added or removed to the cell state is controlled by three gates. An LSTM network computes a mapping from an input sequence $\boldsymbol{x} = (x_1, \ldots, x_t)$ to an output sequence $\boldsymbol{y} = (y_1, \ldots, y_t)$, where initially, the input gate activation vector $\boldsymbol{i}_t$ and the candidate values of the memory cell state $\tilde{c}_t$ are calculated with,

$$\boldsymbol{i}_t = \sigma(\boldsymbol{W}_i \boldsymbol{x}_t + \boldsymbol{U}_i \boldsymbol{h}_{t-1} + \boldsymbol{b}_i); \text{ and} \tag{9}$$

$$\tilde{c}_t = \tanh(\boldsymbol{W}_c \boldsymbol{x}_t + \boldsymbol{U}_c \boldsymbol{h}_{t-1} + \boldsymbol{b}_c) \tag{10}$$

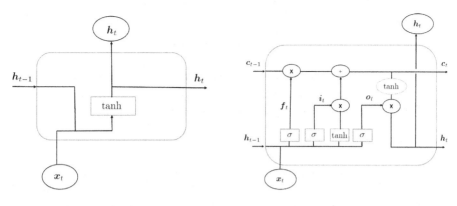

Fig. 3. RNN architecture   **Fig. 4.** LSTM based RNN architecture

respectively, where σ is the logistic sigmoid function. The input gate activation vector helps to store new information in the cell state. $\tilde{c}_t$ is a vector of new candidate values created by each tanh layer to be added to the state. The forget gate activation vector $\boldsymbol{f}_t$ is then calculated using,

$$\boldsymbol{f}_t = \sigma(\boldsymbol{W}_f \boldsymbol{x}_t + \boldsymbol{U}_f \boldsymbol{h}_{t-1} + \boldsymbol{b}_f). \tag{11}$$

The forget gate helps to throw away information from the cell state and it also helps to reset the memory cells. The calculated values of $\boldsymbol{i}_t$, $\tilde{c}_t$ and $\boldsymbol{f}_t$ are then used to calculate the new state of the memory cell $\boldsymbol{c}_t$,

$$\boldsymbol{c}_t = \boldsymbol{i}_t \times \tilde{c}_t + \boldsymbol{f}_t \times \boldsymbol{c}_{t-1}. \tag{12}$$

The cell state works like a conveyor belt which runs through the entire chain. The cell state vector $\boldsymbol{c}_t$ is then used to compute the output gate activation vector $\boldsymbol{o}_t$ via [17,35],

$$\boldsymbol{o}_t = \sigma(\boldsymbol{W}_o \boldsymbol{x}_t + \boldsymbol{U}_o \boldsymbol{h}_{t-1} + \boldsymbol{V}_o \boldsymbol{c}_t + \boldsymbol{b}_o); \text{and} \tag{13}$$

which can then be used to determine the output vector of the LSTM

$$\boldsymbol{h}_t = \boldsymbol{o}_t \tanh(\boldsymbol{c}_t). \tag{14}$$

The output gate helps to compute the output using the cell state along with filtering the cell activations. The weight matrices $\boldsymbol{W}_i$, $\boldsymbol{W}_c$, $\boldsymbol{W}_f$, $\boldsymbol{W}_o$, $\boldsymbol{U}_i$, $\boldsymbol{U}_c$, $\boldsymbol{U}_f$, $\boldsymbol{U}_o$ and $\boldsymbol{V}_o$ are used in Eqs. 9–14 which have to be learnt in the training stage along with $\boldsymbol{b}_i$, $\boldsymbol{b}_c$, $\boldsymbol{b}_f$ and $\boldsymbol{b}_o$ which represent bias vectors. The described LSTM structure does resolve the vanishing gradients issues and has been found to be suitable for long-term dependencies problems [34].

5   Experiments and Results

Multi-step time series forecast analyses on electric load datasets are now performed using the ARIMA model and the LSTM model. Forecast performance of multi-step electric load forecasting of both models are also compared.

5.1 Dataset

Three datasets were obtained from the Open Power System Data on electric load for the Great Britain (GB), Poland (PL) and Italy (IT) [36]. The datasets consist of data of electric load from 2010-02-01 to 2014-01-31 each comprising of 35064 data points with a sampling frequency of 60 min. The sampling frequency of the time series is too granular so downsampling of time series data from 60 min to 1 day was performed using resampling technique [37] for datasets.

5.2 ARIMA Model Formulation

The Dickey-Fuller (ADF) test [4,32] was applied and it was found that the time series data of all three datasets is non-stationary whereas it is important to have the time series data to be stationary for application of the ARIMA forecasting model [4,32]. Seasonal differencing was applied on the three datasets to make the time series stationary as outlined in [4,32]. It was found that before doing the first order differencing the ADF Statistic was found to be more than the 1% critical value and the p-value was close to 0.05 so the null hypothesis of the ADF test could not be rejected. Whereas after the seasonal difference ADF Statistic to be found it was less than the 1% critical value and the p-value was much lower than 0.05 which suggests that the d parameter of the ARIMA model should at least be a value of 1 for better performance. Next, for the other two parameters p and q of the ARIMA model, the Autocorrelation Function (ACF) and Partial Autocorrelation Function (PACF) were computed and plotted following the procedure as outlined in [4,32]. Examining the ACF and PACF plots, the parameters p and q were identified for all three datasets. From the ACF and PACF plots, the 'p' and 'q' values were identified as follows:

– GB Electric Load: $p = 2$ and $q = 2$.
– PL Electric Load: $p = 2$ and $q = 1$.
– IT Electric Load: $p = 2$ and $q = 1$.

Following this, ARIMA $(2,1,2)$, ARIMA $(2,1,1)$ and ARIMA $(2,1,1)$ models were used for the GB Electric Load, PL Electric Load and IT Electric Load datasets respectively.

5.3 LSTM Model Formulation

Selected datasets need to be transformed before fitting in an LSTM model [38]. The transformation of the datasets has been done in three steps. In the first step, the selected time series datasets were divided into input and output components to enable supervised learning to be undertaken. In the Python ecosystem, this was achieved using the shift() function of pandas where, the Previous time steps $(t - n)$ used as an input and current time step (t) used as an output for the observed data [37]. For the second step, the selected time series datasets were then transformed to a stationary time series data set as it is easier to model and it was expected to produce a better forecast. This has been achieved by applying seasonal differencing of the data as the seasonal trend is visible in the actual data [38]. For the third step, the time series data sets were re-scaled to values between −1 and 1, this is because LSTM model requires data to be within the scale of the activation function of the network. This is achieved using the 'MinMaxScaler class' from scikit-learn Python library [37,38].

Data shape needs to be reshaped as the LSTM requires the input to be in a matrix form with the dimensions: [samples, time steps, features]. The original data set is a 2D array [samples, features] which required transforming to a 3D array [samples, timesteps, features] [38]. This was achieved by fixing the time steps at a value of 1. Simple LSTM have been designed where a network structure consisting of 1 hidden layer with 1 LSTM unit, then an output layer with a linear activation and N output values. Value of N was varied according to the number of time steps over which a forecast was required. The network also represented the 'Root Mean Squared Error (RMSE)' as a loss function and the 'ADAM algorithm' [39] as an optimizer. LSTM is stateful in designed network and the network was fitted with 1 epochs for simplicity. Batch size of the network was set to 1, which is also known as online learning for both training and prediction. All these parameters were used to configure the LSTM model in preparation for forecasting.

5.4 Forecast Performance of the Models

The ARIMA and LSTM models were developed using the Python ecosystem [37]. Statsmodel library was used to fit the ARIMA model by calling ARIMA() along with the p, d, and q parameters [41]. Then fit() and predict() functions were called to train the model and make predictions respectively. SciPy environment with Keras deep learning library using the TensorFlow backend was used for the LSTM model [40]. To perform out of sample forecasting using the ARIMA and LSTM models, the datasets were split into train and test. The training data were used to train the models and the test data were used for performance characterization. Ten different ranges of time points were selected for training, simulating a sweeping action through the data sets across time. This sweeping action helps to determine the potential applicability of the models to a real time prediction scenario. Furthermore, it enables the mean error to be calculated across these different prediction scenarios to determine the generalised performance of the discussed models.

The sampling frequency of the data was a day. For example, if 1 day ahead forecast was required then single-step forecasting was performed and the value of N was 1. However, to forecast 10 days ahead multi-step forecasting was performed and the value of N was 10. To measure the performance of the ARIMA and LSTM models, the RMSE performance measurement technique was used. RMSE can be expressed and calculated using

$$\bar{e} = \sqrt{\frac{\sum_{i=1}^{m} (y_i - \widehat{y}_i)^2}{m}} \tag{15}$$

where, y_i are actual values, $\widehat{y}_i$ are forecast values and m is the number of target output data. Forecast performance of both models on the three datasets are provided in Table 1.

Table 1. Multi-step forecast response of GB, PL and IT electric load

Forecast step (N)	GB		PL		IT	
	ARIMA model	LSTM model	ARIMA model	LSTM model	ARIMA model	LSTM model
1	5269	4009	2154	1416	4615	4087
2	4245	2829	1913	1218	4374	3176
3	4471	2950	2135	1991	5548	3798
4	4821	3491	2328	1788	6441	4887
10	4587	3765	2461	1922	6263	4876

5.5   Discussion

The RMSE performance indicator on various forecast steps of the ARIMA and LSTM models using the three datasets are plotted in Figs. 5, 6 and 7.

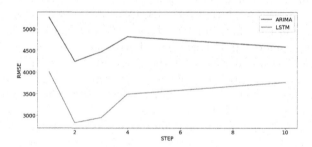

Fig. 5. LSTM and ARIMA model performance on predicting the GB load data.

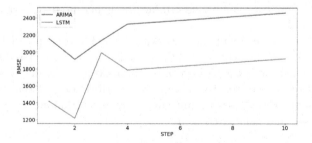

Fig. 6. LSTM and ARIMA model performance on predicting the PL load data.

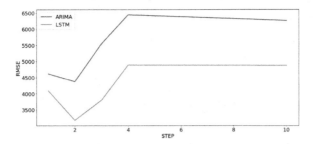

Fig. 7. LSTM and ARIMA model performance on predicting the IT load data.

These show that for every forecast step LSTM provides better performance compared to the ARIMA model for each dataset. It can be concluded that in forecasting single-step and multi-step ahead of selected electric load data sets, LSTM model outperforms ARIMA model.

6   Conclusions

Multi-step forecasting of electric load has been presented in this paper. The experimental results obtained with electric load datasets on the performance of ARIMA and LSTM model are also compared. The comparative analysis revealed that LSTM model outperformed ARIMA model in forecasting multi-step ahead of electric load.

References

1. Mandal, P., Haque, A.U., Meng, J., Srivastava, A.K., Martinez, R.: A novel hybrid approach using wavelet, firefly algorithm and fuzzy ARTMAP for day-ahead electricity price forecasting. IEEE Trans. Power Syst. **28**(2), 1041–1051 (2013)
2. Du Preez, J., Witt, S.F.: Univariate versus multivariate time series forecasting: an application to international tourism demand. Int. J. Forecasting **19**(3), 435–451 (2003)
3. Nataraja, C., Gorawar, M., Shilpa, G., Harsha, J.S.: Short term load forecasting using time series analysis: a case study for Karnataka, India. Int. J. Eng. Sci. Innov. Technol. **1**, 45–53 (2012)
4. Masum, S., Liu, Y., Chiverton, J.: Comparative analysis of the outcomes of differing time series forecasting strategies. In: 13th International Conference on Natural Computation, Fuzzy Systems and Knowledge Discovery. IEEE Press (2017)
5. Wang, J.J., Wang, J.Z., Zhang, Z.G., Guo, S.P.: Stock index forecasting based on a hybrid model. Omega **40**(6), 758–766 (2012)
6. Meyler, A., Kenny, G., Quinn, T.: Forecasting Irish inflation using ARIMA models. Published in Central Bank and Financial Services Authority of Ireland Technical Paper Series, vol. 3, p. 148 (1998)
7. Tabachnick, B.G., Fidell, L.S.: Using Multivariate Statistics, 4th edn. Pearson Education, Upper Saddle River (2001)

8. Kam, K.M.: Stationary and non-stationary time series prediction using state space model and pattern-based approach. The University of Texas at Arlington (2014)
9. Lineesh, M., Minu, K., John, C.J.: Analysis of nonstationary nonlinear economic time series of gold price: a comparative study. In: International Mathematical Forum, vol. 5, no. 34, pp. 1673–1683. Citeseer (2010)
10. Liu, K., Subbarayan, S., Shoults, R.R., Manry, M.T., Kwan, C., Lewis, F.L., Naccarino, J.: Comparison of very short-term load forecasting techniques. IEEE Trans. Power Syst. **11**(2), 877–882 (1996)
11. Hagan, M.T., Behr, S.M.: The time series approach to short term load forecasting. IEEE Trans. Power Syst. **PWRS-2**(3), 785–791 (1987)
12. Yang, H.T., Huang, C.M., Huang, C.L.: Identification of ARMAX model for short term load forecasting: an evolutionary programming approach. IEEE Trans. Power Syst. **11**(1), 403–408 (1996)
13. Espinoza, M., Joye, C., Belmans, R., Moor, B.D.: Short-term load forecasting, profile identification, and customer segmentation: a methodology based on periodic time series. IEEE Trans. Power Syst. **20**(3), 1622–1630 (2005)
14. Mandal, J.K., Sinha, A.K., Parthasarathy, G.: Application of recurrent neural network for short term load forecasting in electric power system. In: IEEE International Conference on Neural Networks, vol. 5, pp. 2694–2698 (1995)
15. Senjyu, T., Takara, H., Uezato, K., Funabashi, T.: One-hour-ahead load forecasting using neural network. IEEE Trans. Power Syst. **17**(1), 113–118 (2002)
16. Bengio, Y., Simard, P., Frasconi, P.: Learning long-term dependencies with gradient descent is difficult. IEEE Trans. Neural Netw. **5**(2), 157–166 (1994)
17. Hochreiter, S., Schmidhuber, J.: Long short-term memory. Neural Comput. **9**(8), 1735–1780 (1997)
18. Sak, H., Senior, A., Beaufays, F.: Long short-term memory recurrent neural network architectures for large scale acoustic modelling. In: Fifteenth Annual Conference of the International Speech Communication Association (2014)
19. Marino, D.L., Amarasinghe, K., Manic, M.: Building energy load forecasting using deep neural networks. In: 42nd Annual Conference of the IEEE Industrial Electronics Society (IECON), pp. 7046–7051 (2016)
20. Wu, W., Chen, K., Qiao, Y., Lu, Z.: Probabilistic short-term wind power forecasting based on deep neural networks. In: International Conference on Probabilistic Methods Applied to Power Systems (PMAPS), pp. 1–8 (2016)
21. Ho, S.L., Xie, M., Goh, T.N.: A comparative study of neural network and Box-Jenkins ARIMA modeling in time series prediction. Comput. Ind. Eng. **42**, 371–375 (2002)
22. Fu, R., Zhang, Z., Li, L.: Using LSTM and GRU neural network methods for traffic flow prediction. In: 31st Youth Academic Annual Conference of Chinese Association of Automation (YAC), pp. 324–328 (2016)
23. Cao, Q., Ewing, B., Thompson, M.: Forecasting wind speed with recurrent neural networks. Eur. J. Oper. Res. **221**, 148–54 (2012)
24. Ma, X., Tao, Z., Wang, Y., Yu, H., Wang, Y.: Long short-term memory neural network for traffic speed prediction using remote microwave sensor data. Transp. Res. Part C: Emerg. Technol. **54**, 187–197 (2015)
25. Tian, Y., Pan, L.: Predicting short-term traffic flow by long short-term memory recurrent neural network. In: IEEE International Conference on Smart City, Chengdu, pp. 153–158 (2015)
26. Cheng, H., Tan, P.N.: Semi-supervised learning with data calibration for long-term time series forecasting. In: Proceedings of the 14th ACM SIGKDD International Conference on Knowledge Discovery and Data Mining, p. I-9. ACM (2008)

27. Molaei, S.M., Keyvanpour, M.R.: An analytical review for event prediction system on time series. In: 2nd International Conference on Pattern Recognition and Image Analysis (IPRIA), pp. 1–6 (2015)
28. Minaei-Bidgoli, B., Lajevardi, S.B.: Correlation mining between time series stream and event stream. In: Fourth International Conference on Networked Computing and Advanced Information Management, Gyeongju, pp. 333–338 (2008)
29. Soyiri, I.N., Reidpath, D.D.: An overview of health forecasting. Environ. Health Prev. Med. **18**(1), 1–9 (2013)
30. Ben Taieb, S., Bontempi, G., Atiya, A.F., Sorjamaa, A.: A review and comparison of strategies for multi-step ahead time series forecasting based on the NN5 forecasting competition. Expert Syst. Appl. **39**(8), 7067–7083 (2012)
31. An, N.H., Anh, D.T.: Comparison of strategies for multi-step ahead prediction of time series using neural network. In: International Conference on Advanced Computing and Applications (ACOMP), pp. 142–149 (2015)
32. George, E.P.B., Gwilym, M.J., Gregory, C.R., Greta, M.L.: Time Series Analysis: Forecasting and Control. Wiley Publisher, New Jersey (2015)
33. Medsker, L., Jain, L.: Recurrent Neural Networks, Design and Applications. CRC Press LLC, Boca Raton (2001)
34. Graves, A.: Neural networks. In: Graves, A. (ed.) Supervised Sequence Labelling with Recurrent Neural Networks. SCI, vol. 385, pp. 15–35. Springer, Heidelberg (2012). https://doi.org/10.1007/978-3-642-24797-2_3
35. Olah, C.: Understanding LSTM networks. http://colah.github.io/posts/2015-08-Understanding-LSTMs
36. Open power system data. https://data.open-power-system-data.org/timeseries/
37. McKinney, W.: Python for Data Analysis. O'Reilly, Sebastopol (2013)
38. Lewis, N.D.: Deep Time Series Forecasting with Python. Create Space Independent Publishing Platform (2016)
39. Kingma, D.P., Ba, J.L.: ADAM: a method for stochastic optimization. In: ICLR, pp. 1–15 (2015)
40. Chollet, F.: Keras, GitHub repository (2015). https://github.com/keras-team/keras
41. Seabold, S., Josef, P.: Statsmodels: econometric and statistical modeling with Python. In: The Proceedings of the 9th Python in Science Conference (2010)

Deep Q-Network Using Reward Distribution

Yuta Nakaya and Yuko Osana[✉]

Tokyo University of Technology,
1404-1, Katakura, Hachioji, Tokyo 192-0982, Japan
osana@stf.teu.ac.jp

Abstract. In this paper, we propose a Deep Q-Network using reward distribution. Deep Q-Network is based on the convolutional neural network which is a representative method of Deep Learning and the Q Learning which is a representative method of reinforcement learning. In the Deep Q-Network, when the game screen (observation) is given as an input to the convolutional neural network, the action value in Q Learning for each action is output. This method can realize learning that acquires a score equal to or higher than that of a human in plural games. The Q Learning learns using the greatest value in the next action, so a positive reward is propagated. However, since negative rewards can not be of greatest value, they are not propagated in learning. Therefore, by distributing negative rewards in the same way as Profit Sharing, the proposed method learn to not take wrong actions. Computer experiments were carried out, and it was confirmed that the proposed method can learn with almost the same speed and accuracy as the conventional Deep Q-Network. Moreover, by introducing reward distribution, we confirmed that learning can be performed so as not to acquire negative reward in the proposed method.

Keywords: Deep Q-Network · Reward distribution · Profit Sharing

1 Introduction

In recent years, as a method which shows better performance than the conventional method in the field of image recognition and speech recognition, the Deep Learning has been drawing attention. Deep Learning is a hierarchical neural network with many layers, and its representative model is the Convolutional Neural Network (CNN) [1] and the Deep Belief Network (DBN) [2].

Also, various studies on reinforcement learning are being conducted as learning methods to acquire appropriate policies through interaction with the environment [3]. In reinforcement learning, learning can proceed by repeating trial and error even in an unknown environment by appropriately setting rewards.

The Deep Q-Network [4] is based on the convolutional neural network which is a representative method of Deep Learning and the Q Learning [5] which is a representative method of reinforcement learning. In the Deep Q-Network, when

L. Rutkowski et al. (Eds.): ICAISC 2018, LNAI 10841, pp. 160–169, 2018.
https://doi.org/10.1007/978-3-319-91253-0_16

the game screen (observation) is given as an input to the convolutional neural network, the action value in Q Learning for each action is output. This method can realize learning that acquires a score equal to or higher than that of a human in plural games. The Q Learning learns using the greatest value in the next action, so a positive reward is propagated. However, since negative rewards can not be of greatest value, they are not propagated in learning.

In this paper, we propose a Deep Q-Network using reward distribution. The proposed method learn to not take wrong actions, by distributing negative rewards in the same way as Profit Sharing [6].

2 Deep Q-Network

Here, we explain the Deep Q-Network [4] that is the basis of the proposed method. The Deep Q-Network is based on the convolutional neural network [1] and the Q Learning [5]. In the Deep Q-Network, when the game screen (observation) is given as an input to the convolutional neural network, the action value in Q Learning for each action is output. This method can realize learning that acquires a score equal to or higher than that of a human in plural games.

2.1 Structure

The structure of Deep Q-Network is shown in Fig. 1. As seen in Fig. 1, the Deep Q-Network is a model based on the convolutional neural network, consisting of three convolution layers and two fully connected layers. The play screen of the game (observation) is input to the convolutional neural network, and the action value for each action corresponding to the observation is outputted. For the first to fourth layers, rectified linear function is used as an output function. The number of neurons in the last finally connected layer which is the output layer is the same as the number of actions that can be taken in the problem to be handled. Since the problem learned by Deep Q-Network can be regarded as a regression problem to learn the relationship between each observation and the action value of each action in the observation, the output function of the output layer is an identity mapping function.

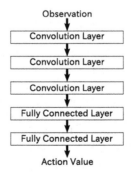

Fig. 1. Structure of Deep Q-Network.

2.2   Learning

Since the action value in Q Learning is used as the output, the following error function used in learning is given by

$$E = \frac{1}{2} \left(r_\tau + \gamma \max_{a' \in C^A(o_{\tau+1})} q(o_{\tau+1}, a') - q(o_\tau, a_\tau) \right)^2 \tag{1}$$

where r_τ is the reward at the time τ, $C^A(o_{\tau+1})$ is the set of actions that an agent can take at the observation $o_{\tau+1}$, γ is the discount factor, $q(o_\tau, a_\tau)$ is the value of taking action a_τ at observation o_τ.

When the game screen o_τ is given to the Deep Q-Network, the value of all actions in observation o_τ is output in the output layer. Based on the output action value, action is determined by the ε-greedy method. In the ε-greedy method, one action is selected randomly with the probability ε $(0 \leq \varepsilon \leq 1)$, the action whose value is highest with the probability of $1 - \varepsilon$.

The probability to select the action a in observation o_τ, $P(o_\tau, a)$ is given by

$$P(o_\tau, a) = \begin{cases} (1 - \varepsilon) + \dfrac{\varepsilon}{|C^A|} & \left(\text{if } a = \operatorname*{argmax}_{a' \in C^A} q(o_\tau, a') \right) \\ \dfrac{\varepsilon}{|C^A|} & \text{(otherwise)} \end{cases} \tag{2}$$

where, $|C^A|$ is the number of action types that the agent can take, which is the same as the number of neurons in the output layer of the Deep Q-Network.

The selected action a_τ is executed, and the state transits to the next state $o_{\tau+1}$. Also, by taking the action a_τ, the reward r_τ is given based on the score, game state and so on.

Learning is unstable merely by approximating the action value of Q Learning using the convolutional neural network, so in the learning of the Deep Q-Network, some ideas called Experience Replay, Fixed Target Q-Network, Reward Clipping are introduced.

3   Deep Q-Network Using Reward Distribution

Here, we explain the proposed Deep Q-Network using reward distribution.

The proposed method is based on the Deep Q-Network, and updates the action value in Q Learning as similar as in the conventional Deep Q-Network. In addition, when a negative reward is given, the negative reward is distributed using the method which is used in the Profit Sharing [6].

The Q Learning learns using the greatest value in the next action, so a positive reward is propagated. However, since negative rewards can not be of greatest value, they are not propagated in learning. Therefore, by distributing negative rewards in the same way as Profit Sharing, the proposed method learn to not take wrong actions.

The distribution of negative rewards is realized by

$$r_\tau = \frac{1}{(|C^{s_\tau}| + 1)^{\tau_r - \tau}} \tag{3}$$

where r_τ is the (negative) reward at the time τ, $|C^{s_\tau}|$ is the number of actions that an agent can take at the time τ, τ_r is the time that the negative reward is obtained.

In the conventional Profit Sharing, reward distribution is done after the episode is over. In contrast, in the proposed model, hen the reward is obtained, the reward is distributed as it goes back in time.

4   Computer Experiment Results

To demonstrate the effectiveness of the proposed method, computer experiments were conducted on three games of Atari 2600 (Pong, Breakout, and Asterix). The results are shown below.

4.1   Experimental Conditions

Table 1 shows the conditions for the convolutional neural networks in the proposed method and the conventional Deep Q-Network. To the convolution neural network, the play screen of the game of four frames is given as an input. The Atari 2600 game screen is an RGB image whose size is 210×160 pixels. In this experiment, the RGB image is converted to a grayscale image, and it is reduced to 110×84 pixels, and the center 84×84 pixels are cut out, and 4 frames are combined into one image and used as input.

Since the output of the output layer is the action value, the number of neurons in the output layer is the same as the number of types of actions that can be taken in the learning game.

Table 2 shows the other conditions. Action is selected by ε-greedy.

Table 1. Experimental conditions for convolutional neural network.

Layer	Filter size	Stride e	Size of output	Output function
Input	—	—	$84 \times 84 \times 4$	—
Convolution 1	8×8	4	$20 \times 20 \times 32$	ReLU
Convolution 2	4×4	2	$9 \times 9 \times 64$	ReLU
Convolution 3	3×3	1	$7 \times 7 \times 64$	ReLU
Fully connected 1	—	—	512	ReLU
Fully connected 2	—	—	The number actions	Identity function

Table 2. Other conditions.

The number of learning episodes	E	$10^3 \sim 5 \times 10^3$
Initial value of ε	ε_{ini}	1
Decrease amount of ε	ε_r	$1/10^6$
Minimum value of ε	ε_{min}	0.1
ε in evaluation	ε'	0.05
Size of replay memory	D_{max}	10^5
Size of mini batch	M	32
Discount factor	γ	0.99
Interval of target network	T_{update}	10^4

4.2 Score Transition

(1) Pong

Figure 2 shows the score transition of the conventional Deep Q-Network and the proposed Deep Q-Network using reward distribution.

In Pong (table tennis game), the final score is calculated by subtracting the opponent's score from player's score, and if it exceeds 0, the player wins. The lowest score is −21 and the highest score is 21. In this experiment, the number of learning episodes was 1000, and the number of evaluation episodes was 112.

In the learning of the conventional Deep Q-Network, the score is often close to −21 until about 15 episodes. From that point the score will gradually grow until about 70 episodes. The mean of the scores of the first 10 episodes was −20.9, whereas the average of the scores of the last 10 episodes was 14.7. The number of steps taken throughout the learning episode was 2408647, and the number of steps taken for the entire evaluation episode was 363930.

In the learning of the proposed method, the score is often close to −21 until about 15 episodes as similar as the conventional method. From that point the score will gradually grow until about 65 episodes. The mean of the scores of the first 10 episodes was −20.8, whereas the average of the scores of the last 10 episodes was 13.5. The number of steps taken throughout the learning episode was 3016489, and the number of steps taken for the entire evaluation episode was 477353.

As shown in this figure, it can be seen that there is no big difference in the score transition between the proposed method and the conventional method. However, the number of steps increases in the proposed method. This is because the time until the enemy takes points in the proposed method is longer than the conventional method.

(2) Breakout

Figure 3(a) shows the score transition of the conventional Deep Q-Network. Breakout is a block breaking game. In this experiment, the number of learning episodes was 5000, and the number of evaluation episodes was 556.

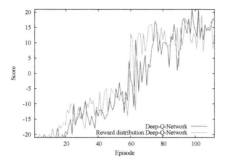

Fig. 2. Score transition (Pong).

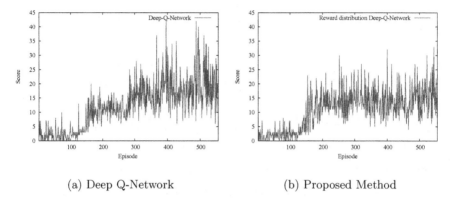

(a) Deep Q-Network   (b) Proposed Method

Fig. 3. Score transition (Breakout).

In the learning of the conventional Deep Q-Network, the score is often close to 0 until about 100 episodes. From that point the score will gradually grow until about 300 episodes. The mean of the scores of the first 10 episodes was 2.7, whereas the average of the scores of the last 10 episodes was 17.6. The number of steps taken throughout the learning episode was 1964570, and the number of steps taken for the entire evaluation episode was 332824.

Figure 3(b) shows the score transition of the proposed Deep Q-Network using reward distribution. In this experiment, the number of learning episodes was 5000, and the number of evaluation episodes was 556.

In the learning of the proposed method, the score is often close to 0 until about 120 episodes as similar as the conventional method. From that point the score will gradually grow until about 200 episodes. The mean of the scores of the first 10 episodes was 1.6, whereas the average of the scores of the last 10 episodes was 14.8. The number of steps taken throughout the learning episode was 2221370, and the number of steps taken for the entire evaluation episode was 180375.

Figure 4 shows the score transition of the conventional Deep Q-Network and the proposed Deep Q-Network using reward distribution. Figure 4(b) shows

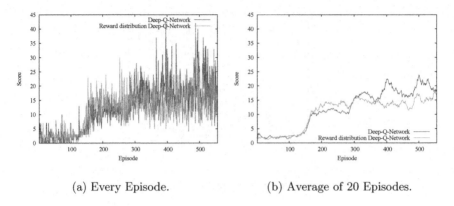

(a) Every Episode.　　　　　　　　(b) Average of 20 Episodes.

Fig. 4. Score transition comparison (Breakout).

average of 20 episodes. As seen in this figure, we can confirmed that the learning converges more quickly in the proposed method. The final score in the proposed method is slightly higher than in the conventional method. In addition, the number of steps of the proposed method is larger than that of the conventional method, and it can be seen that the number of steps up to the game over is increased.

(3) Asterix

Figure 4(a) shows the score transition of the conventional Deep Q-Network. In this experiment, the number of learning episodes was 5000, and the number of evaluation episodes was 556. In the initial episode of learning in the conventional Deep Q-Network, the score goes up and down in the range of 100 to 500. In contrast, in the second half episode of learning, the score goes up and down in the range of 200 to 600. The mean of the scores of the first 10 episodes was 240, whereas the average of the scores of the last 10 episodes was 385. The number of steps taken throughout the learning episode was 1467501, and the number of steps taken for the entire evaluation episode was 191082.

Figure 4(b) shows the score transition of the proposed Deep Q-Network using reward distribution. In the initial episode of learning in the proposed method, the score goes up and down in the range of 100 to 500 as similar as the conventional method. In contrast, in the second half episode of learning, the score goes up and down in the range of 300 to 700. The mean of the scores of the first 10 episodes was 305, whereas the average of the scores of the last 10 episodes was 480. The number of steps taken throughout the learning episode was 1554778, and the number of steps taken for the entire evaluation episode was 180375.

Figure 5 shows the score transition of the conventional Deep Q-Network and the proposed Deep Q-Network using reward distribution. Figure 5(b) shows average of 20 episodes. As seen in this figure, we can confirmed that the final score in the proposed method is higher than in the conventional method.

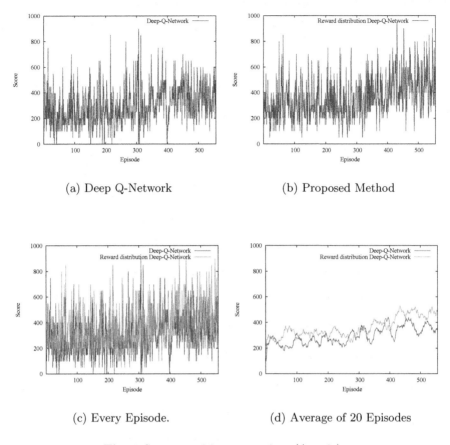

(a) Deep Q-Network   (b) Proposed Method

(c) Every Episode.   (d) Average of 20 Episodes

Fig. 5. Score transition comparison (Asterix).

4.3   Transition of Acquired Negative Reward

Since the proposed method distributes negative rewards, it is thought that if negative learning is properly performed, negative rewards are less likely to be acquired as learning progresses.

(1) Pong

Figure 6(a) shows the transition of the number of acquisitions of negative reward for every 100 thousand steps in the learning for Pong. From this figure, it can be seen that the number of acquisitions of negative compensation is about 85% in the proposed method, compared to the conventional Deep Q-Network which does not distribute compensation.

(2) Breakout

Figure 6(b) shows the transition of the number of acquisitions of negative reward for every 100 thousand steps in the learning for Breakout. From this figure, it

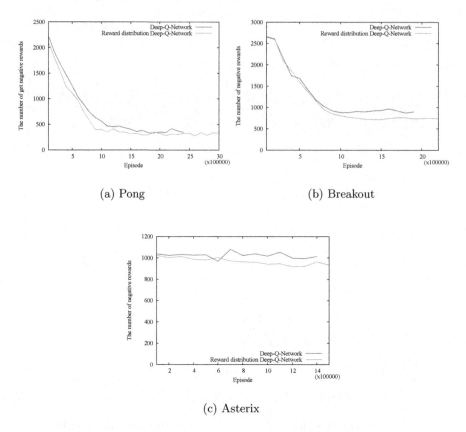

(a) Pong   (b) Breakout

(c) Asterix

Fig. 6. Transition of acquired negative reward (Pong).

can be seen that the number of acquisitions of negative compensation is about 90% in the proposed method, compared to the conventional Deep Q-Network which does not distribute compensation.

(3) Asterix

Figure 6(c) shows the transition of the number of acquisitions of negative reward for every 100 thousand steps in the learning for Asterix. From this figure, it can be seen that the number of acquisitions of negative compensation is about 85% in the proposed method, compared to the conventional Deep Q-Network which does not distribute compensation.

From these results, we confirmed that the proposed method learns not to acquire negative reward for all games.

5   Conclusion

In this paper, we have proposed the Deep Q-Network using reward distribution. The proposed method is based on the conventional Deep Q-Network [4]. In the

Deep Q-Network, when the game screen (observation) is given as an input to the convolutional neural network, the action value in Q Learning for each action is output. This method can realize learning that acquires a score equal to or higher than that of a human in plural games. The Q Learning learns using the greatest value in the next action, so a positive reward is propagated. However, since negative rewards can not be of greatest value, they are not propagated in learning. Therefore, by distributing negative rewards in the same way as Profit Sharing, the proposed method learn to not take wrong actions.

We carried out a series of computer experiments, and confirmed that the proposed method has following features.

(1) The proposed method can learn with almost the same speed and accuracy as the conventional Deep Q-Network.
(2) Final score in the proposed method is higher than in the conventional method for some games.
(3) By introducing reward distribution, the learning can be performed so as not to acquire negative reward in the proposed method.

References

1. LeCun, Y., Bottou, L., Bengio, Y., Haffner, P.: Gradient-based learning applied to document recognition. Proc. IEEE **86**(11), 2278–2324 (1998)
2. Hinton, G.E., Osindero, S., Teh, Y.: A fast learning algorithm for deep belief nets. Neural Comput. **18**, 1527–1544 (2006)
3. Sutton, R.S., Barto, A.G.: Reinforcement Learning: An Introduction. The MIT Press, Cambridge (1998)
4. Mnih, V., et al.: Human-level control through deep reinforcement learning. Nature **518**, 529–533 (2015)
5. Watkins, C.J.C.H., Dayan, P.: Technical note: Q-learning. Mach. Learn. **8**, 55–68 (1992)
6. Grefenstette, J.J.: Credit assignment in rule discovery systems based on genetic algorithms. Mach. Learn. **3**, 225–245 (1988)

Motivated Reinforcement Learning Using Self-Developed Knowledge in Autonomous Cognitive Agent

Piotr Papiez and Adrian Horzyk[✉]

AGH University of Science and Technology,
Mickiewicza Av. 30, 30-059 Krakow, Poland
ppapiez13@gmail.com, horzyk@agh.edu.pl

Abstract. This paper describes the development of a cognitive agent using motivated reinforcement learning. The conducted research was based on the example of a virtual robot, that placed in an unknown maze, was learned to reach a given goal optimally. The robot should expand knowledge about the surroundings and learn how to move in it to achieve a given target. The built-in motivation factors allow it to focus initially on collecting experiences instead of reaching the goal. In this way, the robot gradually broadens its knowledge with the advancement of exploration of its surroundings. The correctly formed knowledge is used for effective controlling the reinforcement learning routine to reach the target by the robot. In such a way, the motivation factors allow the robot to adapt and control its motivated reinforcement learning routine automatically and autonomously.

Keywords: Neural networks
Environment adapted reinforcement learning
Knowledge-based learning and supervision · Motivated learning
Knowledge development · Cognitive systems · Cognitive robot

1 Introduction

Robots enter the lives of people and begin to play an increasingly important role in it. With the increasing requirements for robots, the difficulty of programming them is increasing. Robots which are programmed to perform some specific moves, over which the programmer has total control, are too limited. With help comes artificial intelligence [8,18] that allows robots to learn automatically and even autonomously if they have some built-in motivation factors which can push them to aim targets which satisfy their adequately designed needs or help them avoid punishment [13,17]. Thanks to it, robots can learn what actions they should do to achieve the goal.

© Springer International Publishing AG, part of Springer Nature 2018
L. Rutkowski et al. (Eds.): ICAISC 2018, LNAI 10841, pp. 170–182, 2018.
https://doi.org/10.1007/978-3-319-91253-0_17

One of the ways of learning is supervised learning. It requires a training dataset consisting of observation and action pairs, based on which the robot is learned. Unfortunately, this approach is problematic. The main inconvenience of supervised learning is that it requires the expert knowledge of the teacher [9] who serves important information about the directions of the learning process and defines the goals of it. The training dataset should be created taking into account all representative observations and assigning them appropriate actions. It is difficult to require the programmer to know what movement the robot should do in a given situation. In real biological intelligent systems [6], our intelligence is automatically supported with knowledge [5,7], so we would like to implement similar solutions in artificial intelligent systems [3] on the basis of associations [4] which control our learning and thinking processes in the brain [1] and form various kinds of memories [2]. Taking into account that the robot is in an unfamiliar environment, supervised learning becomes difficult to use. The robot should learn and adapt to any new environment on its own like people [10].

In this paper, it was decided to focus on the aspect of learning the robot without external help. For this purpose, motivated reinforcement learning is used. Thanks to this approach, there is no need to create a learning dataset, which would require the involvement of human intelligence. Instead, the internal needs of the robot are defined, which drive its system of motivations and actions to achieve the desired goal [20].

The paper aims to develop new trends and motivated learning strategies for reinforcement learning algorithms [9,11,13]. In the beginning, the robot does not know the maze, so it must try to perform random actions and draw conclusions from them as the living beings do. It is possible thanks to defined reward function, through which it knows whether the movement it made was correct. For example, the robot knows that falling into a wall is unfavorable but does not know where the walls are. In this way, through the interaction with the environment, the robot knows what movements should be performed in specific situations. The use of this technique allows the cognitive robot to learn independently, automatically, and autonomously. Thanks to this, it discovers how to behave optimally in an unfamiliar environment [12].

Reinforcement learning was expanded by motivated learning [13]. The robot will depend not only on the external rewards but also on satisfying internal motivations. The external reward returned by the environment serves to satisfy the primitive goals on the basis of which the internal system of motivation and goals is built. Thanks to this, the robot may be not guided by the maximization of the external reward but by internal motivations leading to complex concepts and intelligent behavior. The use of motivation for learning can serve to build better knowledge about the environment [13].

To study the described learning method, an example problem was created, which the robot had to face. On this example, motivated reinforcement learning was implemented which is described in more detail in the following chapters. The description of the project carried out as part of the research is as follows:

The work aims to create a new method for a virtual robot (agent) that allows it to move in a maze for a set purpose with the use of motivated reinforcement learning. It should be able to reach a predetermined location in the maze from any starting point in an optimal way after finishing the training process which runs together with the exploration of the maze and formation of the knowledge on this basis [3]. At the starting point in the maze, the robot does not know the surroundings. It also does not know whether there exists the way to the target location, and if so, which way is optimal. The only available information is the coordinates of the target and the current coordinates of the robot. The robot's primary task is to explore surroundings and expand knowledge about it because the knowledge plays a fundamental role in every intelligent behavior [3,5,7]. The second task is to designate the shortest way to the target that bypasses other objects or obstacles [11].

2   Robot's Reinforcement Learning Using Motivation

The main goal of any learning process is to change internal parameters of the model to satisfy defined requirements. The most wanted training routines should be able to adapt the robot automatically to the surrounding and its needs [20]. The robot should be able to use its capabilities (e.g. the ability to move) to explore surroundings and find the way how to satisfy its needs, not to use any external support, control, or an external training dataset. If we want the robot to learn to solve the tasks itself, we need to define on what basis its parameters change to control its behavior satisfactorily. We will use a specific combination of motivated and reinforcement learning to satisfy these assumptions [13]. It is similar to reinforcement learning, but in addition to external rewards, it also defines an internal motivation system. Thanks to this, it can set goals for itself, not rely only on the reinforcement returned by the environment. Remembering that at the beginning it does not have any knowledge about the surrounding environment, it should have appropriate cognitive skills allowing it to develop independently by the gained experience. We can define cognitive abilities [1] on the basis of perception, memory [2], motivation [13,17], needs [20], anticipation, planning, association [4], learning, and reasoning [8,9].

2.1   Interaction with the Surroundings

The robot needs sensors to be able to collect data from the environment. They are just as important to it as senses for people. Without senses, we would be able to do some limited movements that we already know, but we certainly could not learn new ones. For this, we need certain feedback after each move, which allows us to assess the correctness of the taken action and draw conclusions for the future. Perception also allows the cognitive robot to observe the signals from the environment on the basis of which decisions are made.

In the presented experiments, the robot has built-in sensors that provide it with all the information it needs to interact with the environment. These are distance sensors that can stop the robot from colliding with obstacles and encoders informing about the current position of the robot and the direction to the target location/point. The robot interprets the room in which it found itself as a square grid. It can move on it in four directions: forward, backward, right, and left. Thanks to its set of sensors, the robot can find out what its current position is and where it fell into an obstacle. However, how can it know that falling into an obstacle is a bad move and that it should avoid it? It is possible thanks to the defined rewards.

2.2 External Rewards and Reinforcement Learning

How do we know that touching hot things is not a good idea? We were not born with this kind of knowledge, but this is an experience that we collected in a childhood. A child does not know what a hot pot means. In the same way, a robot does not know what it means to fall into the wall. Both behaviors are destructive and should be avoided. After the first touch, the child knows that it should not repeat this kind of movements because it caused great pain.

The absence of external rewards and punishment play the role of pain. After each movement made by the robot, a reinforcement is returned to it, which is an assessment of the made action. Thanks to it, the robot can determine which movement was beneficial and which was not. The ability to assess movement is crucial in the learning process. The learning based on external rewards is known as reinforcement learning [14].

Thanks to the interaction with the surroundings, the robot collects new experiences which are used for learning. Such a learning process aims to maximize the expected external rewards. However, reinforcement learning suffers from the credit assignment problem. It means that the returned external reward not always is a proper evaluation of actions. Moreover, we need to explicitly define the assessment function on the basis of the expert knowledge. Thanks to motivated learning, the robot thoroughly learns the surroundings which, together with the expert, automatically defines rewards and punishments.

2.3 Internal Rewards and Motivation Factors

Using motivated learning, we can also define internal rewards and punishments [13]. In this way, we give the robot another cognitive trait: motivation. It enables to create a better system of rewards and thus to return a more accurate rating. It is another aspect based on human behavior. Human is not guided in life by mere primitive stimuli such as the need for food or avoiding pain. Important factors also come from internal stimuli, which might be even stronger than external ones. In this way, the robot can create the motivation to meet internal needs, which can be more important than maximizing the external rewards or minimizing the external punishments. With the help of internal stimuli, the robot can also develop curiosity, which can mobilize to broaden and deepen its knowledge.

A robot using motivated learning has its own internal needs that it tries to satisfy. They may depend on external rewards sent by the surroundings. Maximizing external rewards may be one of the robot's goals, but it does not have to be the most important. The priority of the needs of robot varies depending on the current motivation to satisfy them. The robot learns to meet its internal needs through interaction with the surroundings. Performing actions, it observes what internal rewards it receives. Thanks to this, it gets to know what movements affect the satisfaction of the specific needs.

2.4   Motivation to Achieve the Goal

To learn to reach a specific location by the robot, we have defined a need for it. Thanks to this, the robot has the motivation to strive for this goal when it is away from it. This need is met when the robot reaches the assumed location.

The key to learning is the correct internal rating returned for the taken actions. The evaluation used in the project uses an external reward to avoid encountered obstacles. Moreover, it uses the current position of the robot and returns a positive value when the robot reaches its final position. So that the robot learns to reach the goal with the shortest route, a small penalty for the empty field was also defined. In this way, having in mind the total reward, the robot will try to determine the route running through the smallest number of fields which reflect its effort. The reward-punishment system motivating the robot to achieve the target location was defined for this research as follows [11]:

- Reaching the goal: the reward equals 1,
- Falling into an obstacle or an object: the punishment is defined as -1,
- Moving to an empty field: the small punishment is represented by -0.05.

After creation of the system of rewards and punishments, the robot can learn by interacting with the surroundings, what exact reward or punishment awaits it for the movement from each field in a specific direction.

However, the robot choosing an action is not guided by the immediate reward but is trying to predict what move will be more beneficial in the context of the whole route. It is interested in the sum of the rewards collected for the entire episode, that is from the current location to the final state. As the final state, we mean falling into an obstacle or reaching a destination. In the final state the robot returns to the starting location using a software module that was implemented in the separate project [15] and tries again to reach the target position.

How can the robot predict what the maximum reward for an entire episode possible to achieve after moving in a specific direction is? How can the collected experience be used to learn the robot to perform the correct movements? The answer to these questions is given by the deep Q-learning algorithm.

2.5   Deep Q-Learning

To satisfy the need of reaching the goal, the robot is guided by the Q function (1) defined in the deep Q-learning algorithm [16].

$$Q(S_t, A_t) \leftarrow R_{t+1} + \gamma \max_a Q(S_{t+1}, a) \tag{1}$$

It requires the current location (S_t) of the robot and the direction (A_t) of the movement that the robot wants to make, and it returns the anticipated future reward for the entire episode being an assessment of the chosen direction (Fig. 1A). In this way, the robot checks all possible movement's directions and chooses the one in which according to the Q function it has the chance to get the largest total reward for the whole episode.

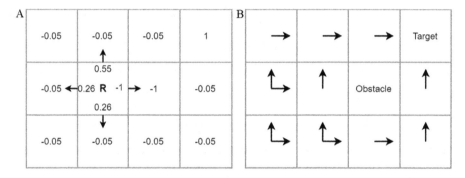

Fig. 1. An example of a room: (A) with rewards for each field and values returned by the Q function, (B) with selected best directions based on the Q function.

Using the Q function, it knows what direction to choose in each field (Fig. 1B). However, using the coefficient γ smaller than 1 rewards more distant in time are less taken into account. This project assumed $\gamma = 0.8$. The Q function should first be learned in order to return correct values. The learning is based on the experiences collected by the robot. After each made a move and the collected reward (R_{t+1}), the function is updated using the equation (1).

Here, the feedforward neural network plays the role of the Q function. At the input, it takes the current coordinates of the robot, and at the output gives four ratings. Each of them is assigned to a specific direction of movement. The used neural network consists of two hidden layers, each of them with 25 neurons. Due to the use of the neural network, the algorithm also uses backward propagation which is used to learn the neural network. The learning coefficient used in this algorithm is 0.05 [11]. At the beginning of learning, following the Q function may usually lead to wrong decisions. It is only after the appropriate number of learned experiences that the network starts to guide the robot better and better. The question is what the sense of following a non-trained function is? Is it not possible to make better use of the initial phase?

2.6   Motivation to Explore and Curiosity

It is well-known that the human curiosity can be a strong motivation to gain new experiences. The opportunity to learn something new affects people favorably because expanding knowledge gives us a better view of the situation, which results in better prediction and a decision-making process.

Before the robot begins to correctly choose the subsequent directions of movements on the way to the target location, it is worth using this time for effective learning. Therefore, the robot was motivated to explore the surroundings in a situation where there is no sufficient knowledge about it. In the initial phase, this motivation exceeds the motivation to reach the goal.

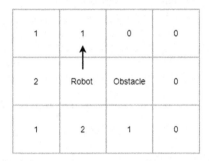

Fig. 2. An example of a room with the number of visits of each field after the first few episodes and with the marked next robot's movement.

Exploration Algorithm:

1. Take from memory information about neighboring fields.
2. Move to the field that is not a wall or obstacle and has the least number of visits. If more than one field meets this condition, then draw one out of them.
3. If the movement was impossible, because the field was a wall or an obstacle, save the field in memory as unavailable and go to step 1.
4. Increase the number of visits to the current field by 1. If the number of visits to any available field in memory is less than 3 then go to step 1.

During exploration, curiosity drags the robot to the places where it was the smallest number of times (Fig. 2). To make it possible, the robot must have another cognitive trait - memory and the ability to remember [1,2]. Thanks to it, it can determine how many times it was in a given place and where it must go to satisfy its curiosity in the best way. With each move, the robot selects the field in which it was the least number of times. It avoids fields that it already knows that are a wall or an obstacle. The motivation for such strategy disappears when each available field stored in the memory has been visited at least three times be. Three times is necessary because the robot can be surrounded by up to three unvisited possible fields (except the starting field) because it has already arrived from a certain field. If it was in the given field three times, according to the algorithm, he could choose one of the unvisited fields each time. Next, the robot has the confidence that every field in the room has been visited.

After visiting each field, the robot begins to satisfy the need for exploration differently. It does this with the help of ϵ-greedy policy [16,19]. It consists in selecting with a certain probability a random direction. At the beginning, the motivation for exploration is strong, and the probability is 100%. Along with the movements and growing knowledge about the surroundings, this motivation decreases, and the motivation to reach the goal increases. Therefore, the robot increasingly chooses the direction selected using the Q function instead of a random direction.

2.7   Experience Replay

Memory, which stores the robot experiences, is a very important part of the robot cognition system [1]. It is used not only to implement curiosity but also to experience replay. The robot stores in its memory all the experiences that it has learned. They consist of the position of the robot before the movement, the chosen direction of movement, the position of the robot after the movement and the obtained reward (Fig. 3).

The robot uses the saved experience to relearn. Every now and then it takes random experiences from memory and uses them to learn the Q function (1) [16,19]. Hence, the robot can still use the knowledge gained during exploration of the surroundings. It has been experimentally found that good results are obtained for taking from memory about 60 experiences, which is about 20% of various experiences that can be obtained in the experimental room.

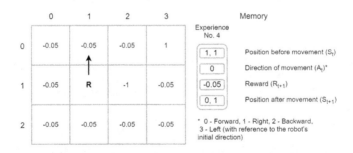

Fig. 3. An example of a fourth robot's movement and its representation in the memory.

2.8   Final Algorithm Controlling the Robot's Movements

1. Initialize the neural network, the starting coordinates as $(0, 0)$, the ϵ as 1, number of iterations (t) as 0 and maximal number of iterations (t_max).
2. Make a move towards the field with the least number of visits and increase t by 1. Finish if t is equal to t_max.

3. Save in memory the experience consisting of S_t, A_t, R_{t+1} and S_{t+1}.
4. Increase the number of visits to the S_t position or mark as a wall or an obstacle. If any field in the memory, which is not a wall or an obstacle, has a number of visits less than 3, go to point 2.
5. With the probability of ϵ move in a random direction. Otherwise, enter the current coordinates for the neural network input and make a move in the direction with the highest rating. Increase t by 1. Finish if t is equal to t_max.
6. Save in memory the experience consisting of S_t, A_t, R_{t+1} and S_{t+1}.
7. Take 60 memory experiences [S, A, R, S'] and use them to improve the neural network using the backpropagation algorithm. As a set value for A direction, use:
   (a) R   if S' is the final state,
   (b) $R + \gamma \max N(S')$ otherwise, where N(S') is the highest rating returned from neural network for position S'.
8. If the robot fell into a wall or an obstacle, it returns to the starting position.
9. If ϵ is greater than 0.01, decrease ϵ by 0.001 and go to point 5.

3   Simulation and Experimental Results

3.1   Virtual Environment

The research was carried out in the created virtual environment using Unity3D library (Fig. 4). The virtual room allows testing whether the robot always chooses one of the shortest routes, even if it is more difficult to overcome obstacles. One of the more difficult positions to learn is shown in Fig. 4. A properly taught robot should choose the route shown in Fig. 4A because it performs fewer moves (11) than according to the route shown in Fig. 4B. However, it must make more turns around the obstacles.

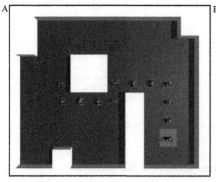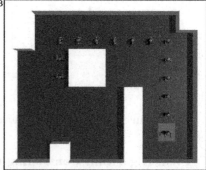

Fig. 4. The virtual room with robot and with the shown routes: (A) the shorter but more difficult route (more turns around), (B) the longer but easier route.

3.2 Robot Rating

To be able to estimate whether the robot correctly learned to move around the room, two indicators were defined. The first (Rating1) is the sum of the fields in which the robot chooses the wrong direction. The second indicator (Rating2) takes into account the exact values that are obtained at the output of the neural network and is expressed by the following formula:

$$Rating2 = \sum_{F}(O - T)^2 \tag{2}$$

where F - set of all available fields, O - vector of values returned by the neural network, T - vector of correct values.

3.3 Learning Rate

The learning coefficient in the backpropagation algorithm affects the rapidity of weight changes in neurons. It must be properly selected. The best result was achieved for a coefficient equal to 0.05 (Fig. 5).

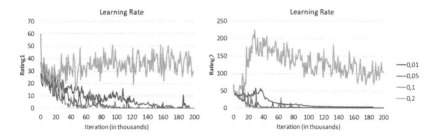

Fig. 5. Ratings of various learning rate values.

3.4 Neural Network Structure

The first and last layers are set due to the number of entries and exits from the network. The number of hidden layers and neurons contained in them must be chosen so that the neural network can quickly learn how to solve the given problem. The selection of the appropriate structure of the neural network is very important. In the case of this project, two hidden layers were applied and found to be enough to solve this problem optimally.

3.5 Motivation to Explore

What impact on the learning process have the exploration methods? The motivation for exploration discussed in this article is very important because it significantly speeds up the learning process (Fig. 7.) and is also one of the most

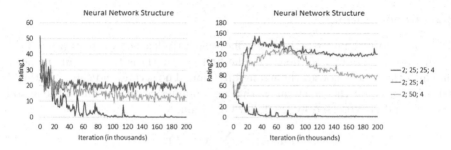

Fig. 6. Ratings of various neural network structures.

important contributions described in this paper. A good method of exploration provides a large number of experiences on the basis of which the robot learns. This is particularly important in this project. If the robot is to be able to reach a given location from any place, it must visit every place in the room. Therefore, it was crucial motivating the robot to go to unvisited places (second method).

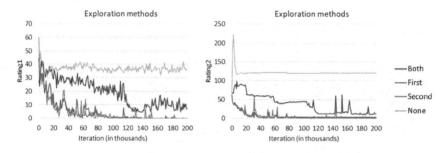

Fig. 7. The impact of exploration methods on learning. First - ϵ-greedy policy, second - going to unvisited places.

4   Conclusions and Final Remarks

The completed project implemented the assumptions and objectives set at the outset. The developed software module allowed the robot to learn how to reach the designated location from any location in the room in an optimal way, i.e. with the smallest possible number of steps. An important aspect of this project was the robot's initial state. It did not know the room in which it found itself. All information, by which it learned, it had to collect first. For this purpose, the robot was given cognitive qualities, and learning was based on the introduced motivated reinforcement learning. By defining the motivation for exploration, the robot can abandon the initial phase of following the neural network and instead gather experience. Thanks to this, the robot's learning was based on a large database of experiences collected during interactions with the surroundings. As shown in the studies, this had a key impact on the outcome of the project. Reinforcement learning was not enough because it did not allow full understanding of the surroundings. Only motivated learning enabled the robot to learn effectively. First, it collected the experience and formed its knowledge about the surroundings, and after that, it learned from it and adapted its moving routines using a neural network. The project used two methods to explore the area: choosing random directions and going to unvisited places. The first method (ϵ-greedy policy), thanks to randomness, could allow selecting the directions omitted in the second method. Going to unvisited places allowed to reach each field but could impose only certain directions in these fields. Both methods were therefore complementary.

Acknowledgments. This work was supported by AGH 11.11.120.612 and the grant from the National Science Centre DEC-2016/21/B/ST7/02220.

References

1. Duch, W.: Brain-inspired conscious computing architecture. J. Mind Behav. **26**, 1–22 (2005)
2. Horzyk, A., Starzyk, J.A., Graham, J.: Integration of semantic and episodic memories. IEEE Trans. Neural Netw. Learn. Syst. **28**(12), 3084–3095 (2017)
3. Horzyk, A.: How does generalization and creativity come into being in neural associative systems and how does it form human-like knowledge? Neurocomputing **144**, 238–257 (2014)
4. Horzyk, A.: Artificial Associative Systems and Associative Artificial Intelligence, pp. 1–276. EXIT, Warsaw (2013)
5. Horzyk, A.: Human-like knowledge engineering, generalization, and creativity in artificial neural associative systems. In: Skulimowski, A.M.J., Kacprzyk, J. (eds.) Knowledge, Information and Creativity Support Systems: Recent Trends, Advances and Solutions. AISC, vol. 364, pp. 39–51. Springer, Cham (2016). https://doi.org/10.1007/978-3-319-19090-7_4
6. Kalat, J.W.: Biological Grounds of Psychology. PWN, Warsaw (2006)
7. Larose, D.T.: Discovering Knowledge from Data: Introduction to Data Mining. PWN, Warsaw (2006)
8. Rutkowski, L.: Techniques and Methods of Artificial Intelligence. PWN, Warsaw (2012)
9. Tadeusiewicz, R.Y.: New trends in neurocybernetics. Comput. Methods Mater. Sci. **10**(1), 1–7 (2010)
10. Smart, W.D.: Making reinforcement learning work on real robots. Ph.D. dissertation, Department of Computer Science, Brown University, Providence, RI (2002)
11. Papiez, P.: Use of a neural network to move a robot to the designated target position in a maze using reinforcement learning. Engineering dissertation, AGH University of Science and Technology in Krakow, Supervised and Promoted by A. Horzyk (2018)
12. Kober, J., Bagnell, J.A., Peters, J.: Reinforcement learning: a survey. Int. J. Robot. Res. **32**, 1238–1274 (2013)
13. Starzyk, J.A.: Motivated learning for computational intelligence. In: Igelnik, B. (ed.) Computational Modeling and Simulation of Intellect: Current State and Future Perspectives (Chap. 11), pp. 265–292. IGI Publishing, Hershey (2011)

14. Sutton, R.S., Barto, A.G.: Reinforcement Learning: An Introduction. MIT Press, Cambridge (1998)
15. Janowski, M.: Use of a neural network to move a robot back to the starting position in a maze using supervised learning. Engineering dissertation, AGH University of Science and Technology in Krakow, Supervised and Promoted by A. Horzyk (2018)
16. Mnih, V., Kavukcuoglu, K., Silver, D., Graves, A., Antonglou, I., Wierstra, D., Riedmiller, M.: Playing Atari with Deep Reinforcement Learning. arXiv preprint arXiv:1312.5602 (2013)
17. Starzyk, J.A., Graham, J., Horzyk, A.: Trust in motivated learning agents. In: 2016 IEEE SSCI, pp. 1–8. Inc. 57 Morehouse Lane Red Hook, NY, 12571, USA. IEEE Xplore (2016). https://doi.org/10.1109/SSCI.2016.7850027, ISBN 978-1-5090-4239-5
18. Tadeusiewicz, R.: Introduction to intelligent systems. In: Wilamowski, B.M., Irvin, J.D. (eds.) The Industrial Electronics Handbook Intelligent Systems, pp. 1–12. CRC Press, Boca Raton (2011)
19. Matiisen, T.: Demystifying Deep Reinforcement Learning (2015). http://neuro.cs.ut.ee/demystifying-deep-reinforcement-learning
20. Tadeusiewicz, R., Horzyk, A.: Man-machine interaction improvement by means of automatic human personality identification. In: Saeed, K., Snášel, V. (eds.) CISIM 2014. LNCS, vol. 8838, pp. 278–289. Springer, Heidelberg (2014). https://doi.org/10.1007/978-3-662-45237-0_27

Company Bankruptcy Prediction
with Neural Networks

Jolanta Pozorska[1][(⊠)] and Magdalena Scherer[2]

[1] Faculty of Mechanical Engineering and Computer Science,
Institute of Mathematics, Częstochowa University of Technology,
al. Armii Krajowej 21, 42-200 Częstochowa, Poland
jolanta.pozorska@im.pcz.pl
[2] Faculty of Management, Częstochowa University of Technology,
al. Armii Krajowej 19, 42-200 Częstochowa, Poland
magdalena.scherer@wz.pcz.pl

Abstract. Bankruptcy prediction is a very important issue in business financing. Raising availability of financial data makes it more and more viable. We use large data concerning the health of Polish companies to predict their possible bankruptcy in a relatively short period. To this end, we utilize feedforward neural networks.

Keywords: Bankruptcy prediction · Neural networks

1 Introduction

Understanding and prediction of corporate bankruptcies have been a subject of interest of researchers for decades. Similar to other issues in economics and finance, such a prediction is not a trivial task as it requires the domain expert knowledge and past data that is as rich as possible. Usually, companies are evaluated on the basis of many factors by a mathematical model created on the basis of past operations [1]. Moreover, the task of creating the model is difficult because of the data imbalance caused by the uneven distribution of successful and bankrupt companies. With more and more data available, researchers had to find new, more sophisticated methods for the prediction model construction. They moved from simple econometrics and data mining models to more sophisticated machine learning methods. In [22] Shin et al. explore possibilities of using support vector machines (SVM, [24]) to detect possible corporate failures. They claim that neural networks have too many parameters to adjust and the gradient descent optimization tends to converge to a local minimum not assuring perfect generalization. Moreover, neural networks require large datasets. That is why they find SVM to be a better choice. du Jardin in [7] combined expert knowledge and classifier bagging and boosting ensembling [5] methods to predict bankruptcy. They built financial profiles using Kohonen maps. In [18] many data mining methods are compared on various bankruptcy data. In [17] the authors preprocess financial data with principal component analysis and classify them

© Springer International Publishing AG, part of Springer Nature 2018
L. Rutkowski et al. (Eds.): ICAISC 2018, LNAI 10841, pp. 183–189, 2018.
https://doi.org/10.1007/978-3-319-91253-0_18

with support vector machines with the radial basis function kernels. They predicted bankruptcy of the Korean companies. A good review of various techniques for bankruptcy prediction is provided in [12] including their advantages and disadvantages. Market and macroeconomic variables are added to the prediction model on more than twenty thousand companies in [23]. The authors claim that financial crises revealed that the prediction at the micro level is not sufficient. In [6,9,11] the fuzzy rule base [3,13,21] is used, while in [2] neural networks are used. Fuzzy rule base was used also for waste production in [19] and neural networks in [20] for sales prediction. The rough set theory is a very good tool for dealing with imperfect data. Rough sets were used for bankruptcy prediction in [4,10,16]. The authors of [14] used decision trees for prediction. It is also possible to use statistical methods for function estimation [8].

In this paper, we use neural networks to predict the bankruptcy of Polish companies. Neural networks require a large amount of training data to accurately model phenomena from data. The paper is organized as follows. Section 2 describes neural networks. Numerical experiments performed on the Polish company bankruptcy dataset [25] are presented in Sect. 3.

2   Methodology of the Research

To predict company bankruptcy we use artificial neural networks (ANNs). The inspiration for their construction were natural neurons connected by synapses and the entire nervous system with the brain. ANNs are mathematical structures implemented by software or hardware models. Artificial neural networks can be used in a broad spectrum of data processing issues, such as pattern classification, prediction, denoising, compression and image and sound recognition, or automation. Neural networks have the ability to process incomplete data and to provide approximate results. They enable fast and efficient processing of large amounts of data. They are resistant to errors and damage.

The basic element of the neural network is the neuron. The neuron model has n inputs, $x_1, \ldots, x_n$ are input signals, $w_0, \ldots, w_n$ are synaptic weights, y is the output value, w_0 is bias and f is the activation function. The operation of the neuron can be described using the formula $y = f(s)$, where $s = \sum_{i=0}^{n} x_i w_i$. The input signals $x_0, \ldots, x_n$ are multiplied by the corresponding weights $w_0, \ldots, w_n$. The resulting values are summed to produce the signal s. The signal is then subjected to an activation function that is usually nonlinear to create many layers. There are many models of neural networks. The neural network division can be made taking into account the following factors: learning method, the direction of signal propagation in the network, type of activation function, type of input data and method of interconnection between neurons. Neural networks consist of interconnected neurons. Depending on how these connections are made, various types of neural networks are distinguished: feedforward, recurrent, convolutional and cellular networks. In feedforward networks, the flow of signals is always in one direction, from the input to the output. Neuron outputs from one layer are neuron inputs in the next layer. In recurrent networks, some of the output signals are simultaneously input signals. In networks of this type, the activation of the network by the input signal causes the activation of some or all of the neurons in the, so-called, network relaxation process. Therefore, in order to validate the operation of the network, a stability condition should be added. The stimulated network must reach a stable state where the baseline values of the neurons remain constant and this process should take place at a finite time. On the other hand, in cellular neural networks, each neuron is connected to neighbouring neurons. Most commonly used neural architecture, both in research and commercial models, are perceptron networks. These are unidirectional networks where neurons are grouped in at least two layers. The first layer is called the input layer and the last layer is the output layer. There may be one or more hidden layers between these layers. Signals are passed from the input layer to the output layer, without feedback to the previous layers. A multi-layer neural network has input signals $x_1, \ldots, x_n$. In a general case, we can have more than one output signals, $y_1, \ldots, y_n$ and several layers denoted by k. The error Q at the network output is defined by

$$Q\left(t\right) = \sum_{i=1}^{N_L} \left(\varepsilon_i^{(L)}\right)^2 (t) = \sum_{i=1}^{N_L} \left(d_i^{(L)}\left(t\right) - y_i^{(L)}\left(t\right)\right)^2, \tag{1}$$

where t is the iteration number, d is a desired value, y is the output of ith neuron defined by

$$y_i^{(k)}\left(t\right) = f\left(s_i^{(k)}\left(t\right)\right), \ \ s_i^{(k)}\left(t\right) = \sum_{j=0}^{N_{k-1}} w_{ij}^{(k)}\left(t\right) x_j^{(k)}\left(t\right), \tag{2}$$

where w is a weight. The backpropagation algorithm propagates the error toward the network input, thus the error in hidden layers is defined as a sum of error in the next layer's neurons weighted by corresponding weights

$$Q_i^{(k)}\left(t\right) = \begin{cases} d_i^{(L)}\left(t\right) - y_i^{(L)}\left(t\right) & dla \ k = L \\ \sum_{m=1}^{N_{k+1}} \delta_m^{(k+1)}\left(t\right) w_{mi}^{(k+1)}\left(t\right) \ dla \ k = 1, \ldots, L-1 \end{cases}, \tag{3}$$

where δ is defined as multiplication of errors in the next layer and the derivative of the activation function

$$\delta_i^{(k)}\left(t\right) = \varepsilon_i^{(k)}\left(t\right) f'\left(s_i^{(k)}\left(t\right)\right), \tag{4}$$

where

$$\varepsilon_i^{(k)}\left(t\right) = \sum_{m=1}^{N_{k+1}} \delta_m^{(k+1)}\left(t\right) w_{mi}^{(k+1)}\left(t\right), \ \ \ k = 1, \ldots, L-1. \tag{5}$$

Finally, we obtain the formula for weight modification in iteration t

$$w_{ij}^{(k)}\left(t+1\right) = w_{ij}^{(k)}\left(t\right) + 2\eta\delta_i^{(k)}\left(t\right) x_j^{(k)}\left(t\right), \tag{6}$$

where η is the learning coefficient responsible for the convergence speed. The number of neurons in each layer is important in the operation of the network.

Too many neurons increase the learning process. In addition, if the number of learning samples in relation to the network size is small, the network can be overtrained and thus lose the ability to generalize knowledge. In this case, the network will learn the learning dataset "by heart" and will probably correctly map only the samples that were included in it. Therefore, after learning the network, we should check the correctness of its operation. For this purpose, a test dataset consisting of samples that were not present in the network learning process is used. Only after testing is it possible to tell whether the network has been properly trained and works properly. There are two methods of learning neural networks: supervised learning and unsupervised learning. Network learning involves enforcing a specific neural network response to the input signals. That is why a very important moment in research is the right choice of the learning method. Supervised teaching, also called learning with a teacher, involves modifying weights so that the output signals are as close as possible to the desired values. Training data includes both input signal groups and desired values for responding to these signals. A special case of supervised learning is reinforcement learning, where the network is trained not to give exact values of the desired output signals, only the information or whether it responds correctly. Unattended learning, called non-teacher learning, is a self-parsing study of dependence in a test set by a neural network. During training, the network receives no information about the desired response. Training data contains only a set of input signals. Networks with such action are called self-organizing or self-associative. Neural networks can learn a broad spectrum of problems on the basis of data. They are better than traditional computer architectures in tasks that people perform naturally, such as image recognition or generalization of knowledge. Advances in computer technology and network learning algorithms have resulted in a steady increase in the complexity of tasks solved by neural networks. New architectures are also emerging, such as convolutional neural networks being able to classify hundreds of image classes. Neural networks are used to solve different problems. However, every problem requires a proper network adaptation. An appropriate network topology, the number of neurons in layers, and the number of network layers must be selected. Next, we need to prepare a training and testing set. The network must be trained learn first and then the correct operation of the network must be verified. In the next section, we use artificial multilayer perceptron network to predict corporate bankruptcy.

3   Experimental Results

We used Polish companies bankruptcy data dataset [25]. The data contain 10,503 objects with 64 features concerning company health (state) such as net profit, total liabilities, working capital, current assets, sales, gross profit, net profit, operating expenses, total costs, etc. The information was collected from Emerging Markets Information Service in the period of 2000–2012, which is a database containing information on emerging markets around the world. The last column indicates the bankruptcy status after, respectively, 1, 2, ..., and 5 years. We used the same two-layer neural network presented in Fig. 1 for all the experiments. We removed objects with missing values existing in the datasets and trained the networks with the Levenberg–Marquardt algorithm [15] for 300 epochs. Errors during training are shown in Fig. 2. The root mean square errors for respective datasets are shown in Table 1.

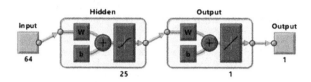

Fig. 1. Neural network structure used in the experiments.

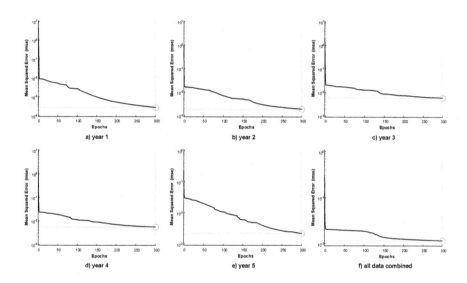

Fig. 2. The root mean square errors for respective datasets. The last plot is for the combined dataset for all five years.

Table 1. Root mean square error for every year bankruptcy data and all the year combined

	Year 1	Year 2	Year 3	Year 4	Year 5	All five years
RMSE	0.016	0.045	0.077	0.074	0.046	0.108

The presented experiments proved that our method is effective and correctly predicts bankruptcy.

4 Conclusion

In the paper, we used feedforward neural networks to predict possible company bankruptcy. We used large datasets with 64 input variables regarding corporate health. Unfortunately, many cases contained missing values, and we were forced not to include them in the training. Future research would involve methods that are able to deal with incomplete data. Even though we removed missing cases, we were able to achieve very high accuracy.

References

1. Altman, E.I., Hotchkiss, E.: Corporate Financial Distress and Bankruptcy: Predict and Avoid Bankruptcy, Analyze and Invest in Distressed Debt, vol. 289. Wiley, Hoboken (2010)
2. Atiya, A.F.: Bankruptcy prediction for credit risk using neural networks: a survey and new results. IEEE Trans. Neural Networks **12**(4), 929–935 (2001)
3. Bartczuk, Ł., Łapa, K., Koprinkova-Hristova, P.: A new method for generating of fuzzy rules for the nonlinear modelling based on semantic genetic programming. In: Rutkowski, L., Korytkowski, M., Scherer, R., Tadeusiewicz, R., Zadeh, L.A., Zurada, J.M. (eds.) ICAISC 2016. LNCS (LNAI), vol. 9693, pp. 262–278. Springer, Cham (2016). https://doi.org/10.1007/978-3-319-39384-1_23
4. Bioch, J., Popova, V.: Bankruptcy prediction with rough sets. ERIM Report Series Research in Management ERS-2001-11-LIS, Erasmus Research Institute of Management (ERIM), ERIM is the joint research institute of the Rotterdam School of Management, Erasmus University and the Erasmus School of Economics (ESE) at Erasmus University Rotterdam, February 2001
5. Bologna, G., Hayashi, Y.: Characterization of symbolic rules embedded in deep dimlp networks: a challenge to transparency of deep learning. J. Artif. Intell. Soft Comput. Res. **7**(4), 265–286 (2017)
6. de Andrés, J., Landajo, M., Lorca, P.: Forecasting business profitability by using classification techniques: a comparative analysis based on a Spanish case. Eur. J. Oper. Res. **167**(2), 518–542 (2005)
7. du Jardin, P.: A two-stage classification technique for bankruptcy prediction. Eur. J. Oper. Res. **254**(1), 236–252 (2016)
8. Galkowski, T., Pawlak, M.: Nonparametric estimation of edge values of regression functions. In: Rutkowski, L., Korytkowski, M., Scherer, R., Tadeusiewicz, R., Zadeh, L.A., Zurada, J. (eds.) ICAISC 2016. LNCS (LNAI), vol. 9693, pp. 49–59. Springer, Cham (2016). https://doi.org/10.1007/978-3-319-39384-1_5
9. Gorzalczany, M.B., Piasta, Z.: Neuro-fuzzy approach versus rough-set inspired methodology for intelligent decision support. Inf. Sci. **120**(1), 45–68 (1999)
10. Greco, S., Matarazzo, B., Slowinski, R.: A new rough set approach to multicriteria and multiattribute classification. In: Polkowski, L., Skowron, A. (eds.) RSCTC 1998. LNCS (LNAI), vol. 1424, pp. 60–67. Springer, Heidelberg (1998). https://doi.org/10.1007/3-540-69115-4_9
11. Jensen, R., Cornelis, C.: Fuzzy-rough nearest neighbour classification and prediction. Theor. Comput. Sci. **412**(42), 5871–5884 (2011). Rough Sets and Fuzzy Sets in Natural Computing
12. Kumar, P.R., Ravi, V.: Bankruptcy prediction in banks and firms via statistical and intelligent techniques - a review. Eur. J. Oper. Res. **180**(1), 1–28 (2007)

13. Łapa, K., Cpałka, K., Galushkin, A.I.: A new interpretability criteria for neuro-fuzzy systems for nonlinear classification. In: Rutkowski, L., Korytkowski, M., Scherer, R., Tadeusiewicz, R., Zadeh, L.A., Zurada, J.M. (eds.) ICAISC 2015. LNCS (LNAI), vol. 9119, pp. 448–468. Springer, Cham (2015). https://doi.org/10. 1007/978-3-319-19324-3_41

14. Lin, F.Y., McClean, S.: A data mining approach to the prediction of corporate failure. Knowl.-Based Systems **14**(3–4), 189–195 (2001)

15. Marquardt, D.W.: An algorithm for least-squares estimation of nonlinear parameters. J. Soc. Ind. Appl. Math. **11**(2), 431–441 (1963)

16. Mckee, T.E.: Developing a bankruptcy prediction model via rough sets theory. Int. J. Intell. Syst. Account. Financ. Manag. **9**(3), 159–173 (2000)

17. Min, J.H., Lee, Y.C.: Bankruptcy prediction using support vector machine with optimal choice of kernel function parameters. Expert Syst. Appl. **28**(4), 603–614 (2005)

18. Olson, D.L., Delen, D., Meng, Y.: Comparative analysis of data mining methods for bankruptcy prediction. Decis. Support Syst. **52**(2), 464–473 (2012)

19. Scherer, M.: Waste flows management by their prediction in a production company. J. Appl. Math. Comput. Mech. **16**, 135–144 (2017)

20. Scherer, M.: Multi-layer neural networks for sales forecasting. J. Appl. Math. Comput. Mech. **17**, 61–68 (2018)

21. Scherer, R.: Multiple Fuzzy Classification Systems. Springer, Heidelberg (2012). https://doi.org/10.1007/978-3-642-30604-4

22. Shin, K.S., Lee, T.S., Kim, H.J.: An application of support vector machines in bankruptcy prediction model. Expert Syst. Appl. **28**(1), 127–135 (2005)

23. Tinoco, M.H., Wilson, N.: Financial distress and bankruptcy prediction among listed companies using accounting, market and macroeconomic variables. Int. Rev. Financ. Anal. **30**, 394–419 (2013)

24. Villmann, T., Bohnsack, A., Kaden, M.: Can learning vector quantization be an alternative to SVM and deep learning? - recent trends and advanced variants of learning vector quantization for classification learning. J. Artif. Intell. Soft Comput. Res. **7**(1), 65–81 (2017)

25. Zikeba, M., Tomczak, S.K., Tomczak, J.M.: Ensemble boosted trees with synthetic features generation in application to bankruptcy prediction. Expert Syst. Appl. **58**, 93–101 (2016)

Soft Patterns Reduction for RBF Network Performance Improvement

Pawel Rozycki[1(✉)], Janusz Kolbusz[1], Oleksandr Lysenko[2],
and Bogdan M. Wilamowski[3]

[1] University of Information Technology and Management in Rzeszow,
Rzeszow, Poland
{prozycki,jkolbusz}@wsiz.rzeszow.pl
[2] National Technical University of Ukraine "Igor Sikorsky Kyiv Polytechnic
Institute", Kyiv, Ukraine
allias@hotmail.com
[3] Auburn University, Auburn, AL 36849-5201, USA
wilambm@auburn.edu

Abstract. Successful training of artificial neural networks depends primarily on used architecture and suitable algorithm that is able to train given network. During training process error for many patterns reach low level very fast while for other patterns remains on relative high level. In this case already trained patterns make impossible to adjust all trainable network parameters and overall training error is unable to achieve desired level. The paper proposes soft pattern reduction mechanism that allows to reduce impact of already trained patterns which helps in getting better results for all training patterns. Suggested approach has been confirmed by several experiments.

Keywords: RBF network training improvement · ErrCor
Error Correction · Soft patterns reduction

1 Introduction

Artificial intelligence is going to create an almost as perfect system as a human being who would be able to learn by interacting with the environment. The milestone in the spread of artificial neural networks was to replace the process of designing the solution by replacing it with the learning process with the development of a learning algorithm. The first such algorithm teaching an artificial neural network was the EBP algorithm (Error Back Propagation) [1]. Artificial neural network trained using this learning algorithm, usually in combination with MLP (Multi Layer Perceptron) topology, was able to solve many problems that until now were not solvable in an algorithmic way [2]. However, this turned out

This work was supported by the National Science Centre, Krakow, Poland, under-grant No. 2015/17/B/ST6/01880.

© Springer International Publishing AG, part of Springer Nature 2018
L. Rutkowski et al. (Eds.): ICAISC 2018, LNAI 10841, pp. 190–200, 2018.
https://doi.org/10.1007/978-3-319-91253-0_19

to be ineffective for more complex problems, and searches aimed at developing new algorithms for teaching artificial neural networks as well as development of new artificial neural network architectures were begun. For neural networks to be able to effectively solve problems whose nature is non-linear and often multidimensional, it is necessary to use unconventional methods.

Effective learning of neural networks requires understanding the neural network architecture and its impact on the learning process and the functioning of the target system that would benefit from a neural network based solution. The main impact on the functioning of the artificial neural network have: network architecture, way of connecting neurons in architecture and the learning algorithm itself [3]. Connections in the neural network architecture play an important role. For example, an MLP architecture consisting of ten neurons is able to solve the problem of Parity–9 [4], at the same time reaching its limit. However, if the MLP architecture will be modified and the connections will be also created between non-adjacent layers, like in BMLP (Bridged MLP) architecture, the network is able to solve the problem of parity–1023 [4,5]. The use of appropriate architecture has a significant impact on solving the problem. Maintaining the same number of neurons can increase the bandwidth of the network, even one hundred times depending on the architecture [6–8].

Going further along this line of reasoning, we come to the conclusion that by expanding the architecture of the neural network with further layers of the network, we can solve more and more complicated problems. However, there is a problem of the lack of an appropriate network learning algorithm with such a deep, deep architecture. Known network learning algorithms, such as EBP or LM, are not suitable for training networks in such an architecture. What is more, nowadays the amount of data processed by artificial neural networks can be very big. Accordingly, this negatively affects the time of their training. In fact, the LM algorithm is not able to train data sets with a large number of patterns, limiting it to solving relatively small problems. This is observable especially for big, deep networks trained by huge amount of data. Therefore, reducing the amount of data necessary for training can accelerate this process.

This limitation in the number of training patterns applies to many known network learning algorithms, making them virtually unusable for use with very complex problems. Ways of solving this type of problem were proposed in [9–11], where the training set is reduced in several general ways by random selection of subset patterns (a), selection of the most important patterns according to the given cost function (b) and by creating clusters (c). An exemplary solution to this problem by removing trained patterns during the training process in a similar way as the elimination of distant states described in [12] was presented in the article [13] where training set has been reduced by elimination of already trained patterns. This paper continues this last idea introduces the concept of soft patterns reduction that instead of removing the trained patterns, allows to reduce their impact on the training process.

The rest of paper is organized as follows. Section 2 contains the description of proposed approach, Sect. 3 shows results of experiments that confirm of proposed method for RBF networks with Error Correction (ErrCor) algorithm, and finally Sect. 4 contains some conclusions and remarks.

2   Concept of Soft Patterns Reduction

Usually, during training process achieving a low level of error occurs fairly quickly for many patterns. The use of all patterns on the subsequent set of iterations of the training is impractical because their influence on the level of error is different. Already trained patterns hold network in steady state preventing futher efficient training with the rest of patterns. The solution to this problem can be a reduction in the number of patterns used to train the network at a stage where the current level of error is almost unchanged. This method was proposed in authors' article [13] where have been presented several methods for removing of already trained patterns. That allowed to achieve a good results, that means better then these reached without pattern reduction, but in most cases in lower training time. The disadvantage of approach presented in [13] is that already trained patterns are removed from training set that can lead to instability of training process. On the other hand, it is possible to reduce the impact of patterns that make a negative contribution to training process, in this case already trained patterns. To reduce the importance of the pattern during the learning process, the SRR (*Soft Remove Rate*) is used, which set the level of the desired neglect of specific data. During training, each pattern for which the error is less than the defined CT (*Cut-off Threshold*) is detected and a value SRR is used as rate of this pattern during the gradient and the Hessian calculation. In consequence of this procedure, the effect of pattern on error level will be decreased.

The procedure of trained patterns detection setting SRR should be done periodically during training process. Moreover, some mechanism for restore pattern to training set under some conditions can be implemented.

The proposed approach has been implemented and tested for the Error Correction algorithm [14,15] and the shallow RBF network.

2.1   Error Correction Algorithm

Error Correction algorithm (*ErrCor*) is constructive training method that is used to train RBF networks shown in Fig. 1 that are build using RBF units with Gaussian activation function defined by (1).

$$\varphi_h(x_p) = \exp\left(-\frac{\|x_p - c_h\|^2}{\sigma_h}\right) \tag{1}$$

where: c_h and σ_h are the center and width of RBF unit h, respectively. $\|\cdot\|$ represents the computation of Euclidean Norm. The output of such network is given by:

$$O_p = \sum_{h=1}^{H} w_h \varphi_h(x_p) + w_o \tag{2}$$

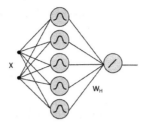

Fig. 1. RBF network architecture

where: H is the number of RBF units, w_h presents the weight on the connection between RBF unit h and network output. w_0 is the bias weight of output unit.

The main idea of the *ErrCor* algorithm is to increase the number of RBF units one by one adjusting all RBF units in network by training them after adding of each unit. The new unit is initially set to compensate largest error in the current error surface and after that all units are trained changing both centers and widths as well as output weights, so *ErrCor* algorithm is not only constructive that allow to achieve a networks with proper number of units but also deterministic.

The update rule of the *ErrCor* training algorithm is given by

$$\Delta_{t+1} = \Delta_t (Q_t - \mu I)^{-1} g_t \tag{3}$$

where Δ refers to all the parameters need to be adjusted, Q is the quasi Hessian matrix, μ is the learning coefficient; I is the identity matrix, and g is the gradient vector.

As can be found in [15], to reduce computation complexity, quasi Hessian matrix Q and gradient vector g are transformed into the sum of subquasi-Hessian matrixes and sum of subgradient vectors as follows

$$Q = \sum_{p=1}^{P} q_p \qquad g = \sum_{p=1}^{P} \eta_p \tag{4}$$

where P is the number of training patterns, q_p is subquasi-Hessian matrixes and η_p is subgradient vectors that be obtained through the calculations of Jacobian row

$$q_p = j_p^T j_p \qquad \eta_p = j_p^T e_p \tag{5}$$

where Jacobian rows j_p are calculated as

$$j_p = \left[\frac{\partial e_p}{\partial c_j} \dots \frac{\partial e_p}{\partial \sigma_j} \dots \frac{\partial e_p}{\partial w_j} \right] \tag{6}$$

where e_p is an error between the desired output and actual output of the p–th pattern. More details of algorithm can be found in [14,15].

To reduce impact of given pattern on training process both Q and g in Eq. 4 can be modified by the factor in range $<0, 1)$

$$Q = \sum_{p=1}^{P} q_p srr_p \qquad g = \sum_{p=1}^{P} \eta_p srr_p \tag{7}$$

where srr_p is SRR value for trained patterns that impact should be reduced and 1 value for the rest of patterns.

Trained patterns are detected periodically by check if their actual error is below given CT value. If actual error is over given RT (Restore Threshold) value the pattern is again treated as untrained. In the proposed solution this procedure is prepared every N added RBF units.

The pseudo code of the Error Correction algorithm with soft pattern reduction is shown in Fig. 2. More important training parameters are: N_{MAX} – maximal number of RBFs used in network ($ErrCor$ as constructive algorithm starts with one RBF unit and in each main loop next RBF unit is added. Number of units that can be added is limited by this parameter): $max\_iterations$ – maximal number of iterations determine what is the max number of execute the loop for training of network after adding each RBF unit; $desired\_R\_error$ – the level of relative change of training error for which the network in given configuration is treated as trained as good as possible; if these value is reached the training process for given number of RBFs is broken and next RBFs is added;

```
1: set error vector err as desired outputs
2: for n ← 1 to N_MAX do                                          ▷ for all new RBF units
3:      find err_max as maximum of absolute value of err
4:      create a new RBF unit with the center c_h at the location of pattern with
        err_max by setting width σ_h and corresponding weight w_h to 1
5:      calculate MSE(iter = 1)
6:      for iter ← 2 to max_iteration do
7:          if N new RBFs are added then ▷ mark patterns as trained or untrained
8:              for patterns abs(err) < CT set srr = SRR
9:              for patterns abs(err) > RT set srr = 1
10:         end if
11:         calculate quasi-Hessian matrix Q and gradient vector g (Eq.7)
12:         update network parameters (Eq.3)
13:         calculate MSE(iter)
14:         while error is not reduced do
15:             adjust the parameter using the LM scheme
16:         end while
17:         if abs( (MSE(iter-1)-MSE(iter)) / MSE(iter-1) ) < desired_R_error then
18:             break
19:         end if
20:     end for
21:     calculate actual err
22:     if MSE(n) < desired_error then
23:         break
24:     end if
25: end for
```

Fig. 2. ErrCor algorithm with soft patterns reduction

desired_error – the level of desired error that determine successful training; if the network reaches this level of error the training process is broken even if number of RBFs is below N_{MAX}.

3   Experimantal Results

To evaluate proposed approach some experiments have been prepared. Modified ErrCor algorithm have been used for several two dimensional benchmark approximation problems: Peaks function, Schwefel function and Schaffer function.

All experiments have been done with Matlab software with the following training parameters: $N_{MAX} = 20$, *max_iterations* $= 100$, *desired_R_error* $= 0.0001$, *desired_error* $= 10^{-12}$, $RT = 2CT$ and $N = 6$. For all functions have been randomly generated three sets of 2977 training and 744 validating patterns. The measured values were Root Mean Square Error for training ($RMSE_T$) and validating ($RMSE_V$) data sets and training time $time_{TR}$. Achieved results have been compared to reached with original ErrCor algorithm [14,15].

3.1   Peaks Function

First experiment has been done for Peaks function given by Example 8 for three different ranges of $<x, y>$: $<-2, 2>$ (A), $<-3, 3>$ (B) and $<-5, 5>$ (C). Surfaces of these functions with $<x, y>$ normalized to range $<-1, 1>$ and value z normalized to range $< 0, 1 >$ are shown in Fig. 3a–c respectively.

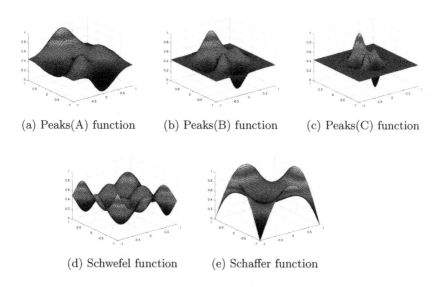

(a) Peaks(A) function   (b) Peaks(B) function   (c) Peaks(C) function

(d) Schwefel function   (e) Schaffer function

Fig. 3. Surfaces of functions used in experiments

Table 1. Results for Peaks(A) function

CT	SRR	1			2			3		
		$RMSE_T$	$RMSE_v$	$time_{TR}$ [s]	$RMSE_T$	$RMSE_v$	$time_{TR}$ [s]	$RMSE_T$	$RMSE_v$	$time_{TR}$ [s]
Original		0.000116	0.000121	55.43	0.000147	0.000145	62.66	0.000037	0.000037	63.22
0.00001	0.9	0.000075	0.000073	72.46	0.000155	0.000149	64.47	0.000019	0.000019	72.56
0.00005		0.000093	0.000100	62.29	0.000049	0.000047	61.42	0.000018	0.000018	63.03
0.00009		0.000098	0.000102	60.95	0.000038	0.000039	59.29	0.000022	0.000023	62.91
0.0007		0.000053	0.000055	61.52	0.000096	0.000100	56.89	0.000043	0.000045	61.20
0.0003		0.000081	0.000088	61.55	0.000045	0.000046	59.68	0.000023	0.000024	58.24
0.001		0.000061	0.000061	63.24	0.000074	0.000074	57.49	0.000038	0.000039	59.27
0.003		0.000040	0.000041	56.84	0.000053	0.000052	57.95	0.000050	0.000055	60.56
0.005		0.000042	0.000044	59.12	0.000136	0.000158	57.07	0.000081	0.000081	56.29
0.00001	0.7	0.000064	0.000065	65.47	0.000148	0.000147	65.55	0.000044	0.000046	66.05
0.00005		0.000053	0.000057	59.17	0.000114	0.000113	53.93	0.000020	0.000020	63.46
0.00009		0.000019	0.000020	62.45	0.000118	0.000124	55.79	0.000049	0.000052	59.96
0.0007		0.000047	0.000046	61.18	0.000058	0.000057	63.56	0.000069	0.000067	60.67
0.0003		0.000048	0.000042	63.01	0.000221	0.000237	56.42	0.000030	0.000032	61.53
0.001		0.000086	0.000090	58.49	0.000103	0.000115	63.72	0.000026	0.000028	58.10
0.003		0.000049	0.000050	64.43	0.000076	0.000075	64.58	0.000021	0.000021	65.05
0.005		0.000026	0.000025	62.69	**0.0000014**	**0.000014**	65.84	0.000040	0.000044	67.10
0.00001	0.3	0.000059	0.000058	69.93	0.000168	0.000161	63.92	**0.000010**	**0.000010**	67.78
0.00005		0.000036	0.000037	61.02	0.000125	0.000125	69.61	0.000046	0.000049	70.18
0.00009		0.000055	0.000060	60.27	0.000040	0.000042	72.70	0.000045	0.000048	69.17
0.0007		0.000056	0.000053	73.90	0.000074	0.000074	72.23	0.000069	0.000074	65.02
0.0003		0.000099	0.000103	68.84	0.000080	0.000081	72.00	0.000068	0.000073	72.50
0.001		0.000092	0.000098	61.98	0.000182	0.000195	59.84	0.000060	0.000060	70.23
0.003		**0.000015**	**0.000015**	69.61	0.000201	0.000202	67.17	0.000024	0.000024	67.88
0.005		0.000099	0.000111	69.75	0.000082	0.000084	68.80	0.000056	0.000056	70.12

$$z(x,y) = -\frac{1}{30} e^{\left(-1-6x-9x^2-9y^2\right)} +$$
$$- \left(0.6x - 27x^3 - 243y^5\right) e^{\left(-9x^2-9y^2\right)}$$
$$+ \left(0.3 - 1.8x + 2.7x^2\right) e^{\left(-1-6y-9x^2-9x^2\right)} \tag{8}$$

Results achieved for these functions with different values of parameters CT and SRR have been presented in Tables 1, 2 and 3. As can be observed in most cases achieved results are better for algorithm with proposed procedure. Best of them are bolded. Note that in the best cases $RMSE_T$ and $RMSE_V$ are even 10 times better than original ErrCor algorithm, e.g. for first data set of Peaks(A) function with $CT = 0.003$ and $SRR = 0.3$ the $RMSE_V$ reaches 0.000015 while original ErrCor only 0.000121, or for second data set of the same function with $CT = 0.005$ and $SRR = 0.7$ where $RMSE_V$ is 0.000014 and result for original algorithm is 0.000145. Also for third data set the best achieved result is over 3 times lower. In the case of the Peaks(B) function for all data sets best reached results are more then two or even three times better than results achieved for ErrCor and most of the others results also are better. The best result reached for Peaks(C) function is over eight times better for first data set and almost

Table 2. Results for Peaks(B) function

		1			2			3		
CT	SRR	$RMSE_T$	$RMSE_V$	$time_{TR}$ [s]	$RMSE_T$	$RMSE_v$	$time_{TR}$ [s]	$RMSE_T$	$RMSE_v$	$time_{TR}$ [s]
Original		0.000050	0.000053	55.72	0.000059	0.000064	56.80	0.000045	0.000045	58.18
0.00001	0.9	0.000035	0.000037	64.61	0.000049	0.000052	63.36	0.000023	0.000026	66.91
0.00005		0.000039	0.000038	57.51	0.000050	0.000053	55.22	0.000051	0.000049	59.19
0.00009		0.000038	0.000041	56.48	0.000051	0.000055	56.62	0.000134	0.000142	57.86
0.0007		**0.000021**	**0.000020**	57.43	0.000060	0.000061	57.70	0.000061	0.000065	61.80
0.0003		0.000073	0.000075	56.65	0.000038	0.000039	58.23	0.000032	0.000033	54.38
0.001		0.000102	0.000107	56.15	0.000061	0.000063	53.54	0.000041	0.000045	60.82
0.003		0.000079	0.000075	58.61	0.000030	0.000033	60.68	0.000025	0.000027	58.37
0.005		0.000074	0.000079	54.07	0.000023	0.000025	59.02	0.000061	0.000064	59.41
0.00001	0.7	0.000035	0.000037	62.76	0.000049	0.000053	60.83	0.000018	0.000019	63.40
0.00005		0.000034	0.000035	54.65	0.000101	0.000104	59.05	0.000024	0.000025	60.56
0.00009		0.000047	0.000050	59.22	0.000038	0.000040	59.06	0.000034	0.000038	60.80
0.0007		0.000042	0.000042	57.89	**0.000012**	**0.000014**	57.31	0.000104	0.000109	55.30
0.0003		0.000044	0.000044	53.77	0.000032	0.000034	59.28	0.000030	0.000034	55.89
0.001		0.000021	0.000021	53.36	0.000026	0.000027	59.93	0.000044	0.000048	57.14
0.003		0.000040	0.000041	59.38	0.000074	0.000076	59.45	**0.000016**	**0.000017**	60.57
0.005		0.000025	0.000024	63.61	0.000046	0.000043	60.41	0.000023	0.000024	63.49
0.00001	0.3	0.000073	0.000070	63.78	0.000057	0.000060	58.87	0.000023	0.000025	64.73
0.00005		0.000039	0.000039	61.48	0.000057	0.000059	64.67	0.000079	0.000084	58.09
0.00009		0.000062	0.000065	63.83	0.000043	0.000045	67.14	0.000059	0.000059	67.84
0.0007		0.000061	0.000063	63.07	0.000033	0.000035	60.68	0.000042	0.000043	68.00
0.0003		0.000049	0.000049	67.56	0.000084	0.000082	67.44	0.000036	0.000038	63.88
0.001		0.000042	0.000044	59.05	0.000045	0.000049	59.59	0.000071	0.000071	64.94
0.003		0.000037	0.000042	65.21	0.000036	0.000035	66.90	0.000049	0.005105	66.43
0.005		0.000059	0.000059	63.84	0.000062	0.000068	61.71	0.000032	0.000030	70.23

four and two times better for second and third data set, respectively. What is important these results has been achieved in the similar training time as original ErrCor.

3.2 Schwefel Function

Second experiment has been prepared for Schwefel function given by Eq. 9 shown in Fig. 3d

$$z\left(x,y\right) = 2 * 418.9829 - xsin\left(\sqrt{|x|}\right) - ysin\left(\sqrt{|y|}\right) \tag{9}$$

Results reached for these function with different values of parameters CT and SRR have been presented in Table 4. As for Peaks functions in this case obtained results are better then these achieved with original ErrCor in most cases. The best results are two times better, eg. for first data set with $CT = 0.1$ and $SRR = 0.3$ where $RMSE_V$ reached 0.00741 while value for original algorithm is 0.01843, and 0.00606 versus 0.01007 for $CT = 0.07$ and $SRR = 0.7$ with third data set. Results for second data set are not so good like in other experiments but also in this case some results are better.

Table 3. Results for Peaks(C) function

		1			2			3		
CT	SRR	$RMSE_T$	$RMSE_V$	$time_{TR}$ [s]	$RMSE_T$	$RMSE_V$	$time_{TR}$ [s]	$RMSE_T$	$RMSE_V$	$time_{TR}$ [s]
Original		0.000239	0.000245	129.22	0.000085	0.000092	137.98	0.000105	0.000105	130.46
0.000001	0.9	0.000137	0.000141	140.88	0.000092	0.000099	148.31	0.000080	0.000089	156.61
0.00005		0.000097	0.000099	165.05	0.000041	0.000044	140.97	0.000091	0.000092	148.28
0.00009		0.000091	0.000098	130.90	0.000032	0.000036	144.95	0.000065	0.000070	135.81
0.0007		0.000064	0.000075	131.13	0.000048	0.000056	140.42	**0.000056**	**0.000057**	139.40
0.0003		0.000061	0.000067	142.81	0.000030	0.000032	141.58	0.000067	0.000074	140.27
0.003		0.000068	0.007522	145.79	0.000038	0.000040	158.30	0.000081	0.000087	137.24
0.005		0.000053	0.000058	152.10	0.000044	0.000048	156.82	0.000078	0.000089	155.20
0.001		0.000090	0.000097	126.29	0.000028	0.000029	143.77	0.000127	0.000143	131.91
0.000001	0.7	0.000074	0.000086	134.79	0.000076	0.000081	151.03	0.000080	0.000089	142.80
0.00009		0.000038	0.000040	138.19	0.000034	0.000039	135.76	0.000061	0.000070	148.53
0.00005		**0.000029**	**0.000029**	161.77	0.000025	0.000028	156.47	0.000110	0.000114	157.20
0.0007		0.000090	0.000104	137.78	0.000058	0.000059	149.11	0.000052	0.000065	120.35
0.0003		0.000136	0.000160	133.93	0.000051	0.000057	149.62	0.000062	0.000065	131.09
0.005		0.000110	0.000125	150.79	**0.000023**	**0.000025**	159.28	0.000100	0.000115	137.49
0.003		0.000122	0.000126	158.18	0.000091	0.000105	148.57	0.000074	0.000083	152.11
0.001		0.000072	0.000079	125.58	0.000041	0.000042	134.82	0.000112	0.000120	138.75
0.000001	0.3	0.000104	0.000112	156.45	0.000096	0.000105	150.39	0.000080	0.000089	164.03
0.00009		0.000034	0.000037	130.54	-	-	139.26	0.000100	0.000120	136.26
0.00005		0.000112	0.000125	151.21	0.000061	0.000068	142.61	0.000061	0.000065	144.70
0.0007		0.000046	0.000051	143.18	0.000060	0.000070	147.83	0.000096	0.000094	124.17
0.0003		0.000064	0.000064	134.36	0.000041	0.000045	148.15	0.000060	0.000064	137.99
0.005		0.000096	0.000109	157.51	0.000064	0.000068	158.31	0.000180	0.000191	155.02
0.003		0.000107	0.000115	118.52	0.000059	0.000061	157.26	0.000080	0.000091	124.32
0.001		0.000047	0.000052	114.32	0.000047	0.000052	149.09	0.000104	0.000116	126.67

3.3 Shaffer Function

Similar experiment has been prepared for Shaffer function given by Eq. 10 shown in Fig. 3e

$$z(x, y) = 0.5 + \frac{sin^2\left(x^2 - y^2\right) - 0.5}{\left[1 + 0.001\left(x^2 - y^2\right)\right]^2} \tag{10}$$

Results achieved for these function with different values of parameters CT and SRR have been presented in Table 5 and again, some of them are a bit better than original ErrCor. Note, that this function is relatively most simple and original version of ErrCor was able to reach very low $RMSE_T$ and $RMSE_V$ results so proposed method does not contribute much due to for most patterns algorithm was able to train well. The best result reached for $CT = 0.00007$ and $SRR = 0.3$ with second data set is two times better.

Table 4. Results for Schwefel function

CT	SRR	$RMSE_T$	$RMSE_V$	$time_{TR}$ [s]	$RMSE_T$	$RMSEv$	$time_{TR}$ [s]	$RMSE_T$	$RMSEv$	$time_{TR}$ [s]
		1			2			3		
Original		0.01588	0.01843	49.18	0.00603	0.00617	55.01	0.01026	0.01007	59.60
0.0003	0.9	0.00704	0.00810	49.41	0.00531	0.00576	49.46	0.01115	0.01197	49.34
0.0007		0.00751	0.00892	47.88	**0.00531**	**0.00574**	53.87	0.01123	0.01130	50.54
0.005		0.00912	0.01029	57.07	0.00796	0.00819	52.21	0.01017	0.00995	64.25
0.007		0.00713	0.00849	43.33	0.00873	0.00935	58.64	0.00988	0.01066	64.12
0.009		0.00718	0.00908	48.29	0.01029	0.01182	41.32	0.00958	0.00965	51.10
0.03		0.00834	0.00985	41.42	0.01316	0.01417	52.38	0.00818	0.00839	49.66
0.07		0.00812	0.00934	49.87	0.00647	0.00706	47.48	0.01159	0.01256	52.50
0.1		0.00855	0.00931	50.24	0.00795	0.00872	51.66	0.00906	0.00917	46.16
0.0003	0.7	0.00912	0.01031	53.72	0.00663	0.00672	43.96	0.01053	0.01136	51.63
0.0007		0.01041	0.01200	49.35	0.00752	0.00821	53.69	0.01068	0.01045	53.92
0.005		0.00678	0.00782	50.65	0.00837	0.00842	54.02	0.00659	0.00707	61.75
0.007		0.00926	0.01006	48.47	0.00834	0.00922	53.68	**0.00574**	**0.00606**	51.29
0.009		0.00964	0.01062	46.83	0.01285	0.01407	45.57	0.00834	0.00843	43.78
0.03		0.00720	0.00897	47.64	0.00816	0.00888	41.20	0.00944	0.00989	59.97
0.07		0.00899	0.01049	54.72	0.00770	0.00778	63.96	0.00995	0.01029	36,45
0.1		0.01022	0.01171	49.09	0.01060	0.01068	52.90	0.01252	0.01292	36.58
0.0003	0.3	**0.00660**	**0.00741**	47.47	0.00797	0.00838	43.46	0.00933	0.00847	50.69
0.0007		0.00838	0.00976	50.65	0.00761	0.00795	51.74	0.01017	0.00994	56.89
0.005		0.00895	0.00982	58.26	0.01086	0.01109	35.39	0.01528	0.01489	39.20
0.007		0.00687	0.00728	42.44	0.01540	0.01566	36.98	0.00973	0.01023	44.08
0.009		0.01035	0.01064	42.66	0.01098	0.01104	40.99	0.01962	0.02086	42.82
0.07		-	-	57.45	0.00681	0.00742	37.39	0.00621	0.00681	48.37
0.03		0.01082	0.01144	21.41	0.00591	0.00603	40.39	-	-	66.49
0.1		0.00692	0.00778	36.80	0.00967	0.01014	49.27	-	-	61.28

Table 5. Results for Schaffer function

CT	SRR	$RMSE_T$	$RMSE_V$	$time_{TR}$ [s]	$RMSE_T$	$RMSEv$	$time_{TR}$ [s]	$RMSE_T$	$RMSEv$	$time_{TR}$ [s]
		1			2			3		
Original		0.0000308	0.0000320	112.61	0.0000432	0.0000510	123.95	0.0000388	0.0000409	117.91
0.00005		0.0000512	0.0000564	130.05	0.0000474	0.0000528	129.39	**0.0000356**	**0.0000391**	119.01
0.00007		0.0000512	0.0000564	131.34	0.0000455	0.0000527	118.79	0.0000361	0.0000391	111.420
0.00009		**0.0000308**	**0.0000319**	119.49	0.0000455	0.0000528	118.39	0.0000356	0.0000391	116.29
0.0001	0.9	0.0000477	0.0000534	130.29	0.0000617	0.0000659	118.06	0.0000389	0.0000411	127.66
0.0005		0.0000368	0.0000392	122.06	0.0000339	0.0000396	126.71	0.0000390	0.0000411	132.53
0.0007		0.0000487	0.0000546	128.25	0.0000425	0.0000507	125.75	0.0000389	0.0000410	128.08
0.00005		0.0000516	0.0000577	119.05	0.0000424	0.0000509	130.74	**0.0000356**	**0.0000391**	117.07
0.00007		0.0000509	0.0000561	134.16	0.0000336	0.0000391	133.28	0.0000388	0.0000408	127.33
0.00009		0.0000402	0.0000419	158.41	0.0000337	0.0000391	159.66	0.0000389	0.0000410	120.57
0.0003	0.7	0.0000468	0.0000524	150.19	0.0000636	0.0000669	117.98	0.0000391	0.0000412	125.33
0.0007		0.0000447	0.0000472	143.33	0.0000567	0.0000608	132.07	0.0000391	0.0000415	155.48
0.001		0.0000378	0.0000410	135.01	0.0000797	0.0000855	123.67	0.0000519	0.0000561	118.12
0.003		0.0000607	0.0000672	121.16	0.0000617	0.0000592	113.44	0.0000398	0.0000435	127.85
0.005		0.0000291	0.0000323	130.31	0.0000395	0.0000423	131.29	0.0000397	0.0000435	119.49
0.00005		0.0000436	0.0000486	133.99	0.0000453	0.0000517	130.39	0.0000361	0.000039	115.06
0.00007		0.0000434	0.0000482	118.88	**0.0000254**	**0.0000251**	110.48	0.0000682	0.0000726	100.62
0.00009		0.0000369	0.0000408	129.41	0.0000337	0.0000391	135.3	0.0000361	0.0000391	107.86
0.0003	0.3	0.0000381	0.0000415	138.83	0.0000433	0.0000434	126.15	0.0000401	0.0000425	133.23
0.0007		0.0000543	0.0000561	141.78	0.0000406	0.0000385	109.94	0.0000595	0.0000629	112.24
0.001		0.0000413	0.0000459	121.65	0.0000586	0.000059	112.43	0.0000496	0.0000487	113.9
0.003		0.0000655	0.0000696	126.14	0.0000538	0.0000557	105.71	0.0000676	0.0000681	99.04
0.005		0.0000639	0.0000674	104.94	0.0000387	0.0000378	122.38	0.0000384	0.0000386	129.86

4   Conclusions

The proposed training method has been verified with several approximation problems. This confirm that reduction of impact of selected, in this case trained, patterns on training process, can be justified or even desirable. Presented approach has been successfully implemented in ErrCor algorithm improving its effectiveness. However, some issues are still open. One of the most important is what parameters should be used for training to achieve best results. All the more that no clear relationship was observed between main parameters CT, SRR and reached result. Further work in these area will be also focused on applying presented approach in different training algorithms.

References

1. Wilamowski, B.M., Yu, H.: Neural network learning without backpropagation. IEEE Trans. Neural Netw. **21**(11), 1793–1803 (2010)
2. Wilamowski, B.M.: Neural network architectures and learning algorithms, how not to be frustrated with neural networks. IEEE Ind. Electron. Mag. **3**(4), 56–63 (2009)
3. Hunter, D., Yu, H., Pukish, M.S., Kolbusz, J., Wilamowski, B.M.: Selection of proper neural network sizes and architectures-a comparative study. IEEE Trans. Ind. Inf. **8**, 228–240 (2012)
4. Hohil, M.E., Liu, D.: Solving the N-bit parity problem using neural networks. Neural Netw. **12**, 1321–1323 (1999)
5. Wilamowski, B.M.: Challenges in applications of computational intelligence in industrial electronics. In: ISIE 2010, pp. 15–22, 4–7 July 2010
6. Fahlman, S.E., Lebiere, C.: The cascade-correlation learning architecture. In: Advances in Neural Information Processing Systems 2, pp. 524–532. Morgan Kaufmann, San Mateo (1990)
7. Lang, K.L., Witbrock, M.J.: Learning to tell two spirals apart. In: Proceedings of the 1988 Connectionists Models Summer School. Morgan Kaufman (1998)
8. Wilamowski, B.M., Korniak, J.: Learning architectures with enhanced capabilities and easier training. In: 19th IEEE International Conference on Intelligent Engineering Systems (INES 2015), 03–05 September, pp. 21–29 (2015)
9. Nguyen, G.H., Bouzerdoum, A., Phung, S.L.: Efficient supervised learning with reduced training exemplars. In: 2008 IEEE International Joint Conference on Neural Networks (IEEE World Congress on Computational Intelligence), Hong Kong, pp. 2981–2987 (2008)
10. Lozano, M.T.: Data reduction techniques in classification processes. Ph.D. Dissertation, Universitat Jaume I, Spain (2007)
11. Chouvatut, V., Jindaluang, W., Boonchieng, E.: Training set size reduction in large dataset problems. In: 2015 International Computer Science and Engineering Conference (ICSEC), Chiang Mai, pp. 1–5 (2015)
12. Kolbusz, J., Rozycki, P.: Outliers elimination for error correction algorithm improvement. In: CS&P Proceedings 24th International Workshop Concurrency, Specification & Programming, (CS&P 2015), vol. 2, pp. 120–129 (2015)
13. Rozycki, P., Kolbusz, J., Lysenko, O., Wilamowski, B.M.: Neural network training improvement by patterns removing. Artif. Intell. Soft Comput. ICAISC **2017**, 154–164 (2017)
14. Yu, H., Reiner, P., Xie, T., Bartczak, T., Wilamowski, B.M.: An incremental design of radial basis function networks. IEEE Trans. Neural Netw. Learn. Syst. **25**(10), 1793–1803 (2014)
15. Xie, T.: Growing and learning algorithms of radial basis function networks. Ph.D. Dissertation, Auburn University, USA (2013)

An Embedded Classifier for Mobile Robot Localization Using Support Vector Machines and Gray-Level Co-occurrence Matrix

Fausto Sampaio, Elias T. Silva Jr$^{(\boxtimes)}$, Lucas C. da Silva,
and Pedro P. Rebouças Filho

Computer Science Department, Federal Institute of Education,
Science and Technology of Ceará, Fortaleza, CE, Brazil
faustos@ppgcc.ifce.edu.br, lucas.costa@lit.ifce.edu.br,
{elias,pedrosarf}@ifce.edu.br

Abstract. Computer vision applications have been largely incorporated into robotics and industrial automation, improving quality and safety of processes. Such systems involve pattern classifiers for specific functions that, many times, demand high processing time and large data memory. Robotics applications usually deal with restricted resources platforms, in order to preserve battery time and to reduce weight and costs. To assist those applications, this paper presents an investigation on GLCM (Gray Level Co-occurrence Matrix) features and image size for an SVM (Support Vector Machines) classifier that can reduce computer resources utilization while preserving high classifier accuracy. Experimental results show a computing time on the embedded platform of 80.5 ms, with an accuracy above to 99%, to classify images of 80×60 pixels.

Keywords: Computer vision · Neural network applications
Robotic applications · GLCM · SVM

1 Introduction

Current studies of computer vision for mobile robots navigation are growing more and more [15]. This is due to the fact that human perception is primarily visual. Therefore, understand and use this type of information is essential to allow human-robot interaction. The visual properties of a place can be used to determine its functional category (e.g., kitchen, office). In addition, visual reference points are commonly used by human beings to plan paths and describe places. These characteristics have motivated researchers to employ visual information in the task of mobile robots localization.

Some computer vision systems have been developed for mobile robot localization, and some combinations of classifiers and feature extractors are employed with very good accuracy results. However, those pattern classifiers, many times, demand high processing time and large data memory.

© Springer International Publishing AG, part of Springer Nature 2018
L. Rutkowski et al. (Eds.): ICAISC 2018, LNAI 10841, pp. 201–213, 2018.
https://doi.org/10.1007/978-3-319-91253-0_20

Robotics applications usually deal with restricted resources platforms, in order to preserve battery life and to reduce weight and costs. Therefore, good results in accuracy must be followed by a low use of resources, when the classifier is deployed in an embedded computer. The design of this type of system is complex since it involves usually neglected concepts for general purpose computing.

To collaborate with those applications, this paper presents an investigation on a patterns classifiers made of GLCM (Gray Level Co-occurrence Matrix) combined with SVM (Support Vector Machines) that can reduce computer resources utilization while preserving the high accuracy of the classifier.

This work evaluates the restrictions to embed a computer vision system for image classification. The proposed approach in this work is based on SVM, as a machine learning classifier, operating together with GLCM, as a feature extractor from digital images. The entire process was tested on a workstation and on an embedded platform, where a final implementation was evaluated.

The aim is to offer empirical results and a reflection on the impact on the classifier accuracy of (1) the features extracted from the image, (2) the image size, and (3) the number of its gray levels. Behind this study is an investigation of the relationships between the characteristics of the image classification process and the consumption of resources in an embedded computer.

This paper is organized as follow: Sect. 2 describe the state of the art, presenting some related works as well as some attempts to embed SVM and GLCM. Section 3 describes the proposed approach and results obtained in a Matlab workstation. Section 4 shows the results of the classifier in an embedded platform. Section 5 concludes the paper discussing the obtained results.

2   Related Works

Mobile Robots Localization and Navigation. An approach for navigation and localization of mobile robots on topological maps was proposed in [12]. The combination of SVM and GLCM obtained an accuracy above 98%. The case study was a robot equipped with one camera that communicated with a microcomputer by radio.

The work of [1] presents a comparison of mobile robots localization methods using Artificial Neural Networks based on images. The robot was equipped with a camera connected to a laptop to execute the algorithms.

Both publications left as future work: (1) to evaluate the complexity of the algorithms, (2) to apply feature selection techniques as well as (3) to investigate the influence of the resolution of the images. All these questions seek to reduce computational resources to enable a computer vision application to be deployed in a mobile and autonomous robot. Therefore, this work aims to fill these gaps, investigating optimizations on GLCM combined with the SVM.

Embedded SVM Classifiers. Many researchers have used SVM in several applications, but few have attempted to embed this classifier.

In [9] a method was proposed to recognize people in aerial images from a Small UAV using Pattern Recognition Systems. SVM was tested combined with the Histogram of Oriented Gradients, among others. The authors pointed out that the implementation was not optimized for the hardware capabilities of the target platform.

The work of [10] presented a study to classify the diseases of mango leaves for agriculture. The SVM classifier was implemented with the OpenCV library, but the use of a hardware to embed the entire system was left as future work.

Embedded GLCM Feature Extractor. In [11], the authors state that the main disadvantage of using GLCM is that it requires a long computing time, especially for very large images. Thus, they proposed an FPGA-based architecture to calculate the CM and extract 6 features. With the same objective, many other works have tried FPGA implementations. It is noted that these solutions converge to suppress some of the GLCM's features, but they do not justify their choices. Most of them do not associate GLCM with a classifier, so they do not evaluate classifier hit rate.

Working in another approach, this research proposes to investigate methods to choose the best set of features while preserving the correctness (accuracy) of the classifier.

3   Proposed Approach

The chosen computer vision application uses GLCM as a feature extractor and SVM (RBF kernel) as the classifier. In order to optimize computer resources usage, three questions guided this research:

- What is the impact of each feature calculation on computing time performance? A process of feature selection and reduction will be conducted.
- Does the image size have an important impact on the classifier accuracy? Aiming an input image size selected according to the accuracy.
- How does the size of Co-Occurrence Matrix affect the classifier accuracy? This is a study on memory consumption, since CM can be really large.

3.1   Data Base

The base consists of a bank of 600 images obtained in the real environment with a conventional camera [12]. A topological map was elaborated to aid in the recognition and navigation of the robot. 6 nodes were defined to represent the localizations in the apartment. From these nodes, 15 classes were named to represent strategic points and indicate the orientation within a given localization. For each class, 40 images were obtained with dimensions of 4000×3000 pixels.

3.2 Feature Extractor and the Pattern Recognition

Extracting features is an essential task of the classification process. The images are converted into feature vectors to be used in the training of some pattern classifier.

The method of feature extraction used in this work consists in representing textures with second order statistics using a Gray Level Co-occurrence Matrix (GLCM) [6]. A Co-occurrence Matrix describes the amount of combinations of gray levels in an image with a certain direction and distance between the neighboring pixels. The size of this matrix (N_g) is equivalent of the amount of gray levels considered in the image. In this work, $N_g = 256$ was initially adopted.

After the feature extraction, the classification process is followed; SVM (Support Vector Machine), in this work. SVM are classifiers that are based on statistical learning theory, which takes into account the structural error minimization, calculated for the training vectors [3], and not only the minimization of the mean squared error (MSE).

In order to train the SVM classifier, the images set of each class (15) was randomly divided into the following proportions: 30 images (or 75%) for the training set and 10 for the test set. It was performed 50 independent training and test runs, using RBF (Radial Basis Function) *kernel*, with $C = 8$ and $\sigma = 2$.

A workstation with Matlab R2015a software (hereafter called WS1) was used to perform the initial evaluations of the classifier, before the entire process is deployed on an embedded platform. WS1 is an Intel Core i5 (5th Gen), 2.2 GHz, 8 GB of RAM. The remainder of the process, which consists of classifying on the embedded system, will be explained in Sect. 4.

3.3 Feature Selection and Reduction

The GLCM [7,13,14] provides a set of 24 statistical measures, which can be extracted from a single image. Table 1 lists all of these measures, which have been adopted as features in this work.

Initially, all 24 features were extracted from all 600 images of indoor environments. Column C1 on Table 2 presents the accuracy of the SVM classifier for each class using all 24 features extracted from the images in their original size (4000 × 3000). The accuracy and standard deviation values come from 50 runs. All experiments use RBF *kernel*, with $C = 8$ and $\sigma = 2$.

Even with a good average global accuracy, around 99%, the use of a vector with 24 features does not seem to be a good solution considering the possible restrictions of computational resources in embedded platforms. Thus, two approaches were applied in reducing the number of features: (1) Features selection based on correlation (CFS). (2) Study on the algorithm complexity of features computation.

Feature Selection Based on Correlation (CFS). CFS [5] is an algorithm that evaluates the performance of a subset of features by considering the individual predictive capacity of each feature along with the degree of redundancy

Table 1. GLCM's features.

Reference	#	Features
Haralick et al. [7]	14	ASM, Contrast, Correlation, Sum of Squares, IDM, Sum Average, IMC I, IMC II, Sum Variance, Sum Entropy, Entropy, Diff. Variance, Diff. Entropy, MCC
Soh and Tsatsoulis [13]	6	Homogeneity, Autocorrelation, Dissimilarity, Cluster Shade, Cluster Prominence, Maximum Probability
Wang et al. [14]	4	Sum Mean, Cluster Tendency, Difference Mean, Inertia
-	**24**	-

Table 2. Classifiers accuracy for each feature set using original image size.

Class	Feature set			
	C1 (24 features)	C2 (17 features)	C3 (10 features)	C4 (14 features)
1	100.00 ± 00.00	100.00 ± 00.00	100.00 ± 00.00	99.92 ± 00.21
2	99.81 ± 00.30	99.74 ± 00.32	99.72 ± 00.32	99.58 ± 00.51
3	100.00 ± 00.00	100.00 ± 00.00	100.00 ± 00.00	99.64 ± 00.45
4	100.00 ± 00.00	100.00 ± 02.75	100.00 ± 00.00	100.00 ± 00.00
5	99.81 ± 00.30	99.90 ± 00.24	99.83 ± 00.28	100.00 ± 00.00
6	99.72 ± 00.41	99.88 ± 00.25	99.38 ± 00.53	99.64 ± 00.41
7	99.95 ± 00.18	100.00 ± 00.00	100.00 ± 00.00	99.60 ± 00.39
8	99.85 ± 00.28	99.86 ± 00.27	99.82 ± 00.29	99.41 ± 00.55
9	100.00 ± 00.00	99.95 ± 00.22	99.85 ± 00.40	99.64 ± 00.49
10	99.69 ± 00.41	99.55 ± 00.57	99.65 ± 00.57	99.47 ± 00.68
11	99.83 ± 00.31	99.78 ± 00.31	99.40 ± 00.60	99.83 ± 00.28
12	99.45 ± 00.45	99.37 ± 00.44	99.36 ± 00.52	97.69 ± 01.30
13	100.00 ± 00.00	100.00 ± 00.00	100.00 ± 00.00	100.00 ± 00.00
14	99.45 ± 00.43	99.29 ± 00.54	99.28 ± 00.51	97.60 ± 01.31
15	98.88 ± 00.91	98.79 ± 00.87	98.83 ± 00.88	99.19 ± 00.92
Average	$\mathbf{99.76 \pm 00.14}$	$\mathbf{99.74 \pm 00.14}$	$\mathbf{99.68 \pm 00.13}$	$\mathbf{99.42 \pm 00.25}$

between them. In this work the CFS was used to select which features contribute the most to making a correct prediction. The expectation is to reduce the number of features used by the SVM classifier, also reducing the computational cost for both: features extraction and classification. Among the 24, CFS selected 17 features: ASM, Contrast, Correlation, Variance, Sum Average, Sum Entropy, Entropy, Difference Variance, Difference Entropy, IMC I, IMC II, MCC, Maximum Probability, Cluster Shade, Cluster Prominence, Dissimilarity and

Table 3. List of features grouped by time complexity.

Running time	Features
$\Theta(n^2)$	ASM, Contrast, IDM, Entropy, Homogeneity, Sum Mean, Maximum Probability, Dissimilarity, Difference Mean, Autocorrelation
$\Theta(n^2 + n)$	Correlation, Inertia
$\Theta(n^2 + 2n)$	Sum Average, Sum Entropy, Difference Variance, Difference Entropy
$\Theta(n^2 + 4n - 2)$	Sum Variance
$\Theta(2n^2)$	Sum of Squares, IMC I, IMC II, MCC, Cluster Tendency, Cluster Shade, Cluster Prominence

Difference Mean. The accuracy of the classifier using the set of features obtained by the CFS is in Table 2, column C2.

Analyzing the results, the mean value of the classifier's accuracy was the same 99% (C1). By checking class by class, one can see very close results obtained using all 24 features. This demonstrates the efficiency of CFS to choosing features for this pattern recognition application. However, considering the goal to minimize the costs of the target embedded system, a complexity study for all 24 features calculations was performed.

Study on Features Complexity. The order of growth of the running time of an algorithm allows to compare the relative performance of alternative algorithms [2]. The complexity study is concerned with how the running time of an algorithm increases with the size of the input n. The notations that are usually used to describe the running time of an algorithm are called asymptotic notation. Usually, an algorithm that is asymptotically more efficient will be the best choice for all but very small inputs.

In this work, the asymptotic notation Θ was used to describe the running time of the algorithms of all 24 features, considering the worst case performances. The value of the input n is the size of the CM, being necessary to go through the entire matrix to extract a feature. In the Table 3 all 24 features are presented according to its time complexity, based on the equations from [4]. Analyzing all values for the feature calculations, the lowest time complexity found was $\Theta(n^2)$.

Note that some features selected by the CFS, like *Cluster Prominence*, have a computation time of $\Theta(2n^2)$, greater than $\Theta(n^2)$, which can be very time-consuming in a limited resources platform.

It should also be noted that, although there are 10 features in the $\Theta(n^2)$ group, they are all calculated with only two nested loops. That is, traversing the CM only once. Therefore, these features were used for a new classifier accuracy analysis. The results are in column C3 on Table 2. Column C4 shows the results of accuracy using the 14 Haralick's Texture Descriptor (Table 1), which were the same as those used in [12].

It is noted by the results on Table 2 that the classifier accuracy for the 4 feature sets is very similar. However, the C3 stands out due to the lower consumption of computational resources.

3.4 Selecting the Input Image Size

Does the image size have an important impact on the classifier accuracy? To answer this question, some tests were carried out with different resizing of the image samples. This study will also help to choose the ideal image size to be used on the embedded platform.

The image resizing was done using a process of reducing the sampling rate of a signal, known as decimation. It produces an approximation of the sequence that would have been obtained by sampling the signal at a lower density, keeping the first sample in each sequence of M samples [8]. In this work, a signal is equivalent to an image and the samples correspond to the image pixels.

To illustrate that, an input image I (6×8), after a decimation of $M = 3$, will result in an image J (2×3). The higher the value of M, the smaller the new image. In this example, only lines 1 and 4 of the input appear in the output image, as well as columns 1, 4, and 7.

An experiment was carried out to evaluate the impact of image size on the classifier accuracy. Image scaling was used with $1 \leq M \leq 1000$. Considering only the M values that produce variations in image size, 163 different values were used for M in total. For each M values, 50 realizations (training and test) of SVM classifier was performed, resulting in the average accuracy of that resolution.

The Fig. 1 shows the average accuracy for different value of M, using the C3 set of features. As the image size decreases to $M = 100$, the accuracy decreases smoothly, without very abrupt variations. The use of smaller size images leads the classifier to make many mistakes, indicating that resizing is compromising recognition information.

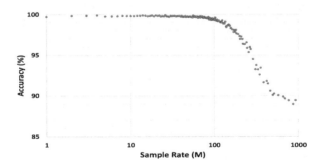

Fig. 1. SVM classifier accuracies for resized images using $1 \leq M \leq 1000$.

The Table 4 shows the accuracy for some selected resized images, using the C3 features. Note that the classifier accuracy remains above 99% for $M \leq 50$, and a small penalty occurs in accuracy (2%) for images as little as 40×30.

Table 4. Classifier accuracies for some resized images.

Class	M = 1(4000 × 3000)	M = 10(400 × 300)	M = 50(80 × 60)	M = 100(40 × 30)
1	99.92 ± 00.21	100.00 ± 00.00	99.37 ± 00.56	99.65 ± 00.47
2	99.58 ± 00.51	100.00 ± 00.00	99.76 ± 00.36	98.26 ± 00.99
3	99.64 ± 00.45	100.00 ± 00.00	99.86 ± 00.35	95.63 ± 01.26
4	100.00 ± 00.00	100.00 ± 00.00	99.87 ± 00.26	99.37 ± 00.64
5	100.00 ± 00.00	99.94 ± 00.23	98.86 ± 00.78	97.29 ± 01.12
6	99.64 ± 00.41	99.96 ± 00.20	99.36 ± 00.53	99.46 ± 00.47
7	99.60 ± 00.39	99.68 ± 00.41	98.85 ± 00.62	97.49 ± 00.95
8	99.41 ± 00.55	100.00 ± 00.00	99.71 ± 00.41	97.68 ± 00.90
9	99.64 ± 00.49	99.64 ± 00.49	99.12 ± 00.70	95.17 ± 01.30
10	99.47 ± 00.68	99.86 ± 00.30	99.23 ± 00.55	99.33 ± 00.75
11	99.83 ± 00.28	99.50 ± 00.44	98.69 ± 00.87	97.90 ± 01.23
12	97.69 ± 01.30	99.85 ± 00.33	99.49 ± 00.52	94.28 ± 01.59
13	100.00 ± 00.00	100.00 ± 00.00	99.99 ± 00.09	97.26 ± 01.20
14	97.60 ± 01.31	99.50 ± 00.62	98.29 ± 00.89	96.86 ± 00.95
15	99.19 ± 00.92	99.64 ± 00.55	97.69 ± 01.04	96.19 ± 01.59
Average	**99.42 ± 00.25**	**99.84 ± 00.11**	**99.21 ± 00.25**	**97.45 ± 00.35**

3.5 Memory Consumption

Aiming to anticipate the memory required in the embedded platform, an analytical study was developed to estimate the memory consumption for GLCM and SVM data structures.

Analytical Study. The study was applied in order to obtain an Estimated Value of Memory Consumption (EVMC) of the GLCM and SVM. The EVMC measurements in *bits* are defined according to the equations below.

Notation:

- N_g: Number of CM gray levels;
- b: Number of *bits* referring to the type of data used;
- w: Number of features;
- n: Number of classes in the problem;
- V: $\sum_{i=1}^{n} v_i$, sum of the number of support vectors v_i of n classes;

Equations:

$$EVMC_{(GLCM)} = (N_g^2 + w).b \tag{1}$$

$$EVMC_{(SVM)} = (V.w + V + n).b \tag{2}$$

Based on the above equations, Table 5 shows the estimated memory consumption in kilobytes (kB) for the GLCM and SVM using resized images. For

Table 5. Memory estimation of GLCM and SVM for different image sizes.

Technique	$M = 1(4000 \times 3000)$	$M = 10(400 \times 300)$	$M = 50(80 \times 60)$	$M = 100(40 \times 30)$
GLCM	256.039 kB			
SVM	12.734 kB	11.445 kB	13.281 kB	18.672 kB
(C3:GLCM)	($V = 326$)	($V = 293$)	($V = 340$)	($V = 478$)
Total (kB)	**268.773**	**267.484**	**269.320**	**274.711**

the GLCM extractor the 10 features of C3 were used ($N_g = 256$). V Represents the Total Support Vectors.

Memory consumption using GLCM, regardless of image size, is directly related to the size of the CM and the number of features selected. The size of CM refers to the amount of gray levels considered in the images. For a CM considering all 256 levels of gray and using the 10 features of C3, a consumption of approximately 256.039 kB was obtained. This takes into account that the data type of all elements of the normalized CM and all features are floating point ($b = 32$ *bits*). The CM elements are 32 bits because they are already normalized.

Regarding the memory consumption of the SVM classifier (after its training process with the GLCM extractor), the main consumption factor is the amount of support vectors V, necessary for pattern recognition. Table 5 shows that for reduced images, the amount of memory used by the SVM classifier increases. This can be justified by the loss of information when resizing images. Consequently, the model will need a greater representation of support vectors in an attempt to separate the patterns.

It is worth mentioning that the memory consumption values of the SVM classifier were extracted from models generated in WS1 with the Matlab tool. It was chosen the models that obtained the best accuracy in the 50 rounds for different sizes of images. The method covered in the multiclass classification was the One-Against-All and all elements of the support vectors are floating point ($b = 32$ *bits*).

The results imply that CM's size (or dimensions) has an important impact on memory consumption for the GLCM technique. So, a new question arises: How does CM's size affect the classifier accuracy? Would it be possible to reduce the CM's impact on the memory usage without sacrificing SVM performance? The next section will dive into this subject.

Co-occurrence Matrix (CM) Reduction

All the experiments performed so far have used a CM matrix that considers 256 levels of gray (N_g), which are found in 8-bit resolution images. However, knowing that the amount of gray levels corresponds to the size of this matrix, and therefore its size, a new experiment was proposed to compare results for different values of N_g. In this study, the CM matrix was reduced by an N factor, which represents the value of the grayscale fraction of the image. For example, for $N = 1$ and $N = 2$, the size of the CM matrix, as well as the amount of gray levels considered in the image, will result, respectively, in 256×256 and 128×128.

Another classifier accuracy evaluation was performed, at this time changing the gray levels. The graph on Fig. 2(a) presents the classifier hit rates for the following configuration: resized images with $1 \leq M \leq 100$, C3 feature set, and CM reduction by $N = \{1, 2, 4, 8, 16\}$. For each combination of N and M values, a hit rate was measured, resulting in 100 values for each of the 5 dispersion groups. Again, 50 realizations (training and test) for each combination.

The results on Fig. 2(a) indicates that moderated reduction in the CM doesn't cause significant penalty in the classifier accuracy. Consequently, this aspect creates an opportunity for a strong reduction in memory footprint. In this application, it is possible to use CM with dimensions of 32×32 ($N_g = 32$), for example, maintaining accuracy above 99%, and reducing memory consumption to only 16.644 kB (considering GLCM and SVM memory usage).

4   Classifier Evaluation on an Embedded Platform

The target embedded platform uses OpenCV. So, another workstation was set up with the OpenCV 2.4.13 library (WS2), working as an intermediate validation for tests performed on the WS1. RBF (Radial Basis Function) kernel was used with $C = 8$ and $\sigma = 2$. The SVM classifier was trained and validated on the WS2 for 50 rounds, presenting results identical to those of WS1.

Embedded platform runs the following steps: (1) CM creation, (2) feature extraction (C3 set), and (3) SVM classification. Raspberry Pi 3 Model B was chosen due to its suitability for a robotic system. The application was implemented using the OpenCV library with support for Python programming language. The test images were placed on non-volatile micro SD memory card. The support vectors, which represent the model defined in the classifier learning process (training), were deployed in the embedded environment with OpenCV's own resources.

In order to compare the SVM classifier results on a workstation with the results on the embedded platform, the following experiment was performed. All 50 test data sets, including the model output labels and support vectors used in WS2, were stored in files and later loaded and run on the embedded platform. For each image processed by the embedded platform, it was checked whether the output corresponded to the prediction made in WS2. In this experiment, it was found that the results of the embedded platform were equal to the WS2 station.

To measure the execution time on the embedded platform, an experiment was performed using a set of resized images from class 1 (stored on the memory card). The classification time doesn't change as a function of sample class, as well as the feature extraction time. Figure 2(b) displays the total computation times obtained on the embedded platform for all resized images. In total, 100 measurements were performed on each image size to determine the computation time on the embedded platform. The time measurements were obtained using the *time.clock*() available in the Python language. Note that those times can be improved using compiled languages. The interpreted version of Python was used just to show the relationship among accuracy, image size and computation time.

The graph on Fig. 2(b) shows a strong reduction in computing time, even with small reductions in image size. This indicates that the time can be greatly

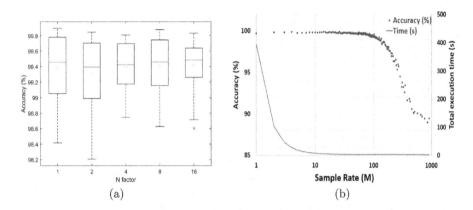

(a) (b)

Fig. 2. (a) Classifier accuracies for resized Co-occurrence Matrix. (b) Computing time for resized images on embedded platform ($N_g = 256$).

Table 6. Performance of the embedded SVM classifier ($M = 50$ and $N_g = 32$).

Class	Metrics	
	Ac(%)	$F1$-Score
1	100.00 ± 00.00	100.00 ± 00.00
2	99.69 ± 00.35	97.65 ± 02.86
3	98.20 ± 01.04	87.60 ± 01.29
4	98.20 ± 01.04	85.18 ± 07.47
5	100.00 ± 00.00	100.00 ± 00.00
6	100.00 ± 00.00	100.00 ± 00.00
7	100.00 ± 00.00	100.00 ± 00.00
8	99.69 ± 00.35	97.75 ± 02.53
9	99.11 ± 00.74	93.72 ± 04.88
10	100.00 ± 00.00	100.00 ± 00.00
11	100.00 ± 00.00	100.00 ± 00.00
12	99.20 ± 00.61	93.75 ± 05.08
13	100.00 ± 00.00	100.00 ± 00.00
14	98.68 ± 00.77	89.83 ± 06.73
15	100.00 ± 00.00	100.00 ± 00.00
Average	$\mathbf{99.52 \pm 00.19}$	$\mathbf{96.37 \pm 01.49}$

improved without having to lose much in the classifier accuracy. On the other hand, it is noticed that since $M = 10(400 \times 300)$ there is no significant reduction in time, not justifying the use of smaller images unless it is necessary to save memory. Therefore, an image size of 80×60 ($M = 50$) is a good choice.

Finally, an SVM classifier was evaluated applying all optimizations. The parameters are: C3 feature set, Image size 80×60, and 32 gray levels; RBF *kernel*, with $C = 8$ and $\sigma = 2$. Table 6 shows the accuracy for each class as well the F1 Score. The execution time in the embedded platform is 80.5 ms and the memory usage is 4.039 kB for GLCM and 12.605 kB for SVM, resulting 16.644 kB.

5   Conclusions

This research investigated three aspects of a computer vision application, made of a combination of GLCM (Gray Level Co-occurrence Matrix) and SVM (Support Vector Machine): (1) The impact of each feature calculation on the execution time; The size of the (2) input image and of the (3) Co-Occurrence Matrix on the classifier accuracy. All these facets affect significantly the resources consumed in the target platform, like memory, clock frequency, and power. The case study is the recognition of internal locations for localization and navigation of a mobile robot using topological map.

A feature selection criterion was proposed based on the study of time complexity of the algorithms. Other investigations related to GLCM+SVM were not concerned about evaluating computational costs. The average classifier accuracy using the proposed rule was kept similar to other results previously published (99.5%).

Input image size affects memory usage and computation time. For each application, there is a range of image resolution that results in maximum accuracy. Experiments demonstrate that it is not necessary to use the larger resolution available, but only the one that can get good accuracy. The evaluated application started with images as large as 4000×3000 pixels, and 80×60 pixels proved to be more than enough.

GLCM can have a large memory footprint due to its co-occurrence matrix, which size depends on the number of gray levels. GLCM+SVM previous investigations usually adopt 256 levels. However, the number of gray levels required for a good pattern classification is not that high, being 32 levels a good value for an environment recognition application. This is an interesting optimization since a simple reduction from 256 to 128 levels results in a reduction from 256 kB to 128 kB.

The SVM memory footprint depends basically on the number Support Vectors (SV) selected by the training process. Additionally, the size of those vector depends on the number of selected features, that was optimized "for free" when they were reduced to decrease computation time. The literature is full of propositions to reduce the number of SV, but this work didn't use any, leaving that for a future investigation.

Finally, this investigation started with a computer vision application that uses images with a size of 4000×3000 pixels [12]. The classifier would consume 268 kB of memory, taking 396 s to run in a Raspberry Pi. At the end, it was verified that it is possible to obtain the same accuracy using 16.6 kB of memory and concluding classification in 80.5 ms.

Acknowledgment. The authors would like to thank the sponsorship from FUNCAP and CAPES via Grant No. 05/2014 FUNCAP/CAPES.

References

1. Bessa, J., Almeida Barroso, D., Rocha Neto, A., Alexandria, A.: Global location of mobile robots using artificial neural networks in omnidirectional images. IEEE Lat. Am. Trans. **13**, 3405–3414 (2015)
2. Cormen, T.H., Leiserson, C.E., Rivest, R.L., Stein, C.: Introduction to Algorithms, 2nd edn. The MIT Press, Cambridge (2001)
3. Cortes, C., Vapnik, V.: Support-vector networks. Mach. Learn. **20**(3), 273–297 (1995)
4. dGB Earth Sciences: Texture directional: a multi-trace attribute that returns textural information based on a statistical texture classification (2015)
5. Hall, M.A.: Correlation-based feature subset selection for machine learning. Ph.D. thesis, University of Waikato, Hamilton, New Zealand (1998)
6. Haralick, R.M.: Statistical and structural approaches to texture. Proc. IEEE **67**, 786–804 (1979)
7. Haralick, R.M., Shanmugam, K., Dinstein, I.: Textural features for image classification. IEEE Trans. Syst. Man Cybern. **SMC–3**, 610–621 (1973)
8. Lyons, R.G.: Understanding Digital Signal Processing, 3rd edn. Prentice-Hall, Upper Saddle River (2010)
9. Oliveira, D.C.D., Wehrmeister, M.A.: Towards real-time people recognition on aerial imagery using convolutional neural networks. In: IEEE 19th International Symposium on Real-Time Distributed Computing (ISORC), pp. 27–34, May 2016
10. Sethupathy, J., Veni, S.: Opencv based disease identification of mango leaves. Int. J. Eng. Technol. Sci. Res. - IJETSR **8**(5), 1990–1998 (2016)
11. Siéler, L., Tanougast, C., Bouridane, A.: A scalable and embedded fpga architecture for efficient computation of grey level co-occurrence matrices and haralick textures features. Microprocess. Microsyst. **34**(1), 14–24 (2010)
12. da Silva, S.P.P., Marinho, L.B., Almeida, J.S., Rebouças Filho, P.P.: A novel approach for mobile robot localization in topological maps using classification with reject option from structural co-occurrence matrix. In: Felsberg, M., Heyden, A., Krüger, N. (eds.) CAIP 2017. LNCS, vol. 10424, pp. 3–15. Springer, Cham (2017). https://doi.org/10.1007/978-3-319-64689-3_1
13. Soh, L.K., Tsatsoulis, C.: Texture analysis of SAR sea ice imagery using gray level co-occurrence matrices. IEEE Trans. Geosci. Remote Sens. **37**(2), 780–795 (1999)
14. Wang, H., Guo, X.H., Jia, Z.W., Li, H.K., Liang, Z.G., Li, K.C., He, Q.: Multi-level binomial logistic prediction model for malignant pulmonary nodules based on texture features of CT image. Eur. J. Radiol. **74**, 124–129 (2010)
15. Wang, Z.L., Cai, B.G., Yi, F.Z., Li, M.: Reviews on planar region detection for visual navigation of mobile robot under unknown environment. In: Lee, G. (ed.) Advances in Automation and Robotics, pp. 593–601. Springer, Heidelberg (2012). https://doi.org/10.1007/978-3-642-25646-2_77

A New Method for Learning RBF Networks by Utilizing Singular Regions

Seiya Satoh[1] and Ryohei Nakano[2(✉)]

[1] National Institute of Advanced Industrial Science and Technology, 2-4-7 Aomi, Koto-ku, Tokyo 135-0064, Japan
seiya.satoh@aist.go.jp
[2] Chubu University, 1200 Matsumoto-cho, Kasugai, Aichi 487-8501, Japan
nakano@cs.chubu.ac.jp

Abstract. The usual way to learn radial basis function (RBF) networks consists of two stages: first, select reasonable weights between input and hidden layers, and then optimize weights between hidden and output layers. When we learn multilayer perceptrons (MLPs), we usually employ the stochastic descent called backpropagation (BP) algorithm or 2nd-order methods such as pseudo-Newton method and conjugate gradient method. Recently new learning methods called singularity stairs following (SSF) methods have been proposed for learning real-valued or complex-valued MLPs by making good use of singular regions. SSF can monotonically decrease training error along with the increase of hidden units, and stably find a series of excellent solutions. This paper proposes a completely new method for learning RBF networks by introducing the SSF paradigm, and compares its performance with those of existing learning methods.

Keywords: Neural networks · RBF networks · Learning method
Singular region · Reducibility mapping

1 Introduction

In the search space of a multilayer perceptron, whether it is real-valued or complex-valued, there exist singular regions throughout which the gradient is zero; thus any search methods using gradient information cannot move in the regions [5,8]. There exist singular regions in the search space of radial basis function (RBF) networks as well.

It is common for learning RBF networks to employ two-stage learning; first, select appropriate weights between input and hidden layers, and then optimize weights between hidden and output layers [2]. Moreover, the EM algorithm [3] was used considering RBF networks as a mixture model [6].

In the learning of MLPs, however, whole weights are updated all at once by using the backpropagation algorithm or pseudo-Newton method. Recently new learning methods called singularity stairs following (SSF) methods have been

© Springer International Publishing AG, part of Springer Nature 2018
L. Rutkowski et al. (Eds.): ICAISC 2018, LNAI 10841, pp. 214–225, 2018.
https://doi.org/10.1007/978-3-319-91253-0_21

proposed for learning real-valued or complex-valued MLPs by making good use of singular regions [9, 10]. SSF can decrease training error monotonically as the number of hidden units increases, and can stably find a series of solutions better than those obtained by existing learning methods.

This paper proposes a completely new method for learning RBF networks by applying the SSF paradigm to the RBF framework and compares its performance with those of the two-stage learning and pseudo-Newton method.

2 Background

2.1 RBF Networks

$\mathrm{RBF}(J)$ denotes an RBF network with J hidden units and one output unit. In $\mathrm{RBF}(J)$ model, let $\boldsymbol{w}_j^{(J)} = (w_{j1}^{(J)}, \cdots, w_{jK}^{(J)})^{\mathrm{T}}$ be weights between input and hidden unit $j (= 1, \cdots, J)$, and let $v_j^{(J)}$ be a weight between hidden unit j and the single output unit. When Gaussian basis function is adopted and μ-th data point $\boldsymbol{x}^\mu = (x_1^\mu, \cdots, x_K^\mu)^{\mathrm{T}}$ is given as input, the output of $\mathrm{RBF}(J)$ can be defined as below. Here σ_j is a parameter of Gaussian basis function at hidden unit j.

$$f_J^\mu = v_0^{(J)} + \sum_{j=1}^{J} v_j^{(J)} \exp\left(-\frac{||\boldsymbol{x}^\mu - \boldsymbol{w}_j^{(J)}||^2}{2(\sigma_j^{(J)})^2}\right) \tag{1}$$

The whole parameter vector of $\mathrm{RBF}(J)$ is given below:

$$\boldsymbol{\theta}^{(J)} = \left(v_0^{(J)}, \cdots, v_J^{(J)}, \left(\boldsymbol{w}_1^{(J)}\right)^{\mathrm{T}}, \cdots, \left(\boldsymbol{w}_J^{(J)}\right)^{\mathrm{T}}, \sigma_1^{(J)}, \cdots, \sigma_J^{(J)}\right)^{\mathrm{T}}.$$

Given training data $\{(\boldsymbol{x}^\mu, y^\mu), \ \mu = 1, \cdots, N\}$, the target function of $\mathrm{RBF}(J)$ learning is given as follows.

$$E_J = \frac{1}{2} \sum_{\mu=1}^{N} (\delta_J^\mu)^2, \qquad \delta_J^\mu \equiv f_J^\mu - y^\mu \tag{2}$$

2.2 Existing Methods for Learning RBF Networks

Most ways to learn RBF networks can be classified into two kinds: one is two-stage learning and the other is one-stage learning.

In the former, under the condition that $\sigma_1^{(J)}, \cdots, \sigma_J^{(J)}$ are fixed to a certain common constant such as $1/\sqrt{2}$ throughout learning, first, select reasonable $\boldsymbol{w}_1^{(J)}, \cdots, \boldsymbol{w}_J^{(J)}$, and then, optimize $v_0^{(J)}, \cdots, v_J^{(J)}$ [2].

In the latter, whole weights are optimized at the same time by using gradient-based methods or the EM algorithm. Gradient-based methods can be classified into the stochastic descent and 2nd-order method such as pseudo-Newton method.

Two-Stage Learning. At the first stage of two-stage learning, one can apply clustering to explanatory data $\{x^\mu\}$ to get J clusters, and then treat all the centroids as $w_1^{(J)}, \cdots, w_J^{(J)}$.

This paper, however, employs function newrb supported in Neural Network Toolbox of MATLAB R2015b. The newrb algorithm gradually expands RBF networks by increasing the number J of hidden units one by one. At each cycle of model expansion, data point x^μ that has the largest value of output error $|f_{J-1}^\mu - y^\mu|$ is added as w_J. Basis function parameters $\{\sigma_j\}$ are not optimized in this learning.

The general flow of the newrb algorithm is shown below. Let $J_{\max}$ be the maximum number of hidden units.

---

newrb algorithm

1: $v_0 \leftarrow (1/N) \sum_{\mu=1}^N y^\mu$
2: **for** $J = 1, \cdots, J_{\max}$ **do**
3:     Compute outputs f_{J-1}^μ, $\mu = 1, \cdots, N$.
4:     $i \leftarrow \arg\max_{\mu \in \{1, \cdots, N\}} |f_{J-1}^\mu - y^\mu|$
5:     $w_J^{(J)} \leftarrow x^i$
6:     With $w_1^{(J)}, \cdots, w_J^{(J)}$ fixed, optimize $v_0^{(J)}, \cdots, v_J^{(J)}$.
7: **end for**

One-Stage Learning. One-stage learning can be classified into gradient-based and mixture-based [6]. Among many gradient-based learning methods, this paper employs the pseudo-Newton method with the BFGS update [4] since it has more powerful learning ability than the stochastic descent. Here we consider two kinds of methods: one is simply called BFGS, and the other is called BFGS(σ). The former optimizes all the weights except $\{\sigma_j\}$, while the latter optimizes both all the weights and $\{\sigma_j\}$ at the same time.

2.3   Singular Regions of RBF Networks

The search space of MLP, whether it is real-valued or complex-valued, has a continuous region where input-output equivalence holds and the gradient is zero [5,8]. Such regions can be generated by reducibility mapping $\alpha\beta$ or γ [9,10]. We call such a region a singular region. The search space of RBF networks has also singular regions, which are generated only by γ reducible mapping.

Below we explain how to generate a singular region of RBF(J) based on the optimal solution of RBF($J-1$). The optimal solution must satisfy the following, where E_{J-1} denotes the target function of RBF($J-1$) learning, and $\theta^{(J-1)}$ denotes the whole parameter vector of RBF($J-1$).

$$\frac{\partial E_{J-1}}{\partial \theta^{(J-1)}} = 0 \qquad (3)$$

This can be broken down element-wise as follows, where $j = 1, \cdots, J - 1$.

$$\frac{\partial E}{\partial v_0^{(J-1)}} = \sum_{\mu=1}^{N} \delta_{J-1}^{\mu} = 0 \tag{4}$$

$$\frac{\partial E}{\partial v_j^{(J-1)}} = \sum_{\mu=1}^{N} \delta_{J-1}^{\mu} \exp\left(-\frac{||\boldsymbol{x}^{\mu} - \boldsymbol{w}_j^{(J-1)}||^2}{2(\sigma_j^{(J-1)})^2}\right) = 0 \tag{5}$$

$$\frac{\partial E}{\partial \boldsymbol{w}_j^{(J-1)}} = \sum_{\mu=1}^{N} \delta_{J-1}^{\mu} v_j^{(J-1)} \exp\left(-\frac{||\boldsymbol{x}^{\mu} - \boldsymbol{w}_j^{(J-1)}||^2}{2(\sigma_j^{(J-1)})^2}\right)$$
$$\frac{\boldsymbol{x}^{\mu} - \boldsymbol{w}_j^{(J-1)}}{\left(\sigma_j^{(J-1)}\right)^2} = 0 \tag{6}$$

$$\frac{\partial E}{\partial \sigma_j^{(J-1)}} = \sum_{\mu=1}^{N} \delta_{J-1}^{\mu} v_j^{(J-1)} \exp\left(-\frac{||\boldsymbol{x}^{\mu} - \boldsymbol{w}_j^{(J-1)}||^2}{2(\sigma_j^{(J-1)})^2}\right)$$
$$\frac{||\boldsymbol{x}^{\mu} - \boldsymbol{w}_j^{(J-1)}||^2}{\left(\sigma_j^{(J-1)}\right)^3} = 0 \tag{7}$$

Let the optimal solution $\widehat{\boldsymbol{\theta}}^{(J-1)}$ of RBF$(J-1)$ be

$$\left(\widehat{v}_0^{(J-1)}, \cdots, \widehat{v}_J^{(J-1)}, \left(\widehat{\boldsymbol{w}}_1^{(J-1)}\right)^{\mathrm{T}}, \cdots, \left(\widehat{\boldsymbol{w}}_J^{(J-1)}\right)^{\mathrm{T}}, \widehat{\sigma}_1^{(J-1)}, \cdots, \widehat{\sigma}_J^{(J-1)}\right)^{\mathrm{T}}.$$

Now apply reducibility mapping γ to the optimal solution $\widehat{\boldsymbol{\theta}}^{(J-1)}$ to get singular region $\widehat{\boldsymbol{\Theta}}_\gamma^{(J)}$. Here $m \in \{2, \cdots, J\}$.

$$\widehat{\boldsymbol{\theta}}^{(J-1)} \xrightarrow{\gamma} \widehat{\boldsymbol{\Theta}}_\gamma^{(J)}$$

$$\widehat{\boldsymbol{\Theta}}_\gamma^{(J)} \equiv \{\boldsymbol{\theta}^{(J)} \mid v_0^{(J)} = \widehat{v}_0^{(J-1)}, v_1^{(J)} + v_m^{(J)} = \widehat{v}_{m-1}^{(J-1)},$$
$$\boldsymbol{w}_1^{(J)} = \boldsymbol{w}_m^{(J)} = \widehat{\boldsymbol{w}}_{m-1}^{(J-1)}, \sigma_1^{(J)} = \sigma_m^{(J)} = \widehat{\sigma}_{m-1}^{(J-1)},$$
$$v_j^{(J)} = \widehat{v}_{j-1}^{(J-1)}, \boldsymbol{w}_j^{(J)} = \widehat{\boldsymbol{w}}_{j-1}^{(J-1)}, \sigma_j^{(J)} = \widehat{\sigma}_{j-1}^{(J-1)},$$
$$\text{for } j \in \{2, \cdots, J\} \backslash \{m\}, \quad m = 2, \cdots, J\} \tag{8}$$

Note that $v_1^{(J)}$ and $v_m^{(J)}$ cannot be determined uniquely since they only have the following constraint.

$$v_1^{(J)} + v_m^{(J)} = \widehat{v}_{m-1}^{(J-1)} \tag{9}$$

This equation can be rewritten using parametric variable q.

$$v_1^{(J)} = q \, \widehat{v}_{m-1}^{(J-1)}, \qquad v_m^{(J)} = (1 - q) \, \widehat{v}_{m-1}^{(J-1)} \tag{10}$$

3   SSF for RBF Networks

A new search paradigm called Singularity Stairs Following (SSF) stably finds a series of excellent solutions by making good use of singular regions. So far, two series of SSF methods have been proposed: The latest versions are SSF1.4 [9] for real-valued MLPs and C-SSF1.3 [10] for complex-valued MLPs.

By applying the SSF paradigm to RBF networks, we have a completely new learning method called RBF-SSF. RBF-SSF optimizes $\{\sigma_j\}$ as well.

3.1   General Flow of RBF-SSF

The general flow of RBF-SSF is shown below. Let $J_{\max}$ be the maximum number of hidden units.

General flow of RBF-SSF

1: Get the best solution of RBF($J =1$) by repeating search with different initial values.
2: **for** $J = 2, \cdots, J_{\max}$ **do**
3:    Select starting points from the singular region obtained by applying reducibility mapping γ to the best solution of RBF($J - 1$).
4:    Calculate Hessian matrix at each starting point, calculate eigenvalues and eigenvectors of the Hessian, and select eigenvectors corresponding to each negative eigenvalue.
5:    Repeat search predefined times from the starting points in the direction and in the opposite direction of a negative eigenvector selected at Step 4, and get the best solution of RBF(J).
6: **end for**

In the following experiments, the starting points from the singular region at the above Step 3 are obtained by setting q to $0.5, 1.0, 1.5$ in Eq. (10). These three points correspond to interpolation, boundary, and extrapolation. Since m is changed in the range of $2, \cdots, J$, the total number of starting points in the search of RBF(J) amounts to $3 \times (J - 1)$.

3.2   Techniques for Making SSF Faster

As described above, the number of starting points of RBF(J) gets large as J gets large. Moreover, the dimension of the search space also gets large as J gets large. Thus, if we perform search for every negative eigenvalue for large J, the number of searches gets very large, and consequently the processing time becomes huge. Hence, we introduce two accelerating techniques into RBF-SSF: one is search pruning and the other is to set upper limit to the number of searches.

Search pruning is an accelerating technique which discards a search if the search is found to proceed along much the same route experienced before [9].

Setting upper limit to the number of searches is an accelerating technique which selects the predefined number of starting points based on the preference.

Here preference is given to smaller negative eigenvalues since such eigenvalues indicate bigger drop from a search point. In the following experiments the upper limit is set to be 100.

4 Experiments

The performance of RBF-SSF was compared with three other learning methods. Table 1 shows the settings of these four learning methods. BFGS and BFGS(σ) are repeated 100 times changing initial values of weights. Note that both BFGS(σ) and RBF-SSF adapt $\{\sigma_j\}$ through learning.

Table 1. Learning methods.

Learning method	Description $(j = 1, \cdots, J)$
newrb	A method of two-stage learning
	Included in Neural Network Toolbox version 9.1
	$\sigma_j^{(J)}$: fixed to be $1/\sqrt{2}$
BFGS	$v_j^{(J)}, \boldsymbol{w}_j^{(J)}$: optimized by BFGS
	The initial value of $v_j^{(J)}$: 0
	The initial value of $\boldsymbol{w}_j^{(J)}$: randomly selected from $\boldsymbol{x}^{(1)}, \cdots, \boldsymbol{x}^{(N)}$
	$\sigma_j^{(J)}$: fixed to be $1/\sqrt{2}$
BFGS(σ)	$v_j^{(J)}, \boldsymbol{w}_j^{(J)}, \sigma_j^{(J)}$: optimized by BFGS($\sigma$)
	The initial value of $v_j^{(J)}$: 0
	The initial value of $\boldsymbol{w}_j^{(J)}$: randomly selected from $\boldsymbol{x}^{(1)}, \cdots, \boldsymbol{x}^{(N)}$
	The initial value of $\sigma_j^{(J)}$: $1/\sqrt{2}$
RBF-SSF	$v_j^{(J)}, \boldsymbol{w}_j^{(J)}, \sigma_j^{(J)}$: optimized by RBF-SSF

The following three datasets were used: engine behavior dataset, building energy dataset, and Parkinsons telemonitoring dateset [7]. The first two are included in MATLAB Neural Network Toolbox. The third was obtained from UCI ML Repository [1]. Each dataset was normalized as follows, where y_{mean} and y_{std} are the average and standard deviation of $\{y^\mu\}$ respectively.

$$\widetilde{x}_k^\mu \leftarrow \frac{x_k^\mu}{\max_\mu(|x_k^\mu|)}, \qquad \widetilde{y}^\mu \leftarrow \frac{y^\mu - y_{\mathrm{mean}}}{y_{\mathrm{std}}} \qquad (11)$$

4.1 Experiments Using Engine Behavior Dataset

The problem for this dataset is to estimate an engine torque from two variables: fuel use and speed. The number N of data points is 1199; 90 % was used for training, and the remaining 10 % was used for test. The number J of hidden units for newrb was changed as $10, 20, \cdots, 200$, while J of BFGS, BFGS(σ), and RBF-SSF was changed as $1, 2, \cdots, 40$.

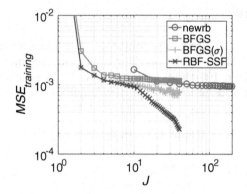

Fig. 1. Training error for engine behavior dataset.

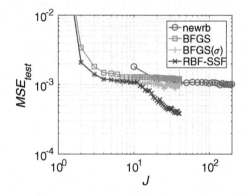

Fig. 2. Test error for engine behavior dataset.

Figures 1 and 2 show minimum training and test errors of each method respectively. Each error is shown in mean squared error (MSE). RBF-SSF and newrb monotonically decreased training error. The final training and test errors of RBF-SSF were the smallest among the four methods, a few times smaller than those of the other three methods. The final training and test errors of BFGS(σ) were smaller than those of BFGS, which means the adaptation of $\{\sigma_j\}$ worked to some extent.

Figure 3 shows the processing time spent by each method at each J. The total time (h:min:sec) spent by newrb, BFGS, BFGS(σ), and RBF-SSF were

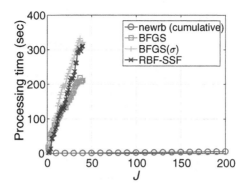

Fig. 3. Processing time for engine behavior dataset.

00:00:05, 01:30:56, 02:08:15, and 01:48:40 respectively. The fastest was newrb and the slowest was BFGS(σ). BFGS(σ) was 4/3 times slower than BFGS since BFGS(σ) had to optimize $\{\sigma_j\}$ aside from weights.

4.2   Experiment Using Building Energy Dataset

The problem for this dataset is to estimate the energy use of a building from 14 variables such as time and weather conditions. The number N of data points is 4208; 90% was used for training, and the remaining 10% was used for test. The number J of hidden units for newrb was changed as $10, 20, \cdots, 400$, while J of BFGS, BFGS(σ), and RBF-SSF was changed as $1, 2, \cdots, 30$.

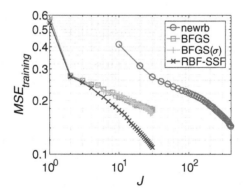

Fig. 4. Training error for building energy dataset.

Figures 4 and 5 show minimum training and test errors of each method respectively. Again, RBF-SSF and newrb monotonically decreased training error. It is clear that the final training and test errors of RBF-SSF were the smallest among

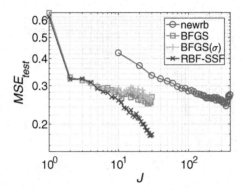

Fig. 5. Test error for building energy dataset.

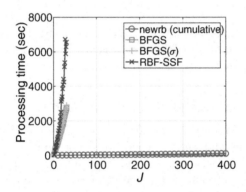

Fig. 6. Processing time for building energy dataset.

the four. The final training and test errors of BFGS(σ) were much the same as those of BFGS, which means the adaptation of $\{\sigma_j\}$ did not work well here. The final training error at $J = 400$ of newrb was smaller than that of BFGS. The best test errors of newrb at $J = 350$ and BFGS(σ) at $J = 29$ were much the same.

Figure 6 shows the processing time spent by each method at each J. The total time (h:min:sec) spent by newrb, BFGS, BFGS(σ), and RBF-SSF Were 00:01:34, 10:19:41, 10:51:10, and 21:56:55 respectively. The fastest was newrb and the slowest was RBF-SSF. RBF-SSF spent the longest time since it had to compute the Hessian repeatedly and the processing time gets even larger as the number of input units gets large.

4.3   Experiment Using Parkinsons Telemonitoring Dataset

The problem for this dataset is to estimate motor UPDRS (unified Parkinson's disease rating score) based on 18 variables such as voice measures. The number N of data points is 5875; 90% was used for training, and the remaining 10% was used for test. The number J of hidden units for newrb was changed as $10, 20, \cdots , 350,$ while J of BFGS, BFGS(σ), and RBF-SSF was changed as $1, 2, \cdots , 20.$

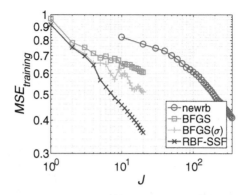

Fig. 7. Training error for Parkinsons telemonitoring dataset.

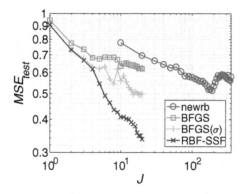

Fig. 8. Test error for Parkinsons telemonitoring dataset.

Figures 7 and 8 show minimum training and test errors of each method respectively. Again, RBF-SSF and newrb decreased training error monotonically. It is clear the final training and test errors of RBF-SSF were the smallest. The final training and test errors of BFGS(σ) were clearly smaller than those of BFGS, which means the adaptation of $\{\sigma_j\}$ worked well here. The final training error at $J = 400$ of newrb was obviously smaller than that of BFGS. The best test errors of newrb at $J = 190$ and BFGS(σ) at $J = 15$ were much the same.

Figure 9 shows the processing time spent by each method at each J. The total time (h:min:sec) spent by newrb, BFGS, BFGS(σ), and RBF-SSF Were 00:03:27, 10:42:19, 10:49:15, 19:03:27 respectively. Again the fastest was newrb, and the slowest was RBF-SSF. RBF-SSF spend the longest time due to the repeated computation of the Hessian.

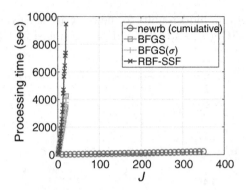

Fig. 9. Processing time for Parkinsons telemonitoring dataset.

4.4   Findings Obtained from Experiments

The following findings can be made from our experiments.

(1) Training Errors
    RBF-SSF could decrease training error monotonically and in a striking man-
    ner, much smaller than those of the other three methods. BFGS(σ) clearly
    exceeded BFGS in training error for two datasets out of three, which means
    the adaptation of $\{\sigma_j\}$ is a meaningful step. The two-stage learning newrb
    decreased training error monotonically, but the decreasing rate is relatively
    small, and it needs hundreds of hidden units to seriously decrease training
    error.
(2) Test Errors
    The tendency of test error decreasing curves by four methods were much
    the same as that of training error. RBF-SSF could decrease test error to
    a remarkable extent, much smaller than those of the other three methods.
    RBF-SSF showed a negative gradient at the largest J for every dataset,
    which indicates the optimal model for each dataset may have $J(> J_{max})$.
    This may suggest the presumable tendency that RBF networks are unlikely
    to cause the overfitting. BFGS(σ) clearly exceeded BFGS in test error for two
    datasets out of three, which means the adaptation of $\{\sigma_j\}$ is a meaningful
    step. Test error of newrb showed U-shaped curve as J got large for two
    datasets out of three. This indicates the overfitting happened in this type of
    learning.
(3) Processing Time
    Two-stage learning newrb was overwhelmingly fast since it only solves linear
    problems repeatedly. On the other hand, the other three methods required
    considerable amount of time since they use BFGS updates so many times.
    BFGS(σ) was slower than BFGS because BFGS(σ) has to adapt $\{\sigma_j\}$ aside
    from weights. Moreover, RBF-SSF has to compute the Hessian repeatedly;
    thus RBF-SSF spent the longest time.

5   Conclusion

This paper proposed a completely new method for learning RBF networks by applying the SSF paradigm to the RBF framework, and compared its performance with those of three existing methods. The proposed RBF-SSF could decrease training and test errors to a remarkable extent, much smaller than those of the other three methods. However, it required the longest processing time.

In the future, we plan to accelerate RBF-SSF even more and apply it to applications which strongly require the advantages of RBF networks.

Acknowledgment. This work was supported by Grants-in-Aid for Scientific Research (C) 16K00342.

References

1. UCI Machine Learning Repository (1996). http://archive.ics.uci.edu/ml/
2. Bishop, C.M.: Neural Networks for Pattern Recognition. Clarendon Press, Oxford (1995)
3. Dempster, A.P., Laird, N.M., Rubin, D.B.: Maximum-likelihood from incomplete data via the EM algorithm. J. R. Stat. Soc. Ser. B **39**, 1–38 (1977)
4. Fletcher, R.: Practical Methods of Optimization, 2nd edn. Wiley, Hoboken (1987)
5. Fukumizu, K., Amari, S.: Local minima and plateaus in hierarchical structure of multilayer perceptrons. Neural Netw. **13**(3), 317–327 (2000)
6. Làzaro, M., Santamarìa, I., Pantaleòn, C.: A new EM-based training algorithm for RBF networks. Neural Netw. **16**, 69–77 (2003)
7. Little, M.A., McSharry, P.E., Roberts, S.J., Costello, D.A.E., Moroz, I.M.: Exploiting nonlinear recurrence and fractal scaling properties for voice disorder detection. Biomed. Eng. OnLine **6**, 23 (2007)
8. Nitta, T.: Local minima in hierarchical structures of complex-valued neural networks. Neural Netw. **43**, 1–7 (2013)
9. Satoh, S., Nakano, R.: Multilayer perceptron learning utilizing singular regions and search pruning. In: Proceedings of International Conference on Machine Learning and Data Analysis, pp. 790–795 (2013)
10. Satoh, S., Nakano, R.: A yet faster version of complex-valued multilayer perceptron learning using singular regions and search pruning. In: Proceedings of 7th International Joint Conference on Computational Intelligence (IJCCI), NCTA, vol. 3, pp. 122–129 (2015)

Cyclic Reservoir Computing with FPGA Devices for Efficient Channel Equalization

Erik S. Skibinsky-Gitlin, Miquel L. Alomar, Christiam F. Frasser, Vincent Canals, Eugeni Isern, Miquel Roca, and Josep L. Roselló[✉]

Electronics Engineering Group, Department of Physics, University of Balearic Islands, 07122 Palma de Mallorca, Spain
`j.rossello@uib.es`

Abstract. The reservoir computation (RC) is a recurrent neural network architecture that is very suitable for time series prediction tasks. Its implementation in specific hardware can be very useful in relation to software approaches, especially when low consumption is an essential requirement. However, the hardware realization of RC systems is expensive in terms of circuit area and power dissipation, mainly due to the need of a large number of multipliers at the synapses. In this paper, we present an implementation of an RC network with cyclic topology (simple cyclic reservoir) in which we limit the available synapses' weights, which makes it possible to replace the multiplications with simple addition operations. This design is evaluated to implement the equalization of a non-linear communication channel, and allows significant savings in terms of hardware resources, presenting an accuracy comparable to previous works.

1 Introduction

Many artificial neural networks (ANNs) applications require the use of specific hardware [1]. These devices can take advantage of the inherent ANNs parallelism that may be beneficial in terms of performance, reliability and cost [2]. Hardware neural networks (HNNs) are used in real-life applications as computer vision [3], image search [4], data mining [5], control of machines and industrial processes [6], distributed sensory networks [7], portable medical applications [8], handwriting and speech recognition systems [9], etc. Different works have been done to develop efficient HNN [10–12], however when a large neuron count is needed, the increased number of synapses' multipliers limits the possibility of implementing massive networks [13]. To solve this problem, the use of approximate multipliers [14] and stochastic computing [13] can be considered as possible solutions to reduce the area requirements at the cost of a lower accuracy.

A widely used neural architecture suited for time-series processing is the Reservoir computing (RC) scheme [15]. In RC systems, the synaptic weights are randomly chosen and kept fixed. The fixed network is called the reservoir and is connected to a readout output layer that can be trained in a simple way by using a multilinear regression analysis. This technique has been applied to robot

© Springer International Publishing AG, part of Springer Nature 2018
L. Rutkowski et al. (Eds.): ICAISC 2018, LNAI 10841, pp. 226–234, 2018.
https://doi.org/10.1007/978-3-319-91253-0_22

control [16], image/video processing [17], or financial forecasting [18]. Fast and efficient hardware designs implementing RC systems can be interesting for many applications, which require real-time intensive data processing and/or the use of low-power devices to ensure a long battery lifetime. The RC system have been implemented in hardware in different ways [15,19,20]. Most of them are sequentially operated to reduce the total resources, which compromises the processing speed. In this work we present a hardware implementation of RC systems with low hardware resources. The proposed solution presents a reduced number of logic gates when compared to conventional designs. This application is used to implement the equalization of a wireless communication channel. Communication systems are aimed at efficiently sending information from a transmitter to a receiver using an available channel. The processing of the received data is necessary as the channel is always responsible for distorting the transmitted signals [21]. Here, we focus on the digital compensation of the nonlinear channel at the receiver side. Summarizing, the problem consists in designing an equalizer to cancel the distortions introduced by the physical environment used for transmission, thus enabling the correct recovery of the original information [22]. The obtained results are compared with different previously-published implementations showing competitive results.

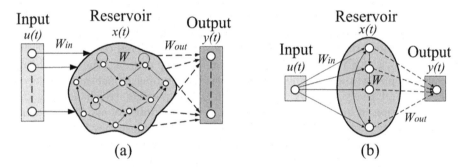

Fig. 1. (a–b) Reservoir computing architectures: (a) general architecture of RC where all connections ($\boldsymbol{W_{in}}$ and $\boldsymbol{W}$) are random and fixed except those from the reservoir to the output units ($\boldsymbol{W_{out}}$), which are trained for each specific task; (b) reservoir with simple cyclic topology and a single input

2   Methodology

2.1   FPGA Implementation of Cyclic Reservoir Systems

A reservoir computing system (Fig. 1a) comprises a total of N neurons, each one providing a value $x_{k,i}$, where $i \in \{1, 2, ..., N\}$ is the neuron index, k is the evolution during time ($k \in \{1, 2, ..., L\}$) and L the total number of samples taken from the reservoir. The time evolution of internal nodes is described by a state matrix with dimensions $L \times N$ (denoted as $\boldsymbol{X}$). At time k the state is defined by the *kth* row. The reservoir state $[\boldsymbol{x(k)}]$ is updated using a nonlinear function evaluating a weighted sum of the reservoir inputs [M inputs $\boldsymbol{u(k)} = (u_1(k), u_2(k), ..., u_M(k))$], and also from the signals coming from the rest of the reservoir (N signals $\boldsymbol{x(k-1)}$). The next expression then applies:

$$\boldsymbol{x(k)} = f\left[\boldsymbol{W_{in}} \boldsymbol{u(k)} + \boldsymbol{W} \boldsymbol{x(k-1)}\right] \tag{1}$$

where f is the activation function $f : \mathbb{R}^N \to \mathbb{R}^N$, $\boldsymbol{W_{in}}$ and $\boldsymbol{W}$ are two $N \times M$ and $N \times N$ weight matrices respectively. In a second phase, a total of Q outputs [$\boldsymbol{\hat{y}(k)} = (\hat{y}_1(k), \hat{y}_2(k), ..., \hat{y}_Q(k))$] are estimated at the output layer performing a linear combination of the reservoir states:

$$\boldsymbol{\hat{y}(k)} = \boldsymbol{W_{out}} \boldsymbol{x(k)} \tag{2}$$

where $\boldsymbol{W_{out}}$ is a $Q \times N$ matrix obtained using a multilinear regression. For this, the expected outputs must be considered [$\boldsymbol{y(k)} = (y_1(k), y_2(k), ..., y_Q(k))$]. Defining $\boldsymbol{Y}$ as the feature matrix of $L \times Q$ to be approximated by the network (ground truth composed by L rows and Q columns), we have that $\boldsymbol{W_{out}}$ may be estimated as:

$$\boldsymbol{W_{out}^T} = \left(\boldsymbol{X^T X}\right)^{-1} \boldsymbol{X^T Y} \tag{3}$$

Therefore, in the RC scheme, both matrices $\boldsymbol{W_{in}}$ and $\boldsymbol{W}$ are taken fixed while $\boldsymbol{W_{out}}$ is trained using (3) that corresponds to the Moore-Penrose pseudo-inverse. The activation function f that we use in this work is a piece-wise function with three linear regions so the RC method can be understood to be an Echo State Network (ESNs). In the case of using spiking neurons, the RC scheme is called Liquid State Machine (LSMs) [23]. The number of neurons employed in ESNs is relatively high ($N > 50$), which makes particularly challenging the implementation in hardware. Considering the large number of synapses present in the circuitry, the use of digital multipliers of each synapse may severely limit the maximum possible value of N. To avoid the use of multipliers we limit the weights values to a sum of powers of two so that shift registers can be employed to perform the product inside the neural synapses instead of multipliers. The output layer of the RC network is implemented using the dedicated embedded multipliers integrated in the FPGA device. The overall area impact of the output layer is low since the number of multipliers to be used is relatively reduced if compared with the number of synapses. The network topology of the RC system used in this work is a simple cyclic architecture (Fig. 1b) [24] that maximizes the packing efficiency. The connections between internal units have been selected with a constant value r whereas the inputs are weighted to both positive $v = |v|$ or negative $v = -|v|$ with the same probability and with the same absolute value. Both parameters r and v are optimized numerically.

In Fig. 2 we show a RC design and the cyclic reservoir when one input is processed. The fixed-point two's complement notation is assumed. The first input [$u(t)$] refers to the external input and the second [$x_{i-1}(t-1)$] to the state of a neighboring neuron at the previous time step. Two different resolutions of n and m bits are considered for the input and the weights respectively (v and r).

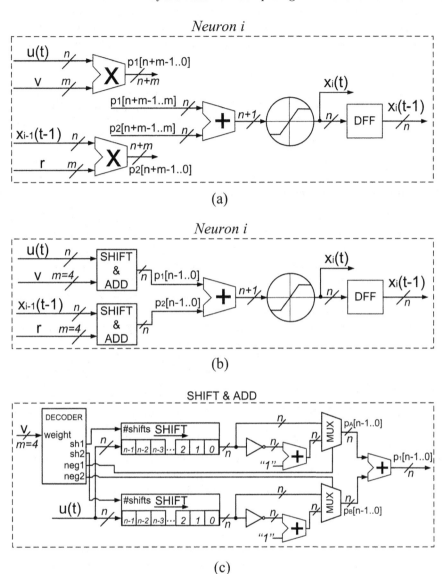

Fig. 2. (a–c) Neuron design: (a) general circuit design of the neuron; (b) reduced implementation scheme when the weight resolution is limited to few bits ($m = 4$) and the multipliers are replaced by simple shift-and-add blocks; (c) description of the shift-and-add block

The multiplier's output is truncated to n bits taking the most significant of the result, but a higher or lower resolution could be employed depending on the desired accuracy. A simple piece-wise linear approximation with three segments [10] is used for the implementation of the activation function, due to its simple implementation. The scheme of Fig. 2a can be simplified to that of Fig. 2b when the weight resolution is limited to a few bits, more specifically $m = 4$. In this case, the multipliers are substituted by shift-and-add blocks. The shift-and-add block is depicted in Fig. 2c showing the multiplication of the input signal $[u(t)]$ by the corresponding weight (v) with a pair of shift registers and an adder. Some additional circuitry is included to perform the negation if necessary. A decoder configures the shift registers (with the number of required shifts, $sh1$ and $sh2$) and controls the activation of the negations ($neg1$ and $neg2$) as a function of the weight value (v).

2.2   Nonlinear Channel Equalization

A representation of a communication system using an equalizer is illustrated in Fig. 3. The transmitter communicates the signal $d(t)$ that can be a given set of discrete values. It is received and demodulated into the analog signal $s(t)$, which is a disturbed version of $d(t)$ due to linear superposition of adjacent symbols ($q(t)$, intersymbol interference), nonlinear distortion and noise ($\nu(t)$). Signal $s(t)$ is sent to the equalizer to recover a delayed version of the sequence $d(t-\tau)$, where τ is the propagation delay associated with the channel. The signal $\hat{y}(t)$ is finally converted into a symbol sequence ($\hat{d}(t-\tau)$). The equalizer is adapted during the training so the reservoir output $\hat{y}(t)$ can restore $s(t)$ to $d(t-\tau)$. Using the training data ($s(t)$ as input and $y(t) = d(t-\tau)$ as desired output), the equalizer is able to minimize the error $\epsilon(t) = d(t-\tau) - \hat{y}(t)$. Once the training has been completed, the weights are fixed and used to estimate the sequence. Reservoir Computing systems represent an attractive equalization solution since they are nonlinear and recurrent. The RC approach has been proposed for the nonlinear channel equalization task using models of diverse complexity [22,24–26]. The channel model used here as a wireless transmission system is the one presented in [27]. The input to the channel is an independent and identically distributed random sequence $d(t)$ with values $\{-3, -1, 1, 3\}$.

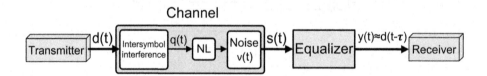

Fig. 3. Schematic of a wireless communication system with a channel equalizer

The functionality of the equalizer is to predict future values based on previously observed for the equalization of a nonlinear channel [28]. The processing is divided in two steps:

- Part of the time-series is used for off-line training using MATLAB. The optimum v and r values and output layer weights ($\boldsymbol{W_{out}}$) are selected. Then, the network is generated using VHDL.

– The rest of the time-series is digitized to 16 bits two's complement that is inserted in the on-chip RAM memory of the FPGA for its processing.

The VHDL code is composed of three parts: RAM memory (containing the input to be processed), the reservoir (constructed using the programmable logic elements of the FPGA), and the output layer (using the dedicated multipliers of the FPGA).

Therefore, the equalization task is a classification problem where deviations from the target signal are acceptable as long as the values are kept within the correct boundaries. The training is performed off-line with a set of 6000 points. Once the network is trained, a VHDL file is generated to test the network with a set of 4000 points digitized to 16 bits two's complement. The quality for the equalization process is measured as the fraction of symbols incorrectly classified with respect the total number of symbols (symbol error rate, SER).

3   Results

We evaluate the performance of the model and compare with previously published works. The area used by the proposed methodology represents an 86% of reduction if compared with an standard digital realization (with full digital multipliers instead of the shift and add blocks). In Table 1 we show the comparison of the logic elements used by the proposed methodology with a conventional digital solution when implemented in an ALTERA FPGA device (Cyclone IV EP4CE115F297C7N). The speed of the system is $f = 1.8 \cdot 10^6$ predictions per second with a power dissipation lower than 1.5 W.

Table 1. Spent logic elements of the Cyclone IV (EP4CE115F297C7N) FPGA for different sizes of the reservoir using a conventional digital solution and the proposed design

Neurons	$N = 50$	$N = 100$	$N = 200$
This work	2.497 (2.2%)	4.947 (4.3%)	9.847 (8.6%)
Conventional	19.147 (16.7%)	37.631 (32.9%)	74.544 (65.1%)

The performance of the channel equalization task is evaluated in terms of the symbol error rate (SER), which is the fraction of misclassified symbols. A sequence of 6000 training and 4000 test symbols is used while the reservoir width is set to $N = 27$. The overall results are shown in Table 2, where the Symbol Error Rate (SER) for four different signal to noise ratio values are considered ($SNR = \{16\,dB, 20\,dB, 24\,dB, 28\,dB\}$). The proposed design is able to decrease the SER value with respect untreated data. These values are similar with other hardware solutions [29]. We compare the obtained with both measurements taken from experimental settings (denoted as Hard. in the table) and purely numerical studies (Soft.). The results provided in [30] do not take into account the intrinsic problems (as system or quantization noise) that is present in experimental settings as in [29] and this work.

Table 2. Symbol error rates for different signal to noise ratio values.

Method	N	16 dB	20 dB	24 dB	28 dB
(Untreated data)		0.167	0.131	0.114	0.098
This work	27	0.0345	0.0095	0.007	0.0042
[29] (Hard.)	246	0.05	0.015	0.006	0.003
[30] (Soft.)	50	0.025	0.006	0.0007	0.00055

4   Conclusions

We have presented an efficient digital implementation of reservoir computing systems for a channel equalization application. The proposed implementation is based on the fact that the reservoir weights can be limited to fixed quantities without decreasing performance too much. The synapses' weights are restricted to combinations of powers of two (sums and subtractions) which allow the use of shift-and-add blocks. The validity of the proposed approach has been demonstrated for the equalization of a nonlinear channel. The proposed model provide considerable advantages with respect conventional digital designs in terms of hardware resources, thus improving the speed and low power characteristics. The accuracy observed is similar than previous models and the proposed design enables the implementation of large networks optimizing hardware [1,2,13,14] with a processing speed of the order of MHz. At the same time, the power dissipation is optimized and the accuracy is similar to other hardware solutions. Therefore, it has been shown that using low-resolution weights for the reservoir do not reduce the system's performance considerably at the time that a substantial hardware reduction is obtained.

Acknowledgment. This work has been partially supported by the Spanish Ministry of Economy and Competitiveness (MINECO), the Regional European Development Funds (FEDER), and the Comunitat Autonoma de les Illes Balears under grant contracts TEC2014-56244-R, TEC2017-84877-R and a fellowship (FPI/1513/2012) financed by the European Social Fund (ESF) and the Govern de les Illes Balears (Conselleria d'Educació, Cultura i Universitats).

References

1. Baptista, D., Abreu, S., Freitas, F., Vasconcelos, R., Morgado-Dias, F.: A survey of software and hardware use in artificial neural networks. Neural Comput. Appl. **23**(3–4), 591–599 (2013)
2. Misra, J., Saha, I.: Artificial neural networks in hardware: a survey of two decades of progress. Neurocomputing **74**(1–3), 239–255 (2010)

3. Amir, M.F., Kim, D., Kung, J., Lie, D., Yalamanchili, S., Mukhopadhyay, S.: NeuroSensor: a 3D image sensor with integrated neural accelerator. In: 2016 SOI-3D-Subthreshold Microelectronics Technology Unified Conference, S3S 2016 (2017)
4. Krizhevsky, A., Sutskever, I., Hinton, G.E.: ImageNet classification with deep convolutional neural networks. Adv. Neural Inf. Process. Syst. 25(NIPS2012), 1–9 (2012)
5. Morro, A., Canals, V., Oliver, A., Alomar, M.L., Galan-Prado, F., Ballester, P.J., Rossello, J.L.: A stochastic spiking neural network for virtual screening (2017)
6. Li, H., Zhang, D., Foo, S.Y.: A stochastic digital implementation of a neural network controller for small wind turbine systems. IEEE Trans. Power Electron. 21(5), 1502–1507 (2006)
7. Chauhan, A., Semwal, S., Chawhan, R.: Artificial neural network-based forest fire detection system using wireless sensor network. In: 2013 Annual IEEE India Conference (INDICON), pp. 1–6 (2013)
8. Raghunathan, S., Gupta, S.K., Ward, M.P., Worth, R.M., Roy, K., Irazoqui, P.P.: The design and hardware implementation of a low-power real-time seizure detection algorithm. J. Neural Eng. 6(5), 056005 (2009)
9. Lee, M., Hwang, K., Park, J., Choi, S., Shin, S., Sung, W.: FPGA-based low-power speech recognition with recurrent neural networks. In: IEEE Workshop on Signal Processing Systems, SiPS: Design and Implementation, pp. 230–235 (2016)
10. Basterretxea, K., Tarela, J.M., del Campo, I.: Digital design of sigmoid approximator for artificial neural networks. Electron. Lett. 38(1), 35–37 (2002)
11. Baptista, D., Morgado-Dias, F.: Low-resource hardware implementation of the hyperbolic tangent for artificial neural networks. Neural Comput. Appl. 23(3–4), 601–607 (2013)
12. Carrasco-Robles, M., Serrano, L.: Accurate differential $\tanh(nx)$ implementation. Int. J. Circuit Theory Appl. 37(5), 613–629 (2009)
13. Nedjah, N., De MacEdo Mourelle, L.: Reconfigurable hardware for neural networks: binary versus stochastic. Neural Comput. Appl. 16(3), 249–255 (2007)
14. Lotrič, U., Bulić, P.: Applicability of approximate multipliers in hardware neural networks. Neurocomputing 96, 57–65 (2012)
15. Lukoševičius, M., Jaeger, H., Schrauwen, B.: Reservoir computing trends. KI - Künstliche Intell. 26(4), 365–371 (2012)
16. Antonelo, E.A., Schrauwen, B.: On learning navigation behaviors for small mobile robots with reservoir computing architectures. IEEE Trans. Neural Netw. Learn. Syst. 26(4), 763–780 (2015)
17. Jalalvand, A., Wallendael, G.V., Walle, R.V.D.: Real-time reservoir computing network-based systems for detection tasks on visual contents. In: Proceedings - 7th International Conference on Computational Intelligence, Communication Systems and Networks, CICSyN 2015, pp. 146–151 (2015)
18. Lin, X., Yang, Z., Song, Y.: Short-term stock price prediction based on echo state networks. Expert Syst. Appl. 36(3 PART 2), 7313–7317 (2009)
19. Alomar, M.L., Canals, V., Perez-Mora, N., Martínez-Moll, V., Rosselló, J.L.: FPGA-based stochastic echo state networks for time-series forecasting. Comput. Intell. Neurosci. 2016 (2016)
20. Alomar, M.L., Soriano, M.C., Escalona-Morán, M., Canals, V., Fischer, I., Mirasso, C.R., Rosselló, J.L.: Digital implementation of a single dynamical node reservoir computer. IEEE Trans. Circuits Syst. II Express Briefs 62(10), 977–981 (2015)
21. Benedetto, S., Biglieri, E.: Principles of Digital Transmission: With Wireless Applications. Kluwer Academic Publishers, Norwell (1999)

22. Boccato, L., Lopes, A., Attux, R., Von Zuben, F.J.: An echo state network architecture based on Volterra filtering and PCA with application to the channel equalization problem. In: Proceedings of the International Joint Conference on Neural Networks, pp. 580–587 (2011)

23. Rossello, J.L., Alomar, M.L., Morro, A., Oliver, A., Canals, V.: High-density liquid-state machine circuitry for time-series forecasting. Int. J. Neural Syst. **26**(5), 1550036 (2016)

24. Rodan, A., Tiño, P.: Minimum complexity echo state network. IEEE Trans. Neural Netw. **22**(1), 131–144 (2011)

25. Jaeger, H., Haas, H.: Harnessing nonlinearity: predicting chaotic systems and saving energy in wireless communication. Science **304**(5667), 78–80 (2004)

26. Bauduin, M., Smerieri, A., Massar, S., Horlin, F.: Equalization of the non-linear satellite communication channel with an Echo state network. In: IEEE Vehicular Technology Conference, vol. 2015 (2015)

27. Mathews, V.J., Lee, J.: Adaptive algorithms for bilinear filtering. In: Proceedings of SPIE - The International Society for Optical Engineering, vol. 2296, pp. 317–327 (1994)

28. Nguimdo, R.M., Verschaffelt, G., Danckaert, J., Van Der Sande, G.: Simultaneous computation of two independent tasks using reservoir computing based on a single photonic nonlinear node with optical feedback. IEEE Trans. Neural Netw. Learn. Syst. **26**(12), 3301–3307 (2015)

29. Ortín, S., Soriano, M.C., Pesquera, L., Brunner, D., San-Martín, D., Fischer, I., Mirasso, C.R., Gutiérrez, J.M.: A unified framework for reservoir computing and extreme learning machines based on a single time-delayed neuron. Scientific Reports 5 (2015)

30. Vinckier, Q., Duport, F., Smerieri, A., Haelterman, M., Massar, S.: Autonomous bio-inspired photonic processor based on reservoir computing paradigm. In: 2016 IEEE Photonics Society Summer Topical Meeting Series, SUM 2016. pp. 183–184 (2016)

Discrete Cosine Transform Spectral Pooling Layers for Convolutional Neural Networks

James S. Smith[1](✉) and Bogdan M. Wilamowski[1,2](✉)

[1] ECE Department, Auburn University, Auburn, AL, USA
jss0036@auburn.edu, wilam@ieee.org
[2] University of IT and Management, Rzeszow, Rzeszów, Poland

Abstract. Pooling operations for convolutional neural networks provide the opportunity to greatly reduce network parameters, leading to faster training time and less data overfitting. Unfortunately, many of the common pooling methods such as max pooling and mean pooling lose information about the data (i.e., they are lossy methods). Recently, spectral pooling has been utilized to pool data in the spectral domain. By doing so, greater information can be retained with the same network parameter reduction as spatial pooling. Spectral pooling is currently implemented in the discrete Fourier domain, but it is found that implementing spectral pooling in the discrete cosine domain concentrates energy in even fewer spectra. Although Discrete Cosine Transforms Spectral Pooling Layers (DCTSPL) require extra computation compared to normal spectral pooling, the overall time complexity does not change and, furthermore, greater information preservation is obtained, producing networks which converge faster and achieve a lower misclassification error.

1 Introduction

In the surging big data era, convolutional neural networks (CNNs) are used with tremendous success as a deep learning technique to process grid-like data such as images [2]. A CNN is a supervised learning technique that applies kernel operators (convolution) as a feature extraction method to data that is then fed into a deep, fully connected Multilayer Perceptron (MLP) neural network. Two high payoff areas to improve convolutional neural networks are computational efficiency and prediction accuracy.

Methods and operations on data in CNNs are implemented as individual layers. Each layer takes as input the output from the previous layer and feeds its processed output into the next layer. The initial layer is the input data and the final layer is class predictions. To train a convolutional neural network, a forward pass is conducted on the entire network starting with the initial layer and ending with the final layer. Loss is calculated and a gradient is passed backwards to each

This work was supported by the National Science Centre, Krakow, Poland, under grant No. 2015/17/B/ST6/01880.

© Springer International Publishing AG, part of Springer Nature 2018
L. Rutkowski et al. (Eds.): ICAISC 2018, LNAI 10841, pp. 235–246, 2018.
https://doi.org/10.1007/978-3-319-91253-0_23

layer in reverse order. During this process, each layer updates its parameters and propagates the gradient backwards to the next layer.

Convolutional layers, used to extract features from data, are the key distinguishing layer that separates CNNs from other artificial neural networks. In a convolutional layer, several kernels are convolved with the input data, with the results being fed into the next layer. The kernel parameters are updated as the gradient is propagated in the backward pass. Several other layers are regularly included in CNNs to improve performance. Rectified Linear Units (ReLUs) are used to avoid the diminishing gradient phenomenon during the backpropagation pass of the training process [4]. Dropout is method used to prevent networks from overfitting data [11]. Pooling layers, such as the most commonly used max pooling layer, are used to down-sample data as a means for parameter reduction and over-training avoidance [10]. Pooling layers are especially important for CNNs because the convolution layer passes a large amount of data forward.

Recently, several innovations have been used to improve efficiency of the convolution process using domain transforms such as the Fast-Fourier Transform (FFT) [7]. A more efficient FFT-based convolution process is implemented in [3] by using the overlap-and-add method. Rippel et al. proposed frequency-domain pooling operations (spectral pooling) in [9] by utilizing the FFT. The work in [5] further extends this concept by mapping the operations and parameters of the entire convolution into the frequency domain.

This paper's proposed network reduces training time while retaining prediction accuracy of current state-of-the-art CNN techniques by thoroughly examining and comparing various implementations of two spectral based techniques, 2-dimensional convolution operations using the FFT and spectral pooling using both the FFT and Discrete Cosine Transformation (DCT). The purpose of these experiments is to find the most effective methods that create a practical network that can be used for common applications such as computer vision, health care, weather forecasting, and natural language, image, and video processing.

DCT has been successfully used in computationally intelligence as both a function approximation method [8] and a network compression method [12]. The network compression method has many similarities to this paper's approach; however, this work does not implement the method in a layer-based approach to interact with standard CNNs and is not compared to similar pooling methods. Rather than subsampling data to improve network generalization, the work gives a practical CNN implementation to compress large network parameters.

This paper is organized as follows. Section 2 reviews spectral representation of images. Section 3 introduces the DCT spectral pooling technique. Section 4 empirically compares different spectral methods for convolution and pooling. Finally, Sect. 5 gives final remarks.

2   Spectral Representation

The spectral domain offers two advantages for training CNNs. The first advantage is that the convolution operation can be sped considerably by performing the convolution in the spectral domain as element-wise multiplication. This is given as the convolution theorem:

$$\mathscr{F}\{f * g\} = \mathscr{F}\{f\} \cdot \mathscr{F}\{g\} \tag{1}$$

The second advantage is that the spectral domain can be used for parameter reduction while preserving great amounts of energy. The reason for this is that energy is often concentrated in only a portion of the entire spectra of the transformed data. Not all of the transformed coefficients are needed to represent the original data for practical purposes.

2.1 Spectral Convolution

The cause for convolution in the spectral domain being faster than convolution in the spatial domain is that the transform to the spectral domain can take advantage of the radix-2 based Cooley-Tukey FFT algorithm [1]. By using the FFT, the time complexity for the respective convolution techniques assuming a image and a $l_1 \times l_2$ image and a $m_1 \times m_2$ filter are:

$$T_{\text{spatial}} = O\left(l_1 l_2 m_1 m_2\right) \tag{2}$$

$$T_{\text{spectral}} = O\left(n_1 n_2 log\left(n_1 n_2\right)\right) \tag{3}$$

where:

$$n_i = l_i + m_1 + 1 \tag{4}$$

Note that the time complexity for spectral convolution is not always better than that of spatial convolution. For very small filter sizes (such as a few pixels), the time complexity for spatial convolution can be advantageous. The primary reason for this is that the FFT algorithm requires both the image and filter to be zero-padded to the dimensions given in (4). Furthermore, it may be faster to use spatial convolution for small image sizes given the overhead computations that must be accounted for in spectral convolution. With these drawbacks being considered, spectral convolution is still powerful because of it will outperform spatial convolution for large, practical-sized problems.

In the case of dealing with large images accompanied by small filters, the overlap-add (OA) convolution method can be used to prevent padding the filter to size $n_1 \times n_2$. This is not practical for the case of Fourier-spectral pooling, but it is practical for using spectral convolution without Fourier-spectral pooling. This method will also be used in the later presented DCT-based spectral pooling.

In order to compare 2d convolution methods, Fig. 1 shows computation time for images of increasing sizes using spatial convolution, spectral convolution implemented with a FFT, and the OA method using Matlab commands. The kernel dimensions for each image is 5% of the image dimensions to demonstrate typical convolution times seen in CNNs, where the filter size is typically considerably smaller than the image size. Pay careful attention to the log scale on the time axis; the spatial convolution grows exponentially faster than the other two methods. OA is more efficient than using the FFT (keep in mind the log scale); however, this advantage is dimensioned as the filter size increases.

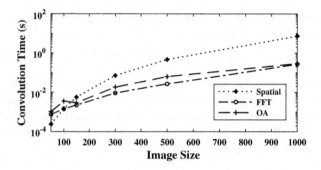

Fig. 1. Training time vs image size (in pixels) for various convolution techniques with filter dimensions as 5% of image dimensions

2.2   Spectral Pooling

Spectral pooling was developed by Rippel et al. [9] to take advantage of the powerful frequency domain properties. Spectral pooling applies a low-pass filter to the spectral representation of a image-kernel product. Simply put, spectral pooling is truncating data in the frequency domain, as described in Algorithm 1. Assume an image with c color channels and a desired dimension size of $p_1 \times p_2$

Algorithm 1. Spectral Pooling

Input: $X \in c \times m_1 \times m_2$ image
Output: $Y \in c \times p_1 \times p_2$
   compute FFT of X
   truncate X to size $c \times p_1 \times p_2$
   return IFFT of X

Spectral pooling is desirable because it can be combined with the convolution theorem to achieve faster results. Specifically, one can combine convolution and pooling by performing the truncation step immediately after the element-wise multiplication in the frequency domain. This is described in the full spectral pooling convolutional layer in Algorithm 2, where k_1 is the total number of images in a batch and k_2 is the number of filters. The CNN can be further optimized by considering the kernels to be in frequency domain initially. That is, there would be no need to transform the kernels at each pass and the kernels would be updated with the transformed propagated gradient. This paper does not include this implementation, but it is a theoretically valid method with considerable potential advantages for convolutional layers with larger filter sizes.

Spectral pooling can retain much more information about the image compared to the commonly used pooling operators such as max pooling and average pooling, which reduce dimensionality by applying the max and average operator, respectively, over small sub matrices contained in each image. Spectral pooling also allows much more variability in pooling amount given that the truncation amount does not have to be a factor of the image size. Another pooling method that utilizes a different type of transformation will be presented in the next section.

Algorithm 2. Spectral Pooling Convolution Layer

Input: $X \in k_1 \times c \times m_1 \times m_2$ image matrix, $F \in k_2 \times c \times l_1 \times l_2$ filter matrix
Output: $Y \in (k_1 \cdot k_2) \times c \times p_1 \times p_2$
  $n_1 = l_1 + m_1 - 1$
  $n_2 = l_2 + m_2 - 1$
  zero-pad filters and images to $k_i \times c \times n_1 \times n_2$
  for all images i **do**
    compute and store FFT of i
  end for
  for all filters f **do**
    compute FFT of f
    for all images i **do**
      multiply FFT(i) with FFT(f)
      truncate result to size $c \times p_1 \times p_2$
      store truncated results
    end for
  end for

A few implementation details must be observed to guarantee correct implementation of this method. One must be careful that the FFT shifts the zero-frequency components to the center so that the cropped spectra can be symmetric about the origin. Otherwise, the results will not represent the original data. In order to calculate the gradient in the backwards pass, the passed gradient from the layer ahead should be zero padded to the original image size (pre-truncation) and then applied the same operator that would be applied in a standard convolutional layer.

3   Discrete Cosine Transform Spectral Pooling

As an alternative to the spectral pooling algorithm, a DCT Spectral Pooling Layer (DCTSPL) is presented to compare to the Fourier-based spectral pooling layer. The DCT gives a spectral representation of data using only cosine functions instead of complex sinusoids. As a result, only real-valued spectral coefficients must be stored; furthermore, more energy is concentrated in even fewer spectra compared to the Fourier representation. It is because of this that the DCT is commonly used for lossy data compression [13].

In order to visualize the motivation behind both Fourier (as done in [9]) and cosine based spectral pooling, the previously discussed pooling methods are compared by applying max pooling, mean pooling, DFT pooling (implemented with a FFT), and DCT pooling to a 2048 × 2048 sized image of a bird on the gulf coast. Figure 2 shows the reconstructed images from the subsampling pooling methods while Figs. 3 and 4 show the spectral representations of the bird image using DFT and DCT, respectively. Figure 2 shows that, for varying degrees of pixel subsampling, the spectral pooling techniques greatly outperform the spatial pooling techniques in terms of information preservation. This can be explained by closely examining Figs. 3 and 4. Observe that most spectral power is focused in the center of the DFT spectrum and in the upper left corner for the DCT spectrum. The significance of this is that the image can be greatly preserved even by truncating many of the spectral components, which is the motivation behind spectral pooling.

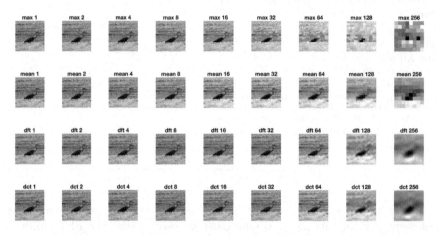

Fig. 2. Image subsampling for image of bird using max pooling (first row), mean pooling (second row), Discrete Fourier Transform (third row) Discrete Cosine Transform (fourth row)

A drawback of the DCT is that no existing radix-2 algorithm exists for its computation (such as the FFT algorithm for the DFT). Instead, the DCT can be implemented by using FFTs, which is slower than performing a single FFT but much faster than the vanilla DCT implementation. A more significant drawback for the purpose of the paper is that convolution cannot be performed using the DCT unless the filter is both real and symmetric. Therefore, the DCT convolutional layer must separate the convolution and pooling processes. The extra computation cost is not too severe given that the OA method can be used for convolution.

The DCT based convolutional layers is given as Algorithm 3. When truncating the spectral representation for DCT, the first p_i desired indexes are kept (this is different than DFT, as shown in Figs. 3 and 4). Compared to the Fourier-based convolutional layer, the DCT convolutional layer is roughly double the number of operations. However, it is still of the same time complexity, and its information preservation makes up for what it loses in training time.

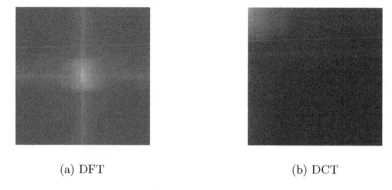

(a) DFT   (b) DCT

Fig. 3. Spectral representation of bird image using (a) Discrete Fourier Transform and (b) Discrete Cosine Transform

---

Algorithm 3. Spectral Pooling Convolution Layer

Input: $X \in k_1 \times c \times m_1 \times m_2$ image matrix, $F \in k_2 \times c \times l_1 \times l_2$ filter matrix
Output: $Y \in (k_1 \cdot k_2) \times c \times p_1 \times p_2$
  $n_1 = l_1 + m_1 - 1$
  $n_2 = l_2 + m_2 - 1$
  zero-pad filters and images to $k_i \times c \times n_1 \times n_2$
  for all filters f **do**
    for all images i **do**
      convolve i with f using OA method
      compute DCT
      truncate result to size $c \times p_1 \times p_2$
      store truncated results
    end for
  end for

---

4   Experiment Results

In order to compare the presented implementations of spectral convolution and pooling, experiments were conducted on both the original (experiment 1) and a modified version (experiment 2) of the MNIST benchmark classification dataset [6]. The original MNIST dataset is composed of 70,000 hand-drawn grey-scale digits of 28×28 pixels in 10 classes ranging from 0 to 9 (60,000 are used for training and 10,000 are used for testing). The second experiment stack 16 identical digits in the same image to create images of 112×112 pixels. The purpose of this is to effectively demonstrate the power of spectral-based techniques because, as previously mentioned, the advantage grows exponentially with image size (with an upfront additional computation cost that outweighs the benefit for small image size).

The CNN was implemented with standard Matlab statements in order to facilitate a rapid development and visualization process. A vanilla CNN library was developed from scratch to enable a fair comparison of new layers (i.e., to

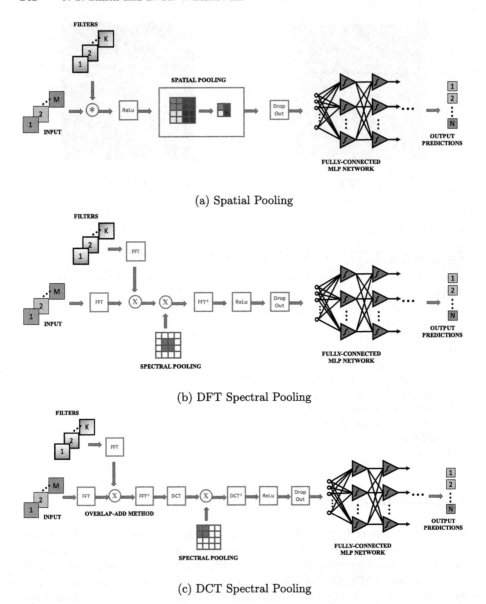

(a) Spatial Pooling

(b) DFT Spectral Pooling

(c) DCT Spectral Pooling

Fig. 4. Convolution neural network architecture implementation for various pooling methods

not compare a vanilla custom layer to a highly optimized layer from a language such as Tensorflow). The spectral convolution and pooling layers were developed based on the given algorithms, using optimized Matlab functions when available. A full-working network trainer software was built on top of the library and used for these experiments.

Table 1. Training results for various pooling methods

Pooling	Stride	Size	Truncation (%)	Reduction (%)	Training time	Classification accuracy (%)
Max	2	2	–	75	182.3	79.0
Mean	2	2	–	75	199.5	78.0
FFT	–	–	75	75	290.6	78.6
DCT	–	–	75	75	10002.2	83.8
Max	4	4	–	93.75	4236.5	70.4
Mean	4	4	–	93.75	4251.1	72.7
FFT	–	–	93.75	93.75	2438.5	73.4
DCT	–	–	93.75	93.75	4056.5	82.5

It is important to observe that this preliminary work is implemented with basic Matlab statements and not all processes are optimized; therefore, the important measurement for this work is classification rate. Many operations in the spectral layers include computationally expensive loops that will be avoided in a future tensor-based implementation. The purpose of these experiments is to show the powerful classification abilities of these layers, given that the theoretical speedup factors have already been presented.

All experiments were implemented with the networks in Fig. 4 with experiment variations being confined to the convolution and pooling layers. The network was trained with the stochastic gradient decent algorithm (batch size = 256) with a momentum rate of 0.9. A probability of 0.05% was used in the dropout layers. The convolutional layer contained 32 kernels of size 3×3. For each experiment trial, the network is trained on 50 epochs of training data and then evaluated on testing data; final results are averaged from ten trials total. It is important to note that hyper-parameters for these experiments were tuned with simple grid searches for each strategy, but this tuning is not presented in these results because it is irrelevant to the purpose of this paper. Rather, these experiments are to compare spectral techniques on a network holding all other settings the same.

The first experiment uses pooling techniques that result in data reduction of 75% (i.e., the number of elements being produced by the pooling layer is 75% of the number of elements being fed into the layer) and the second experiment uses pooling techniques that result in data reduction of 93.75%. Note that the networks are identical if the reduction is 0%. Both experiments were evaluated on classification accuracy and training time. The layer parameters and results are given in Table 1 while the learning curves are given in Fig. 5.

When comparing results, the FFT spectral pooling is the fastest method as the problem size scales up (experiment 2) while the DCT is the most accurate method. The DCT is able to converge faster and to a higher classification rate than the other methods, especially in the second experiment. For this size data,

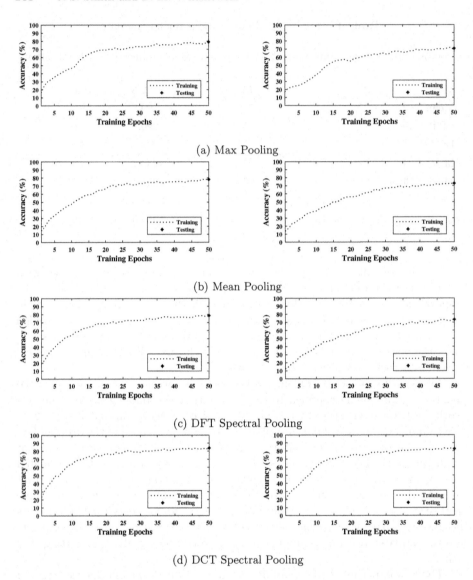

(a) Max Pooling

(b) Mean Pooling

(c) DFT Spectral Pooling

(d) DCT Spectral Pooling

Fig. 5. Training curves for (a) max pooling, (b) mean pooling, (c) FFT spectral pooling, and (d) DCT spectral pooling for experiment 1 (left column) and experiment 2 (right column)

the training times of the spatial and spectral networks are comparable; however, as discussed earlier in Fig. 1, the training time of the spatial networks will grow exponentially compared to that of the spectral networks. Furthermore, it was earlier shown that the DCT network will take longer than the FFT network, but will grow at a linear rate in comparison. This means that the high relative training time is not a major concern; rather, the high classification rate for large layer reduction is a significant achievement.

5 Final Remarks

This paper presents a discrete cosine transform spectral pooling layer (DCTSPL) to be used for subsampling large layer data while retaining maximum information. Convolution times for practical image and filter size combinations are investigate to conclude that spectral convolution outperforms spatial convolution, even for very small filter sizes. The advantage of spectral pooling over max and mean pooling for information retention is visually demonstrated. Finally, experimental results demonstrate that the DCTSPL achieved higher classification rates on a practical dataset compared to networks using the same amount data reduction. Whereas other pooling methods take a significant cut to classification accuracy for data reduction through the pooling layer, the DCT is able to retain greater accuracy.

The natural next step for this work is to continue testing the four pooling techniques on even more datasets using a wider variety of network parameters. It could be that the DCTSPL does not always outperform competing strategies and therefore it should be determined when it is most practical to use. When this layer is further optimized for a tensor-based library, it could be highly desired for problems where a sharp reduction in training time and memory costs is desired for only a small loss in accuracy.

References

1. Cooley, J.W., Tukey, J.W.: An algorithm for the machine calculation of complex Fourier series. Math. Comput. **19**(90), 297–301 (1965)
2. Goodfellow, I., Bengio, Y., Courville, A.: Deep Learning. MIT Press, Cambridge (2016)
3. Highlander, T., Rodriguez, A.: Very efficient training of convolutional neural networks using fast Fourier transform and overlap-and-add. arXiv:1601.06815 [cs], January 2016
4. Jarrett, K., Kavukcuoglu, K., Ranzato, M., LeCun, Y.: What is the best multi-stage architecture for object recognition? In: 2009 IEEE 12th International Conference on Computer Vision, pp. 2146–2153, September 2009
5. Ko, J.H., Mudassar, B., Na, T., Mukhopadhyay, S.: Design of an energy-efficient accelerator for training of convolutional neural networks using frequency-domain computation. In: Proceedings of the 54th Annual Design Automation Conference 2017, DAC 2017, pp. 59:1–59:6. ACM, New York (2017)
6. Lecun, Y., Bottou, L., Bengio, Y., Haffner, P.: Gradient-based learning applied to document recognition. Proc. IEEE **86**(11), 2278–2324 (1998)
7. Mathieu, M., Henaff, M., LeCun, Y.: Fast training of convolutional networks through FFTs. arXiv:1312.5851 [cs], December 2013

8. Olejczak, A., Korniak, J., Wilamowski, B.M.: Discrete cosine transformation as alternative to other methods of computational intelligence for function approximation. In: Rutkowski, L., Korytkowski, M., Scherer, R., Tadeusiewicz, R., Zadeh, L.A., Zurada, J.M. (eds.) ICAISC 2017. LNCS (LNAI), vol. 10245, pp. 143–153. Springer, Cham (2017). https://doi.org/10.1007/978-3-319-59063-9_13

9. Rippel, O., Snoek, J., Adams, R.P.: Spectral representations for convolutional neural networks. arXiv:1506.03767 [cs, stat], June 2015

10. Scherer, D., Müller, A., Behnke, S.: Evaluation of pooling operations in convolutional architectures for object recognition. In: Diamantaras, K., Duch, W., Iliadis, L.S. (eds.) Part III, ICANN 2010. LNCS, vol. 6354, pp. 92–101. Springer, Heidelberg (2010). https://doi.org/10.1007/978-3-642-15825-4_10

11. Srivastava, N., Hinton, G., Krizhevsky, A., Sutskever, I., Salakhutdinov, R.: Dropout: a simple way to prevent neural networks from overfitting. J. Mach. Learn. Res. **15**, 1929–1958 (2014)

12. Wang, Y., Xu, C., You, S., Tao, D., Xu, C.: CNNpack: packing convolutional neural networks in the frequency domain. In: Lee, D.D., Sugiyama, M., Luxburg, U.V., Guyon, I., Garnett, R. (eds.) Advances in Neural Information Processing Systems, vol. 29, pp. 253–261. Curran Associates, Inc., Red Hook (2016)

13. Watson, A.B.: Image compression using the discrete cosine transform. Math. J. **4**(1), 81–88 (1994)

Extreme Value Model for Volatility Measure in Machine Learning Ensemble

Ryszard Szupiluk and Paweł Rubach[(✉)]

Warsaw School of Economics, Al. Niepodleglosci 162, 02-554 Warsaw, Poland
{rszupi,prubac}@sgh.waw.pl

Abstract. This paper presents a method of model aggregation using multivariate decompositions where the main problem is to properly identify the components that carry noise. We develop a volatility measure which uses generalized extreme value decomposition. It is applied to destructive and constructive latent component classification. A practical experiment was conducted in order to validate the effectiveness of the introduced method.

Keywords: Prediction · Blind separation · Ensemble models · Noise detection
Theta noise measure

1 Introduction

The popularity of predictor aggregation approaches is a natural consequence of the development of modeling methods in such disciplines as data mining, machine learning, artificial intelligence or nonparametric modeling. The main idea of model aggregation is to combine the knowledge derived from different existing models. However, this seemingly natural concept is not fully realized by the most popular aggregation methods such as bagging or boosting, in which a standard approach means that essentially the same models are aggregated and the only difference is related to the learning process that is accomplished using different subsets of training data [1–3]. Such methods do not fully realize the basic idea of model aggregation, in which we assume that we compose combinations of diverse sources of information.

These restrictions do not apply to the aggregation approach based on Blind Signal Separation (BSS) methods [4, 5]. In this approach, we treat the set of prediction results generated by different models as one multidimensional variable. We assume that this variable contains hidden constructive and destructive components. We expect that these hidden underlying components have a "physical nature", and that the elimination of such a destructive component should result in improved forecasts measured by different criteria. This aspect is one of the key differences from the "pure" mathematical optimization of the selected single criterion of error typically used in other aggregation methods. In this wide framework, we propose novel extensions associated with the identification of destructive components. This method is addressed for data with time structure e.g. signals or time series.

© Springer International Publishing AG, part of Springer Nature 2018
L. Rutkowski et al. (Eds.): ICAISC 2018, LNAI 10841, pp. 247–256, 2018.
https://doi.org/10.1007/978-3-319-91253-0_24

2 Prediction Results Improvement

Let us assume the existence of a set of m prediction results such as x_i, where $i = 1, \ldots, m$. These results will be collected in one multidimensional variable: $\mathbf{x} = [x_1, \ldots, x_m]^T$. We can say that a given result is a combination of hidden constructive components $\hat{s}_j, j = 1, \ldots, r$, and destructive components that make the predicted results differ from target $\tilde{s}_l, l = 1, \ldots, h$. In the case of a linear mixing scheme we have

$$\mathbf{x}(k) = \mathbf{A}\mathbf{s}(k), \qquad (1)$$

where k is the observation number or time index, the matrix $\mathbf{A} = [a_{ij}] \in R^{m \times n}$ represents the mixing system and the vector $\mathbf{s}(k) = [\hat{s}_1(k), \ldots, \hat{s}_r(k), \tilde{s}_{r+1}(k). \ldots, \tilde{s}_{r+h}(k)]^T$ represents the set of latent base components where $n = r + h$. For simplicity we ordered components in blocks of destructive and constructive ones. We also assume that $m = n$, matrix $\mathbf{A}$ is nonsingular and $E\{\mathbf{s}\} = \mathbf{0}$, where $E\{.\}$ is the expected value operator. We can see that by identifying the mixing system $\mathbf{A}$ and the base components $\mathbf{s}$ and next by eliminating the destructive components (assuming respectively $\tilde{s}_l = 0$) we get

$$\hat{\mathbf{x}}(k) = \mathbf{A}[\hat{s}_1(k), \ldots, \hat{s}_r(k), 0_{r+1}(k) \ldots, 0_n(k)]^T, \qquad (2)$$

where $\hat{\mathbf{x}} = [\hat{x}_1, \ldots, \hat{x}_m]^T$ is the improved version of the prediction results x. We can also look at the whole process as an attempt to separate unknown base components mixed in an unknown system. This leads directly to the type of multivariate decompositions used in Blind Signal Separation or Blind Source Separation (BSS) [6, 7]. Consequently, the problem can be represented as looking for an inverse transformation to (1) such that

$$\mathbf{y}(k) = \mathbf{W}\mathbf{x}(k) \approx \mathbf{s}(k), \qquad (3)$$

where the matrix $\mathbf{W} = \mathbf{A}^{-1}$ is the inversion of the $\mathbf{A}$ matrix. Many solutions used for the identification of $\mathbf{W}$ use different forms of decorrelation (linear, nonlinear, time delay etc.) resulting in different numerical algorithms. There are many BSS methods of finding $\mathbf{W}$ such as Independent Component Analysis (ICA), Smooth Component Analysis (SmCA) and AMUSE or SOBI algorithms [4, 6, 7].

As a result, the whole procedure takes the following form:

1. Make predictive models and gather their results in one variable.
2. Transform (decompose) the results of the blind separation prediction by obtaining latent base components (sources).
3. Identify the destructive components and then eliminate them (replace them with zeros) to obtain an improved set of base components.
4. For corrected tapered components, make a reverse transformation to the separating transformation from step 2.
5. Check the quality of the solutions obtained, in case the improvement is unsatisfactory repeat the whole procedure while changing the decomposition method or component system classification.

It should also be noted that the described process is the basic version of BSS aggregation. It can be significantly expanded by adding many stages of different decompositions as well as by replacing (2) nonlinear transformations with neural networks [4].

However, even in the basic form, the abovementioned conceptual BSS aggregation scheme is an idealized one. Assuming its proper implementation, we could have expected that the destructive components would have some physical character – according to BSS nature. We understand thereby that in this separation process the distortions related to specific distorting factors such as noise in the training data, not fully adequate predictive models or optimization inaccuracies will be identified. As a result, the elimination of such "physical" factors should lead to improved prediction according to all assessment criteria. This approach significantly differs from other methods where distortions are defined *a priori*, as mathematical white or color noise with given characteristics [9]. Of course, in fact, our ideal model in its pure form can rarely be fully realized.

In practice, assumptions about the static and linear forms of the mixing model and the choice of a particular type of separation method, e.g. ICA, may not be fully adequate. Actual components may be mixed in a non-linear or dynamic manner, and therefore, the most appropriate methods of distinction may be the ones based on non-negative, smoothness, or sparse criteria. Consequently, the obtained components may not be pure white or color noise. As a result, the standard methods of noise identification based on correlation, spectral or R/S analysis are not always effective. Therefore, there is a need for the development of new methods [5, 8].

3   Volatility as Destructive Components Indicator

We will now propose a method of identifying destructive components for cases when the data contains a time structure (signals, time series). The analyzed characteristic will be the variability (it can also be understood as smoothness) understood as a kind of difference in successive values of the signal in relation to the maximum range of fluctuations. In general, we can define it as the following expression

$$\Theta^{(p,q)}(y) = \frac{1}{N} \frac{\sum_{k=q+1}^{N} |y(k) - y(k-q)|^p}{I((\max(y) - \min(y))^p)}, \tag{4}$$

where

$$I(u) = \begin{cases} u \text{ for } u \neq 0 \\ 1 \text{ for } u = 0 \end{cases}, \tag{5}$$

and q is a chosen nonnegative integer value, and p some positive real value.

The interpretation of the Theta measure (4) is relatively simple and intuitive. We obtain the average signal q order differences to the power of p with respect to the square of the fluctuation range in the same power. The function $I(u)$ acts as a zero indicator and prevents division by zero for a constant signal. For $q = 1$ the value reaches its maximum when the signal changes are equal to the range of variation, and the minimum value is obtained for a constant signal. For expression (4) there are also deeper theoretical consequences, for example for $p = 1$ assuming that $u(k) = y(k + 1) - y(k)$ and approximating the expression $|y|$ with the function $\log(\cosh(y))$ we obtain $1/N\{\sum \log(\cosh(u))\} \approx E\{\log(\cosh(u))\}$ effectively the volatility measure (4) for $p = q = 1$ can be expressed as:

$$\Theta^{(1,1)}(y) \approx E\{\log(\cosh(u))\}I((\max(y) - \min(y))^p)^{-1}. \tag{6}$$

The function $E\{\log(\cosh(u))\}$ is widely used in the context of machine learning based on the maximization (or minimization) of entropy and negentropy [7]. Consequently, the $\Theta^{(p,q)}$ measure fits well with the broadly understood theory of information and machine learning.

To illustrate the performance of the proposed measure $\Theta^{(2,1)}$ its value characteristics of the noise level for the case of a sinusoid with Gaussian noise are presented in Fig. 1. The noise level is illustrated using the Signal-to-noise ratio (SNR).

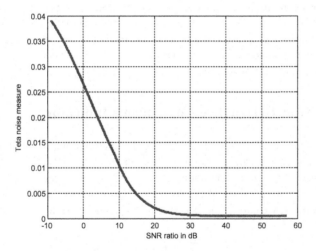

Fig. 1. Theta noise measure vs. SNR for the sinusoid signal with Gaussian noise.

One of the main constraints associated with the measure (4) is its high sensitivity to properly determining the extreme values of signals. This means that the results may be strongly dependent on the choice of a particular learning sample causing potential difficulties with the correct generalization of the modeling process. To overcome this sensitivity to direct observations, in place of extreme empirical values, the model extreme values determined from generalized extreme value distribution (GEVD) will be used [10, 11]. The probability density function, the generalized extreme value distribution, with the position parameter μ, the scale parameter σ, and the shape parameter $\gamma \neq 0$ are given as

$$f(z) = \frac{1}{\sigma}\left(1 + \gamma\frac{z - \mu}{\sigma}\right)^{-1-\frac{1}{\gamma}} \exp\left(-\left(1 + \gamma\frac{z - \mu}{\sigma}\right)^{-1/\gamma}\right), \tag{7}$$

where $1 + \gamma\frac{z - \mu}{\sigma} > 0$, and $\gamma > 0$ (distribution type II) or $\gamma < 0$ (distribution type III). For $\gamma = 0$ (distribution type I) GEVD is given as

$$f(z) = \frac{1}{\sigma} \exp\left(-\exp\left(\frac{z - \mu}{\sigma}\right) - \frac{z - \mu}{\sigma}\right). \tag{8}$$

Distribution parameters (8) can be estimated using the moment method which leads to $\tilde{\mu} = \pi^{-1}s\sqrt{6}$ and $\sigma = \bar{z} - 0.5772\tilde{\mu}$ where $\bar{z}$ and s are the mean and standard deviation of the sample.

The method of determining the signal's volatility using GEVD is as follows:

1. From the observed signal, we randomly draw (using the bootstrap method) samples and determine their minimums as the value of the variable z,
2. Assuming that z is the realization of the distribution of GEVD $f(z)$, we find the parameter μ_-,
3. We determine their maxima for a set of randomly selected samples and find the parameter μ_+ of the function $f(-z)$,
4. We regulate y to y_{reg} by adjusting the extreme values to the interval $[\mu-, \mu+]$,
5. We calculate $\Theta^{(p,q)}(y)$ for the regularized signal.

$$\Theta^{(p,q)}(y_{reg}) = \frac{1}{N}\frac{\sum\limits_{k=q+1}^{N}|y(k) - y(k - q)|^p}{I\left((\mu_+ - \mu_-)^p\right)}. \tag{9}$$

The aforementioned algorithm and the Theta signal noise measure have been assessed using a practical experiment that is described in the following section.

4   Practical Experiment

This experiment presents the effectiveness of the proposed method for improving the prediction of short-term electricity consumption. For this purpose, six MLP neural models optimized with the Levenberg-Marquardt (LM) algorithm were used. Each of the models was first used for prediction separately and the prediction error was measured using MSE and MAPE. These error rates are presented in Table 1.

Table 1. Prediction error rates for primary models.

Comp.	x_1	x_2	x_3	x_4	x_5	x_6
MSE (10^{-4})	8.5616	8.3066	6.9451	8.2068	6.2407	8.8686
MAPE (10^{-2})	2.1454	2.1408	1.9121	2.1108	1.8243	2.1855

The next step was the decomposition process realized using the AMUSE RS Blind Signal Separation algorithm, the result of which, are the 6 base components $y_1 \ldots y_6$. These components are now analyzed using the proposed Theta noise measure (4) to identify among them the ones which have a negative influence on the prediction and eliminate them (by replacing them with zeros). Table 2 contains the computed values of $\Theta^{(2,1)}$ and $\Theta^{(4,1)}$ for each component $y_1 \ldots y_6$.

Table 2. Theta noise measure (4) for each base component.

	y_1	y_2	y_3	y_4	y_5	y_6
$\Theta^{(2,1)}(y)(10^{-2})$	0.40	0.78	0.86	1.18	1.25	1.11
$\Theta^{(4,1)}(y)(10^{-2})$	0.00759	0.0415	0.0363	0.0653	0.0695	0.0822

The results show that the first component y_1 stands out as having not only the lowest value of the Theta noise measures but also a significantly lower value than any other base component. This suggests that the values of y_1 have a lower noise than other base components and should influence the final prediction stronger than any other component. This hypothesis has been confirmed by the experiment. Several combinations of elimination of individual components were analyzed and in all of the cases where y_1 was chosen as one of the components to be eliminated, the final prediction error rates increased. This is shown in form of highly positive values for all the combinations involving elimination of y_1 in Tables 3 and 4. These tables contain the MAPE and MSE prediction error rates achieved for cases when pairs of base components y_x and y_z identified by the row number x and column number z are eliminated. The cases shown on the diagonals of Tables 3 and 4 represent the elimination of only one base component.

Table 3. MAPE average error reduction rate in case of the elimination of every one of the base components (on diagonal) and every combination of 2 base components.

	y_1	y_2	y_3	y_4	y_5	y_6
y_1	4792.90%	4792.72%	4792.93%	4793.04%	4792.87%	4792.67%
y_2	4792.72%	−1.39%	−5.42%	−2.77%	−2.76%	−2.86%
y_3	4792.93%	−5.42%	−4.08%	−5.39%	−5.51%	−5.65%
y_4	4793.04%	−2.77%	−5.39%	−1.24%	−2.56%	−2.60%
y_5	4792.87%	−2.76%	−5.51%	−2.56%	−1.34%	−2.63%
y_6	4792.67%	−2.86%	−5.65%	−2.60%	−2.63%	−1.39%

The last 3 components (y_4, y_5, y_6) have a substantially larger value of the Theta noise measure (seen in Table 2) both computed for $p = 2$ and $p = 4$; therefore, there is a greater probability to find noise or destructive components among those 3. An arbitrary decision was made to eliminate 2 out of those 3 base components. All possible combinations were analyzed (Tables 3 and 4) and it turned out that all of them lead to smaller average MSE and MAPE prediction error rates than the ones computed for original signals. For the chosen elimination of y_5 and y_6 the error rates for the initial prediction as well as the "corrected" ones, measured using MSE and MAPE respectively, are depicted in Fig. 2. The average error rate reductions achieved using this method amount to 2.63% using MAPE and 5.53% using MSE respectively.

Table 4. MSE average error reduction rate in case of the elimination of every one of the base components (on diagonal) and every combination of 2 base components.

	y_1	y_2	y_3	y_4	y_5	y_6
y_1	154860.00%	154857.75%	154852.80%	154857.99%	154857.94%	154857.12%
y_2	154857.75%	−2.55%	−10.04%	−4.86%	−4.91%	−5.73%
y_3	154852.80%	−10.04%	−7.49%	−9.80%	−9.85%	−10.67%
y_4	154857.99%	−4.86%	−9.80%	−2.31%	−4.66%	−5.48%
y_5	154857.94%	−4.91%	−9.85%	−4.66%	−2.36%	−5.53%
y_6	154857.12%	−5.73%	−10.67%	−5.48%	−5.53%	−3.18%

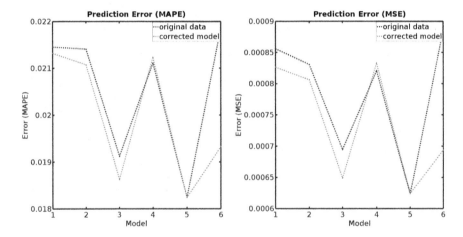

Fig. 2. Prediction error rates for original and "corrected" data for the elimination of y_5 and y_6.

The best improvements of error rates for individual base components in the analyzed example reach 11.57% in case of MAPE and 21.75% in case MSE.

The last part of the experiment involved testing the performance of the Theta noise measure (4) for different values of the q parameter which is responsible for the delay between compared values of the input data, that is base components $y_1 \ldots y_6$ in our case.

Figure 3 demonstrates the values of Theta for each base component as well as its average computed for $p = 2$ depending on the value of the delay q. Figure 4 is similar, however, it depicts the Thetas calculated for $p = 4$.

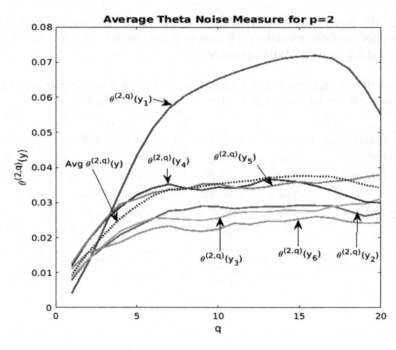

Fig. 3. Theta noise measure for $p = 2$ each base component $y_1 \ldots y_6$ depending on the delay q.

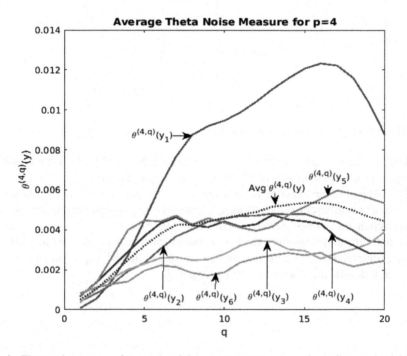

Fig. 4. Theta noise measure for $p = 4$ each base component $y_1 \ldots y_6$ depending on the delay q.

As one can clearly see in both figures, the values of the Theta noise measure (4) computed for the first base component y_1 differ significantly from those calculated for all other components $y_2 \ldots y_6$. Taking into account the previous results of the experiment that show evidence of y_1 being definitely constructive for the prediction, one should note that the proposed Theta noise measure performs well as a means of identifying constructive and destructive components. Additionally, the Figs. 3. and 4. show that good results may be obtained by manipulating the measure's both parameters – p as well as q.

5   Conclusions

This paper presents a method of improving prediction results with blind separation decompositions in the context of noise detection. The proposed GEVD model generalizes our basis volatility measure. Destructive components, usually identified with noise, may have characteristics far removed from the mathematical definition of iids (identically independent distributed). In many cases, the components that are actually destructive in terms of statistical characteristics differ only slightly from the constructive components. Moreover, the evaluation of a single component can be misleading because of its interaction with other base components. As a result, a component rated under the *a priori* criteria as noise or destructive may interact with other components and become constructive. Consequently, in principle, the only way to uniquely determine the nature of individual components is to test the prediction while eliminating all possible combinations of individual components. Of course, in the case of a large number of models, and consequently a large number of possible base components, the combinatorial test is impossible, leading to the choice of statistical *a priori* criteria. One such new concept of noise evaluation was proposed in this paper. The results of the experiment show the performance of the volatility measure introduced. The experiment provided evidence that the proposed Theta noise measure enables the identification of constructive and destructive components based on both the measure's p exponent as well as the q parameter responsible for the delay.

References

1. Breiman, L.: Bagging predictors. Mach. Learn. **24**, 123–140 (1996)
2. Drucker, H.: Improving regressors using boosting techniques. In: ICML, vol. 97, pp. 107–115 (1997)
3. Hoeting, J., Madigan, D., Raftery, A., Volinsky, C.: Bayesian model averaging: a tutorial. Stat. Sci. **14**, 382–417 (1999)
4. Szupiluk, R., Wojewnik, P., Ząbkowski, T.: Prediction improvement via smooth component analysis and neural network mixing. In: Kollias, S., Stafylopatis, A., Duch, W., Oja, E. (eds.) ICANN 2006. LNCS, vol. 4132, pp. 133–140. Springer, Heidelberg (2006). https://doi.org/10.1007/11840930_14
5. Szupiluk, R., Wojewnik, P., Zabkowski, T.: Noise detection for ensemble methods. In: Rutkowski, L., Scherer, R., Tadeusiewicz, R., Zadeh, L.A., Zurada, J.M. (eds.) ICAISC 2010. LNCS (LNAI), vol. 6113, pp. 471–478. Springer, Heidelberg (2010). https://doi.org/10.1007/978-3-642-13208-7_59

6. Comon, P., Jutten, C.: Handbook of Blind Source Separation: Independent Component Analysis and Applications. Academic Press, Cambridge (2010)
7. Hyvärinen, A.: Independent component analysis: recent advances. Philos. Trans. R. Soc. A **371**, 20110534 (2013)
8. Szupiluk, R., Wojewnik, P., Zabkowski, T.: The noise identification method based on divergence analysis in ensemble methods context. In: Dobnikar, A., Lotrič, U., Šter, B. (eds.) ICANNGA 2011. LNCS, vol. 6594, pp. 206–214. Springer, Heidelberg (2011). https://doi.org/10.1007/978-3-642-20267-4_22
9. Vasegi, S.V.: Advanced Signal Processing and Digital Noise Reduction. Wiley, Chichester (2008)
10. Evans, M., Hastings, N., Peacock, B.: Statistical Distributions, 3rd edn. Wiley, Hoboken (2000)
11. McFadden, D.: Modeling the choice of residential location. Transp. Res. Rec. **673**, 72–77 (1978)

Deep Networks with RBF Layers
to Prevent Adversarial Examples

Petra Vidnerová[(✉)] and Roman Neruda

Institute of Computer Science, Czech Academy of Sciences,
Pod vodárenskou věží 2, Praha 8, Prague, Czech Republic
{petra,roman}@cs.cas.cz

Abstract. We propose a simple way to increase the robustness of deep neural network models to adversarial examples. The new architecture obtained by stacking deep neural network and RBF network is proposed. It is shown on experiments that such architecture is much more robust to adversarial examples than the original one while its accuracy on legitimate data stays more or less the same.

Keywords: Adversarial examples · RBF networks
Deep neural networks · Convolutional networks

1 Introduction

Deep neural networks (DNN) and convolutional neural networks (CNN) enjoy high interest nowadays. They have become the state-of-art methods in many fields of machine learning, and have been applied to various problems, including image recognition, speech recognition, and natural language processing [1].

In the area of pattern recognition, deep and convolutional neural networks achieved several human-competitive results [2–4]. Concerning these results, there is a question if these methods achieve similar capabilities to human vision, such as a generalization. This paper deals with a property of machine learning models that demonstrates a difference. Let us have a classifier and an image, correctly classified by the classifier as one class (for example an image of a hand-written digit five). It is possible to slightly change the image, so as for human eyes, there is almost no difference, but the classifier classifies the image as something completely else (such as digit zero).

This counter-intuitive property of neural networks was first described in [5]. It relates to the stability of a neural network with respect to small perturbation of their inputs. Such perturbed examples are known as *adversarial examples*. The adversarial examples differ only slightly from correctly classified examples drawn from the data distribution, but they are classified incorrectly by the classifier learned on the data. Not only they are classified incorrectly, they can often be classified as a class of our choice.

The vulnerability to adversarial examples is not only the case of deep neural network models, but spreads through all machine learning methods, including

© Springer International Publishing AG, part of Springer Nature 2018
L. Rutkowski et al. (Eds.): ICAISC 2018, LNAI 10841, pp. 257–266, 2018.
https://doi.org/10.1007/978-3-319-91253-0_25

shallow architectures (like SVMs) or decision trees. Networks with local units, RBF networks, are known to be more robust to adversarial examples. In this paper we examine the way of using RBF layers in deep architecture to protect the architecture from adversarial examples. We propose the new architecture obtained by stacking the deep architecture and an RBF network. We show that such a model is much less vulnerable to adversarial examples than the original model, while its accuracy remains almost the same.

This paper is organized as follows. First, in Sect. 2 we explain how adversarial examples work and review related work. Then, Sect. 3 introduces the new architecture. Section 4 deals with the results of our experiments. Finally, Sect. 5 concludes our paper.

2   Adversarial Examples and Related Work

The adversarial examples were first introduced in [5]. The paper shows that having a trained network it is possible to arbitrarily change the network prediction by applying an imperceptible non-random perturbation to an input image. Such perturbations are found by optimizing the input to maximize the prediction error. The box-constrained Limited-memory Broyden-Fletcher-Goldfarb-Shanno algorithm (L-BFGS) is used for this optimization.

On some data sets, such as ImageNet, the adversarial examples are so close to the original examples that they are indistinguishable by human eye. In addition, the authors state that adversarial examples are relatively robust, and they generalize between neural networks with varied number of layers, activations, or trained on different subsets of the training data. In other words, if we use one neural network to generate a set of adversarial examples, these examples are also misclassified by another neural network even when it was trained with different hyperparameters, or when it was trained on a different subset of examples.

Paper [6] suggests that it is the linear behaviour in high-dimensional spaces that is sufficient to cause adversarial examples (for example, a linear classifier exhibits this behaviour, too). The authors propose a fast method of generating adversarial examples (adding small vector in the direction of the sign of the derivation).

Let us have a linear classifier and let x and $\tilde{x} = x + \eta$ be input vectors. The classifier should assign x and $\tilde{x}$ same classes as long as $||\eta||_\infty \leq \varepsilon$, where ε is the precision of features.

Consider the dot product between weight vector w and input vector $\tilde{x}$:

$$w^\top \tilde{x} = w^\top x + w^\top \eta.$$

Adding η to the input vector, the activation increases by $w^\top \eta$. We can maximize this increase by $\eta = \varepsilon sign(w)$. If n is the dimension and m is the average magnitude of w, the activation grows by εmn. Note that $||\eta||_\infty$ does not grow with n, but the change in activation caused by perturbation η does grow linearly. It is possible to make many infinitesimal changes to the input that add up to a large change of the activation. Therefore, a simple linear model can have adversarial examples if its input has sufficient dimensionality.

The above observation can be generalized to nonlinear models [6]. Let θ be the parameters of a model, x an input, y the targets for x, and $J(\theta, x, y)$ the cost function. If we linearize the cost function around the θ, we obtain an optimal perturbation: $\eta = \varepsilon\ \mathrm{sgn}(\bigtriangledown_x J(\theta, x, y))$. This represents an efficient way of generating adversarial examples and it is referred to as the *fast gradient sign method* (FGSM). See Fig. 1 for adversarial images crafted by FGSM on MNIST data set [7] for CNN.

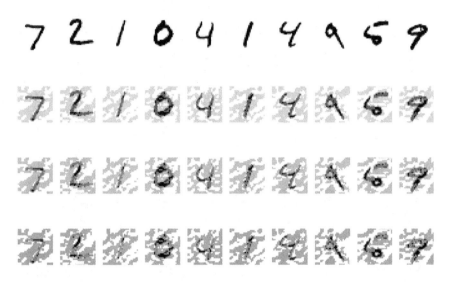

Fig. 1. Original test examples and corresponding adversarial examples crafted by FGSM with ϵ 0.2, 0.3, and 0.4.

Other results of fooling deep and convolutional networks can be found in [8]. This paper studies the generation of images looking as noise or regular patterns by evolutionary algorithms. To generate regular patterns the authors use Compositional pattern-producing network (CPPN), that has similar structure as neural networks. It takes indexes (x, y) as an input and outputs pixel value. Nodes are functions like Gaussian, sinus, sigmoid, linear. The CPPNs are created by evolutionary algorithms, and the resulting images are regular patterns that are classified as desired images from the training set with high confidence.

In [9] another class of crafting algorithms is proposed, and in [10] a black-box strategy to adversarial attacks is described.

In our paper [11], we examine a vulnerability to adversarial examples throughout variety of machine learning methods. We propose a genetic algorithm for generating adversarial examples. Though the evolutionary search for adversarial examples is slower than techniques described in [5,6], it enables us to obtain adversarial examples without the access to model's weights. Thus, we have a unified approach for a wide range of machine learning models, including not only neural networks, but also support vector machine classifiers (SVMs), decision trees, and possibly others. The only thing this approach needs is the possibility to query the classifier to evaluate a given example. See Fig. 2 for adversarial images crafted by our genetic algorithm for CNN.

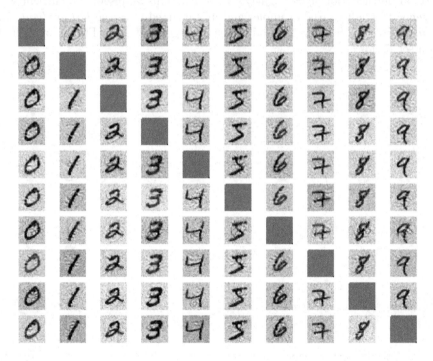

Fig. 2. Adversarial examples crafted by GA. Images on the first line are all classified as zero by the target CNN, images on the second line as one, etc.

The question of how to make the neural networks robust to adversarial examples is dealt with in [12]. The authors tried several methods, from noise injection and Gaussian blur, using autoencoder, to method they call *deep contractive network* (that applies a regularization term penalizing large changes of activation in respect to change of input to the cost function). However, the methods cure the adversarial examples only to some extend.

Another attempt to prevent adversarial examples is proposed in [13], based on distillation, i.e. training another network based on outputs produced by target network.

3   Deep Networks with RBF Layers

RBF networks [14–18] are neural networks with one hidden layer of RBF units and a linear output layer.

By an *RBF unit* we mean a neuron with multiple real inputs $\boldsymbol{x} = (x_1, \ldots, x_n)$ and one output y. Each unit is determined by n-dimensional vector $\boldsymbol{c}$ which is called *centre*. It can have additional parameter $\beta > 0$ that determines its width. The output y is computed as:

$$y = \varphi(\xi); \quad \xi = \beta ||\boldsymbol{x} - \boldsymbol{c}||^2 \tag{1}$$

where $\varphi : \mathbb{R} \to \mathbb{R}$ is suitable activation function, typically Gaussian $\varphi(z) = e^{-z^2}$.

Thus, the network computes the following function $\boldsymbol{f} : \mathbb{R}^n \to \mathbb{R}^m$:

$$f_s(\boldsymbol{x}) = \sum_{j=1}^{h} w_{js} \varphi \left(\beta_j \parallel \boldsymbol{x} - \boldsymbol{c_j} \parallel \right), \tag{2}$$

where $w_{ji} \in \mathbb{R}$ and f_s is the output of the s-th output unit.

The history of RBF networks can be traced back to the 1980s, particularly to the study of interpolation problems in numerical analysis. It is where the radial basis functions were first introduced, in the solution of the real multivariate interpolation problem [19,20].

The RBF networks benefit from a rich spectrum of learning possibilities. The study of these algorithms together with experimental results was also published in our papers [21,22].

With the boom of deep learning the popularity of RBF networks vanishes. However, we show that they can bring advantages when combined with deep neural networks.

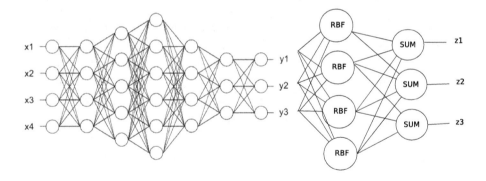

Fig. 3. Deep neural network architecture followed by RBF network.

We introduce new deep architecture that is defined as a concatenation of a feedforward deep neural network and an RBF network (see Fig. 3). Let us have a deep neural network DN that realizes a function $f_{DN} : \mathbb{R}^n \to \mathbb{R}^m$ and a RBF network RBF that realizes a function $f_{RBF} : \mathbb{R}^m \to \mathbb{R}^m$. Then feeding the outputs of DN to inputs of RBF we get a network implementing function: $f : \mathbb{R}^n \to \mathbb{R}^m$, where

$$f(\boldsymbol{x}) = f_{RBF}(f_{DN}(\boldsymbol{x})).$$

For classification tasks we can add softmax activation function to the output layer of RBF network.

The training procedure is the following:

1. train the *DN* by any appropriate learning algorithm
2. set the centers of *RBF* randomly, drawn from uniform distribution on $(0, 1.0)$
3. set the parameters β to the constant value
4. initialize the weights of RBF output layer to random small values
5. retrain the whole network DNRBF (by back propagation).

While the DN part of the network is already trained, it is usually sufficient to train the whole stacked network only for few epochs.

4   Experimental Results

For our experiments we use the FGSM implemented in Cleverhans library [23]. To implement deep neural networks we use Keras [24] and our RBF layer implementation [25]. The scripts used for experiments can be found at [26].

Table 1. Accuracies on legitimate test examples and adversarial examples for MLP and MLPRBF with various initial widths. Average accuracies over 30 runs of learning algorithm.

Model	Legitimate samples				Adversarial samples			
	Mean	Std	Min	Max	Mean	Std	Min	Max
MLP	**98.35**	**0.12**	**98.04**	**98.59**	**1.95**	**0.41**	**1.30**	**2.86**
MLPRBF(0.01)	97.62	2.43	88.44	98.65	2.56	2.09	1.16	10.71
MLPRBF(0.1)	88.61	8.56	69.91	98.36	10.04	6.45	1.71	23.10
MLPRBF(1.0)	98.23	0.10	98.08	98.48	81.77	7.84	64.18	94.06
MLPRBF(2.0)	**98.19**	**0.14**	**97.91**	**98.38**	**89.21**	**5.03**	**66.28**	**94.83**
MLPRBF(3.0)	98.18	0.14	97.88	98.45	81.66	4.38	70.13	87.23
MLPRBF(5.0)	97.64	2.09	89.34	98.36	69.47	13.26	13.01	81.95
MLPRBF(10.0)	80.94	11.82	58.57	98.33	21.49	16.32	2.48	65.11

We have two target architectures—MLP (two dense hidden layers with 512 ReLU units each, dense output layer of 10 softmax units) and CNN (two convolutional layers with 32 3×3 filters, ReLU activation, 2×2 max pooling layer, dense layer with 128 ReLU units, dense output layer of 10 softmax units).

These two architectures were trained 30 times by RMSProp for 20 and 12 epochs for MLP and CNN respectively. We obtained 98.35% average accuracy for MLP and 98.97% average accuracy for CNN on test data, but only 1.95% (MLP) and 8.49% (CNN) on adversarial data drawn by FGSM from test set.

Table 2. Accuracies on legitimate test examples and adversarial examples for CNN and CNNRBF with various initial widths. Average accuracies over 30 runs of learning algorithm.

Model	Legitimate samples				Adversarial samples			
	Mean	Std	Min	Max	Mean	Std	Min	Max
CNN	**98.97**	**0.07**	**98.84**	**99.13**	**8.49**	**3.52**	**3.11**	**16.43**
CNNRBF(0.01)	98.36	1.73	89.12	99.01	15.60	4.28	10.26	28.44
CNNRBF(0.1)	94.19	8.21	58.88	98.92	18.58	6.42	6.01	31.29
CNNRBF(1.0)	98.83	0.13	98.46	99.04	57.09	9.23	33.39	78.99
CNNRBF(2.0)	**98.85**	**0.13**	**98.38**	**99.09**	**74.57**	**7.69**	**53.07**	**84.67**
CNNRBF(3.0)	98.82	0.14	98.55	99.10	68.65	7.77	44.36	80.13
CNNRBF(5.0)	98.74	0.11	98.49	98.94	62.35	7.04	48.03	77.04
CNNRBF(10.0)	97.86	2.24	89.33	98.84	64.71	8.32	46.61	79.89

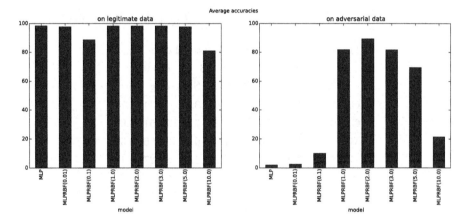

Fig. 4. Accuracies on legitimate and adversarial data for MLP and MLPRBF with various initial widths.

To each of the 30 trained networks we added the RBF network and retrained the whole new networks for 3 epochs. We found that the results depend on the parameters β of the Gaussians, therefore we tried several initial setups. The best results were obtained with initial β 2.0, and on adversarial data they were 89.21% for MLPRBF and 74.57% for CNNRBF. The complete results can be found in Tables 1, 2 and Figs. 4, 5. It shows that adding RBF network to the deep network may significantly decrease the vulnerability to adversarial examples.

In addition, Table 3 lists the average accuracies on adversarial data crafted with FGSM with different values of ϵ.

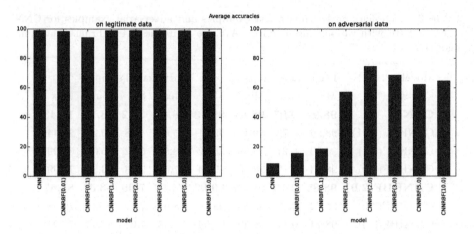

Fig. 5. Accuracies on legitimate and adversarial data for CNN and CNNRBF with various initial widths.

Table 3. Accuracies on adversarial data crafted by FGSM with different ϵ.

Model	Accuracy on adversarial data					
	$\epsilon = 0.2$		$\epsilon = 0.3$		$\epsilon = 0.4$	
	Avg	Std	Avg	Std	Avg	Std
CNN	33.85	7.58	8.49	3.52	4.34	1.71
CNNRBF	76.88	6.25	74.57	7.69	73.51	8.08
MLP	3.01	0.69	1.95	0.41	1.66	0.38
MLPRBF	90.14	4.82	89.21	5.03	88.27	5.14

5 Conclusion

In this paper we dealt with the problem of adversarial examples. We have proposed the new deep architecture that is obtained by stacking a feedforward deep neural network and an RBF network. Only a few learning epochs for the whole stacked network are needed to retrain, and to obtain the accuracy close to the accuracy of original deep neural network. We have shown that the new stacked network is much less vulnerable to adversarial examples than the original one.

Acknowledgments. This work was partially supported by the Czech Grant Agency grant GA18-23827S and institutional support of the Institute of Computer Science RVO 67985807.

Access to computing and storage facilities owned by parties and projects contributing to the National Grid Infrastructure MetaCentrum provided under the programme "Projects of Large Research, Development, and Innovations Infrastructures" (CESNET LM2015042), is greatly appreciated.

References

1. Goodfellow, I., Bengio, Y., Courville, A.: Deep Learning. MIT Press, Cambridge (2016)
2. Bengio, Y.: Learning deep architectures for AI. Found. Trends Mach. Learn. **2**(1), 1–127 (2009)
3. Hinton, G.E.: Learning multiple layers of representation. Trends Cognit. Sci. **11**, 428–434 (2007)
4. Krizhevsky, A., Sutskever, I., Hinton, G.E.: ImageNet classification with deep convolutional neural networks. In: Bartlett, P., Pereira, F., Burges, C., Bottou, L., Weinberger, K. (eds.) Advances in Neural Information Processing Systems, vol. 25, pp. 1106–1114. Neural Information Processing Systems Foundation (2012)
5. Szegedy, C., Zaremba, W., Sutskever, I., Bruna, J., Erhan, D., Goodfellow, I., Fergus, R.: Intriguing properties of neural networks (2013). arXiv:1312.6199
6. Goodfellow, I.J., Shlens, J., Szegedy, C.: Explaining and harnessing adversarial examples (2014). arXiv:1412.6572
7. LeCun, Y., Cortes, C.: The MNIST database of handwritten digits (2012)
8. Nguyen, A.M., Yosinski, J., Clune, J.: Deep neural networks are easily fooled: high confidence predictions for unrecognizable images. CoRR abs/1412.1897 (2014)
9. Papernot, N., McDaniel, P.D., Jha, S., Fredrikson, M., Celik, Z.B., Swami, A.: The limitations of deep learning in adversarial settings. CoRR abs/1511.07528 (2015)
10. Papernot, N., McDaniel, P.D., Goodfellow, I.J., Jha, S., Celik, Z.B., Swami, A.: Practical black-box attacks against deep learning systems using adversarial examples. CoRR abs/1602.02697 (2016)
11. Vidnerová, P., Neruda, R.: Evolutionary generation of adversarial examples for deep and shallow machine learning models, pp. 43:1–43:7 (2016)
12. Gu, S., Rigazio, L.: Towards deep neural network architectures robust to adversarial examples. CoRR abs/1412.5068 (2014)
13. Papernot, N., McDaniel, P.D., Wu, X., Jha, S., Swami, A.: Distillation as a defense to adversarial perturbations against deep neural networks. CoRR abs/1511.04508 (2015)
14. Moody, J., Darken, C.: Fast learning in networks of locally-tuned processing units. Neural Comput. **1**, 289–303 (1989)
15. Poggio, T., Girosi, F.: A theory of networks for approximation and learning. Technical report, Cambridge, MA, USA (1989) A. I. Memo No. 1140, C.B.I.P. Paper No. 31
16. Broomhead, D., Lowe, D.: Multivariable functional interpolation and adaptive networks. Complex Syst. **2**, 321–355 (1988)
17. Peng, J.X., Li, K., Irwin, G.W.: A novel continuous forward algorithm for RBF neural modelling. IEEE Trans. Autom. Control **52**(1), 117–122 (2007)
18. Fu, X., Wang, L.: Data dimensionality reduction with application to simplifying RBF network structure and improving classification performance. IEEE Trans. Syst. Man Cybern. Part B (Cybern.) **33**(3), 399–409 (2003)
19. Powel, M.: Radial basis functions for multivariable interpolation: a review. In: IMA Conference on Algorithms for the Approximation of Functions and Data, RMCS, Shrivenham, England, pp. 143–167 (1985)
20. Light, W.: Some aspects of radial basis function approximation. In: Approximation Theory, Spline Functions and Applications, pp. 163–190. Kluwer Academic Publishers, Dordrecht (1992)

21. Neruda, R., Kudová, P.: Learning methods for radial basis functions networks. Future Gener. Comput. Syst. **21**, 1131–1142 (2005)
22. Neruda, R., Kudová, P.: Hybrid learning of RBF networks. Neural Netw. World **12**(6), 573–585 (2002)
23. Papernot, N., et al.: cleverhans v2.0.0: an adversarial machine learning library. arXiv preprint arXiv:1610.00768 (2017)
24. Chollet, F.: Keras (2015). https://github.com/fchollet/keras
25. Vidnerová, P.: RBF for keras (2017). https://github.com/PetraVidnerova/rbf_keras
26. Vidnerová, P.: Experiments with deep RBF networks (2017). https://github.com/PetraVidnerova/rbf_tests

Application of Reinforcement Learning to Stacked Autoencoder Deep Network Architecture Optimization

Roman Zajdel and Maciej Kusy[(✉)]

Faculty of Electrical and Computer Engineering, Rzeszow University of Technology,
al. Powstancow Warszawy 12, 35-959 Rzeszow, Poland
{rzajdel,mkusy}@prz.edu.pl

Abstract. In this work, a new algorithm for the structure optimization of stacked autoencoder deep network (SADN) is introduced. It relies on the search for the numbers of the neurons in the first and the second layer of SADN through an approach based on reinforcement learning (RL). The $Q(0)$-learning based agent is constructed, which according to received reinforcement signal, picks appropriate values for the neurons. Considered network, with the architecture adjusted by the proposed algorithm, is applied to the task of MNIST digit database recognition. The classification quality is computed for SADN to determine its performance. It is shown that, using the proposed algorithm, the semi-optimal configuration of the number of hidden neurons can be achieved much faster than the successive exploration of the entire space of layers' arrangement.

Keywords: Stacked autoencoder deep network
Reinforcement learning · Classification quality

1 Introduction

Deep learning (DL) is a class of machine learning methods which are capable of learning complex (deep) features of input data. Majority of DL algorithms utilize neural network architectures arranged in succession. Such models are referred to as deep neural networks (DNN). Deep belief networks [1], deep Boltzmann machines [2], convolutional neural networks [3] or stacked autoencoders [4,5] represent typical DNN architectures. The depth of the networks expressed in terms of a number of hidden layers plays crucial role in the classification quality of DNNs. In literature, various numbers of layers are explored, e.g.: 7 in LeNet–5 [6], 13 in AlexNet [7] or 16 in VGG–16 D [8]. However, one finds the contributions where much greater structures are studied: 100 layers [9] or 152 layers [10]. The other crucial aspect of DNNs' performance is a selection of proper size of a particular layer. General approaches can be pointed out here: growing [11], pruning [12] or simply trial and error methods. Some authors also use RL based techniques in the search of optimal DNN's architectures [13].

© Springer International Publishing AG, part of Springer Nature 2018
L. Rutkowski et al. (Eds.): ICAISC 2018, LNAI 10841, pp. 267–276, 2018.
https://doi.org/10.1007/978-3-319-91253-0_26

In this paper, only one type of DNN architecture, namely, the stacked autoencoder is investigated. Considered SADN is applied in the task of the MNIST database [14] digits recognition. The network with only two hidden layers is utilized due to its very good classification quality results. The algorithm is proposed, which by means of RL, allows the network to find the semi-optimal number of the neurons in both layers. The $Q(0)$-learning method is adjusted for this purpose.

This paper consists of the following parts. Section 2 outlines the structure and the operation of the SADN. In Sect. 3, the fundamentals of one of the RL algorithms, i.e., $Q(0)$-learning are highlighted. In Sect. 4, the algorithm for an automatic construction of the SADN's architecture is introduced. Section 5 presents the results obtained by the proposed algorithm and the discussion on considered approach. The work is concluded in Sect. 6.

2 Stacked Autoencoder Deep Network

A SADN is a neural network composed of multiple layers of autoencoders stacked on top of each other. This network allows features of input patterns

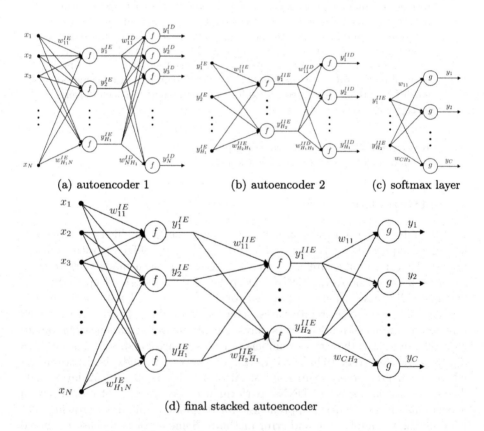

(a) autoencoder 1   (b) autoencoder 2   (c) softmax layer

(d) final stacked autoencoder

Fig. 1. The architecture of the stacked autoencoder.

to be extracted in succession through a number of layers. The first layer of the SADN learns first-order features (e.g., edges of an image); the second layer learns second-order features (e.g., contour of an image); higher layers learn higher-order features (more primitive elements).

The structure of two layer SADN is depicted in Fig. 1(d). The arrangement of the architecture of this network consists in removing decoding layers, i.e., output layers of autoencoders 1 and 2 shown in Fig. 1(a) and (b), respectively. The remaining layers of autoencoders are connected to each other followed by an output layer. The hidden layers are usually activated by means of a sigmoid functions (f) while the output layer is a softmax classifier (g). The procedure of discarding of decoding layers is strictly related to the multi-stage training process of SADN. At the beginning, the autoencoder 1 is trained using the input data set composed of $x_1, \ldots, x_N$ features. Next, the autoencoder 2 is trained on the output $\mathbf{y}^{IE}$ of the hidden layer of autoencoder 1. Then, the softmax layer (Fig. 1(c)) undergoes training using the output $\mathbf{y}^{IIE}$ of autoencoder 2 as its input. Afterwards, all layers are joined together to form a deep network (see Fig. 1(d)). The final stacked autoencoder is fine tuned by retraining it on the training data in a supervised fashion. The conjugate gradient algorithm is utilized in all stages of SADN's training process.

3   Reinforcement Learning

3.1   Introduction

The aim of the reinforcement learning (RL) is to make the agent learn to perform some task by means of the trial and error interaction with an unknown environment. The interaction between the agent and the environment is performed until the terminal state is reached. The agent uses its sensors to discover the environment. On the basis of current sensory input, it selects and performs an action in the environment. Based on the result of the action's use, the agent receives a reward [15]. The goal of RL is to maximize the discounted sum of future reinforcements r_t obtained in long term in any time step t, what is usually formalized as $\sum_{t=0}^{\infty} \gamma^t r_t$, where $\gamma \in [0, 1]$ is the discount rate.

RL method is modeled by a Markov decision process, which is defined as the quadruple $\langle S, A, P^{a_t}_{s_t s_{t+1}}, r_t \rangle$ where S is a set of states, A is a set of actions, $P^{a_t}_{s_t s_{t+1}}$ stands for the probability of the transition to the state $s_{t+1} \in S$ after execution of the action $a_t \in A$ in the state $s_t \in S$.

3.2 $Q(0)$-learning

The $Q(0)$-learning proposed by Watkins [16] is one of the most often used RL algorithms. $Q(0)$-learning computes the Q–table consisting of the elements $Q(s, a)$ which represent the expected reward that agent can receive in state s after it performs action a. The Q–table is updated for actual state-action pair (s_t, a_t) according to the following formula [17]

$$Q_{t+1}(s_t, a_t) = Q_t(s_t, a_t) + \alpha \left(r_t + \gamma \max_a Q_t(s_{t+1}, a) - Q_t(s_t, a_t) \right), \quad (1)$$

where $\alpha \in (0, 1]$ is the learning rate. In (1), the maximization operator refers to the action value a which may be performed in next state s_{t+1}. The formula (1) will be used as the basis of the algorithm for the SADN structure optimization presented in Sect. 4.

4   Proposed Approach

In this section, the RL based algorithm for determining the semi-optimal number of hidden layer neurons in SADN is proposed. By semi-optimal, it is assumed that: (i) the solution is given in the limited number of neurons' configurations in the first and second layer; (ii) the number of original interactions between the agent and the environment, thus the number of unrepeatable neurons' configurations, is finite and additionally smaller than their total number. Taking (i) and (ii) into account, the obtained solution is the best in terms of maximization of SADN's quality indicator.

The proposed algorithm is shown in Algorithm 1. It begins with the initialization of the array of states $\mathbf{H}$ in step $\mathbf{2}$. Each element of $\mathbf{H}$ stores the label of a state which corresponds to the combination of a pair of neurons' numbers H_1 and H_2 in two hidden layers of SADN. Figure 2 presents the graphical interpretation of $\mathbf{H}$. As shown, the array of states $\mathbf{H}$ is initialized in the way that $H_1 = \{20, 30, \ldots, 200\}$ while $H_2 = \{10, 20, \ldots, 190\}$ subject to $H_1 > H_2$, what gives 190 unique states. Gray cells in the figure denote available states which are labeled by subsequent natural numbers. For example, for $s = 1$ and $s = 28$, the corresponding pairs of H_1 and H_2 are $\{20, 10\}$ and $\{80, 70\}$, respectively. Step $\mathbf{3}$ defines the set of actions, that represent possible modification values for neurons in hidden layer. For the example from Fig. 2, $a_{H_1} = a_{H_2} = 10$. This allows H_1 and H_2 to change their values by ∓ 10. The start-up state s_1 is the first state which is then used to initialize the set of unique states $\mathbf{U}$ (step $\mathbf{5}$). In the next step, the number of unique visited states is computed. Thereafter, for the initial state s_1, the pair of hidden neurons' number $\{H_1, H_2\}$ is retrieved from the array of states $\mathbf{H}$. Step $\mathbf{8}$ is used to initialize the action value function Q with zeros. In step $\mathbf{9}$, the SADN training process is conducted on the training set and classification quality is determined on the test set by computing the following indicator

$$q(s_t) = 1 - \frac{1}{|\mathbf{X}^{\text{test}}|} \sum_{x_i^{\text{test}} \in \mathbf{X}^{\text{test}}} \delta \left[y(s_t, x_i^{\text{test}}) \neq t_i^{\text{test}} \right], \tag{2}$$

where $\mathbf{X}^{\text{test}}$ denotes the test set, $y(s_t, x_i^{\text{test}})$ is SADN's output obtained for testing case x_i^{test} at the state s_t, and t_i^{test} is the target of x_i^{test}. In (2), $\delta [\cdot] = 1$ when $y(x_i^{\text{test}}) \neq t_i^{\text{test}}$ and 0, otherwise. Since the aim of the algorithm is to find the semi-optimal configuration of the SADN in terms of $H_{1,\max}$ and $H_{2,\max}$ coded by a state label, therefore, in steps $\mathbf{10}$–$\mathbf{13}$, the semi-optimal state label $s_{\max}$ and the corresponding network quality $q_{\max}$ are established. In RL, states can be visited multiple times, hence, the number of states visited at least once ($u_{\max}$) is adopted to be the algorithm's stopping criterion. In step $\mathbf{15}$, the main loop

---

1 Initialize:
2   array of states $\mathbf{H}$
3   action set $A = \{-a_{H_1}, a_{H_1}, -a_{H_2}, a_{H_2}\}$
4   start state $s_1 = 1$
5   set of unique states $\mathbf{U} = \{s_1\}$
6   number of unique visited states $u = |\mathbf{U}|$
7   numbers of hidden layer neurons $\{H_1, H_2\} \leftarrow \mathbf{H}(s_1)$
8      $Q(s, a) = 0$ for each $s \in S$ and $a \in A$
9 Calculate classification quality $q(s_1)$ of SADN
10 $s_{\max} = s_1$
11 $q_{\max} = q(s_1)$
12 $H_{1,\max} = H_1$
13 $H_{2,\max} = H_2$
14 Assume maximum number of unique visited states $u_{\max}$
15 **while** $u <= u_{\max}$ **do**
16 | Choose action a_t for s_t using policy derived from Q_t
17 | Apply action a_t
18 | Observe next state s_{t+1}
19 | **if** $s_{t+1} \notin \mathbf{U}$ **then**
20 | | Get numbers of hidden layer neurons $\{H_1, H_2\} \leftarrow \mathbf{H}(s_{t+1})$
21 | | Calculate classification quality $q(s_{t+1})$ of SADN
22 | | $\mathbf{U} = \mathbf{U} \cup \{s_{t+1}\}$
23 | | $u = |\mathbf{U}|$
24 | | **if** $q(s_{t+1}) > q_{\max}$ **then**
25 | | | $s_{\max} = s_{t+1}$
26 | | | $q_{\max} = q(s_{t+1})$
27 | | | $H_{1,\max} = H_1$
28 | | | $H_{2,\max} = H_2$
29 | | | Determine label of unique visited state $l = u$
30 | | **end**
31 | **end**
32 | **if** $s_{t+1} \neq s_t$ **then**
33 | | $r_t = q(s_{t+1}) - q(s_t)$
34 | **else**
35 | | $r_t = -1$
36 | **end**
37 | $\Delta_t = r_t + \gamma \max_{a'} Q_t(s_{t+1}, a') - Q_t(s_t, a_t)$
38 | Update $Q_{t+1}(s_t, a_t) = Q_t(s_t, a_t) + \alpha \Delta_t$
39 **end**

Algorithm 1. The $Q(0)$-learning based algorithm for providing the semi-optimal number of hidden layer neurons in SADN's structure.

begins. Here, the action a_t is selected in the state s_t based on the policy represented by Q_t (step **16**). The action is applied to the environment expressed by the array of states what results in the change of the current state to the next state s_{t+1} (step **18**).

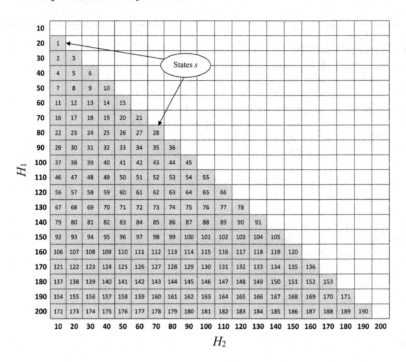

Fig. 2. Array of states as the form of coding of the hidden neurons' number H_1 and H_2.

Step **19** verifies whether the next state is unique ($s_{t+1} \notin \mathbf{U}$). If the condition is fulfilled, the number of neurons in both hidden layers is taken from $\mathbf{H}$. The training and testing process is performed for the SADN composed of new number of hidden neurons H_1 and H_2 and the network's quality is determined. Then, the state s_{t+1} is added to the set of unique states $\mathbf{U}$. Steps **24–30** aim at providing semi-optimal SADN's configuration.

The next stage of the algorithm (steps **32–36**) is responsible for calculation of the reinforcement signal as follows

$$
r_t = \begin{cases} q(s_{t+1}) - q(s_t) & \text{if } s_{t+1} \neq s_t \\ -1 & \text{if } s_{t+1} = s_t \end{cases} .
\tag{3}
$$

The difference in the SADN's classification quality between the states s_{t+1} and s_t defines the reinforcement signal when $s_{t+1} \neq s_t$. This permits realization of the strategy for rewarding or punishing the agent when, by the change of H_1 and H_2, the increase or decrease of quality takes place, respectively. If, as a consequence of applying an action, no change of the states occurs, the reinforcement signal is set to the value of -1. In this way, the agent is punished for "staying in the same place" and the lack of exploration of the SADN's structure in the sense of changing the number of hidden neurons. In steps **37** and **38**, the temporal differences error is computed and the action value function is updated according to (1), respectively. The algorithm terminates when the maximum number of unique states is reached.

5 Empirical Results and Discussion

This section presents the classification quality results provided by SADN with the structure optimized by means of Algorithm 1. The input data set used in the experiments is taken from the MNIST database repository [14]. It is composed of the 5,000 training and 5,000 testing digits from the set $\{0, 1, \ldots, 9\}$ each normalized into 28×28 pixel image. In Fig. 3, four exemplary training and testing digits are shown.

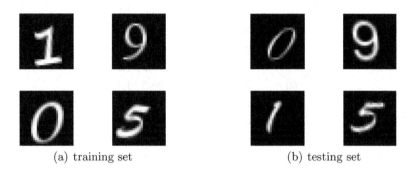

(a) training set   (b) testing set

Fig. 3. Selected digits from MNIST data set.

Because of the size of the considered images, the number of SADN's input neurons is $N = 784$. The number of epochs is set to 400 in all network's training stages which are highlighted in Sect. 2. The SADN is trained and tested 10 times for the same number of hidden neurons H_1 and H_2. The number of unique visited states is assumed to be $u_{max} = 50$ while the RL parameters are as follows: greedy parameter $\epsilon = 0.2$, discount rate $\gamma = 0.95$, learning rate $\alpha = 0.01$.

The obtained results are presented in Table 1. In two left-most columns, the number of neurons in the first and the second layer of the network is indicated. Third column shows the label l of the unique visited state for which the highest averaged quality q_{max} is attained. The next column includes the label of the state from the array $\mathbf{H}$ that would be visited if the search were performed starting from the first state sequentially up to the state s_{max}. For example, for $H_1 = 160$ and $H_2 = 100$, the SADN's classification quality is equal $q_{max} = 98.3\%$. This result is achieved at $s_{max} = 115$, which is determined after visiting $l = 46$ unique states. This means that the algorithm requires 69 state visits less in comparison to the case when all states need to be checked. If $l \geqslant s_{max}$, there is no gain in applying RL to find semi-optimal structure of SADN. In the current study, there are only two such cases.

In Fig. 4, the states visited by the agent controlled by Algorithm 1 are marked gray. One can observe that within an exploration of the SADN's structure configuration, the agent searches the semi-optimal values of H_1 and H_2. For three initial states $\{1, 2, 3\}$, the obtained classification qualities are visualized in the

Table 1. Exemplary SADN's structure configurations ($H_{1,\max}$, $H_{2,\max}$), unique (l) and visited ($s_{\max}$) state labels, and averaged classification quality results (in %) along with standard deviation (in %) obtained by the proposed algorithm.

$H_{1,\max}$	$H_{2,\max}$	l	$s_{\max}$	$q_{\max}$	sd
110	70	25	52	98.7	4.5
140	80	45	86	98.7	1.0
180	50	45	141	98.7	1.6
130	70	45	73	98.5	2.2
90	60	26	34	98.4	1.7
160	100	46	115	98.3	1.9
110	40	38	49	98.3	5.1
80	60	31	27	98.2	1.2
160	50	51	110	98.2	1.4
70	60	21	21	97.9	6.3
200	40	47	175	97.5	8.1
160	30	51	108	97.3	2.2
140	20	43	80	96.5	7.4

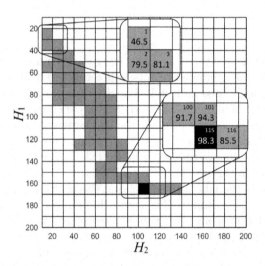

Fig. 4. All 50 states visited by the agent starting from the state s_1. For initial states and the states in the neighborhood of $s_{\max}$, the values of classification quality are provided along with the state label.

enlargement. The black color is used for the state labeled 115 for which the highest classification quality $q_{\max}$ is received.

In order to make current study more comprehensive, an additional comparative analysis is performed in the considered classification task. Four data classifiers are utilized: single decision tree (SDT), probabilistic neural network (PNN), support vector machine (SVM) and multilayer perceptron (MLP).

The utilized classifiers are parametrized as follows. For SDT, the quality of splits is evaluated with entropy. The depth of the tree is set to 10. The tree is pruned with respect to minimum cross-validation error. In the case of PNN, the smoothing parameter is represented by means of a vector corresponding to each attribute. The network is trained using the conjugate gradients method. SVM is simulated with a radial basis kernel function. An optimal grid search for model's parameters $C \in [10^{-1}, 10^4]$ and $sc \in [0.5, 50]$ is applied. Finally, MLP is composed of two hidden layers with sigmoidal activation functions and various number of neurons. The conjugate gradient algorithm is applied to train the network. For SDT, PNN, SVM and MLP, the following qualities (in %) are obtained: 86.4 ± 0.5, 92.9 ± 1.1, 99.7 ± 0.1 and 94.9 ± 2.4, respectively. Thus, SVM achieves the highest classification performance among all regarded models. As shown, the results obtained by means of proposed approach are much better than the outcomes of PNN, MLP and SDT, and narrowly worse than the one provided by SVM (quality margin equal 1.0%). It is necessary to point out that the SVM's quality is received by optimal search method while the proposed solution is only semi-optimal.

6   Conclusions

This paper proposed the procedure for the automatic design of the architecture of stacked autoencoder deep network. The idea was based on the search for the semi-optimal number of neurons in two hidden layers of the network. For this purpose, one of the reinforcement learning algorithms, i.e., $Q(0)$-learning was used. The important element of the proposed approach was the definition of the reinforcement signal which rewarded or penalized an agent depending on the performance of the network. The efficiency of SADN was evaluated on the MNIST database composed of 5000 training and testing 28×28 pixel digits by computing classification quality. The proposed algorithm provided the possibility of finding semi-optimal configuration of hidden neurons after significantly less iterations which would be required if the search throughout entire neurons' space were performed. A better solution can be obtained by increasing the number of maximum unique visited states, the neurons in each hidden layer or the layers of SADN.

The current work is a preliminary study on designing deep network architectures with the use of RL based algorithms in the task of image recognition. In future, the authors will focus on automatic architecture search for a broader range of classifiers, including both SADN and other models such as deep belief networks or convolution neural networks. More complex data sets will be applied.

Acknowledgements. The work was supported by Rzeszow University of Technology, Department of Electronics Fundamentals Grant for Statutory Activity (DS 2018).

References

1. Hinton, G.E., Osindero, S., Teh, Y.: A fast learning algorithm for deep belief nets. Neural Comput. **18**(7), 1527–1554 (2006)
2. Salakhutdinov, R., Hinton, G.E.: Deep Boltzmann machines. In: International Conference on Artificial Intelligence and Statistics, Clearwater Beach, USA, pp. 448–455 (2009)
3. LeCun, Y., et al.: Handwritten digit recognition with a back-propagation network. In: Touretzky, D.S. (ed.) Advances in Neural Information Processing Systems, vol. 2, pp. 396–404. Morgan-Kaufmann, Burlington (1990)
4. Kang, X., Li, C., Li, S., Lin, H.: Classification of hyperspectral images by Gabor filtering based deep network. IEEE J. Sel. Top. Appl. Earth Obs. Remote Sens. (2017). https://doi.org/10.1109/JSTARS.2017.2767185
5. Chen, Y., Lin, Z., Zhao, X., Wang, G., Gu, Y.: Deep learning-based classification of hyperspectral data. IEEE J. Sel. Top. Appl. Earth Obs. Remote Sens. **7**(6), 2094–2107 (2014)
6. LeCun, Y., Bottou, L., Bengio, Y., Haffner, P.: Gradient-based learning applied to document recognition. Proc. IEEE **86**(11), 2278–2324 (1998)
7. Krizhevsky, A., Sutskever, I., Hinton, G.E.: Imagenet classification with deep convolutional neural networks. In: Pereira, F., Burges, C.J.C. (eds.) Advances in Neural Information Processing Systems, vol. 25, pp. 1097–1105. Curran Associates Inc., Red Hook (2012)
8. Simonyan, K., Zisserman, A.: Very deep convolutional networks for large-scale image recognition. arXiv preprint arXiv:1409.1556 (2014)
9. Szegedy, C., Liu, W., Jia, Y., Sermanet, P., Reed, S., Anguelov, D., Erhan, D., Vanhoucke, V., Rabinovich, A.: Going deeper with convolutions. In: IEEE Conference on Computer Vision and Pattern Recognition, pp. 1–9. IEEE Press, Boston (2015)
10. He, K., Zhang, X., Ren, S., Sun, J.: Deep residual learning for image recognition. In: Proceedings of the IEEE Conference on Computer Vision and Pattern Recognition, pp. 770–778. IEEE Press, Washington (2016)
11. Bodenhausen, U., Manke, S.: Automatically structured neural networks for handwritten character and word recognition. In: Gielen, S., Kappen, B. (eds.) ICANN 1993. Springer, London (1993). https://doi.org/10.1007/978-1-4471-2063-6_283
12. LeCun, Y., Denker, J.S., Solla, S.A.: Optimal brain damage. In: Advances in Neural Information Processing Systems, vol. 2, pp. 598–605 (1990)
13. Baker, B., Gupta, O., Naik, N., Raskar, R.: Designing neural network architectures using reinforcement learning. In: International Conference on Learning Representations, Toulon, France (2017). https://openreview.net/pdf?id=S1c2cvqee
14. LeCun, Y., Cortes, C., Burges, C.J.C.: The MNIST database of handwritten digits (1998). http://yann.lecun.com/exdb/mnist/
15. Lanzi, P.: Adaptive agents with reinforcement learning and internal memory. In: Sixth International Conference on the Simulation of Adaptive Behavior, pp. 333–342. The MIT Press, Cambridge (2000)
16. Watkins, C.: Learning from delayed Rewards. Ph.D. thesis. Cambridge University, Cambridge (1989)
17. Sutton, R.S., Barto, A.G.: Reinforcement Learning: An Introduction. MIT Press, Cambridge (1998)

Evolutionary Algorithms and Their Applications

An Optimization Algorithm Based on Multi-Dynamic Schema of Chromosomes

Radhwan Al-Jawadi[1,2(✉)] and Marcin Studniarski[3]

[1] Faculty of Mathematics, Informatics and Mechanics, University of Warsaw,
Banacha 2, 02-097 Warsaw, Poland
radwanyousif@mimuw.edu.pl, radwanyousif@yahoo.com
[2] Technical College of Mosul, Mosul, Iraq
[3] Faculty of Mathematics and Computer Science, University of Łódź, Banacha 22,
90-238 Łódź, Poland
marstud@math.uni.lodz.pl

Abstract. In this work, a new efficient evolutionary algorithm to enhance the global optimization search is presented, which applies double populations, each population divided into several groups. The first population is original and the second one is a copy of the first one but with different operators are applied to it. The operators used in this paper are dynamic schema, dynamic dissimilarity, dissimilarity, similarity and a random generation of new chromosomes. This algorithm is called Multi-Dynamic Schema with Dissimilarity and Similarity of Chromosomes (MDSDSC) which is a more elaborate version of our previous DSC and DSDSC algorithms. We have applied this algorithm to 20 test functions in 2 and 10 dimensions. Comparing the MDSDSC with the classical GA, DSC, DSDSC and, for some functions, BA and PSO algorithms, we have found that, in most cases, our method is better than the GA, BA and DSC.

Keywords: Dynamic schema
Dissimilarity and Similarity of Chromosomes · Double population

1 Introduction

The idea of double population in evolutionary algorithms was use to improve the search for optimal solution and also to increase the diversity of a population. In [1] the authors have used a double population with Swarm Optimization Algorithm for optimization problems, in [2] a dual-population genetic algorithm was presented, which employs two populations, where the main population was used to find a good solution to the given problem and the second population was used to evolve and provide controlled diversity to the main population.

In this paper, a new evolutionary algorithm for solving optimization problems called Multi-Dynamic Schema with Dissimilarity and Similarity of Chromosomes

© Springer International Publishing AG, part of Springer Nature 2018
L. Rutkowski et al. (Eds.): ICAISC 2018, LNAI 10841, pp. 279–289, 2018.
https://doi.org/10.1007/978-3-319-91253-0_27

(MDSDSC) is presented. This algorithm is complementary to our previous algorithms called Dynamic Schema with Dissimilarities and Similarities of Chromosomes (DSDSC) [3] and Dissimilarity and Similarity of Chromosomes (DSC) [4]. In the MDSDSC algorithm a new technique is used, that is the double population of chromosomes working together to improve the efficiency of optimization and increase the chance to reach the best solution, where the first population is the original one and the second one is a copy of the first one but different types of operations are applied to it.

Briefly, the algorithm aims at finding the optimal solution by fixing the highest bits of a chromosome (i.e., fixing the highest bits of all variables $(x_1, \ldots x_n)$ which are contained in the chromosome) and changing the lower bits at the same time, thus the algorithm focuses on the searching for optimal solution in a small area.

2   Methodology

The MDSDSC starts with a random population (P0) of M elements representing a number of solutions to the problem. This population is sorted, then a new population (P1) is formed which is a copy of a part of (P0), each population (P0, P1) is divided into several equal groups and some different operators are applied to these groups (see Table 1).

Briefly, the MDSDSC creates new chromosomes by exploring dissimilarity, similarity, dynamic schema and dynamic dissimilarity. These operators are described as follows:

2.1   Dissimilarity Operator

For the first two chromosomes (A, B) in a group, check each pair of corresponding bits: if the two bits are equal, put a star (*) in the second (B) chromosome; otherwise leave this bit unchanged in the second chromosome. Then randomly put 0 or 1 in the bits with stars (*). Compare this new second chromosome with the third chromosome in the group and so on (see Table 2).

2.2   Similarity Operator

For the first two chromosomes (A, B) in a group, check each pair of corresponding bits: if the two bits are not equal, put a star (*) in the second (B) chromosome; otherwise leave this bit unchanged in the second chromosome. Then randomly put 0 or 1 in the bits with stars (*). Compare this new second chromosome with the third one and so on (see Table 3).

2.3   Dynamic Schema Operator

The dynamic schema operator is applied onto two chromosomes (A, B). This operator works as follows (see Table 4):

Table 1. Populations (P0) and (P1) and the seven groups of chromosomes.

Original groups of chromosomes (P0)		Copy groups of chromosomes (P1)	
Ch_1	G1: To the first group the dynamic dissimilarity operator is applied	Ch_1	G5: To the fifth group the dissimilarity operator is applied
Ch_2		Ch_2	
Ch...		Ch...	
Ch...		Ch...	
$Ch_{M/4}$		$Ch_{M/4}$	
$Ch_{M/4+1}$	G2: To the second group the similarity operator is applied	$Ch_{M/4+1}$	G6: To the sixth group the dynamic dissimilarity operator is applied
$Ch_{M/4+2}$		$Ch_{M/4+2}$	
Ch...		Ch...	
Ch...		Ch...	
$Ch_{M/2}$		$Ch_{M/2}$	
$Ch_{M/2+1}$	G3: To the third group the dynamic schema operator is applied	$Ch_{M/2+1}$	G7: To the seventh group the dynamic schema operator is applied
$Ch_{M/2+2}$		$Ch_{M/2+2}$	
Ch...		Ch...	
Ch...		Ch...	
$Ch_{M/2+M/4}$		$Ch_{M/2+M/4}$	
$Ch_{M/2+M/4+1}$	G4: The fourth group is generated randomly		
$Ch_{M/2+M/4+1}$			
Ch...			
Ch...			
Ch_M			

Table 2. The dissimilarity operator.

Before change								
Chromosome A	1	1	0	0	1	0	1	1
Chromosome B	1	0	1	1	0	0	0	1
Check: putting (*) in the second chromosome								
Chromosome A	1	1	0	0	1	0	1	1
Chromosome B	*	0	1	1	0	*	0	1
After change: put randomly 0 or 1 in (*) bits								
Chromosome A	1	1	0	0	1	0	1	0
Chromosome B	1	0	1	1	0	0	0	0

Table 3. The similarity operator.

Before change									
Chromosome A	1	1	0	0	1	0	1	1	
Chromosome B	1	0	1	1	0	0	0	1	
Check: putting (*) in the second chromosome									
Chromosome A	1	1	0	0	1	0	1	1	
Chromosome B	1	*	*	*	*	0	*	1	
After change: put randomly 0 or 1 in (*) bits									
Chromosome A	1	1	0	0	1	0	1	0	
Chromosome B	1	1	0	1	0	0	1	1	

Table 4. The dynamic schema operator.

Before change: an example for finding a schema from two chromosomes (A,B). Here shadow bits are not destroyed.									
No. of Ch.	m_1					m_2			
	R_1		$m_1 - R_1$			R_2	$m_2 - R_2$		
Ch. A	1	1	0	0	1	0	1	0	1	0
Ch. B	0	1	1	0	0	1	0	0	0	1
Schema	1	1	*	0	*	*	1	0	*	*
After finding the schema: copy it K times.									
$Ch_{M/2+1}$	1	1	*	0	*	*	1	0	*	*
$Ch_{M/2+2}$	1	1	*	0	*	*	1	0	*	*
Ch...	1	1	*	0	*	*	1	0	*	*
Ch...	1	1	*	0	*	*	1	0	*	*
$Ch_{3M/4}$	1	1	*	0	*	*	1	0	*	*
After change: put randomly 0 or 1 in (*) bits.									
$Ch_{M/2+1}$	1	1	1	0	1	0	1	0	0	1
$Ch_{M/2+2}$	1	1	1	0	0	0	1	0	1	1
Ch...	1	1	0	0	1	0	1	0	1	0
Ch...	1	1	0	0	0	1	1	0	0	0
$Ch_{3M/4}$	1	1	1	0	1	1	1	0	1	1

First, divide each chromosome into n parts corresponding to variables $x_1, \ldots, x_n$, the i-th part having length m_i, where m_i is the number of bits for x_i. Next, for each variable x_i, generate a random integer R_i from the set $\{3, \ldots, m_i/2\}$. Define the "gray" part of x_i as the first segment of length R_i of the string corresponding to x_i. Define the "white" part of x_i as the second segment of length $m_i - R_i$ of the same string.

For the "white" parts of both chromosomes, if the two bits are not equal then put a star (*) in the schema, otherwise leave this bit unchanged. After finding the schema, copy it $K = M/4$ times and put it in group (G3), then randomly put 0 or 1 in the positions having (*). The positions marked in gray remain unchanged.

Note. The name "dynamic schema operator" is justified by the fact that the lengths of "gray" and "white" segments of chromosomes may differ from iteration to iteration.

2.4   Dynamic Dissimilarity Operator

The dynamic dissimilarity operator is applied onto the first two chromosomes (A, B) in a group. This operator works similarly to the dynamic schema operator to find the "gray" and "white" parts corresponding to variables $x_1, \ldots, x_n$. The "gray" part of x_i is not destroyed, in the "white" part of x_i, if the two bits are equal, put a star (*) in the second (B) chromosome; otherwise leave this bit unchanged. Then randomly put 0 or 1 in the bits with stars (*) in the second chromosome. Compare this new second chromosome with the third one in the same way and so on (see Table 5).

Table 5. The dynamic dissimilarity operator.

Before change: an example for the chromosomes (A, B). Here shadow bits are not destroyed.									
No. of Ch.	m_1					m_2			
	R_1		$m_1 - R_1$			R_2	$m_2 - R_2$		
Ch. A	1	1	0	0	1	0	1	0	1	0
Ch. B	0	0	1	0	0	1	0	0	1	1
Ch. A	1	1	0	0	1	0	1	0	1	0
Ch. B	1	0	1	*	0	1	1	*	*	1
After change: put randomly 0 or 1 in (*) bits.									
Ch. A	1	1	0	0	1	0	1	0	1	0
Ch. B	1	0	1	1	0	1	0	1	0	1

3   The MDSDSC Algorithm

The following optimization problem is considered:

$$f : \mathbb{R}^n \to \mathbb{R}$$
$$\text{minimize} \mid \text{maximize } f(x_1, \ldots, x_n) \text{ subject to}$$
$$x_i \in [a_i, b_i], i = 1, \ldots, n$$

where $f : \mathbb{R}^n \to \mathbb{R}$ is a given function.

In the algorithm described below, it uses a standard encoding of chromosomes as in the book of Michalewicz [5]. In particular, it uses the following formula to decode a real number $x_i \in [a, \ldots, b_i]$:

$$x_i = a_i + \text{decimal}(1001..001) * \frac{b_i - a_i}{2^{m_i} - 1}$$

where m_i is the length of a binary string and "decimal" represents the decimal value of this string. The value of m_i for each variable depends on the length of the interval $[a, \ldots, b_i]$. To encode a point $(x_1, \ldots x_n)$, a decimal string of length $m = \sum_{i=1}^{n} m_i$ is used.

Let M be a positive integer divisible by 8. The MDSDSC algorithm consists of the following steps:

1. Generate $2M - M/4$ chromosomes, each chromosome representing a point $(x_1, \ldots x_n)$. Divide the chromosomes into two populations (P0) and (P1), where (P0) consists of a four groups (G1, G2, G3, G4), and P1 consists of three groups (G5, G6, G7), each group having $M/4$ chromosomes.
2. Compute the values of the fitness function f for each chromosome in the population (G1, ..., G7).
3. Sort the chromosomes according to the descending (for maximization) or ascending (for minimization) values of the fitness function.
4. Copy the groups (G1, G2) onto (G5, G6), replacing the original chromosomes.
5. Copy C times the first chromosome and put it in C randomly chosen positions in the first half of population (P0), replacing the original chromosomes, where $C = M/8$.
6. Apply the dynamic schema operator for chromosomes A = Ch_1 and B = $Ch_{M}/4$ from populations (P0), (that is, the chromosomes on positions 1 and $M/4$, respectively). Copy this schema $M/4$ times and put it in (G3).
7. Apply the dynamic schema operator for chromosomes A = Ch_1 and B = $Ch_{M}/4$ from populations (P0), (that is, the chromosomes on positions 1 and $M/4$ respectively). Copy this schema $M/4$ times and put it in (G7).
8. Apply the dynamic dissimilarity and similarity operators to groups (G1) and (G2) respectively. Apply the dissimilarity and dynamic dissimilarity operators to groups (G5) and (G6) respectively.
9. All the chromosomes created in Steps 6 to 8 replace the original ones in positions from 2 to $3M/4$ in populations (P0) and (P1). Then randomly generate chromosomes for group (G4).
10. Go to Step 2 and repeat until the stopping criterion is reached.

Notes:

1. The stopping criterion for the algorithm depends on the example being considered, see Sect. 4.
2. An injection technique to the population was suggested in [6] to preserve the diversity of the population. They use fixpoint injection, which means that they introduce new randomly generated chromosomes to the population for certain numbers of generations. A similar strategy has been applied in the MDSDSC algorithm by generating the group (G4) randomly in each iteration.

4   Experimental Results

In this section, we report on computational testing (by using the Matlab R2015b software on a computer having CPU core i5 2.4 MHz with 8 GB RAM) of the

Table 6. Test functions

Function Name	Interval	Function	Global Optimum Min/Max
Easom	$x,y\in$ $[-100,100]$	$f(x,y)=-\cos(x)\cos(y)\exp(-(x-\pi)^2+(y-\pi)^2)$	$f(\pi,\pi)=-1$, min
Matyas	$x,y\in$ $[-10,10]$	$f(x,y)=0.26(x^2+y^2-0.48xy)$	$f(0,0)=0$, min
Beale's	$x,y\in$ $[-4.5,4.5]$	$f(x,y)=(1.5-x-xy)^2+(2.25-x+xy^2)^2$ $+(2.625-x-xy^2)^2$	$f(3,0.5)=0$, min
Booth's	$x,y\in$ $[-10,10]$	$f(x,y)=(x+2y-7)^2+(2x+y-5)^2$	$f(1,3)=0$, min
Goldstein -Price	$x,y\in[-2,2]$	$f(x,y)=(1+(x+y+1)^2(19-14x+3x^2-14y+6xy+3y^2))$ $\cdot(30+(2x-3y)^2(18-32x+12x^2+48y-36xy+27y^2))$	$f(0,-1)=3$, min
Schaffer N.2	$x,y\in$ $[-100,100]$	$f(x,y)=0.5+\dfrac{\sin^2(x^2-y^2)-0.5}{(1+0.001(x^2+y^2))^2}$	$f(0,0)=0$, min
Schwefel's	$x_1,x_2\in$ $[-500,500]$	$f(x)=\sum\limits_{i=1}^{n}x_i\cdot\sin\left(\sqrt{|x_i|}\right)$	$f(1,1)=0$, min
Branins's rcos	$x_1\in[-5,10]$ $x_2\in[0,15]$	$f(x_1,x_2)=a\cdot(x_2-b\cdot x_1^2+c\cdot x_1-d)^2+e\cdot(1-f)\cdot\cos(x_1)+e$ $a=1,\ b=\frac{5.1}{4\cdot\pi^2},\ c=\frac{5}{\pi},\ d=6,\ e=10,\ f=\frac{1}{8\pi}$	$f(\pi,2.275)$ or $f(9.42478,2.475)$ or $f(-\pi,12.275)=$ 0.397887, min
Six-hump camel back	$x_1\in[-3,3]$ $x_2\in[-2,2]$	$f(x_1,x_2)=\left(4-2.1x_1^{\frac{4}{3}}\right)\cdot x_1^2+x_1x_2+\left(-4+4x_2^2\right)\cdot x_2^2$	$f(-0.0898,0.7126)$ $=-1.0316$, min
Shubert	$x_1,x_2\in$ $[-10,10]$	$f(x_1,x_2)=\left(\sum\limits_{i=1}^{5}i\cos[(i+1)x_1+i]\right)\cdot\left(\sum\limits_{i=1}^{5}i\cos[(i+1)x_2+i]\right)$	18 global min $f=-186.7309$, min
Martin and Gaddy	$x_1,x_2\in[0,10]$	$f(x_1,x_2)=(x_1-x_2)^2\cdot\left((x_1+x_2-10)/3\right)^2$	$f(5,5)=0$, min
Zbigniew Michalewicz	$x_1\in[-3,12.1]$ $x_2\in[-4.1,5.8]$	$f(x_1,x_2)=21.5+x_1\cdot\sin(4\pi x_1)+x_2\cdot\sin(20\pi x_2)$	$f(11.631407,5.724824)$ $=38.81208$, max,
Holder table	$x_1,x_2\in$ $[-10,10]$	$f(x_1,x_2)=-\left|\sin(x_1)\cos(x_2)\exp\left(\left|1-\dfrac{\sqrt{x_1^2+x_2^2}}{\pi}\right|\right)\right|$	$f(8.05502,9.66458)$ or $f(8.05502,-9.66458)$ or $f(-8.05502,9.66458)$ or $f(-8.05502,-9.66458)$ $=-19.2085$, min
Drop-wave	$x_1,x_2\in$ $[-4.12,5.12]$	$f(x_1,x_2)=-\dfrac{1+\cos\left(12\sqrt{x_1^2+x_2^2}\right)}{0.5(x_1^2+x_2^2)+2}$	$f(0,0)=-1$, min
Levy N.13	$x_1,x_2\in$ $[-10,10]$	$f(x_1,x_2)=\sin^2(3\pi x_1)+(x_1-1)^2[1+\sin(3\pi x_2)]$ $+(x_2-1)^2[1+\sin(2\pi x_2)]$	$f(1,1)=0$, min
Rastrigin's	$x_i\in$ $[-5.12,5.12]$	$f(x_i)=10\cdot d\sum\limits_{i=1}^{d}[x_i^2-10\cos(2\pi x_i^2)]$	$f(0,0)=0$, min
Rosenbrock	$x_i\in[-2.048,2.048]$	$f(x)=20[100(x_2-x_1^2)+(x_1-1)^2]$	$f(1,1)=0$, min
Sum of squares $d=10$	$x_i\in[-10,10]$	$f(x_i)=\sum\limits_{i=1}^{d}ix_i^2$	$f(0,...,0)=0$, min
Sphere $d=10$	$x_i\in$ $[-5.12,5.12]$	$f(x_i)=\sum\limits_{i=1}^{d}x_i^2$	$f(0,...,0)=0$, min
Sum of different powers $d=10$	$x_i\in[-1,1]$	$f(x_i)=\sum\limits_{i=1}^{d}|x_i|^{i+1}$	$f(0,...,0)=0$, min

MDSDSC algorithm on 17 functions of 2 variables and 3 functions of 10 variables. The test functions are taken from literature. After each test, the result of MDSDSC has been compared with the known global optimum and with the result of a classical GA taken from the respective reference. In Table 6 we mention all 20 functions with optimal solutions. We have applied the algorithm with 80 chromosomes (P0) with the stopping criterion that the difference between our best solution and the known optimal solution is less than or equal a given threshold. This threshold was equal to 0.001 for most two-dimensional functions, 0.01 for the Shubert function, 0.04 for the Michalewicz function, and 0.1 for ten-dimensional functions.

The MDSDSC algorithm has found optimum solutions for some optimization problems (like Easom, Booth's, Schaffer, Schwefel's, Shubert) that the classical

Table 7. The best value of functions for 50 runs of the MDSDSC algorithm.

Function name	Min no. of iterations	Max no. of iterations	Mean no. of iterations for all succ. runs	Mean of the best solution fitness from all runs	Succ. of DSDSC and MDSDSC	Succ. of DSC	Succ. of GA
Easom	4	291	62	−0.99934	100%	100%	0% [7]
Matyas	2	10	5	0.000484	100%	100%	70% [8]
Beale's	2	74	16	0.000497	100%	100%	6% [9]
Booth's	2	48	17	0.000519	100%	100%	0% [8]
Goldstein-Price	2	62	20	3.000497	100%	100%	72% [7]
Schaffer N.2	2	39	14	0.000354	100%	100%	0%
Schwefel's	8	253	65	0.000684	100%	92%	0% [7]
Branins's rcos	2	103	9	0.398415	100%	100%	100%
Six-hump camel back	2	61	8	−1.03112	100%	100%	98% [10]
Shubert	2	169	33	−186.714	100%	100%	0% [7]
Martin-Gaddy	2	11	6	0.000445	100%	100%	1% [7]
Michalewicz	2	546	71	38.81849	100%	100%	73%
Holder table	4	87	24	−19.208	100%	100%	78% [11]
Drop-wave	6	189	44	−0.99959	100%	100%	30%
Levy N.13	4	47	19	0.000527	100%	100%	70%
Rastrigin's	8	133	38	0.000418	100%	100%	100%
Rosenbrock	3	102	24	0.0005517	100%	100%	1% [7]
Sum squares 10-D	38	251	128	0.072773	100%	25%	80%
Sphere 10-D	14	38	23	0.073044	100%	100%	90%
Sum powers 10-D	1	7	4	0.029555	100%	100%	80%

Table 8. The number of function evaluations for 50 runs of the MDSDSC algorithm.

Function name	Min no. of fun. eval.	Max no. of fun. eval.	Average no. of fun. eval.
Easom	560	40740	8680
Matyas	280	1400	700
Beale's	280	10360	2240
Booth's	280	6720	2380
Goldstein–Price	280	8680	2800
Schaffer N.2	280	5460	1960
Schwefel's	1120	32420	9100
Branins's rcos	280	14420	1260
Six-hump camel back	280	8540	1120
Shubert	280	23660	4620
Martin and Gaddy	280	1540	840
Zbigniew Michalewicz	280	76440	9940
Holder table	560	12180	3360
Drop-wave	840	26460	6160
Levy N. 13	560	6580	2660
Rastrigin's	1120	18620	5320
Rosenbrock	420	14280	3360
Sum squares 10-D	10640	70280	35840
Sphere 10-D	3920	10640	6440
Sum of different powers 10-D	280	1960	1120

Table 9. Comparison of BA, PSO and MDSDSC algorithms in terms of average number of functions evaluations and success rate.

Function name	Succ. BA	Fun. Eval. BA	PSO	Fun. Eval. PSO	MDSDSC	Fun. Eval. MDSDSC
Easom	72%	5868	100%	2094	100%	8680
Shubert	0%	—	100%	3046	100%	4620
Schwefel's	85%	5385	86%	3622	100%	9100
Goldstein-Price	7%	9628	100%	1465	100%	2800
Martin-Gaddy	100%	1448	3%	9707	100%	840
Rosenbrock	46%	7197	100%	1407	100%	3360

genetic algorithm cannot solve, as shown in Table 7. Column eight shows 0% success rate with bit string as it has been mentioned in [7,8]. For our algorithm all success rates are 100% with 80 chromosomes in (P0) for all problems.

The MDSDSC algorithm keeps the best solution from each iteration at the first position until it is replaced by a better one.

Note that the average number of iterations to find the best solution was especially high (71) for the Zbigniew Michalewicz function, see Table 7. For 10-dimensional problems we used 160 chromosomes for population (P0).

Table 8 shows the minimum, maximum and average numbers of function evaluations for 50 runs of the MDSDSC algorithm.

Table 9 presents a comparison of the success rate and the number of function evaluations (for two-dimensional functions only) for three algorithms: Bees Algorithm (BA), Particle Swarm Optimization (PSO), and MDSDSC. The results for BA and PSO are taken from [7].

5   Conclusion

The MDSDSC is a new multi-population evolutionary algorithm that uses two populations. This algorithm uses different operators to find the optimal solution, where through the dynamic schema operator the algorithm obtains the best area of solutions, and searches within that area in each iteration as it detects the schema from the best solution in the population. While the dynamic dissimilarity operator performs searching in a wide range of solutions in (G1) and (G6), where the high bits are kept without change and the lower bits are changed. The dissimilarity and similarity operators possess the ability of searching in the whole search space because every bit of a chromosome can be changed by them. The fifth operator generates chromosomes randomly in (G4) to help increasing the diversity and not to stick in a local optimum solution.

We have applied the GA, DSC, DSDSC, MDSDSC algorithms on 20 tested function with 2 and 10 dimensions. The results show the MDSDSC algorithm is superior on the GA and DSC and DSDSC algorithms for most functions.

Through our experiments we found that whenever the function range is small like $(-1, 1)$ or $(-5, 5)$, the solution was obtained faster compared to the larger range $(-500, 500)$.

Acknowledgments. The first author would like to thank the Ministry of Higher Education and Scientific Research (MOHESR), Iraq.

References

1. Wu, Y., Sun, G., Su, K., Liu, L., Zhang, H., Chen, B., Li, M.: Dynamic self-adaptive double population particle swarm optimization algorithm based on Lorenz equation. J. Comput. Commun. **5**(13), 9–20 (2017)
2. Park, T., Ryu, K.R.: A dual-population genetic algorithm for adaptive diversity control. IEEE Trans. Evol. Comput. **14**(6), 865–884 (2010)

3. Al-Jawadi, R.: An optimization algorithm based on dynamic schema with dissimilarities and similarities of chromosomes. Int. J. Comput. Electr. Autom. Control
   Inf. Eng. **7**(8), 1278–1285 (2016)
4. Al-Jawadi, R., Studniarski, M., Younus, A.: A new genetic algorithm based on
   dissimilarities and similarities. Comput. Sci. J. **19**(1), 19 (2018)
5. Michalewicz, Z.: Genetic Algorithms + Data Structures = Evolution Programs.
   Artificial Intelligence, 3rd edn. Springer, Heidelberg (1996). https://doi.org/10.
   1007/978-3-662-03315-9
6. Sultan, A.B.M., Mahmod, R., Sulaiman, M.N., Abu Bakar, M.R.: Maintaining
   diversity for genetic algorithm: a case of timetabling problem. J. Teknol. Malaysia
   44(5), 123–130 (2006)
7. Eesa, A.S., Brifcani, A.M.A., Orman, Z.: A new tool for global optimization
   problems- Cuttlefish algorithm. Int. J. Comput. Electr. Autom. Control Inf. Eng.
   8(9), 1198–1202 (2014)
8. Ritthipakdee, A., Thammano, A., Premasathian, N., Uyyanonvara, B.: An
   improved firefly algorithm for optimization problems. In: ADCONP, Hiroshima,
   no. 2, pp. 159–164 (2014)
9. Iqbal, M.A., Khan, N.K., Mujtaba, H., Baig, A.R.: A novel function optimization
   approach using opposition based genetic algorithm with gene excitation. Int. J.
   Innov. Comput. Inf. Control **7**(7), 4263–4276 (2011)
10. Odili, J.B., Nizam, M., Kahar, M.: Numerical function optimization solutions using
    the African buffalo optimization algorithm (ABO). Br. J. Math. Comput. Sci.
    10(1), 1–12 (2015)
11. Scott, E.O., De Jong, K.A.: Understanding simple asynchronous evolutionary algo-
    rithms. In: United Kingdom ACM FOGA 2015, Aberystwyth, 17–20 January 2015

Eight Bio-inspired Algorithms Evaluated for Solving Optimization Problems

Carlos Eduardo M. Barbosa and Germano C. Vasconcelos[✉]

Center for Informatics, Federal University of Pernambuco, Recife, Brazil
{cemb,gcv}@cin.ufpe.br
http://www.cin.ufpe.br

Abstract. Many bio-inspired algorithms have been proposed to solve optimization problems. However, there is still no conclusive evidence of superiority of particular algorithms in different problems, diverse experimental situations and varied testing scenarios. Here, eight methods are investigated through extensive experimentation in three problems: (1) benchmark functions optimization, (2) wind energy forecasting and (3) data clustering. Genetic algorithms, ant colony optimization, particle swarm optimization, artificial bee colony, firefly algorithm, cuckoo search algorithm, bat algorithm and self-adaptive cuckoo search algorithm are compared, concerning, the quality of solutions according to several performance metrics and convergence to best solution. A bio-inspired technique for automatic parameter tuning was developed to estimate the optimal values for each algorithm, allowing consistent performance comparison. Experiments with thousands of configurations, 12 performance metrics and Friedman and Nemenyi statistical tests consistently evidenced that cuckoo search works efficiently, robustly and superior to the other methods in the vast majority of experiments.

Keywords: Bio-inspired algorithms · Swarm intelligence
Optimization · Automatic parameter tuning

1 Introduction

Optimization problems often occur in a large number of scientific, financial, industrial, and management domains [1–3]. Many everyday problems require optimization with extensive computer time to find optimal solutions. Exact methods are able to obtain optimal solutions but are of limited practical use because usually demand more complex modeling and are only efficient when involving small problem instances, short dimensional spaces and limited number of parameters. Alternately, optimization can be reached by heuristic and meta-heuristic methods which, although no guaranteeing the exact optimal case, usually provide efficient, close to optimal, solutions. Many of the advances in optimization algorithms are inspired by natural systems, given their flexibility to adaptation, seen as a great strategy for solving many complex problems.

© Springer International Publishing AG, part of Springer Nature 2018
L. Rutkowski et al. (Eds.): ICAISC 2018, LNAI 10841, pp. 290–301, 2018.
https://doi.org/10.1007/978-3-319-91253-0_28

In this context, bio-inspired metaheuristic algorithms have increasingly evolved as powerful tools for optimization by providing, approximate solutions, reached in acceptable computational times [3]. Among the bio-inspired algorithms, particularly, Genetic Algorithms (GA) [4], Ant Colony Optimization (ACO) [5], Particle Swarm Optimization (PSO) [6], Artificial Bee Colony (ABC) [7], Firefly Algorithm (FA) [8], Cuckoo Search (CS) [9], Bat Algorithm (BAT) [10] and Self Adaptive Cuckoo Search (SACS) [11] have been applied to several applications [3]. This work evaluates those algorithms regarding quality (solutions reached and convergence speed), in a group of optimization applications with different characteristics (benchmark functions, wind prediction and data clustering), through a vast experimental framework with each algorithm stressed to reach its best performance.

2 Bio-inspired Algorithms

GA is the most popular heuristic evolutionary method and has been applied to problems ranging from cluster analysis [12] to classifier systems [13]. ACO is inspired on the behavior of ants that search the best path between their colony and a food source and has been applied to discrete optimization [5] and wind energy forecasting [2]. PSO is inspired by the dynamics and social behavior of groups of animals, such as schools of fishes and flocks of birds, during their search for food, and has been used to wind energy forecasting [2], neural network training [6], and data clustering [14]. ABC is based on the behavior of bee colonies in their search for food sources, and has been applied to continuous optimization problems [7]. FA is inspired by the bioluminescence process of fireflies, and was applied to the traveling salesman problem, data clustering, digital processing and image compression [8]. CS is based on the parasitic reproduction behavior of some cuckoo bird species, and has been applied to global optimization problems [9], image processing, scheduling, planning, feature selection and forecasting [15, 16]. BAT is inspired by the echolocation process of bats during the flight [10], and has been employed to search strategies and data clustering [17]. And, SACS [11] is a more recent version of CS that adds two mutational rules and a self-adaptive parameter to CS.

3 Automatic Parameter Tuning

Each bio-inspired algorithm has several parameters to adjust, depending on the complexity of the method, to accomplish full potential. Many works employ suggested values or carry out trial-and-error processes with few variations. Here, a more refined parameter examination is conducted based on the use of a bio-inspired algorithm itself for parameter tuning. The technique formulates an objective function to measure the quality of the parameters as a combination of the number of iterations (I) required to reach a given threshold error rate, the minimum function value (E) reached at the end of the maximum number of iterations, and the total number of individuals (N) in the algorithm.

This way, the technique tries to obtain the optimal parameters to reach a high degree of success in minimizing the error and fast convergence, within the smallest computational cost possible, using as few individuals as possible. PSO was used as the optimization algorithm due to its simplicity and success [6,14,18], although other algorithms could also be considered. The objective function is defined as $f_{obj} = IEN$, and the method is referred to as Tuning-PSO. Since PSO is used, 5 parameters are adjusted: swarm size N, inertia weight w, cognitive learning coefficient $c1$, social learning coefficient $c2$ and speed rate v_r. Each particle is optimized within a range of search space values for each variable, as shown in Table 1(a), with the parameter variation limits for all algorithms. After experimentation, the best parameter values found for Tuning-PSO were $N = 20$, $w = 0.729$, $c1 = 2.05$, $c2 = 2.05$, $v_r = 0.1$.

4   Optimization Problems

Benchmark Functions. Six typical optimization functions [19] with different levels of complexity were investigated: Sphere, Rosenbrock, Rastrigin, Griewank, Ackley and Schwefel. Table 1(b) shows details of the functions with respect to search space interval considered, optimal solution $(x*)$ and function value in the optimal case $(f(x*))$. Since the number of dimensions (d) is a relevant issue [20], dimensions 2, 5, 10, 20, 50 and 100 were experimented. Table 1(c) shows the optimal parameter values for each optimization algorithm, for each function. With optimal parameter values found, experiments were performed to a maximum of 10000 iterations, after observing this number guaranteed complete convergence of the algorithms. Mean and standard deviation were measured over 20 runs.

Wind Forecasting. Wind energy forecasting is a relevant real world problem and was considered for comparing the algorithms. The databases of Montana and Texas [21] were employed for training a numeric weather prediction (NWP) model [22], which provides short interval forecasts. The databases (US NREL lab) record values of wind speed and wind energy produced every ten minutes, from 2004 to 2006. The set with the first 8760 (365 days, 24 records per day) samples from Montana database; and the set with the first 8760 samples from Texas database, were used. The first 50% of records were used for training, the next 25% for validation (for preventing overfitting) and the last 25% for testing.

Clustering. The clustering problem consisted of identifying different clusters in 13 datasets [23]: Balance Scale, Bupa, Cancer, Haberman, Hillvalley, Ionosphere, Iris, Pima, Image Segmentation, Sonar, Transfusion, Vehicle and Wine. Its definition is given in [12]. The objective is to minimize the sum of distances between elements of the same cluster. Each cluster must contain at least one element and one element cannot belong to more than one cluster. Datasets were divided using 10-fold cross-validation. Each subset was used for testing and the remaining 9 were used for training, with circular rotation of subsets.

Table 1. Parameter configuration settings for all algorithms investigated

(a) Parameter ranges for all algorithms A, explored by Tuning-PSO

A	Parameter variation ranges
GA	Population size (N): [10,50]. Crossover percentage (pc): [0.1,1.0]. Crossover factor (tc): [0.1,1.0]. Mutation percentage (pm): [0.1,1.0]. Mutation rate (tm): [0.1,1.0]
ACO	Population size (N): [10,50]. Sample size (S): [10,50]. Intensification factor (q): [0.1,1.0]. Distance deviation rate (ζ): [0.1,1.0]
PSO	Population size (N): [10,50]. Inertia weight (w): [0.4,0.9]. Cognitive coefficient $(c1)$: [0.5,2.5]. Social coefficient $(c2)$: [0.5,2.5]. Speed rate (v_r): [0.1,1.0]
ABC	Population size (N): [10,50]. Upper limit of acceleration coefficient (a): [0.2,1.2]
FA	Population size (N): [10,50]. Mutation coefficient (α): [0.1,1.0]. Base value of the coefficient of attraction (β_0): [0.1,3.0]. Coefficient of decay of light intensity (γ): [0.1,1.0]. Reduction rate of the mutation coefficient (α_r): [0.1,1.0]
CS	Population size (N): [10,50]. Probability of abandoning the nest (p_a): [0.1,1.0]
BAT	Population size (N): [10,50]. Coefficient of sound amplitude update (α): [0.1,1.0]. Coefficient of pulse rate update (λ): [0.1,1.0]
SACS	Population size (N): [10,50]

(b) Benchmark functions

Function	Search Space	x*	f(x*)
Sphere	[-5.12,5.12]	[0,...,0]	0
Rosenbrock	[-2.048,2.048]	[1,...,1]	0
Rastrigin	[-5.12,5.12]	[0,...,0]	0
Griewank	[-600,600]	[0,...,0]	0
Ackley	[-32.768,32.768]	[0,...,0]	0
Schwefel	[-500,500]	[420.9687,...,420.9687]	0

(c) Benchmark - best parameter configuration ($d = 2$)

	Sphere	Rosenbrock	Rastrigin	Griewank	Ackley	Schwefel
GA	N: 43 pc: 1.00 tc: 0.26 pm: 0.12 tm: 0.38	N: 40 pc: 0.33 tc: 0.65 pm: 0.61 tm: 0.62	N: 19 pc: 0.81 tc: 0.45 pm: 0.36 tm: 0.72	N: 39 pc: 0.77 tc: 0.25 pm: 0.22 tm: 0.44	N: 18 pc: 1.00 tc: 0.36 pm: 0.45 tm: 0.50	N: 10 pc: 0.72 tc: 0.85 pm: 0.75 tm: 0.13
ACO	N: 10 S: 40 q: 0.10 ζ: 0.51	N: 18 S: 30 q: 0.48 ζ: 0.78	N: 11 S: 45 q: 0.46 ζ: 0.23	N: 26 S: 48 q: 0.13 ζ: 0.25	N: 10 S: 48 q: 0.10 ζ: 0.59	N: 12 S: 22 q: 0.29 ζ: 0.17
PSO	N: 24 w: 0.40 c1: 0.89 c2: 0.82 v_r: 0.62	N: 46 w: 0.44 c1: 1.59 c2: 1.57 v_r: 0.48	N: 10 w: 0.40 c1: 0.94 c2: 1.08 v_r: 0.32	N: 35 w: 0.40 c1: 1.15 c2: 1.00 v_r: 0.56	N: 10 w: 0.40 c1: 0.71 c2: 0.56 v_r: 0.82	N: 10 w: 0.62 c1: 1.39 c2: 1.75 v_r: 0.38
ABC	N: 50 a: 1.18	N: 34 a: 0.37	N: 47 a: 1.09	N: 31 a: 1.07	N: 50 a: 1.13	N: 39 a: 0.29
FA	N: 36 α: 0.54 β_0: 1.75 γ: 0.40 α_r: 0.10	N: 30 α: 0.50 β_0: 1.52 γ: 0.47 α_r: 0.38	N: 11 α: 0.83 β_0: 1.85 γ: 0.34 α_r: 0.10	N: 39 α: 0.97 β_0: 2.32 γ: 0.49 α_r: 0.10	N: 10 α: 0.56 β_0: 1.89 γ: 0.92 α_r: 0.11	N: 10 α: 0.62 β_0: 1.83 γ: 0.62 α_r: 0.46
CS	N: 41 p_a: 0.44	N: 26 p_a: 0.33	N: 42 p_a: 0.63	N: 36 p_a: 0.75	N: 13 p_a: 0.23	N: 10 p_a: 0.27
BAT	N: 37 α: 0.32 λ: 0.53	N: 33 α: 0.85 λ: 0.96	N: 45 α: 0.63 λ: 0.31	N: 27 α: 0.42 λ: 0.82	N: 39 α: 0.50 λ: 0.73	N: 32 α: 0.37 λ: 0.64
SACS	N: 38	N: 40	N: 10	N: 33	N: 11	N: 10

5   Performance Analysis

Benchmark Functions. Table 2(a) and (b) show the results (mean and standard deviation) for 2 and 100 dimensions, respectively (best in bold). CS was the best for all functions, while SACS, FA and PSO equalled CS in only 2 out of 6 functions. In general, CS was the best among all algorithms and functions (32 out of 36 cases), except for 10 and 20 dimensions for Sphere (where SACS and PSO outperformed CS). In specific situations, CS was rivalled by particular algorithms. In 5 dimensions, CS was equalled by SACS and FA in Sphere, and SACS and PSO in Ackley. In 10 and 20 dimensions, CS was equalled by SACS in Ackley. Additionally, CS continued to attain the best solutions in higher dimensions, while the performance of all other algorithms dropped.

Wind Forecasting. Six metrics were employed for performance analysis: MAE, MAPE, MSE, POCID, UTHEIL and ARV [24]. Table 2(c) shows the results over 20 runs (best in bold) in the Montana dataset. CS achieved the best results for all metrics, except MAPE, for which SACS was the best. FA also performed well. Conversely, ACO, PSO and, especially, BAT were consistently inferior in all metrics, except ARV. In the Texas dataset, SACS was the best for all metrics, except POCID, with CS being the best. ACO and ABC also showed attractive results. GA, FA, and, especially, BAT, were consistently inferior in all metrics, except ARV. In general, experiments showed satisfactory results will all algorithms, although there was a clear superiority of CS and SACS in the vast majority of cases. Besides, only CS, SACS and ABC showed values of UTHEIL below 1 in both datasets, meaning only those algorithms were superior to a random walk model. FA in the Montana dataset also reached this performance. In both datasets, MAE, MAPE and MSE values for CS and SACS were quite satisfactory. For POCID, all values were higher than 50%, indicating all models were superior to a coin tossing experiment. For ARV, all values were much lower than 1, tending to 0, a performance much higher than the mean model.

Clustering. Table 2(d) shows the results (best in bold) with respect to accuracy, precision, coverage, F-measure and IRC. CS achieved the lowest distances for all databases, and was the only algorithm to reach the lowest distances in Image Segmentation and Hillvalley. Regarding the 5 metrics, CS attained the best results. In precision and coverage, CS was the best (alone or shared) in 10 out of 13 datasets. In accuracy, CS was the best in 9 out of 13 datasets. In F-measure, CS was the best in 9 out of 13 datasets. And, in IRC, CS was the best in 7 out of 13 datasets. BAT was the worst in all metrics, with worst intra-cluster distance results, in 7 datasets, followed by ACO, in 4 datasets.

6   Statistical Analysis

Benchmark Functions. Friedman test with significance level of 5% was used to detect statistical differences in performance, and Nemenyi post-test was used to find out which pairs of algorithms differed, with 95% confidence. Table 3(a)

Table 2. Results of all algorithms in the different problems and datasets.

(a) Benchmark - $d = 2$

	Sphere	Rosenbrock	Rastrigin	Griewank	Ackley	Schwefel
GA	1.98e-35 (8.84e-35)	2.76e-06 (5.35e-06)	3.14e-04 (1.28e-03)	3.70e-03 (3.79e-03)	6.35e-05 (1.41e-04)	71.06 (70.85)
ACO	2.93e-13 (1.01e-12)	2.89e-05 (2.41e-05)	1.00 (1.37)	2.75e-02 (3.16e-02)	1.61e-02 (7.20e-02)	57.56 (81.81)
PSO	**4.94e-324** (**0**)	**1.23e-32** (**0**)	1.64 (1.22)	1.85e-03 (3.29e-03)	0.44 (1.93)	82.91 (67.66)
ABC	5.48e-21 (5.14e-21)	1.17e-05 (1.16e-05)	1.01e-11 (1.03e-11)	3.77e-04 (3.53e-04)	1.42e-08 (6.28e-09)	1.06e-02 (1.00e-02)
FA	**4.94e-324** (**0**)	9.85e-05 (3.61e-04)	0.55 (0.82)	**4.94e-324** (**0**)	0.64 (1.56)	116.51 (86.44)
CS	**4.94e-324** (**0**)	**1.23e-32** (**0**)	**3.55e-15** (**0**)	**4.94e-324** (**0**)	**2.22e-15** (**0**)	**2.55e-05** (**0**)
BAT	2.34e-09 (4.12e-09)	5.38e-03 (9.05e-03)	3.89e-07 (5.69e-07)	2.25e-03 (3.47e-03)	7.36e-04 (6.14e-04)	3.00e-05 (5.90e-06)
SACS	**4.94e-324** (**0**)	1.42e-32 (8.27e-33)	0.55 (0.99)	4.19e-03 (7.42e-03)	**2.22e-15** (**0**)	17.77 (43.39)

(b) Benchmark - $d = 100$

	Sphere	Rosenbrock	Rastrigin	Griewank	Ackley	Schwefel
GA	8.34 (7.44)	836.15 (221.78)	300.24 (67.36)	14.77 (7.80)	14.60 (1.42)	2.59e+04 (1.50e+03)
ACO	1.42 (2.18)	215.05 (45.66)	301.61 (18.72)	1.98 (3.37)	8.97 (1.59)	1.29e+03 (1.33e+03)
PSO	0.66 (1.28)	78.55 (4.99)	119.19 (26.09)	0.66 (0.62)	10.33 (1.19)	2.24e+04 (2.19e+03)
ABC	1.07 (0.15)	533.86 (104.30)	1.18e+03 (28.83)	16.94 (3.05)	16.58 (0.66)	983.07 (949.82)
FA	4.32e-05 (1.19e-04)	934.17 (676.09)	380.71 (46.99)	9.11 (28.61)	15.54 (0.48)	1.57e+04 (2.15e+03)
CS	**1.49e-250** (**0**)	**1.23e-28** (**0**)	**1.14e-13** (**0**)	**4.94e-324** (**0**)	**2.22e-15** (**0**)	**1.27e-03** (**0**)
BAT	3.59e-04 (7.57e-05)	145.41 (57.39)	0.15 (3.52e-02)	7.61e-02 (2.42e-02)	0.27 (1.90e-02)	1.01 (0.24)
SACS	0.12 (0.30)	147.12 (40.75)	430.69 (95.42)	0.56 (0.47)	14.26 (2.12)	1.41 (1.34)

(c) Wind - Montana

	MAE	MAPE	MSE	POCID	UTHEIL	ARV
GA	2.25 (1.56)	18.88 (12.45)	36.56 (31.32)	86.62 (14.10)	1.71 (1.22)	9.53e-05 (1.82e-05)
ACO	8.12 (6.40)	162.69 (111.23)	232.69 (226.54)	75.94 (11.69)	3.43 (1.64)	3.63e-04 (1.27e-04)
PSO	14.44 (2.10)	283.96 (114.47)	797.13 (332.87)	68.94 (12.46)	3.83 (1.21)	3.79e-04 (1.64e-04)
ABC	2.02 (1.01)	36.87 (19.92)	5.68 (2.64)	86.62 (2.37)	0.47 (0.17)	2.17e-05 (3.05e-05)
FA	1.39 (0.28)	22.23 (6.56)	3.74 (2.54)	86.62 (15.56)	0.38 (0.34)	1.71e-05 (1.58e-05)
CS	**0.83** (2.20e-04)	9.21 (3.67e-04)	**1.39** (1.22e-03)	86.67 (**0**)	0.14 (1.28e-04)	**5.93e-06** (5.32e-09)
BAT	27.17 (13.05)	363.92 (150.96)	3.76e+03 (2.86e+03)	66.35 (11.25)	4.75 (0.98)	5.16e-04 (1.08e-04)
SACS	0.85 (0.21)	**9.19** (**1.28**)	1.45 (0.35)	**86.67** (7.74e-02)	0.14 (3.30e-02)	**5.93e-06** (2.15e-06)

(d) Clustering - Segmentation

	J	Accuracy	Precision	Coverage	F-measure	IRC
GA	1.74e+05 (128.42)	0.89 (1.22e-03)	0.75 (2.14e-03)	0.71 (4.29e-03)	0.72 (3.40e-03)	0.49 (1.31e-02)
ACO	4.66e+05 (9.68e+04)	0.77 (2.01e-03)	0.22 (2.82e-02)	0.20 (7.04e-03)	0.18 (1.06e-03)	2.67e-02 (7.40e-03)
PSO	1.65e+05 (2.83e+03)	0.91 (2.62e-04)	0.75 (5.29e-04)	0.72 (9.18e-04)	0.72 (6.66e-04)	0.50 (4.23e-04)
ABC	2.19e+05 (2.59e+03)	0.89 (5.25e-04)	0.66 (5.14e-03)	0.63 (1.84e-03)	0.64 (2.77e-03)	0.41 (6.72e-03)
FA	2.76e+05 (9.75e+03)	0.87 (7.17e-03)	0.57 (1.91e-02)	0.54 (2.51e-02)	0.53 (2.63e-02)	0.29 (3.34e-02)
CS	**1.64e+05** (**1.31e+03**)	**0.92** (**5.25e-04**)	**0.76** (**1.69e-03**)	**0.74** (**1.84e-03**)	**0.74** (**2.07e-03**)	**0.52** (**3.83e-03**)
BAT	2.62e+05 (1.24e+03)	0.76 (0)	0 (0)	0.14 (0)	3.57e-02 (0)	0 (0)
SACS	1.86e+05 (1.30e+03)	**0.92** (**2.45e-03**)	0.75 (9.14e-03)	0.71 (8.57e-03)	0.72 (7.86e-03)	0.49 (7.89e-03)

Table 3. Results of Friedman test applied to all problems and datasets, according to different metrics.

(a) Benchmark

d	Benchmark - p-value					
	Sphere	Rosenbrock	Rastrigin	Griewank	Ackley	Schwefel
2	0	0	1.50e-10	2.22e-16	0	8.73e-10
5	0	0	0	0	1.88e-12	0
10	0	0	0	0	7.68e-14	0
20	0	0	0	0	0	0
50	0	0	0	0	0	0
100	0	0	0	0	0	0

(b) Wind

Dataset	Wind Forecasting - p-value					
	MAE	MAPE	MSE	POCID	UTHEIL	ARV
Montana	0	0	0	0	0	0
Texas	0	0	0	1.72e-13	0	0

(c) Clustering

Dataset	Clustering - p-value					
	J	Accuracy	Precision	Coverage	F-measure	IRC
balance	0	2.22e-16	0	0	2.22e-16	2.22e-16
bupa	0	2.22e-16	2.22e-16	2.22e-16	2.22e-16	4.44e-16
cancer	0	1.69e-11	1.69e-11	2.51e-09	1.69e-11	1.69e-11
haberman	0	1.23e-08	1.25e-02	1.25e-02	1.25e-02	5.12e-02
hillvalley	0	0	0	0	0	0
ionosphere	0	0	1.23e-12	1.03e-12	4.98e-13	0
iris	0	4.44e-16	4.44e-16	4.44e-16	4.44e-16	4.44e-16
pima	0	2.22e-16	2.22e-16	2.22e-16	2.22e-16	2.22e-16
segmentation	0	0	0	0	0	0
sonar	0	0	0	0	0	0
transfusion	0	2.27e-08	2.27e-08	2.27e-08	2.27e-08	2.27e-08
vehicle	0	5.63e-14	3.25e-10	0	2.27e-08	3.79e-11
wine	0	3.57e-09	1.29e-11	1.29e-11	1.29e-11	2.27e-08

shows the results of Friedman test. Table 4(a) shows the result of Nemenyi test for 2 dimensions (significant differences indicated with 1). For Sphere and Rosenbrock, there were differences between CS and {GA, ACO, ABC} in all dimensions, except for GA in Sphere for 2 dimensions, and for ACO in Sphere for 50 dimensions. For Rastrigin, CS showed differences to ABC and FA in all dimensions, except for ABC in 2 dimensions. For Griewank, CS differed from GA and ABC, except for GA in 10 dimensions. For Ackley, CS outperformed GA, PSO and ABC in 10 dimensions, ABC in 2 and 5 dimensions, and GA in 2 dimensions. For Schwefel, CS was superior to all algorithms in all dimensions, except to SACS and BAT.

Wind Forecasting. For both Montana and Texas datasets, and all metrics, Friedman test (Table 3(b)) confirmed statistical difference in the results. Concerning Montana dataset, CS and SACS outperformed the other algorithms in all metrics, except POCID, being statistically superior to ACO, PSO and BAT. The exception was CS and SACS compared to each other, in which case there

Table 4. Results of Nemenyi test applied to all problems and datasets, according to different metrics.

(a) Benchmark

		GA	ACO	PSO	ABC	FA	CS	BAT
Sphere	ACO	1						
	PSO	1	1					
	ABC	1	0	1				
	FA	1	1	0	1			
	CS	1	1	0	1	0		
	BAT	1	1	1	0	1	1	
	SACS	1	1	0	1	0	0	1
Rosenbrock	ACO	0						
	PSO	1	1					
	ABC	0	0	1				
	FA	1	1	1	1			
	CS	1	1	0	1	1		
	BAT	1	0	1	1	1	1	
	SACS	1	1	0	1	1	0	1
Rastrigin	ACO	1						
	PSO	1	0					
	ABC	0	1	1				
	FA	0	1	1	0			
	CS	1	1	1	1	1		
	BAT	1	0	0	1	0	1	
	SACS	0	1	1	0	0	1	1
Griewank	ACO	1						
	PSO	1	1					
	ABC	0	1	1				
	FA	1	1	0	1			
	CS	1	1	1	1	0		
	BAT	0	1	1	0	1	1	
	SACS	1	1	0	1	1	1	1
Ackley	ACO	1						
	PSO	1	0					
	ABC	0	1	1				
	FA	0	1	1	0			
	CS	1	1	1	1	1		
	BAT	0	1	1	1	0	1	
	SACS	1	0	0	1	1	0	1
Schwefel	ACO	1						
	PSO	0	1					
	ABC	0	1	0				
	FA	1	0	1	1			
	CS	1	1	1	1	1		
	BAT	0	1	0	0	1	0	
	SACS	1	1	1	1	1	1	1

(b) Wind - Montana

		GA	ACO	PSO	ABC	FA	CS	BAT
MAE	ACO	0						
	PSO	1	1					
	ABC	0	0	1				
	FA	0	0	1	0			
	CS	1	1	1	1	1		
	BAT	1	0	1	0	0	1	
	SACS	1	1	1	1	1	0	1
MAPE	ACO	0						
	PSO	0	1					
	ABC	0	0	0				
	FA	0	0	1	0			
	CS	1	1	1	1	1		
	BAT	0	0	1	1	0	1	
	SACS	1	1	1	1	1	0	1
MSE	ACO	0						
	PSO	1	1					
	ABC	0	0	1				
	FA	0	0	1	0			
	CS	1	1	1	1	1		
	BAT	0	0	1	1	0	1	
	SACS	1	1	1	1	1	0	1
POCID	ACO	0						
	PSO	0	1					
	ABC	0	0	0				
	FA	0	0	1	0			
	CS	0	1	1	0	0		
	BAT	1	0	1	1	0	1	
	SACS	0	1	1	0	0	0	1
UTHEIL	ACO	0						
	PSO	1	1					
	ABC	0	0	1				
	FA	0	0	1	0			
	CS	1	1	1	1	1		
	BAT	0	0	1	0	0	1	
	SACS	1	1	1	1	1	0	1
ARV	ACO	0						
	PSO	1	1					
	ABC	0	0	1				
	FA	0	0	1	0			
	CS	1	1	1	1	1		
	BAT	0	0	1	0	0	1	
	SACS	1	1	1	1	1	0	1

(c) Clustering - Segmentation

		GA	ACO	PSO	ABC	FA	CS	BAT
J	ACO	1						
	PSO	0	1					
	ABC	0	1	1				
	FA	1	0	1	0			
	CS	0	1	0	1	1		
	BAT	1	0	1	0	0	1	
	SACS	0	1	0	0	1	1	1
Accuracy	ACO	1						
	PSO	0	1					
	ABC	0	0	1				
	FA	0	0	1	0			
	CS	1	1	0	1	1		
	BAT	1	0	1	1	0	1	
	SACS	0	1	0	0	1	0	1
Precision	ACO	1						
	PSO	0	1					
	ABC	0	0	1				
	FA	1	0	1	0			
	CS	0	1	0	0	1		
	BAT	1	0	1	0	0	1	
	SACS	0	1	0	1	1	0	1

		GA	ACO	PSO	ABC	FA	CS	BAT
Coverage	ACO	1						
	PSO	0	1					
	ABC	0	0	1				
	FA	0	0	1	0			
	CS	1	1	0	1	1		
	BAT	1	0	1	1	0	1	
	SACS	0	1	0	0	1	0	1
F-measure	ACO	1						
	PSO	0	1					
	ABC	0	0	1				
	FA	0	0	1	0			
	CS	1	1	0	1	1		
	BAT	1	0	1	1	0	1	
	SACS	0	1	0	0	1	0	1
IRC	ACO	1						
	PSO	0	1					
	ABC	0	0	1				
	FA	0	0	1	0			
	CS	1	1	0	1	1		
	BAT	1	0	1	1	0	1	
	SACS	0	1	0	0	0	0	1

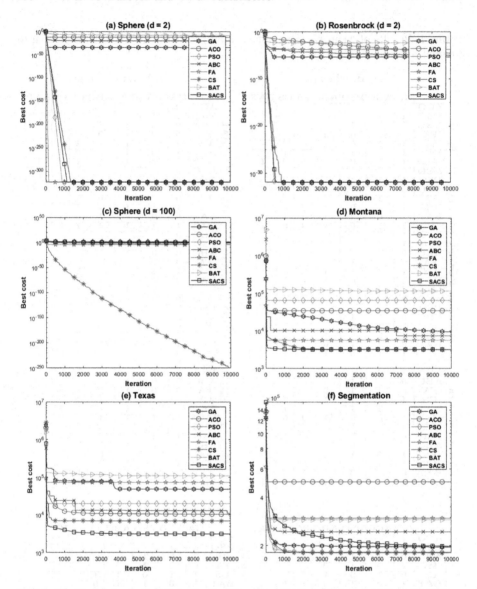

Fig. 1. Convergence analysis of all algorithms in the benchmark function, wind prediction and segmentation problems.

was no significant difference. In Texas dataset, SACS was superior to all algorithms, except CS in MAPE; CS and PSO in POCID; and CS and ABC in ARV. CS was also superior to all algorithms, except SACS in MAPE and MSE; and GA, FA and BAT in the other metrics. Table 4(b) shows the results of Nemenyi tests for the Montana dataset (significant differences indicated with 1).

Clustering. Friedman (Table 3(c)) and Nemenyi tests showed significant differences in all metrics, for all datasets, except for IRC in the Haberman dataset. Table 4(c) shows the results of Nemenyi test for the Segmentation database (significant differences indicated with 1). CS outperformed all algorithms except SACS and PSO in accuracy, coverage, F-measure and IRC; except GA and PSO in J; and except SACS, GA, PSO and ABC in precision. Conversely, BAT and ACO showed the worst results in many pairs for all metrics.

7   Convergence Analysis

Benchmark Functions. An analysis was conducted to assess convergence of the algorithms. Figure 1(a), (b) and (c) show 3 out of the 36 convergence graphs, for Sphere and Rosenbrock in 2 dimensions, and for Sphere in 100 dimensions, respectively. CS presented better convergence than all other algorithms, in 31 out of 36 cases. Only in particular cases, other algorithms converged faster. PSO reached better convergence than both CS and SACS in Sphere and Rosenbrock, in 2 dimensions. The same occurred with Sphere in 20 dimensions. Likewise, FA outperformed CS and SACS for Sphere, in 2 and 5 dimensions, and for Griewank in 2 dimensions.

Wind Forecasting. Figure 1(d) and (e) show convergence for Montana and Texas datasets. In Montana dataset, SACS and CS converged rapdily to smaller errors, followed by FA, ABC and GA, in this order. BAT, PSO and ACO possibly got trapped in local minima with errors far from the best solutions. BAT presented the worst performance. In Texas dataset, SACS converged faster than the other algorithms, followed by CS, ACO, and ABC, in this order. BAT and FA got trapped in local minima, with BAT presenting the worst performance.

Clustering. Figure 1(f) shows convergence of the algorithms for the Segmentation dataset. CS converged faster than the other algorithms, typically reaching solutions in less than 1000 iterations, followed by FA, GA, SACS, ABC, BAT, PSO and ACO.

8   Conclusion

This work analyzed a subset of bio-inspired algorithms in three optimization problems with distinct characteristics. To explore the full potential of the algorithms, an automatic tuning technique, Tuning-PSO, based on a bio-inspired algorithm itself, was proposed to parameterize the bio-inspired algorithms. Among the algorithms investigated, cuckoo search (CS) and its variation, the self adaptive cuckoo search (SACS) have shown, by far, the best performance when compared to all other algorithms in the three problems. Cuckoo search is a simple method with only two parameters to be estimated: population size and probability of abandoning the nest. The extensive experimental investigation, together with its simplicity, provided strong evidences to make cuckoo search a solid and strong candidate for solving optimization problems. Another work being currently carried out is examining the reasons for the efficiency of CS with respect to its main parameters, aiming at explaining their impact on performance.

References

1. Pousinho, H.M.I., Mendes, V.M.F., da Silva Catalo, J.P.: A hybrid PSO–ANFIS approach for short-term wind power prediction in Portugal. Energy Convers. Manag. **52**(1), 397–402 (2011)
2. Rahmani, R., Yusof, R., Seyedmahmoudian, M., Mekhilef, S.: Hybrid technique of ant colony and particle swarm optimization for short term wind energy forecasting. J. Wind Eng. Ind. Aerodyn. **123**, 163–170 (2013)
3. Talbi, E.-G., Melab, N., Cahon, S.: Handbook of Bioinspired Algorithms and Applications (2006)
4. Holland, J.H.: Algoritmos genéticos. Investigación y Ciencia **192**, 38–45 (1992)
5. Dorigo, M., Di Caro, G., Gambardella, L.M.: Ant algorithms for discrete optimization. Artif. Life **5**(2), 137–172 (1999)
6. Eberhart, R.C., Kennedy, J., et al.: A new optimizer using particle swarm theory. In: Proceedings of the Sixth International Symposium on Micro Machine and Human Science, New York, NY, vol. 1, pp. 39–43 (1995)
7. Karaboga, D.: An idea based on honey bee swarm for numerical optimization. Technical report, Technical report-tr06, Erciyes University, Engineering Faculty, Computer Engineering Department (2005)
8. Yang, X.-S.: Firefly algorithm, stochastic test functions and design optimisation. Int. J. Bio-inspired Comput. **2**(2), 78–84 (2010)
9. Yang, X.-S., Deb, S.: Cuckoo search via levy flights. In: World Congress on Nature and Biologically Inspired Computing, NaBIC 2009, pp. 210–214. IEEE (2009)
10. Yang, X.-S.: A new metaheuristic bat-inspired algorithm. In: González, J.R., Pelta, D.A., Cruz, C., Terrazas, G., Krasnogor, N. (eds.) NICSO 2010. SCI, vol. 284, pp. 65–74. Springer, Heidelberg (2010). https://doi.org/10.1007/978-3-642-12538-6_6
11. Li, X., Yin, M.: Modified cuckoo search algorithm with self adaptive parameter method. Inf. Sci. **298**, 80–97 (2015)
12. Hruschka, E.R., Ebecken, N.F.F.: A genetic algorithm for cluster analysis. Intell. Data Anal. **7**(1), 15–25 (2003)
13. Booker, L.B., Goldberg, D.E., Holland, J.H.: Classifier systems and genetic algorithms. Artif. Intell. **40**(1), 235–282 (1989)
14. Van der Merwe, D.W., Engelbrecht, A.P.: Data clustering using particle swarm optimization. In: The 2003 Congress on Evolutionary Computation, CEC 2003, vol. 1, pp. 215–220. IEEE (2003)
15. Yang, X.-S., Deb, S.: Cuckoo search: recent advances and applications. Neural Comput. Appl. **24**(1), 169–174 (2014)
16. Barbosa, C.E.M., Vasconcelos, G.C.: Cuckoo search optimization for short term wind energy forecasting. In: 2016 IEEE Congress on Evolutionary Computation (CEC), pp. 1765–1772. IEEE (2016)
17. Jensi, R., Wiselin Jiji, G.: MBA-LF: a new data clustering method using modified bat algorithm and levy flight. ICTACT J. Soft Comput. **6**(1), 1093–1101 (2015)
18. Zhao, B.: An improved particle swarm optimization algorithm for global numerical optimization. In: Alexandrov, V.N., van Albada, G.D., Sloot, P.M.A., Dongarra, J. (eds.) ICCS 2006. LNCS, vol. 3991, pp. 657–664. Springer, Heidelberg (2006). https://doi.org/10.1007/11758501_88

19. Jamil, M., Yang, X.-S.: A literature survey of benchmark functions for global optimisation problems. Int. J. Math. Model. Numer. Optim. **4**(2), 150–194 (2013)
20. Friedman, J.H.: An overview of predictive learning and function approximation. In: Cherkassky, V., Friedman, J.H., Wechsler, H. (eds.) From Statistics to Neural Networks. NATO ASI Series, vol. 136, pp. 1–61. Springer, Heidelberg (1994). https://doi.org/10.1007/978-3-642-79119-2_1
21. U.S. Department of Energy. National renewable energy laboratory (2016). https://www.nrel.gov/wind/. Accessed 24 Apr 2016
22. Phillips, N.A.: Numerical weather prediction. Adv. Comput. **1**, 43–90 (1960)
23. Center for Machine Learning and Intelligent Systems. UCI machine learning repository (2016). https://archive.ics.uci.edu/ml/datasets.html. Accessed 04 June 2016
24. Zhao, X., Wang, S., Li, T.: Review of evaluation criteria and main methods of wind power forecasting. Energy Procedia **12**, 761–769 (2011)

Robotic Flow Shop Scheduling with Parallel Machines and No-Wait Constraints in an Aluminium Anodising Plant with the CMAES Algorithm

Carina M. Behr and Jacomine Grobler[(✉)]

Department of Industrial and Systems Engineering,
University of Pretoria, Pretoria, South Africa
jacomine.grobler@gmail.com

Abstract. This paper proposes a covariance matrix adaptation evolution strategy (CMAES) based algorithm for a robotic flow shop scheduling problem with multiple robots and parallel machines. The algorithm is compared to three popular scheduling rules as well as existing schedules at a South African anodising plant. The CMAES algorithm statistically significantly outperformed all other algorithms for the size of problems currently scheduled by the anodising plant. A sensitivity analysis was also conducted on the number of tanks required at critical stages in the process to determine the effectiveness of the CMAES algorithm in assisting the anodising plant to make business decisions.

Keywords: Robotic flow shop scheduling
Covariance matrix adaptation evolution strategy

1 Introduction

Anodising is an electrolytic reaction used to produce a layer of aluminium oxide on an aluminium alloy. Anodising production lines often consist of a series of chemical processes with material needing to be moved from station to station by means of overhead cranes - a complex production scheduling problem. Developing an optimization algorithm to solve a problem of this nature has both practical and academic significance. Firstly, effective production scheduling has a major impact on the cost of production, utilization of resources and customer satisfaction. Secondly, the optimization problem is a discrete-valued optimization problem with a number of complicated constraints, which is challenging to solve by means of an evolutionary algorithm.

In this paper, a scheduling problem at a South African anodising plant is formulated as a robotic flow shop problem with multiple robots and parallel machines with no-wait constraints. A covariance matrix adaptation evolution

© Springer International Publishing AG, part of Springer Nature 2018
L. Rutkowski et al. (Eds.): ICAISC 2018, LNAI 10841, pp. 302–312, 2018.
https://doi.org/10.1007/978-3-319-91253-0_29

strategy (CMAES) [1] based scheduling algorithm is developed to solve the problem. The CMAES based algorithm is benchmarked against a number of scheduling heuristics on real data from the anodising plant. The results are also compared against the existing schedules generated by the production personnel. The CMAES algorithm statistically significantly outperformed all the benchmark algorithms and existing schedules for problems where around 35 jobs needed to be scheduled on 13 stations. As problem size increased the performance of the CMAES algorithm worsened. A sensitivity analysis was also conducted on the number of tanks required at critical stages in the process.

This paper is significant because, to the best of the authors' knowledge, it describes the first CMAES based algorithm for solving a robotic flow shop problem.

The rest of the paper is organized as follows: Sect. 2 describes the actual scheduling problem in more detail and provides an overview of existing approaches used in literature for solving similar problems. Section 3 provides a brief introduction to CMAES and describes the scheduling algorithm in more detail. The experimental setup and results are described in Sect. 4 before the CMAES as tool for business decision making is evaluated in Sect. 5. Finally, the paper is concluded in Sect. 6.

2   Robotic Flow Shop Problems

The robotic flow shop problem with parallel processing stations and multiple robots requires n jobs to be processed by s stations, one operation on each station. All jobs complete processing in the same sequence with standard processing times [2]. The purpose is to determine the sequence in which the jobs should be processed to minimize the makespan (total time to complete all required jobs). The cranes in the anodising process (modelled as robots) have machine availability constraints, which imply that a crane can only be utilized to move material between two stations if it is not already in use for another move. A move time is calculated for each job to be transported between two stations as the time required for the crane to move to the pickup location and the actual time required for the crane to move the job to the next station.

Figure 1 illustrates the specific anodising process and its stations. The brackets indicate parallel stations. The hot etch, anodising and sealing stations are critical stations or stations with a no-wait scenario, which means the cranes need to pick up jobs from these stations the moment the processing time has been completed (no-wait constraints). The remaining stations may wait for the crane to become available before a pickup is made.

Since the cranes cannot cross over each other, the two cranes are each pre-allocated stations to service with the first crane servicing the first half of the stations and the second crane the second half of the stations. A detailed formulation of this problem can be found in [3] and will thus not be repeated here for the sake of conciseness.

A number of researchers have already solved variations of the robotic flow shop scheduling problem. The first variations focused largely on solving problems

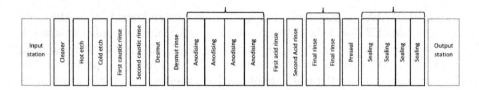

Fig. 1. Anodising stations.

where only one robot was available to move products between stations [4,5]. The problems were soon expanded to consider multiple robots [6,7] and parallel machines [3,8], where a stage has more than one machine which can be used to process a job.

Three types of part pickup criteria have been considered, namely no-wait [9, 10], interval [11,12], and free pickup. Additional complexities such as reentrance [13], fixed robot routes [14] and sequence-dependent setup-times [4] can also be found.

With regard to solution strategies, formulating problems as mixed integer linear programmes which are solved exactly by software packages such as CPLEX [3,6,11,13], is common for small sized problems. Larger problems are solved by heuristics [8,14] and meta-heuristics such as genetic algorithms [5], ant colony optimization [16] and cuckoo search [4].

Two main conclusions can be reached from analyzing the existing robotic flow shop literature. Firstly, there is no evidence of previous use of CMAES for solving a robotic flow shop scheduling problem. Secondly, the number of jobs typically considered in literature are significantly less than the number of jobs which need to be considered for the anodising plant described in this paper. Zhou et al. [11], for example, focused on scheduling 3 jobs on 24 stages (resulting in 72 operations) and Li et al. [3] focused on scheduling 3 jobs on 20 stages (resulting in 60 operations). The problems considered in this paper ranges from 455 operations to 1300 operations. The next section will discuss the CMAES based algorithm developed to solve these problems in more detail.

3   CMAES for Robotic Flow Shop Scheduling

Lei and Wang [15] proved that robotic flow shop problems with time window constraints, such as no-wait constraints, are np-complete. The more complicated problem addressed in this paper can thus not be solved to optimality in polynomial time. Approximation algorithms such as heuristic rules or evolutionary algorithms are thus logical options. CMAES was selected as basis for the scheduling algorithm due to its excellent performance versus more well known evolutionary algorithms such as genetic algorithms (GAs) and differential evolution algorithms on continuous optimization problems [1]. Furthermore, CMAES has only one parameter to tune and requires significantly fewer individuals per iteration compared with, for example a GA, to obtain satisfactory performance.

3.1 Background on CMAES

CMAES is a stochastic, non-linear optimization algorithm. The CMAES algorithm consists of four main phases, namely solution generation, selection and recombination, covariance matrix update, and step size update. During the first generation phase, a population of solutions is generated at each iteration according to a multivariate normal distribution such that

$$x_i(t+1) \sim N(m(t), \sigma_{CMA}^2(t))C(t), \tag{1}$$

where $x_i(t+1)$ is the ith candidate solution at iteration $t+1$, $N(m(t), \sigma_{CMA}^2(t))$ denotes a normal distribution with mean $m(t)$ and standard deviation $\sigma_{CMA}(t)$. The mean of the CMAES population at time t is denoted by $m(t)$, σ_{CMA} denotes the step size of the algorithm at time t, and $C(t)$ is the covariance matrix at time t. After the solutions are evaluated and sorted, selection and recombination takes place by adjusting the mean of the population as follows:

$$m(t+1) = \sum_{k=1}^{n_s} w_k x_k, \tag{2}$$

where n_s is the population size and w_k is the kth recombination weight in the CMAES algorithm.

The covariance matrix, $C(t)$, is then updated as:

$$C(t+1) = (1 - c_{cov})C(t) + \frac{c_{cov}}{\mu_{cov}} p_{c_{CMA}} p_{c_{CMA}}^T + c_{cov}\left(1 - \frac{1}{\mu_{cov}}\right)$$
$$\times \sum_{k=1}^{n_s} w_k \left(\frac{x_k(t+1) - m(t)}{\sigma_{CMA}(t)}\right)\left(\frac{x_k(t+1) - m(t)}{\sigma_{CMA}(t)}\right)^T, \tag{3}$$

where

$$\mu_{cov} \geq 1, \tag{4}$$
$$\mu_{cov} = \mu_{eff}, \text{and} \tag{5}$$
$$c_{cov} \approx \min(\mu_{cov}, \mu_{eff}, n_x^2)/n_x^2. \tag{6}$$

The symbol c_{cov} denotes the learning rate for the covariance matrix update, μ_{eff} denotes the variance effective selection mass and μ_{cov} denotes the parameter which weighs between the rank-one update and rank-μ update. The rank-one update uses only the previous iteration to estimate the covariance matrix where the rank-μ update uses all previous iterations. n_x is the number of problem dimensions.

The CMAES algorithm makes use of cumulative step-size adaptation. A cumulative path is used which is a combination of all the steps an algorithm has made with the importance of a step decreasing exponentially with time [17]. Two evolution paths are used in the CMAES algorithm, the anisotropic evolution path, $p_{c_{CMA}}$, associated with the covariance matrix and the isotropic evolution path, p_σ, associated with the step size. $p_{c_{CMA}}$ is calculated as follows:

$$p_{c_{CMA}} = (1 - c_{c_{CMA}})p_{c_{CMA}} + \sqrt{c_{c_{CMA}}(2 - c_{c_{CMA}})\mu_{eff}}\left(\frac{m(t+1) - m(t)}{\sigma_{CMA}(t)}\right),$$
(7)

where μ_{eff} is given by

$$\mu_{eff} = \left(\sum_{k=1}^{n_s} w_k^2\right)^{-1}$$
(8)

and $c_{c_{CMA}}$ is the backward time horizon of the anisotropic evolution path.

Finally, the step size, $\sigma_{CMA}(t+1)$, is updated as follows:

$$\sigma_{CMA}(t+1) = \sigma_{CMA}(t)\exp\left(\frac{c_\sigma}{d_\sigma}\left(\frac{\|p_\sigma(t+1)\|}{E\|N(0,I)\|} - 1\right)\right),$$
(9)

where d_σ is the damping parameter in the CMAES algorithm, $\frac{1}{c_\sigma}$ is the backward time horizon of the isotropic evolution path, p_σ:

$$p_\sigma = (1 - c_\sigma)p_\sigma + \sqrt{c_\sigma(2 - c_\sigma)\mu_{eff}}C(t)^{-0.5}\left(\frac{m(t+1) - m(t)}{\sigma_{CMA}(t)}\right).$$
(10)

The Covariance Matrix Adaption Evolution Strategy (CMAES) algorithm has already been used for a number of real world problems such as to optimise the layout of a three column wind farm [18] and for the optimization of irrigation scheduling [19].

3.2   The CMAES-Based Robotic Flow Shop Scheduling Algorithm

Since the CMAES algorithm operates in a continuous space, a mapping mechanism is required to convert the candidate solutions ($x_i(t+1)$) to valid schedules [20]. Each candidate solution is interpreted as a set of priorities for the jobs to be scheduled. This prioritized list of jobs is then provided as input to a scheduling algorithm described in Fig. 2. The algorithm provides the makespan associated with the schedule of each candidate solution which is then returned to the CMAES algorithm.

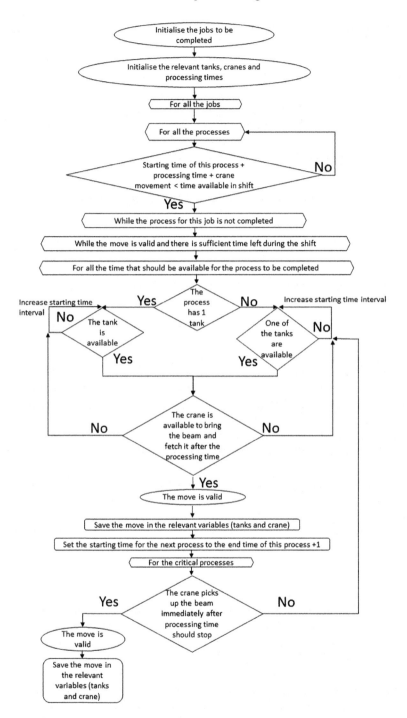

Fig. 2. Flowchart of the algorithm used to calculate the fitness function of each individual.

4   Empirical Evaluation

The evaluation of the CMAES scheduling algorithm was conducted on 5 datasets derived from real world data from the South African anodising plant. The datasets range in size from 34 jobs (corresponding to the current planning horizon) to 100 jobs which needs to be produced on 13 processing stages. These datasets are available for comparison purposes from the corresponding author.

The parameter settings used for the CMAES algorithm is based on recommendations from literature [1] and is listed in Table 1.

Table 1. The CMAES algorithm parameters.

Parameter	Value used
Maximum iterations	$\frac{1e3(n_x+5)^2}{\sqrt{n_s}}$
Search interval ($[LB, UB]$)	$[-100, 100]$
Population size (n_s)	$4 + floor(3 \log n_x)$
Sigma	$0.35(UB - LB)$

For the first three datasets, the actual production schedule as planned and executed by the anodising plant could be obtained. These schedules, referred to as the "as-is" schedules can be used as a baseline to compare the performance of the CMAES algorithm. Furthermore, another three production scheduling rules were used for benchmarking purposes. These rules are the current state-of-the-art algorithms developed specifically for the anodising plant's problem. The rules are FIFO (First In, First Out), Priority (where priority orders are completed first and then the as-is schedule from the plant is followed for the rest of the orders) and SPT (shortest processing time) (all 10 μm orders, followed by 15 μm and then 25 μm).

The results of the comparison between the CMAES scheduling algorithm, the benchmarking algorithms and the as-is schedules are recorded in Table 2. Due to the stochastic nature of the CMAES algorithm, all CMAES results were recorded over 30 independent simulation runs. Throughout the rest of this section, μ and σ respectively denote the mean and standard deviation associated with the makespan of the algorithm.

A statistical analysis was conducted to validate the results obtained. For every dataset, a Mann-Whitney U test at 95% significance was performed to compare the CMAES algorithm results to the benchmarking algorithms. The results in bold indicate that a statistical significant performance improvement was obtained. As can be seen, the CMAES algorithm outperforms the other algorithms for the first three datasets. Up to 115 min can be saved per day by the anodising plant if they utilize the CMAES algorithm instead of their existing scheduling procedures. With regard to the larger two datasets, the CMAES performs statistically similar to the SPT heuristic, indicating that further design

Table 2. Comparative results of the CMAES scheduling algorithm to the benchmark algorithms and as-is schedules.

Dataset	Operations	CMAES			FIFO	SPT	Priority	As-is
		μ	σ	Time (s)	C	C	C	C
07-Aug	34	**457.23**	11.596	412.16	603	571	600	572
22-Aug	34	**470.6**	13.051	415.72	556	546	579	560
23-Aug	42	**544.3**	10.639	649.57	643	608	727	658
50	50	627	0	2276.28	NA	627	NA	798
100	100	1177	0	5985.65	NA	1177	NA	1621

improvements will need to be made to the CMAES algorithm if a larger scheduling horizon is considered in future.

The time required to obtain a solution for each problem was also recorded in Table 2. A computer running Windows 7 Enterprise with an Intel(R) Core(TM) i7-4710MQ CPU @ 2.50 GHz with 8 Gb RAM was used to obtain the results. Running an algorithm overnight to schedule the next morning's production is considered acceptable in industry and thus the computational time required to obtain a solution is not an issue.

5   Using the CMAES Algorithm for Business Decisions

From the results obtained, it is clear that the CMAES algorithm could add value by reducing the total time required to produce material. This section investigates the use of the CMAES algorithm to assist the anodising plant in making business decisions. A sensitivity analysis was conducted on the number of tanks used at two critical stages in the manufacturing process. Three scenarios were developed. Scenario 1 involved the addition of two anodising tanks to the existing four tanks at the anodising station. Scenario 2 involved adding two sealing tanks to the existing four sealing tanks at the sealing station. Scenario 3 involved adding both additional anodising and sealing tanks to the existing production line. The CMAES algorithm was used to solve the three scenarios and the results were recorded in Table 3.

A statistical analysis was again conducted. For every comparison of scenarios, a Mann-Whitney U test was performed (using the two sets of 30 data points of the two scenarios being compared) and if the existing scenario statistically significantly outperformed the second scenario, a win was recorded. If no statistical difference could be observed a draw was recorded. If the second scenario resulted in statistically significantly better results than the as-is scenario, a loss was recorded for the as-is scenario. As an example, (5-0-0) in row 1 column 1, indicates that the "as-is" scenario significantly outperformed Scenario 1 for five of the datasets. No draws or losses were recorded (Table 4).

Interestingly, the results show that adding additional tanks does not have a significant impact on the schedule. One possible explanation could be that the

Table 3. Sensitivity analysis of the CMAES algorithm results to additional tanks.

Dataset	As-is setup		Scenario 1		Scenario 2		Scenario 3	
	μ	σ	μ	σ	μ	σ	μ	σ
07-Aug	457.23	11.596	524.23	1.2507	536.23	11.227	458.03	12.076
22-Aug	470.6	13.051	496.93	4.0338	507.37	14.39	451	0
23-Aug	544.3	10.639	628.73	0.82768	648.53	5.8648	545.23	11.596
50	627	0	630.37	3.81	649.9	3.2308	627	0
100	1177	0	1182.5	5.0596	1243.3	4.4347	1177	0

Table 4. Hypotheses analysis of the impact of additional tanks on scheduling performance.

Scenario 1	Scenario 2	Scenario 3	TOTAL
5-0-0	5-0-0	0-4-1	10-4-1

process bottleneck is now simply moved to another resource, such as the cranes, which are required to move all jobs between stations. Future work can focus on a more in-depth analysis of the system bottlenecks.

6   Conclusion

This paper described the development of a CMAES based algorithm for a robotic flow shop scheduling problem with multiple robots and parallel machines. Data from a South African anodising plant was used to compare the CMAES algorithm to three popular scheduling rules as well as existing schedules used at the plant. The CMAES algorithm statistically significantly outperformed all other algorithms for problems with around 500 operations. An example analysis was also conducted to show how the algorithm can be used to evaluate various system parameters.

Future research opportunities lie in the improvement of algorithm performance on larger problems, benchmarking the CMAES algorithm against other meta-heuristic algorithms, more in-depth tuning of the CMAES parameters, and a more thorough analysis of the system bottlenecks.

Acknowledgements. This work is based on the research supported wholly or in part by the National Research Foundation of South Africa (Grant Number 109273). The authors would also like to thank the University of Twente for their financial support.

References

1. Auger, A., Hansen, N.: A restart CMA evolution strategy with increasing population size. In: Proceedings of the 2005 IEEE Congress on Evolutionary Computation, pp. 1769–1776 (2005)
2. Tang, L.X., Liu, P.: Two-machine flow shop scheduling problems involving a batching machine with transportation or deterioration consideration. Appl. Math. Model. **33**, 1187–1199 (2009)
3. Li, X., Chan, F.T.S., Chung, S.H.: Optimal multi-degree scheduling of multiple robots without overlapping in robotic flow shops with parallel machines. J. Manufact. Syst. **36**, 62–75 (2015)
4. Majumder, A., Laha, D.: A new cuckoo search algorithm for 2-machine robotic cell scheduling problem with sequence-dependent setup times. Swarm Evol. Comput. **28**, 131–143 (2016)
5. Lim, J.M.: A genetic algorithm for a single hoist scheduling in the printed-circuit-board electroplating line. Comput. Ind. Eng. **33**(3–4), 789–792 (1997)
6. Che, A., Lei, W., Feng, J., Chu, C.: An improved mixed integer programming approach for multi-hoist cyclic scheduling problem. IEEE Trans. Autom. Sci. Eng. **11**(1), 302–309 (2014)
7. Che, A., Chu, C.: Optimal scheduling of material handling devices in a PCB production line: problem formulation and a polynomial algorithm. Math. Probl. Eng. **2008** (2008). Article ID 364279
8. Geismar, H.N., Pinedo, M., Sriskandarajah, C.: Robotic cells with parallel machines and multiple dual gripper robots: a comparative overview. IIE Trans. **40**(12), 1211–1227 (2008)
9. Che, A., Chu, C.: Multi-degree cyclic scheduling of a no-wait robotic cell with multiple robots. Eur. J. Oper. Res. **199**(1), 77–88 (2009)
10. Che, A., Chabrol, M., Gourgand, M., Wang, Y.: Scheduling multiple robots in a no-wait re-entrant robotic flow shop. Int. J. Prod. Econ. **135**(1), 199–208 (2012)
11. Zhou, Z., Che, A., Yan, P.: A mixed integer programming approach for multi-cyclic robotic flow shop scheduling with time window constraints. Appl. Math. Model. **36**(8), 3621–3629 (2012)
12. Dawande, M., Geismar, H.N., Pinedo, M., Sriskandarajah, C.: Throughput optimization in dual-gripper interval robotic cells. IIE Trans. **42**(1), 1–15 (2009)
13. Li, X., Fung, R.Y.: A mixed integer linear programming solution for single hoist multi-degree cyclic scheduling with reentrance. Eng. Optim. **46**(5), 704–723 (2014)
14. Kats, V., Levner, E.: Parametric algorithms for 2-cyclic robot scheduling with interval processing times. J. Sched. **14**(3), 267–279 (2011)
15. Lei, L., Wang. T.: A proof: the cyclic hoist scheduling problem is NP-complete. Graduate School of Management, Rutgers University, Working Paper, pp. 89–116 (1989)
16. Elmi, A., Topaloglu, S.: Cyclic job shop robotic cell scheduling problem: ant colony optimization. Comput. Ind. Eng. **111**, 417–432 (2017)
17. Chotard, A., Auger, A., Hansen, N.: Cumulative step-size adaptation on linear functions. In: Coello, C.A.C., Cutello, V., Deb, K., Forrest, S., Nicosia, G., Pavone, M. (eds.) PPSN 2012. LNCS, vol. 7491, pp. 72–81. Springer, Heidelberg (2012). https://doi.org/10.1007/978-3-642-32937-1_8
18. Saravanan, A.J., Karthikeyan, C.P., Samuel, A.A.: Optimization of exogenous and endogenous variables for a three column wind farm using CMAES. Appl. Mech. Mater. **573**, 777–782 (2014)

19. Belaqziz, S., Mangiarotti, S., Le Page, M., Khabba, S., Er-Raki, S., Agouti, T., Drapeau, L., Kharrou, M.H., El Adnani, M., Jarlan, L.: Irrigation scheduling of a classical gravity network based on the covariance matrix adaptation - evolutionary strategy algorithm. Comput. Electron. Agric. **102**, 64–72 (2014)
20. Grobler, J., Engelbrecht, A.P., Kok, S., Yadavalli, V.S.S.: Metaheuristics for the multi-objective FJSP with sequence-dependent set-up times, auxiliary resources and machine down time. Ann. Oper. Res. **180**(1), 165–196 (2010)

Migration Model of Adaptive Differential Evolution Applied to Real-World Problems

Petr Bujok[(✉)]

Department of Informatics and Computers, University of Ostrava,
30. dubna 22, 70103 Ostrava, Czech Republic
petr.bujok@osu.cz
http://prf.osu.eu/kip/

Abstract. Ten variants of migration model are compared with six adaptive differential evolution (DE) algorithms on real-world problems. Two parameters of migration model are studied experimentally. The results of experiments demonstrate the superiority of the migration models in first stages of the search process. A success of adaptive DE algorithms employed by migration model is systematically influenced by the studied parameters. The most efficient algorithm in the comparison is proposed migration model P15x50. The worst performing algorithm is adaptive DE.

Keywords: Differential evolution · Migration model
Migration frequency · Sub-population size · Experimental study
Real-world problems

1 Introduction

In many areas of research and industry arises necessity to solve various problems. Each problem to be solved is represented by an objective function. These functions should be very simple, but in prevalent cases, the computational costs to solve more complex problems are non-negligibly high. An essential feature of the real problems is their dimensionality that substantially influences the total computational costs of the solution process.

In last decades, a favorite kind of optimization techniques called evolutionary algorithms (EAs) is rapidly developed. An essence of this fact is that EAs are very effective even in the solution of complex problems with higher dimensionality. Popularity of EAs is given also by its simplicity because the main idea is to use randomness to reduce time demands and keep the quality of the provided solution.

One of the most widely studied and applied representatives of EAs is differential evolution (DE). Differential evolution is very popular optimization technique used to search the solution of problems in many areas of research and industry. It was introduced by Storn and Price in 1995, and its popularity could be also measured by more than 7500 citations of the original paper [14].

© Springer International Publishing AG, part of Springer Nature 2018
L. Rutkowski et al. (Eds.): ICAISC 2018, LNAI 10841, pp. 313–322, 2018.
https://doi.org/10.1007/978-3-319-91253-0_30

Although DE algorithm is very useful optimizer there are even real complex problems which DE is not able to solve. In last decades, many researchers of optimization tasks developed adaptation mechanisms of DE control parameters to increase the possibility of finding the global solution of the tasks. Beside the adaptive DE variants, many other principles increasing algorithm efficiency are included in original DE. A comprehensive review of state-of-the-art in a case of DE algorithms are in [7,8].

The main motivation for this experimental study is applied the existing parallel migration model [4] to real-world problems. These problems are defined by various dimensionality, therefore it is necessary to study two main control parameters of migration model to find out the proper settings. Although the experiments of parallel migration model are performed on single CPU computer, the efficiency measured by the number of the objective function evaluations should detect some good and poor migration settings compare to the original adaptive DE variants. Given results and conclusions of this experiment could provide potential ways how to set parameters of migration model and solve complex problems in physical parallel computational grids.

There are several existing works focused on parallel DE algorithm and real problems. In a first, a multi-objective maximization problem of an amount of ethylene and propylene in a petrochemical industry process is solved by parallel DE in [16]. Several settings A short-term hydro scheduling problem of power plants generator is solved by the parallel model of self-adaptive DE [9]. Authors of [12] study many parameters of migration model of DE. The state-of-the-art of parallel evolutionary algorithms containing deep analysis of various parallel models, several parallel programming frameworks and hardware for distributed computing is summed in [10].

In this paper, a migration model of adaptive DE variants is applied to real-world optimization problems CEC 2011. The main goal is to show how several various settings of migration of individuals influence the efficiency of the proposed model. The performance of the migration model is also compared with the original adaptive DE variants. The remaining part of the paper is organized as follows. A brief review of adaptive state-of-the-art DE variants used in this paper is explained in Sect. 2. Details of the migration model are described in Sect. 3. A test suite of real-world problems CEC 2011 with the experiment settings is introduced in Sect. 4. Results of the experimental comparison of the migration model are presented in Sects. 5 and some remarks are provided in Sect. 6.

2 Adaptive Variants of Differential Evolution

Although DE is efficient optimizer there exist many various adaptive variants of DE which are widely used to solve optimization problems. In this experimental study, four state-of-the-art adaptive DE algorithms [1,11,13,18] have been compared experimentally with DE algorithm using composite trial vector generation strategies and control parameters (CoDE) [17] and with a variant of competitive DE [15]. The main purpose to select this sixth of adaptive DE is to tie up to the previous successful experimental study [4] and show that simple idea should provide significant increasing of the performance. Necessary to note that more sophisticated adaptive DE variants arisen latter should be also implemented in presented migration model. A very brief familiarization of used adaptive DE follows. More details are provided in the original references.

Self-adaptive *jDE* algorithm proposed by Brest et al. [1] uses the DE/rand-/1/bin strategy with an evolutionary self-adaptation of F and CR. Differential Evolution with Strategy adaptation (*SaDE*) [13] uses four strategies which are preferred according to its success rate in the previous LP generations. *JADE* variant of adaptive differential evolution [18] extends the original DE concept with three different improvements – current-to-pbest mutation, a new adaptive control of parameters F and CR, and archive. Next adaptive DE variant using Ensemble of Parameter values and mutation Strategies (*EPSDE*), has mutation strategies and the values of control parameters F and CR stored in pools and was proposed in [11]. In competitive DE H strategies (in this study *b6e6rl* uses $H = 12$ strategies) are used with their control-parameter values held in the pool and each strategy is preferred according to its success rate in the preceding step. DE algorithm with three well-studied composite trial vector generation strategies simultaneously applied on each point in population and control parameters (*CoDE*) has been recently presented [17].

3   Migration Model of DE

There are several scenarios to distribute operations of DE algorithm to achieve better results. A migration model used in this experiment is widely-used because of its simplicity and possibility of exchange information between parallel processes [2,3,5]. Migration model is controlled by several parameters which setting is provided as follows. A comprehensive state-of-the-art of distributed evolutionary algorithms is available in [10].

In general migration model there are k islands and each contains one sub-population, P_j, $j = 1, 2, \ldots, k$. Each island is linked only to a special island called *mainland* and individuals can migrate only between the linked islands. This topology is called *star*. A pseudocode of experimentally compared parallel migration model is shown in Algorithm 1. Sub-populations has the same size, N_p is an input parameter. Each sub-population is developed independently by one of the six adaptive DE algorithm described above until the moment to migration is reached. Necessary to note that in one studied model only b6e6rl variant is performed in all sub-populations and also mainland population (the name of this variants is in an index). The migration from the islands to the mainland occurs after performing a given number of generations *nde* (studied input parameter of the algorithm). In the migration model used here, the individual with the least function value of the ith sub-population ($\boldsymbol{x}_{best,i}$) replaces the ith individual ($\boldsymbol{x}_i^m$) of the mainland population, and N_{rnd} other randomly chosen points of the sub-population (except $\boldsymbol{x}_{best,i}$) overwrite N_{rnd} individuals of mainland population on places corresponding to kth sub-population, N_{rnd} is also an input parameter.

In this study, this parameter is not deeply studied, and it is set to $N_{rnd} = 4$. Thus, $N_{rnd} + 1$ individuals from each island are copied to central mainland sub-population. It is obvious the size of the mainland sub-population N_m should be set up to $N_m \geq k \times (N_{rnd} + 1)$. If $N_m = k \times (N_{rnd} + 1)$, the mainland sub-population is refreshed completely in each epoch, and the elitism of the migration model is ensured. In order to satisfy this condition, the input parameter N_{rnd} was set up to $N_{rnd} = 4$ and $N_m = 6 \times 5 = 30$ in all the experiments.

After finishing the migration of the selected individuals from the islands to the mainland, the search process continues applying a DE variant on the mainland until the stopping condition for the current epoch (1) is reached. In the proposed migration model, competitive *b6e6rl* as the most reliable in preliminary experiments was chosen for the mainland sub-population. Only in the one case of migration model setting, the fastest JADE variant was performed on mainland sub-population (the name of JADE algorithm is in the index of the model). The stopping condition for the mainland, and the current epoch was formed as follows:

$$f_{\max} - f_{\min} < 1 \times 10^{-6} \quad \text{OR} \quad FES_m > 10^{(epoch-1)} \times 2 \times nde \times N_m, \quad (1)$$

where $f_{\max}$ and $f_{\min}$ are the worst and the best objective function values of the mainland sub-population, respectively, and FES_m is the number of function evaluations in the mainland population during this epoch. Notice that in early epochs the evolution on the mainland tends to stop due to the given limit of allowed function evaluations (after $2 \times nde$ generations in the first epoch) while in late epochs due to the small difference of the function values in the mainland population. The whole search process of the algorithm is stopped after the pre-defined number of objective function evaluations given by used benchmark set of CEC 2011.

Algorithm 1. Migration Model of Adaptive Differential Evolution

---

initialize mainland population, and sub-populations P_i, $i = 1, 2, \ldots, k$
$epoch = 1$
while stopping condition not reached **do**
    for $i = 1, 2, \ldots, k$ **do**
        perform nde generations of ith sub-population by ith adaptive DE
        migrate the best point and N_{rnd} points randomly chosen from P_i to mainland
    end for
    while stopping condition (1) not reached **do**
        develop mainland sub-population by a adaptive DE variant
    end while
    for $i = 1, 2, \ldots, k$ **do**
        migrate $1 + N_{rnd}$ points from the mainland to ith island
    end for
    $epoch = epoch + 1$
end while

---

4   Experimental Settings

The test suite of 22 real-world problems selected for CEC 2011 competition in Special Session on Real-Parameter Numerical Optimization [6] is used as one benchmark in the experimental comparison. The functions in the benchmark differ in the computational complexity, and in the dimension of the search space which varies from $D = 1$ to $D = 240$, the dimensionality of most problems exceeds $D = 20$. For each algorithm and problem, 25 independent runs were carried out. The run of the algorithm stops if the prescribed number of objective function evaluations $MaxFES = 150000$ is reached. The partial results of the algorithms after reaching one third and two-thirds of $MaxFES$ were also recorded. The point in the terminal population with the smallest function value is the solution of the problem found in the run. The population size of original adaptive DE variants $N = 100$ was used in all CEC 2011 problems. The sub-populations size of the proposed migration models is studied parameter set to $N_p = 10, 15, 45$ and 90. The number of generations performed on all k islands before migration of individuals is second analysed parameter and its values are $nde = 5, 10, 20,$ and 50. The migration models are labelled based on using explicit N_p, and nde settings 'P+N_p+x+nde', i.e. 'P15x5'. The other control parameters are set up according to a recommendation of authors in their original papers. All the algorithms are implemented in Matlab 2010a, and all computations were carried out on a standard PC with Windows 7, Intel(R) Core(TM)i7-4790 CPU 3.6 GHz, 16 GB RAM.

5   Results and Discussion

In this experimental study six original adaptive DE variants are compared with ten newly proposed migration models. Because the lack of place, detailed results are not provided. In Table 2 medians of all algorithms on all real-world problems are presented. The best value is underlined and printed bold, algorithms on second place has median printed bold, and median of algorithm on third place is only underlined. We can see that the best performing algorithms are among the original DE variants, and also migration models. The least median value is not the best testified statistics. Therefore in the last four rows of this table numbers of significant best, second, third and last positions based on Kruskal-Wallis test are printed. Each migration model in study is able to be competitive with the original DE variants. The most migration models are never the worst performing except P10x5 and P15x5. Combination of small sub-population size and frequent migration causes worse results.

Kruskal-Wallis non-parametric one-way ANOVA test was applied to each test problem. It was found that the performance of the algorithms in comparison differs significantly, the null hypothesis on the same performance is rejected in all the problems with achieved significance level $p < 0.000005$. Multiple comparison was then applied using Kruskal-Wallis z method (Dunn's test). The details of the significant positions of the Kruskal-Wallis test for each problem are in Table 1.

Table 1. Significantly best, second, third and worst performing algorithms from Kruskal-Wallis test.

Fun	D	1st	2nd	3rd	last
T01	6	p45x5	p15x50$_{b6e6}$	b6e6rl	CoDE
T02	30	p15x50$_{jade}$	p15x50	p15x20	CoDE
T03	1	No significant difference			
T04	1	No significant difference			
T05	30	p15x50$_{jade}$	p15x10	p15x20	EPSDE
T06	30	JADE	SaDE	p15x50	EPSDE
T07	20	p15x10	p15x5	p10x5	EPSDE
T08	7	No significant difference			
T09	126	SaDE	b6e6rl	jDE	CoDE
T10	12	b6e6rl	EPSDE	SaDE	p10x5
T11.1	120	p15x5	p15x50$_{b6e6}$	p45x5	b6e6rl
T11.2	240	p45x5	p15x50$_{b6e6}$	p15x20	CoDE
T11.3	6	b6e6rl	b6e6rl	b6e6rl	p15x5
T11.4	13	b6e6rl	SaDE	p90x50	p10x5
T11.5	15	b6e6rl	EPSDE	p90x5	JADE
T11.6	40	CoDE	EPSDE	p90x50	p10x5
T11.7	140	CoDE	EPSDE	p90x50	jDE
T11.8	96	CoDE	EPSDE	b6e6rl	jDE
T11.9	96	JADE	EPSDE	b6e6rl	jDE
T11.10	96	CoDE	EPSDE	b6e6rl	jDE
T13	26	p15x50	p10x5	p15x5	CoDE
T14	22	p90x5	p15x10	p45x5	CoDE

We can see that in three problems all algorithms perform similarly. The best performing original algorithms are b6e6rl, JADE, and CoDE. The best performing migration models are P15x50$_{jade}$ and P45x5. The worst performing algorithms are CoDE, jDE, EPSDE, and P10x5. The CoDE variant provides good performance only for problems T11.1–T11.10. These problems are very similar, and therefore CoDE is able to solve 'one kind' of CEC 2011 problems.

For better comparison of 16 algorithms on 22 problems in three equidistant stages the Friedman test on medians is applied, and results are depicted in Fig. 1. The test was carried out on medians of minimal function values at three stages of the search, namely after $FES = 50,000$, $100,000$, and $150,000$. The null hypothesis on the equivalent efficiency of the algorithms was rejected at the all stages of the search with $p < 5 \times 10^{-6}$. The algorithms in this table are ordered from left to right with respect to their mean rank from Friedman test at the finish of the search, i.e. after reaching $MaxFES = 150,000$.

Table 2. Medians of all functions from 25 independent runs, and count of significant positions for all algorithms based on Kruskal-Wallis test.

Fun	b6e6rl	code	epsde	jade	jde	sade	p10x5	p15x5	p45x5	p90x5	p15x10	p15x20	p15x50	p90x50	p15x50jade	p15x50b6e6
T01	**0**	2.69825	8.01E-01	3.53E-02	2.94E-01	1.02276	**0**	**0**	**0**	2.84E-08	**0**	**0**	**0**	1.67E-02	**1.83E-27**	**0**
T02	-20.5239	-11.6394	-12.2997	-23.4393	-20.7562	-20.4201	-22.997	-24.2639	-24.1439	-23.7111	-24.5733	-24.766	**-26.3623**	-17.2066	-25.7944	-25.021
T03	1.15E-05	1.15149E-05	1.15E-05	1.15E-05	1.15E-05	1.15E-05	1.15E-05	1.15E-05	1.15E-05	1.15E-05	1.15E-05	1.15E-05	1.15E-05	1.15E-05	1.15E-05	1.15E-05
T04	0	0	0	0	0	0	0	0	0	0	0	0	0	0	0	0
T05	-34.6563	-29.8468	-29.3899	-35.4652	-34.3078	-33.8831	**-35.97**	-35.856	-35.2389	**-35.9186**	**-35.97**	**-35.97**	-35.5547	-33.2848	**-36.7803**	-35.0104
T06	-27.8868	-21.6885	-20.9441	**-29.0552**	-27.3214	-28.5373	-23.0059	-23.0059	-27.9904	-23.273	-27.4298	-27.4298	-29.124	-22.2496	-27.4298	-23.0059
T07	1.31442	1.39838	1.46226	1.13173	1.34771	1.27682	8.58E-01	**8.09E-01**	1.19482	1.1708	**8.00E-01**	8.46E-01	8.48E-01	1.30385	1.05281	8.70E-01
T08	220	220	220	220	220	220	220	220	220	220	220	220	220	220	220	220
T09	**1800.09**	2719.55	2385.26	2290.08	2058.85	1783.97	2112.92	2170.66	2143.57	2088.81	2064.67	2341.87	1996.4	2172.28	2313.56	2101.07
T10	**-21.6444**	-21.5793	**-21.6444**	-21.4385	-21.5014	-21.6437	-21.4382	-21.4354	-21.6427	-21.5628	-21.4433	-21.6013	-21.6306	-21.5496	-21.4728	-21.4658
T11.1	87766.4	53034.7	52099.3	52342.8	61001.8	52107.7	52099.4	**51736.8**	52050	51983.1	52001.6	51931.9	52256.6	52147.1	52202.1	**51851.4**
T11.2	1.07E+06	1.07E+06	1.07E+06	1.07E+06	1.07E+06	1.07E+06	1.07E+06	1.07E+06	**1.07E+06**	1.07E+06	1.07E+06	1.07E+06	1.07E+06	1.07E+06	1.07E+06	**1.07E+06**
T11.3	15444.2	15444.2	15444.2	15444.2	15444.2	15444.2	15444.3	15445.7	15444.2	15444.2	15444.2	15444.2	15444.2	15444.2	15444.2	15444.2
T11.4	**18106.4**	18372.8	18326.7	18124.4	18262.1	**18117.3**	18391.3	18354.7	18274.5	18238.7	18282.2	18260.8	18310.2	18161	18297.4	18352.3
T11.5	**32743.3**	32847.6	**32745.9**	32964.9	32839.6	32851.4	32859.8	32818	32795.2	32789	32814.6	32832.4	32807.6	32838	32842.7	32832.3
T11.6	128028	**128703**	127128	127534	133584	130711	134217	133620	130184	129635	132041	132287	131425	127812	131569	132388
T11.7	1.90E+06	**1.89E+06**	1.89E+06	1.91E+06	2.07E+06	1.91E+06	1.93E+06	1.94E+06	1.92E+06	1.91E+06	1.93E+06	1.93E+06	1.92E+06	1.90E+06	1.92E+06	1.93E+06
T11.8	938815	**937003**	938271	938159	986658	940206	945568	945193	943887	940863	944488	944188	943595	939135	943919	944947
T11.9	943960	960844	943887	**943361**	1.44E+06	1.01E+06	1.22E+06	1.15E+06	1.03E+06	978260	1.21E+06	1.15E+06	1.12E+06	962454	1.11E+06	1.16E+06
T11.10	938355	**936501**	937586	939246	980234	940379	946017	945832	945266	940539	945464	943493	944620	938441	943726	942976
T13	19.0466	20.4819	19.6354	17.5794	18.1192	16.7843	**15.1013**	15.2185	15.2915	17.016	15.2853	15.1563	**14.3166**	18.3744	15.7274	15.5208
T14	14.0444	21.14	16.7738	18.6502	14.6702	18.4972	14.4482	14.2952	14.2502	**11.8391**	12.0035	13.6729	14.3056	15.7403	16.0027	15.4958
1st	3	4	0	2	0	1	0	1	2	1	1	0	1	0	2	0
2nd	1	0	7	0	0	2	2	1	0	0	2	0	1	0	0	3
3rd	4	0	0	0	1	1	1	1	2	1	0	3	1	3	0	0
last	1	6	3	1	4	0	3	1	0	0	0	0	0	0	0	0

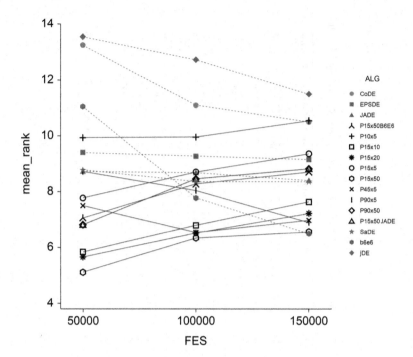

Fig. 1. Mean ranks from Friedman tests at three stages of the search, CEC 2011 test suite, $FES = 50000, 100000, 150000$.

In Fig. 1 the mean rank of all the compared algorithms are provided. We can observe a good performance of most migration models in the first stage. This result is important when a restrict number of evaluations is provided. The worst performing migration model is P10x5, the small sub-populations size ($N_p = 10$), and frequent migration ($nde = 5$) decreases the efficiency. The best performing model is P15x50, therefore the model with JADE on mainland, and model with b6e6rl on all sub-populations employs these settings. The performance of P90x5 variant is increased with increasing number of function evaluations. This conclusion means better performance for algorithms with 'casual' population size in DE. Although the best population size, and the good number of generations before migration is combined, unfortunately the performance of P90x50 is rather worse. When the number of function evaluations increases, the performance of migration models decreases whereas the performance of the original DE is rather increasing. The best performing P15x50 is slightly outperformed in the last stage by b6e6rl.

Further success of adaptive DE variants in migration model is analyzed. Each sub-population is developed by one DE, and the number of successfully generated individuals of each sub-population is weighted to % and depicted in Fig. 2. The most successful original algorithm is EPSDE, but the efficiency decreases with increasing sub-populations size N_p. The higher efficiency of EPSDE variant in migration model in most experiments could be caused by a huge number of small positive changes. Further research will be focused on an analysis of the diversity of the sub-populations during the search process.

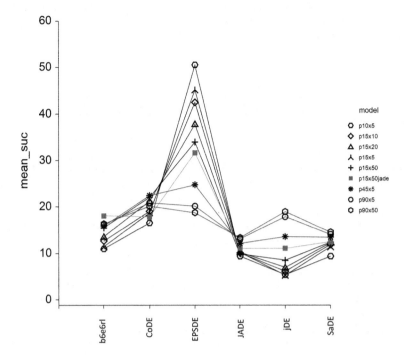

Fig. 2. Success of adaptive DE variants in migration models (%).

6 Conclusion

The results of the experimental comparison of ten various migration adaptive DE models with six popular adaptive DE algorithms demonstrate clearly very good efficiency of most of the migration models. The best performing algorithm in comparison is migration model P15x50, worst performing is adaptive jDE. Most of the migration models perform better than the original DE in the first stage of CEC 2011. This fact is important for real problems restricted by small function evaluation amount. The most successful adaptive DE in migration models is EPSDE, the efficiency of this variant is decreased with increasing sub-populations size. Although migration model of adaptive DE provide very good performance, further study of another migration parameters is fundamental for next research.

References

1. Brest, J., Greiner, S., Boškovič, B., Mernik, M., Žumer, V.: Self-adapting control parameters in differential evolution: a comparative study on numerical benchmark problems. IEEE Trans. Evol. Comput. **10**, 646–657 (2006)
2. Bujok, P.: Synchronous and asynchronous migration in adaptive differential evolution algorithms. Neural Netw. World **23**(1), 17–30 (2013)

3. Bujok, P.: Hierarchical topology in parallel differential evolution. In: Dimov, I., Fidanova, S., Lirkov, I. (eds.) NMA 2014. LNCS, vol. 8962, pp. 62–69. Springer, Cham (2015). https://doi.org/10.1007/978-3-319-15585-2_7
4. Bujok, P., Tvrdík, J.: Parallel migration model employing various adaptive variants of differential evolution. In: Rutkowski, L., Korytkowski, M., Scherer, R., Tadeusiewicz, R., Zadeh, L.A., Zurada, J.M. (eds.) EC/SIDE -2012. LNCS, vol. 7269, pp. 39–47. Springer, Heidelberg (2012). https://doi.org/10.1007/978-3-642-29353-5_5
5. Bujok, P., Tvrdík, J.: New variants of adaptive differential evolution algorithm with competing strategies. Acta Electronica at Informatica 15(2), 49–56 (2015)
6. Das, S., Suganthan, P.N.: Problem definitions and evaluation criteria for CEC 2011 competition on testing evolutionary algorithms on real world optimization problems. Jadavpur University, India and Nanyang Technological University, Singapore, Technical report (2010)
7. Das, S., Mullick, S.S., Suganthan, P.: Recent advances in differential evolution - an updated survey. Swarm Evol. Comput. 27, 1–30 (2016)
8. Das, S., Suganthan, P.N.: Differential evolution: a survey of the state-of-the-art. IEEE Trans. Evol. Comput. 15, 27–54 (2011)
9. Glotic, A., Glotic, A., Kitak, P., Pihler, J., Ticar, I.: Parallel self-adaptive differential evolution algorithm for solving short-term hydro scheduling problem. IEEE Trans. Power Syst. 29(5), 2347–2358 (2014)
10. Gong, Y.J., Chen, W.N., Zhan, Z.H., Zhang, J., Li, Y., Zhang, Q., Li, J.J.: Distributed evolutionary algorithms and their models: a survey of the state-of-the-art. Appl. Soft Comput. 34, 286–300 (2015)
11. Mallipeddi, R., Suganthan, P.N., Pan, Q.K., Tasgetiren, M.F.: Differential evolution algorithm with ensemble of parameters and mutation strategies. Appl. Soft Comput. 11, 1679–1696 (2011)
12. Penas, D., Banga, J., González, P., Doallo, R.: Enhanced parallel differential evolution algorithm for problems in computational systems biology. Appl. Soft Comput. 33, 86–99 (2015)
13. Qin, A.K., Huang, V.L., Suganthan, P.N.: Differential evolution algorithm with strategy adaptation for global numerical optimization. IEEE Trans. Evol. Comput. 13(2), 398–417 (2009)
14. Storn, R., Price, K.V.: Differential evolution - a simple and efficient heuristic for global optimization over continuous spaces. J. Glob. Optim. 11, 341–359 (1997)
15. Tvrdík, J.: Self-adaptive variants of differential evolution with exponential crossover. Analele West Univ. Timisoara Ser. Math.-Inform. 47, 151–168 (2009). http://www1.osu.cz/~tvrdik/
16. Wang, X., Tang, L.: Multiobjective operation optimization of naphtha pyrolysis process using parallel differential evolution. Ind. Eng. Chem. Res. 52(40), 14415–14428 (2013)
17. Wang, Y., Cai, Z., Zhang, Q.: Differential evolution with composite trial vector generation strategies and control parameters. IEEE Trans. Evol. Comput. 15, 55–66 (2011)
18. Zhang, J., Sanderson, A.C.: JADE: adaptive differential evolution with optional external archive. IEEE Trans. Evol. Comput. 13, 945–958 (2009)

Comparative Analysis Between Particle Swarm Optimization Algorithms Applied to Price-Based Demand Response

Diego L. Cavalca[1], Guilherme Spavieri[2], and Ricardo A. S. Fernandes[1,2(✉)]

[1] Graduate Program in Computer Science,
Federal University of Sao Carlos, Sao Carlos, Brazil
diego.cavalca@dc.ufscar.br, ricardo.asf@ufscar.br
[2] Department of Electrical Engineering, Federal University of Sao Carlos,
Sao Carlos, Brazil

Abstract. Demand-side management is a useful and necessary strategy in the context of smart grids, as it allows to reduce electricity consumption in periods of increased demand, ensuring system reliability and minimizing resources wastage. In its range of activities, Demand Response programs have received great attention in recent years due to their potential impact measured in several studies. In this work, different approaches of the Particle Swarm Optimization algorithm are applied to the autonomous and distributed demand response optimization model based on energy price. In addition, a stochastic mechanism is proposed to mitigate the structural bias problem that such algorithm presents, boosting its application in the analyzed problem. Results provided by computational simulations show that the proposed approach contributes significantly to reduce the energy consumption costs in relation to tariff variations, as well as minimizing the use of residential equipment during peak hours of a group of consumers.

Keywords: Particle Swarm Optimization · Demand response
Smart grid

1 Introduction

Until the 1970s, power distribution utilities planned their energy delivery capacity according to demand growth, once energy consumption was highly predictable. However, since the 1980s, due to economic, political, social and technological factors, demand has started to become less and less predictable [1]. Currently, the use of renewable sources has increased such volatility.

In this context, Demand-Side Management (DSM) was conceived as the planning, implementation and monitoring of activities that aim to influence the use of electricity in a way that produces desired changes in the load curve of utilities [2]. The DSM covers a set of actions for load management, which include, for example, the adoption of variable tariffs, measures for rational energy use,

© Springer International Publishing AG, part of Springer Nature 2018
L. Rutkowski et al. (Eds.): ICAISC 2018, LNAI 10841, pp. 323–332, 2018.
https://doi.org/10.1007/978-3-319-91253-0_31

renewable sources, energy efficiency, and Demand Response (DR). In this way, DSM mitigates the risks that can compromise the efficiency, reliability and stability of the power system, since it allows the relief of the power grid during peak hours.

According to [3], with the advances in data communication technologies and energy metering infrastructure, an adequate scenario is presented for the efficient management of energy resources. Therefore, in [4], the authors demonstrate that recent literature presents several studies that discuss the need of automatic load management and consumer behaviour analysis for DR programs to enable DSM actions. In addition, the authors argue that sustainable management of energy resources can be achieved through cooperation between the utility and its consumers, balancing the benefits between them.

Following this context, this paper addresses the DSM issue as a DR optimization problem, which aims to obtain the residential load scheduling that minimizes the cost of the energy consumed in face of tariff variations, respecting consumer habits as well as the characteristics of their electrical equipment. Another, but no less important goal concerns the reduction of the peak consumption that could be generated by a group of consumers. Thus, the reduction of this peak of consumption maximizes the reliability of the power system.

2   Demand Response and Load Scheduling

Demand Response methods refer to mechanisms that aim to manage the consumption patterns of end consumers in response to generation, supply, environmental, economic, and other conditions [5]. In this way, six load modulation strategies are defined by means of DR programs: peak reduction, valley filling, strategic growth, load shifting, strategic energy conservation, and flexible load curve.

Based on the ability of consumers to respond to an action by the system operator due to the change in energy prices, the potential impact of the DR is estimated to reduce peak demand. There are several ways to implement price-based DR programs, such as the Time-of-Use (TOU) pricing model, where the utility establishes prices that vary according to predefined time periods, which can include hours of the day, days of the week or seasons of the year, for example.

From the advances on Information and Communication Technologies and the consequent introduction of the Smart Grid concept, together with the DSM mechanisms, consumers play an important role in the energy scenario, since they can manage their consumption in an appropriate manner, selecting a preferred supplier and scheduling the operation of each residential load. Obviously, the scheduling of loads is not a trivial task, since it should consider a large set of information, objectives and constraints governed by mutual benefit among all agents of the system. Therefore, the DSM as an optimization problem leads to the development of decision support methods that meet the objectives of the utility and its consumers.

3   Autonomous and Distributed Modelling

From mathematical models that represent the loads and profile of residential consumers, it becomes possible to maximize the advantages of joining DR programs. Therefore, the choice of these models considers different factors related to the energy consumption.

Thus, this paper considers an autonomous and distributed model, published by [6], which was idealized for price-based DR programs. In this model, it is assumed that a group of nearby consumers, connected on the same power grid, has a bidirectional communication with the utility and interact with each other. It is also assumed that each consumer has a device called Energy Consumption Scheduler (ECS), which is responsible to obtain measurements and manage the flow of information between consumers. In addition, all communication between them and the utility is carried out via LAN (Local Area Network).

Considering the minimization of the cost for energy consumption, the Peak-to-Average Ratio (PAR) reduction is also part of the objective function of this model. The PAR corresponds to the ratio between the maximum demand and the average consumption demand, which reflects how much demand is concentrated in the peak period [3], being an important element that contributes to the energy price.

In the autonomous and distributed model, η denotes the group of consumers, where the number of consumers is $N \doteq |\eta|$. For each consumer $n \in \eta$, the total energy consumption at the time $h \in H \doteq \{1, ..., H\}$ is denoted by l_n^h, where $H = 24$. Aiming to maintain the generalization of the model, the time discretization considered is one hour. Therefore, the daily consumption profile for the consumer n is denoted by $l_n \doteq \{l_n^1, ..., l_n^H\}$. Based on these definitions, the total consumption at each hour of the day ($h \in H$) considering all consumers can be calculated as:

$$L_h \doteq \sum_{n \in \eta} l_n^h. \tag{1}$$

Peak and average daily consumption can be calculated respectively by:

$$L_{peak} = \max_{h \in H} L_h, \tag{2}$$

$$L_{avg} = \frac{1}{H} \sum_{h \in H} L_h. \tag{3}$$

Therefore, the PAR of load demand is given by:

$$PAR = \frac{L_{peak}}{L_{avg}} = \frac{H \max_{h \in H} L_h}{\sum_{h \in H} L_h}. \tag{4}$$

For each consumer $n \in \eta$, A_n denotes the set of electrical devices present in the residence. For each load/appliance $a \in A_n$, it was defined a power consumption planning vector $x_{n,a} \doteq \{x_{n,a}^1, ..., x_{n,a}^H\}$, where the scalar $x_{n,a}^h$ represents the planned energy consumption by the consumer n for the load/appliance a at time h. Thus, the total hourly consumption of each consumer is obtained as follows:

$$l_n^h \doteq \sum_{a \in A_n} x_{n,a}^h. \tag{5}$$

In this model, the objective of the ECS of each consumer is to calculate, through the optimization algorithm, the best power consumption planning vector $(x_{n,a})$ for each residential load. Thus, it can be defined the daily consumption profile of the consumer. Clearly, the definition of feasible planning should consider the preferences and needs of consumers throughout the day, as well as the operating characteristics of each load. It is important to mention that the objective is not to change the total amount of energy consumed, but rather to manage and allocate the loads efficiently to reduce the total daily energy cost paid by the consumer as well as reduce consumption at times of peak demand.

For this purpose, the consumer should define the beginning $\alpha_{n,a} \in H$ and end $\beta_{n,a} \in H$ of time interval in which each load/appliance can be turned on, so that $\alpha_{n,a} < \beta_{n,a}$. The definition of this operating window imposes time constraints on the planning vector and the total energy consumption previously determined must be carried out within the established range, such that $\sum_{h=a_{n,a}}^{\beta_{n,a}} x_{n,a}^h = E_{n,a}$ and $x_{n,a}^h = 0, \forall h \in H \backslash H_{n,a}$. Thus, $H_{n,a} \doteq \{\alpha_{n,a}, ..., \beta_{n,a}\}$ is the operating window of each device defined a priori. For each load/appliance, this time interval must be greater than the interval required to perform its function completely.

Therefore, it can be noticed that the daily energy consumed by all loads/appliances is equal to the sum of the total consumption of each load/appliance of each consumer. In this sense, the energy balance ratio will be maintained:

$$\sum_{h \in H} L_h \doteq \sum_{n \in \eta} \sum_{a \in A_n} E_{n,a}. \tag{6}$$

In general, the operation of certain equipment is not as flexible as the changes in the schedule. Therefore, the ECS does not impact on the consumption planning of such equipment. Thus, for each load/appliance $a \in A_n$, the minimum energy consumption in standby mode $(\gamma_{n,a}^{min})$ and the consumption relative to the maximum power $(\gamma_{n,a}^{max})$ are defined, so that $\gamma_{n,a}^{min} \le x_{n,a}^h \le \gamma_{n,a}^{max}, \forall h \in H_{n,a}$.

Thus, given the assumptions established in the autonomous and distributed model, the problem of minimizing the diary energy bill can be expressed as follows:

$$\min_{x_{n,a}, \forall n \in \eta, \forall a \in A_n} \sum_{h=1}^{H} C_h \left(\sum_{n \in \eta} \sum_{a \in A_n} x_{n,a}^h \right), \tag{7}$$

where C_h is the function that defines the energy cost in the hour h. This model considers a previously known and strictly convex function for the energy cost.

In addition to the cost, the reduction of PAR is also necessary to obtain efficient consumption planning [6]. In this way, it is possible to represent the PAR of the energy consumption planning vectors as:

$$\frac{H \max_{h \in H} \left(\sum_{n \in \eta} \sum_{a \in A_n} x_{n,a}^h \right)}{\sum_{n \in \eta} \sum_{a \in A_n} E_{n,a}}. \tag{8}$$

Therefore, having prior knowledge of all consumers' needs, the solution to the problem results in an efficient planning of energy consumption with respect to PAR:

$$\min_{x_{n,a}, \forall n \in \eta, \forall a \in A_n} \frac{H \max_{h \in H} \left(\sum_{n \in \eta} \sum_{a \in A_n} x_{n,a}^h \right)}{\sum_{n \in \eta} \sum_{a \in A_n} E_{n,a}}. \tag{9}$$

Given the planning vectors, the terms H and $\sum_{n \in \eta} \sum_{a \in A_n} E_{n,a}$ are constants. Consequently, they can be removed from Eq. 9, so that the following equivalent problem can be determined:

$$\min_{x_{n,a}, \forall n \in \eta, \forall a \in A_n} \max_{h \in H} \left(\sum_{n \in \eta} \sum_{a \in A_n} x_{n,a}^h \right) \tag{10}$$

According to the described mathematical modeling, the problem discussed in this paper can be solved by a variety of optimization approaches, such as metaheuristics. In this sense, the application of the PSO algorithm is feasible to solve the established price-based Demand Response problem, given its algorithmic simplicity and its ability to solve these category of problems.

4 Particle Swarm Optimization

4.1 Classical and Linear Decreasing Weight PSO

The PSO algorithm was proposed by [7] as an evolutionary computational optimization technique where the solution of a problem, or a particle, is found within a swarm containing a fixed number of particles. With its coordinates, each particle has a record of its best-known fitness, called *pBest*, and the best overall fitness of the swarm, *gBest*. Therefore, the swarm always moves towards the best solutions found.

The position of a particle is determined based on its previous position, $P_i(X_1, ..., X_n)$, and by its velocity, $V_i(X_1, ..., X_n)$, so that $\{X_1, ..., X_n\}$ are the coordinates of the particle. Therefore, according to [7], the movement of the swarm is governed by the equations:

$$V_i^{(t+1)} = \omega * V_i^t + \phi_1 r_1 \left(X_{pBest_i}^t - X_i^t \right) + \phi_2 r_2 \left(X_{gBest_i}^t - X_i^t \right), \tag{11}$$

$$X_i^{(t+1)} = X_i^t + V_i^{(t+1)}, \tag{12}$$

where $V_i^{(t+1)}$ is the velocity coordinate at the next iteration; V_i^t is the current coordinate of the velocity; $X_i^{(t+1)}$ is the coordinate of the position at the next iteration; X_i^t is the current coordinate of the position in the iteration t; $X_{pBest_i}^t$ and $X_{gBest_i}^t$ are the best coordinates for a particle and for the swarm, respectively; ϕ_1 and ϕ_2 are factors for local and global exploration, respectively (namely cognitive and social parameters); and finally r_1 and r_2 are random numbers uniformly distributed between 0 and 1, which insert a stochastic characteristic in the process of exploration of the search space.

It should be noticed that the inertia factor ω in Eq. 11 is not part of the original PSO, since it was proposed by [8] in a new approach called Linear Decreasing Weight PSO (LDW-PSO). The authors proved that this factor significantly increases the performance of the algorithm in relation to classical PSO in many cases. In the LDW-PSO, ω is the inertia factor, which usually decreases linearly from 0.9 to 0.4. According to the authors, the value of the inertia factor can be obtained through the equation:

$$\omega = \omega_{max} - \frac{\omega_{max} - \omega_{min}}{iter_{max}} * iter, \tag{13}$$

where, ω_{max} represents the maximum value that the inertia factor can obtain, ω_{min} is the minimum value, $iter_{max}$ is the maximum number of iterations that the PSO will execute, and $iter$ refers to the current iteration of the algorithm. According to [8], a large value for the inertia factor guarantees the global exploration of the search space. Otherwise, a small value helps with local explorations.

4.2 Proposed PSO Based on Stochastic Population Mechanism

A potential deficiency of meta-heuristics, including the PSO algorithm, referred to as structural bias, is discussed by [9]. According to the authors, a heuristic algorithm is structurally biased when it is more likely to visit some parts of the search space than the others. This behavior is not justified by the objective function. The authors further suggest that the structural bias has a greater impact on the exploration efficiency of the search space according to the level of difficulty inherent in solving the problem. In addition, the adoption of an effective particle sampling strategy increases the chance of PSO convergence.

Thus, assuming the autonomous and distributed modelling and the exposed objectives, a stochastic mechanism for the generation of particles is proposed (Algorithm 1), which aims to potentiating the application of LDW-PSO to the DR problem. The main idea of this mechanism is to generate individuals in function of the energy tariff in order to stochastically define viable candidate solutions that present better fitness and to explore more efficiently the search space of the problem. Therefore, this paper design a new PSO algorithm, which will be treated in this paper as SPM-PSO (Stochastic Population Mechanism PSO).

5   Results and Discussions

Based on the autonomous and distributed model presented, this paper aims to conduct an effective comparison regarding the performance of the PSO algorithms proposed in Sect. 4, highlighting the particularities of these algorithms when applied to solve the autonomous and distributed model of DR problem.

Thus, to represent a complete cycle of consumption behavior in a residence, a planning horizon of 24 h was analyzed. The data considered were generated by the *Load Profile Generator* software. The simulations were performed considering a set three residences, inhabited by 3 persons (2 adults and 1 child). In addition, each residence has 40 electrical load/appliances, which have their own operating configurations.

Alg. 1: Run for each load/appliance $a \in A_n$

1:	Initialize $x_{n.a}$ as a *vector of zeros*
2:	Calculate the *minimum duration* λ (in hours) of the operation based on planned consumption $E_{n.a}$ and $\gamma_{n.a}^{máx}$
3:	**If** $\lambda > 0$ **then**
4:	**If** $\left|H_{n.a}\right| > \lambda$ **then**
5:	Extract the *vector of peak hours* of energy cost $\mathcal{P}$, based on C_h
6:	Create the *auxiliary vector* δ, where each dimension δ_i represent the probability of the time $h \in H_{n.a}$ allocate part of the load $E_{n.a}$, containing equal probabilities
7:	For each $\rho_i \in \mathcal{P}$, *to penalize* δ_i by 90% its value, decreasing the probability of $H_{n.a_i}$ being used in planning
8:	*Rebalance* vector δ, potentializing off-peak times, such that $\sum \delta_i = 1$
9:	Define *stochastically the hours* that will allocate the load $E_{n.a}$, taking into account δ and λ, resulting in the vector $\mathcal{E}$
10:	**Otherwise**
11:	Define *planned schedule vector* for load allocation $\mathcal{E} = \mathcal{H}_{n.a}$ (non-flexible load)
12:	**End**
13:	Respecting the constraints, *generating a random* vector of load C
14:	*Allocate* the load $c_i \in C$ in the time $e_i \in \mathcal{E}$ of the planning vector $x_{n.a}$
15:	**End**

Aiming to minimize the cost of energy consumption and the PAR, factors represented by Eqs. 7 and 10, the objective function considered in the implemented algorithms was described as:

$$\lambda_{cost} \sum_{h \in H} \left(C_h \sum_{n \in \eta} \sum_{a \in A_n} x_{n,a}^h \right) + \lambda_{PAR} \max_{h \in H} \left(\sum_{n \in \eta} \sum_{a \in A_n} x_{n,a}^h \right), \qquad (14)$$

where h, H, n, η, a, A_n, C_h, and $x_{n,a}^h$ have the meanings already expressed above. λ_{cost} and λ_{PAR} are used to weight the impact of the minimization of each factor of Eq. 14, both having value 1 in the realized tests.

Each PSO was tested 10 times, using a swarm of 50 individuals and limited to 1000 iterations. The parameters ϕ_1 and ϕ_2 were both defined equal to 2.05. For both the LDW-PSO and the SPM-PSO, ω_{max} equal to 0.9 and ω_{min} equal to 0.4. For function C_h, the TOU tariff was adopted so that the kWh between 6 pm. and 8 pm. is R\$0.50. For the remaining hours the cost is R\$0.35. The individual was considered as a vector of dimension 2880, which represents 40 loads/appliances for 3 residences in 24 h of the day.

In the subsequent analysis, only the best result of each PSO will be considered among all the tests performed. Table 1 presents the summary of the obtained results.

The *Fitness* line represents the final value obtained for the objective function, the *Cost* line is the amount paid for the energy consumption (first term of the objective function), *Peak Consumption* line is the quantity of load (in kW) scheduled in the peak, i.e., for the second term of the objective function, and finally the line *PAR* is the ratio between the maximum and average demands.

Table 1. Summary of the performances reached by each PSO algorithm.

Parameter	Classical PSO	LDW-PSO	SPM-PSO
Fitness	28.901	28.675	23.526
Cost	22.115	22.086	21.414
Peak Consumption	6.786	6.589	2.112
PAR	3.075	3.093	4.127

As can be seen, SPM-PSO presented the highest efficiency in relation to the objective function, resulting in a *Fitness* of 23.526. In contrast, the classical PSO and LDW-PSO reached 28.675 and 28.901, respectively. Separating the first and second terms of Eq. 14, it is possible to obtain the total *Cost* and *Peak Consumption* for a period of 24-hours. In this sense, the SPM-PSO presented light reduction of the cost of energy consumption. However, the energy consumption for peak times was drastically reduced by using the proposed SPM-PSO when compared to the alternative approaches. Thus, it is important to mention that this reduction is expected by utilities in order to relieve the load of the power grid. This behavior is evident in Fig. 1, which shows the optimized load profile of the best solution obtained by the classical PSO, the LDW-PSO and the

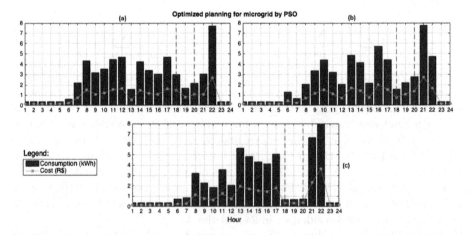

Fig. 1. Optimized load profile obtained by the algorithms: (a) classical PSO; (b) LDW-PSO; (c) SPM-PSO.

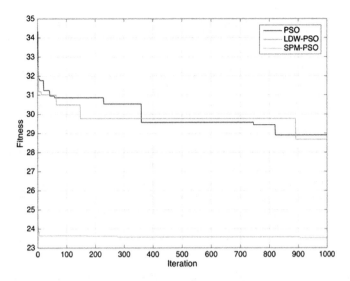

Fig. 2. Convergence analysis of each PSO (classical PSO, LDW-PSO and SPM-PSO).

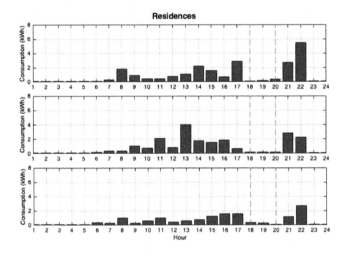

Fig. 3. Optimized individual residence consumption planning.

proposed SPM-PSO, where the red vertical lines delimit the period where the energy presents high costs (peak times).

Regarding the convergence of the best solution found by each algorithm (Fig. 2), SPM-PSO presented an advantage over all the others, since the stochastic mechanism acts to distribute the loads for periods in which the energy has the lowest cost, avoiding peak times. It is worth mentioning that the proposed stochastic mechanism is used in the initialization of the swarm. This choice is justified by the need to measure the impact of this mechanism on the convergence process of the algorithm within feasible search space, i.e., the objective is to verify if the proposed mechanism is efficient in the context of the analyzed problem. Thus, this result corroborates with [9], since the classical PSO and LDW-PSO algorithms are susceptible to have structural bias, necessitating strategies to mitigate such undesirable effect, especially in the case of the DR problem, which fitness has strong correlation with an external function (energy tariff C_h).

Therefore, based on the best solution obtained by SPM-PSO, it is possible to observe the optimized daily load profile for each residence (Fig. 3).

The residences have optimized plans that are different from each other, having as a common characteristic the avoidance of energy consumption at times when the tariff presents the highest cost, respecting the individual preferences of the consumers. Therefore, the precepts discussed in this paper are evidenced in Fig. 3, in which SPM-PSO optimizes load scheduling efficiently in face of energy price variations.

Acknowledgements. This paper was supported by FAPESP (grant number 2015/12599-0), CNPq (grant number 420298/2016-9) and CAPES.

References

1. Gellings, C.W.: The concept of demand-side management for electric utilities. Proc. IEEE **73**(10), 1468–1470 (1985)
2. Gellings, C.W.: Evolving practice of demand-side management. J. Mod. Power Syst. Clean Energy **5**(1), 1–9 (2017)
3. Batchu, R., Pindoriya, N.M.: Residential demand response algorithms: state-of-the-art, key issues and challenges. Lect. Notes Inst. Comput. Sci. Soc. Inf. Telecommun. Eng. **154**(2), 18–32 (2015)
4. Haider, H.T., See, O.H., Elmenreich, W.: A review of residential demand response of smart grid. Renew. Sustain. Energy Rev. **59**, 166–178 (2016)
5. Albadi, M.H., El-Saadany, E.F.: Demand response in electricity markets. In: IEEE PES General Meeting, pp. 1–5 (2007)
6. Mohsenian-Rad, A.H., Wong, V.W.S., Jatskevich, J., Schober, R., Leon-Garcia, A.: Autonomous demand side management based on game-theoretic energy consumption scheduling for the future smart grid. IEEE Trans. Smart Grid **1**(3), 320–331 (2010)
7. Eberhart, R., Kennedy, J.: A new optimizer using particle swarm theory. In: Proceedings of the Sixth International Symposium on Micro Machine and Human Science, pp. 39–43 (1995)
8. Shi, Y., Eberhart, R.C.: Empirical study of particle swarm optimization. In: Proceedings of the 1999 Congress on Evolutionary Computation, vol. 3, pp. 1945–1950 (1999)
9. Kononova, A.V., Corne, D.W., De Wilde, P., Shneer, V., Caraffini, F.: Structural bias in population-based algorithms. Inf. Sci. **298**, 468–490 (2015)

Visualizing the Optimization Process for Multi-objective Optimization Problems

Bayanda Chakuma and Mardé Helbig[(✉)]

Department of Computer Science, University of Pretoria, Pretoria, South Africa
mhelbig@cs.up.ac.za

Abstract. Visualization techniques used to visualize the optimization process of multi-objective evolutionary algorithms (MOEAs) have been discussed in the literature, predominantly in the context of aiding domain experts in decision making and in improving the effectiveness of the design optimization process. These techniques provide the decision maker with the ability to directly observe the performance of individual solutions, as well as their distribution in the approximated Pareto-optimal front. In this paper a visualization technique to study the mechanics of a MOEA, as it is solving multi-objective optimization problems (MOOPs), is discussed. The visualization technique uses a scatterplot animation to visualize the evolutionary process of the algorithms search, focusing on the changes in the population of non-dominated solutions obtained for each generation. The ability to visualize the optimization process of the algorithm provides the means to evaluate the performance of the algorithm, as well as visually observing the trade-offs between objectives.

Keywords: Visualization · Scatterplot
Multi-objective optimization · Multi-objective evolutionary algorithms

1 Introduction

Multi-objective optimization (MOO) considers problems that require two to three objective functions to be optimized simultaneously [1]. Most of these problems characterize real-world design problems where the essence is to find a balanced design to satisfy multiple, possibly conflicting objectives at the same time [2]. When dealing with MOO problems (MOOPs), the aim is to find a set of solutions representing the trade-offs or good compromises among the objective functions being evaluated. The set of optimal trade-off solutions is called a Pareto-optimal set (POS) in the decision variable space and it comprises all the non-dominated solutions that are considered to be equally good among themselves. The solutions in the POS are such that there is no feasible solution that can be improved further in terms of a single objective without causing a simultaneous degradation in at least one other objective [3]. The corresponding solutions in the objective space is referred to as the Pareto-optimal front (POF).

© Springer International Publishing AG, part of Springer Nature 2018
L. Rutkowski et al. (Eds.): ICAISC 2018, LNAI 10841, pp. 333–344, 2018.
https://doi.org/10.1007/978-3-319-91253-0_32

Optimization using multi-objective evolutionary algorithms (MOEAs) to solve MOOPs has become a very popular research topic [3]. This calibre of algorithms has also gained recognition in a wide variety of disciplines, due to their stochastic nature and flexibility in handling multiple objectives when compared to other conventional optimization methods [3]. In the case of a MOEA its performance is influenced by the genetic operators used, such as mutation, crossover and population size [4,5]. Although MOEAs are recommended for solving MOOPs and finding the POS, there is still a lack of quantitative techniques to evaluate the performance of the algorithms' search and to compare the influence of different genetic operators and their parameters predominantly in the perspective of algorithm designers.

The objective of this research is to investigate a visualization technique to analyse the search process of a MOEA as it is solving MOOPs, focusing on the evolutionary process of the non-dominated set of solutions for each generation. This in turn will provide insight into the mechanics of how generations of solutions evolve over time and to answer questions related to the performance of the algorithm depending on the genetic parameters used.

The rest of the paper is organised as follows: Sect. 2 provides background information on visualization techniques proposed in literature. The proposed method, as well as the data used for visualization, is discussed in Sect. 3. Section 4 discusses the experimental setup, along with the results obtained for each experiment. Lastly the conclusions and future work are discussed in Sect. 5.

2   Background

This section provides background information required for the rest of the paper. Section 2.1 discusses visualization techniques and Sect. 2.2 discusses visualization techniques used specifically for low-dimensional spaces.

2.1   Visualization Techniques

Visualizing the search process and the results of an algorithm's search has been addressed in literature in the context of aiding domain experts in decision making [6], and in improving the effectiveness of a design optimization process. Being able to visualize the optimization process in real-time, i.e. effectively monitoring and representing the optimization progress of an algorithm's search, can lead to an increase in productivity and in the effectiveness of decision making for system designers [6,7].

Visualization techniques to visualize the optimization process and the set of found optimal solutions have been developed and refined over the years [8]. For the purpose of this paper methods that can be applied to visualizing the POF are discussed. The first group of visualization methods encompasses two similar approaches: the scatter plot method and the bubble chart method [9]. Both methods are robust visualization methods used to show the distribution and the shape of the POF. The scatter plot is however limited to showing two to three objectives in a low-dimensional space, while the bubble chart is limited to five objectives. The main advantage of the bubble chart is that all the information can be represented in a single plot [10].

The second group uses the parallel coordinate system in high-dimensional objective spaces. Parallel coordinates [11] have an advantage of representing the independence between objectives. The drawback is its inability to represent the shape of the POF. Similar to parallel coordinates, heatmaps [12] also show independence between objectives, using colour to show the value of objectives. One disadvantage of heatmaps is that they cannot clearly show trade-offs between objectives. Also, when the number of solutions are increased, the number of colours also increase, making it visually difficult to distinguish between a large number of colours.

The third group categorizes visualization methods based on mapping [10]. These methods use sophisticated mapping techniques for dimension reduction to the two dimensional (2D) space. Sammon mapping and neuroscale both aim to minimize the stress function, which in turn preserves the local distance [8]. Other methods in this group include radial coordinate visualization (RadVis), isomaps [11,13], principal component analysis (PCA) and self-organizing maps (SOMs) [14,15]. All these visualization methods are discussed and compared in [8,10].

The last group were proposed specifically for visualizing the POF and the progress of the search process of algorithms. In order to observe whether enough information could be acquired from the visualization to improve the genetic operators and their parameters, SOMs were used for visualizing the search process of a MOEA [14]. Using SOMs to visualise the POF have two disadvantages: the maps have incorrect points that represent non Pareto-optimal solutions that in some cases proved to be infeasible; and the coverage of the maps for the edge region of the Pareto optimal solution is not satisfactory or good [16]. Therefore, a visualization method using SOMs-neural gas (SOM-NG), an improved SOM visualization that incorporates neural gas, was proposed that introduces a learning parameter using differential evolution (DE) and that forms maps by considering the similarities on the input space.

Taghavi and Pimentel [17] proposed the Multi-Objective Design space eXploration (VMODEX) visualization tool that enables designers to easily understand how a MOEA explores the design space, where the optimum design points are located, how design parameters influence each objective and the relationship between objectives [17]. The drawback of this visualization is that it uses a DSE tree for its visualization and the progress of the algorithms' search is not clearly presented. Lastly, three-dimensional virtual reality facilities were used to present multi-dimensional POFs, allowing zooming and rotating of the POF to better understand and compare the trade-off among solutions [18].

2.2   Low-dimensionality Visualization Techniques

Most of the visualization techniques reviewed above are primarily proposed to view data in a high-dimensional objective space. This is with the exception of the scatterplot approach, which has been acclaimed for visualizing data in low-dimensional objective spaces, while exhibiting some disadvantages when applied to high-dimensional data [8]. Using the scatterplot approach, each objective function can be visualised on one axis, with the information about the location, range and shape of the approximation POF clearly visible. Other approaches that were proposed for visualizing data in a low-dimensional objective space include a visualization technique for analysing and comparing non-dominated solutions by calculating and visualizing the distance between the solutions, as well as their distribution [19]. Contrary to how most visualization techniques behave when visualizing the optimization process of the algorithms, where the focus is on the fitness values of the first generation and the last, a genetic algorithm visualization system that uses a combination of coverage, contour and distance maps to visualize the fitness of solutions for every generation in the evolutionary process was proposed in [20]. This approach facilitates the analysis of the coverage of the search space, as well as the convergence behaviour of the solutions.

3   Proposed Approach

The visualization technique proposed in this paper attempts to combine the ideas of the simple scatter plot approach discussed in [8] and the video approach proposed in [20]. This section discusses the proposed method, as well as the test functions used for the evaluation of the algorithm.

3.1   Scatterplot

The scatterplot method is extremely useful for visualizing large quantities of data and showing correlations between variables. In the context of this paper, it shows a good representation of the location, range and the shape of the POF, which enables the trade-offs and or conflicts between objectives to be observed. It also simplifies the process of choosing a preferred solution from the POS. However, this visualization technique also has limitations, i.e. when it is over populated it is difficult to assess the quality of the approximation set, as well as to monitor the progress or the convergence of an optimization run. To compensate for this limitation, the video technique is introduced. For every optimization run, a scatterplot is produced representing the state of the search process. In addition, the scale of each frame is adjusted for the values of each generation being visualised. This information is then mapped to a video format, producing an animation of the search process. The adjustment of solution parameters is observed with every changing frame.

3.2   Animation

There are many characteristics linked to the animation technique [21], e.g. rotation. For the purpose of the proposed method, the transition between data graphics is considered. According to [22], animated transitions can be used to represent stages or the changes in the data. One of the benefits of using animated transitions is to facilitate cognition. The usage of animated transitions can also improve graphical perception of the changes in data graphics. In this paper animation plays an important role in representing the solutions as moving dots. Each frame visualises the state of the evolutionary process. When the frames are combined the evolution of the generations of solutions can be observed over time. This can assist in showing the convergence behaviour of a MOEA and the characteristics of the test functions.

Most vocalization techniques [19,23] focus on the data produced after the algorithm has completed the optimization process, considering only the fitness values from the first and the last generation. This is very useful when used as a performance measure, for example visualizing the distribution of the solutions in the non-dominated set. However, with regards to the search process it only gives a feel of the progress of the search, but not the actual features of the evolution as it is taking place. Methods using data generated from every generation [12,20] provide the current state of evolutionary process, but some of them visualize the data using chromosomes instead of the individuals' position in the search space. The proposed method takes into account all the data or solutions produced for every generation and visualises the objective values of each individual in the solution set.

3.3   Visualizing the State of the Algorithm

The proposed method investigates the behaviour of NSGA-II [24], focusing mainly on the relationships generated between individuals during the search process, i.e. observing the influence of genetic operators and their parameters throughout the search process. The motivation behind this is to aid algorithm designers to understand the features of the algorithm's search process, especially when changes to the parameters of generic operators are made. Furthermore, this will assist domain experts when choosing their preferred solutions for their designs.

NSGA-II is a relatively old algorithm, but is still used as a benchmark to which other MOEAs are compared against. The algorithm's main advantages are that it has a relatively low computational complexity, because of its fast non-dominated solution ranking technique, and it also provides good diversity of solutions when two or three objectives are optimized.

The first part of the visualization covers the key features of the algorithm, with each frame of the animation representing a population of solutions. For each frame the scale is adjustable to ensure that the objective values of each solution are represented accordingly for each generation. The rank level of solutions (based on the level of non-domination) is provided through colour, with blue representing the rank of the most preferred solutions in each generation i.e. blue = rank 1, green = rank 2. The effect of the crowding distance of the solutions can also be viewed in each frame. Regions of high density can be viewed in areas where solutions are overlapping and clustered, while low density is seen in areas with few solutions.

The second part of the visualization covers known features of the Deb-Thiele-Laumanns-Zitzler (DTLZ) group of test functions [25], visualizing how the NSGA-II algorithm performs on these functions using the hypervolume [26] performance measure.

4   Experimental Setup and Results

In this section the influence of genetic parameters on the performance of NSGA-II is studied and assessed using the DTLZ test functions [25]. To visually observe the influence of genetic operators on the search process of the algorithm, genetic parameters are varied and the changes in the non-dominated solutions of each generation are represented as an animation of moving dots in a two-dimensional plane. For the purpose of this paper only the influence of the population size on the algorithm's performance is considered to demonstrate the proposed visualization method.

For each parameter value being investigated, the algorithm has one independent run, consisting of iterations equal to the number of generations used. It is important to note that the stochastic nature of evolutionary algorithms necessitates that their evaluation be conducted using multiple independent simulation runs to obtain good statistical information. However, the focus of the paper is to observe the phenomenon of the search process using visualization techniques and not to compare the statistical information. With that said, the hypervolume, which is used to measure the convergence and diversity of solutions, is calculated for each run. The number of generations is kept constant at 300, with the population size varying from 100, 200 to 500, respectively. Crossover rate and the mutation probability are kept constant. All the simulations were ran using Distributed Evolutionary Algorithms in Python (DEAP) [27].

4.1   Experiment I

For the first experiment, the DTLZ3 test function was used to study the performance of NSGA-II when changing the size of the population for each algorithm run. Table 1 shows the parameter values used for this experiment [24]. For this experiment two objectives were used and results reported are non-dominated solutions obtained for each generation in the optimization process.

Table 1. Parameter values used in Experiment I

#Generations	Population size	Crossover rate	Mutation probability (m_p)
300	100, 200, 500	1-m_p	0.2

Figures 1 to 2 illustrate the transition of the algorithm over 300 generations. Each figure illustrates the state of the algorithm after 100 and 300 generations respectively. The horizontal axis represents values of $f_1(\mathbf{x})$, while the vertical axis represents values for $f_2(\mathbf{x})$. The grid or the two dimensional plane on which the non-dominated solutions are plotted, is adjusted for every generation to accommodate the values of $f_1(\mathbf{x})$ and $f_2(\mathbf{x})$ for each generation.

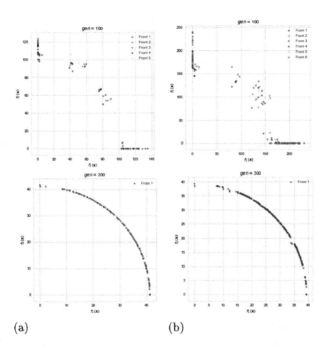

(a)   (b)

Fig. 1. NSGA-II after 100 and 300 generations for (a) population size 100 and (b) population size 200 (Color figure online)

Non-dominated Solutions. Figure 1(a) and (b) (generation 300) show that almost all the solutions of the POF were evenly distributed between both objective functions. However, looking at both Fig. 1(a) and (b), solutions in areas with high values of $f_2(\mathbf{x})$ and low values of $f_1(\mathbf{x})$ were not evenly spread and lacked diversity. Looking at solutions from the 100th generation to the last generation, the search was effective, i.e. the solutions improved towards the end of the optimization process. Figure 2 shows that the solutions found at generation 300 were not optimal solutions. From generation 299 to the final generation the algorithm seemed to have stagnated and stopped showing any signs of real improvement. This might indicate that one of the genetic parameters responsible for diversity needs to be improved.

Effects of Varying Population Size. From the simulations conducted with varying population size, the results showed that a small population size was most optimal compared to the other population sizes. From Figs. 1 to 2 it can

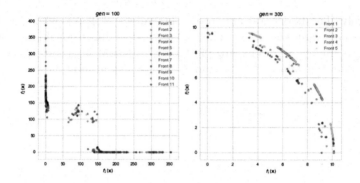

Fig. 2. NSGA-II after 100 and 300 generations for population size 500

be seen that a population size of 100 showed very few gaps, i.e. areas on the POF containing no solutions. A smaller population lead to faster convergence, but resulted in less solutions. Increasing the population size also increased the time it took for the solutions to reach optimality. The time recorded was 10.45 s, 35.4 s, and 3 min 13 s for population sizes of 100, 200, and 500 respectively. From the observed results, it is clear that the population size had a significant influence on the evolutionary process, i.e. a larger population size presented a larger pool of solutions, giving decision makers more solutions to choose from. However, even though a larger population size produced more solutions in certain areas on the POF, it also lead to more gaps on the POF.

Results of Visualization. The proposed visualization method is an animation of moving dots illustrating the optimization process of NSGA-II. Since only two objective functions are used, the non-dominated solutions are plotted on a two-dimensional plane. The visualization method can also be extended to three objective functions, which in turn requires the non-dominated solutions to be visualized in a three-dimensional plane. From the generated videos, NSGA-II can be observed, starting with an initial random population that is sorted and ranked according to non-domination. The initial population is at generation 0 for all the videos. The sorting and rank assignment of the non-dominated solutions can be observed over time. The rank is used to identify fronts the solution belongs to, with the fronts indicated in enumerated colours. The fronts are enumerated in ascending order with one being the true POF, indicated in blue. The following front is indicated in green and the last front in orange.

The spread of solutions throughout the evolutionary process can also be observed with areas of high density of non-dominated solutions and areas with low density of solutions. Initially the solutions are widely separated (generations 0 to 9) but from generation 10 clear clustering and crowding of solutions can be seen. As stated in [25], the DTLZ3 test function introduces multiple local POFs and the algorithm is occasionally attracted to one of them. A population size of 500 showed this behaviour with the algorithm getting stuck in one of the local POFs and never reaching optimality.

Figure 3 shows the hypervolume plots of the varied population sizes respectively. Referring to Fig. 3 (for all population sizes' illustrations), solutions obtained in later generations produced a higher hypervolume measure than solutions in generations 10 to 50. Comparing solutions or the hypervolume across the three population size variations (100, 200, and 500) leads to the conclusion that NSGA-II with a smaller population size produced the most optimal solutions.

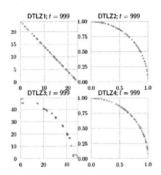

Fig. 3. Hypervolume for population size 100, 200 and 500 (Color figure online)

4.2   Experiment II

Experiment two deals with using the proposed visualization method to visualize the features of the DLZ benchmarks. The number of iterations are equal to the number of generations for the simulation run and all parameters are fixed.

Results of Visualization. Figures 4 to 6, illustrate the transition of NSGA-II over 1000 generations, for DTLZ1 to DTLZ4, and DTLZ7. Figure 4 illustrates the results obtained during the optimization run in the range of 40 to 50 generations, Fig. 4(b) represents the range of generations from 400–500 and Fig. 5 represents the range of generations from 900 to 1000 for DTLZ1 to DTLZ4 respectively. Figure 5 illustrates these ranges for the DTLZ7 test function.

From the obtained results, the most prominent feature that can be observed is the variety of POF shapes [25]. DTLZ2 to 4 have the same POF shape, i.e. concave, while DTLZ1 has a linear shaped POF and DTLZ7 has a discontinuous POF.

The second feature that is observed is the challenge brought about by the existence of multiple local POFs in the DTLZ3 test function. The existence of these local POFs presents a difficulty for the algorithm when searching towards the global POF, as can be seen in Fig. 4(b). Figure 5 shows the ability of NSGA-II to maintain solutions in disconnected portions of the POF.

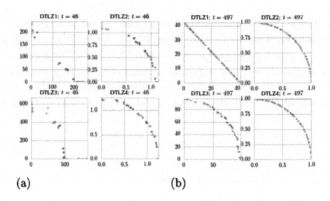

Fig. 4. Generation range (a) 40 to 50; and (b) 400 to 500

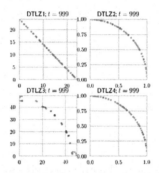

Fig. 5. Generation range 900 to 1000

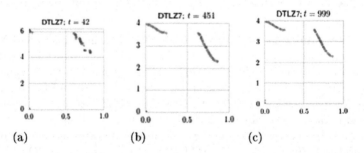

Fig. 6. Solutions found by NSGA-II after (a) 42, (b) 451 and (c) 999 generations

5   Conclusion

In this paper a visualization method of using animated non-dominated solutions to understand the search process of NSGA-II when solving MOOPs was investigated. First the effects of varying the population size on the performance of the algorithm were investigated using the DTLZ3 test function. The results verified that using a small population size for the optimization process results in optimum solutions. Next, the features of different DTLZ test functions were assessed using the visualization method. The illustrations showed that the shape function of the test functions was the most noticeable feature.

In future, the research will be extended to investigate the optimization process of dynamic problems. Another possible feature for future work is the tracking feature that tracks individuals (solutions) that are propagated through the evolutionary process and visualizes all the generations each individual is present or survives.

Acknowledgements. This work is based on the research supported by the National Research Foundation (NRF) of South Africa (Grant Number 46712). The opinions, findings and conclusions or recommendations expressed in this article is that of the author(s) alone, and not that of the NRF. The NRF accepts no liability whatsoever in this regard.

References

1. Coello Coello, C.: Evolutionary multi-objective optimization: a historical view of the field. IEEE Comput. Intell. Mag. **1**(1), 28–36 (2006)
2. Konak, A., Coit, D., Smith, A.: Multi-objective optimization using genetic algorithms: a tutorial. Reliab. Eng. Syst. Saf. **91**(9), 992–1007 (2006)
3. Coello Coello, C., Lamont, G., van Veldhuizen, D.: Evolutionary Algorithms for Solving Multi-Objective Problems, vol. 5. Springer, New York (2007). https://doi.org/10.1007/978-0-387-36797-2
4. Deb, K.: Multi-Objective Optimization Using Evolutionary Algorithms, vol. 16. Wiley, Hoboken (2001)
5. Jones, D., Mirrazavi, S., Tamiz, M.: Multi-objective meta-heuristics: an overview of the current state-of-the-art. Eur. J. Oper. Res. **137**(1), 1–9 (2002)
6. Messac, A., Chen, X.: Visualizing the optimization process in real-time using physical programming. Eng. Optim. **32**(6), 721–747 (2000)
7. Jones, C.: Visualization and optimization. ORSA J. Comput. **6**(3), 221–257 (1994)
8. Tuar, T., Filipi, B.: Visualization of Pareto front approximations in evolutionary multiobjective optimization: a critical review and the prosection method. IEEE Trans. Evol. Comput. **19**(2), 225–245 (2015)
9. Ashby, M.: Multi-objective optimization in material design and selection. Acta Mater. **48**(1), 359–369 (2000)
10. He, Z., Yen, G.: Visualization and performance metric in many-objective optimization. IEEE Trans. Evol. Comput. **20**, 386–402 (2016)
11. Inselberg, A., Dimsdale, B.: Parallel coordinates: a tool for visualizing multidimensional geometry. In: Proceedings of the Conference on Visualization, pp. 361–378 (1990)
12. Pryke, A., Mostaghim, S., Nazemi, A.: Heatmap visualization of population based multi objective algorithms. In: Obayashi, S., Deb, K., Poloni, C., Hiroyasu, T., Murata, T. (eds.) EMO 2007. LNCS, vol. 4403, pp. 361–375. Springer, Heidelberg (2007). https://doi.org/10.1007/978-3-540-70928-2_29
13. Kudo, F., Yoshikawa, T.: Knowledge extraction in multi-objective optimization problem based on visualization of Pareto solutions. In: Proceedings of the IEEE Congress on Evolutionary Computation, pp. 1–6, June 2012

14. Yamashiro, D., Yoshikawa, T., Furuhashi, T.: Efficiency of search performance through visualizing search process. In: Proceedings of the IEEE International Conference in Systems, Man and Cybernetics, pp. 1114–1119, October 2006

15. Lotif, M.: Visualizing the population of meta-heuristics during the optimization process using self-organizing maps. In: Proceedings of the IEEE Congress on Evolutionary Computation, pp. 313–319, July 2014

16. Kobayashi, Y., Okamoto, T., Koakutsu, S.: A Pareto optimal solution visualization method using SOM-NG with learning parameter optimization. In: Proceedings of the IEEE International Conference on Systems, Man, and Cybernetics, pp. 4525–4531 (2016)

17. Taghavi, T., Pimentel, A.: VMODEX: a visualization tool for multi-objective design space exploration. In: Proceedings of the International Conference on Field-Programmable Technology, pp. 357–360 (2010)

18. Madetoja, E., Ruotsalainen, H., Monkkonen, V., Hamalainen, J., Deb, K.: Visualizing multi-dimensional Pareto-optimal fronts with a 3D virtual reality system. In: Proceedings of the International Multiconference on Computer Science and Information Technology, pp. 907–913 (2008)

19. Ang, K., Chong, G., Li, Y.: Visualization technique for analyzing non-dominated set comparison. In: Proceedings of the Asia-Pacific Conference on Simulated Evolution and Learning, vol. 1 (2002)

20. Shine, W., Eick, C.: Visualizing the evolution of genetic algorithm search processes. In: Proceedings of the IEEE International Conference on Evolutionary Computation, pp. 367–372, April 1997

21. Elmqvist, N., Dragicevic, P., Fekete, J.: Rolling the dice: multidimensional visual exploration using scatterplot matrix navigation. IEEE Trans. Vis. Comput. Graph. **14**(6), 1141–1148 (2008)

22. Heer, J., Robertson, G.: Animated transitions in statistical data graphics. IEEE Trans. Vis. Comput. Graph. **13**(6), 1240–1247 (2007)

23. Pohlheim, H.: Visualization of evolutionary algorithms-set of standard techniques and multidimensional visualization. In: Proceedings of the Annual Conference on Genetic and Evolutionary Computation, vol. 1, pp. 533–540, July 1999

24. Deb, K., Pratap, A., Agarwal, S., Meyarivan, T.: A fast and elitist multiobjective genetic algorithm: NSGA-II. IEEE Trans. Evol. Comput. **6**(2), 182–197 (2002)

25. Deb, K., Thiele, L., Laumanns, M., Zitzler, E.: Scalable multi-objective optimization test problems. In: Proceedings of the 2002 Congress on Evolutionary Computation, pp. 825–830, May 2002

26. Fleischer, M.: The measure of Pareto optima applications to multi-objective metaheuristics. In: Fonseca, C.M., Fleming, P.J., Zitzler, E., Thiele, L., Deb, K. (eds.) EMO 2003. LNCS, vol. 2632, pp. 519–533. Springer, Heidelberg (2003). https://doi.org/10.1007/3-540-36970-8_37

27. Fortin, F., Rainville, F., Gardner, M., Parizeau, M., Gagné, C.: DEAP: evolutionary algorithms made easy. J. Mach. Learn. Res. **13**, 2171–2175 (2012)

Comparison of Constraint Handling Approaches in Multi-objective Optimization

Rohan Hemansu Chhipa and Mardé Helbig$^{(\boxtimes)}$

Department of Computer Science, University of Pretoria, Pretoria, South Africa
mhelbig@cs.up.ac.za

Abstract. When considering real-world optimization problems the possibility of encountering problems having constraints is quite high. Constraint handling approaches such as the penalty function and others have been researched and developed to incorporate an optimization problem's constraints into the optimization process. With regards to multi-objective optimization, in this paper the two main approaches of incorporating constraints are explored, namely: Penalty functions and dominance based selection operators. This paper aims to measure the effectiveness of these two approaches by comparing the empirical results produced by each approach. Each approach is tested using a set of ten benchmark problems, where each problem has certain constraints. The analysis of the results in this paper showed no overall statistical difference between the effectiveness of penalty functions and dominance based selection operators. However, significant statistical differences between the constraint handling approaches were found with regards to specific performance indicators.

Keywords: Constrained multi-objective optimization
Pareto dominance · Penalty functions

1 Introduction

In many real-world problems one has to find a set of optimal solutions, since the problem has multiple objectives that could possibly be conflicting [1–3]. It should be noted that there exists a fundamental difference between a single-objective and a multi-objective problem. When solving a single-objective problem the final result of the optimization process is a single solution which optimizes the single objective function. However, when solving a multi-objective optimization problem (MOOP) there cannot exist a single solution that optimizes all objectives [3]. Thus a multi-objective problem tends to result in a set of solutions where one solution is not necessarily better than another solution within that set, i.e. each solution effectively represents some form of trade-off between the objectives [3].

There are two common approaches for solving a MOOP. The first approach is the simplest and involves combining each objective function into a single objective function [2,3]. This effectively converts the problem into a single-objective

© Springer International Publishing AG, part of Springer Nature 2018
L. Rutkowski et al. (Eds.): ICAISC 2018, LNAI 10841, pp. 345–362, 2018.
https://doi.org/10.1007/978-3-319-91253-0_33

optimization problem. The manner in which this can be achieved is by taking the weighted sum of each objective function to produce a scalar value, which represents the fitness of the entire solution. Using this approach it is possible to assign priorities to each of the objectives where an objective with a larger weight has a higher priority than an objective with a smaller weight. However, it is important to place emphasis on the selection of the weights for each objective as even the smallest change in weight values can result in a different solution. Deb further states that unless a reliable set of objective weights is available, any solution found using this approach would be highly subjective to that particular user [2,3].

The second approach does not combine the objective functions in any way and because of this the fitness of a solution is no longer quantified by a single, scalar value. Using this approach each solution will have a vector of fitness values, where each value within this vector was produced by an objective function and through this approach it becomes possible to find a set of trade-off optimal solutions [3]. Each solution within this set should exhibit the following property: suppose there are two minimization objectives f_1 and f_2, each solution within the trade-off optimal set is optimized according to the two functions. However suppose there exists two solutions S_1 and S_2 such that: $[f_1(S_1) < f_1(S_2)$ and $f_2(S_2) < f_2(S_1)]$.

Neither solution is better than the other, and both are some form of trade-off between the objectives. It is these types of solutions that the trade-off optimal solution set should consist of, i.e. non-dominated solutions.

Constraints can be found in most real-world optimization problems [2]. Constraint handling is important in any optimization process due to the fact that the introduction of constraints to a problem can allow the optimization algorithm to find two different types of solutions, namely: feasible and infeasible solutions [3]. Thus, it is important to choose an effective constraint handling approach for the problem at hand to ensure that the final set of solutions produced is a set of non-dominated solutions where every solution within the set is feasible, i.e. does not violate any constraint. The main focus of this paper is on the comparison of two main constraint handling approaches available for MOOPs, namely: (i) penalty functions and (ii) dominance based selection operators. Both of these approaches will be discussed in more detail in Sect. 2.

As pointed out by Konak et al., constraints in multi-objective optimization have not been adequately researched [2], and thus the main focus of this paper is to compare the two main constraint handling approaches for constrained MOOPs to determine whether there exists a statistical significant difference in effectiveness of each approach. In addition, the diversity and feasibility of the final set of solutions from each approach will be observed and finally the results will be compared to determine if such a statistical significant difference exists.

This paper does not propose any new constraint handling approach nor does it propose any new method of optimisation, but instead develops insight into the effectiveness of current constraint handling approaches. The chosen set of benchmark problems is assumed to be sufficient in terms of complexity and is suitable for comparing the two approaches.

The rest of the paper is organized as follows: Sect. 2 presents a literature survey of constraint handling approaches. The experimental setup is discussed in Sect. 3. Section 4 presents the results and finally the conclusions are discussed in Sect. 5.

2 Background

Mathematical approaches have been proposed to solve multi-objective optimization problems. However, many mathematical programming techniques are susceptible to the shape of the Pareto front and struggle to find solutions when the PF is concave or disconnected, i.e. the Pareto front consists of disconnected pieces. Many approaches also require the calculation of the derivative, which is not always feasible. Most mathematical approaches also only produce one solution per run. Therefore, to obtain multiple solutions, the algorithm has to be run multiple times [4,5]. A good discussion of mathematical approaches can be found in [6].

Since multi-objective optimization problems typically have conflicting objectives, a single solution does not exist and the goal of a multi-objective algorithm is to find the set of optimal trade-off solutions. An evolutionary algorithm is capable of finding multiple solutions from a single run. Therefore, it is well suited to find the set of optimal trade-off solutions. Furthermore, these algorithms do not require the calculation of gradients [4].

When considering real world optimization problems there always exists a possibility that a particular problem has constraints that need to be satisfied [2]. Konak et al. states four simple approaches to address constraints: (i) discarding any solutions that violate any constraints, also known as the "death penalty", (ii) "repairing" each solution, (iii) creating operators that always result in feasible solutions and finally (iv) penalty functions, where each solution is penalized with respect to the degree in which they violated the constraints [2]. These methods can be applied to both single and multi-objective problems, however the manner in which they are implemented might vary.

When considering any constrained multi-objective problems the set of solutions for these problems can be classified as either feasible or infeasible. A solution is feasible if and only if it satisfies all given constraints, thus if a constraint is violated then that solution is classified as an infeasible solution. If an infeasible solution is found during the optimization process one of the four approaches outlined by Konak et al. [2] can be used to handle that particular solution. Ideally, the result of any optimization algorithm for a constrained problem should be a set of solutions that are both the global optima and feasible.

Penalty functions are the most common approach for constraint handling [7,8]. It involves reducing the overall fitness of a solution, where the reduction in fitness is directly proportional to the degree in which the solution violates the constraints [9]. Le Riche et al. and Coello Coello discuss the dangers of having a penalty that is too high or too low. They stated that if the penalty is too high it will constrain solutions to a feasible region and will effectively reduce the exploration of the search space. However, in the event that the penalty is too low the individuals might spend most of their time exploring infeasible regions, reducing the overall quality of the solution. Thus, the penalty needs to be kept at an optimal level [7,9]. Penalty functions were implemented in various different ways [7]: (i) static penalties, where the penalties applied to solutions remained constant throughout the optimization process, (ii) dynamic penalties, where the penalty applied depends on the current generation, (iii) annealing penalties, based on simulated annealing [10], where at the start of the optimization process the penalties applied are small and then they increase as the process progresses, i.e. start in a "hot" state (low penalties) and over time the temperature decreases (high penalties), (iv) death penalties in which infeasible solutions are discarded, risking reducing the diversity of the population and in the event that all the solutions within a population are infeasible then the entire population will be discarded and the process will stagnate [7].

One approach of applying penalty functions to a multi-objective optimization problem would be converting it into a single-objective problem using the weighted sum approach discussed earlier. If this is done then any solution will only have one overall fitness value, which then a penalty value could be added to. However, as pointed out by Deb, weight selection is not a simple task and can be highly subjective [3]. Thus, to avoid the issue of weight selection Deb discusses a second approach where a problem's objectives are kept separate and the penalty is applied to each fitness value. However, the penalties for each objective must be scaled to ensure the penalty being applied and objective's magnitudes are the same [3]. An issue with penalty functions is the high number of parameters that need to be tuned [7]. A different approach with fewer parameters is discussed next.

Coello Coello stated that a number of constraint strategies were prone to having a large number of fitness function evaluations and solutions that have complex encodings [11]. An alternative to the penalty function approach would be to use dominance based operators, which use the concept of Pareto dominance. In this approach a solution's dominance over another solution is taken into consideration. The Niched Pareto Genetic Algorithm (NPGA), proposed by Horn et al. [12], is one such approach. The NPGA uses Pareto dominance tournaments for its selection, which works as follows: two solutions are chosen at random from the population along with a comparison set also chosen at random. If one solution dominates every other solution within the comparison set and the other solution does not, then the dominant solution is chosen. In the event that both solutions are non-dominated or dominated, the tie is broken by selecting the solution that has the least number of solutions in its niche [13]. NPGA also takes fitness into account when checking the dominance of a solution, which in turn can have a high computational cost [11].

As an alternative, Coello Coello proposed another approach, that is based on NPGA, but uses neither niching nor fitness values in dominance checking. Instead of niching, the algorithm uses a selection ratio (S_r), which is used to control the diversity of the population [11,14]. In this approach all feasible solutions are given a high rank, while infeasible solutions are given a lower rank. The algorithm works by searching for solutions that are non-dominated and have a low constraint violation, and assigning a rank to them. Dominated solutions are given a lower rank. Higher ranked solutions are given preference in selection, however dominated and infeasible solutions are also given a chance to allow for diversity within the population [11]. Unlike NPGA, this approach is only applicable to single-objective problems.

The NPGA and its single objective variant fall under the genetic algorithm, i.e. they can be applied when a genetic algorithm is been used to optimize a problem. However, the use of dominance based comparisons for constraints can be extended to other optimization algorithms as well, the OMOPSO proposed by Sierra and Coello Coello [15] is one such algorithm. The OMOPSO works by selecting multiple leaders and adjusting the rest of the solutions based on the selected set of leaders. Sierra and Coello Coello [15] stated that the set of leaders must be a set of non-dominated solutions. This set is formed by using the dominance relation with regards to the objectives, however the selection process can be further augmented by using the dominance relation with regards to the constraints as well, thus the set of leaders produced will be feasible. Similarly the constraint dominance relation can also be applied to the select operators in Differential Evolution (DE).

3   Experimental Setup

The main focus of this paper is to gather sufficient empirical data to determine if there exists a statistical significant difference between different constraint handling approaches across multiple optimization algorithms. The algorithms, along with their parameters, used in this paper are listed in Sect. 3.1. The performance measures are discussed in Sect. 3.2. Section 3.3 discusses the different libraries and frameworks used for the experiment and finally Sect. 3.4 discusses the experimental procedure.

3.1   Algorithms

For this study three optimization algorithms were implemented, namely: NSGA-II [16], OMOPSO [15] and GDE3 [17]. For NSGA-II a population size of 100 was used, along with SBX crossover with a crossover probability of 0.9 and polynomial mutation with a mutation probability of $\frac{1}{n}$, where n is the number of parameters to be optimized. The distribution index for the mutation and crossover was set to 20 [16]. Similarly, OMOPSO was also run with a population size of 100, along with Uniform and Non-uniform mutation, both with a probability of $\frac{1}{n}$, where n is once again the number of parameters to be optimized. Finally, GDE3 was also run with a population size of 100, while using the standard operators for crossover and mutation.

3.2   Performance Measures

The following performance measures were used to compare the different constraint approaches:

- **Generational Distance (GD)** - This measure takes the solutions produced by an optimization algorithm and computes the average Euclidean distance between those solutions and the true Pareto front [18]. A lower GD value implies that the solutions are *closer* to the true Pareto front and are therefore better.
- **Inverted generational distance (IGD)** - This measure is the inverse of the GD measure with a key difference. The IGD calculates the minimum Euclidean distance instead of the average distance [18]. A low IGD value is desired.
- **Hypervolume (HV)** - This measure can be used to determine the accuracy and diversity of a set of solutions. HV can be used to find the size of the objective space that is covered by a set of solutions i.e. diversity [18]. A large HV value is desired.
- **Spread** - The spread performance measure is used to quantify the distribution of the final set of solutions [18]. A low spread value indicates that the solutions are clustered towards a certain area on the Pareto front. Thus a high spread value is desired.

3.3   Implementation

A Java framework was used to implement the algorithms and collect all necessary information with regards to the optimization process. JMetal 5 [19] was chosen as it provides all the necessary algorithms and allows for easily creating new operators and objective functions. JMetal also includes a feature that allows for carrying out experimental studies. The study feature allows for configuring an algorithm to solve a certain number of problems, a predefined number of times. Once the experiments are complete, the JMetal framework calculates all the specified performance indicators and places the results in latex tables and/or R scripts.

3.4   Procedure

The experiment consisted of implementing each constraint handling approach for each of the algorithms, and then testing each algorithm on all ten benchmark functions. The following algorithms were used: NSGA-II, OMOPSO and GDE3. Each algorithm, for a given constraint handling approach, and benchmark function, was run 30 times in an attempt to minimize any random effects. The benchmark functions were taken from the 2009 CEC problem set [20], where every selected problem had constraints.

The constraint handling approaches implemented are: Dominance based tournament selection (which will be referred to as Pareto dominance from this point on) along with Static and Dynamic Penalty functions under the penalty function category. The purpose of this study is to determine if there exists a

Table 1. Pareto dominance vs. Static penalties

Function class	Result	NSGA-II				OMOPSO				GDE3			
		GD	IGD	HV	Spread	GD	IGD	HV	Spread	GD	IGD	HV	Spread
Single constraint	Wins	1	0	2	1	1	0	7	1	0	0	1	1
	Losses	1	3	0	0	6	4	0	4	2	1	0	0
	Draws	6	5	6	7	1	4	1	3	6	7	7	7
Double constraint	Wins	0	0	1	0	1	1	1	1	0	0	2	0
	Losses	0	1	0	0	1	1	1	1	0	1	0	0
	Draws	2	1	1	2	0	0	0	0	2	1	0	2
Double objectives	Wins	0	0	1	0	2	1	5	2	0	0	3	1
	Losses	0	1	0	0	4	2	1	3	1	1	0	0
	Draws	7	6	6	7	1	4	1	2	6	6	4	6
Triple objectives	Wins	1	0	2	1	0	0	3	0	0	0	0	0
	Losses	1	3	0	0	3	3	0	2	1	1	0	0
	Draws	1	0	1	2	0	0	0	1	2	2	3	3

statistical significant difference between the two constraint handling approaches. Therefore, the results of the Pareto dominance approach was compared with the results of both the static and dynamic penalty approaches.

Pair-wise Mann-Whitney U tests were used to determine the difference in performance for each constraint handling approach for the different problems. If the Mann-Whitney U test indicated a statistical significant difference, the average performance measure values were compared. The algorithm with the best average value was awarded a win and the other algorithm a loss.

4   Results

All the results for the experiments are in presented in Tables 1, 2, 3, 4, 5 and 6. Table 1 summarizes all the wins, losses and draws for the Pareto dominance approach, compared to the Static Penalty approach. Similarly, Table 2 provides a summarized view of the wins, losses and draws for the Pareto dominance approach compared to the Dynamic Penalty approach. Tables 3, 4, 5 and 6 contain all the values gathered for each performance measure for each algorithm and constraint handling approach.

With regards to Tables 1 and 2 the problems were grouped into four different classes depending on their characteristics. The four classes are Single Constraint, Double Constraint, Double Objectives and Triple Objectives where Single constraint consists of problems 1 to 5 and 8 to 10, Double constraint consists of problems 6 and 7, Double objectives consists of the problems 1 to 7 and finally Triple objectives consists of problems 8 to 10. With regards to Tables 1 and 2 the following observations were made for each performance measure.

4.1   GD

With regards to NSGA-II, for the single constraint function class the Pareto dominance approach tied with the static penalty approach, where both approaches performed better for 12.5% of the problems, while 75% of the remaining problems

Table 2. Pareto dominance vs. Dynamic penalties

Function class	Result	NSGA-II				OMOPSO				GDE3			
		GD	IGD	HV	Spread	GD	IGD	HV	Spread	GD	IGD	HV	Spread
Single constraint	Wins	0	0	5	0	5	5	6	3	2	1	5	1
	Losses	2	3	0	3	2	2	2	5	1	4	0	3
	Draws	6	5	3	5	1	1	0	0	5	3	3	4
Double constraint	Wins	0	0	1	1	2	1	1	0	0	0	1	1
	Losses	0	1	0	0	0	0	0	2	0	1	0	0
	Draws	2	1	1	1	0	1	1	0	2	1	1	1
Double objectives	Wins	0	0	3	1	4	3	6	2	2	1	3	2
	Losses	1	1	0	1	2	2	0	5	0	2	0	1
	Draws	6	6	4	5	1	2	1	0	5	4	4	4
Triple objectives	Wins	0	0	3	0	3	3	1	1	0	0	3	0
	Losses	1	3	0	2	0	0	2	2	1	3	0	2
	Draws	2	0	0	1	0	0	0	0	2	0	0	1

showed no significant difference in performance. However, the dynamic penalty approach, when being compared to the Pareto dominance approach, performed significantly better for 25% of the single constraint problems, while the remaining 75% showed no significant difference. For both the double constraints and double objective classes there were no statistical significant difference in performance for the static penalty approach. However, the dynamic penalties performed better on 14.3% and 33.3% of the respective classes.

For OMOPSO, the Pareto dominance approach, when compared to the static penalty approach, did not achieve better GD scores for any of the four function classes. For the single constraint class, Pareto dominance performed better for 12.5% of the problems, while the static penalty approach performed better for 75% of the problems. There was no significant difference in performance for the remaining 12.5%. In the double objectives class, Pareto dominance performed better for 28.6% of the problems, while static penalty performed significantly better on 54.1% of the problem set. For the final triple objective class, Pareto dominance did not outperform static penalty for any problems within that set. However, Pareto dominance achieved significantly better results when compared to the dynamic penalty approach. In the single constraint class, Pareto dominance performed better for 62.5% of the problems, while dynamic penalty produced better results for only 25%. Similarly, for double objectives and triple constraints, Pareto dominance achieved better results for all problems within those sets.

In the instance of GDE3 it can be seen that Pareto dominance performed worse than static penalty. This can be seen in Table 1 where Pareto dominance achieved no wins across all four classes. However, it can be seen in Table 2 that Pareto dominance achieved better results against dynamic penalty. For the single constraint class Pareto dominance performed well for 25% of the problems while dynamic penalty only performed well for 12.5% of the problems. There existed no significant difference in the results for the remaining 62.5%. A similar result was seen for the double objectives class where Pareto dominance performed well for 28.6% of the problems while there was no significant difference in the performance of the remaining 71.4%.

Table 3. Mean GD and standard deviation for each algorithm and constraint handling approach.

	NSGA-II	OMOPSO	GDE3
Pareto dominance			
P_1	$4.15\mathrm{e}{-}03_{2.0\mathrm{e}{-}03}$	$7.04\mathrm{e}{-}06_{2.8\mathrm{e}{-}06}$	$1.54\mathrm{e}{-}03_{1.4\mathrm{e}{-}04}$
P_2	$1.96\mathrm{e}{-}02_{2.0\mathrm{e}{-}02}$	$1.10\mathrm{e}{-}04_{5.2\mathrm{e}{-}05}$	$3.92\mathrm{e}{-}03_{8.4\mathrm{e}{-}04}$
P_3	$1.55\mathrm{e}{-}01_{6.1\mathrm{e}{-}02}$	$1.55\mathrm{e}{-}02_{3.9\mathrm{e}{-}03}$	$2.37\mathrm{e}{-}02_{3.3\mathrm{e}{-}03}$
P_4	$7.79\mathrm{e}{-}03_{7.3\mathrm{e}{-}03}$	$7.41\mathrm{e}{-}04_{7.1\mathrm{e}{-}04}$	$3.41\mathrm{e}{-}03_{4.3\mathrm{e}{-}04}$
P_5	$6.67\mathrm{e}{-}02_{8.4\mathrm{e}{-}02}$	$2.84\mathrm{e}{-}03_{2.5\mathrm{e}{-}03}$	$2.08\mathrm{e}{-}02_{3.3\mathrm{e}{-}03}$
P_6	$2.82\mathrm{e}{-}03_{1.8\mathrm{e}{-}03}$	$3.54\mathrm{e}{-}04_{1.8\mathrm{e}{-}04}$	$5.11\mathrm{e}{-}03_{1.6\mathrm{e}{-}03}$
P_7	$8.11\mathrm{e}{-}02_{1.1\mathrm{e}{-}01}$	$2.69\mathrm{e}{-}03_{2.5\mathrm{e}{-}04}$	$1.06\mathrm{e}{-}02_{4.9\mathrm{e}{-}03}$
P_8	$1.49\mathrm{e}{+}01_{2.0\mathrm{e}{+}01}$	$1.68\mathrm{e}{-}03_{1.4\mathrm{e}{-}04}$	$3.99\mathrm{e}{-}03_{8.2\mathrm{e}{-}04}$
P_9	$3.05\mathrm{e}{-}01_{1.4\mathrm{e}{-}01}$	$2.24\mathrm{e}{-}03_{1.8\mathrm{e}{-}05}$	$3.93\mathrm{e}{-}03_{8.1\mathrm{e}{-}04}$
P_{10}	$4.90\mathrm{e}{+}01_{2.5\mathrm{e}{+}01}$	$5.18\mathrm{e}{-}03_{3.1\mathrm{e}{-}04}$	$7.30\mathrm{e}{-}01_{4.2\mathrm{e}{-}01}$
Static penalty			
P_1	$3.61\mathrm{e}{-}03_{2.5\mathrm{e}{-}03}$	$8.75\mathrm{e}{-}06_{6.1\mathrm{e}{-}06}$	$1.62\mathrm{e}{-}03_{1.6\mathrm{e}{-}04}$
P_2	$2.47\mathrm{e}{-}02_{1.4\mathrm{e}{-}02}$	$1.07\mathrm{e}{-}04_{5.1\mathrm{e}{-}05}$	$3.97\mathrm{e}{-}03_{8.6\mathrm{e}{-}04}$
P_3	$1.59\mathrm{e}{-}01_{5.7\mathrm{e}{-}02}$	$1.46\mathrm{e}{-}02_{1.1\mathrm{e}{-}03}$	$2.36\mathrm{e}{-}02_{3.4\mathrm{e}{-}03}$
P_4	$7.75\mathrm{e}{-}03_{9.7\mathrm{e}{-}03}$	$5.62\mathrm{e}{-}04_{4.9\mathrm{e}{-}04}$	$3.59\mathrm{e}{-}03_{1.8\mathrm{e}{-}03}$
P_5	$2.96\mathrm{e}{-}02_{3.9\mathrm{e}{-}02}$	$1.20\mathrm{e}{-}03_{1.2\mathrm{e}{-}03}$	$2.11\mathrm{e}{-}02_{2.6\mathrm{e}{-}03}$
P_6	$2.91\mathrm{e}{-}03_{2.2\mathrm{e}{-}03}$	$3.23\mathrm{e}{-}04_{1.2\mathrm{e}{-}04}$	$4.01\mathrm{e}{-}03_{1.4\mathrm{e}{-}03}$
P_7	$3.72\mathrm{e}{-}02_{4.8\mathrm{e}{-}02}$	$3.70\mathrm{e}{-}03_{1.5\mathrm{e}{-}04}$	$7.84\mathrm{e}{-}03_{4.1\mathrm{e}{-}03}$
P_8	$5.34\mathrm{e}{-}01_{3.9\mathrm{e}{-}01}$	$1.55\mathrm{e}{-}03_{9.1\mathrm{e}{-}05}$	$4.04\mathrm{e}{-}03_{1.5\mathrm{e}{-}03}$
P_9	$3.37\mathrm{e}{-}01_{9.2\mathrm{e}{-}02}$	$2.23\mathrm{e}{-}03_{1.1\mathrm{e}{-}05}$	$3.89\mathrm{e}{-}03_{6.5\mathrm{e}{-}04}$
P_{10}	$5.55\mathrm{e}{-}01_{4.7\mathrm{e}{-}01}$	$2.74\mathrm{e}{-}03_{3.5\mathrm{e}{-}05}$	$1.43\mathrm{e}{-}02_{6.0\mathrm{e}{-}03}$
Dynamic penalty			
P_1	$4.45\mathrm{e}{-}03_{3.2\mathrm{e}{-}03}$	$5.94\mathrm{e}{-}06_{3.8\mathrm{e}{-}06}$	$1.55\mathrm{e}{-}03_{1.5\mathrm{e}{-}04}$
P_2	$1.77\mathrm{e}{-}02_{1.3\mathrm{e}{-}02}$	$1.32\mathrm{e}{-}04_{6.1\mathrm{e}{-}05}$	$4.60\mathrm{e}{-}03_{6.6\mathrm{e}{-}04}$
P_3	$1.86\mathrm{e}{-}01_{1.1\mathrm{e}{-}01}$	$8.06\mathrm{e}{-}03_{1.8\mathrm{e}{-}03}$	$2.13\mathrm{e}{-}02_{3.5\mathrm{e}{-}03}$
P_4	$1.14\mathrm{e}{-}02_{1.2\mathrm{e}{-}02}$	$1.48\mathrm{e}{-}03_{1.6\mathrm{e}{-}04}$	$3.44\mathrm{e}{-}03_{2.0\mathrm{e}{-}04}$
P_5	$2.78\mathrm{e}{-}02_{4.9\mathrm{e}{-}02}$	$1.27\mathrm{e}{-}03_{1.1\mathrm{e}{-}04}$	$2.03\mathrm{e}{-}02_{4.1\mathrm{e}{-}03}$
P_6	$4.40\mathrm{e}{-}03_{3.5\mathrm{e}{-}03}$	$1.10\mathrm{e}{-}03_{1.7\mathrm{e}{-}05}$	$1.73\mathrm{e}{-}03_{3.7\mathrm{e}{-}04}$
P_7	$4.09\mathrm{e}{-}02_{6.8\mathrm{e}{-}02}$	$2.74\mathrm{e}{-}03_{2.3\mathrm{e}{-}04}$	$8.93\mathrm{e}{-}03_{4.4\mathrm{e}{-}03}$
P_8	$7.36\mathrm{e}{-}01_{5.2\mathrm{e}{-}01}$	$1.77\mathrm{e}{-}03_{1.6\mathrm{e}{-}04}$	$2.81\mathrm{e}{-}03_{1.3\mathrm{e}{-}03}$
P_9	$3.25\mathrm{e}{-}01_{3.8\mathrm{e}{-}02}$	$2.26\mathrm{e}{-}03_{6.9\mathrm{e}{-}07}$	$2.82\mathrm{e}{-}03_{1.9\mathrm{e}{-}04}$
P_{10}	$3.64\mathrm{e}{-}01_{3.6\mathrm{e}{-}01}$	$8.96\mathrm{e}{-}03_{0.0\mathrm{e}{+}00}$	$4.20\mathrm{e}{-}03_{8.0\mathrm{e}{-}04}$

Table 4. Mean IGD and standard deviation for each algorithm and constraint handling approach.

	NSGA-II	OMOPSO	GDE3
Pareto dominance			
P_1	$4.15\mathrm{e}{-}03_{2.0\mathrm{e}-03}$	$5.88\mathrm{e}{-}05_{2.0\mathrm{e}-05}$	$2.10\mathrm{e}{-}03_{3.9\mathrm{e}-05}$
P_2	$1.96\mathrm{e}{-}02_{2.0\mathrm{e}-02}$	$4.75\mathrm{e}{-}03_{4.2\mathrm{e}-03}$	$1.71\mathrm{e}{-}02_{1.5\mathrm{e}-02}$
P_3	$1.55\mathrm{e}{-}01_{6.1\mathrm{e}-02}$	$2.08\mathrm{e}{-}01_{1.3\mathrm{e}-01}$	$2.89\mathrm{e}{-}01_{5.2\mathrm{e}-02}$
P_4	$7.79\mathrm{e}{-}03_{7.3\mathrm{e}-03}$	$6.50\mathrm{e}{-}03_{4.9\mathrm{e}-03}$	$2.47\mathrm{e}{-}02_{5.8\mathrm{e}-02}$
P_5	$6.67\mathrm{e}{-}02_{8.4\mathrm{e}-02}$	$4.56\mathrm{e}{-}02_{8.6\mathrm{e}-02}$	$8.83\mathrm{e}{-}02_{1.4\mathrm{e}-01}$
P_6	$2.82\mathrm{e}{-}03_{1.8\mathrm{e}-03}$	$8.23\mathrm{e}{-}05_{6.5\mathrm{e}-05}$	$6.39\mathrm{e}{-}03_{8.8\mathrm{e}-03}$
P_7	$8.11\mathrm{e}{-}02_{1.1\mathrm{e}-01}$	$4.00\mathrm{e}{-}03_{4.6\mathrm{e}-03}$	$2.54\mathrm{e}{-}02_{6.6\mathrm{e}-02}$
P_8	$1.49\mathrm{e}{+}01_{2.0\mathrm{e}+01}$	$4.11\mathrm{e}{-}01_{1.8\mathrm{e}-02}$	$3.48\mathrm{e}{-}01_{3.6\mathrm{e}-01}$
P_9	$3.05\mathrm{e}{-}01_{1.4\mathrm{e}-01}$	$9.12\mathrm{e}{-}02_{1.2\mathrm{e}-02}$	$3.67\mathrm{e}{-}01_{5.9\mathrm{e}-02}$
P_{10}	$4.90\mathrm{e}{+}01_{2.5\mathrm{e}+01}$	$4.99\mathrm{e}{-}01_{6.3\mathrm{e}-02}$	$6.85\mathrm{e}{+}01_{4.0\mathrm{e}+01}$
Static penalty			
P_1	$3.61\mathrm{e}{-}03_{2.5\mathrm{e}-03}$	$6.81\mathrm{e}{-}05_{2.1\mathrm{e}-05}$	$2.09\mathrm{e}{-}03_{3.0\mathrm{e}-05}$
P_2	$2.47\mathrm{e}{-}02_{1.4\mathrm{e}-02}$	$4.93\mathrm{e}{-}03_{3.6\mathrm{e}-03}$	$1.77\mathrm{e}{-}02_{1.6\mathrm{e}-02}$
P_3	$1.59\mathrm{e}{-}01_{5.7\mathrm{e}-02}$	$1.17\mathrm{e}{-}01_{9.9\mathrm{e}-03}$	$2.79\mathrm{e}{-}01_{4.0\mathrm{e}-02}$
P_4	$7.75\mathrm{e}{-}03_{9.7\mathrm{e}-03}$	$1.40\mathrm{e}{-}03_{5.7\mathrm{e}-04}$	$8.06\mathrm{e}{-}03_{1.7\mathrm{e}-02}$
P_5	$2.96\mathrm{e}{-}02_{3.9\mathrm{e}-02}$	$2.85\mathrm{e}{-}03_{6.3\mathrm{e}-03}$	$6.07\mathrm{e}{-}02_{1.1\mathrm{e}-01}$
P_6	$2.91\mathrm{e}{-}03_{2.2\mathrm{e}-03}$	$4.61\mathrm{e}{-}05_{1.7\mathrm{e}-05}$	$3.86\mathrm{e}{-}03_{3.2\mathrm{e}-03}$
P_7	$3.72\mathrm{e}{-}02_{4.8\mathrm{e}-02}$	$9.46\mathrm{e}{-}03_{1.1\mathrm{e}-02}$	$3.14\mathrm{e}{-}03_{6.1\mathrm{e}-03}$
P_8	$5.34\mathrm{e}{-}01_{3.9\mathrm{e}-01}$	$2.76\mathrm{e}{-}01_{1.6\mathrm{e}-02}$	$5.55\mathrm{e}{-}01_{6.2\mathrm{e}-01}$
P_9	$3.37\mathrm{e}{-}01_{9.2\mathrm{e}-02}$	$7.50\mathrm{e}{-}02_{1.1\mathrm{e}-02}$	$3.72\mathrm{e}{-}01_{1.9\mathrm{e}-02}$
P_{10}	$5.55\mathrm{e}{-}01_{4.7\mathrm{e}-01}$	$1.60\mathrm{e}{-}01_{1.1\mathrm{e}-02}$	$3.27\mathrm{e}{-}01_{1.7\mathrm{e}-01}$
Dynamic penalty			
P_1	$4.45\mathrm{e}{-}03_{3.2\mathrm{e}-03}$	$7.83\mathrm{e}{-}05_{2.1\mathrm{e}-05}$	$2.10\mathrm{e}{-}03_{3.9\mathrm{e}-05}$
P_2	$1.77\mathrm{e}{-}02_{1.3\mathrm{e}-02}$	$3.59\mathrm{e}{-}03_{8.2\mathrm{e}-04}$	$1.58\mathrm{e}{-}02_{8.2\mathrm{e}-04}$
P_3	$1.86\mathrm{e}{-}01_{1.1\mathrm{e}-01}$	$7.96\mathrm{e}{-}02_{1.5\mathrm{e}-02}$	$3.00\mathrm{e}{-}01_{5.7\mathrm{e}-02}$
P_4	$1.14\mathrm{e}{-}02_{1.2\mathrm{e}-02}$	$1.94\mathrm{e}{-}02_{7.0\mathrm{e}-03}$	$8.73\mathrm{e}{-}03_{8.3\mathrm{e}-04}$
P_5	$2.78\mathrm{e}{-}02_{4.9\mathrm{e}-02}$	$2.43\mathrm{e}{-}03_{1.1\mathrm{e}-03}$	$9.97\mathrm{e}{-}02_{1.1\mathrm{e}-01}$
P_6	$4.40\mathrm{e}{-}03_{3.5\mathrm{e}-03}$	$3.81\mathrm{e}{-}04_{2.8\mathrm{e}-04}$	$4.41\mathrm{e}{-}03_{3.4\mathrm{e}-03}$
P_7	$4.09\mathrm{e}{-}02_{6.8\mathrm{e}-02}$	$3.72\mathrm{e}{-}02_{6.0\mathrm{e}-02}$	$1.60\mathrm{e}{-}03_{1.6\mathrm{e}-03}$
P_8	$7.36\mathrm{e}{-}01_{5.2\mathrm{e}-01}$	$7.45\mathrm{e}{-}01_{2.7\mathrm{e}-02}$	$6.59\mathrm{e}{-}01_{7.1\mathrm{e}-01}$
P_9	$3.25\mathrm{e}{-}01_{3.8\mathrm{e}-02}$	$2.99\mathrm{e}{-}01_{2.1\mathrm{e}-02}$	$3.72\mathrm{e}{-}01_{2.1\mathrm{e}-02}$
P_{10}	$3.64\mathrm{e}{-}01_{3.6\mathrm{e}-01}$	$1.27\mathrm{e}{+}00_{2.4\mathrm{e}-01}$	$1.00\mathrm{e}{+}00_{5.3\mathrm{e}-01}$

Table 5. Mean HV and standard deviation for each algorithm and constraint handling approach.

	NSGA-II	OMOPSO	GDE3
Pareto dominance			
P_1	$4.15\mathrm{e}{-03}_{2.0\mathrm{e}-03}$	$5.00\mathrm{e}{-01}_{4.2\mathrm{e}-05}$	$4.87\mathrm{e}{-01}_{4.9\mathrm{e}-04}$
P_2	$1.96\mathrm{e}{-02}_{2.0\mathrm{e}-02}$	$5.35\mathrm{e}{-01}_{2.6\mathrm{e}-03}$	$4.17\mathrm{e}{-01}_{1.9\mathrm{e}-02}$
P_3	$1.55\mathrm{e}{-01}_{6.1\mathrm{e}-02}$	$5.60\mathrm{e}{-02}_{1.7\mathrm{e}-02}$	$0.00\mathrm{e}{+00}_{0.0\mathrm{e}+00}$
P_4	$7.79\mathrm{e}{-03}_{7.3\mathrm{e}-03}$	$4.84\mathrm{e}{-01}_{3.1\mathrm{e}-02}$	$3.60\mathrm{e}{-01}_{2.4\mathrm{e}-02}$
P_5	$6.67\mathrm{e}{-02}_{8.4\mathrm{e}-02}$	$4.43\mathrm{e}{-01}_{5.7\mathrm{e}-02}$	$4.67\mathrm{e}{-02}_{2.5\mathrm{e}-02}$
P_6	$2.82\mathrm{e}{-03}_{1.8\mathrm{e}-03}$	$6.35\mathrm{e}{-01}_{4.3\mathrm{e}-03}$	$5.52\mathrm{e}{-01}_{9.0\mathrm{e}-03}$
P_7	$8.11\mathrm{e}{-02}_{1.1\mathrm{e}-01}$	$5.23\mathrm{e}{-01}_{1.2\mathrm{e}-02}$	$2.49\mathrm{e}{-01}_{2.0\mathrm{e}-01}$
P_8	$1.49\mathrm{e}{+01}_{2.0\mathrm{e}+01}$	$2.08\mathrm{e}{-01}_{1.8\mathrm{e}-02}$	$6.41\mathrm{e}{-02}_{3.3\mathrm{e}-02}$
P_9	$3.05\mathrm{e}{-01}_{1.4\mathrm{e}-01}$	$2.51\mathrm{e}{-01}_{8.7\mathrm{e}-03}$	$5.56\mathrm{e}{-02}_{2.5\mathrm{e}-02}$
P_{10}	$4.90\mathrm{e}{+01}_{2.5\mathrm{e}+01}$	$3.27\mathrm{e}{-02}_{4.5\mathrm{e}-03}$	$1.94\mathrm{e}{-04}_{1.0\mathrm{e}-03}$
Static penalty			
P_1	$3.61\mathrm{e}{-03}_{2.5\mathrm{e}-03}$	$5.00\mathrm{e}{-01}_{4.0\mathrm{e}-05}$	$4.87\mathrm{e}{-01}_{4.2\mathrm{e}-04}$
P_2	$2.47\mathrm{e}{-02}_{1.4\mathrm{e}-02}$	$5.36\mathrm{e}{-01}_{2.5\mathrm{e}-03}$	$4.17\mathrm{e}{-01}_{1.9\mathrm{e}-02}$
P_3	$1.59\mathrm{e}{-01}_{5.7\mathrm{e}-02}$	$6.56\mathrm{e}{-02}_{1.6\mathrm{e}-02}$	$0.00\mathrm{e}{+00}_{0.0\mathrm{e}+00}$
P_4	$7.75\mathrm{e}{-03}_{9.7\mathrm{e}-03}$	$5.02\mathrm{e}{-01}_{2.3\mathrm{e}-02}$	$3.67\mathrm{e}{-01}_{7.0\mathrm{e}-02}$
P_5	$2.96\mathrm{e}{-02}_{3.9\mathrm{e}-02}$	$4.70\mathrm{e}{-01}_{3.6\mathrm{e}-02}$	$4.92\mathrm{e}{-02}_{2.0\mathrm{e}-02}$
P_6	$2.91\mathrm{e}{-03}_{2.2\mathrm{e}-03}$	$6.36\mathrm{e}{-01}_{2.7\mathrm{e}-03}$	$5.60\mathrm{e}{-01}_{7.6\mathrm{e}-03}$
P_7	$3.72\mathrm{e}{-02}_{4.8\mathrm{e}-02}$	$4.91\mathrm{e}{-01}_{1.0\mathrm{e}-02}$	$3.18\mathrm{e}{-01}_{2.1\mathrm{e}-01}$
P_8	$5.34\mathrm{e}{-01}_{3.9\mathrm{e}-01}$	$2.29\mathrm{e}{-01}_{1.1\mathrm{e}-02}$	$7.91\mathrm{e}{-02}_{4.5\mathrm{e}-02}$
P_9	$3.37\mathrm{e}{-01}_{9.2\mathrm{e}-02}$	$2.55\mathrm{e}{-01}_{9.1\mathrm{e}-03}$	$6.07\mathrm{e}{-02}_{2.4\mathrm{e}-02}$
P_{10}	$5.55\mathrm{e}{-01}_{4.7\mathrm{e}-01}$	$1.67\mathrm{e}{-01}_{6.6\mathrm{e}-03}$	$2.33\mathrm{e}{-02}_{4.7\mathrm{e}-02}$
Dynamic penalty			
P_1	$4.45\mathrm{e}{-03}_{3.2\mathrm{e}-03}$	$5.00\mathrm{e}{-01}_{2.9\mathrm{e}-05}$	$4.87\mathrm{e}{-01}_{4.7\mathrm{e}-04}$
P_2	$1.77\mathrm{e}{-02}_{1.3\mathrm{e}-02}$	$5.94\mathrm{e}{-01}_{2.6\mathrm{e}-03}$	$4.72\mathrm{e}{-01}_{1.3\mathrm{e}-02}$
P_3	$1.86\mathrm{e}{-01}_{1.1\mathrm{e}-01}$	$1.16\mathrm{e}{-01}_{4.5\mathrm{e}-02}$	$0.00\mathrm{e}{+00}_{0.0\mathrm{e}+00}$
P_4	$1.14\mathrm{e}{-02}_{1.2\mathrm{e}-02}$	$5.39\mathrm{e}{-01}_{2.0\mathrm{e}-02}$	$4.01\mathrm{e}{-01}_{2.9\mathrm{e}-03}$
P_5	$2.78\mathrm{e}{-02}_{4.9\mathrm{e}-02}$	$5.35\mathrm{e}{-01}_{2.1\mathrm{e}-02}$	$4.72\mathrm{e}{-02}_{3.3\mathrm{e}-02}$
P_6	$4.40\mathrm{e}{-03}_{3.5\mathrm{e}-03}$	$6.65\mathrm{e}{-01}_{7.4\mathrm{e}-04}$	$5.96\mathrm{e}{-01}_{7.5\mathrm{e}-03}$
P_7	$4.09\mathrm{e}{-02}_{6.8\mathrm{e}-02}$	$5.25\mathrm{e}{-01}_{9.5\mathrm{e}-03}$	$2.41\mathrm{e}{-01}_{2.5\mathrm{e}-01}$
P_8	$7.36\mathrm{e}{-01}_{5.2\mathrm{e}-01}$	$1.60\mathrm{e}{-01}_{1.7\mathrm{e}-02}$	$1.79\mathrm{e}{-01}_{8.4\mathrm{e}-02}$
P_9	$3.25\mathrm{e}{-01}_{3.8\mathrm{e}-02}$	$2.83\mathrm{e}{-01}_{8.6\mathrm{e}-06}$	$2.04\mathrm{e}{-01}_{2.7\mathrm{e}-02}$
P_{10}	$3.64\mathrm{e}{-01}_{3.6\mathrm{e}-01}$	$0.00\mathrm{e}{+00}_{0.0\mathrm{e}+00}$	$5.23\mathrm{e}{-02}_{4.4\mathrm{e}-02}$

Table 6. Mean Spread and standard deviation for each algorithm and constraint handling approach.

	NSGA-II	OMOPSO	GDE3
Pareto dominance			
P_1	$5.26\mathrm{e}{-01}_{1.1\mathrm{e}-01}$	$6.64\mathrm{e}{-01}_{2.7\mathrm{e}-02}$	$2.67\mathrm{e}{-01}_{2.0\mathrm{e}-02}$
P_2	$1.46\mathrm{e}{+00}_{4.9\mathrm{e}-02}$	$1.20\mathrm{e}{+00}_{2.3\mathrm{e}-02}$	$1.36\mathrm{e}{+00}_{7.7\mathrm{e}-02}$
P_3	$1.32\mathrm{e}{+00}_{8.1\mathrm{e}-02}$	$1.08\mathrm{e}{+00}_{1.1\mathrm{e}-01}$	$8.47\mathrm{e}{-01}_{6.7\mathrm{e}-02}$
P_4	$1.50\mathrm{e}{+00}_{1.1\mathrm{e}-01}$	$1.05\mathrm{e}{+00}_{1.8\mathrm{e}-01}$	$1.32\mathrm{e}{+00}_{1.6\mathrm{e}-01}$
P_5	$1.34\mathrm{e}{+00}_{2.0\mathrm{e}-01}$	$1.17\mathrm{e}{+00}_{1.1\mathrm{e}-01}$	$1.16\mathrm{e}{+00}_{1.8\mathrm{e}-01}$
P_6	$8.02\mathrm{e}{-01}_{1.3\mathrm{e}-01}$	$8.27\mathrm{e}{-01}_{2.0\mathrm{e}-02}$	$9.17\mathrm{e}{-01}_{1.8\mathrm{e}-01}$
P_7	$1.34\mathrm{e}{+00}_{1.7\mathrm{e}-01}$	$1.22\mathrm{e}{+00}_{4.3\mathrm{e}-02}$	$1.51\mathrm{e}{+00}_{3.6\mathrm{e}-01}$
P_8	$9.81\mathrm{e}{-01}_{8.2\mathrm{e}-02}$	$1.33\mathrm{e}{+00}_{9.9\mathrm{e}-02}$	$8.88\mathrm{e}{-01}_{1.9\mathrm{e}-01}$
P_9	$8.26\mathrm{e}{-01}_{7.1\mathrm{e}-02}$	$8.09\mathrm{e}{-01}_{1.6\mathrm{e}-02}$	$6.84\mathrm{e}{-01}_{7.9\mathrm{e}-02}$
P_{10}	$1.00\mathrm{e}{+00}_{4.6\mathrm{e}-13}$	$1.12\mathrm{e}{+00}_{5.7\mathrm{e}-02}$	$9.80\mathrm{e}{-01}_{8.0\mathrm{e}-02}$
Static penalty			
P_1	$5.03\mathrm{e}{-01}_{1.1\mathrm{e}-01}$	$6.59\mathrm{e}{-01}_{2.0\mathrm{e}-02}$	$2.60\mathrm{e}{-01}_{2.2\mathrm{e}-02}$
P_2	$1.47\mathrm{e}{+00}_{6.4\mathrm{e}-02}$	$1.29\mathrm{e}{+00}_{1.5\mathrm{e}-02}$	$1.36\mathrm{e}{+00}_{8.1\mathrm{e}-02}$
P_3	$1.32\mathrm{e}{+00}_{8.4\mathrm{e}-02}$	$1.13\mathrm{e}{+00}_{9.5\mathrm{e}-02}$	$8.77\mathrm{e}{-01}_{8.1\mathrm{e}-02}$
P_4	$1.46\mathrm{e}{+00}_{1.2\mathrm{e}-01}$	$7.07\mathrm{e}{-01}_{1.1\mathrm{e}-01}$	$1.43\mathrm{e}{+00}_{1.0\mathrm{e}-01}$
P_5	$1.34\mathrm{e}{+00}_{2.0\mathrm{e}-01}$	$8.73\mathrm{e}{-01}_{1.0\mathrm{e}-01}$	$1.13\mathrm{e}{+00}_{1.8\mathrm{e}-01}$
P_6	$8.24\mathrm{e}{-01}_{1.0\mathrm{e}-01}$	$7.96\mathrm{e}{-01}_{1.4\mathrm{e}-02}$	$8.90\mathrm{e}{-01}_{1.2\mathrm{e}-01}$
P_7	$1.32\mathrm{e}{+00}_{1.7\mathrm{e}-01}$	$1.30\mathrm{e}{+00}_{8.8\mathrm{e}-02}$	$1.63\mathrm{e}{+00}_{3.1\mathrm{e}-01}$
P_8	$1.03\mathrm{e}{+00}_{1.1\mathrm{e}-01}$	$1.31\mathrm{e}{+00}_{6.3\mathrm{e}-02}$	$8.18\mathrm{e}{-01}_{1.3\mathrm{e}-01}$
P_9	$8.28\mathrm{e}{-01}_{6.6\mathrm{e}-02}$	$7.91\mathrm{e}{-01}_{1.2\mathrm{e}-02}$	$6.82\mathrm{e}{-01}_{4.9\mathrm{e}-02}$
P_{10}	$1.09\mathrm{e}{+00}_{1.5\mathrm{e}-01}$	$6.93\mathrm{e}{-01}_{1.5\mathrm{e}-02}$	$9.19\mathrm{e}{-01}_{1.6\mathrm{e}-01}$
Dynamic penalty			
P_1	$5.34\mathrm{e}{-01}_{1.3\mathrm{e}-01}$	$4.78\mathrm{e}{-01}_{3.3\mathrm{e}-03}$	$2.61\mathrm{e}{-01}_{2.2\mathrm{e}-02}$
P_2	$1.31\mathrm{e}{+00}_{1.1\mathrm{e}-01}$	$6.36\mathrm{e}{-01}_{1.8\mathrm{e}-02}$	$1.20\mathrm{e}{+00}_{8.0\mathrm{e}-02}$
P_3	$1.29\mathrm{e}{+00}_{9.8\mathrm{e}-02}$	$1.21\mathrm{e}{+00}_{7.5\mathrm{e}-02}$	$8.62\mathrm{e}{-01}_{8.0\mathrm{e}-02}$
P_4	$1.50\mathrm{e}{+00}_{1.2\mathrm{e}-01}$	$1.26\mathrm{e}{+00}_{9.4\mathrm{e}-03}$	$1.44\mathrm{e}{+00}_{5.8\mathrm{e}-02}$
P_5	$1.24\mathrm{e}{+00}_{1.9\mathrm{e}-01}$	$8.18\mathrm{e}{-01}_{1.7\mathrm{e}-01}$	$1.23\mathrm{e}{+00}_{1.8\mathrm{e}-01}$
P_6	$8.98\mathrm{e}{-01}_{1.2\mathrm{e}-01}$	$7.34\mathrm{e}{-01}_{1.8\mathrm{e}-01}$	$9.45\mathrm{e}{-01}_{8.7\mathrm{e}-02}$
P_7	$1.35\mathrm{e}{+00}_{2.0\mathrm{e}-01}$	$1.12\mathrm{e}{+00}_{6.8\mathrm{e}-02}$	$1.88\mathrm{e}{+00}_{4.0\mathrm{e}-02}$
P_8	$9.81\mathrm{e}{-01}_{1.4\mathrm{e}-01}$	$1.10\mathrm{e}{+00}_{2.1\mathrm{e}-02}$	$7.94\mathrm{e}{-01}_{1.5\mathrm{e}-01}$
P_9	$7.45\mathrm{e}{-01}_{6.8\mathrm{e}-02}$	$8.39\mathrm{e}{-01}_{1.2\mathrm{e}-02}$	$6.12\mathrm{e}{-01}_{5.0\mathrm{e}-02}$
P_{10}	$8.98\mathrm{e}{-01}_{1.2\mathrm{e}-01}$	$9.74\mathrm{e}{-01}_{4.0\mathrm{e}-02}$	$6.73\mathrm{e}{-01}_{7.1\mathrm{e}-02}$

4.2 IGD

For the IGD values, the Pareto dominance approach for NSGA-II performed significantly worse than both the static and dynamic penalty approaches across all four function classes. With regards to Tables 1 and 2 it can be seen that the Pareto dominance approach did not a achieve any wins while both the static and dynamic penalty approaches performed better for 37.5%, 50%, 14.3% and 100% of the problems for the single constraint, double constraint, double objectives and triple objective classes respectively.

The OMOPSO also performed poorly when comparing the Pareto dominance results to the static penalty results. For the single constraint class the static penalty approach performed better for 50% of the problem set while the remaining 50% showed no significant difference. For the double objectives class the Pareto dominance approach performed better for 14.3% of the problems while the static penalty approach performed better for 28.6% of the problems. The remaining 57.1% showed no significant difference in performance. For the final triple objective class the static penalty approach completely outperformed Pareto dominance by achieving better results for all problems within that set. With regards to Table 2 it can be seen that Pareto dominance performed well when compared to dynamic penalty. With the single constraint class the Pareto dominance approach performed significantly better for 62.5% of the problems while dynamic penalty only performed better for 25% of the problem set. With regards to the double objective class, Pareto dominance performed better for 42.9% of the problems and dynamic penalty for 28.6% of the problems. For the final triple objective class, dynamic penalty was completely outperformed by Pareto dominance which performed significantly better for all the problems within that problem set.

The GDE3 achieved similar results than NSGA-II. Both the static and dynamic penalty approaches slightly outperformed the Pareto dominance approach for all function classes. The static penalty approach performed better for 12.5%, 50%, 14.3% and 33.3% of the problems for the single constraint, double constraint, double objective and triple objective classes respectively. However, for each of those respective classes 87.5%, 50%, 57.1% and 66.7% of the problems showed no significant difference in performance, which is a slight improvement over both NSGA-II and OMOPSO.

4.3 HV

For this quality indicator the Pareto dominance approach for all three algorithms performed extremely well against the static and dynamic penalty approaches. For NSGA-II it can be seen in both Tables 1 and 2 that Pareto dominance had zero losses against both static and dynamic penalties. However there exists a high number of problems for the static penalty where there were no significant differences between the two approaches. For the comparison between Pareto dominance and static penalty, the former performed significantly better for 25%, 50%, 14.3% and 66.7% of the problems for the single constraint, double constraint, double objective and triple objective classes respectively, while the remaining 75%, 50%, 85.7% and 33.3% of the problems of each respective class showed no significant difference in performance. A similar observation can be made with regards to dynamic penalties, where Pareto dominance performed better for 62.5%, 50%, 42.9% and 100% of the problems for each respective class. Thus it can be seen that Pareto dominance outperformed dynamic penalties to a greater extent than static penalties.

The Pareto dominance approach for OMOPSO also produced good results and suffered very few losses and draws with static and dynamic penalties approach. With regards to Table 1 it can be seen that for the single constraint class Pareto dominance achieved better result for 87.5% of the problems while only 12.5% showing no significant difference. For the double constraint class both approaches performed equally well for 50% of the problems. For the double objectives class Pareto dominance performed extremely well by producing better results for 71.4% of the problems, while static penalty only produced better results for 14.2%. For the final class, triple objectives, Pareto dominance produced better results for all the problems within that class. Similarly, with regards to the dynamic penalty approach, for the single constraint class the Pareto dominance approach performed well for 75% of the problems while the dynamic penalty approach produced better values for remaining 25%. For the double constraint class Pareto dominance produced better results for 50% of the problems while the remaining 50% showed no difference in the performance. In the double objectives class Pareto dominance achieved better results for 85.7% of the problems while the remaining 14.3% showed no significant difference. The only instance of where the Pareto dominance lost to the static and dynamic penalty approaches is for the final triple objective class in Table 2, where the dynamic penalty approach produced better results for 66.7% of the problems while Pareto dominance produced better results for the remaining 33.3%.

For the remaining GDE3 algorithm, Pareto dominance did not suffer any losses against static and dynamic penalties. However, there was a significant amount of problems where there was no significant difference in the results of the different approaches. With regards to Table 1 it can be seen that Pareto dominance produced better results for 12.5%, 100%, 42.9% and 0% of the problems for the classes single constraint, double constraint, double objectives and triple objectives respectively, while the remaining 87.5%, 0%, 57.1% and 100% of the problems for each respective class showed no significant difference in performance. Finally a similar scenario can be seen for dynamic penalties where once again Pareto dominance suffered no losses against dynamic penalties. With regards to Table 2 it can be seen that Pareto dominance produced better results for 62.5%, 50%, 42.9% and 100% of the problems for each respective class while the remaining 37.5%, 50%, 57.1% and 0% of the problems for each respective class showed no significant difference in terms of their performance.

4.4   Spread

For the final performance measure, the Pareto dominance and static penalty approach for NSGA-II showed no significant difference in their performance. The static penalty approach had no wins and the Pareto dominance approach only produced better results for 14.3% and 33.3% of problems for the single constraint and triple objectives class respectively. For the remaining problems across the remaining classes there existed no significant difference in performance. The Pareto dominance and dynamic penalty showed similar results, however the dynamic penalties produced slightly better results for the single constraint and triple objectives class where it performed better on 37.5% and 66.7% of the problems. The remaining 60%, 50%, 71.4% and 33.3% of problems for single constraint, double constraint, double objectives and triple objectives respectively showed no significant difference.

For OMOPSO the penalty approaches appear to have outperformed the Pareto dominance approach. With regards Table 1 it can be seen that Pareto dominance suffered more losses than the other two algorithms. For the single constraint class Pareto dominance only performed well for 12.5% of the problems while static penalties performed well for 50% of the problems and the remaining 37.5% showed no difference. This can be observed once again for the double objectives class where Pareto dominance performed well on 28.6% of the problems, while static penalties performed well on 42.9%. With regards to Table 2 it be seen that Pareto dominance was completely outperformed by dynamic penalties. Pareto dominance only achieved better results for 37.5%, 0%, 28.6% and 33.3% of the problems for the single and double constraint as well the double and triple objective classes respectively, the dynamic penalty approach achieved better results for 62.5%, 100%, 71.4% and 66.7% of the problems for the respective classes.

With regards to the GDE3 it can be seen that both Pareto dominance and static penalty approach did not outperform each other. The only two instances where Pareto dominance produced better results were for the single constraint and double objective class where it performed better for 12.5% and 14.3% of the problems. The remaining problems across all classes showed no significant difference. For the dynamic penalties, the Pareto dominance approach did not perform well for the single constraint and triple objective classes.

By analyzing the performance measures one can begin to draw conclusions with regards to the effectiveness of each constraint handling approach for each algorithm. By looking at the data in Tables 1 and 2, the following observations can be made:

- The Pareto dominance approach achieved the best HV values when compared to both the static and dynamic penalty approach. Thus the Pareto dominance approach discovers solutions that are more diverse and cover most of the objective space.
- The OMOPSO implementation of Pareto dominance produced better solutions than the dynamic penalty implementation.

- With regards to OMOPSO, both static and dynamic penalties produced better values for Spread. Thus, the solutions produced were more distributed when compared to the solutions produced by the Pareto dominance approach.
- With regards to NSGA-II both the static penalty and Pareto dominance approach performed equally with regards to GD. However, the dynamic penalty approach showed a slight advantage over Pareto dominance in terms of performance.

5   Conclusion

There currently exist two main approaches to constraint handling in multiobjective optimization problems, namely: penalty functions and dominance based selection operators (selection based on Pareto dominance). The type of constraint handling approach chosen for a specific problem can have a significant impact on the final set of solutions that will be produced. Thus, it is important to select a suitable constraint handling approach when optimizing for a particular problem.

This paper conducted an experimental study to compare three constraint handling methods, namely: dominance based tournament selection (Pareto dominance) and static and dynamic penalty functions. These three methods were applied to NSGA-II, OMOPSO and GDE3. Each algorithm was run on a benchmark set consisting of 10 problems. Based on the statistical observations made, this paper concludes there was no overall significant difference between the two constraint handling approaches i.e. dominance based selection and penalty functions. However, there were instances where a constraint handling method produced significantly better results for a particular performance measure. Over all the functions, constraint handling approaches and optimization algorithms, the best HV values were produced by the Pareto dominance approach. Similarly, with regards to OMOPSO, the static and dynamic penalty approaches showed significantly better spread values when compared to spread values produced by the Pareto dominance approach.

Avenues for further research may involve expanding the benchmark problem set to include more problems of different classes i.e. concave, convex or disjoint Pareto fronts. The list of algorithms may also be expanded, as well as the constraint handling approaches, i.e. including different constraint handling approaches and different penalty function implementations.

Acknowledgements. This work is based on the research supported by the National Research Foundation (NRF) of South Africa (Grant Number 46712). The opinions, findings and conclusions or recommendations expressed in this article is that of the author(s) alone, and not that of the NRF. The NRF accepts no liability whatsoever in this regard.

References

1. Zitzler, E., Thiele, L.: Multiobjective optimization using evolutionary algorithms— a comparative case study. In: Eiben, A.E., Bäck, T., Schoenauer, M., Schwefel, H.-P. (eds.) PPSN 1998. LNCS, vol. 1498, pp. 292–301. Springer, Heidelberg (1998). https://doi.org/10.1007/BFb0056872
2. Konak, A., Coit, D.W., Smith, A.E.: Multi-objective optimization using genetic algorithms: a tutorial. Reliab. Eng. Syst. Saf. **91**(9), 992–1007 (2006)
3. Deb, K.: Multi-objective Optimization Using Evolutionary Algorithms. Wiley, Hoboken (2005)
4. Coello Coello, C.A.: Evolutionary multi-objective optimization: a historical view of the field. IEEE Comput. Intell. Mag. **1**(1), 28–36 (2006)
5. Miettinen, K.: Nonlinear Multiobjective Optimization. Kluwer Academic Publishers, Boston (1999)
6. Marler, R., Arora, J.: Survey of multi-objective optimization methods for engineering. Struct. Multidiscip. Optim. **26**(6), 369–395 (2004)
7. Coello Coello, C.A.: Theoretical and numerical constraint-handling techniques used with evolutionary algorithms: a survey of the state of the art. Comput. Methods Appl. Mech. Eng. **191**(11), 1245–1287 (2002)
8. Coello Coello, C.A.: A survey of constraint handling techniques used with evolutionary algorithms. Lania-RI-99-04, Laboratorio Nacional de Informática Avanzada (1999)
9. Le Riche, R., Knopf-Lenoir, C., Haftka, R.T.: A segregated genetic algorithm for constrained structural optimization. In: ICGA, pp. 558–565. Citeseer (1995)
10. Joines, J.A., Houck, C.R.: On the use of non-stationary penalty functions to solve nonlinear constrained optimization problems with GA's. In: Proceedings of the First IEEE Conference on Evolutionary Computation, IEEE World Congress on Computational Intelligence, pp. 579–584. IEEE (1994)
11. Coello Coello, C.A., Montes, E.M.: Constraint-handling in genetic algorithms through the use of dominance-based tournament selection. Adv. Eng. Inform. **16**(3), 193–203 (2002)
12. Horn, J., Nafpliotis, N., Goldberg, D.E.: A niched Pareto genetic algorithm for multiobjective optimization. In: Proceedings of the First IEEE Conference on Evolutionary Computation, IEEE World Congress on Computational Intelligence, pp. 82–87. IEEE (1994)
13. Zitzler, E., Thiele, L.: Multiobjective evolutionary algorithms: a comparative case study and the strength Pareto approach. IEEE Trans. Evol. Comput. **3**(4), 257–271 (1999)
14. Abraham, A., Jain, L.: Evolutionary multiobjective optimization. In: Abraham, A., Jain, L., Goldberg, R. (eds.) Evolutionary Multiobjective Optimization, pp. 1–6. Springer, London (2005). https://doi.org/10.1007/1-84628-137-7_1
15. Sierra, M.R., Coello Coello, C.A.: Improving PSO-based multi-objective optimization using crowding, mutation and ∈-dominance. In: Coello Coello, C.A., Hernández Aguirre, A., Zitzler, E. (eds.) EMO 2005. LNCS, vol. 3410, pp. 505–519. Springer, Heidelberg (2005). https://doi.org/10.1007/978-3-540-31880-4_35
16. Deb, K., Pratap, A., Agarwal, S., Meyarivan, T.: A fast and elitist multiobjective genetic algorithm: NSGA-II. IEEE Trans. Evol. Comput. **6**(2), 182–197 (2002)
17. Kukkonen, S., Lampinen, J.: GDE3: the third evolution step of generalized differential evolution. In: The 2005 IEEE Congress on Evolutionary Computation, vol. 1, pp. 443–450. IEEE (2005)

18. Riquelme, N., Von Lücken, C., Baran, B.: Performance metrics in multi-objective optimization. In: Computing Conference (CLEI), 2015 Latin American, pp. 1–11. IEEE (2015)
19. Jmetal 5. https://jmetal.github.io/jMetal/. Accessed 03 May 2017
20. Zhang, Q., Zhou, A., Zhao, S., Suganthan, P.N., Liu, W., Tiwari, S.: Multiobjective optimization test instances for the CEC 2009 special session and competition. University of Essex, Colchester, UK and Nanyang technological University, Singapore, Special Session on Performance Assessment of Multi-objective Optimization Algorithms, Technical report 264 (2008)

Genetic Programming for the Classification of Levels of Mammographic Density

Daniel Fajardo-Delgado[1](✉), María Guadalupe Sánchez[1],
Raquel Ochoa-Ornelas[1], Ismael Edrein Espinosa-Curiel[2], and Vicente Vidal[3]

[1] Department of Systems and Computation, ITCG, 49100 Ciudad Guzmán, Mexico
{dfajardo,msanchez,raqueoo}@itcg.edu.mx
[2] Department of Computer Science, CICESE-UT3, 63173 Tepic, Mexico
ecuriel@cicese.mx
[3] Department of Informatics Systems and Computing,
Universitat Politècnica de València, 46022 Valencia, Spain
vvidal@dsic.upv.es

Abstract. Breast cancer is the second cause of death of adult women in Mexico. Some of the risk factors for breast cancer that are visible in a mammography are the masses, calcifications, and the levels of mammographic density. While the first two have been studied extensively through the use of digital mammographies, this is not the case for the last one. In this paper, we address the automatic classification problem for the levels of mammographic density based on an evolutionary approach. Our solution comprises the following stages: thresholding, feature extractions, and the implementation of a genetic program. We performed experiments to compare the accuracy of our solution with other conventional classifiers. Experimental results show that our solution is very competitive and even outperforms the other classifiers in some cases.

Keywords: Breast cancer · Levels of mammographic density
Genetic programming

1 Introduction

Breast cancer is a health world problem and an epidemic priority in several developed countries [1]. In Mexico, this type of cancer is the second cause of death among women from 30 to 54 years old [2] and, according to Franco-Marina et al. [3], this mortality will be increasing in the next years. Thus, opportune and effective treatments is a priority to reverse this tendency.

A variety of factors determine the risk of breast cancer in women. Some of these factors relate to the family heritage or to the clinical history of the patient [4]. In spite of the variety of procedures for the early detection, which includes the auto-exploration and clinical tests, mammography is the only one that which offer enough opportune diagnosis for this type of cancer in Mexico [5].

© Springer International Publishing AG, part of Springer Nature 2018
L. Rutkowski et al. (Eds.): ICAISC 2018, LNAI 10841, pp. 363–375, 2018.
https://doi.org/10.1007/978-3-319-91253-0_34

Some of the main risk factors for breast cancer that are visible from a mammography image include the detection and classification of masses, calcifications, and the levels of mammographic density. While the detection of masses and calcifications have been widely studied through the use of digital mammographies, this is not the case for the classification of the levels of mammographic density [6] (in spite of the fact that it is one of the main risk factors for breast cancer in early stages [7]).

The levels of mammographic density reflect the mammary tissue composition. In a mammography, the mammary epithelium and stroma appear white while the fat appears dark. The Breast Imaging Reporting and Data System (BI-RADS) of the American College of Radiology (ACR) is a tool intended to standardize the terms used to report the findings on a mammogram [8]. According to the BI-RADS classification system, there are four categories for the mammographic densities: (T.I) fatty, (T.II) scattered areas of fibroglandular density, (T.III) heterogeneously dense, which may obscure masses, and (T.IV) extremely dense, which lowers sensitivity.

In this paper, we address the automatic classification problem for the levels of mammographic density by using genetic programming. This work is inspired by the initial study of Burling-Claridge et al. [9], who were the first to deal with this problem under an evolutionary approach. Similarly to them, our study only focuses on binary classification. The motivation of our work is twofold. First, we broaden the scope of the initial study of Burling-Claridge et al. through the use of different feature extraction methods. Second, we rather simplified the methodology proposed by them without necessarily the use of transfer learning phase. We also explore the use of others operators to produce new classification trees. Our experimental results shows that our solution outperforms the results reported in [9] for certain situations.

We organize the rest of this paper as follows. Section 2 presents the previous work of this research. Section 3 describes the methodology used in this work. Section 4 shows the experimental results obtained from the comparison of the proposed genetic program with other classifiers. Finally, Sect. 5 presents some concluding remarks.

2   Previous Work

The Computer-Aided Diagnosis (CAD) for the classification of the levels of mammographic density has been of great interest and it still is being a fertile area of research. A diversity of the methods and algorithms related to the automatic classification problem in mammographies include: wavelets [10], statistical methods [11], computer vision [12], artificial neural networks [13], and evolutionary classifiers [9]. From these classifying methods, the last one, respectively, is the least explored.

Burling-Claridge et al. [9] propose an evolutionary solution intended only for binary classifications of mammographic density levels. The aim of their study was to investigate whether genetic programming and learning classifier systems can perform well enough for this problem. They mainly use two methods to extract the features of the mammographies: statistical analysis and Local Binary Patterns (LBP). Both of them are similar to the proposed by Reyad et al. [14], with the difference that they consider the whole breast segment of the image rather than subsections of regions of interest. On the other hand, they implement a two staged method of transfer learning in their genetic program, reusing the extracted knowledge of one stage into the other. This process requires of n executions of their genetic program for each stage of the knowledge transfer, which involves a lot of computational resources to produce good results.

Although our work has been inspired by the initial study of Burling-Claridge et al. in [9], our approach differs in the following ways:

- We do not use the knowledge transfer process to improve our results, our solution consists of only one execution of the genetic program.
- The set of primitives (functions and terminals) used by our genetic program produces classification trees that outputs boolean values rather than scalar values.
- Similarly to Reyad et al. [14], our proposed solution applies the multiresolution analysis and the LBP method to subsections of regions of interest in the image instead of the whole breast segment.
- We also use the multiresolution analysis proposed in [14]; however, they applied their feature extraction methods for breast mass classification but not for the classification of the mammography density.

3   Methodology

Our proposed solution for the automatic classifications for the levels of mammographic density consists of the following three stages: thresholding, feature extractions, and the implementation of the genetic program. In this section, we describe each of these stages in more detail.

3.1   Thresholding

This stage consists of applying a thresholding method to automatically delimit the region of interest (ROI) of the mammographies. Thresholding is a widely used tool for image segmentation, which consists in identifying the different homogeneous components of an image. We use the Otsu method [15] for this aim, a non-parameterized and a supervised thresholding method that seeks to maximize the separability of resulting classes in a grayscale image. The result is a binary mask for each mammography image that allows discarding unimportant information such as the black background of the images. Figure 1 shows an example of the preprocessing of a mammography image of the INbreast dataset before and after applying the Otsu method.

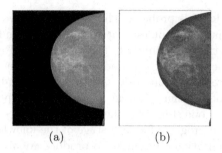

(a)   (b)

Fig. 1. Example of the preprocessing of a mammography image of the INbreast dataset; (a) the original image; (b) the image after applying the Otsu method

3.2 Feature Extractions

In the area of image pattern recognition, the process of feature extractions is one of the most important steps that influences both the design and performance of a classifier [16]. In this context, an image represents a mosaic of different texture regions, and the characteristics of the image are associated with those regions. According to Reyad et al. [14], mammographies are among the most difficult medical images to analyze. This is because of the low contrast and level of intensity of this type of images which can be easily affected by the breast tissue, blood vessels, and noise. In the present work, we use three approaches proposed by [14] to extract the features of the mammographies: statistical features, Local Binary Patterns (LBP), and multiresolution analysis. For the first two, respectively, we divide the ROI into 3×3 blocks (as shown in Fig. 2) to extract the features for each block.

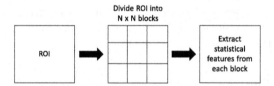

Fig. 2. Block diagram for the statistical features and local binary patterns

Statistical Features. Texture analysis based on statistical properties of intensity histogram is a common approach in image pattern recognition. In particular, we use the following six statistical features: mean, standard deviation, smoothness, skewness, uniformity, and entropy. Table 1 shows these features based on the intensity histogram of a region. In this table, the expression of the n-th order moments about the mean is $\mu_n = \sum_{i=0}^{L-1}(Z_i - m)^m p(z_i)$, where Z_i is a random variable that indicates intensity, $p(z_i)$ is the histogram of the intensity levels in a region, L is the number of possible intensity levels, and m is the mean intensity.

Table 1. Descriptors of texture based on the intensity histogram of a region

Moment	Expression	Measure of texture
Mean	$m = \sum_{i=0}^{L-1} z_i p(z_i)$	A measure of average intensity
Standard deviation	$\sigma = \sqrt{\mu_2(z)} = \sqrt{\sigma^2}$	A measure of average contrast
Smoothness	$R = 1 - \frac{1}{1+\sigma^2}$	Measures the relative smoothness of the intensity in a region
Skewness	$\mu_3 = \sum_{i=0}^{L-1} (Z_i - m)^3 p(z_i)$	Measures the skewness of a histogram
Uniformity	$U = \sum_{i=0}^{L-1} p^2(z_i)$	Measures the uniformity of intensity in the histogram
Entropy	$e = -\sum_{i=0}^{L-1} p(z_i) \, log_2 \, p(z_i)$	A measure of randomness

Local Binary Patterns. The local binary pattern (LBP) [17] is a local texture feature used in a wide pattern recognition applications. This feature provides a high discriminative power and tolerance to changes in illumination. Let c be a pixel of a mammography image, and let P be the set of neighboring pixels of c within a radius of distance R. We compute the LBP code of c through Eq. (1), where we compare the graylevel value of c, denoted as g_c, with the graylevel values of g_P from the set of neighboring pixels. In Eq. (1), $s(x) = 1$ if $x \geq 0$, and $s(x) = 0$ if $x < 0$.

$$LBP_{P,R} = \sum_{P=0}^{P-1} s(g_P - g_c)2^P \tag{1}$$

Assuming that c has the coordinates $(0,0)$, then the coordinates of P are $(-R\sin(\frac{2\pi P}{P}), R\cos(\frac{2\pi P}{P}))$. Figure 3 shows examples from the set of neighboring pixels for different P and R values.

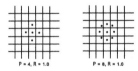

P = 4, R = 1.0   P = 8, R = 1.0

Fig. 3. Circularly symmetric neighbor pixels for different P and R

We compute the LBP code for each pixel $c_{i,j}$ with coordinates (i,j) within a $N \times M$ mammography image. The LBP histogram is built through Eq. (2), where K is the maximum value of the LBP code.

$$H(k) = \sum_{i=1}^{N} \sum_{j=1}^{M} (LBP_{P,R}(c_{i,j}), k), k \in [0, K] \tag{2}$$

Multiresolution Analysis. Another important technique for image texture analysis is the multiresolution analysis [18]. This method consists of decomposing a mammography image into different frequency sub-bands through an iterative and recursive process. This process generates a collection of decreasing resolution images arranged in the shape of a pyramid. The base of the pyramid contains the highest resolution approximation of the image, while the top contains the lowest one (see Fig. 4).

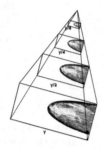

Fig. 4. Pyramid of images for the multiresolution analysis

In the present study, we use the two-dimensional Discrete Wavelet Transform (DWT) of Daubechies [19] to represent the image texture in terms of spectral spatial/frequency components. A mammography image is decomposed into a sub-band of low-frequency coefficients (A), and into three sub-bands of high-frequency coefficients: vertical (V), horizontal (H) and diagonal (D) high-frequency sub-bands. This process is repeated for more decomposition levels applied to the low-frequency sub-band (A). In this work, we use the DWT at its last level. The lowest sub-band is converted to a feature vector corresponding to the ROI and passed directly as a feature vector. Figure 5 illustrates a wavelet multiresolution decomposition at level 2.

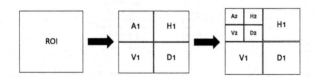

Fig. 5. Wavelet multiresolution decomposition at level 2

3.3 Genetic Programming

Genetic programming is an evolutionary technique with a great potential for classification [20]. This technique, proposed by Koza in [21], uses the principles of natural biological evolution to automatically adapt programs to solve problems.

The programs, as individuals of a population, produce offsprings by means of reproduction and mutation operators in each generation, in such a way that the best heritage traits preserve and improve progressively.

In the implementation of the genetic program, we consider the traditional tree structure as the representation for each individual of the population. The leaves of a program tree are elements from the set of terminals $T = \{$IN0, IN1, IN2,...$\}$ that consists of the features extracted from the mammographies pre-computed in stage 2 (see Sect. 3.2). On the other hand, each internal node of a program tree is an element from the set of functions $F = \{+, -, \times, \div, \wedge, \vee, \neg, \leq, =, if\_then\_else\}$ that consists of arithmetical, logical and conditional operations that can be applied to the terminals. We consider a 'protected' division $\div$, which prevents the generation of exceptions in the program when we divide a number by zero. The evaluation of each program tree returns a boolean value to represent two different classes.

The following pseudo-code shows an outline of the structure of the proposed genetic program:

1. (Initialization) Generate a set of μ random program trees generated through the 'ramped half-and-half' initialization method.
2. (Fitness evaluation) Evaluate each program tree in terms of the sum of coincidences of correctly classified images; i.e. $TP + TN$, where TP and TN denote the number of true positive and true negative predictions, respectively.
3. **while** number of generation is less than τ
   (a) (Parent selection) Randomly sample three individuals with replacement from the population and selects the one with the best fitness.
   (b) (Crossover) Recombine all the pairs of parents to generate λ offsprings. For each pair of parents, a random crossover point (which may be a node or edge in the tree) is chosen to exchange the secondary branches of that point derived from that point between them.
   (c) (Mutation) Randomly select a mutation point in the program tree and replace it with a random sub-tree generated by the 'full' initialization method.
   (d) (Fitness evaluation) Evaluate each program tree in terms of the sum of coincidences of correctly classified images.
   (e) (Survival selection) Create a new generation of μ individuals using deterministic fitness-based replacement.
   end_while
4. **return** the best individual of the population.

In Step 1, the well-known 'ramped half-and-half' initialization method ensures a variety of sizes and shapes of the computer trees in the initial population. Steps 2 and 3d compute a fitness measure for each individual in the population. Since the fitness of each individual is in terms of how well it classifies the mammographies, we seek to maximize the fitness measure. Step 3a consists of the well-known tournament selection method with size three. Step 3b combines all pairs of parents to obtain a set of new individuals who 'inherit' the desirable traits of their parents. Step 3c introduces a level of variability in the population through the random modification of one or more individuals. Step 3e builds a new generation by selecting the best μ individuals from the union of the μ individuals of the previous generation and the λ offsprings. Step 3 iterates until it completes τ generation and returns the best program tree in the last population (Step 4).

4   Experimental Results

With the aim to test the performance of our proposed solution, we performed experimental simulations to compare the accuracy of our solution with other conventional classifiers. For all the experiments, we performed 30 runs for each classifier. We executed the experiments on a 3.6 GHz Intel Core i7 (Mac) of four cores, with 8 GB of RAM, and under the operating system OS X 10.13.2. We used Python 3.6 to implement the classification method in all its stages.

Table 2 shows the configuration set used by the genetic program to build each binary classifier. We selected this configuration set by the execution of preliminary trials considering the tradeoff between time and efficiency.

Table 2. Set of configurations for the genetic programs.

Population size	$\mu = 1024$
Recombination rate	0.90
Mutation rate	0.29
Parent selection	Tournament with size 3
Survivor selection	Fitness-based $(\mu + \gamma)$
Completion criteria	$\tau = 100$ generations

We use the set of mammographies from the databases INbreast [22] and MIAS [23]. INbreast has a total of 410 images corresponding to 115 clinical cases, from which 90 are two image sets (MLO and CC) corresponding to each breast, and the other 25 correspond to women with two images from a different perspective of the same breast. On the other hand, MIAS database consists of a total of 322 mammographies with three levels of mammographic density, depending on the character of background tissue: fatty, fatty-glandular, and dense.

Table 3 shows the class count for the levels of mammographic density of INbreast and MIAS databases (corresponding to the columns INbreast_1 and MIAS, respectively). We grouped the images of each database into only two classes, considering for the first one the classes T.I and T.II and for the second one the classes T.III and T.IV (columns MIAS_Binary and INbreast_Binary_1 of Table 3). With the aim to compare the results of our solution with those

Table 3. Class counts for the MIAS and INbreast databases.

Class	Levels of mammographic density				BI-RADS	
	MIAS	MIAS_Binary	INbreast $_1$	INbreast_Binary$_1$	INbreast $_2$	INbreast_Binary$_2$
T.I	106	210	137	283	67	287
T.II	104		146		220	
T.III	112	112	99	127	23	123
T.IV	N.A		28		100	
Total	322	322	410	410	410	410

reported in [9], we also use the class count for INbreast according to the BI-RADS assessment categories (columns INbreast$_2$ and INbreast_Binary$_2$ of Table 3).

We compare the accuracy of our evolutionary solution (Our_GP) regarding other well-known classifiers: Gaussian Naive Bayes (NB), Decision Trees (DT), K-Nearest Neighbor (KNN), and Support Vector Machines (SVM). We also compare Our_GP with the Genetic Program (GP) and the Learning Classifier System (LCS) proposed by [9]. We performed a statistical test analysis of the results. Since the data is not normally distributed, we applied the Wilcoxon rank sum test to determine if there exists a statistically significant difference among the outputs.

Table 4 shows the accuracy of all classifiers when they use the statistical features. Despite that our proposed solution (Our_GP) has not the best results, the difference compared to the other classifiers is not significant. It is important to emphasize that Our_GP presents the lowest variability for MIAS and the second lowest for INbreast.

Table 4. Our_GP compared to other classifiers for statistical features.

		Dataset	
		MIAS_Binary	INbreast_Binary $_2$
Our_GP	Training (%)	84.70 ± 0.03	75.46 ± 0.02
	Testing (%)	71.16 ± 0.03	64.84 ± 0.02
GP [9]	Training (%)	76.82 ± 0.53	72.46 ± 0.25
	Testing (%)	72.19 ± 1.24	69.40 ± 0.79
LCS [9]	Training (%)	$\mathbf{94.48 \pm 0.50}$	$\mathbf{85.33 \pm 0.24}$
	Testing (%)	$\mathbf{73.44 \pm 3.18}$	68.57 ± 2.26
NB	Testing (%)	67.07 ± 0.73	67.61 ± 0.83
DT	Testing (%)	72.71 ± 1.34	69.37 ± 0.54
K-NN	Testing (%)	71.30 ± 1.14	59.11 ± 1.02
SVM	Testing (%)	72.95 ± 0.59	$\mathbf{70.00 \pm 0.00}$

Table 5. Our_GP compared to other classifiers for LBP features.

		Dataset	
		MIAS_Binary	INbreast_Binary$_2$
Our_GP	Training (%)	93.09 ± 0.02	79.95 ± 0.02
	Testing (%)	**79.82 ± 0.02**	65.61 ± 0.02
GP [9]	Training (%)	79.08 ± 1.49	73.87 ± 0.32
	Testing (%)	70.69 ± 2.47	68.75 ± 0.96
LCS [9]	Training (%)	**95.56 ± 0.35**	**93.93 ± 0.50**
	Testing (%)	65.57 ± 3.11	69.83 ± 2.18
NB	Testing (%)	62.86 ± 0.70	51.99 ± 1.28
DT	Testing (%)	66.74 ± 2.23	68.19 ± 1.23
K-NN	Testing (%)	71.50 ± 1.05	58.80 ± 1.06
SVM	Testing (%)	74.53 ± 0.79	**70.84 ± 0.24**

Table 6. Our_GP compared to other classifiers for statistical and LBP features.

		Dataset	
		MIAS_Binary	INbreast_Binary$_2$
Our_GP	Training (%)	89.97 ± 0.03	88.95 ± 0.03
	Testing (%)	**76.82 ± 0.03**	**71.31 ± 0.15**
GP [9]	Training (%)	78.80 ± 1.03	73.91 ± 0.33
	Testing (%)	71.31 ± 2.37	69.05 ± 1.00
LCS [9]	Training (%)	**95.22 ± 0.17**	**92.98 ± 0.38**
	Testing (%)	66.58 ± 2.92	69.84 ± 2.18
NB	Testing (%)	63.27 ± 0.70	52.15 ± 1.23
DT	Testing (%)	68.29 ± 2.24	66.89 ± 1.47
K-NN	Testing (%)	72.86 ± 1.08	57.76 ± 0.99
SVM	Testing (%)	74.92 ± 1.23	70.90 ± 0.24

Table 5 shows the results of using LBP features across all classifiers. Similarly to the statistical features, Our_GP presents the lowest variability for the LBP features. Our_GP obtains the highest precision over all classifiers for MIAS. Regarding to INbreast, Our_GP is still competitive.

Table 6 shows the results of using statistical and LBP features across all classifiers. Similarly to the previous two, Our_GP presents the lowest variability for the LBP features. Our_GP obtains the highest precision over all classifiers for MIAS and INbreast.

Table 7 shows the results by using the coefficients of the multiresolution analysis as the feature vector. These values are lower than the obtained by the other feature extraction methods.

Table 7. Our_GP results by using the coefficients of the multiresolution analysis.

		Dataset	
		MIAS_Binary	INbreast_Binary$_2$
Our_GP	Training (%)	85.38 ± 0.03	81.50 ± 0.03
	Testing (%)	66.78 ± 0.02	60.60 ± 0.02

Table 8. Our_GP results by using the classification of the levels of mammographic density for INbreast_Binary$_1$.

		INbreast_Binary$_1$
Statistical features	Training (%)	80.09 ± 0.02
	Testing (%)	59.48 ± 0.02
LBP features	Training (%)	79.47 ± 0.01
	Testing (%)	59.43 ± 0.02
Statistical and LBP features	Training (%)	79.74 ± 0.02
	Testing (%)	59.54 ± 0.01
Multiresolution features	Training (%)	$\mathbf{81.52 \pm 0.03}$
	Testing (%)	$\mathbf{60.67 \pm 0.03}$

Finally, Table 8 shows the results by using the classification of the levels of mammographic density for the INbreast_Binary$_1$ of Table 3. The results show that Our_GP has the highest precision when it uses the feature vector of the multiresolution analysis.

The experimental results show that the efficiency of Our_GP varies according to the feature extraction method. Our_GP obtains the bests results when it uses the combination of the statistical features with the LBP for MIAS and INbreast databases; however, this is not the case when it uses such feature vectors independently. We conjecture that this behavior occurs because of the random variability of chosen elements from the set of terminals for each program tree.

5 Conclusions

In this work, we present a new evolutionary solution for the automatic classification of the levels of mammographic density. Similarly to [9], we do not use a previous detected ROI within the mammographies for the classification. Instead of that, we apply a thresholding process to the complete image with the aim to include the whole breast as the ROI.

Experimental results show that our proposed solution is, in general, very competitive. Even for some cases of the MIAS database, it outperforms the results of the other classifiers. We notice that our solution presents the lowest level of variability with respect to the other classifiers. On the other hand, we use the multiresolution analysis as a feature extraction method for our evolutionary approach. Although the results by using this method show a lower precision than those reported by the other feature extraction methods, we believe that is because of the decomposition down to the last level of the DWT.

Acknowledgments. This work was supported by PRODEP under grant 511-6/17-8931 (ITCGUZ-CA-7), and by the TecNM under the projects 6055.17-P and 6307.17-P. Additionally, it was partially funded by the Spanish Ministry of Economy and Competitiveness under grant TIN2015-66972-C5-4-R co-financed by FEDER funds.

References

1. Knaul, F.M., Nigenda, G., Lozano, R., Arreola-Ornelas, H., Langer, A., Frenk, J.: Breast cancer in Mexico: a pressing priority. Reprod. Health Matter **16**(32), 113–123 (2008)
2. Lozano, R., Knaul, F., Gómez-Dantés, H., Arreola-Ornelas, H., Méndez, O.: Trends in mortality of breast cancer in Mexico, 1979–2006, observatory of health. Work Document. Competitiveness and Health, Mexican Foundation for the Health (2008). (in Spanish)
3. Franco-Marina, F., Lazcano-Ponce, E., López-Carrillo, L.: Breast cancer mortality in Mexico: an age-period-cohort analysis. Pub. Health Mex. **51**, s157–s164 (2009). (in Spanish)
4. Tyrer, J., Duffy, S.W., Cuzick, J.: A breast cancer prediction model incorporating familial and personal risk factors. Stat. Med. **23**(7), 1111–1130 (2004)
5. Brandan, M.E., Villaseñor, Y.: Detection of breast cancer: state of the mammography in Mexico. Cancerology **1**(3), 14–162 (2006). (in Spanish)
6. Byrne, C.: Studying mammographic density: implications for understanding breast cancer. JNCI-J. Natl. Cancer Inst. **89**(8), 531–532 (1997)
7. Wolfe, J.: Breast patterns as an index of risk for developing breast cancer. Am. J. Roentgenol. **126**(6), 1130–1137 (1976)
8. Obenauer, S., Hermann, K., Grabbe, E.: Applications and literature review of the BI-RADS classification. Eur. Radiol. **15**(5), 1027–1036 (2005)
9. Burling-Claridge, F., Iqbal, M., Zhang, M.: Evolutionary algorithms for classification of mammographie densities using local binary patterns and statistical features. In: IEEE Congress on Evolutionary Computation (CEC), pp. 3847–3854 (2016)
10. Qian, W., Li, L., Clarke, L.P.: Image feature extraction for mass detection in digital mammography: influence of wavelet analysis. Med. Phys. **26**(3), 402–408 (1999)
11. Chan, T.F., Golub, G.H., LeVeque, R.J.: Updating formulae and a pairwise algorithm for computing sample variances. In: Caussinus, H., Ettinger, P., Tomassone, R. (eds.) COMPSTAT 1982 5th Symposium held at Toulouse 1982, pp. 30–41. Springer, Heidelberg (1982). https://doi.org/10.1007/978-3-642-51461-6_3
12. Polakowski, W.E., Cournoyer, D.A., Rogers, S.K., DeSimio, M.P., Ruck, D.W., Hoffmeister, J.W., Raines, R.A.: Computer-aided breast cancer detection and diagnosis of masses using difference of Gaussians and derivative-based feature saliency. IEEE Trans. Med. Imaging **16**(6), 811–819 (1997)
13. Li, L., Clark, R.A., Thomas, J.A.: Computer-aided diagnosis of masses with full-field digital mammography. Acad. Radiol. **9**(1), 4–12 (2002)
14. Reyad, Y.A., Berbar, M.A., Hussain, M.: Comparison of statistical, LBP, and multi-resolution analysis features for breast mass classification. J. Med. Syst. **38**(9), 100 (2014)

15. Otsu, N.: A threshold selection method from gray-level histograms. IEEE Trans. Syst. Man Cybern. **9**(1), 62–66 (1979)
16. Ding, S., Zhu, H., Jia, W., Su, C.: A survey on feature extraction for pattern recognition. Artif. Intell. Rev. **37**(3), 169–180 (2012)
17. Ojala, T., Pietikainen, M., Harwood, D.: A comparative study of texture measures with classification based on featured distributions. Pattern Recognit. **29**(1), 51–59 (1996)
18. Gonzalez, R.C., Woods, R.E.: Image processing. Digital Image Process. **2** (2007)
19. Daubechies, I., et al.: Ten Lectures on Wavelets. CBMS-NSF Regional Conference Series in Applied Mathematics, vol. 61. SIAM, Philadelphia (1991)
20. Espejo, P.G., Ventura, S., Herrera, F.: A survey on the application of genetic programming to classification. IEEE Trans. Syst. Man Cybern. **40**(2), 121–144 (2010)
21. Koza, J.R.: Genetic Programming: On the Programming of Computers by Means of Natural Selection. MIT Press, Cambridge (1992)
22. Moreira, I.C., Amaral, I., Domingues, I., Cardoso, A., Cardoso, M.J., Cardoso, J.S.: Inbreast: toward a full-field digital mammographic database. Acad. Radiol. **19**(2), 236–248 (2012)
23. Suckling, J., Parker, J., Dance, D., Astley, S., Hutt, I., Boggis, C., Ricketts, I., Stamatakis, E., Cerneaz, N., Kok, S., et al.: The mammographic image analysis society digital mammogram database. Exerpta Medica. Int. Congr. Ser. **1069**, 375–378 (1994)

Feature Selection Using Differential Evolution for Unsupervised Image Clustering

Matheus Gutoski[1], Manassés Ribeiro[2], Nelson Marcelo Romero Aquino[1],
Leandro Takeshi Hattori[1], André Eugênio Lazzaretti[1],
and Heitor Silvério Lopes[1(✉)]

[1] Graduate Program in Electrical and Computer Engineering,
Federal University of Technology – Paraná (UTFPR), Av Sete de Setembro 3165,
Curitiba, PR 80230-901, Brazil
matheusgutoski@gmail.com, hslopes@utfpr.edu.br
[2] Catarinense Federal Institute of Education, Science and Technology–(IFC),
Rod. SC 135 km 125, Videira, SC 89560-000, Brazil

Abstract. Due to the accelerated growth of unlabeled data, unsupervised classification methods have become of great importance, and clustering is one of the main approaches among these methods. However, the performance of any clustering algorithm is highly dependent on the quality of the features used for the task. This work presents a Differential Evolution algorithm for maximizing an unsupervised clustering measure. Results are evaluated using unsupervised clustering metrics, suggesting that the Differential Evolution algorithm can achieve higher scores when compared to other feature selection methods.

Keywords: Differential evolution · Feature selection · Image clustering

1 Introduction

Classification of unlabeled data is an important and challenging task. The importance is mainly due to the huge volume of unlabeled data being created every second by surveillance cameras, Internet searches, and many other sources [1]. Although supervised classification methods have been highly successful, real-world scenarios seldom make unlabeled data available. As a consequence, a great challenge arises for developing robust unsupervised feature extraction techniques that can represent the data in meaningful ways. The problem becomes even more difficult as volume and dimensionality increases, which is a trend in real-world scenarios [2].

When it comes to analyzing unlabeled data, one of the most used techniques is clustering. The goal of a clustering algorithm is to partition the data such that similar instances are grouped together, and distinct instances are well separated. Many clustering algorithms have been proposed over the years, with K-Means [3] being one of the most popular due to its simplicity and performance.

© Springer International Publishing AG, part of Springer Nature 2018
L. Rutkowski et al. (Eds.): ICAISC 2018, LNAI 10841, pp. 376–385, 2018.
https://doi.org/10.1007/978-3-319-91253-0_35

However, the performance of clustering algorithms strongly depends on the quality of the features extracted from the data. A poor feature set usually leads to poor classification results. In image processing, the usual procedure for classification is to extract meaningful features from images, creating a more informative representation than the raw array of pixels. Image feature extractors can be handcrafted, such as Histogram of Oriented Gradients (HOG) [4], or learned, such as in Convolutional Neural Networks (CNNs) [5]. Despite their great success, CNNs fall under the supervised learning category. In supervised learning, ground truth information guides the network into learning features that are discriminative between each of the classes. However, this is not the case in handcrafted feature extractors. Image features obtained in unsupervised ways may not be discriminative enough to correctly classify images in complex scenarios. Furthermore, some of the features may be irrelevant to the classification process.

Eliminating irrelevant features is necessary to improve the performance of classifiers. This role is fulfilled by feature selection methods. Moreover, selecting a subset of important features can speed up the classification process, reduce computational complexity [1] and improve the classification performance [6].

Many methods have been developed for performing the feature selection task. In [7], a Fast Clustering-Based Feature Subset Selection Algorithm (FAST) was used to select the most relevant features from data using class label information. An unsupervised feature selection method was introduced by [8], where feature saliency is introduced and estimated by an expectation-maximization algorithm. [9] also proposed another unsupervised feature selection method that combines cluster analysis and sparse structural analysis. Bio-inspired feature selection methods have also been explored in the literature. In [10], feature selection is performed using Ant Colony Optimization. In [11], a combination of Genetic Algorithms and Particle Swarm Optimization achieve great results on a supervised feature selection task. [12] performs feature selection and classification simultaneously using a Genetic Programming approach, reaching interesting results.

Inspired by the potential of bio-inspired algorithms, we propose a Differential Evolution (DE) approach for selecting a subset of features in an unsupervised way. The objective function of the algorithm is an unsupervised clustering measure.

Clustering results are evaluated using unsupervised clustering measures. In a fully unsupervised scenario, supervised measures cannot be computed, since they require human-made annotations. Contrariwise, unsupervised measures analyze cluster results by, for instance, computing cluster density and distance between centroids. We show that the proposed DE approach outperforms other methods in regard to unsupervised measures.

The contributions of this paper are: (i) a Differential Evolution algorithm for selecting features from a large dimensional space with an unsupervised clustering measure as the objective function; (ii) a comparison between four feature selection and dimensionality reduction techniques applied to an unsupervised image classification problem; (iii) a semantic analysis of the cluster meaning after classification.

The paper is organized as follows: Sect. 2 describes the theoretical aspects of the methods employed in this work, Sect. 3 presents the method, Sect. 4 presents the experimental results and analysis, and Sect. 5 provides the final remarks along with a discussion and future works.

2   Theoretical Aspects

2.1   Feature Selection

Feature selection fulfills the role of eliminating irrelevant features from data. This process leads to a smaller dimensionality and in some cases can improve the accuracy of a classifier by eliminating noisy features that could be misleading [6].

Feature selection methods are traditionally divided into four groups: filter, wrapper, embedded, and hybrid methods [7]. Filter methods provide a feature ranking by means of mathematical analysis of the data, without the need to build a classifier [8], making them computationally efficient. Wrapper methods select feature subsets based on the results of a classifier. Such methods can be computationally expensive, as they require a classification process for each evaluation. Embedded methods combine feature selection and classification in a single process, such as Neural networks [7]. Hybrid methods combine filter and wrapper methods by doing a two-step feature selection, where the filter is applied first, and the wrapper method comes in sequence [7].

The main feature selection approach used in this work is a wrapper method. DE is a meta-heuristic from the field of Evolutionary Computation that follows the Darwinian evolution principles. The evolution process is guided by an objective function. In this case, we employ the Calinski-Harabasz (CH) coefficient, which is an unsupervised clustering measure, as the function to be maximized.

2.2   Differential Evolution

Differential Evolution (DE) is an evolutionary computation method introduced by Storn and Price [13]. The algorithm performs global optimization by minimizing complex functions such as nonlinear and non-differentiable functions [13]. The algorithm became popular due to its simplicity. It is currently widely used in diverse areas [14, 15].

Similar to other evolutionary algorithms, DE evolves a population of possible solutions (vectors) using genetic operations such as crossover and mutation, along with selection techniques to choose the best solutions that will generate new solutions (the next generation). The mutation and crossover rates are parameters that have to be set in order to run the algorithm. Other parameters include population size, number of generations and selection method. The main difference between DE and other evolutionary algorithms is the way DE creates a new population. It is based on using the scaled differences of randomly selected vectors of the current population [14].

According to [16], the DE algorithm is usually described using the form DE/x/y/z, where x describes the differential mutation base, y describes the number of vector differences added to the base vector and z describes the crossover method. The most common definition of the DE algorithm is the (DE/rand/1/bin), which describes random mutation, single vectorial difference and binomial crossover [17]. Different methods for the crossover and mutation operations have been discussed in [16].

Algorithm 1 presents the pseudo-code of the DE algorithm [17]. The inputs to the algorithms are: $NP=$ number of individuals of the population; $CR=$ crossover probability; $F=$ weighting factor; $NV=$ number of variables (length of the vectors).

Algorithm 1. Pseudo-code of the DE algorithm.

---

function DE(NP, CR, F, NV);
Generate randomly the initial population (NP individuals);
$x \leftarrow random(NP, NV)$;
Compute the *fitness* for all individuals of the population;
$fitness_x \leftarrow f(x)$;
while *stopping criterion=FALSE* **do**
    for $i = 1$ **to** NP **do**
        $v_i^{G+1} \leftarrow$ mutation(x_i^G, F);
        $u_i^{G+1} \leftarrow$ crossover(x_i^G, v_i^{G+1}, CR);
    $fitness_u \leftarrow f(u)$;
    for $i = 1$ **to** NP **do**
        if $fitness_u(i) > fitness_x(i)$ **then**
            $x_i^{G+1} \leftarrow u_i^{G+1}$;
        else
            $x_i^{G+1} \leftarrow x_i^G$;
    Update *stopping criterion*;

2.3 Unsupervised Cluster Evaluation Metrics

Unsupervised metrics evaluate clustering based on internal information. One of those metrics is the Calinski-Harabasz score [18], which is defined by the ratio of between and within cluster dispersions. The index is defined by Eq. 1, where k is the number of clusters, B_k is the between cluster dispersion, W_k is the within cluster dispersion, and n is the number of points in the data.

$$CH(k) = \frac{B_k}{W_k} \times \frac{n - k}{k - 1}. \tag{1}$$

The between cluster dispersion matrix can be calculated as shown in Eq. 2:

$$B_k = \sum_{i=1}^{k} n_i \left\| \mathbf{c_i} - \mu \right\|^2, \tag{2}$$

where k is the number of groups, n_i is the number of observations in group i, $\mathbf{c_i}$ is the center of cluster i, μ is the data mean, and $\left\| \mathbf{c_i} - \mu \right\|$ is the euclidean norm. The within cluster dispersion matrix can be calculated as shown in Eq. 3:

$$W_k = \sum_{i=1}^{k} \sum_{\mathbf{x} \in C_i} \|\mathbf{x} - \mathbf{c_i}\|^2, \tag{3}$$

where k is the number of groups, $\mathbf{x}$ is a point, C_i is the cluster i and $\mathbf{c_i}$ is the center of cluster i. The Calinski-Harabasz index produces higher values when clusters are well defined, hence maximizing it is desirable.

The Silhouette score another metric used in this work. The index calculates the *fitness* of each data point to its cluster assignment. The index ranges from -1 to $+1$, where higher values represent more compact cluster assignments. The final score is defined by the silhouette mean of all data samples. The score can be calculated for individual data points following Eq. 4:

$$s_i = \frac{b_i - a_i}{\max(a_i, b_i)}, \tag{4}$$

where a_i is the mean of samples assigned to the same group, and b_i is the mean distance between the sample i and every other point in the nearest cluster.

3   Method

3.1   Overview

The proposed method is illustrated in Fig. 1. First, an input image is converted to grayscale and resized. Next, features are extracted from the images using the HOG algorithm. DE is then applied in order to select the best features according to the CH clustering measure. Next, clustering is done by using the K-Means algorithm. We evaluate the feature selection methods using two unsupervised clustering metrics: CH and Silhouette. Moreover, for the sake of comparison, we present results using two other feature selection methods: Highest Variance (VAR) and the well known Principal Component Analysis (PCA). Whilst the former is not a feature selection method, it is often used as a dimensionality reduction technique. We also present the clustering results using the Full Data, which is the original HOG feature vector. Implementation was performed using the scikit-learn and Inspyred python packages.

3.2   Preprocessing

Before the feature extraction process, the image data was reshaped to 64×128 pixels in order to allow the HOG feature extractor to operate with its default parameters. The images were also converted to grayscale.

3.3   Feature Extraction and Selection Setup

Using the default parameters, the HOG feature extractor generates 3780 features per image. In order to evaluate different feature selection methods, each algorithm was applied separately to the HOG data, selecting the top 100 features.

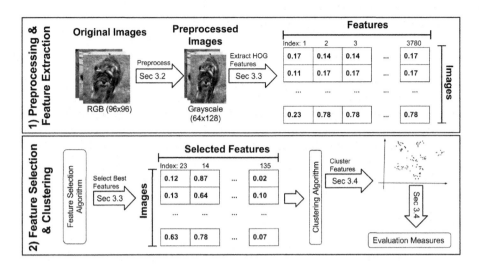

Fig. 1. Overview of the proposed method.

The only parameters required by the PCA and Variance filters are the n number of components to select.

The individuals of the DE algorithm are initialized with a random vector of length 100. Floating point values range from 1 to 3780, matching the length of the HOG feature vector. The lower and upper limits are ensured by the Bounding Function 5.

$$x = \begin{cases} 1 & \text{if } x < 1 \\ 3780 & \text{if } x > 3780 \\ x & \text{otherwise} \end{cases} \quad (5)$$

The decoding process takes the integer part of each value, where each element selects the index of a feature in the original HOG vector. Individuals are penalized in case of re-occurrence of any index by reducing their fitness by 10% per repetition.

The objective function maximizes the CH index (see Eq. 1). The DE (DE/rand/1/bin) parameters were set as follows: population size: 30; heuristic crossover rate: 0.75; Gaussian mutation rate: 0.01; generations: 2,000; Elitism: 0; selection by tournament with 2 candidates and 1 winner. The result reported is the best of 5 runs with different random seeds.

3.4 Clustering and Evaluation

Clustering is done with K-Means algorithm using the K-Means++ initialization method. The algorithm runs 10 times with different random seeds. The best run is then selected based on the least sum of squared distances from each point to its centroid. Evaluation is done using the Silhouette and CH coefficients.

4   Experimental Results and Analysis

4.1   Dataset Description

The STL-10[1] dataset contains 96×96 pixels color image data divided into 10 classes: airplane, bird, car, cat, deer, dog, horse, monkey, ship, truck. The dataset was developed for unsupervised and weakly supervised learning algorithms, and is divided as follows: unsupervised train: 100, 000 unlabeled instances containing data from all 10 classes and some variations; supervised train: contains 500 labeled images per class; test: 800 images per class. In this work, we employ only the Test set, since our method does not include a training phase.

4.2   Evaluation Measures

We perform the evaluation using the measures discussed in Sect. 2.3. Since the clustering algorithms require the number of groups parameters to be set, we propose two experiments. First, we cluster the data into 10 groups, since it is the real number of classes in the STL-10 dataset. Next, we propose a two groups experiment to approach the problem without using any ground truth information.

Experiment #1: 10 Groups. This experiment aims at evaluating the feature selection and clustering methods in a scenario where the number of clusters is set according to a prior ground truth information. The absolute classification performance is not the concern of this experiment. Instead, the relative performance of the different methods was evaluated.

In this experiment, we evolve the DE algorithm and cluster the data in 10 groups, i.e $K = 10$. Table 1 shows the results for this experiment.

Analyzing the results, a few conclusions can be drawn. Regarding the unsupervised measures, DE shows superior results compared to the other methods. The Silhouette score is also slightly higher, however, it shows that clusters are completely overlapped in the feature space.

PCA and *Full Data* presented similar results, as PCA captures most of the variance in 100 features. It also shows that a much smaller subset of features can produce the same results, with fewer computational endeavor. The variance filter performs worse than the other feature selection methods with respect to the CH coefficient, but has slightly better performance than PCA and Full Data regarding the Silhouette coefficient.

Experiment #2: 2 Groups. In this experiment, we also seek to explore the relative performances of the different feature selection methods. However, unlike the previous experiment, we assume no prior knowledge regarding the class distribution.

---

[1] http://cs.stanford.edu/~acoates/stl10/.

Without using any a priori information, we perform this experiment using $K = 2$ as a parameter for K-Means. Results are displayed in Table 1.

Results show that the algorithms form more compact and distant clusters, as shown by the CH and Silhouette scores. As before, the DE algorithm outperforms the other methods. Full Data and PCA presented very similar results, and the VAR filter was outperformed by all methods.

Despite the ground truth containing 10 classes, in a scenario where the user must define the number of groups without any previous knowledge, the 2 groups solution would be preferable over the 10 groups solution. Addressing this, we analyze the contents of each cluster in the two-groups solution searching for semantic patterns in each group. The analysis is found in the following Section.

Table 1. Evaluation of data clustering using K-Means with $K = 10, 2$

Metric	K = 10				K = 2			
	Full data	DE	Variance	PCA	Full data	DE	Variance	PCA
Calinski-Harabasz	188.30	483.59	143.35	188.08	992.21	2,164.37	724.65	992.11
Silhouette	0.007	0.04	0.01	0.006	0.13	0.23	0.12	0.13

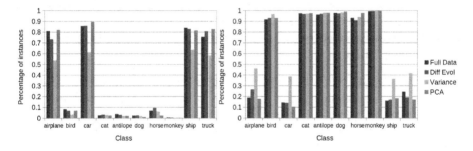

Fig. 2. Histogram of class occurrence per cluster. Cluster 1 (left), Cluster 2 (right)

4.3 Cluster Analysis

In order to better explore the 2 clusters solution, we present a histogram of class occurrences per cluster. Each cluster histogram counts every data point belonging to a class in ground truth level. Figure 2 shows the histograms of clusters 1 and 2. The x-axis shows each class at ground-truth level. The y-axis shows the percentage of instances belonging to a given class in the current cluster.

The chart shows a clear indication that HOG features form clusters with semantic meaning. Cluster 1 (left) shows that vehicles such as airplanes, cars, ships, and trucks were clustered together, whilst in cluster 2 (right) there is a clear predominance of animal classes such as birds, cats, antelopes, dogs, horses and monkeys.

Regarding the feature selection methods, there is no noticeable difference between DE, PCA and Full Data. The VAR filter, however, appears not to be as consistent regarding the animal/vehicle separation observed in other methods.

5   Conclusion

This paper presented a DE approach for performing feature selection in an unsupervised clustering task. The effectiveness of the DE algorithm was compared to other two simple feature selection methods. The evaluation was done by using two unsupervised clustering metrics. A qualitative analysis showing the contents of each cluster was also presented.

Results have shown that the DE feature selection outperformed other methods in terms of unsupervised metrics. Moreover, we have shown that HOG features are sufficiently discriminative between animal and vehicle classes on the STL-10 dataset.

Whilst clustering can be used as a classification method and evaluated as such, it is also true that being an unsupervised method, there is no single correct solution to the classification problem. For a clustering result to match the desired classification, it is necessary that the features are discriminative enough to differ between the target classes. Hence, future works should aim at developing methods for extracting semantic features from images.

Unlabeled image clustering is a difficult task. Separating data into groups with meaningful semantic information in an unsupervised way has proven to be one of the biggest challenges in the fields of Machine Learning, Deep Learning, and Computer Vision.

Acknowledgements. Author M. Gutoski and L.T. Hattori would like to thank CAPES for the scholarship; Author M. Ribeiro would like to thank the Catarinense Federal Institute of Education, Science and Technology and IFC/CAPES/Prodoutoral for the scholarship; Author N. Aquino would like to thank the Organization of the American States, the Coimbra Group of Brazilian Universities and the Pan American Health Organization; author H. S. Lopes would like to thank to CNPq for the research grant number 440977/2015-0.

References

1. Jain, A.K.: Data clustering: 50 years beyond K-means. Pattern Recogn. Lett. **31**(8), 651–666 (2010)
2. Gantz, J.F.: The diverse and exploding digital universe: an updated forecast of worldwide information growth through 2011. Technical report, IDC (2008)
3. Lloyd, S.P.: Least squares quantization in PCM. IEEE Trans. Inf. Theory **28**(2), 129–137 (1982)
4. Dalal, N., Triggs, B.: Histograms of oriented gradients for human detection. In: Proceedings of the IEEE Conference on Computer Vision and Pattern Recognition (CVPR), Piscataway, NJ, vol. 1, pp. 886–893. IEEE Press (2005)

5. Krizhevsky, A., Sutskever, I., Hinton, G.E.: ImageNet classification with deep convolutional neural networks. In: Proceedings of the 25th International Conference on Neural Information Processing Systems, (USA), vol. 1, pp. 1097–1105. Curran Associates Inc. (2012)
6. Chandrashekar, G., Sahin, F.: A survey on feature selection methods. Comput. Electr. Eng. **40**(1), 16–28 (2014)
7. Song, Q., Ni, J., Wang, G.: A fast clustering-based feature subset selection algorithm for high-dimensional data. IEEE Trans. Knowl. Data Eng. **25**(1), 1–14 (2013)
8. Law, M.H., Figueiredo, M.A., Jain, A.K.: Simultaneous feature selection and clustering using mixture models. IEEE Trans. Pattern Anal. Mach. Intell. **26**(9), 1154–1166 (2004)
9. Li, Z., Liu, J., Yang, Y., Zhou, X., Lu, H.: Clustering-guided sparse structural learning for unsupervised feature selection. IEEE Trans. Knowl. Data Eng. **26**(9), 2138–2150 (2014)
10. Tabakhi, S., Moradi, P., Akhlaghian, F.: An unsupervised feature selection algorithm based on ant colony optimization. Eng. Appl. Artif. Intell. **32**(1), 112–123 (2014)
11. Ghamisi, P., Benediktsson, J.A.: Feature selection based on hybridization of genetic algorithm and particle swarm optimization. IEEE Geosci. Remote Sens. Lett. **12**(2), 309–313 (2015)
12. Nag, K., Pal, N.R.: A multi-objective genetic programming-based ensemble for simultaneous feature selection and classification. IEEE Trans. Cybern. **46**(2), 499–510 (2016)
13. Storn, R., Price, K.: Differential evolution - a simple and efficient heuristic for global optimization over continuous spaces. J. Glob. Optim. **11**(4), 341–359 (1997)
14. Das, S., Suganthan, P.N.: Differential evolution: a survey of the state-of-the-art. IEEE Trans. Evol. Comput. **15**(1), 4–31 (2011)
15. Hattori, L.T., Lopes, H.S., Lopes, F.M.: Evolutionary computation and swarm intelligence for the inference of gene regulatory networks. Int. J. Innov. Comput. Appl. **7**(4), 225–235 (2016)
16. Lin, C., Qing, A., Feng, Q.: A comparative study of crossover in differential evolution. J. Heuristics **17**(6), 675–703 (2011)
17. Krause, J., Lopes, H.S.: A comparison of differential evolution algorithm with binary and continuous encoding for the MKP. In: Proceedings of the BRICS Congress on Computational Intelligence and 11th Brazilian Congress on Computational Intelligence, pp. 381–387 (2013)
18. Caliński, T., Harabasz, J.: A dendrite method for cluster analysis. Commun. Stat.-Theory Methods **3**(1), 1–27 (1974)

A Study on Solving Single Stage Batch Process Scheduling Problems with an Evolutionary Algorithm Featuring Bacterial Mutations

Máté Hegyháti[(✉)] , Olivér Ősz , and Miklós Hatwágner

Széchenyi István University, Győr 9026, Hungary
hegyhati@sze.hu
http://it.sze.hu

Abstract. The short term scheduling of batch processes is an active research field of chemical engineering, that has been addressed by many different techniques over the last decades. These approaches, however, are unable to solve long-term scheduling problems due their size, and the vast number of discrete decisions they entail. Evolutionary algorithms already proved to be efficient for some classes of large scheduling problems, and recently, the utilization of bacterial mutations has shown promising results on other fields.

In this paper, an evolutionary algorithm featuring bacterial mutation is introduced to solve a case study of a single stage product scheduling problem. The solution performance of the algorithm was compared to a method from the literature. The results indicate that the proposed approach can find the optimal solution under relatively short execution times.

Keywords: Single stage product scheduling
Bacterial Evolutionary Algorithm

1 Introduction and Literature Review

Batch process scheduling is a widely researched area of chemical engineering. The utilization of batch units is preferable for high value products with changing market demands. These units can be shared among several production processes to fulfill similar steps. Designing batch processes, however, has its own challenges, as the decisions must be made for both the assignment and scheduling of the production steps.

Over the last two or three decades, many approaches have been developed to solve various classes of batch process scheduling problems [7,11]. Most of the published methods rely on a Mixed-Integer Linear Programming formulation of the problem. Among these, different branches of modeling techniques were introduced [3]. Combinatorial tools, such as the S-graph framework [17], or

© Springer International Publishing AG, part of Springer Nature 2018
L. Rutkowski et al. (Eds.): ICAISC 2018, LNAI 10841, pp. 386–394, 2018.
https://doi.org/10.1007/978-3-319-91253-0_36

state space enumeration techniques, e.g., Linear Priced Timed Automata [15] or Timed Placed Petri Nets [5] were also developed and adopted for various industrial case studies. Aside from a few modeling errors [2, 8, 18], the aforementioned techniques are all exact methods guaranteeing the global optimal solution for short term scheduling problems.

Due to the large number of discrete decisions these problems have, the aforementioned approaches cannot be expected to provide optimal schedules for the long term planning of batch processes. This issue can be addressed with a number of different approaches, e.g., parallelization, heuristics, or expert-based restriction of the search space. Starting from the last, it is a common industrial practice, to assume, that the production plan is cyclic in nature, and thus only the most fitting cycle time, and the schedule of one cycle should be planned [10]. Parallelization is also a reasonable direction, considering the direction of recent hardware development. There has been many papers on the parallelization of MILP solvers [16], and the other approaches are suitable for parallel execution as well [19].

Evolutionary algorithms are widely used in engineering and research to solve complex design, search and optimization problems. One member of the populous family of these algorithms is the Bacterial Evolutionary Algorithm (BEA) [12]. The name of the method expresses the fact that it imitates the evolution of bacteria in nature. It is a direct descendant of Pseudo-Bacterial Genetic Algorithm [14] which can be considered a modified Genetic Algorithm [6]. BEA was originally designed to optimize the parameters of fuzzy systems [13], but has been applied successfully to solve other problems as well, e.g. the Traveling Salesman Problem [4] or the Permutation Flow Shop Problem [1]. It can find near-optimal, quasi-optimal solutions of even a non-continuous, non-linear, multi-modal or high-dimensional problem. The relatively easy implementation of this effective and robust algorithm made it popular among researchers.

BEA works with a collection of possible solutions, the so-called population. In most cases it is generated randomly at start, and improved during the consecutive generations. These generations are developed by using the two main operators: bacterial mutation and gene transfer.

Bacterial mutation independently modifies the elements of the current population, called individuals or bacteria. All genes of a bacterium are mutated in a randomized order, only one at a time. Usually two-three copies of the bacterium, the so-called clones are used to test the effect of a mutated gene value (allele). The modified values of the current gene are inserted into the clones, and then they are evaluated. If the change leads to a better fitness value, the value is preserved in the bacterium.

The second operator is the gene transfer. It also contains randomized activities, but its primary goal is to combine the already existing genetic information of the population in order to find better solutions. First, it sorts the population based on the fitness of bacteria. After that, it selects one bacterium from the better (superior) half of the population, and another one from the worse (inferior) half. Then it selects one or some genes, and copies the values at these positions of the superior bacterium into the inferior one. After the modifications, the bacterium must be re-evaluated and the population sorted again. If the gene transfer raised the fitness of the inferior bacterium with the necessary extent, the bacterium becomes a member of the superior half, and will be a potential source of gene data during the consecutive gene transfers.

These two operators are performed repeatedly to create the series of generations until a termination condition is fulfilled. A limit on the number of generations, wall clock time, convergence speed, etc. can serve as stop criterion, or a combination of them. Finally, the best bacteria of the last population are considered result.

2   Problem Definition

The problems investigated in this paper are two case studies introduced by Kopanos et al. [9]. In these problems, single-stage batch processes need to be scheduled, i.e., the product of each order is produced via a single production step, which is often referred to as a task. For each task, several production units may be suitable, however, it should be assigned to exactly one of them, i.e., multiple units cannot work on a single task in parallel. Similarly, a unit may work on only one task at a time, and preemption is not allowed.

A strict deadline is given for each order, which the completion times of the tasks must not exceed. The goal of the problem is to minimize the total earliness, that results in proportional storage costs. Note, that due to the strict deadlines, the problem may be infeasible as well.

An additional industrial restriction of the problem is that between executing two tasks, a unit may require cleaning, retooling or other maintenance. The time required for these jobs is called setup time, and consists of two parts: sequence-dependent and sequence-independent. Sequence-dependent setup times are given for each task pair, and may be asymmetric for the task order. Sequence-independent setup time depends only on the equipment unit but the problem can be generalized so it would depend on the task as well. Sequence-dependent setup time is independent from the unit but similarly, it can be extended to be unit-specific as well.

In later sections, P and U will denote the set of products and units, respectively. The set of units suitable for a product $p \in P$ is $U_p \subseteq U$, and similarly, the set of products, that can be performed by a unit $u \in U$ is $P_u \subseteq P$.

3   The Proposed Approach

The presented approach is an evolutionary algorithm, whose mutation follows the search trait of BEAs. The size of the population is fixed over all generations, and the initial population is generated randomly. For each new generation, a BEA-like mutation is applied to all individuals. Next, several new individuals are generated randomly, and mutated immediately. Some parents are selected for a simple crossover operation to generate additional individuals for the new generation. In this study a very simple, scattered crossover operation was applied, that randomly selects one of the parents' allele for each gene. These new individuals are also mutated immediately. Last, the size of the population is reduced to its fixed size. A fixed number of best individuals are kept, and the remaining ones are selected with a probability based on their fitness value.

3.1 Representation of Schedules and the Mutation

An individual representing a schedule has $|P|$ genes, which can take any value from the $[0, U)$ domain. The whole part of an allele indicates to which unit the corresponding task is assigned. The production order of the tasks assigned to the same units is determined by the increasing order of the fractional part of their alleles. This representation can be represented as a circle, that is divided into $|U|$ parts, the genes are dots around the circle, and the production sequence of a schedule is determined by their order.

This representation is not unambiguous, i.e., several individuals may represent the same schedule, although their alleles are different. However, these genes are suitable for a bacterial mutation, by generating several random values in a given neighborhood of the current allele for the local copies. Changing the allele may not alter the schedule at all, or it changes the place of the corresponding product in the production order of the assigned unit, or the product can also be moved to another unit like this.

Note, that if the new assignment of a product is infeasible, i.e., $p \in P$ is assigned to a new unit $u \in U \backslash U_p$, then the allele is further decreased or increased by 1, until a suitable unit is found.

3.2 Fitness Function

The fitness value of an individual equals to the cost of the best schedule with that production order, if it is feasible, otherwise it is punished by a great penalty. The best schedule for a production order can easily be generated by the following approach:

1. The task of the product assigned as last to a unit is timed such that its completion time equals the product deadline.
2. The task belonging to the previous product is pushed to the rightmost starting allowed by the next task, the setup times, and the processing time.
3. If this task violates the deadline of the corresponding product, it is shifted back, so that the completion time equals the deadline.
4. Repeat from step 2, until there is no other products left.

After the schedule is generated in this manner, the total earliness can easily be accumulated. If there are any tasks, that should start earlier than 0, an additional proportional penalty is added to the fitness value of the individual.

3.3 Fine Tuning

The proposed algorithm can be fine-tuned via several parameters:

population the fixed size of the population
localpool the number of modified alleles (clones) generated locally for mutation
range the range from where the modified alleles are selected during mutation
bestkeep the number of best individuals that are always kept for the next
    generation
new the number of the randomly generated new solutions in each generation
crossover the number of new individuals generated with crossovers.

4 Computational Results

The case studies presented by Kopanos et al. [9] contain 20 and 25 products
respectively, and 4 units. The processing times of the 20 tasks are the same as

Table 1. Processing times of the problem

Products	Processing time (days)				Deadline
	u1	u2	u3	u4	
P1	1.538	–	–	1.194	15
P2	1.500	–	–	0.789	30
P3	1.607	–	–	0.818	22
P4	–	–	1.564	2.143	25
P5	–	–	0.736	1.017	20
P6	5.263	–	–	3.200	30
P7	4.865	–	3.025	3.214	21
P8	–	–	1.500	1.440	26
P9	–	–	1.869	2.459	30
P10	–	1.282	–	–	29
P11	–	3.750	–	3.000	30
P12	–	6.796	7.000	5.600	21
P13	11.25	–	–	6.716	30
P14	2.632	–	–	1.527	25
P15	5.000	–	–	2.985	24
P16	1.250	–	–	0.783	30
P17	4.474	–	–	3.036	30
P18	–	1.492	–	–	30
P19	–	3.130	–	2.687	13
P20	2.424	–	1.074	1.600	19
P21	7.317	–	3.614	–	30
P22	–	–	0.864	–	20
P23	–	–	3.624	–	12
P24	–	–	2.667	4.000	30
P25	5.952	–	3.448	4.902	17
Setup	0.180	0.175	0.000	0.237	

Table 2. Setup times between consecutive products in Example 1

	P1	P2	P3	P4	P5	P6	P7	P8	P9	P10	P11	P12	P13	P14	P15	P16	P17	P18	P19	P20
P1	–	0.3	0.8	1.5	0.6	0.5	2.0	1.1	0.0	–	0.5	1.0	0.2	0.8	0.7	0.5	1.8	–	2.5	0.3
P2	0.2	–	1.3	0.9	2.5	0.2	0.8	2.5	0.4	–	0.6	2.5	0.5	0.2	0.6	0.0	1.1	–	0.8	2.5
P3	0.5	0.9	–	0.5	0.7	0.4	1.5	0.4	0.9	–	0.2	1.5	0.8	0.7	0.0	2.0	0.6	–	0.5	1.3
P4	1.1	0.7	0.2	–	0.8	2.0	0.9	0.0	1.3	–	1.5	1.0	1.8	0.6	1.3	0.6	1.5	–	1.0	0.5
P5	0.5	1.0	0.0	1.3	–	0.5	2.0	1.3	0.9	–	0.4	0.3	2.0	1.0	2.0	0.7	0.2	–	0.3	0.9
P6	0.2	0.0	1.3	1.0	1.0	–	0.7	1.3	0.8	–	0.7	0.6	0.5	0.7	0.5	2.0	0.9	–	1.1	0.5
P7	0.9	0.5	1.1	0.0	1.4	0.6	–	4.0	0.5	–	0.5	0.8	0.3	0.4	1.1	0.5	1.5	–	0.9	1.5
P8	1.5	2.0	0.4	1.3	0.5	0.9	0.7	–	0.9	–	0.4	1.8	0.6	1.5	0.6	0.5	0.7	–	0.9	1.1
P9	2.5	0.6	0.5	0.8	0.6	1.8	0.6	0.2	–	–	2.0	1.5	2.0	0.6	0.9	1.3	1.8	–	0.7	0.8
P10	–	–	–	–	–	–	–	–	–	–	1.0	1.3	–	–	–	–	–	0.0	0.8	–
P11	0.8	1.0	1.3	0.8	1.1	0.4	2.5	0.9	2.0	0.0	–	0.8	1.0	2.5	1.5	0.6	0.8	2.5	1.3	0.6
P12	0.2	0.7	0.6	0.3	0.9	0.3	0.5	0.2	0.4	0.4	0.2	–	2.0	1.1	0.9	0.2	2.0	–	0.6	0.5
P13	0.9	0.8	1.3	1.1	1.3	0.6	0.4	1.5	0.5	–	0.4	1.8	–	0.0	1.8	0.8	0.6	–	2.5	1.0
P14	1.8	1.5	2.0	1.5	0.4	2.5	0.5	0.5	1.1	–	0.6	1.5	0.8	–	0.5	0.5	0.0	–	1.1	1.5
P15	1.5	0.9	1.3	0.9	0.6	0.1	0.2	1.1	0.3	–	1.3	0.5	0.4	0.6	–	1.3	1.0	–	1.3	1.0
P16	1.3	2.0	1.5	0.5	0.4	0.9	1.8	0.6	0.7	–	1.5	2.0	0.6	0.4	0.8	–	0.9	–	0.5	0.2
P17	0.7	0.7	0.9	0.8	1.4	0.6	0.8	1.0	0.6	–	0.9	0.4	0.5	0.9	2.0	1.3	–	–	0.7	1.1
P18	–	–	–	–	–	–	–	–	–	0.0	0.8	1.3	–	–	–	–	–	–	1.3	–
P19	0.6	0.5	1.1	0.5	0.4	1.4	0.9	0.4	0.6	0.4	2.5	0.0	0.7	0.7	0.5	1.3	0.7	0.2	–	2.0
P20	0.7	0.5	2.0	1.4	0.0	1.1	0.5	0.6	1.4	2.0	0.4	0.9	2.0	0.8	0.7	0.3	0.5	–	0.8	–

of the first 20 of the 25 tasks from the bigger example. These values are given in Table 1 along with the deadlines of the products and sequence-independent setup times of units. The two examples differ in the sequence-dependent setup times: values for Example 1 are shown in Table 2, and the data for Example 2 can be found in the cited article [9].

The computational tests were executed on an Intel Core i7-6700HQ 2.60 GHz CPU. For more accurate performance comparison, all tests were run on a single thread.

First, the importance of crossover operations was investigated on Example 1. With a population of 100, and parameters $localpoolsize = 5, bestkeep = 30, new = 50$, the number of crossovers were changed between 10, 30, and 50. The results are shown in Fig. 1.

Even though more crossovers resulted in more computation per iteration, the algorithm found the optimal solution faster in most cases because less iterations were necessary.

After values of the other parameters were examined on Example 1 and the best settings were determined, the algorithm was compared with the MILP approach of Kopanos et al. [9]. The MILP model was solved with the IBM ILOG CPLEX 12.7.1 foptimization solver. The best resulted solutions were examined under time limits. Table 3 shows the average results of 5 runs with each parameter set. The optimal objective values for Example 1 and 2 are 17.199 and 29.556 respectively. BEA parameters were $population = 100, localpool = 5, range = 0.5, bestkeep = 30, new = 50, crossover = 50$.

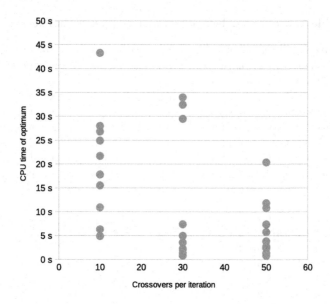

Fig. 1. Effect of increasing the crossover count

Table 3. Comparison results

Problem	Parameters	Time limit (s)	Objective value
Example 1	MILP	60	25.654
	BEA	60	17.199
	MILP	10	33.961
	BEA	10	18.077
Example 2	MILP	300	31.336
	BEA	300	29.556
	MILP	120	41.869
	BEA	120	30.716

The comparison shows that the proposed approach provided better solutions on average, than the MILP model under various time limits. The algorithm found the optimal solution faster than the MILP approach in all cases. Given the heuristic nature of the method, it cannot guarantee optimality but a fast convergence was observed, thus, it can be used to find good approximate solutions where the solution times of exact methods are too long.

5   Concluding Remarks

An evolutionary algorithm using bacterial mutation was presented for minimizing earliness in a single stage product scheduling problem. A case study from the literature was investigated. Different parameter settings of the algorithm were examined, and the efficiency of the approach was compared to a literature method.

In summary, the results showed that the proposed method is able to provide even optimal solutions under short running times and also finds good approximate solutions.

Acknowledgments. This research was supported by the ÚNKP-17-4 New National Excellence Program of the Ministry of Human Capacities. This research was supported by the EFOP-3.6.1-16-2016-00017; "Internationalization, initiatives to establish a new source of researchers and graduates, and development of knowledge and technological transfer as instruments of intelligent specializations at Szechenyi University" grant.

References

1. Balázs, K., Horváth, Z., Kóczy, L.T.: Different chromosome-based evolutionary approaches for the permutation flow shop problem. Acta Polytech. Hung. **9**(2), 115–138 (2012)
2. Ferrer-Nadal, S., Capón-García, E., Méndez, C.A., Puigjaner, L.: Material transfer operations in batch scheduling. A critical modeling issue. Ind. Eng. Chem. Res. **47**, 7721–7732 (2008)
3. Floudas, C.A., Lin, X.: Continuous-time versus discrete-time approaches for scheduling of chemical processes: a review. Comput. Chem. Eng. **28**(11), 2109–2129 (2004)
4. Földesi, P., Botzheim, J., Kóczy, L.T.: Eugenic bacterial memetic algorithm for fuzzy road transport traveling salesman problem. Int. J. Innov. Comput. Inf. Control **7**(5), 2775–2798 (2011)
5. Ghaeli, M., Bahri, P.A., Lee, P., Gu, T.: Petri-net based formulation and algorithm for short-term scheduling of batch plants. Comput. Chem. Eng. **29**(2), 249–259 (2005)
6. Goldberg, D.E.: Genetic Algorithms in Search, Optimization, and Machine Learning. Addison-Wesley, Boston (1989)
7. Hegyhati, M., Friedler, F.: Overview of Industrial Batch Process Scheduling. Chem. Eng. Trans. **21**, 895–900 (2010)
8. Hegyháti, M., Majozi, T., Holczinger, T., Friedler, F.: Practical infeasibility of cross-transfer in batch plants with complex recipes: S-graph vs MILP methods. Chem. Eng. Sci. **64**(3), 605–610 (2009)
9. Kopanos, G.M., Lainez, J.M., Puigjaner, L.: An efficient mixed-integer linear programming scheduling framework for addressing sequence-dependent setup issues in batch plants. Ind. Eng. Chem. Res. **48**(13), 6346–6357 (2009)
10. Levner, E., Kats, V., De Pablo, D.A.L., Cheng, T.C.: Complexity of cyclic scheduling problems: a state-of-the-art survey. Comput. Ind. Eng. **59**(2), 352–361 (2010)
11. Mendez, C.A., Cerda, J., Grossmann, I.E., Harjunkoski, I., Fahl, M.: State-of-the-art review of optimization methods for short-term scheduling of batch processes. Comput. Chem. Eng. **30**(6–7), 913–946 (2006)
12. Nawa, N.E., Furuhashi, T.: A study on the effect of transfer of genes for the bacterial evolutionary algorithm. In: 1998 Proceedings of the Second International Conference on Knowledge-Based Intelligent Electronic Systems, KES 1998, vol. 3, pp. 585–590. IEEE (1998)

13. Nawa, N.E., Furuhashi, T.: Fuzzy system parameters discovery by bacterial evolutionary algorithm. IEEE Trans. Fuzzy Syst. **7**(5), 608–616 (1999)
14. Nawa, N.E., Hashiyama, T., Furuhashi, T., Uchikawa, Y.: A study on fuzzy rules discovery using pseudo-bacterial genetic algorithm with adaptive operator. In: 1997 IEEE International Conference on Evolutionary Computation, pp. 589–593. IEEE (1997)
15. Panek, S., Engell, S., Subbiah, S., Stursberg, O.: Scheduling of multi-product batch plants based upon timed automata models. Comput. Chem. Eng. **32**(1–2), 275–291 (2008)
16. Ralphs, T., Shinano, Y., Berthold, T., Koch, T.: Parallel solvers for mixed integer linear optimization. Technical report 16T–014-R3, 16–74, ISE, Lehigh University and Zuse Institute Berlin (ZIB) (2016)
17. Sanmarti, E., Holczinger, T., Puigjaner, L.L., Friedler, F., Sanmartí, E., Puigjaner, L.L., Holczinger, T., Friedler, F.: Combinatorial framework for effective scheduling of multipurpose batch plants. AIChE J. **48**(11), 2557–2570 (2002)
18. Shaik, M.A., Floudas, C.A.: Novel unified modeling approach for short-term scheduling. Ind. Eng. Chem. Res. **48**(6), 2947–2964 (2009)
19. Smidla, J., Heckl, I.: S-graph based parallel algorithm to the scheduling of multipurpose batch plants. Chem. Eng. Trans. **21**(1994), 937–942 (2010)

Observation of Unbounded Novelty in Evolutionary Algorithms is Unknowable

Eric Holloway$^{(\boxtimes)}$ and Robert Marks

Department of Electrical and Computer Engineering, Baylor University,
Waco, TX 76706, USA
{Eric_Holloway,Robert_Marks}@baylor.edu

Abstract. Open ended evolution seeks computational structures whereby creation of unbounded diversity and novelty are possible. However, research has run into a problem known as the "novelty plateau" where further creation of novelty is not observed. Using standard algorithmic information theory and *Chaitin's Incompleteness Theorem*, we prove no algorithm can detect unlimited novelty. Therefore observation of unbounded novelty in computer evolutionary programs is nonalgorithmic and, in this sense, unknowable.

1 Artificial Life and Endless Novelty

Evolutionary emergence is inspirational in two fields of computer science: artificial life and evolutionary computing [1]. The goal of the more theoretical field of artificial life is creation of conditions that lead to unbounded novelty explosion seemingly apparent in biology [2–4]. Evolutionary computing attempts to harness these innovative powers to solve engineering problems.

But conditions for unbounded novelty have yet to be discovered. All approaches hit a point where novelty ceases to be produced [5]. Many think the root cause of the "novelty plateau" is the reliance on an objective fitness function which causes evolution to become stuck on local minima [6] or hit the fitness upper bound [7]. Open ended evolution seeks to bypass the plateau by removing the fitness function [8–10]. The fitness function is replaced with a type of swarm intelligence [11–13] where each individual in the swarm makes its own decisions which may or may not cause it to survive. The environment is infused with rich information so the swarm can both grow in novelty and complexity and then contribute its own information to future generations [14,15].

Finding the necessary and sufficient conditions to produce boundless novelty is a goal of current artificial life research [6,16,17]. It is impossible to observe an artificial life simulation forever to know whether it can produce boundless novelty. To claim a simulation produces boundless novelty, the claim must be proven from initial conditions. Establishing such "conditions" is equivalent to foundational "axioms" in the development of mathematical disciplines [18,19]. These unbounded evolutionary axioms are then considered to be the basis for a proof or demonstration of ever increasing unbounded novelty.

L. Rutkowski et al. (Eds.): ICAISC 2018, LNAI 10841, pp. 395–404, 2018.
https://doi.org/10.1007/978-3-319-91253-0_37

Insofar as the conditions are the same as axioms, we prove axioms for unbounded open evolution do not exist in the sense that ever increasing novelty cannot be observed. The proof assumes there are no inconsistencies, such as false positives and false negatives, in the observation process. Even in the absence of inconsistencies, the bound on novelty for a set of axioms can still be very high. The observable novelty plateau for a set of conditions can thus be highly elevated – but not unbounded.

1.1   Identifying Unlimited Novelty

Computational open ended evolution assumes that bitstrings of ever increasing novelty are computable. Indeed, bitstrings of increasing novelty will be contained in the trivial enumeration:

Listing 1.1. An algorithm that generates an endless amount of novel bitstrings, to illustrate endless novelty production is easily accomplished.

```
for i in range(∞): print binary(i)
```

Listing all binary bitstrings will generate all novel bitstrings. Lacking is a method to identify which bitstrings contain increasing novelty. Doing so for an unlimited number of novel bitstrings is not possible due to *Chaitin's incompleteness theorem* [20, 21].

2   Prefix Free Complexity and Algorithmic Information

Algorithmic information theory [20, 22–25], the study of the mathematics of algorithms, is a useful tool for insight into artificial life and evolutionary computing [8, 26–29]. We here apply it to the analysis of open ended evolution.

Kolmogorov complexity [20] is the length of the shortest program y^* that generates a bitstring x when executed on a universal Turing machine, $\mathcal{U}$,

$$K(x) = \min_{y:\, \mathcal{U}(y)=x} \ell(y). \tag{1}$$

where $\ell(y)$ is the length, typically measured in bits, of the program y. Minimization is over all y programs to find the shortest program y^* that outputs x so that

$$K(x) = \ell(y^*).$$

Chaitin refers to y^* as the *elegant program* for x [30]. We use prefix free (a.k.a. self-delimiting) code [31, 32] to simplify analysis. Without prefix free Kolmogorov complexity, the chain rule of Kolmogorov complexity is [33]

$$K(a, b) = K(a|b) + K(b) + O(\log K(a, b)). \tag{2}$$

We can understand the source of the logarithmic term in (2) by analyzing the set generating program p, which generates the set $\{a, b\}$ from the elegant program for b (denoted b^*) and the elegant program that generates a given b (denoted a_b^*). To generate $\{a, b\}$ from the elegant programs a_b^* and b^*, the set generator program p will need to know the size of the length of the smaller elegant program. Since size of the length of the smaller elegant program is at most a logarithm of the length, the logarithmic term is added to (2).

Ming and Vitányi show [33] that if we use $K(a|b, K(b))$ with prefix free coding, then the logarithmic accuracy term becomes a constant,

$$K(a, b) = K(a|b, K(b)) + K(b) + c.$$

The result can be further simplified by observing that an elegant program for b gives us both b and $K(b)$. Knowing $K(b)$ we find the first short program b^* that generates b and $K(b)$ by running all programs of length $K(b)$. The shortest program that halts and outputs b is b^*, and we know this procedure will halt since we know $K(b)$. We can therefore replace $\{b, K(b)\}$ with b^*. The complexity expression where this is done is denoted

$$Kc(a|b) := K(a|b^*).$$

2.1   Defining the Set Generating Program

The constant c in

$$Kc(a, b) = Kc(a|b) + Kc(b) + c \tag{3}$$

is the size of the program p_c needed to generate the set $\{a, b\}$ from elegant programs b^* and the elegant program that generates a given b^* (denoted $a_{b^*}^*$). We rewrite (3) using the notation[1]

$$Kc(a, b) \underset{c}{=} Kc(a|b) + Kc(b). \tag{4}$$

To understand how the set generation program p_c works, assume we have the elegant program b^* for b, and the elegant program $a_{b^*}^*$ that generates a when given b^* as input. If b^* does not allow the construction of a program $a_{b^*}^*$ that is shorter than the elegant program a^* for a, then $a_{b^*}^*$ is a^*.

The concatenation of b^*, $a_{b^*}^*$ and stopcode s gives the input[2] $i = b^* a_{b^*}^* s$ for the calculation program p_c. The set generation program p_c's operation is as follows.

---

[1] Some authors use the notation "$\underset{c}{\overset{+}{=}}$" in lieu of "$\underset{c}{=}$" [31,34].

[2] Even though i is not prefix free (i.e. it will halt once b^* is executed, leaving $a_{b^*}^*$ unread), the program that is run on the Turing machine $\mathcal{U}$ is still prefix free because i is appended to p_c, so the full execution on $\mathcal{U}$ is $\mathcal{U}(p_c i)$. Since p_c is prefix free, then so is $p_c i$, as it will only halt once the entire string is read.

1. p_c executes i on a Turing machine $\mathcal{U}$ until $\mathcal{U}(i)$ halts. Since b^* is prefix free, that means $\mathcal{U}(i)$ has output b.
2. The remaining portion of i is run with the previous portion of i as input, which equates to running $a_{b^*}^*$ with b^* as input, which produces a.
3. Once the stopcode s is encountered, p_c collects b and a from the multi-part execution of i, and outputs the set $\{a, b\}$.

2.2   Kolmogorov Complexity is not Computable

No program can tell us what the Kolmogorov complexity is for every bit-string [30,35]. That there is no such program can be easily shown using proof by contradiction.

Assume that there is such a program, the True Kolmogorov Complexity Printer (TKCP). We can then use it in the program in Listing 1.2 to produce a contradiction.

Listing 1.2. Code showing a Klomogorov Complexity printer results in a contradiction. len(self) means the length of this program, including all code for the subfunctions such as TKCP.

```
def contradiction ():
    for i in range (∞):
        bs = binary (i)
        if TKCP (bs) > len (self):
            return bs
```

The contradiction program iterates through all integers until it finds one that has a binary encoding with a greater Kolmogorov complexity than the size of contradiction (which includes the code for all subfunctions such as TKCP). Once the program finds such an integer, it outputs the binary encoding of this integer bs^*. Thus, we have a program with a smaller size than the Kolmogorov complexity for bs^* which nevertheless outputs bs^*. Yet, the definition of Kolmogorov complexity is the length of the shortest program that outputs bs^*. This is a contradiction. Therefore, TKCP cannot exist.

2.3   Chaitin's Incompleteness Theorem

Chaitin's incompleteness theorem is based on a similar argument but deals with axioms and proofs instead of programs. The theorem similarly shows an axiomatic system cannot prove the Kolmogorov complexity of a bitstring above a certain limit, and that this limit is dependent on the size of the axioms plus a constant for the length of the proof assembling program.

Listing 1.3. A program demonstrating there is a limit to proving lower bounds on Kolmogorov complexity. `len(self)` means the length of this program, including all code for the subfunctions.

```
def contradiction(axioms):
    L = len(self) + len(axioms)
    for i in range(∞):
        bs = binary(i)
        goal = "K("+bs+")>"+str(L)
        for proof in all_proofs(axioms):
            if proves(proof, goal):
                return bs
```

Listing 1.3 shows again we end up with a contradiction, since a program that is shorter than the Kolmogorov complexity of bs^* generates bs^*. The difference between this listing and the previous is the function accepts a set of axioms it can use to prove a lower bound on the bitstring's Kolmogorov complexity.

3   Limits on Identifying Novelty

3.1   Definitions of Novelty

There are many different approaches to defining novelty. In a biological setting, novelty refers to new aspects in an organism that not homogonous or homologous with ancestral organisms [36] and fulfill unique functions [37]. The field of information retrieval defines novelty as a new information nugget in a user's interest set that is also contained within the document set [38] and sentences that contain information not contained in previous sentences [39]. In the field of computer generated art, novelty is defined using Dorin and Korb's definition [40] which characterizes a system S_2 that can reliably generate patterns that cannot be created by another system S_1, where S_1 is the existing worldview of an audience.

Within the fields of anomaly [41], fault [42], and outlier detection [43], the systems are generally trained or designed to recognize the typical behavior, and flag atypical behavior. Atypical behavior is not necessarily novel, since it may be well understood as in fault detection where a variety of failure states are derived ahead of time. Novelty is atypical behavior that is unanticipated, and these disciplines provide different techniques for defining the typical region, ranging from rule based, to statistical to complex nonlinear regression models such as neural networks [44–46].

A digital organism can be measured as novel with reference to an existing population. Novelty is measured by a distance function from the rest of the population where the distance function is domain specific [47]. Selecting for novelty is also used in evolutionary computation [48]. In the case of digital life and evolutionary computation, the typical population changes as the algorithm progresses, but there is still the desire to find individuals that do not fit the latest typical population.

3.2   Commonalities of Novelty

In all these domains, novelty has common characteristics. Novelty is

- defined relative to a typical population,
- measured using a distance, and
- unanticipated.

Within a computational domain, everything can be represented by a bitstring. We can measure a bitstring's distance from a population in many ways. A large distance signifies something outside of the typical population. We can characterize the typical population by the smallest program that can produce the population, the size of which is the Kolmogorov complexity of the population.[3] When a member is added to the population that requires the program size to increase, then the member is atypical albeit not necessarily novel. Note if a member is novel, then it is atypical, thus it will increase the population's Kolmogorov complexity.

3.3   Necessary Condition for Novelty

We can measure atypical information in bitstring b_N by the conditional Kolmogorov complexity in reference to the smallest program that generates the current population $\{b_1, b_2, \ldots, b_{N-1}\}$. If the bitstring b_N contains new information then the conditional complexity is positive.

$$Kc(b_N | b_1, b_2, \ldots, b_{N-1}) > 0. \tag{5}$$

Since Kolmogorov complexity corresponds to the length of a program in bits, its measure is restricted to positive integers. Therefore (5) can equivalently be written as

$$Kc(b_N | b_1, b_2, \ldots, b_{N-1}) \geq 1. \tag{6}$$

Randomness also meets this definition, which makes (6) a necessary but not sufficient definition for novelty. For the purpose of this proof, however, a necessary condition is all that is needed.

The distinction between necessary and sufficient conditions is an important qualification, since many bitstrings are random, and their addition will most likely increase the $Kc(\cdot)$ without increasing the true novelty in the population. It may even be true that every new bitstring added to a population increases $Kc(\cdot)$. In either case, whether the new bitstring is entirely random or if it contains true novelty, $Kc(\cdot)$ will be increased. Increasing $Kc(\cdot)$ alone is therefore not sufficient to indicate the addition of novelty. However, we will see that regardless of the source of $Kc(\cdot)$ increase, randomness or novelty, the fact that $Kc(\cdot)$ increases when novelty is added makes it impossible to detect novelty beyond a limit.

---

[3] While Kolmogorov complexity is an exact metric, and has to account for both meaningful structure and random noise in the population, the Kolmogorov sufficient statistic can be used to measure just the meaningful structure in the population.

3.4 Unbounded Novelty Detection is Nonalgorithmic

Assume there exists a set of axioms that can identify novelty when novelty is added to a population. When a new bitstring is generated, we use the axioms to derive a proof that the new bitstring is novel. Note these axioms are not directly calculating the conditional Kolmogorov complexity of the new bitstring. The axioms are only being used to prove the new bitstring b_N is novel with respect to prior observations. However, based on the argument in Sect. 3.3, proving b_N is novel indirectly entails (6), namely that the conditional complexity of b_N is positive.

If we have a set of axioms that always identifies novelty added to the population, then when novelty is added, (6) states the conditional complexity of b_N is at least one bit. With this knowledge, we can use the Kolmogorov complexity chain rule [33], as defined in (4), to estimate a lower bound on the population's joint Kolmogorov complexity, substituting 1 whenever we detect novelty and 0 otherwise. Assuming we've identified M instances of novelty during the generation of N bitstrings, we can lower bound the joint complexity using (6).

$$
\begin{aligned}
Kc(b_1, b_2, \ldots, b_N) &\underset{c}{=} Kc(b_1) + Kc(b_2|b_1) + \ldots \\
&\quad + Kc(b_N|b_1, b_2, \ldots, b_{N-1}) \\
&\underset{c}{\geq} M.
\end{aligned}
\tag{7}
$$

Since c is the size of the set generating program p_c, as detailed in Sect. 2.1, and thus is positive, then by removing c we can make the inequality exact

$$
Kc(b_1, b_2, \ldots, b_N) \geq M.
\tag{8}
$$

As proven in Sect. 2.3, Chaitin's incompleteness theorem states we cannot prove $K(b) > \mathcal{L}$ for some $\mathcal{L}$ that is based on the proof axioms. However, if we can detect unbounded novelty, then M in (8) becomes arbitrarily big, and at some point we can prove $Kc(b_1, b_2, \ldots, b_N) \geq M > \mathcal{L}$, which is a contradiction.

Therefore, we can only detect novelty a finite number of times. Furthermore, this limit is not much larger than the size of the axioms as can be inferred from the foundation of Chaitin's incompleteness theorem.

Additionally, this argument works for novelty generating algorithms. If we have an algorithm that we know always creates novel bitstrings, then the conditional complexity in (6) is always greater than zero, and eventually the algorithm will create an M that is greater than $\mathcal{L}$ thereby contradicting Chaitin's incompleteness theorem.

4 Conclusion

The great novelty and diversity resulting from biological evolution suggests there is an algorithm that can produce the same. Finding this algorithm is the goal of evolutionary computation and artificial life research. Yet, it is not sufficient to produce endless novelty, but also identify novelty when it occurs.

However, a reliable method of identifying an endless amount of novelty would also imply the ability to calculate a lower bound of arbitrary size on Kolmogorov complexity. Since every axiomatic system has a limit to the lower bound it can calculate, due to Chaitin's incompleteness theorem, the reliable method of novelty detecting introduces a contradiction. This same contradiction results if we have a program we know only generates novel bitstrings.

As such, we must conclude there is no reliable method of identifying an endless degree of novelty, nor of producing only novel bitstrings. We can only reliably detect novelty to a finite amount, and not significantly more than the Kolmogorov complexity of the axioms used for detection.

The bound on novelty observation can be high, so our analysis does not specifically preclude observation of novelty in the evolutionary process - but does prove there are limitations. There is also the assumption of consistency. If the novelty requirement is relaxed to also allow inconsistencies such as the occurrence of false positives and false negatives, then observing endless novelty might still be possible such as labeling all bitstrings generated by the algorithm in Listing 1.1 as "novel".

References

1. Mitchell, M., Forrest, S.: Genetic algorithms and artificial life. Artif. life **1**(3), 267–289 (1994)
2. Huneman, P.: Determinism, predictability and open-ended evolution: lessons from computational emergence. Synthese **185**(2), 195–214 (2012)
3. Komosinski, M., Rotaru-Varga, A.: From directed to open-ended evolution in a complex simulation model. Artif. Life **7**, 293–299 (2000)
4. Sayama, H.: Seeking open-ended evolution in swarm chemistry. In: 2011 IEEE Symposium on Artificial Life (ALIFE), pp. 186–193. IEEE (2011)
5. Li, J., Storie, J., Clune, J.: Encouraging creative thinking in robots improves their ability to solve challenging problems. In: Proceedings of the 2014 Annual Conference on Genetic and Evolutionary Computation, pp. 193–200. ACM (2014)
6. Soros, L., Stanley, K.O.: Identifying necessary conditions for open-ended evolution through the artificial life world of chromaria. Artif. life **14**, 793–800 (2014)
7. Basener, W.F.: Exploring the concept of open-ended evolution. In: Biological Information: New Perspectives, pp. 87–104. World Scientific (2012)
8. Bedau, M.A., McCaskill, J.S., Packard, N.H., Rasmussen, S., Adami, C., Green, D.G., Ikegami, T., Kaneko, K., Ray, T.S.: Open problems in artificial life. Artif. life **6**(4), 363–376 (2000)
9. Ruiz-Mirazo, K., Peretó, J., Moreno, A.: A universal definition of life: autonomy and open-ended evolution. Orig. Life Evol. Biosph. **34**(3), 323–346 (2004)
10. Ruiz-Mirazo, K., Umerez, J., Moreno, A.: Enabling conditions for open-ended evolution. Biol. Philos. **23**(1), 67–85 (2008)
11. Bonabeau, E., Dorigo, M., Theraulaz, G.: Swarm Intelligence: from Natural to Artificial Systems, vol. 1. Oxford University Press, Oxford (1999)
12. Ewert, W., Marks, R.J., Thompson, B.B., Yu, A.: Evolutionary inversion of swarm emergence using disjunctive combs control. IEEE Trans. Syst. Man Cybern. Syst. **43**(5), 1063–1076 (2013)

13. Roach, J., Ewert, W., Marks, R.J., Thompson, B.B.: Unexpected emergent behaviors from elementary swarms. In: 2013 45th Southeastern Symposium on System theory (SSST), pp. 41–50. IEEE (2013)
14. Roach, J.H., Marks, R.J., Thompson, B.B.: Recovery from sensor failure in an evolving multiobjective swarm. IEEE Trans. Syst. Man Cybern. Syst. **45**(1), 170–174 (2015)
15. Taylor, T.: Exploring the concept of open-ended evolution. In: Proceedings of the 13th International Conference on Artificial life, pp. 540–541 (2012)
16. Jakobi, N.: Encoding scheme issues for open-ended artificial evolution. In: Voigt, H.-M., Ebeling, W., Rechenberg, I., Schwefel, H.-P. (eds.) PPSN 1996. LNCS, vol. 1141, pp. 52–61. Springer, Heidelberg (1996). https://doi.org/10.1007/3-540-61723-X_969
17. Channon, A.: Three evolvability requirements for open-ended evolution. In: Artificial Life VII Workshop Proceedings, Portland, OR, pp. 39–40 (2000)
18. Mueller, I.: Euclid's elements and the axiomatic method. Br. J. Philos. Sci. **20**(4), 289–309 (1969)
19. Shenoy, P.P., Shafer, G.: Axioms for probability and belief-function propagation. In: Yager, R.R., Liu, L. (eds.) Classic Works of the Dempster-Shafer Theory of Belief Functions. STUDFUZZ, vol. 219, pp. 499–528. Springer, Heidelberg (2008). https://doi.org/10.1007/978-3-540-44792-4_20
20. Chaitin, G.J.: Algorithmic information theory. IBM J. Res. Dev. **21**(4), 350–359 (1977)
21. Raatikainen, P.: On interpreting Chaitin's incompleteness theorem. J. Philos. Log. **27**(6), 569–586 (1998)
22. Chaitin, G.J.: Information, Randomness & Incompleteness: Papers on Algorithmic Information Theory, vol. 8. World Scientific, Singapore (1990)
23. Grünwald, P.D., Vitányi, P.M., et al.: Algorithmic information theory. In: Handbook of the Philosophy of Information, pp. 281–320 (2008)
24. Seibt, P.: Algorithmic Information Theory. Springer, Heidelberg (2006). https://doi.org/10.1007/978-3-540-33219-0
25. Van Lambalgen, M.: Algorithmic information theory. J. Symb. Log. **54**(4), 1389–1400 (1989)
26. Chaitin, G.: Proving Darwin: Making Biology Mathematical. Vintage, New York (2012)
27. Chaitin, G.J.: Toward a mathematical definition of life. In: Information, Randomness & Incompleteness: Papers on Algorithmic Information Theory, pp. 86–104. World Scientific (1987)
28. Gecow, A.: The purposeful information. On the difference between natural and artificial life. Dialogue Univers. **18**(11/12), 191–206 (2008)
29. Pattee, H.H.: Artificial life needs a real epistemology. In: Morán, F., Moreno, A., Merelo, J.J., Chacón, P. (eds.) ECAL 1995. LNCS, vol. 929, pp. 21–38. Springer, Heidelberg (1995). https://doi.org/10.1007/3-540-59496-5_286
30. Chaitin, G.J.: The Unknowable. Springer Science & Business Media, Heidelberg (1999)
31. Bennett, C.H., Gács, P., Li, M., Vitanyi, P., Zurek, W.H.: Information distance. arXiv preprint arXiv:1006.3520 (2010)
32. Calude, C.S.: Algorithmic randomness, quantum physics, and incompleteness. In: Margenstern, M. (ed.) MCU 2004. LNCS, vol. 3354, pp. 1–17. Springer, Heidelberg (2005). https://doi.org/10.1007/978-3-540-31834-7_1
33. Ming, L., Vitányi, P.M.: Kolmogorov complexity and its applications. Algorithms Complex. **1**, 187 (2014)

34. Vitányi, P.M., Li, M.: Minimum description length induction, Bayesianism, and Kolmogorov complexity. IEEE Trans. Inf. Theory **46**(2), 446–464 (2000)
35. Wallace, C.S., Dowe, D.L.: Minimum message length and Kolmogorov complexity. Comput. J. **42**(4), 270–283 (1999)
36. Muller, G.B., Wagner, G.P.: Novelty in evolution: restructuring the concept. Ann. Rev. Ecol. Syst. **22**(1), 229–256 (1991)
37. Pigliucci, M.: What, if anything, is an evolutionary novelty? Philos. Sci. **75**(5), 887–898 (2008)
38. Li, X., Croft, W.B.: An information-pattern-based approach to novelty detection. Inf. Process. Manag. **44**(3), 1159–1188 (2008)
39. Zhao, L., Zhang, M., Ma, S.: The nature of novelty detection. Inf. Retr. **9**(5), 521–541 (2006)
40. Kowaliw, T., Dorin, A., McCormack, J.: An empirical exploration of a definition of creative novelty for generative art. In: Korb, K., Randall, M., Hendtlass, T. (eds.) ACAL 2009. LNCS (LNAI), vol. 5865, pp. 1–10. Springer, Heidelberg (2009). https://doi.org/10.1007/978-3-642-10427-5_1
41. Chandola, V., Banerjee, A., Kumar, V.: Anomaly detection: a survey. ACM Comput. Surv. (CSUR) **41**(3), 15 (2009)
42. Venkatasubramanian, V., Rengaswamy, R., Yin, K., Kavuri, S.N.: A review of process fault detection and diagnosis: part I: quantitative model-based methods. Comput. Chem. Eng. **27**(3), 293–311 (2003)
43. Hodge, V., Austin, J.: A survey of outlier detection methodologies. Artif. Intell. Rev. **22**(2), 85–126 (2004)
44. Pimentel, M.A., Clifton, D.A., Clifton, L., Tarassenko, L.: A review of novelty detection. Sig. Process. **99**, 215–249 (2014)
45. Reed, R., Marks, R.J.: Neural Smithing: Supervised Learning in Feedforward Artificial Neural Networks. MIT Press, Cambridge (1999)
46. Thompson, B.B., Marks, R.J., Choi, J.J., El-Sharkawi, M.A., Huang, M.Y., Bunje, C.: Implicit learning in autoencoder novelty assessment. In: 2002 Proceedings of the 2002 International Joint Conference on Neural Networks, IJCNN 2002, vol. 3, pp. 2878–2883. IEEE (2002)
47. Lehman, J., Stanley, K.O.: Abandoning objectives: evolution through the search for novelty alone. Evol. Comput. **19**(2), 189–223 (2011)
48. Mouret, J.B.: Novelty-based multiobjectivization. In: Doncieux, S., Bredèche, N., Mouret, J.B. (eds.) New Horizons in Evolutionary Robotics. SCI, vol. 341, pp. 139–154. Springer, Heidelberg (2011). https://doi.org/10.1007/978-3-642-18272-3_10

Multi-swarm Optimization Algorithm Based on Firefly and Particle Swarm Optimization Techniques

Tomas Kadavy$^{(\boxtimes)}$ ⓘ, Michal Pluhacek ⓘ, Adam Viktorin ⓘ,
and Roman Senkerik ⓘ

Faculty of Applied Informatics, Tomas Bata University in Zlin,
T. G. Masaryka 5555, 760 01 Zlin, Czech Republic
{kadavy,pluhacek,aviktorin,senkerik}@utb.cz

Abstract. In this paper, the two hybrid swarm-based metaheuristic algorithms are tested and compared. The first hybrid is already existing Firefly Particle Swarm Optimization (FFPSO), which is based, as the name suggests, on Firefly Algorithm (FA) and Particle Swarm Optimization (PSO). The secondly proposed hybrid is an algorithm using the multi-swarm method to merge FA and PSO. The performance of our developed algorithm is tested and compared with the FFPSO and canonical FA. Comparisons have been conducted on five selected benchmark functions, and the results have been evaluated for statistical significance using Friedman rank test.

Keywords: Firefly Algorithm · Particle Swarm Optimization
Hybridization · Multi-swarm

1 Introduction

The swarm-based algorithms, in general, are quite popular amongst researchers, as the number of published papers on this topic shows over the years [1]. Nowadays, one of the favorite technique in metaheuristic optimization lies in the

T. Kadavy—This work was supported by the Ministry of Education, Youth and Sports of the Czech Republic within the National Sustainability Programme Project no. LO1303 (MSMT-7778/2014), further by the European Regional Development Fund under the Project CEBIA-Tech no. CZ.1.05/2.1.00/03.0089 and by Internal Grant Agency of Tomas Bata University under the Projects no. IGA/CebiaTech/2018/003. This work is also based upon support by COST (European Cooperation in Science & Technology) under Action CA15140, Improving Applicability of Nature-Inspired Optimisation by Joining Theory and Practice (ImAppNIO), and Action IC1406, High-Performance Modelling and Simulation for Big Data Applications (cHiPSet). The work was further supported by resources of A.I.Lab at the Faculty of Applied Informatics, Tomas Bata University in Zlin (ailab.fai.utb.cz).

© Springer International Publishing AG, part of Springer Nature 2018
L. Rutkowski et al. (Eds.): ICAISC 2018, LNAI 10841, pp. 405–416, 2018.
https://doi.org/10.1007/978-3-319-91253-0_38

ensemble method. Several research papers prove this claim, for example [2,3]. The ensemble method is based on the idea that combination of multiple algorithms could assimilate the positives of them. And also, eliminate their disadvantages. However, achieving this ideal state is a difficult task. The similar results could be obtained with another useful mechanic, the so-called multi-swarm [4,5].

The most typical representative of an optimization algorithm with a long history and many modifications [6–8] is Particle Swarm Optimization (PSO) [9]. The PSO has, apart from other weaknesses, a strong tendency for a premature convergence to a local minimum. Despite this drawback and PSOs age (proposed 1995), it is still prevalent and used in many applications [10–12]. In recent years, another swarm-based algorithm proves its usefulness and is becoming quite popular. This mentioned novel optimization technique is a Firefly Algorithm (FA) which was introduced in [13]. Since then, like PSO, many extensions and modifications were proposed for this powerful optimization algorithm [14–16].

In this paper, attempt to merge these two exciting and years proven algorithms together is presented. Similar research was already done with impressive results in the application field [17]. The new proposed algorithm depends solely on the previously mentioned technique called multi-swarm. In short, the resulted multi-swarm hybrid algorithm consists of two independent swarms. The first swarm represents the FA algorithm, and it should introduce the exploration phase to the whole hybrid. The second swarm serves as the exploitation part of the hybrid, and it is represented by PSO. Thus, the weakness of PSO, which is a previously mentioned premature convergence, could serve as a benefit to this newly proposed hybrid.

The proposed optimization algorithm is tested and statistically evaluated on selected well-known benchmark functions. It is also compared with other hybridized algorithm FFPSO, with the classical version of FA. The rest of the paper is structured as follows. Brief descriptions of PSO, FA and FFPSO, are in Sects. 2 and 3. Section 3 also covers a description of the proposed hybrid algorithm. In Sect. 4, the benchmark functions are defined, and the parameter settings of tested algorithms are shown as well. The results and conclusion sections follow.

2   Particle Swarm Optimization

This swarm-based algorithm is one of current main representative on a field of similar swarm intelligence algorithms. This Particle Swarm Optimization (PSO) was first published by Ebenhart and Kennedy in [9]. This algorithm mimics the social behavior of swarming animals in nature. Despite the fact that its quite long time from its first appearance, its still plenty used across many optimization problems.

In every step of the algorithm, the new positions of particles are calculated based on previous positions and velocities. The new position of a particle is checked if it still lies in the space of possible solutions. The position of particle x is then calculated according to the formula (1).

$$x_i' = x_i + v_i \tag{1}$$

Where x_i' is a new position of particle i, x_i is the previous old position of a particle and v is the velocity of a particle. The velocity of a particle v is calculated according to (2).

$$v_i' = w \cdot v_i + c_1 \cdot r_1 \cdot (pBest_i - x_i) + c_2 \cdot r_2 \cdot (gBest - x_i) \tag{2}$$

Where w is inertia weight, $c1$ and $c2$ are learning factors, and $r1$ and $r2$ are random numbers of unimodal distribution in the range $<0,1>$. The $pBest$ is a personal best found solution so far and the $gBest$ is the best-obtained solution by the whole population.

3   Firefly Algorithm

This optimization nature-based algorithm was developed and introduced by Yang in [13]. The fundamental principle of this algorithm lies in simulating the mating behavior of fireflies at night when fireflies emit light to attract a suitable partner. In the first subsection, the classical version of FA is presented. In the second subsection, the existent hybrid FA is briefly described. Finally, the newly proposed hybrid is described in the third subsection.

3.1   A Canonical Version of Firefly Algorithm

The main idea of Firefly Algorithm (FA) is that the objective function value that is optimized is associated with the flashing light of these fireflies. The movement of one firefly towards another one is defined by Eq. (3). Where x_i' is a new position of firefly i, x_i is the current position of firefly i and x_j is a selected brighter firefly (with better objective function value). The α is a randomization parameter and $sign$ simply provides random direction -1 or 1.

$$x_i' = x_i + \beta \cdot (x_j - x_i) + \alpha \cdot sign \tag{3}$$

The brightness I of a firefly is computed by the Eq. (4). This equation of brightness consists of three factors mentioned in the rules above. On the objective function value, the distance between two fireflies and the last factor is the absorption factor of a media in which fireflies are.

$$I = \frac{I_0}{1 + \gamma r^m} \tag{4}$$

Where Io is the objective function value, the γ stands for the light absorption parameter of a media in which fireflies are, and the m is another user-defined coefficient, and it should be set $m \geq 1$. The variable r is the Euclidian distance (5) between the two compared fireflies.

$$r_{ij} = \sqrt{\sum_{k=1}^{d} (x_{i,k} - x_{j,k})^2} \tag{5}$$

Where r_{ij} is the Euclidian distance between fireflies x_i and x_j. The d is current dimension size of the optimized problem.

The attractiveness β (6) is proportional to brightness I as mentioned in rules above and so these equations are quite similar to each other. The β_0 is the initial attractiveness defined by the user, the γ is again the light absorption parameter, and the r is once more the Euclidian distance. The m is also the same as in Eq. (4).

$$\beta = \frac{\beta_0}{1 + \gamma r^m} \tag{6}$$

3.2 Firefly Particle Swarm Optimization

This hybrid algorithm, as the name suggests, is composed of two previously mentioned popular algorithms. The first is freshly described FA and the second is very popular PSO. The Firefly Particle Swarm Optimization (FFPSO) [17] was introduced in late 2015 by authors Kora and Rama Krishna. The main principle remains the same as in standard FA, but the equation for firefly motion (3) is slightly changed according to PSO movement (2) and is newly computed as (7) while still using the full attractions as in canonical FA.

$$x_i' = w \cdot x_i + c_1 \cdot e^{-r_{px}^2} \cdot (pBest_i - x_i) + c_2 \cdot e^{-r_{gx}^2} \cdot (gBest - x_i) + \alpha \cdot sign \tag{7}$$

Where w, $c1$, and $c2$ are variables transferred from PSO and their values often depends on the user. Also, the $pBest$ and $gBest$ are variables originally belonging to PSO algorithm. They both represent the memory of best position where $pBest$ is best position of each particle and $gBest$ is globally achieved best position so far. The remaining variables rpx (8) and rgx (9) are distances between particle xi and $pBesti$ and $gBest$.

$$r_{px} = \sqrt{\sum_{k=1}^{d} \left(pBest_{i,k} - x_{i,k}\right)^2} \tag{8}$$

$$r_{gx} = \sqrt{\sum_{k=1}^{d} \left(gBest_k - x_{i,k}\right)^2} \tag{9}$$

The pseudocode below shows the fundamentals of FFPSO operations.

3.3 Proposed Hybrid Algorithm

The new proposed hybrid algorithm is composed of FA and PSO, similar to [17], but in a different manner. The inner structure of the hybrid algorithm is based on multi-swarm approach [4,5]. The proposed algorithm contains two or more different swarms that share selected properties or knowledge and can cooperate.

The proposed hybrid algorithm used in this study contains precisely two swarms. The first swarm is represented by a typical FA and the second swarm is

---

Algorithm 1. FFPSO

1: FFPSO initialization
2: **while** terminal condition not met **do**
3:     **for** $i = 1$ to all fireflies **do**
4:         **for** $j = 1$ to all fireflies **do**
5:             **if** $I_j < I_i$ **then**
6:                 calculate r_{px} and r_{gx}
7:                 move x_i to x_j
8:                 evaluate x_i
9:             **end if**
10:         **end for**
11:     **end for**
12:     record the best firefly
13: **end while**

PSO. Each of them has a unique collection of particles and the communication between swarms is provided by PSO global memory feature called *gBest*. The PSOs *gBest* is updated if better value is found by firefly swarm, or in a classical way by PSO swarm. After each generation/iteration, the swarms are evaluated sequentially. Firefly swarm is evaluated first, and PSO swarm second. The firefly swarm movement is computed by Eq. (3). The PSO swarm works precisely the same as described in a previous Sect. 2. The basic idea behind this structure is that the firefly swarm, with appropriate setting, should serve as an exploration provider and the PSO swarm is used for its exploitation abilities.

The pseudocode below shows the fundamentals of hybrid algorithm.

4 Experimental Setup

The experiments were performed on a simple set of five well-known benchmark functions consisting of unimodal and multimodal types:

1. Sphere function (f_1) (10),
2. Rosenbrock function (f_2) (11),
3. Rastrigin function (f_3) (12),
4. Schwefel function (f_4) (13),
5. Ackley function (f_5) (14).

$$f(x)_1 = \sum_{i=1}^{d} x_i^2 \tag{10}$$

$$f(x)_2 = \sum_{i=1}^{d-1} \left[100 \cdot \left(x_{i+1} - x_i^2 \right)^2 + \left(1 - x_i \right)^2 \right] \tag{11}$$

$$f(x)_3 = 10 \cdot d + \sum_{i=1}^{d} \left[x_i^2 - 10 \cdot \cos \left(2\pi x_i \right) \right] \tag{12}$$

Algorithm 2. Proposed hybrid algorithm

1: Initialization
2: **while** terminal condition not met **do**
3:    **for** $i = 1$ to all FA members **do**
4:       **for** $j = 1$ to all FA members **do**
5:          **if** $I_j < I_i$ **then**
6:             move x_i to x_j
7:          **end if**
8:       **end for**
9:    **end for**
10:    **for** $i = 1$ to all PSO members **do**
11:       move x_i
12:    **end for**
13:    **for** $i = 1$ to all FA members **do**
14:       **if** $f(x_i) < gBest$ **then**
15:          $gBest = x_i$
16:       **end if**
17:    **end for**
18:    **for** $i = 1$ to all PSO members **do**
19:       **if** $f(x_i) < pBest$ **then**
20:          $pBest = x_i$
21:          **if** $pBest < gBest$ **then**
22:             $gBest = pBest$
23:          **end if**
24:       **end if**
25:    **end for**
26:    record the gBest
27: **end while**

$$f(x)_4 = 418.9829 \cdot d - \sum_{i=1}^{d} \left(x_i \cdot \sin \sqrt{|x_i|} \right) \tag{13}$$

$$f(x)_5 = -20 \cdot e^{-0.2 \cdot \sqrt{d^{-1} \cdot \sum_{i=1}^{d} x_i^2}} - e^{d^{-1} \cdot \sum_{i=1}^{d} \cos(2 \cdot \pi \cdot x_i)} + 20 + e \tag{14}$$

For this experiment, the unshifted and nonrotated version of functions were used. The selected tested dimensions were 2, 5, 10 and 15. The maximal number of the evaluation was set as $2000 \cdot d$ (dimension size). The number of particles (*NP*) was set to 40 for all dimension sizes. Every test function was repeated for 30 independent runs and the results were statistically evaluated.

The tested and compared algorithms with their settings are displayed in Table 1. Parameters for FA and FFPSO were set to recommended values according to algorithms authors [13,17]. These parameters were simply passed on the proposed hybrid without any additional change. As can be seen in Table 1, the proposed hybrid algorithm was tested with 3 different modifications. These modifications only affect the relative sizes of algorithm subpopulations. The NP_{PSO} stand for PSO subswarm population size, and the NP_{FA} is the number of population in FA subswarm.

Table 1. Parameters of tested algorithms.

Name	Parameters		Description
FA	$\alpha = 0.5, \beta_0 = 0.2, \gamma = 1$		A classical Firefly Algorithm
FFPSO	$\alpha = 0.5, \beta_0 = 0.2, \gamma = 1, \ w = 0.729,$ $c_1 = c_2 = 1.49445$		A hybrid of the Firefly Algorithm and Particle Swarm Optimization
Hybrid1	$NP_{FA}{=}10,$ $NP_{PSO}{=}30$	$\alpha = 0.5, \beta_0 = 0.2,$ $\gamma = 1, \ w = 0.729,$ $c_1 = c_2 = 1.49445$	A proposed hybrid algorithm
Hybrid2	$NP_{FA}{=}20,$ $NP_{PSO}{=}20$		
Hybrid3	$NP_{FA}{=}30,$ $NP_{PSO}{=}10$		

5   Results

The results of performed experiments are given in this section. Firstly, the results overviews and comparisons are presented in Table 2, which contain the simple statistic like mean and std. dev. values. Further, examples of convergence behavior of the compared methods are given in Figs. 1, 2, 3 and 4.

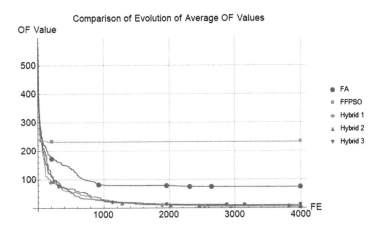

Fig. 1. Convergence graph for f_4 dimension 2.

Furthermore, we present the Friedman ranks with critical distance evaluated according to the Nemenyi Critical Distance post-hoc test for multiple comparisons. The visual outputs of multiple comparisons with rankings are given in Figs. 5 and 6. All Friedman rank test hypothesis are relevant with p-value lower than 0.05. The dashed line represents the critical distance from the best-performed algorithm (the lowest mean rank). The critical distance (CD) value for this experiment has been calculated as 2.72802; according to the definition given in (15) and value $q_a = 2.72802$; using k = 5 compared algorithms and a number of data sets N = 5 (5 tested functions).

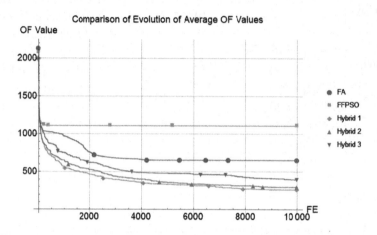

Fig. 2. Convergence graph for f_4 dimension 5.

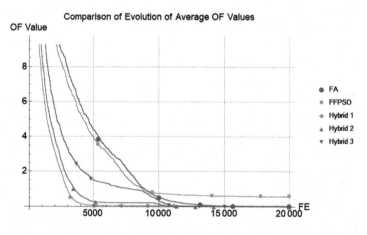

Fig. 3. Convergence graph for f_5 dimension 10.

From the results, it is noticeable, that the proposed hybrid function, all three tested settings, achieved better results compared to other tested algorithms. Nevertheless, with the given results, supported by convergence figures, the proposed

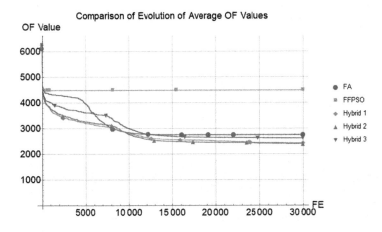

Fig. 4. Convergence graph for f_4 dimension 15.

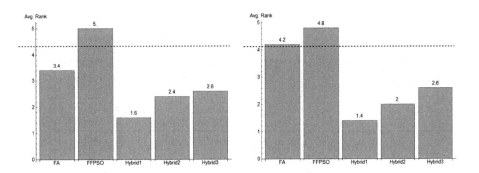

Fig. 5. Friedman rank test for dimension size 2 (left) and 5 (right).

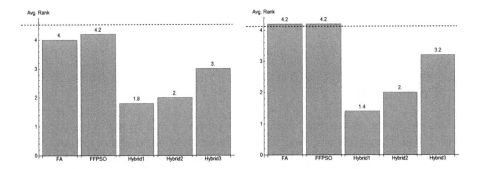

Fig. 6. Friedman rank test for dimension size 10 (left) and 15 (right).

hybrid suggesting noticeable potential. The results presented by Friedman rank test also showing the impact of population sizes in each subswarm.

$$CD = q_a \sqrt{\frac{k(k+1)}{6N}} \tag{15}$$

Table 2. Statistical results (mean. and std. dev.)

Dimension 2										
F	FA		FFPSO		Hybrid 1		Hybrid 2		Hybrid 3	
f_1	6.3E-10	6.2E-10	8.4E-04	9.2E-04	1.2E-11	1.9E-11	4.5E-11	1.2E-10	1.7E-10	2.4E-10
f_2	6.2E-08	1.3E-07	2.7E-04	2.1E-01	1.9E-05	5.8E-05	1.3E-05	5.6E-05	3.2E-07	6.6E-07
f_3	1.3E-01	3.4E-01	2.2E-01	2.4E-01	1.5E-07	3.2E-07	6.6E-02	2.E-01	3.3E-02	1.8E-01
f_4	7.3E01	7.2E01	2.3E02	1.0E02	4.3E00	2.1E01	4.5E00	2.1E01	9.5E00	3.0E01
f_5	4.3E-04	2.3E-04	1.0E-01	6.0E-02	3.1E-05	2.5E-05	5.7E-05	1.6E-04	1.8E-04	1.6E-04
Dimension 5										
F	FA		FFPSO		Hybrid 1		Hybrid 2		Hybrid 3	
f_1	3.1E-08	1.1E-08	2.2E-02	1.0E-02	6.3E-18	1.3E-17	4.7E-17	7.5E-17	6.9E-15	1.8E-14
f_2	1.1E00	1.8E00	3.9E00	6.7E-01	5.7E-01	7.5E-01	6.1E-01	7.4E-01	5.6E-01	6.7E-01
f_3	3.6E00	2.1E00	3.2E00	2.1E00	1.2E00	1.0E00	1.1E00	9.5E-01	1.8E00	1.6E00
f_4	6.5E02	1.4E02	1.1E03	1.1E02	2.6E02	1.3E02	3.0E02	1.1E01	3.9E02	1.4E02
f_5	1.8E-03	4.4E-04	3.8E-01	1.9E-01	2.4E-08	2.3E-08	9.4E-08	1.3E-07	2.8E-06	6.3E-06
Dimension 10										
F	FA		FFPSO		Hybrid 1		Hybrid 2		Hybrid 3	
f_1	2.0E-07	5.7E-08	7.0E-02	2.3E-02	5.4E-24	9.2E-24	7.0E-22	1.5E-21	1.6E-13	8.8E-13
f_2	9.9E00	2.3E01	1.1E01	1.6E01	3.0E00	1.8E00	3.2E00	1.7E00	4.3E00	2.1E00
f_3	1.1E01	4.4E00	5.0E00	4.1E00	6.2E00	3.4E00	7.1E00	3.4E00	7.4E00	3.1E00
f_4	1.6E03	3.1E02	2.6E03	2.4E02	1.2E03	2.4E02	1.3E03	2.6E02	1.4E03	2.4E02
f_5	3.6E-03	8.2E-04	5.7E-01	2.1E-01	3.8E-02	2.1E-01	1.3E-07	4.2E-07	1.5E-05	3.1E-05
Dimension 15										
F	FA		FFPSO		Hybrid 1		Hybrid 2		Hybrid 3	
f_1	5.4E-07	1.4E-07	9.0E-01	1.8E-02	2.0E-26	4.2E-26	5.1E-23	1.6E-22	6.8E-09	3.6E-08
f_2	1.6E01	1.6E01	1.6E01	2.2E00	7.8E00	3.4E00	8.3E00	3.2E00	8.9E00	3.7E00
f_3	1.9E01	6.7E00	6.0E00	5.9E00	1.3E01	4.1E00	1.3E01	4.2E00	1.8E01	5.8E00
f_4	2.7E03	4.0E02	4.5E03	2.8E02	2.4E03	3.5E02	2.4E03	2.7E02	2.6E03	3.5E02
f_5	4.7E-03	9.2E-04	5.7E-01	1.6E-01	1.1E-07	4.5E-07	2.1E-07	3.2E-07	3.3E-04	5.4E-04

6 Conclusion

In this study, the new hybrid algorithm is proposed and tested. This proposed hybrid algorithm was compared to one similar hybridized algorithm and with the standardized version of FA. Also, the three different population ratios of the proposed hybrid algorithm were analyzed. According to analyzed statistical data, the proposed hybrid algorithm outperformed the FFPSO on all dimensions with only one exception. However, on all dimension settings, the proposed hybrid algorithm was constantly achieving the lowest rank in Friedman rank test. The ratio between population sizes of subswarms also has a noticeable impact on given results. Based on the presented results, the more particles belong to PSO; the better results hybridized algorithm could provide. This observation correlates with the fundamental principle of this method. It means that PSO subswarm serves as an exploitation provider.

The results of this study suggesting that more research is needed. In this paper, the parameters of the hybrid were set according to a recommendation of authors of the FA and FFPSO algorithms on which the hybrid is based on. However, with their hybridization, the formerly recommended parameters may no longer be sufficient. Definitely, several tunings and tests are needed to find the optimal parameters setting. The finding of optimal parameters for the proposed hybrid is related to a selection of more robust benchmark functions in future research (e.g. CEC benchmark function set). Also, design of the proposed hybrid algorithm offering several possible changes that could affect its performance. The sizes of used swarms, the number of swarms itself or design of another communication method between the swarms. These are just a small number of tasks for future research.

References

1. Pluhacek, M., Senkerik, R., Viktorin, A., Kadavy, T., Zelinka, I.: A review of real-world applications of particle swarm optimization algorithm. In: Duy, V., Dao, T., Zelinka, I., Kim, S., Phuong, T. (eds.) AETA 2017. LNEE. Springer, Cham (2018). https://doi.org/10.1007/978-3-319-69814-4_112018
2. Du, W., Li, B.: Multi-strategy ensemble particle swarm optimization for dynamic optimization. Inf. Sci. **178**(15), 3096–3109 (2008)
3. Wang, H., Wu, Z., Rahnamayan, S., Sun, H., Liu, Y., Pan, J.: Multi-strategy ensemble artificial bee colony algorithm. Inf. Sci. **20**(279), 587–603 (2014)
4. Blackwell, T., Branke, J.: Multi-swarm optimization in dynamic environments. In: Raidl, G.R., et al. (eds.) EvoWorkshops 2004. LNCS, vol. 3005, pp. 489–500. Springer, Heidelberg (2004). https://doi.org/10.1007/978-3-540-24653-4_50
5. Liang, J.J., Suganthan, P.N.: Dynamic multi-swarm particle swarm optimizer with local search. IEEE (2005)
6. Lynn, N., Suganthan, P.N.: Heterogeneous comprehensive learning particle swarm optimization with enhanced exploration and exploitation. Swarm Evol. Comput. **1**(24), 11–24 (2015)
7. Nepomuceno, F.V., Engelbrecht, A.P.: A self-adaptive heterogeneous PSO for real-parameter optimization. IEEE (2013)
8. Zhan, Z.-H., Zhang, J., Li, Y., Shi, Y.-H.: Orthogonal learning particle swarm optimization. TEVC **15**(6), 832–847 (2011)
9. Eberhart, R., Kennedy, J.: A new optimizer using particle swarm theory. In: Proceedings of the Sixth International Symposium on Micro Machine and Human Science, MHS 1995, pp. 39–43. IEEE (1995)
10. Allahverdi, A., Al-Anzi, F.S.: A PSO and a Tabu search heuristics for the assembly scheduling problem of the two-stage distributed database application. Comput. Oper. Res. **33**(4), 1056–1080 (2006)

11. Assareh, E., Behrang, M.A., Assari, M.R., Ghanbarzadeh, A.: Application of PSO (Particle Swarm Optimization) and GA (Genetic Algorithm) techniques on demand estimation of oil in Iran. Energy **35**(12), 5223–5229 (2010)
12. Rudek, M., Canciglieri Jr., O., Greboge, T.: A PSO application in skull prosthesis modelling by superellipse. ELCVIA Electron. Lett. Comput. Vis. Image Anal. **12**(2), 1–12 (2013). https://doi.org/10.5565/rev/elcvia.514
13. Yang, X.-S.: Nature-Inspired Metaheuristic Algorithms. Luniver Press, Frome (2010)
14. Gandomi, A.H., Yang, X.S., Talatahari, S., Alavi, A.H.: Firefly algorithm with chaos. Commun. Nonlinear Sci. Numer. Simul. **18**(1), 89–98 (2013)
15. Yang, X.: Firefly algorithm, Lévy flights and global optimization. In: Bramer, M., Ellis, R., Petridis, M. (eds.) Research and Development in Intelligent Systems XXVI, pp. 209–218. Springer, London (2010). https://doi.org/10.1007/978-1-84882-983-1_15
16. Farahani, S.M., Abshouri, A.A., Nasiri, B., Meybodi, M.R.: A Gaussian firefly algorithm. Int. J. Mach. Learn. Comput. **1**(5), 448 (2011)
17. Kora, P., Rama Krishna, K.S.: Hybrid firefly and particle swarm optimization algorithm for the detection of bundle branch block. Int. J. Cardiovasc. Acad. **2**(1), 44–48 (2016)

New Running Technique for the Bison Algorithm

Anezka Kazikova$^{(\boxtimes)}$, Michal Pluhacek, Adam Viktorin, and Roman Senkerik

Faculty of Applied Informatics, Tomas Bata University in Zlin,
T. G. Masaryka 5555, 760 01 Zlin, Czech Republic
{kazikova,pluhacek,aviktorin,senkerik}@utb.cz

Abstract. This paper examines the performance of the Bison Algorithm with a new running technique. The Bison Algorithm was inspired by the typical behavior of bison herds: the swarming movement of endangered bison as the exploitation factor and the running as the exploration phase of the optimization.

While the original running procedure allowed the running group to scatter throughout the search space, the new approach proposed in this paper preserves the initial formation of the running group throughout the optimization process.

At the beginning of the paper, we introduce the Bison Algorithm and explain the new running technique procedure. Later the performance of the adjusted algorithm is tested and compared to the Particle Swarm Optimization and the Cuckoo Search algorithm on the IEEE CEC 2017 benchmark set, consisting of 30 functions. Finally, we evaluate the meaning of the experiment outcomes for future research.

Keywords: Bison Algorithm · Running technique · Swarm algorithms

1 Introduction

The swarm algorithms simulate the behavior of animal swarms to optimize various problems. These may be either continuous (where the minimum of a multi-modal function is searched for) or discrete like the Travelling Salesman Problem, where we seek the optimum trace of a salesman, who ought to visit certain cities with the lowest traveling cost [1,2].

The inspiration of the intelligent swarms is vast. The algorithms simulate the behavior of fish [3], fireflies [4], cuckoos [5], wolves [6] and many others. The Bison Algorithm was developed by the author et al. in 2017 [7]. It divides the population into two groups. The first one simulates the behavior of the endangered herd by swarming the weak ones in between of the strongest. The second group simulates the running behavior of the bison herd. The algorithm proved to be able to solve nontrivial continuous minimization problems. However, the original running procedure seemed to have some inconveniences. In multiple cases, the running group got scattered over the search space, leaving the algorithm with lone running individuals rather than with a wholesome running group as intended.

© Springer International Publishing AG, part of Springer Nature 2018
L. Rutkowski et al. (Eds.): ICAISC 2018, LNAI 10841, pp. 417–426, 2018.
https://doi.org/10.1007/978-3-319-91253-0_39

In Sect. 2, the original Bison Algorithm is described and a new running technique that fixes the mentioned problem is proposed. In Sect. 3, the experiments are designed to compare the performance of the adjusted algorithm with the original Bison Algorithm and to compare its performance to other optimization algorithms such as the Particle Swarm Optimization and the Cuckoo Search [8] on the set of benchmark functions from IEEE CEC 2017 [9]. The results are also presented in Sect. 3. Finally, in Sect. 4, we discuss the results of the experiments and their meaning for future research.

2 Bison Algorithm

The inspiration of the Bison Algorithm comes from the typical behavior patterns of bison herds. When bison are endangered by predators, they form a circle, with the outline of the strongest individuals, protecting the calves and the weak ones inside. Bison are also famous for their tireless running performance, which makes them a great exploration model [10].

2.1 Bison Algorithm Definition

The Bison Algorithm is defined by two groups of bison: the swarming and the running group. The first exploits the search space by moving the weaker solutions closer to the center of the better ones, while the second explores it by a continuous run. The original algorithm divides the groups by the objective function value: the fitter solutions belong to the swarming group, while the worse ones to the running group [7]. Both groups can, therefore, exchange their members.

Algorithm 1. Pseudocode of the original Bison Algorithm

```
1.   Initialization:
     objective function: f(x) = (x₁, x₂, ..., x_dim)
     generate the swarming group randomly
     generate the running group around the worst bison
     generate the run direction vector (Eq. 5)
2.   for every migration round do
3.       compute the center of the elite bison group
4.       for every bison in the swarming group do
5.           compute new position candidate x_new (Eq. 4)
6.           if f(x_new) < f(x_old) then move to the x_new
7.       end for
8.       adjust the run direction vector (Eq. 6)
9.       for every bison in the running group do
10.          move to the new position (Eq. 7)
11.      end for
12.      check boundaries
13.      sort the population by objective function value
14.  end for
```

Swarming Behavior. The swarming movement starts by computing the center of the best solutions. This study uses the ranked center model (Eqs. 1, 2). Then, each member of the swarming group proposes a new solution closer to the center (Eq. 3). The swarming bison moves, if the quality of the proposed solution improves.

$$weight = (10, 20, 30, \ldots, 10 \cdot s) \tag{1}$$

$$ranked\ center = \sum_{i=1}^{s} \frac{weight_i \cdot x_i}{\sum_{j=1}^{s} weight_j} \tag{2}$$

$$x_{new} = x_{old} + (center - x_{old}) \cdot random(0, overstep)_{dim} \tag{3}$$

where s is the elite group size, x_i, x_{new} and x_{old} represent the ith, the current and the previous solutions respectively and dim is a dimension.

Running Behavior. The running movement is based on the run direction vector, which is generated during the initialization of the algorithm (Eq. 4) and slightly altered in every iteration (Eq. 5). Runners move without elitism (Eq. 6).

$$run\ direction = random(\frac{ub - lb}{45}, \frac{ub - lb}{15})_{dim} \tag{4}$$

$$run\ direction = run\ direction \cdot random(0.9, 1.1)_{dim} \tag{5}$$

$$x_{new} = x_{old} + run\ direction \tag{6}$$

where *run direction* is the run direction vector, ub and lb are the upper and lower boundaries of the search space, x_{new} and x_{old} represent the current and the previous solutions respectively and dim is a dimension.

So far, the description of the original and the newly proposed running technique is the same. The difference occurs when a member of the running group finds a better solution than the last member of the swarming group. In the original algorithm, the whole population was sorted by the solution quality after every iteration. Therefore both groups could switch their members easily. However, adding a formerly swarming bison to the running group usually caused a scattering of the running group, as demonstrated in Fig. 1. To keep them together, the running group was during the initialization generated around the worst member of the swarming group.

Proposition of the New Running Behavior. The new running technique used in this paper, promotes the successful runners into the swarming group, while leaves the running group untouched. The swarm group members with worse solutions are abandoned, and the new swarming group is sorted by the objective function value.

This approach allows duplicities in the population, however the diversity is not endangered, as the running group keeps on moving throughout the whole optimization process and must be different the very next iteration. The new running also enables a more intelligent way of the initialization: the running group may now be generated around the best swarmer instead of the worst one.

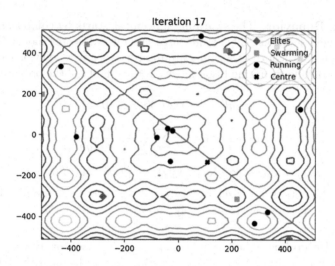

Fig. 1. The scattered running group caused by the original running technique

Bison Algorithm with the New Running Behavior. The newly proposed Bison Algorithm implements the same swarming principles and parameters as the original Bison Algorithm, while changes the rules of successful runners and the initial generation of the running group. The changed algorithm is described in pseudocode Algorithm 2.

Algorithm 2. Pseudocode of the Bison Algorithm with the new running technique

```
1.   Initialization:
       objective function: f(x) = (x₁, x₂, ..., x_dim)
       generate the swarming group randomly
       generate the running group around the best bison
       generate the run direction vector (Eq. 5)
2.   for every migration round do
3.       compute the center of the elite bison group
4.       for every bison in the swarming group do
5.           compute new position candidate x_new (Eq. 4)
6.           if f(x_new) < f(x_old) then move to the x_new
7.       end for
8.       adjust the run direction vector (Eq. 6)
9.       for every bison in the running group do
10.          move to the new position (Eq. 7)
11.      end for
12.      check boundaries
13.      if f(x_runner) < f(x_swarmer)
14.          then place x_runner to the swarming group
15.      sort the swarming group by f(x) values
16. end for
```

Parameters and Out of Bounds Behavior. The parameters are described in Table 1. The *swarm group size* determines the number of bison performing the swarming movement. The *elite group size* indicates the number of the fittest solutions used to compute the center for the swarming movement. The *overstep* defines the maximum length of the swarming movement. The running group is the complement of the swarming group to the population. Crossing the borders means moving to the other side of the exceeded dimension (Fig. 2).

Table 1. Parameters of the Bison Algorithm

Parameter	Recommended value
Population size	50
Elite group size	20
Swarm group size	40
Center computation	Ranked
Overstep	3.5–4.1

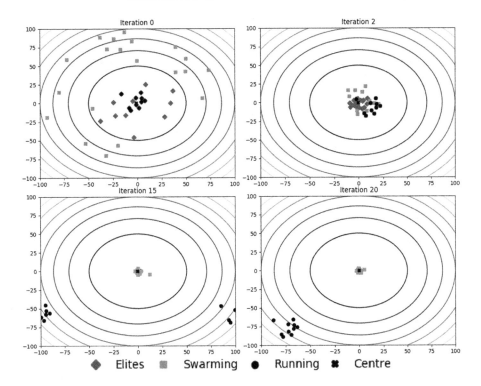

Fig. 2. Bison Algorithm with the new running technique on 2D Rastrigin function

3   Experiments and Results

The Bison Algorithm was tested on 30 minimization problems of the IEEE CEC 2017 benchmark [9]. In accordance with the CEC 2017 benchmark instructions, the algorithms were tested on 51 independent runs, each consisting of $n = 10000 \cdot dimensions$ evaluations of the objective function. The parameter configurations of the examined optimization algorithms are described in Table 2.

First, we tested the new running approach against the original version. Figure 3 shows the mean results of the Bison Algorithm using both the mentioned running techniques for 50 dimensions on a logarithmic scale. Both of the lines are quite similar with a very slight deviation around the F2, F6, F14, and F15. However, after closer examination, these results (and results for 10 and 30

Table 2. Parameter settings for the experiments

Bison Algorithm		PSO		Cuckoo Search	
Population	50	Population	50	Population	50
Elite group size	20	v_{max}	6	P_a	0.25
Swarm group size	40	w_{Max}	0.9		
Center computation	Ranked	w_{Min}	0.2		
Overstep	3.5	c_1	2		
		c_2	2		

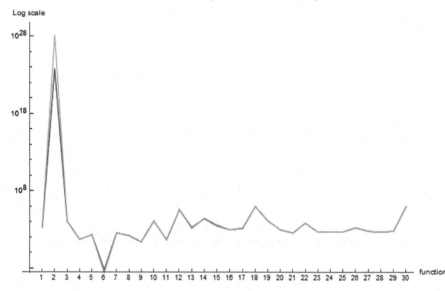

Fig. 3. Comparison of the Bison Algorithm running techniques on 50 dimensions CEC 2017

dimensions, which we also tested) seemed to favor the proposed modification over the entire benchmark set slightly.

The performance of the Bison Algorithm with the new running technique was then compared to the Particle Swarm Optimization Algorithm and the Cuckoo

Table 3. Performance of the Bison Algorithm, PSO, and CS in 30-dimensional CEC 2017

	Bison Algorithm		PSO		Cuckoo Search		Min
	Avg.	Std.	Avg.	Std.	Avg.	Std.	Alg.
f_1	2.09E+03	1.85E+03	3.91E+03	5.11E+03	8.90E+02	7.49E+02	CS
f_2	7.30E+11	5.87E+11	5.02E+23	3.58E+24	2.85E+11	6.21E+11	-
f_3	6.89E+01	8.57E+01	0.00E+00	0.00E+00	3.39E+04	5.76E+03	PSO
f_4	1.06E+01	2.51E+01	9.31E+01	2.89E+01	6.82E+01	1.89E+01	BIA
f_5	7.03E+01	6.11E+01	1.52E+02	2.82E+01	1.37E+02	2.00E+01	BIA
f_6	0.00E+00	0.00E+00	3.06E+01	9.78E+00	4.15E+01	9.07E+00	BIA
f_7	1.78E+02	3.17E+01	1.02E+02	2.21E+01	1.63E+02	2.11E+01	PSO
f_8	6.31E+01	5.85E+01	1.06E+02	1.90E+01	1.36E+02	1.76E+01	BIA
f_9	5.38E+00	9.01E+00	2.21E+03	9.68E+02	3.72E+03	1.14E+03	BIA
f_{10}	6.92E+03	2.79E+02	3.51E+03	6.38E+02	3.74E+03	2.57E+02	PSO
f_{11}	3.29E+01	2.33E+01	1.00E+02	3.55E+01	8.84E+01	1.82E+01	BIA
f_{12}	2.46E+04	1.17E+04	6.57E+05	2.05E+06	1.50E+05	5.39E+04	BIA
f_{13}	1.09E+04	7.72E+03	1.03E+05	6.20E+05	5.91E+03	4.01E+03	-
f_{14}	3.48E+03	1.71E+03	5.08E+03	5.80E+03	1.18E+02	1.84E+01	CS
f_{15}	4.69E+03	2.09E+03	1.06E+04	1.28E+04	9.92E+02	6.81E+02	CS
f_{16}	1.06E+03	4.17E+02	8.49E+02	2.54E+02	9.25E+02	1.35E+02	-
f_{17}	1.43E+02	1.52E+02	5.21E+02	1.90E+02	3.01E+02	8.57E+01	BIA
f_{18}	1.51E+05	1.03E+05	1.53E+05	1.55E+05	5.57E+04	1.50E+04	CS
f_{19}	5.66E+03	3.96E+03	4.86E+03	6.42E+03	3.82E+02	2.91E+02	CS
f_{20}	1.83E+02	1.29E+02	1.83E+02	1.30E+02	4.13E+02	8.57E+01	-
f_{21}	2.58E+02	5.87E+01	3.37E+02	2.86E+01	3.25E+02	4.04E+01	BIA
f_{22}	1.00E+02	6.70E-01	1.98E+03	1.96E+03	8.29E+02	1.52E+03	BIA
f_{23}	3.80E+02	1.39E+01	6.58E+02	1.04E+02	4.86E+02	2.73E+01	BIA
f_{24}	4.52E+02	1.17E+01	6.98E+02	6.72E+01	5.44E+02	4.98E+01	BIA
f_{25}	3.90E+02	8.93E+00	3.90E+02	4.03E+00	3.85E+02	1.27E+00	CS
f_{26}	1.07E+03	6.42E+02	2.07E+03	1.60E+03	1.05E+03	4.45E+02	-
f_{27}	5.31E+02	1.11E+01	5.77E+02	5.06E+01	5.29E+02	7.43E+00	-
f_{28}	3.24E+02	5.47E+01	4.24E+02	4.38E+01	3.87E+02	3.43E+01	BIA
f_{29}	5.54E+02	1.28E+02	9.34E+02	2.20E+02	9.28E+02	7.87E+01	BIA
f_{30}	4.50E+03	7.83E+02	5.19E+03	3.11E+03	1.10E+04	3.43E+03	-

Table 4. Sum of functions, where one algorithm outperformed the others

Dimension	None	Bison Algorithm	PSO	Cuckoo Search
10 D	9	9	2	10
30 D	7	14	3	6
50 D	6	15	4	5

Search. Both of these algorithms were implemented from the EvoloPy optimization library [8]. The following experiments refer to the Particle Swarm Optimization as PSO, the Cuckoo Search as CS and the Bison Algorithm as BIA.

Table 3 shows the average solutions and their standard deviations of BIA, PSO, and CS in 30-dimensional problems. The last column presents the most successful algorithm for solving the examined function, subject to Wilcoxon rank-sum test ($\alpha = 0.05$). The winner is recognized if it outperforms all the other algorithms with statistical significance according to the rank-sum test. Table 4 sums the number of wins, where one optimization algorithm outperformed all the others with statistical significance, for each algorithm in 10, 30 and 50 dimensions.

Figures 4 and 5 show the mean final solution values of the tested algorithms in 30 and 50-dimensional problems. The Bison Algorithm outperformed the others in most of the problems in 30 and 50 dimensions. In 10 dimensions, the Cuckoo Search achieved a slightly better performance.

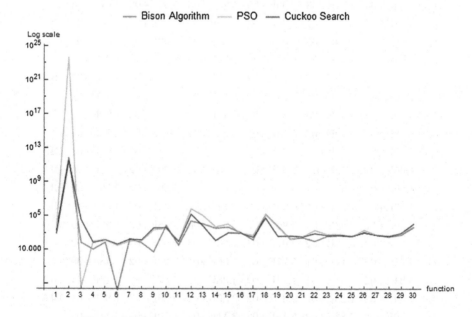

Fig. 4. Mean results of tested algorithms in 30 dimensions

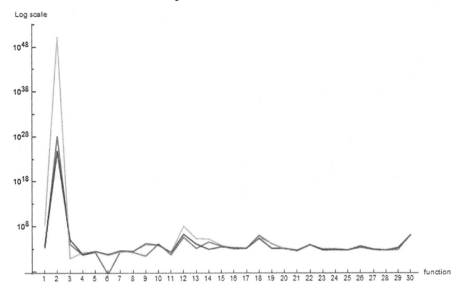

Fig. 5. Mean results of tested algorithms in 50 dimensions

4 Discussion

According to above-presented results, the Bison Algorithm with the modified running behavior outperformed the PSO and CS algorithms in most cases, especially in higher dimensions. The results were subject to Wilcoxon rank-sum test over all pairs to show the significance of the results. In the case of 50 dimensions, the proposed method managed to significantly outperform the other algorithms over the half of the benchmark with six functions being indecisive. It is also worth pointing out that the Bison Algorithm achieves superior performance over all function types, from simple unimodal and basic multimodal functions to complex and hybrid functions.

The main goal of the running group is the exploration of the search space and the diversification of the population. Therefore the success rate of the running group alone is not essential. However, when compared to the original approach, it is clear, that there is still space for future improvements of the running behavior.

5 Conclusion

We propose a redefinition of the Bison Algorithm with a new, more logical and integrated running model. The Bison Algorithm with the new running standard showed promising results in comparison with other optimization algorithms.

The proposed approach opened new ways to explore. Suddenly there is a possibility of implementing more running groups to the Bison Algorithm, which was not achievable before since the groups did not hold together.

However, the running group still seems to have limited impact on the overall performance of the optimization process. Investigating the cause of this will direct our future research.

Acknowledgment. This work was supported by the Ministry of Education, Youth and Sports of the Czech Republic within the National Sustainability Programme Project no. LO1303 (MSMT-7778/2014), further by the European Regional Development Fund under the Project CEBIA-Tech no. CZ.1.05/2.1.00/03.0089 and by Internal Grant Agency of Tomas Bata University under the Projects no. IGA/CebiaTech/2018/003. This work is also based upon support by COST (European Cooperation in Science and Technology) under Action CA15140, Improving Applicability of Nature-Inspired Optimisation by Joining Theory and Practice (ImAppNIO), and Action IC1406, High-Performance Modelling and Simulation for Big Data Applications (cHiPSet). The work was further supported by resources of A.I.Lab at the Faculty of Applied Informatics, Tomas Bata University in Zlin (ailab.fai.utb.cz).

References

1. Ouaarab, A., Ahiod, B., Yang, X.S.: Discrete Cuckoo Search algorithm for the travelling salesman problem. Neural Comput. Appl. **24**(7–8), 1659–1669 (2014)
2. Duan, Y., Ying, S.: A particle swarm optimization algorithm with ant search for solving traveling salesman problem. In: International Conference on Computational Intelligence and Security, Beijing, pp. 137–141 (2009)
3. Kennedy, J., Eberhart, R.: Particle swarm optimization. Proc. IEEE Int. Conf. Neural Netw. **4**, 1942–1948 (1995)
4. Yang, X.-S.: Firefly algorithm, Lévy flights and global optimization. In: Bramer, M., Ellis, R., Petridis, M. (eds.) Research and Development in Intelligent Systems XXVI, pp. 209–218. Springer, London (2010). https://doi.org/10.1007/978-1-84882-983-1_15
5. Yang, X.-S., Deb, S.: Cuckoo Search via Lévy flights. In: Proceedings of World Congress on Nature & Biologically Inspired Computing (NaBIC 2009), pp. 210–214, India, December 2009. IEEE Publications (2009)
6. Mirjalili, S., Mirjalili, S.M., Lewis, A.: Grey wolf optimizer. Adv. Eng. Softw. **69**, 46–61 (2014)
7. Kazikova, A., Pluhacek, M., Senkerik R., Viktorin, A.: Proposal of a new swarm optimization method inspired in Bison behavior. In: Matousek, R. (ed.) Recent Advances in Soft Computing (Mendel 2017). Advances in Intelligent Systems and Computing. Springer, Heidelberg (2017, in press)
8. Faris, H., Aljarah, I., Mirjalili, S., Castillo, P., Merelo, J.: EvoloPy: an open-source nature-inspired optimization framework in Python. In: Proceedings of the 8th International Joint Conference on Computational Intelligence (IJCCI 2016), ECTA, vol. 1, pp. 171–177 (2016)
9. Awad, N.H., Ali, M.Z., Liang, J.J., Qu, B.Y., Suganthan, P.N.: Problem definitions and evaluation criteria for the CEC 2017 special session and competition on single objective bound constrained real-parameter numerical optimization. Technical report, Nanyang Technological University, Singapore (2016)
10. Berman, R.: American Bison (Nature Watch). Lerner Publications, Minneapolis (2008)

Evolutionary Design and Training of Artificial Neural Networks

Lumír Kojecký and Ivan Zelinka$^{(\boxtimes)}$

Department of Computer Science, FEECS, VŠB - Technical University of Ostrava,
17. listopadu 15, 708 33 Ostrava, Poruba, Czech Republic
{lumir.kojecky,ivan.zelinka}@vsb.cz

Abstract. The dynamics of neural networks and evolutionary algorithms share common attributes and based on many research papers it seems to be that from dynamic point of view are both systems indistinguishable. In order to compare them mutually from this point of view, artificial neural networks, as similar as possible to natural one, are needed. In this paper is described part of our research that is focused on the synthesis of artificial neural networks. Since most current ANN structures are not common in nature, we introduce a method of a complex network synthesis using network growth model, considered as a neural network. Synaptic weights of the synthesized ANN are then trained by an evolutionary algorithm to respond to an input training set successfully.

Keywords: Neural network synthesis · Network growth model
Complex network · Evolutionary algorithms

1 Introduction

Our current research is focused on neural and swarm-based algorithms that represent a class of search methods that can be used for solving optimization problems and challenges of learning. Swarm and neural algorithms are different kind of algorithms, distinct in its nature, however, with common attributes of its dynamics (based on parallelism, processing and storing information, exhibit universal behavior like chaos).

Artificial neural networks (ANN) [3,12] have been successfully used as a tool for solving various kind of problems like classification, regression, or forecasting. During the time, several algorithms were developed to train the network in shorter time and with the more accurate response, for example, backpropagation, deep learning [6], or even optimization techniques like well-known genetic algorithms [7].

During the time there have also been many improvements of the traditional feedforward [3] network structure, for example, recurrent neural networks [12], deep neural networks [6], or synthesized multilayer network structures by genetic algorithms [7].

© Springer International Publishing AG, part of Springer Nature 2018
L. Rutkowski et al. (Eds.): ICAISC 2018, LNAI 10841, pp. 427–437, 2018.
https://doi.org/10.1007/978-3-319-91253-0_40

All these approaches, however, provide artificial network structures that are not common in nature. The main aim of this paper is to show our approaches how to synthesize a neural network structure, and how to train the network for the further analysis of its behavior.

In our previous research, we synthesized ANNs using evolutionary algorithms [1,9] and symbolic regression [13], which will be briefly introduced here. We will newly introduce a method of a complex network synthesis using network growth model [2] – the network will be then considered as a neural network with defined inputs and outputs, and finally trained by an evolutionary algorithm to successfully respond to an input training set.

2   Used Methods and Algorithms

For our experiments described here, specific hardware and algorithms have been used. All important information about algorithms used in our experiments is mentioned and referred here. For this study, as initial one, for the first experiments were selected two algorithms. The SOMA [9] (a representative of swarm/evolutionary algorithms) and evolutionary strategies (ES, [1], a representative of simpler ancestor of evolutionary algorithms).

2.1   Self-Organizing Migrating Algorithm

Self-Organizing Migrating Algorithm (SOMA) [9] is a swarm-based optimization algorithm imitating the behavior of cooperating individuals searching the best food source. In SOMA each individual in a population is represented as one point in N-dimensional space. The search process is represented as vector operations between the points. Quality of the food source, individual's fitness, is evaluated by so-called fitness (or cost) function f_{cost}. Purpose of optimization algorithms (i.e., SOMA) is, therefore, searching the global minimum of f_{cost} at a given interval.

SOMA starts with a randomly initialized population of individuals, each evaluated by f_{cost}. At the beginning of each migration loop, an individual with the best fitness is chosen as the leader. All individuals then move towards the leader through space and search their best position with the best fitness on their way. The SOMA migration process is visualized in Fig. 1.

2.2   Evolution Strategies

Evolution Strategies (ES) [1] is one of the first stochastic optimization algorithms, also based on the population of individuals and vector operations between them. In ES, like in SOMA, the individual is represented as one point in N-dimensional space, its quality is evaluated by f_{cost}, and it is used to search the global minimum of f_{cost} at a given interval.

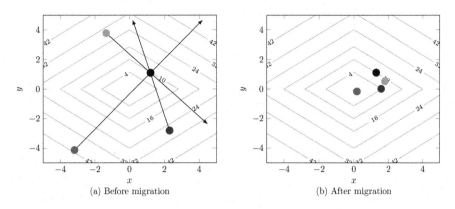

(a) Before migration   (b) After migration

Fig. 1. Illustration of the SOMA migration process

ES also starts with randomly initialized population of μ individuals – parents, each evaluated by f_{cost}. For each parent $(\boldsymbol{x})$ there is generated a number (λ) of offspring $(\boldsymbol{y})$ using Gaussian mutation operator – $\boldsymbol{y} = \mathcal{N}\left(\boldsymbol{x}, \sigma^2\right)$. In each generation, all $\mu \cdot \lambda$ offspring are evaluated by f_{cost} and best μ offspring are selected as parents for a next generation. The ES generation process is visualized in Fig. 2.

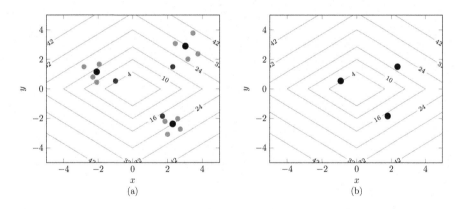

(a)   (b)

Fig. 2. Illustration of the EA generation process for $\mu = 3$ and $\lambda = 5$

3   Experiment Design

In our previous research [4] we have focused on two approaches of an ANN synthesis that will be briefly introduced in following paragraphs – synthesis using Analytic Programming (AP) [11] and synthesis ANN with the arbitrary structure using an evolutionary algorithm (EA). However, both these approaches produce networks with unnatural/artificial structure – tree in case of AP, or random graph in case of EA. Therefore we decided to implement a network growth model to obtain a more natural structure – complex network.

All the synthesized neural networks consist of neurons and synapses between these neurons. The synapse weights are not known at the time of ANN synthesis – it is necessary to estimate them using an evolutionary algorithm [10]. Here, an individual is represented as the vector of all ANN weights, and f_{cost} returns the global network error for the input training set.

3.1 Synthesis Using Analytic Programming

The main principle of AP [13] is the composition of simple elements (like an artificial neuron or weight) into more complex units like ANN. This process can be used for example to synthesize an ANN that properly responds to inputs from a training set. The simple elements are placed in a General Functional Set (GFS), which consists three elements for our purposes – a simple addition function with two arguments, an artificial neuron (parameterized hyperbolic tangent function), and a weighted input for the AN.

At the AP input, there is a set of integers-pointers to the GFS. At the output there is an ANN that consists of the indexed elements – arguments of the elements are substituted in a FIFO order as next indexed elements. The composition process is visualized in Fig. 3. To find the most suitable ANN structure, it is necessary to find the best input number set – the set can be represented as an individual in an evolutionary algorithm.

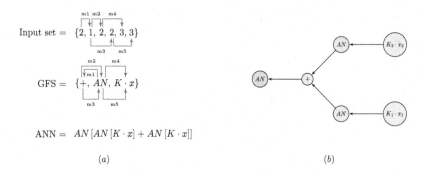

(a)   (b)

Fig. 3. Illustration of the expression synthesis process (a) and its output visualization (b) as a graph. The green nodes represent the network input, while the red node represents the network output (Color figure online)

3.2 Synthesis by Means of Evolutionary Algorithm

In this process, individuals are generated in range $[0, 1]$ like $\begin{bmatrix} 0.87 & 0.15 & 0.51 & 0.77 \end{bmatrix}$. Each element of the vector of dimension n^2 is rounded and the vector is transformed into the $n \times n$ matrix $\begin{bmatrix} 1 & 0 \\ 1 & 1 \end{bmatrix}$. This matrix represents an adjacency matrix of the result graph – neural network.

In the neural network, x nodes, having at least one output connection, are marked as inputs, y nodes, having at least one input connection, are marked as outputs. Unnecessary nodes are then removed – non-input nodes with no input connection and non-output nodes with no output connections. In Fig. 4 there is visualized a sample weighted adjacency matrix and its corresponding graph.

$$A_G = \begin{bmatrix} 0 & 0 & w_1 & 0 \\ w_2 & w_3 & 0 & 0 \\ 0 & w_4 & w_5 & 0 \\ 0 & w_6 & w_7 & w_8 \end{bmatrix}$$

(a)   (b)

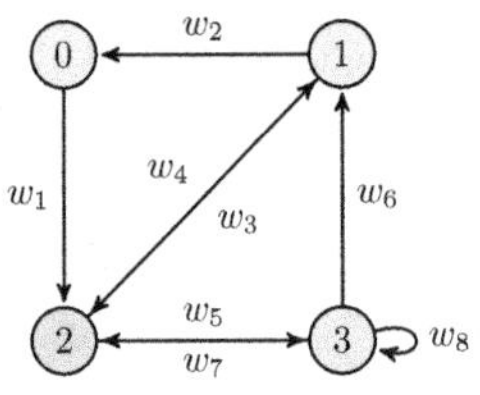

Fig. 4. A weighted adjacency matrix (a) and its network (b). The green nodes (0 and 1) represent the network input, while the red nodes (2) represent the network output (Color figure online)

3.3 Synthesis Using Network Growth Model

To obtain a more natural structure than in the two previous approaches (i.e., complex network), we decided to implement Bianconi-Barabási network growth model [2]. Process in this model starts with a basic small network, where the nodes are randomly connected. When the new node n_i should be connected to the network, one of the current nodes n_r is selected randomly and connected to the new one. With probability p, the new node is connected to one neighbor of n_r. Additionally, with probability $1 - p$, the new node is connected to another randomly selected node. A sample process of attaching new node to the network is shown in Fig. 5.

The process above considers the synthesis of the undirected network. Our implementation, however, deals with directed networks. Considering a randomly chosen node as n_r and the new node as n_i – with probability p, direction of the edge is oriented $n_r \rightarrow n_i$, while with probability $1 - p$, direction of the edge is oriented $n_r \leftarrow n_i$.

In a standard feedforward neural networks [3], the connections between neurons do not form a cycle, and thus information moves only in one direction – from the input neurons, through the hidden neurons, to the output neurons. On the other hand, connections in neural networks created by this network growth model can (and most probably will) form a cycle – a recurrent neural network [12]. In a cycle, the calculated values at the neuron outputs affect inputs for the previous neurons, and therefore it is necessary to iteratively recalculate all the output values until the output is stable.

At the beginning of the synthesis, network with $(x_{inputs} + y_{outputs})$ neurons is made with random connections between the neurons, and x_{inputs} neurons

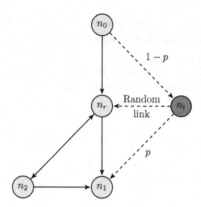

Fig. 5. A sample process of attaching new node n_i to randomly chosen node n_r, its neighbor n_1 with probability p, and another node n_0 with probability $1 - p$.

are chosen as the network input. As was mentioned at the beginning of this paper, weights of the network are estimated by an evolutionary algorithm with f_{cost} returning the global network error for the input training set. In this initial study, we consider networks with only one output ($y_{outputs} = 1$) – we, therefore, evaluate the global network error for each neuron and f_{cost} returns the lowest error.

In some cases, the stability of the neuron outputs cannot be achieved. Some of the output values will not stabilize during the specified time – their values change periodically or non-periodically (i.e., chaotically). Therefore we added a penalization to the f_{cost} – the cost value is increased by the output values difference in time.

The network growth model described above dealt with a uniform selection of any random node in the network. Additionally, we implemented also preferential attachment – nodes/neurons with better fitness are more likely to be attached to a new node n_i.

4   Simulations and Results

ANN synthesis using network growth model was implemented in Open MPI platform, and the simulations were performed on Anselm, x86-64 based supercomputer cluster hosted at national supercomputing center IT4Innovations[1].

4.1   Algorithm Settings

Algorithm settings of used evolutionary algorithms are summarized in Table 1.

As the training data for a synthesized ANN, two kinds of datasets were used: truth table of XOR problem (Dataset 1) and Sample, also linearly inseparable, data (Dataset 2) visualized in Fig. 6.

---

[1] https://www.it4i.cz/?lang=en.

Table 1. Algorithm settings of used evolutionary algorithms.

Parameter SOMA	Value SOMA	Parameter ES	Value ES
Dimension	Number of weights	Dimension	Number of weights
PopSize	Number of weights	Parents	Number of weights
Migrations	20	Generations	20
PRT	0.1	Offspring	50
PathLength	3	σ	$\frac{2}{3}$
Step	0.11		

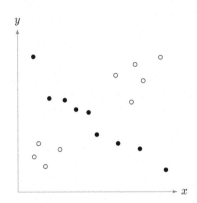

Fig. 6. Sample dataset used as a training data

Both random and preferential attachment was tested with weight estimation using SOMA and ES. As a random number generator, Mersenne Twister [8] was used for all simulations. For each combination of attachment and EA, 50 simulations were performed. For each combination was also used the same random number generator seed sequence.

4.2 Results

In Tables 2 and 3 there are summarized all simulation results. Performance of random attachment and preferential attachment is compared as well as performance of weight estimation by means of SOMA and ES. Each table contains these information:

- Ratio between successful and not successful network synthesis
- In case both approaches successfully synthesized the network there is information, which approaches provided a network with lower global error
- Also, when both approaches successfully synthesized the network there is information, which approaches synthesized the result faster (in fewer iterations – using fewer nodes).

Table 2. Mutual comparison of random attachment and preferential attachment.

Attachment	Simulations	Success	Not success	Better result	Faster result
Random	200	75%	25%	27%	33%
Preferential	200	82%	18%	73%	67%

Table 3. Mutual comparison of weight estimation by means of SOMA and ES.

EA	Simulations	Success	Not success	Better result	Faster result
SOMA	200	82%	18%	50%	73%
ES	200	82%	18%	50%	27%

4.3   Sample Output

In Figs. 7 and 8 there are visualized output networks of 200 and 500 nodes, synthesized by approach introduced in Sect. 3.3. Unlike in Sect. 3.2, the unnecessary nodes are not removed from the network. Keeping these nodes do not have any

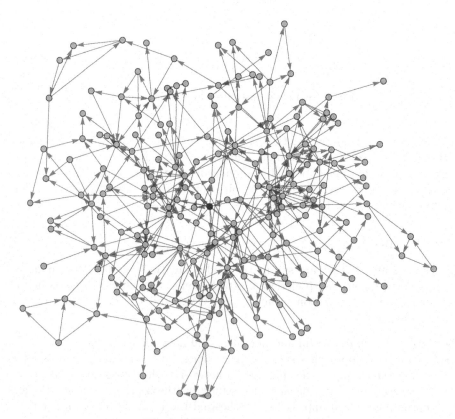

Fig. 7. ANN with 200 neurons.

impact on the ANN function – default input value of neurons with no input is zero, and thus output value will also be zero ($\tanh(0) = 0$). Although omitting this step can have a negative impact on the synthesis speed, these "unnecessary" nodes can be connected to a new node in the future, and create a more powerful network.

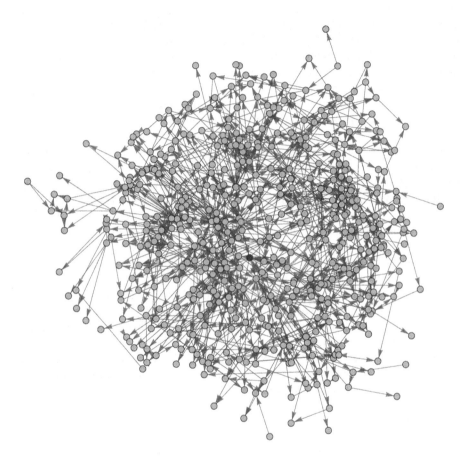

Fig. 8. ANN with 500 neurons.

5   Conclusion

In this paper, we discussed possibilities of artificial neural network synthesis and their training using the evolutionary algorithm. Two our current methods were mentioned – synthesis using Analytic Programming and synthesis using evolutionary algorithms. We proposed another method – synthesis using network growth model, Bianconi-Barabási model in particular. This method was implemented in two regimes – random and preferential attachment of new nodes. The final ANN was trained using SOMA and ES evolutionary algorithms.

In the result summarization tables, we compared the impact of random/preferential node attachment and also impact of different training algorithm. From totally 400 simulations, preferential attachment provided 7% more successfully synthesized networks. When both approaches successfully synthesized the ANN, preferential attachment approach provided ANN with a lower global error in 73% of simulations. Also in 67% of simulations, preferential attachment provided the result faster (in fewer iterations – using fewer nodes).

In all 400 simulations, both SOMA and ES successfully synthesized 82% or networks. When both approaches successfully synthesized the ANN, both SOMA and ES were equal in a number of networks with a lower global error. However, SOMA provided faster result in 73% of simulations.

Based on this information, the best possible algorithm combination is network growth model with preferential node attachment trained by SOMA algorithm. This combination will provide the best quality results in the shortest time (the fastest synthesis took 173 s, while the slowest synthesis time was 2,879 s).

Since the synthesis using network growth model study is in an initial status, we have used only simple test problems with only two inputs (for example XOR). In the future work, we are going to extend these networks for much more inputs (i.e., hundreds). These networks can be used for example for classification of astroinformatics big data [5]. As a random number generator for these simulations was used Mersenne Twister. Due to significant performance and time differences between different random number generators [4] we will also try impact of the generators on this synthesis.

This initial study handles neural networks with fixed input nodes and only one output. In the future work, we will also find the best possible network input as well as the network output, including a selection of the best activation function for each neuron.

Based on verification that ANN can be synthesized in this way, as a future continuation dynamics of such ANNs will be examined and compared with dynamics of selected swarm algorithms.

Acknowledgment. The following grants are acknowledged for the financial support provided for this research: Grant of SGS No. 2018/177, VSB-Technical University of Ostrava and by the European Union's Horizon 2020 research and innovation programme under grant agreement No. 710577.

References

1. Beyer, H.G., Schwefel, H.P.: Evolution strategies-a comprehensive introduction. Nat. Comput. **1**(1), 3–52 (2002)
2. Bianconi, G., Darst, R.K., Iacovacci, J., Fortunato, S.: Triadic closure as a basic generating mechanism of communities in complex networks. Phys. Rev. E **90**(4), 042806 (2014)
3. Hornik, K., Stinchcombe, M., White, H.: Multilayer feedforward networks are universal approximators. Neural Netw. **2**(5), 359–366 (1989)

4. Kojecký, L., Zelinka, I., Prasad, A., Vantuch, T., Tomaszek, L.: Investigation on unconventional synthesis of astroinformatic data classificator powered by irregular dynamics. IEEE Intell. Syst. (in press)
5. Kojeckỳ, L., Zelinka, I., Šaloun, P.: Evolutionary synthesis of automatic classification on astroinformatic big data. Int. J. Parallel Emerg. Distrib. Syst. **32**(5), 429–447 (2017)
6. LeCun, Y., Bengio, Y., Hinton, G.: Deep learning. Nature **521**(7553), 436–444 (2015)
7. Leung, F.H.F., Lam, H.K., Ling, S.H., Tam, P.K.S.: Tuning of the structure and parameters of a neural network using an improved genetic algorithm. IEEE Trans. Neural Netw. **14**(1), 79–88 (2003)
8. Matsumoto, M., Nishimura, T.: Mersenne Twister: a 623-dimensionally equidistributed uniform pseudo-random number generator. ACM Trans. Model. Comput. Simul. (TOMACS) **8**(1), 3–30 (1998)
9. Onwubolu, G.C., Babu, B.: New Optimization Techniques in Engineering, vol. 141. Springer, Heidelberg (2004). https://doi.org/10.1007/978-3-540-39930-8
10. Oplatková, Z.: Metaevolution-synthesis of evolutionary algorithms by means of symbolic regression. Tomas Bata University in Zlín (2008)
11. Vařacha, P.: Neural network synthesis (2011)
12. Williams, R.J., Zipser, D.: A learning algorithm for continually running fully recurrent neural networks. Neural Comput. **1**(2), 270–280 (1989)
13. Zelinka, I.: Analytic programming by means of soma algorithm. In: Proceedings of the 8th International Conference on Soft Computing, Mendel, vol. 2, pp. 93–101 (2002)

Obtaining Pareto Front in Instance Selection with Ensembles and Populations

Mirosław Kordos[1]([⊠]), Marcin Wydrzyński[1], and Krystian Łapa[2]

[1] Department of Computer Science and Automatics,
University of Bielsko-Biala, Bielsko-Biała, Poland
mkordos@ath.bielsko.pl
[2] Częstochowa University of Technology,
Institute of Computational Intelligence, Częstochowa, Poland
krystian.lapa@iisi.pcz.pl

Abstract. Collective computational intelligence can be used in several ways, for example as taking the decision together by some form of a bagging ensemble or as finding the solutions by multi-objective evolutionary algorithms. In this paper we examine and compare the application of the two approaches to instance selection for creating the Pareto front of the selected subsets, where the two objectives are classification accuracy and data size reduction. As the bagging ensemble members we use DROP5 algorithms. The evolutionary algorithm is based on NSGA-II. The findings are that the evolutionary approach is faster (contrary to the popular belief) and usually provides better quality solutions, with some exceptions, were the outcome of the DROP5 ensemble is better.

1 Introduction

Data preprocessing is frequently the most important step in data mining, even more important than choice of the classifier and its parameters, as even the best classifier cannot produce good outcome if the data quality is poor. One part of data preprocessing is data selection. Improving the data quality is one objective of data selection. The other objective is limitation the data size to make the data easier to interpret and analyze and to make the learning process faster. The data selection process can be seen as a two-objective optimization, aimed at data size reduction and classification accuracy improvement. Frequently it happens that a small reduction of data size (when done properly) can achieve both objectives. Although further reduction of data size also reduces the prediction accuracy, it can still be beneficial for data analysis and fast experiments with the learning model.

In this work we want to find a Pareto-front in the compression-accuracy space - that is such a set of solutions, that for any of them there does not exist a solution that can simultaneously improve both: compression and accuracy. The closer is the Pareto front situated to the 0% of selected instances and 100% accuracy point in the accuracy-compression space (upper-left corner in Figs. 2,

© Springer International Publishing AG, part of Springer Nature 2018
L. Rutkowski et al. (Eds.): ICAISC 2018, LNAI 10841, pp. 438–448, 2018.
https://doi.org/10.1007/978-3-319-91253-0_41

4, 5, and 6.) the better. Data selection can be decomposed into feature and instance selection and in our previous work [1] we analyzed the relations and interactions between feature and instance selection. In this work we focus on instance selection and its Pareto-Front.

The first (obvious) observation is that in order to achieve the front we need many solutions. On the other hand most instance selection methods produce only a single solution. There are two ways to obtain the Pareto front. One is to perform the instance selection multiple times with different parameters of the instance selection method, that however can be applied only to these instance selection methods that can be parameterized in this way. The other solution is to use ensembles or populations of instance selection algorithms [2,3,9]. Ensemble methods are widely used in classification and regression problems as they are known to obtain better results than the single best method. The same can be applied to instance selection and the same methods of differentiating the results can be applied (as instance bagging or feature bagging for example) and different voting schemes can be used to move along the Pareto front, as discussed in Sect. 2.

Evolutionary algorithms in contrast use populations and cooperation between the population members (using the crossover operator). There are also known multi-objective evolutionary algorithms, as the NSGA-II [4], which provides additional operators to ensure that the best members of the population occupy possibly uniformly the Pareto front. Using them is simply faster than running many single-objective optimizations with different parameters and it may also produce better results (lower Pareto front), as discussed in Sect. 3.

In Sect. 4 we experimentally compare the Pareto front obtained by the two approaches: a bagging ensemble of DROP5 algorithms and an NGSA-II based evolutionary instance selection.

2   Bagging Ensembles of DROP5 Instance Selection Algorithms

The ensemble methods are proved to be efficient for enhancing the classification process and many solutions are widely used for that purpose [5].

However, the idea of ensembles of instance selection algorithms has been considered only by few authors. The first idea related to ensembles was presented in the all-kNN algorithm, where the several ENN algorithms with different k values were used [6]. In [7] the authors attempted to adapt boosting to instance selection, where the objective was not only to improve the success rate, but also to reduce the data size. The weight given by boosting to each instance defined relevance of the instance, and a statistical test was used to decide whether it can be discarded without affecting classification accuracy.

In [8] classifier ensembles were constructed using weighted instance selection. In [3,9] two variants of bagging of instance selection algorithms were discussed; in the first paper feature based bagging and in the second instance-based bagging. In [10] the authors applied boosting for several instance selection algorithms, as CNN, IB3, DROP3, ICF, approaching the instance selection problem as a two-class classification. They tested several boosting methods, as AdaBoost, FloatBoost, MultiBoost, ReweightBoost and the obtained results confirmed that in most cases the boosting methods showed better performance than the single instance selection algorithm.

The boosting methods, should be used only after applying noise filter or on data that does not require noise filters. Otherwise, the noisy instances would be promoted by the algorithm, as they are usually mis-classified, what can lead to incorrect results. For the same reason in the DROP (Decremental Reduction Optimization Procedure) algorithm noise filter is applied before the condensation part [11].

The ENN (Edited Nearest Neighbor) [11] algorithm is a noise filer. It starts from a whole dataset and check which instances are wrongly classified by k-NN. Each wrongly classified instance is marked for removal. Finally all the instances marked for removal get removed.

The instance selection algorithms from the DROP family methods belongs to the best instance selection methods for classification tasks [12,13]. Two concepts are defined in those methods: nearest neighbors and associates. Nearest neighbors of an instance p are those k instances to which the distance from the instance p is shorter than to the remaining instances. Associates are those instances that have p as one of their k nearest neighbors.

The DROP5 algorithm used in this work was developed based on its predecessors in the following way:

- DROP1 eliminates an instance p, if this does not affect the classification of its associates.
- DROP2 before starting the selection sorts the instances in descending order by the distances from the instance to its nearest enemy (the instance from another class). So first the instances located among the same class instances are processed and later the instances close to the class boundary.
- DROP3 additionally first applies a noise filter that works like the ENN algorithm, removing the instances incorrectly classified by k-NN.
- DROP4 applies a modified version of the noise filter, by additionally verifying if removing an instance does not cause a misclassification of another instance and if it does, the instance will not be removed.
- DROP5 is similar to DROP2, but instead of using the noise filter, it starts the analysis from the instances that are closest to the nearest enemies (those on the class boundary). In this way as a matter of fact the noise filter is applied at the first stage. However, its computational complexity is still $O(n^3)$.

To obtain the Pareto front we used an ensemble of 10 instance selection algorithms, were the differentiation between the ensemble members was achieved by providing to each member a random subset of the data set. Then we analyzed how many members of the ensemble voted for each instance and thus we obtained 10 datasets: the first one containing the instances for which every member voted (however, this was frequently an empty set), the next one for which at least 9 members voted, and so one. The last subset was the biggest and contained the instances for which at least one member voted. We evaluated the classification accuracy for classifiers trained on each of the datasets on the test set. In this way we obtained the Pareto front in the compression-accuracy space. That allows us to select one point on the front depending on the importance we assign to the compression and classification accuracy. The results are presented in the Sect. 4.

3 NSGA-II Based Evolutionary Instance Selection

The way in which standard genetic or evolutionary algorithms work is with very high probability well known to the reader so we will not waste space explaining this, especially that all the details can be easily found in literature [4,14–16].

There are two fundamental differences between the classical and evolutionary-based instance selection algorithms. In the first case a single solution is obtained and some properties of the search must be a priori defined, as for example how to tell the boundary point from the point situated among the same class members, how to define a noisy point and so on. A great advantage of evolutionary-based instance selection is that we do not have to carry about all the definitions and parameters.

Each individual in the genetic population encodes the entire training set and the value at each chromosome position indicates whether the instance is present (value > 0) or not (value $= 0$). In case of instance selection the first value $= 1$. In case of instance weighting it can be a real number between 0 and 1. Most of evolutionary approaches to instance selection define the fitness function in a similar way, as a weighted sum of the achieved compression and classification accuracy on the test set. Thus to obtain a set of solutions the optimization has to be performed several times with various weights. However, this can be avoided using a multi-objective evolutionary algorithms [4,16], as we use in the experimental section.

There have been already some propositions in the literature to use genetic or evolutionary algorithms for instance selections [17–22]. The results obtained by these authors were better in terms of accuracy-compression balance than the results of the best classical algorithms as DROP3, DROP4 and DROP5, however they compared only single points and not the whole Pareto fronts.

It is widely believed that evolutionary optimization can find better solutions that the local search or gradient based-solutions, but its computational cost is much higher. As the first statement is generally true, we show in this work that in the case of instance selection the second statement can be false. The evolutionary instance selection method we use can find the solution in $O(n^2)$ time, while the best classical methods, as the DROP family has the $O(n^3)$ complexity, which for large datasets can be prohibitive. Thus in that case evolutionary methods can be better at both: the obtained results and the evaluation time. We calculate and sort the distance matrix only once at the beginning of the optimization and then reading the classes of the nearest neighbors from the sorted arrays is very fast.

Because the purpose of this work is to obtain not a single solution, but a Pareto front that consists of the set of the best solutions (called also non-dominated solutions), we use a multi-objective evolutionary algorithm based on NSGA-II [4], as shown in the pseudo-code. The two criteria that we use are compression and classification accuracy. We tested 3 different solutions: based on single-objective evolutionary algorithm with changing the weights for accuracy and compression in the fitness function, the SEEA-based solution [4] and the NSGA-II based solution. The last one in most cases produced the best results, so we decided to use it in the final experiments.

Although there are newer multi-objective evolutionary algorithms, NSGA-II often produces the best or at equal results, and due to the fact that the NSGA-II method is well known, we have decided to use it and modify it for the purpose of this study, by adjusting it to binary instance weights, introducing the proper population initialization (the probability of zero is set to the expected compression level) [23] and additional parameters that force the small values to zeros (which rejects the not very useful instances). In [24] the authors investigated the concept of e-dominance with multi-objective evolutionary algorithms and found out that in some cases this concept significantly helps to reduce the execution time. However, e-dominance can also drastically slow down the process for problems where the number of objective vectors is small. We did not use e-dominance, because in our case it did not improved the performance.

We can perform either binary instance selection or instance weighting. With instance weighting each instance is assigned a value between 0 and 1 representing the importance of this instance. While calculating the majority class of an instance k-NN adds the class of its neighbors multiplied by their weights. To calculate the compression we treat as rejected only those instances with zero weights and all others as selected. If during the optimization a given weight in a given individual takes a value below some threshold (e.g. 0.01) it is forced to zero to ensure effective instance rejection. The instance weighting method allows for achieving higher accuracy in the regions of very stronger compression. However, instance weighting is mostly useful for the instance selection process. In the final prediction all the non-zero values can be converted to ones usually with little loss of accuracy. But if we are focused more or accuracy improvement then binary values used during the optimization are usually producing better solutions with low compression but high prediction accuracy, frequently even above the accuracy obtained on the entire dataset (with the exception of situations as in Fig. 6). There is much more to study in the area of instance selection vs instance weighting and the detailed analysis will be conducted in our future research.

4   Experiments and Results

We conducted the experiments on classification datasets from the KEEL Repository [25]. The experiments with DROP5 were performed in RapidMiner using the Data Selection Extension [3,9] with the Weka Instance Selection Module [26], which contained the DROP5 algorithm. The experiments with evolutionary instance selection were performed using a software we created in C# language. We make available all the software, including GPU-based implementation (with

Algorithm 1. NSGA-II

1: $\mathbf{P} := initialization(N, M)$ //$\mathbf{P}$ is population, $\mathbf{F}$ is Pareto front
2: $evaluation(\mathbf{P})$
3: $\mathbf{F} = fast\_nondominated\_sort(\mathbf{P}, N)$
4: $crowding\_distance(\mathbf{F})$
5: **while** $stop\_condition()$ **do**
6:     $\mathbf{P'} = \emptyset$
7:     **for** $i = 1$ **to** N **do**
8:         $parentA = select\_parent(\mathbf{P})$
9:         $parentB = select\_parent(\mathbf{P})$
10:        $child = new\_individual(parentA, parentB)$
11:        $\mathbf{P'} = \mathbf{P'} \cup child$
12:    **end for**
13:    $evaluation(\mathbf{P'})$
14:    $\mathbf{P} = \mathbf{P} \cup \mathbf{P'}$
15:    $\mathbf{F} = fast\_nondominated\_sort(\mathbf{P}, 2N)$
16:    $crowding\_distance(\mathbf{F})$
17:    $\mathbf{P} = selection(\mathbf{P}, \mathbf{F})$
18: **end while**
19: **return** $\mathbf{F}_1$ {list of non dominated individuals}

Nvidia Cuda Toolkit) of evolutionary instance selection and the detailed results of all the experiments so that the interested reader can find much more information at our web page *www.kordos.com/icaisc*2018.

Although the computational time between different implementations could not be compared directly, as the RadidMiner implementations are generally much slower due to several reasons, its dependence on the number of instances can be. In the first case it was almost $O(n^3)$ what agrees with the theoretical complexity of the DROP algorithms and in the second case it was about $O(nlog(n))$, although we are aware that for a datasets much bigger than that used in the experiments it may asymptotically approach $O(n^2)$, which is the complexity of calculating the distance matrix for k-NN. Of course the time can be reduced in both cases by using more advanced approximate calculations of the distance matrix, but this will not change the relation.

All the tests were run in 10-fold crossvalidation. However, first we randomly changed the order of the instances in each dataset and then we used linear sampling in crossvalidation to ensure that both of the methods use exactly the same training and test subsets - so that the comparison is made on exactly the same data. DROP5 produces the same results with each run on the same data. The NSGA-II based optimization while repeated on the same data displays some variability, but mostly in the number of selected instances and the variability of the obtained accuracy is really very low, as can be seen from the figures how closely particular points adhere to the blue line. The accuracy obtained on the whole dataset is shown in the green line. The final classification algorithm and the inner evaluation algorithm was k-NN with optimal k for each dataset (the k allowing for the highest classification accuracy). Also the same k value was used for DROP5. The optimal k was usually 1 for the datasets, where the achievable classification accuracy was about 99% and growing with the decrease of the accuracy, reaching 7 or more for the accuracies below 80%. However, in the experiments, we limited the maximum k to 7. For the NSGA-II based optimization we used $S = 96$ individuals and from $E = 20$ to 200 epochs - more for larger datasets.

We also trained decision trees of the type described in [27] and the MLP neural networks with the VSS algorithm [28] on the selected subsets to verify how other classifiers would work on the data and the classification accuracies were highly correlated to that obtained with the optimal k k-NN. We have also observed some interesting issues about the optimization of instance selection process for particular final classifier models. However, formulating the conclusions and recommendations require further studies, which will be another topic of our future research (Fig. 1).

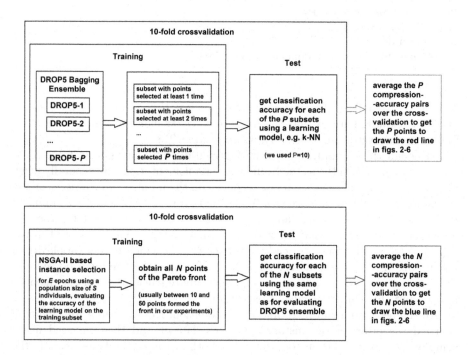

Fig. 1. The experimental setup

5   Conclusions

We discussed the problem of obtaining a Pareto front in the accuracy-compression space in instance selection. We evaluated two approaches: bagging

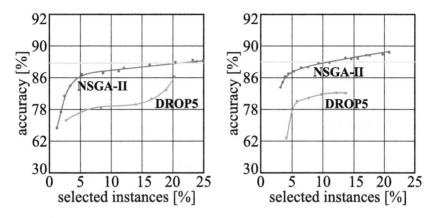

Fig. 2. The Pareto fronts for Balance (left) and Ionosphere (right) datasets. Light green: accuracy without instance selection. (Color figure online)

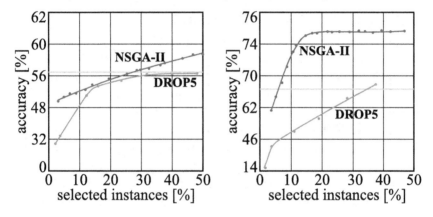

Fig. 3. The Pareto fronts for Yeast (left) and Led7digit (right) datasets. (Color figure online)

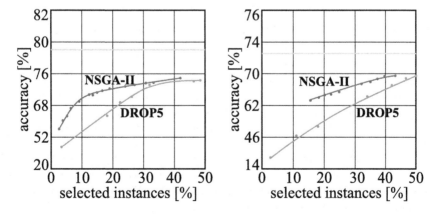

Fig. 4. The Pareto fronts for Bupa (left) and Vehicle (right) datasets. (Color figure online)

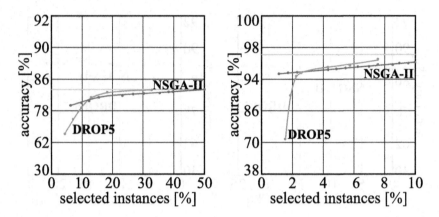

Fig. 5. The Pareto fronts for Magic (left) and Pageblocks (right) datasets. (Color figure online)

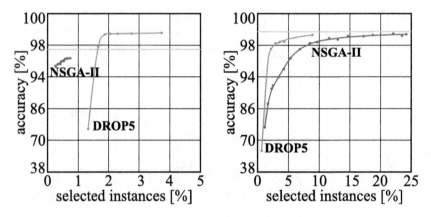

Fig. 6. The Pareto fronts for Twonorm (left) and Penbased (right) datasets. (Color figure online)

of DROP5 algorithms and multi-objective evolutionary optimization based on the NSGA-II algorithm. DROP5 was selected as it was one of the best non-evolutionary instance selection algorithms. The NGSA-II was chosen as one of the best multi-objective evolutionary algorithms. However, we had to adjust it to instance selection problem. The experiments showed the superiority of the evolutionary approach in terms of computational time and also terms of Pareto front position with the exception of the datasets with very high compression (about 1% remaining instances) obtained for high accuracy (above 98%), where the DROP5 ensembles were better. It is the problem with the NSGA-II method that in order to create an equally distributed front, it does not prefer the extreme solutions, so the front gets narrowed from both sides. We are currently working on this problem and we hope it will be solved soon (Fig. 3).

Acknowledgments. This work was supported by the NCN (Polish National Science Center) grant "Evolutionary Methods in Data Selection" No. 2017/01/X/ST6/00202.

References

1. Kordos, M.: Data selection for neural networks. Schedae Informaticae **25**, 153–164 (2017)
2. Arnaiz-González, Á., Blachnik, M., Kordos, M., García-Osorio, C.: Fusion of instance selection methods in regression tasks. Inf. Fusion **30**, 69–79 (2016)
3. Blachnik, M.: Ensembles of instance selection methods based on feature subset. IEEE Proc. Comput. Sci. **35**, 388–396 (2014)
4. Deb, K.: Multi-Objective Optimization using Evolutionary Algorithms. Wiley, Hoboken (2001)
5. Kuncheva, L.I.: Combining Pattern Classifiers: Methods and Algorithms. Wiley, Hoboken (2004)
6. Tomek, I.: An experiment with the edited nearest-neighbor rule. IEEE Trans. Syst. Man Cybern. **6**, 448–452 (1976)
7. Sebban, M., et al.: Stopping criterion for boosting based data reduction techniques: From binary to multiclass problem. J. Mach. Learn. Res. **3**, 863–885 (2002)
8. Garcia-Pedrajas, N.: Constructing ensembles of classifiers by means of weighted instance selection. IEEE Trans. Neural Netw. **20**, 258–277 (2009)
9. Blachnik, M., Kordos, M.: Bagging of instance selection algorithms. In: Rutkowski, L., Korytkowski, M., Scherer, R., Tadeusiewicz, R., Zadeh, L.A., Zurada, J.M. (eds.) ICAISC 2014. LNCS (LNAI), vol. 8468, pp. 40–51. Springer, Cham (2014). https://doi.org/10.1007/978-3-319-07176-3_4
10. García-Pedrajas, N., De Haro-García, A.: Boosting instance selection algorithms. Knowl.-Based Syst. **67**, 342–360 (2014)
11. Wilson, D.R., Martinez, T.R.: Reduction techniques for instance-based learning algorithms. Mach. Learn. **38**, 257–286 (2000)
12. Olvera-López, A., Carrasco-Ochoa, J., Martínez-Trinidad, F., Kittler, J.: A review of instance selection methods. Artif. Intell. Rev. **34**(2), 133–143 (2010)
13. Garcia, S., Derrac, J., Cano, J.R., Herrera, F.: Prototype selection for nearest neighbor classification: Taxonomy and empirical study. IEEE Trans. Pattern Anal. Mach. Intell. **34**(3), 417–435 (2012)
14. Goldberg, D.: Genetic Algorithms in Search, Optimization and Machine Learning. Addison Wesley, Boston (1989)
15. Lobo, F.G., Lima, C.F., Michalewicz, Z.: Parameter Setting in Evolutionary Algorithms. Studies in Computational Intelligence, vol. 54. Springer, Heidelberg (2007). https://doi.org/10.1007/978-3-540-69432-8
16. Konak, A., Coit, D., Smith, A.: Multi-objective optimization using genetic algorithms: A tutorial. Reliab. Eng. Syst. Safety **91**, 992–1007 (2006)
17. Antonelli, M., Ducange, P., Marcelloni, F.: Genetic training instance selection in multiobjective evolutionary fuzzy systems: A coevolutionary approach. IEEE Trans. Fuzzy Syst. **20**(2), 276–290 (2012)
18. Tsaia, C.-F., Eberleb, W., Chu, C.-Y.: Genetic algorithms in feature and instance selection. Knowl.-Based Syst. **39**, 240–247 (2013)
19. Cano, J.R., Herrera, F., Lozano, M.: Using evolutionary algorithms as instance selection for data reduction in KDD: An experimental study. IEEE Trans. Evol. Comput. **7**(6), 561–575 (2003)

20. Cano, J.R., Herrera, F., Lozano, M.: Instance selection using evolutionary algorithms: an experimental study. In: Pal, N.R., Jain, L. (eds.) Advanced Information and Knowledge Processing, pp. 127–152. Springer, London (2004). https://doi.org/10.1007/1-84628-183-0_5

21. Derrac, J., et al.: Enhancing evolutionary instance selection algorithms by means of fuzzy rough set based feature selection. Inf. Sci. **186**, 73–92 (2012)

22. Kordos, M.: Optimization of evolutionary instance selection. In: Rutkowski, L., Korytkowski, M., Scherer, R., Tadeusiewicz, R., Zadeh, L.A., Zurada, J.M. (eds.) ICAISC 2017. LNCS (LNAI), vol. 10245, pp. 359–369. Springer, Cham (2017). https://doi.org/10.1007/978-3-319-59063-9_32

23. Łapa, K., Cpałka, K., Hayashi, Y.: Hybrid initialization in the process of evolutionary learning. In: Rutkowski, L., Korytkowski, M., Scherer, R., Tadeusiewicz, R., Zadeh, L.A., Zurada, J.M. (eds.) ICAISC 2017. LNCS (LNAI), vol. 10245, pp. 380–393. Springer, Cham (2017). https://doi.org/10.1007/978-3-319-59063-9_34

24. Horoba, C., Numann, F.: Benefits and drawbacks for the use of e-dominance in evolutionary multi-objective optimization. In: Genetic and Evolutionary Computation Conference. ACM Press, pp. 641–680 (2008)

25. Alcala-Fdez, J., et al.: KEEL Data-Mining Software Tool and Data Set Repository http://sci2s.ugr.es/keel/datasets.php (2017)

26. Arnaiz-González, Á., Díez-Pastor, J.F., Rodríguez, J.J., García-Osorio, C.: Instance selection for regression: Adapting DROP. Neurocomputing **201**, 66–81 (2016)

27. Kordos, M., Blachnik, M., Perzyk, M., Kozłowski, J., Bystrzycki, O., Gródek, M., Byrdziak, A., Motyka, Z.: A hybrid system with regression trees in steel-making process. In: Corchado, E., Kurzyński, M., Woźniak, M. (eds.) HAIS 2011. LNCS (LNAI), vol. 6678, pp. 222–230. Springer, Heidelberg (2011). https://doi.org/10.1007/978-3-642-21219-2_29

28. Kordos, M., Duch, W.: Variable step search algorithm for MLP training. In: The 8th IASTED International Conference on Artificial Intelligence and Soft Computing, Marbella, pp. 215–220, September 2004

Negative Space-Based Population Initialization Algorithm (NSPIA)

Krystian Łapa[1($\boxtimes$)], Krzysztof Cpałka[1], Andrzej Przybył[1],
and Konrad Grzanek[2,3]

[1] Institute of Computational Intelligence, Częstochowa University of Technology,
Częstochowa, Poland
{krystian.lapa,krzysztof.cpalka,andrzej.przybyl}@iisi.pcz.pl
[2] Information Technology Institute, University of Social Sciences,
90-113 Łódź, Poland
[3] Clark University, Worcester, MA 01610, USA

Abstract. There are many different varieties of population-based algorithms. They are interesting techniques for investigating of the search space of solutions and can be used, among others, to solve optimization problems. They usually start from initialization of a population of individuals, each of which encodes parameters of a single solution to the problem under consideration. After initialization, the preselected individuals are processed in a way that depends on the specifics of the algorithm. Therefore, properly implemented population initialization can significantly improve the algorithm's operation and increase the quality of obtained results. This article describes a new population initialization algorithm. Its characteristic feature is the marginalization of those areas of the search space, in which once localized individuals were assessed as not satisfying. The proposed algorithm is of particular importance for problems in which no information is available that can improve the search procedure (black-box optimization). To test the proposed algorithm simulations were carried out using well-known benchmark functions.

Keywords: Population-based algorithms · Population initialization
Black-box optimization · Negative space

1 Introduction

Population-based algorithms are part of the computational intelligence methods (see e.g. [8,11,23,40]) that can be used in the optimization problems [15,39, 50]. The idea of population-based algorithms relies on an iterative processing of the population of individuals with the purpose of searching for better and better solutions in terms of the adopted fitness function. In such algorithms, each individual from the population encodes parameters of a single solution to the problem under consideration. The modification of the individuals takes place through the use of operators depending on the type of the algorithm. For

© Springer International Publishing AG, part of Springer Nature 2018
L. Rutkowski et al. (Eds.): ICAISC 2018, LNAI 10841, pp. 449–461, 2018.
https://doi.org/10.1007/978-3-319-91253-0_42

example, in the genetic algorithm, a crossover and mutation operators are used [42]. An important step for population-based algorithms is the initialization of the population [22]. Properly selected population supports the operation of the population algorithm, facilitates finding the global optimum and reduces the computational effort related to the execution of the search procedure [4]. This initialization is especially important in case of solving hybrid problems (e.g. choosing the structure and parameters of neuro-fuzzy systems [2,3,13,30,43,45–48,51–55], controllers [34], neural networks [5–7], etc.).

Initialization methods were initially implemented with the purpose of uniform coverage the search space of the considered problem. This was particularly important in Monte Carlo methods [38] and in solving problems with a low number of dimensions. In the paper [36] the authors showed that correct search space coverage requires testing of at least 10 values of parameters for each dimension. Therefore, when using population-based algorithms, it is required to evaluate 10^D individuals of the population (D means the number of dimensions). The evaluation is performed using the fitness function depending on the problem under consideration and it can determine the computational complexity. In the typical optimization problems, the number of dimensions is usually higher than 30. With this assumption, the number of evaluations would be so large that it would significantly limit the use of population algorithms [18]. Hence, to improve the performance of population-based algorithms other methods of population initialization were developed (see e.g. [32,35]), including ways to narrow down the search space (see e.g. [36]). This is especially important because population algorithms can be used to solve wide area of problems, including system identification [17], prediction [24,48], regression [16], classification [37,41] or control problems [29,34].

Various approaches to the classification of initialization methods have been proposed in the literature. In [21] the following criteria are considered: (a) randomness (this group distinguish stochastic and deterministic methods [12,31,33]), (b) compositionality (this group distinguishes a single and multi-step methods [9,28]), and (c) generality (this group distinguishes generic and application specific methods [14,19,36,44]). In [25] a more generic division of the methods was proposed (see Fig. 1) as follows:

– Number generators. This group contains methods used to generate numbers, including both stochastic (see e.g. Lagged Fibonacci Generator [31], Mersenne Twister [9]) and deterministic (see e.g. Halton Sequence [12], Tent Map [22]) methods. In contrast to [21] it consist of methods regardless of their compositionality or generality.
– Number transformations. This group contains methods that can be used for transforming previously generated numbers (any number of generators can be used as a base). It contains space cutting methods with idea of narrowing the search space (see e.g. Center-based Initialization [36]), function transformation methods with idea of concentrating more values in specific areas of search space (see e.g. Box-Muller [49], Hyperbolic Transformation [25], etc.),

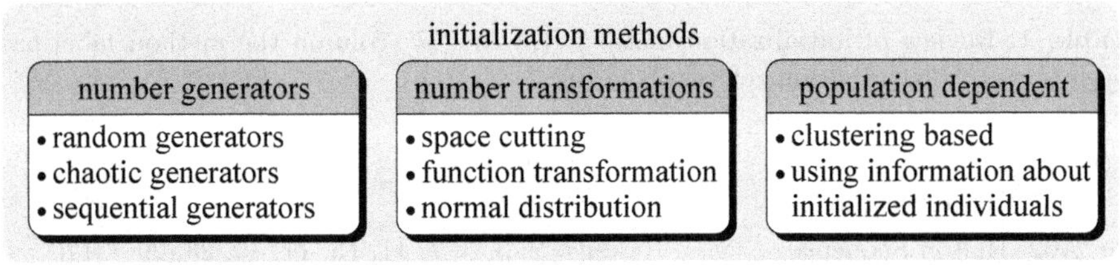

Fig. 1. The division of initial population initialization methods [25].

and normal distribution based methods where normal distribution is used for transformation of values (see e.g. Latin Hypercube Sampling [28]).

- Population dependent. This group contains all methods that rely on individuals in initialization. It is worth mentioning that the use of the individuals is not related to the generation or transformation of numbers, due to that such methods can be used together with any number generators and number transformations [25]. This group contains methods that use clustering of initialized individuals (see e.g. CVT [44]) and methods that use information about already initialized individuals to determine the rest of the initial population (see e.g. AR - Adaptive Randomness [32], SSI - Simple Sequential Inhibition [14]).

In addition to the methods listed above, other worth mentioning initialization techniques are also in use. These include among the others: generation of the initial population of a larger size than used in the population-based algorithm [10], re-initializations in case of achieving convergence of the populations [27], initialization with creating significant differences between individuals [20], etc. As a part of this paper, a wide range of different initialization methods was tested (the details can be found in Table 1). This is especially important because depending on the simulation problem, the use of different initialization methods may give different results [26]. However, the most important element presented in this paper is a new population initialization algorithm based on negative space. Negative space is a part of search space defined by individuals which are not satisfactory in terms of the fitness function and it is determined dynamically during the initialization process. An advantage of the proposed algorithm is the initial rejection of some individuals that are candidates to the population, without the need for additional and time-consuming evaluation. The proposed solution was tested in finding the minims of typical benchmark functions and compared with the results obtained by other initialization methods.

The paper is structured as follows: in the Sect. 2 description of the proposed algorithm is placed, in the Sect. 3 simulation results are presented, while in the Sect. 4 final remarks are included.

Table 1. Review of initialization methods (in the "#" column the method label has been given). Their division refers to the Fig. 1.

#	Method	Parameters	Cite
A01	Halton Sequence	$p = 2, 3$	[12]
A02	Halton Sequence	$p = 2, 3, 5, 7, 11, 13, 17, 19, 23, 29$	[12]
A03	LFG	$\Diamond = +,\ m = 2^{30},\ \mathbf{p} = \begin{bmatrix} 1 & 3 \end{bmatrix}$	[31]
A04	LFG-Knuth-TAOCP	$\Diamond = +,\ m = 2^{30},\ \mathbf{p} = \begin{bmatrix} 37 & 100 \end{bmatrix}$	[31]
A05	LFG-Marsa-LFIB4	$\Diamond = +,\ m = 2^{32},$ $\mathbf{p} = \begin{bmatrix} 55 & 119 & 179 & 256 \end{bmatrix}$	[31]
A06	LFG-Ziff98	$\Diamond = \text{XOR},\ m = 2^{32},$ $\mathbf{p} = \begin{bmatrix} 471 & 1586 & 6988 & 9689 \end{bmatrix}$	[31]
A07	Mersenne Twister 19937	Default	[9]
A08	Mrg32k3a	Default	[9]
A09	Sobol Sequence	Direction vector: new-joe-kuo-6.21201	[12]
A10	Tent Map	$\mu = 1.95$	[22]
B01	Box-Muller	Default	[49]
B02	Center-Based	$C = 70\%$	[36]
B03	Center-Based	$C = 80\%$	[36]
B04	Center-Based	$C = 90\%$	[36]
B05	Exponential Transformation	Default	[25]
B06	Hyperbolic Transformation	Default	[25]
B07	Latin Hypercube Sampling	$N = 2,\ M = 5$	[28]
B08	Latin Hypercube Sampling	$N = 3,\ M = 10$	[28]
B09	Latin Hypercube Sampling	$N = 5,\ M = 12$	[28]
B10	Normal Sampling	$\mu = 0,\ \sigma = 0.2$	[36]
C01	Adaptive Randomness	$k = 3$	[32]
C02	Adaptive Randomness	$k = 10$	[32]
C03	CVT+k-means	$N = 1000,\ iterations = 50$	[44]
C04	CVT+k-means	$N = 5000,\ iterations = 50$	[44]
C05	Dynamic SSI	$\Delta = 0.32,\ \alpha = 0.999,$ $distance = \text{Manhattan}$	[25]
C06	Dynamic SSI	$\Delta = 0.40,\ \alpha = 0.999,$ $distance = \text{Euclidean}$	[25]
C07	Opposite-Based Learning	Default	[35]
C08	Quadratic interpolation	Default	[1]
C09	SSI	$\Delta = 0.27,\ distance = \text{Manhattan}$	[14]
C10	SSI	$\Delta = 0.35,\ distance = \text{Euclidean}$	[14]

2   Description of Proposed Initialization Algorithm

Comments regarding the proposed negative space-based population initialization algorithm (NSPIA) can be summarized as follows:

- It allows eliminating in the initialization process the candidate individuals that are localized in the negative space. The negative space is defined from scratch in every step of the initialization and it is based on the best and worst value of the fitness function in the population. The purpose of such action is to block access to the population of individuals initially assumed as not satisfactory (those that would be very likely to have worse values of the fitness function) and whose evaluation at the stage of the population-based algorithm would not improve the quality of future solutions.
- It allows initializing the population of $Npop$ individuals using $Npop$ evaluation.
- Rejection of certain areas of search space at the stage of population initialization does not exclude finding a solution in them in the operational phase of a population-based algorithm.

Algorithm 1. NSPIA in case of minimization problem

1: Create empty population $\mathbf{P} := \{\emptyset\}$.
2: Get the number of individuals $Npop > 2$ and the threshold value of similarity $\alpha \in (0, 1)$ of individuals in comparison to individuals that define negative space.
3: Create two individuals $\mathbf{X}_1$ i $\mathbf{X}_2$, with random initialized parameters, that encodes the solutions to the considered problem. Evaluate them (set $ff_1 := \mathrm{ff}(\mathbf{X}_1)$ and $ff_2 := \mathrm{ff}(\mathbf{X}_2)$) and add them to the $\mathbf{P}$ (then $|\mathbf{P}| = 2$).
4: **while** $|\mathbf{P}| < Npop$ **do**   ▷ *Create population.*
5:    Set $ffmin := \min\limits_{i=1,2,\dots,|\mathbf{P}|} \{ff_i\}$ and $ffmax := \max\limits_{i=1,2,\dots,|\mathbf{P}|} \{ff_i\}$.
6:    Create candidate individual $\mathbf{X}*$ with random initialized parameters.
7:    **for** $i := 1$ to $|\mathbf{P}|$ **do**   ▷ *Try reject candidate $\mathbf{X}*$.*
8:        **if** $ff_i > 0.5 \cdot (ffmin + ffmax)$ **then**   ▷ *Is $\mathbf{X}_i$ defining negative space?*
9:            **if** $d(\mathbf{X}*, \mathbf{X}_i) < \alpha$ **then**   ▷ *Is $\mathbf{X}*$ close to the $\mathbf{X}_i$?*
10:                Set $p_1 := 1 - \frac{1}{\alpha} \cdot d(\mathbf{X}*, \mathbf{X}_i)$.   ▷ *Set similarity $\mathbf{X}*$ to the $\mathbf{X}_i$.*
11:                Set $p_2 := 1 - \frac{(ffmax - ff_i)}{(0.5 \cdot (ffmax - ffmin))}$.   ▷ *Set normalized similarity of*
                                                                        ▷ $\mathrm{ff}(\mathbf{X}_i)$ *to the $ffmax$.*
12:                Set $p := fp(p_1, p_2)$.   ▷ *Set probability of rejection $\mathbf{X}*$ (see Table 2).*
13:                **if** $p \leq \mathrm{random}(0, 1)$ **then**   ▷ *Should candidate $\mathbf{X}*$ be rejected?*
14:                    **go to** line 6.   ▷ *Reject candidate $\mathbf{X}*$.*
15:                **end if**
16:            **end if**
17:        **end if**
18:    **end for**
19:    Place not rejected candidate $\mathbf{X}*$ at the end of the $\mathbf{P}$.
20:    Evaluate $\mathbf{X}_{|\mathbf{P}|}$ (set $ff_{|\mathbf{P}|} := \mathrm{ff}(\mathbf{X}_{|\mathbf{P}|})$).
21: **end while**

The manner of operation of the proposed population initialization algorithm that uses negative space is presented in the listing 1. After population creation (line 1) and getting algorithm parameters (line 2) algorithm starts. First, to the empty population two individuals are added (line 3). Then the main loop of algorithms begins (line 4) with a purpose to fill the population with new individuals. After determining best and worst values of fitness function from the population (line 5) a new candidate individual is generated (line 6). This individual will be added to the population only when it will not be rejected by any of the individuals (line 7) forming the so-called negative space (line 8). If the candidate individual is close enough to any of individuals forming negative space (line 9) then the attempt to its elimination takes place (line 13). The probability of elimination is increased by the similarity of candidate individual to the negative individual (line 10) and similarity of the negative individual to the worst individual in the population (line 11). Both components are aggregated into the probability of candidate rejection (line 12). It is easy to notice that in the proposed algorithm its harder to reject individuals that differ more from negative space forming individuals. It is also worth emphasizing, that the set of individuals creating the negative space changes with each iteration of the algorithm (line 8). This results in the continuous update of information about the best and worst individuals (line 5), implemented after each population extension (lines: 3, 19 and 20).

3   Simulation Results

The simulations were performed using typical benchmark functions, which are presented in Table 3. For some of these functions, the search space domain was changed in such a way that the minimum of function was not located in the middle of the area (this was a significant facilitation for many initialization algorithms). The populations initialized by different methods (see Tables 1 and 2) have been processed using typical genetic algorithm [42]. The following parameters of this algorithm were set: crossover probability 0.90, mutation probability 0.30, mutation range 0.15, number of iterations 200, number of simulation repetitions 100 (the results were averaged). An overview of how individual initialization algorithms work for a Rosenbrock's valley problem ($D = 2$) is shown in

Table 2. Considered variants of calculating of $d(\cdot)$ and $fp(\cdot)$ (in column "#" a label of variant has been given).

#	α	Distance $d(\cdot)$	Probability $fp(\cdot)$	#	α	Distance $d(\cdot)$	Probability $fp(\cdot)$
N01	0.35	Euclidean	$p_1 \cdot p_2$	N05	0.35	Euclidean	$0.5 \cdot (p_1 + p_2)$
N02	0.40	Euclidean	$p_1 \cdot p_2$	N06	0.40	Euclidean	$0.5 \cdot (p_1 + p_2)$
N03	0.30	Manhattan	$p_1 \cdot p_2$	N07	0.40	Euclidean	$p_1^2 \cdot p_2$
N04	0.35	Manhattan	$p_1 \cdot p_2$	N08	0.45	Euclidean	$p_1^2 \cdot p_2$

Table 3. Considered simulation problems (in column "#" a label of problem has been given, in column D a number of dimension has been given).

#	Function name	D (dimension)	x_i range	Function minimum
F01	Ackley's problem	30	$[-15, 32]$	$0, 0, \ldots, 0$
F02	4th De Jong	30	$[-0.6, 1.3]$	$0, 0, \ldots, 0$
F03	Levy's function	30	$[-10, 10]$	$1, 1, \ldots, 1$
F04	Rastrigin's function	30	$[-2.4, 5]$	$0, 0, \ldots, 0$
F05	Rosenbrock's valley	30	$[-2, 2]$	$1, 1, \ldots, 1$
F06	Sphere model	30	$[-2.5, 5.1]$	$1, 1, \ldots, 1$
F07	Alpine function	60	$[-5.5, 10]$	$0, 0, \ldots, 0$
F08	Inverted cosine wave function	60	$[-1.2, 2.5]$	$0, 0, \ldots, 0$
F09	Pathological function	60	$[-24, 50]$	$0, 0, \ldots, 0$
F10	Schwefel's problem 2.21	60	$[-50, 99]$	$0, 0, \ldots, 0$
F11	Schwefel's problem 2.22	60	$[-4, 10]$	$0, 0, \ldots, 0$
F12	Zakharov's function	60	$[-5.5, 10]$	$0, 0, \ldots, 0$

Fig. 2. This is a pictorial drawing that illustrates well the differences in the way specific algorithms create initial population. As it can be seen on Fig. 2, the initial population obtained on the basis of the initialization algorithm proposed in this paper was located in accordance with the optimum of the benchmark function. The detailed simulations results are presented in Table 4. The simulations conclusions are as follows:

- Depending on the simulation problem the best results were obtained for different initialization methods (see Table 4). This is in line with the conclusions presented in the [26].
- Each benchmark function have different sensitivity to used initialization method (see Table 4). The less sensitive problems are: F05, F06, F10 and F11; and the most sensitive problems were: F01, F03, F07, F09, and F12.
- The following initialization methods gave the most stable results in terms of standard deviation: A04, A06, C08, N01, N02.
- The proposed in this paper negative space-based population initialization algorithm proved to be the most universal and ensured the best results (see Table 4, column *avg*).

Table 4. Simulation results obtained for different initialization methods (see Table 1 and Table 2) (*avg* stands for a normalized and an averaged results for all benchmark functions (see Table 3), *dev* stands for an average standard deviation shown in percentages). For each test function the best 5 results were highlighted and the best result was bolded.

#	F01	F02	F03	F04	F05	F06	F07	F08	F09	F10	F11	F12	avg	dev
A01	2.82	8.4E-5	0.24	24.62	34.46	0.13	31.45	24.77	7.7E-3	12.45	17.47	5.64	0.586	16.2
A02	2.75	8.4E-5	0.19	17.59	38.61	0.13	16.17	17.91	2.4E-3	12.27	18.43	8.25	0.459	15.5
A03	2.88	9.3E-5	0.33	17.43	31.95	0.13	15.62	17.01	4.1E-3	12.36	16.40	3.76	0.352	13.8
A04	2.80	10.2E-5	0.41	**15.87**	32.05	0.14	15.40	15.88	3.9E-3	12.18	16.68	3.37	0.395	12.5
A05	2.73	9.3E-5	0.26	16.39	31.71	0.13	12.31	16.64	2.6E-3	12.26	16.57	2.01	0.266	15.2
A06	2.82	8.2E-5	0.28	17.86	31.50	0.15	15.39	15.86	3.5E-3	12.39	16.36	2.28	0.303	12.9
A08	2.72	7.9E-5	0.32	15.94	31.63	0.14	14.21	15.83	4.0E-3	12.51	16.98	2.54	0.295	14.2
A09	2.77	10.1E-5	0.27	16.97	31.93	0.13	15.55	16.82	2.8E-3	12.47	16.48	9.50	0.369	13.6
A10	2.72	8.2E-5	0.27	20.67	31.83	0.13	21.59	15.76	4.0E-3	12.64	17.18	3.68	0.365	13.8
B01	2.77	8.1E-5	0.32	19.17	32.17	0.14	15.52	14.40	8.2E-3	12.35	16.84	2.51	0.360	13.6
B02	2.75	9.7E-5	0.31	18.07	32.33	0.14	14.73	14.68	7.0E-3	12.16	17.12	1.03	0.354	14.9
B03	2.70	9.0E-5	0.26	17.16	32.18	0.14	14.46	**14.36**	4.5E-3	12.27	16.58	2.44	0.299	15.4
B04	2.69	8.4E-5	0.27	17.01	32.10	0.14	14.35	14.69	3.2E-3	12.35	16.39	2.77	0.278	13.4
B05	2.79	7.3E-5	0.38	17.37	31.90	0.13	14.69	14.91	4.1E-3	12.30	16.56	2.44	0.295	13.2
B06	2.81	7.8E-5	0.28	17.77	32.04	0.13	14.78	14.91	4.2E-3	12.44	16.68	2.91	0.284	15.4
B07	2.86	9.2E-5	0.27	16.01	38.60	0.13	12.13	17.81	**0.7E-3**	12.24	16.60	3.32	0.341	15.9
B08	2.83	9.4E-5	0.30	17.81	31.41	0.13	10.82	16.86	2.2E-3	11.97	16.61	0.34	0.268	15.6
B09	2.80	8.6E-5	0.30	18.08	31.97	0.14	**10.40**	17.03	0.9E-3	11.84	16.75	**0.30**	0.274	15.7
B10	2.81	9.7E-5	0.28	16.84	34.47	0.13	15.03	15.87	3.1E-3	12.21	16.92	2.38	0.329	15.1
C01	2.88	8.9E-5	0.31	16.82	31.58	0.13	14.89	17.12	2.6E-3	12.30	17.11	2.17	0.328	14.2
C02	2.73	11.3E-5	0.19	16.25	36.06	0.15	14.65	17.17	2.4E-3	12.25	17.23	2.29	0.399	15.3
C03	2.80	8.4E-5	0.35	18.93	30.23	0.13	15.01	14.76	5.2E-3	19.09	16.86	1.54	0.391	15.3
C04	2.78	9.4E-5	0.22	17.99	31.56	0.12	16.46	15.15	17.2E-3	12.74	16.58	1.29	0.319	16.9
C05	2.80	9.0E-5	0.32	16.53	33.44	0.14	14.33	16.87	3.3E-3	12.22	16.48	2.07	0.331	14.3
C06	2.73	8.0E-5	0.25	16.77	32.29	0.14	14.89	17.79	3.4E-3	12.27	17.19	3.70	0.312	15.4
C07	2.72	8.5E-5	0.25	16.63	31.81	0.13	14.66	15.84	3.3E-3	**11.82**	17.15	3.29	0.270	13.3
C08	2.72	7.2E-5	0.28	19.04	31.68	0.14	16.39	17.67	2.5E-3	12.27	16.46	4.20	0.308	**12.5**
C09	2.72	8.4E-5	0.24	16.63	32.29	0.14	14.47	18.76	2.8E-3	12.45	16.75	3.09	0.315	15.4
C10	2.77	9.5E-5	0.21	16.73	32.80	0.13	15.07	17.15	4.9E-3	12.15	17.06	4.04	0.330	15.6
N01	2.62	9.3E-5	0.32	16.21	31.93	0.13	14.39	17.13	3.0E-3	12.32	16.65	2.60	0.291	13.1
N02	2.95	**7.1E-5**	0.26	16.37	32.13	0.12	14.93	16.48	2.7E-3	12.28	16.91	3.06	0.250	13.1
N03	2.53	9.7E-5	0.33	17.01	32.20	0.14	14.69	17.04	2.5E-3	12.40	**16.34**	2.69	0.331	13.4
N04	2.66	8.2E-5	0.21	16.04	31.73	**0.11**	14.53	17.28	3.5E-3	12.32	16.52	2.31	**0.191**	14.1
N05	2.68	8.3E-5	0.20	15.97	31.91	0.13	15.03	15.72	2.9E-3	12.50	17.06	4.39	0.262	15.4
N06	3.04	8.5E-5	0.29	16.82	33.81	0.14	14.70	16.11	2.3E-3	12.38	16.77	2.25	0.331	14.1
N07	**1.72**	8.2E-5	0.23	17.26	32.45	0.14	15.04	17.17	3.8E-3	12.34	16.94	2.61	0.237	15.8
N08	2.55	7.6E-5	0.27	16.10	33.28	0.12	14.04	15.45	2.8E-3	12.26	16.54	3.86	0.231	14.2

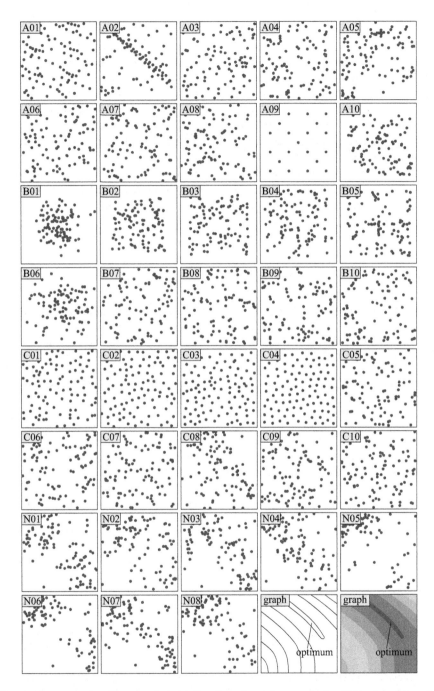

Fig. 2. An example of how each initialization algorithm works for Rosenbrock's valley problem (for $D = 2$) ('graph' shows search space with its optimum).

4   Conclusions

In this paper, a negative space-based population initialization algorithm (NSPIA) was proposed. The idea of the algorithm is the initialization of the population, in which a minimum number of evaluations is performed (equal to the number of initialized individuals). It is based on the negative space, that evolves during the initialization process. In the performed simulations the proposed algorithm proved to be the most universal and ensured the best-averaged results. Moreover, selected variants of the algorithm enabled finding stable solutions in terms of standard deviation.

In the future, it is planned to (a) test the proposed initialization algorithm in combination with different population algorithms and (b) test the proposed initialization algorithm using real test problems in the field of regression and classification.

References

1. Ali, M., Pant, M., Abraham, A.: Unconventional initialization methods for differential evolution. Appl. Math. Comput. **219**(9), 4474–4494 (2013)
2. Bartczuk, Ł., Łapa, K., Koprinkova-Hristova, P.: A new method for generating of fuzzy rules for the nonlinear modelling based on semantic genetic programming. In: Rutkowski, L., Korytkowski, M., Scherer, R., Tadeusiewicz, R., Zadeh, L.A., Zurada, J.M. (eds.) ICAISC 2016. LNCS (LNAI), vol. 9693, pp. 262–278. Springer, Cham (2016). https://doi.org/10.1007/978-3-319-39384-1_23
3. Bartczuk, Ł., Przybył, A., Cpałka, K.: A new approach to nonlinear modelling of dynamic systems based on fuzzy rules. Int. J. Appl. Math. Comput. Sci. **26**(3), 603–621 (2016)
4. Basu, M.: Quasi-oppositional differential evolution for optimal reactive power dispatch. Electr. Power Energy Syst. **78**, 29–40 (2016)
5. Bilski, J., Smoląg, J.: Parallel architectures for learning the RTRN and Elman dynamic neural networks. IEEE Trans. Parallel Distrib. Syst. **26**(9), 2561–2570 (2015)
6. Bilski, J., Wilamowski, B.M.: Parallel learning of feedforward neural networks without error backpropagation. In: Rutkowski, L., Korytkowski, M., Scherer, R., Tadeusiewicz, R., Zadeh, L.A., Zurada, J.M. (eds.) ICAISC 2016. LNCS (LNAI), vol. 9692, pp. 57–69. Springer, Cham (2016). https://doi.org/10.1007/978-3-319-39378-0_6
7. Bilski, J., Wilamowski, B.M.: Parallel Levenberg-Marquardt algorithm without error backpropagation. In: Rutkowski, L., Korytkowski, M., Scherer, R., Tadeusiewicz, R., Zadeh, L.A., Zurada, J.M. (eds.) ICAISC 2017. LNCS (LNAI), vol. 10245, pp. 25–39. Springer, Cham (2017). https://doi.org/10.1007/978-3-319-59063-9_3
8. Bobulski, J.: 2DHMM-based face recognition method. In: Choraś, R.S. (ed.) Image Processing and Communications Challenges 7. AISC, vol. 389, pp. 11–18. Springer, Cham (2016). https://doi.org/10.1007/978-3-319-23814-2_2

9. Bradley, T., Toit, J.D., Tong, R., Giles, M., Woodhams, P.: Parallelization techniques for random numbers generators. In: GPU Computing Gems Emerald Edition, pp. 231–246 (2011)
10. Bramlette, M.F.: Initialization, mutation and selection methods in genetic algorithms for function optimization. In: Proceedings of the Fourth International Conference on Genetic Algorithms, pp. 100–107 (1991)
11. Chang, O., Constante, P., Gordon, A., Singana, M.: A novel deep neural network that uses space-time features for tracking and recognizing a moving object. J. Artif. Intell. Soft Comput. Res. 7(2), 125–136 (2017)
12. Cheng, J., Ruzdzel, M.J.: Computational investigation of low-discrepancy sequences in simulation algorithms for Bayesian networks. In: Proceedings of the Sixteenth Annual Conference on Uncertainty in Artificial Intelligence, pp. 72–81 (2000)
13. Cpałka, K., Rebrova, O., Nowicki, R., Rutkowski, L.: On design of flexible neuro-fuzzy systems for nonlinear modelling. Int. J. Gen Syst 42(6), 706–720 (2013)
14. Diggle, P.J.: Statistical Analysis of Spatial Point Patterns (Mathematics in Biology). Academic Press, Cambridge (1983)
15. Dziwiński, P., Bartczuk, Ł., Tingwen, H.: A method for non-linear modelling based on the capabilities of PSO and GA algorithms. In: Rutkowski, L., Korytkowski, M., Scherer, R., Tadeusiewicz, R., Zadeh, L.A., Zurada, J.M. (eds.) ICAISC 2017. LNCS (LNAI), vol. 10246, pp. 221–232. Springer, Cham (2017). https://doi.org/10.1007/978-3-319-59060-8_21
16. Galkowski, T., Pawlak, M.: Nonparametric Estimation of edge values of regression functions. In: Rutkowski, L., Korytkowski, M., Scherer, R., Tadeusiewicz, R., Zadeh, L.A., Zurada, J.M. (eds.) ICAISC 2016, Part II. LNCS (LNAI), vol. 9693, pp. 49–59. Springer, Cham (2016). https://doi.org/10.1007/978-3-319-39384-1_5
17. Gałkowski, T., Rutkowski, L.: Nonparametric recovery of multivariate functions with applications to system identification. Proc. IEEE 73, 942–943 (1985)
18. Glorot, X., Bengio, Y.: Understanding the difficulty of training deep feedforward neural networks. In: Proceedings of the Thirteenth International Conference on Artificial Intelligence and Statistics, pp. 249–256 (2010)
19. Grefenstette, J.J.: Incorporating problem specific knowledge into genetic algorithms. In: Davis, L. (ed.) Genetic Algorithms and Simulated Annealing, pp. 42–60. Morgan Kaufmann, Los Altos (1987)
20. Iba, H.: Random tree generation for genetic programming. In: Voigt, H.-M., Ebeling, W., Rechenberg, I., Schwefel, H.-P. (eds.) PPSN 1996. LNCS, vol. 1141, pp. 144–153. Springer, Heidelberg (1996). https://doi.org/10.1007/3-540-61723-X_978
21. Kazimipour, B., Li, X., Qi, A.K.: A review of population initialization techniques for evolutionary algorithms. In: Proceedings of 2014 IEEE Congress on Evolutionary Computation (CEC), 6–11 July, pp. 2585–2592 (2014)
22. Kazimipour, B., Li, X., Qin, A.K.: Effects of population initialization on differential evolution for large scale optimization. In: Proceedings of 2014 IEEE Congress on Evolutionary Computation (CEC), 6–11 July, pp. 2404–2411 (2014)
23. Khan, N.A., Shaikh, A.: A smart amalgamation of spectral neural algorithm for nonlinear Lane-Emden equations with simulated annealing. J. Artif. Intell. Soft Comput. Res. 7(3), 215–224 (2017)
24. Liu, H., Gegov, A., Cocea, M.: Rule based networks: an efficient and interpretable representation of computational models. J. Artif. Intell. Soft Comput. Res. 7(2), 111–123 (2017)

25. Łapa, K., Cpałka, K., Hayashi, Y.: Hybrid initialization in the process of evolutionary learning. In: Rutkowski, L., Korytkowski, M., Scherer, R., Tadeusiewicz, R., Zadeh, L.A., Zurada, J.M. (eds.) ICAISC 2017. LNCS (LNAI), vol. 10245, pp. 380–393. Springer, Cham (2017). https://doi.org/10.1007/978-3-319-59063-9_34

26. Maaranen, H., Miettinen, K., Penttinen, A.: On initial populations of a genetic algorithm for continuous optimization problems. J. Glob. Optim. **37**(3), 405–436 (2007)

27. Maresky, J., Davidor, Y., Gitler, D., Aharoni, G.: Selectively destructive re-start. In: Eschelman L.J. (ed.) Proceedings of the 6th International Conference on Generic Algorithms, pp. 144–150. Morgan Kaufmann (1995)

28. McKay, M.D., Beckman, R.J., Conover, W.J.: A comparison of three methods for selecting values of input variables in the analysis of output from a computer code. Technometrics **21**(2), 239–245 (1979)

29. Notomista, G., Botsch, M.: A machine learning approach for the segmentation of driving maneuvers and its application in autonomous parking. J. Artif. Intelli. Soft Comput. Res. **7**(4), 243–255 (2017)

30. Nowicki, R., Scherer, R., Rutkowski, L.: A method for learning of hierarchical fuzzy systems. In: Intelligent Technologies-Theory and Applications, pp. 124–129 (2002)

31. Orue, A.B., Montoya, F., Encinas, L.H.: Trifork, a new pseudorandom number generator based on lagged fibonacci maps. J. Comput. Sci. Eng. **1**(10), 46–51 (2010)

32. Pan, W., Li, K., Wang, M., Wang, J., Jiang, B.: Adaptive randomness: a new population initialization method. Math. Probl. Eng. **2014**, 1–14 (2014)

33. Peng, L., Wang, Y., Dai, G., Cao, Z.: A novel differential evolution with uniform design for continuous global optimization. J. Comput. **7**(1), 3–10 (2012)

34. Przybył, A., Łapa, K., Szczypta, J., Wang, L.: The method of the evolutionary designing the elastic controller structure. In: Rutkowski, L., Korytkowski, M., Scherer, R., Tadeusiewicz, R., Zadeh, L.A., Zurada, J.M. (eds.) ICAISC 2016. LNCS (LNAI), vol. 9692, pp. 476–492. Springer, Cham (2016). https://doi.org/10.1007/978-3-319-39378-0_41

35. Rahnamayan, S., Tizhoosh, H.R., Salama, M.M.A.: A novel population initialization method for accelerating evolutionary algorithms. Comput. Math. Appl. **53**(10), 1605–1614 (2007)

36. Rahnamayan, S., Wang, G.G.: Toward effective initialization for large-scale search spaces. WSEAS Trans. Syst. **3**(8), 355–367 (2009)

37. Riid, A., Preden, J.-S.: Design of fuzzy rule-based classifiers through granulation and consolidation. J. Artif. Intell. Soft Comput. Res. **7**(2), 137–147 (2017)

38. Robert, C.P.: Monte Carlo Methods. Wiley, Hoboken (2004)

39. Rotar, C., Iantovics, L.B.: Directed evolution-a new metaheuristic for optimization. J. Artif. Intelli. Soft Comput. Res. **7**(3), 183–200 (2017)

40. Rutkowski, L.: Non-parametric learning algorithms in time-varying environments. Sig. Process. **182**, 129–137 (1989)

41. Rutkowski, L.: Adaptive probabilistic neural networks for pattern classification in time-varying environment. IEEE Trans. Neural Netw. **15**(4), 811–827 (2004)

42. Rutkowski, L.: Computational Intelligence. Springer, Heidelberg (2008). https://doi.org/10.1007/978-3-540-76288-1

43. Rutkowski, L., Cpałka, K.: Compromise approach to neuro-fuzzy systems. In: Proceedings of the 2nd Euro-International Symposium on Computation Intelligence. Frontiers in Artificial Intelligence and Applications, vol. 76, pp. 85–90 (2002)

44. Saka, Y., Gunzburger, M., Burkardt, J.: Latinized, improved LHS, and CVT point sets in hypercubes. Int. J. Numer. Anal. Model. **4**(3–4), 729–743 (2007)

45. Scherer, R.: Multiple Fuzzy Classification Systems. Springer, Heidelberg (2012). https://doi.org/10.1007/978-3-642-30604-4

46. Scherer, R., Rutkowski, L.: A fuzzy relational system with linguistic antecedent certainty factor. In: Rutkowski, L., Kacprzyk, J. (eds.) Neural Networks and Soft Computing, pp. 563–569. Springer, Heidelberg (2003). https://doi.org/10.1007/978-3-7908-1902-1_86

47. Scherer, R., Rutkowski, L.: Neuro-fuzzy relational classifiers. In: Rutkowski, L., Siekmann, J.H., Tadeusiewicz, R., Zadeh, L.A. (eds.) ICAISC 2004. LNCS (LNAI), vol. 3070, pp. 376–380. Springer, Heidelberg (2004). https://doi.org/10.1007/978-3-540-24844-6_54

48. Scherer, R., Rutkowski, L.: Connectionist fuzzy relational systems. In: Halgamuge, S.K., Wang, L. (eds.) Computational Intelligence for Modelling and Prediction, pp. 35–47. Springer, Heidelberg (2005). https://doi.org/10.1007/10966518_3

49. Shinzato, T.: Box Muller Method (2007)

50. Yang, S., Sato, Y.: Swarm intelligence algorithm based on competitive predators with dynamic virtual teams. J. Artif. Intell. Soft Comput. Res. **7**(2), 87–101 (2017)

51. Zalasiński, M.: New algorithm for on-line signature verification using characteristic global features. In: Wilimowska, Z., Borzemski, L., Grzech, A., Świątek, J. (eds.) Information Systems Architecture and Technology: Proceedings of 36th International Conference on Information Systems Architecture and Technology – ISAT 2015 – Part IV. AISC, vol. 432, pp. 137–146. Springer, Cham (2016). https://doi.org/10.1007/978-3-319-28567-2_12

52. Zalasiński, M., Cpałka, K.: New algorithm for on-line signature verification using characteristic hybrid partitions. In: Wilimowska, Z., Borzemski, L., Grzech, A., Świątek, J. (eds.) Information Systems Architecture and Technology: Proceedings of 36th International Conference on Information Systems Architecture and Technology – ISAT 2015 – Part IV. AISC, vol. 432, pp. 147–157. Springer, Cham (2016). https://doi.org/10.1007/978-3-319-28567-2_13

53. Zalasiński, M., Cpałka, K., Hayashi, Y.: A method for genetic selection of the most characteristic descriptors of the dynamic signature. In: Rutkowski, L., Korytkowski, M., Scherer, R., Tadeusiewicz, R., Zadeh, L.A., Zurada, J.M. (eds.) ICAISC 2017. LNCS (LNAI), vol. 10245, pp. 747–760. Springer, Cham (2017). https://doi.org/10.1007/978-3-319-59063-9_67

54. Zalasiński, M., Cpałka, K., Er, M.J.: Stability evaluation of the dynamic signature partitions over time. In: Rutkowski, L., Korytkowski, M., Scherer, R., Tadeusiewicz, R., Zadeh, L.A., Zurada, J.M. (eds.) ICAISC 2017. LNCS (LNAI), vol. 10245, pp. 733–746. Springer, Cham (2017). https://doi.org/10.1007/978-3-319-59063-9_66

55. Zalasiński, M., Łapa, K., Cpałka, K., Saito, T.: A method for changes prediction of the dynamic signature global features over time. In: Rutkowski, L., Korytkowski, M., Scherer, R., Tadeusiewicz, R., Zadeh, L.A., Zurada, J.M. (eds.) ICAISC 2017. LNCS (LNAI), vol. 10245, pp. 761–772. Springer, Cham (2017). https://doi.org/10.1007/978-3-319-59063-9_68

Deriving Functions for Pareto Optimal Fronts Using Genetic Programming

Armand Maree, Marius Riekert, and Mardé Helbig[(✉)]

Department of Computer Science, University of Pretoria, Pretoria, South Africa
mhelbig@cs.up.ac.za

Abstract. Genetic Programming is a specialized form of genetic algorithms which evolve trees. This paper proposes an approach to evolve
an expression tree, which is an N-Ary tree that represents a mathematical equation and that describes a given set of points in some space.
The points are a set of trade-off solutions of a multi-objective optimization problem (MOOP), referred to as the Pareto Optimal Front (POF).
The POF is a curve in a multi-dimensional space that describes the
boundary where a single objective in a set of objectives cannot improve
more without sacrificing the optimal value of the other objectives. The
algorithm, proposed in this paper, will thus find the mathematical function that describes a POF after a multi-objective optimization algorithm
(MOA) has solved a MOOP. Obtaining the equation will assist in finding
other points on the POF that was not discovered by the MOA. Results
indicate that the proposed algorithm matches the general curve of the
points, although the algorithm sometimes struggles to match the points
perfectly.

Keywords: Multi-objective optimization · Pareto optimal front
Genetic programming

1 Introduction

Optimization is a field in computer science that deals with finding the best
possible value that satisfies some conditions. The best value can either be the
maximum or the minimum of a specified optimization problem.

Real world problems often do not have a single objective that needs to be satisfied, but is rather described by multiple conflicting objectives [1]. For instance,
the price of a car versus comfort. Choosing a cheaper car results in a decrease
in comfort and vice versa. Therefore, these two objectives are in conflict where
improved value in one objective results in a less optimal value in the other
objective.

These problems, referred to as multi-objective optimization problems
(MOOPs), can successfully be solved by computational intelligence (CI) algorithms [2–4]. A popular and efficient CI approach for finding a set of optimal
solutions (a set of points where one objective cannot be optimized more without

© Springer International Publishing AG, part of Springer Nature 2018
L. Rutkowski et al. (Eds.): ICAISC 2018, LNAI 10841, pp. 462–473, 2018.
https://doi.org/10.1007/978-3-319-91253-0_43

sacrificing the other objective, referred to as a Pareto optimal front (POF)), is the genetic algorithm (GA) [5]. GAs, as well as other CI algorithms, have sets of individuals that can traverse the search space [1]. Each individual, or group of individuals, can therefore find a different solution along the POF [1].

A Pareto optimal point (POP) is a point in the search space where there does not exist another point which dominates it [6]. Assuming minimization, a point, $\mathbf{x}^*$, dominates another point, $\mathbf{x}$, if and only if (iff):

$$\forall f \in \mathbf{F}, \forall \mathbf{x} \in \mathbf{X}, \exists g \in \mathbf{F} \text{ such that } f(\mathbf{x}^*) \leq f(\mathbf{x}) \text{ and } g(\mathbf{x}^*) < g(\mathbf{x}) \qquad (1)$$

where $\mathbf{F}$ is the set of objective functions. A complete set of these POPs is then collectively called the POF in the objective space and the Pareto optimal set (POS) in the decision variable space. Figure 1 shows an example of a POF.

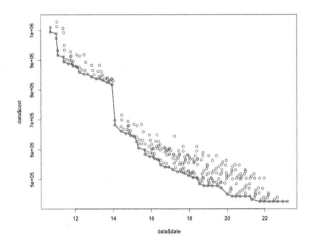

Fig. 1. Pareto optimal front example (https://i.stack.imgur.com/5KLU2.png)

Algorithms that solve a MOOP often produce an incomplete set of POPs, rather than a mathematical formula that represents the entire POF [7]. The objective of this research is thus to find a mathematical formula that describes the points on the POF. This is attempted using genetic programming (GP) to evolve an expression tree. This expression tree can then be used to find other points on the POF that was not found by the optimization algorithm. Furthermore, the expression tree can be used to formulate new MOOPs for which the solution, or POF, is unknown.

2   Background

This section provides background information about the algorithms or approaches used to find equations in Sect. 2.1 and genome representation in Sect. 2.2.

2.1   Approaches

Even though some tools exist to find equations from provided points, such as GeneticSharp [8], not much research has been conducted to find equations of a MOOP. The existing tools are also not well suited for solutions where a POF equation contains summations such as Eq. 2:

$$f(x) = \sum_{i=1}^{5} x^i \qquad (2)$$

Gene expression programming (GEP) is an algorithm that finds a function that describes a given set of data points in a multi-dimensional search space [9]. These function finding algorithms will be adapted for MOOPs and applied to discover the POFs based on provided POPs.

Another technique to represent the trees is to use arrays [9]. Issues can arise when dealing with problems that have long equations and thus require large trees. However, in these cases a tree of unbounded size can be used if the size of the tree is regulated by adding a penalty to larger trees [10]. Another approach is to only grow the tree as soon as no better solution can be found [10].

GP is a specialized version of a GA used for evolving trees [11]. Various techniques are used to evolve tree structures, including various mutation and cross over techniques used in GAs [12]. Through the evolutionary mechanics and the dynamic nature of the algorithm, GP also has the ability to grow solutions to an appropriate size without having to make assumptions about the size of the solution in advance. The pseudo-random components of the algorithm also affords the opportunity to explore areas of the search space which might not be considered by other methods for approximating functions.

Expression trees used in GP consist of two node types. The first type is function nodes, or non-terminal nodes, and the second type is leaf nodes, or terminal nodes [11]. Function nodes refer to mathematical functions, such as addition, subtraction, multiplication, etc. Each of these operations require one or more input parameters, leading to a subtree. Leaf nodes consist of variables and constants, such as x_i or 10, and do not have a subtree.

2.2   Genome Representation

Each individual (genome) is represented through two trees. The trees consist of nodes classified into operators and operands. The list of supported operators are: addition (+), subtraction (−), multiplication (*), division (/), square root (#), power (^) and summation ($).

An individual is represented by two trees corresponding to the left hand side (LHS) and right hand side (RHS) of the equation. In this paper, the LHS RHS consist of only a single variable (y) and a multi-node tree possibly containing the x variable respectively. This paper also only focuses on the two dimensional space, although the principle can be extended to more dimensions.

Figure 2 shows an example of an individual's genome. This specific genome represents the equation $y = x^{x+2}$. In the implementation of the algorithm, nodes are represented by objects and association is used to form the tree structure.

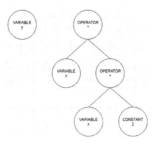

Fig. 2. Tree example

3   Algorithm

This section discusses the specific details of the proposed algorithm, starting with initialization in Sect. 3.1, followed by the description of a single iteration in Sect. 3.2. Sections 3.3, 3.4, and 3.5 discuss the selection, crossover and mutation operators respectively. The stopping conditions are discussed in Sect. 3.6. Some optimization aspects, namely clean up and the introduction of new individuals, are discussed in Sects. 3.7 and 3.8. Finally, Sect. 3.9 addresses the fitness calculation of an individual.

3.1   Initialization

All the points and variables are read in from a file. The user has freedom to choose the variables (a and b, x and y, etc.). The variables are marked as reserved and cannot be used by the algorithm at a later stage.

Next, all the individuals are created. The number of individuals is selected based on a control variable. Upon creation, each individual assigns the y variable to the LHS of the equation and generates a random two-level deep tree, which it assigns to the RHS. This second tree consists of a single operator node at the root and leaf nodes (variable or constant node) as children. The number of children depend on the number of operands the operator requires.

3.2   One Iteration

A single iteration contains a few steps: Firstly a new generation of individuals has to be created. This is done by selecting parents (Sect. 3.3) from the population and breeding them to create new children by means of crossover (Sect. 3.4) and mutation (Sect. 3.5). This breeding process is continued until the new generation size is the same size as the parent generation. Once this is completed, the new generation completely replaces the previous generation. There is however one exception to this rule. In order to preserve the best individual's genes, this individual is slightly mutated and allowed to survive to the next generation. This process is continued until the stopping conditions have been reached (Sect. 3.6).

At specific intervals during the algorithm run, other tasks occur to assist with the performance of the algorithm. The first task is a clean up to reduce tree sizes and the second is the introduction of new individuals. These tasks will be discussed in more detail in Sects. 3.7 and 3.8 respectively.

At the end of each iteration the best individual's fitness is calculated (Sect. 3.9) and returned.

3.3   Selection Operator

The selection of individuals for the crossover process, used to produce new individuals, are done using tournament selection. Tournament selection is the process of selecting a few individuals from the population and placing them in a pool [13]. The two best individuals in this pool are then selected as the parents and will be bred. There is a control parameter called the *selection pressure*. This refers to the likelihood of an individual to be selected [13]. The higher the selection pressure, the less likely weaker individuals are to be selected [13]. This is regulated by the pool size. If the pool size is 100% of the population then the two strongest individuals will always be selected (elitism), thus a high selection pressure. If the pool size is two, then all the individuals have an equal chance of being selected as parents (this is basically random selection). For this study, the pool size was set to 20% of the population of 30 individuals.

3.4   Crossover

The *One-Point Recombination* concept from [9] was adapted to the trees used in this algorithm. A node in each of the parents is selected at random. This node, along with its subtree, is then swapped with the selected node and subtree of the other parent. The two new trees that are formed are the children nodes. Refer to Fig. 3 for an example. In the figure the two top trees are the parents and the two bottom trees are the children.

3.5   Mutation

Two types of mutation were used. The first is a modification and the second is an adaption. The differentiation lies within the degree of change a genome will experience (meaning a minor change in values or a large change in structure). While iterating over all the genes (nodes) in the genome (tree) of an individual there is a chance that the node will be mutated (either modified or adapted). This percentage is a control parameter called the mutation probability and was set to 30% during the experiments. The probability of a node experiencing a modification versus an adaption was set to 50%.

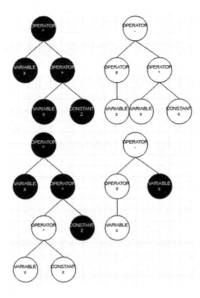

Fig. 3. Crossover example

Modification. This is a simple process which replaces the node (and consequently the entire subtree) with a new random node. The node can either be a constant, a variable or an operator. If the node changes to an operator, a random subtree is generated. This subtree is only two layers deep.

Adaption. The purpose of this mutation type is to allow an individual to fine-tune constant nodes. In other words, only constant nodes are allowed to experience any change and the change can only be based on the node's value. There exists a 50% chance of the value increasing and a 50% chance of the value decreasing. Equations 3 and 4 show how the value is either increased or decreased respectively. The equations update the value based on the difference between the fitness of the current individual (f_{x_i}) and the fitness of the best individual (f_{x_b}).

$$newValue = oldValue + scale(f_{x_i} - f_{x_b})/f_{x_b} \qquad (3)$$

$$newValue = oldValue - scale(f_{x_i} - f_{x_b})/f_{x_b} \qquad (4)$$

3.6 Stopping Conditions

Since these types of algorithms might never find a perfect solution, there is usually some condition that causes the algorithm to stop executing and return the current best solution. For the proposed algorithm, the user can specify a maximum number of iterations where no change has been observed. Thus, the algorithm will stop executing if the fitness of the best individual does not improve after the specified number of iterations have been reached.

3.7   Clean Up

In order to assist the algorithm in avoiding redundant nodes in the trees, a clean up function has been implemented. This is required because the depth of the tree severely impacts the performance of the algorithm.

After a predetermined number of iterations have elapsed, each individual in the population undergoes a clean up, where unnecessary constant nodes are removed as these tend to form frequently in the tree. An example of this is that the equation $y = x + 2 + 2 + 2$ can be written as $y = x + 6$, which would reduce the depth of the tree.

3.8   New Individuals

Stagnation of the gene pool is prevented by introducing new individuals at predetermined intervals during the execution of the algorithm. This process involves killing off the weakest 25% of individuals (called the genocide percentage) and replacing them with completely new individuals. These new individuals are generated the same way as described in the initialization section (Sect. 3.1).

3.9   Fitness Calculation

The fitness of an individual indicates how accurate an individual's expression tree represents the data points provided in the input. This is calculated by substituting the input variables of each point into the tree and calculating the absolute value of the difference between the LHS and the RHS. The fitness is then the summation of this difference for all the points. A value of 0 would thus mean that an equation has been found that describes the points perfectly. The fitness of an individual should thus be minimized.

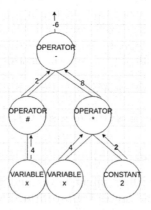

Fig. 4. Fitness example

The fitness calculation works as follows:

- request the root node to calculate its value based on the provided points
- root node asks its children for their fitness
- this process carries on until a leaf node is reached
- if leaf node is a constant it returns its value
- if leaf node is a variable then the corresponding value of the provided point is returned
- operator nodes perform the required operation after receiving the values of their children.

Figure 4 illustrates how the values are sent from the bottom of the tree to the top if a point is provided where $x = 4$.

4   Summation Operator

The summation operator functions differently than the other operators: firstly a recursive process is used calculating its value, and secondly it introduces a new variable that is exclusively available to the nodes in its subtree.

Section 4.1 discusses the maximum and minimum values for the counter variable (discussed in Sect. 4.2) of the summation operator. The calculation of a subtree that contains a summation operator is discussed in Sect. 4.3.

4.1   Start and End Value

For this study, the start and end value for the sigma summation operator was fixed at 1 and 10 respectively. Future research will explore the possibility of evolving these values as if they were normal subtrees.

4.2   Counter Variable

Since multiple summation operators can exist in one subtree (refer to Eq. 5) it is important that the counter variable name has to be managed dynamically. Whenever a new summation operator node is created, it requests a list of variables already used in the parent tree. It then allocates the first available variable following the pattern i_k, where k refers to the kth summation operator in this subtree. In other words, the first summation operator uses the counter variable i_0, the second uses i_1 and so on.

$$\Sigma_{i=1}^{10} \Sigma_{j=1}^{10} i x^j \tag{5}$$

This also changes the way the variable nodes work, since a variable node can be a counter variable for a summation operator. In order to overcome this hurdle, when a variable is created it recursively requests its parent for a list of variables allocated in the subtree. This list contains all the variables in the input (only x is applicable at the current stage of the program) and also all the dynamic counter variables. The variable node will then randomly choose one variable from this list.

4.3   Value Calculation

The value for the subtree needs to be calculated differently compared to the other operators. The operator iteratively requests its subtree to calculate its value and sums each returned result. This means that the variables, along with their values, have to be passed down the tree. The root node starts by passing the values for the current point being evaluated to its children. From this point forward, all the nodes pass the values down the tree so that the variable nodes can obtain their values. On the way down, a summation node may be crossed. At this point the summation node places its counter variable in the list, along with the first value, and passes it down the tree. This process is repeated for each iteration of the summation operator's loop.

5   Results

This section discusses the results that were obtained from the study. The following equations were tested:

$$y = x^2 + 100 \tag{6}$$

$$y = 20 \times e^{-x} + 1 \tag{7}$$

$$y = 50 + \Sigma_{i=1}^{10} x^{\frac{i}{10}} \tag{8}$$

The following control parameters were used:

- Population size: 30
- Maximum no improvement epochs: 100000
- Mutation probability: 0.3
- Modify probability: 0.5
- Genocide percentage: 0.25

The obtained results are presented in Table 1. The figures listed in the table indicates how well or poorly the equation produced by the algorithm matches up with the points that were provided.

The input points are the points in the two dimensional space that were provided as input to the program. The equation that the algorithm managed to calculate is displayed in the "Produced Equation" column. The figure column indicates the figure number that compares the "Input Points" and the "Produced Equation" columns, i.e. a visual illustration of how well the equation matches the provided points.

Looking at the result for Eq. 6 which is shown in Fig. 5(a), it is clear that the algorithm closely matches the given points. The equation is also nearly identical to Eq. 6. The result for Eq. 7 (shown in Fig. 5(b)), matches the curve that the

Table 1. Results for the equations

Equation number	Fitness	Input points (x, y)	Produced equation	Figure
6	10.07	(0, 101) (1, 102) (2, 104) (3, 108) (4, 116) (5, 132) (6, 164)	$y = x^{1.9931} + 99.5904$	5(a)
7	44.03	(−4, 1093) (−3, 403) (−2, 149) (−1, 55) (0, 21) (1, 8) (2, 4) (3, 2) (4, 1)	$y = 1.67^{-(4.3578+2x)}$	5(b)
8	44.03	(0, 50) (10, 94) (20, 123) (30, 150) (40, 176) (50, 201)	$y = 58.7976 + x + \left(\Sigma_{i=1}^{10} x\right)^{0.7097}$	5(c)
Freely plotted points	6.84	(0.1, 100) (1, 10) (2, 5) (3, 3) (4, 1) (5, 0)	$y = \frac{10}{x}$	5(d)

points form, but is offset in the negative x direction. It is interesting to note that this result and also the result for Eq. 8 (Fig. 5(c)) show an Equation that differs in form from the original equation, but the curve that it forms represents the points fairly well. Based on the result in Fig. 5(d), which corresponds to the freely plotted points, it can be seen that as x approaches 0 the curve matches the points well, but accuracy decreases as x increases.

6   Conclusion and Future Work

From the results and analysis of it we can see the algorithm manages to find equations for the points in some situations. It manages to find the general shape of

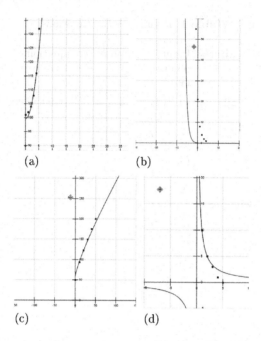

Fig. 5. Results obtained for Eq. 6 (a), Eq. 7 (b), Eq. 8 (c) and Randomly plotted points (d)

the simple equations, although it still struggles finding an equation that matches the points perfectly.

An interesting thing to note is that although it does not necessarily find an equation that has the same structure as the test equation, it can find an equation that still match the points. As more points are provided the number of equations that matches the points decrease. This will of course increase the run time but will assist the algorithm to find the correct equation.

One advantage of the algorithm is the duration of the program. For simple equations like the ones tested here the program exits within a few minutes.

With some additional tweaks to the way the algorithm functions there might be potential for this algorithm to solve more complex equations.

Since this paper only tested some very basic equations in an attempt to find an algorithm that can find curves (and also incorporate the summation operator) the next step would be firstly allow the summation operator to have its parameters adapted dynamically. The algorithm also needs to be tested against more complex equations in order to evaluate its performance and potential even further.

Acknowledgements. This work is based on the research supported by the National Research Foundation (NRF) of South Africa (Grant Number 46712). The opinions, findings and conclusions or recommendations expressed in this article is that of the author(s) alone, and not that of the NRF. The NRF accepts no liability whatsoever in this regard.

References

1. Konak, A., Coit, D.W., Smith, A.E.: Multi-objective optimization using genetic algorithms: a tutorial. Reliab. Eng. Syst. Saf. **91**(9), 992–1007 (2006)
2. Ganesan, T., Elamvazuthi, I., Vasant, P.: Multiobjective design optimization of a nano-CMOS voltage-controlled oscillator using game theoretic-differential evolution. Appl. Soft Comput. **32**, 293–299 (2015)
3. Shirazi, A., Najafi, B., Aminyavari, M., Rinaldi, F., Taylor, R.A.: Thermal-economic-environmental analysis and multi-objective optimization of an ice thermal energy storage system for gas turbine cycle inlet air cooling. Energy **69**, 212–226 (2014)
4. Courteille, E., Mortier, F., Leotoing, L., Ragneau, E.: Multi-objective robust design optimization of an engine mounting system. Technical report, SAE Technical Paper (2005)
5. Deb, K., Sundar, J.: Reference point based multi-objective optimization using evolutionary algorithms. In: Proceedings of the 8th Annual Conference on Genetic and Evolutionary Computation, pp. 635–642. ACM (2006)
6. Marler, R.T., Arora, J.S.: Survey of multi-objective optimization methods for engineering. Struct. Multidiscip. Optim. **26**(6), 369–395 (2004)
7. Wang, F., Lai, X., Shi, N.: A multi-objective optimization for green supply chain network design. Decis. Support Syst. **51**(2), 262–269 (2011)
8. Giacomelli, D.: Geneticsharp. https://github.com/giacomelli/GeneticSharp. Accessed 02 Feb 2018
9. Ferreira, C.: Gene expression programming in problem solving. In: Roy, R., Köppen, M., Ovaska, S., Furuhashi, T., Hoffmann, F. (eds.) Soft Computing and Industry, pp. 635–653. Springer, London (2002). https://doi.org/10.1007/978-1-4471-0123-9_54
10. Banzhaf, W., Nordin, P., Keller, R.E., Francone, F.D.: Genetic Programming: An Introduction, vol. 1. Morgan Kaufmann, San Francisco (1998)
11. Engelbrecht, A.P.: Computational Intelligence: An Introduction. Wiley, Hoboken (2007)
12. Koza, J.R.: Genetic Programming: On the Programming of Computers by Means of Natural Selection, vol. 1. MIT Press, Cambridge (1992)
13. Miller, B.L., Goldberg, D.E., et al.: Genetic algorithms, tournament selection, and the effects of noise. Complex Syst. **9**(3), 193–212 (1995)

Identifying an Emotional State from Body Movements Using Genetic-Based Algorithms

Yann Maret[1,2], Daniel Oberson[2], and Marina Gavrilova[1(✉)]

[1] University of Calgary, Calgary, Alberta, Canada
mgavrilo@ucalgary.ca
[2] The School of Engineering and Architecture of Fribourg, Fribourg, Switzerland

Abstract. Emotions may not only be perceived by humans, but could also be identified and recognized by a machine. Emotion recognition through pattern analysis can be used in expert systems, lie detectors, medical emergencies, as well as during rescue operations to quickly identify people in distress. This paper describes a system capable of recognizing emotions based on the arm movement. Features extracted from 3D skeleton using Kinect sensor are classified by five commonly used machine learning techniques: K nearest neighbors, SVM, Decision tree, Neural Network and Naive Bayes. A genetic algorithm is then invoked to find the best system parameters to achieve the higher recognition rate. The system achieved 98.96% average accuracy on the experimental dataset.

Keywords: Kinect sensor · Genetic algorithm · Machine learning · Biometric
Activity recognition · Image processing · Risk assessment

1 Introduction

Our society has reached a critical mass in a race for automation of traditional human only tasks. From smart fridges that can remind of an expiration date of a product, to self-driving autonomous cars, our population relies more and more on machine learning in an effort to build an optimally run and efficient society. Machine learning has evolved to one of the most sophisticated and widespread methodologies, with applications in security, image processing, medicine, biometrics, commerce and social media [5, 7, 16, 20, 21]. Human emotions remain one of the last frontiers to concur with machine learning [14]. Only very recently researchers started to question whether computer systems can identify human emotions, actions or intent, and to perceive a possible risk [15]. This critical information could be then used in a situation awareness systems to alert the operator, to coordinate a rescue operation by providing immediate help to an individual with more severe injuries, or to evaluate the condition of people waiting in check-out lanes at a supermarket.

This paper develops a Kinect-based system capable of recognize four gestures of a person: friendly, distress, annoyed and neutral, based on the arm movements. The arm movement is tracked using machine learning system, which relies on the body skeleton extracted by Kinect camera. This paper makes two important contributions to the body of literature on machine learning and its applications in a biometric domain. First of all,

© Springer International Publishing AG, part of Springer Nature 2018
L. Rutkowski et al. (Eds.): ICAISC 2018, LNAI 10841, pp. 474–485, 2018.
https://doi.org/10.1007/978-3-319-91253-0_44

it develops a new framework for using Kinect gait sensor to identify a person in distress, an annoyed person, or a friendly person, through observation of a person's upper body arm movement. The classification relies on five supervised and unsupervised machine-learning methods, including KNN, SVM, Decision tree, Neural Network and Naive Bayes methods. Secondly, this paper proposes to use genetic algorithm to arrive to an optimal weight for the ensemble of classifiers, which allows achieving a high recognition of a person's state from the arm movement. The experiments conducted on the data set of eight subjects demonstrate a high recognition performance of the method. The follow up research conducted presently is focused on the embedded algorithm for fast and efficient emotion from body movement recognition in real-time.

2 Literature Review

The recent release of Kinect sensor has created new opportunities to address the problems related to real-time motion analysis, leading to a spike in the interest in gait recognition using Kinect [5]. In addition to different data streams that can be obtained from Kinect (RGB, depth, audio, etc.), it can also construct a 3D virtual skeleton from human body and track it in real-time, rendering the traditional video pre-processing tasks, such as background modeling and silhouette extraction, unnecessary. As a result, some of the recent gait recognition methods found in literature utilize the computationally inexpensive real-time depth sensing and skeleton tracking in order to model the gait signature. One of the early works on Kinect-based gait analysis was conducted in 2012 [17], where the authors utilized features extracted from the lower body skeletal joints to construct the feature descriptor. An alternative approach, developed the same year [18], introduced thirteen biometric features for gait recognition, including height, legs, torso, upper arms, forearms, step- length, and speed. Researchers also studied the differences in the skeletal joint positions between consecutive frames to model a human gait [19].

Kinect sensor can be used not only for biometric gait recognition, but also for activity recognition. Although the tracked skeletal joints obtained from the Kinect are relatively noisy, the computationally inexpensive nature of the Kinect real-time skeleton tracking has contributed to popularity of the recent action recognition methods, that utilize Kinect for efficient motion representation. Commonly used approaches frequently rely on machine learning for feature classification tasks. For example, the extracted posture-based visual words were trained using a hidden Markov model (HMM) to identify 10 different activities in [8]. A method based on accumulated motion energy (AME), which combined both static and dynamic features extracted from body joints, was proposed in [9]. A non-parametric Naïve Bayes Nearest Neighbor (NBNN) classifier for action recognition was proposed in [10]. New features based on Joint-Relative Angle (JRA) and Join-Relative Distance (JRD) were trained using three different classifiers–SVM, Decision trees, and K-NN in [6].

Despite Kinect versatility, it has been rarely used outside gait and activity recognition. While there has been a surge in a number of action recognition systems developed based on Kinect sensor technology, recognizing emotion or performing crowd observation from the body movement has been considered in the scientific discourse only very recently [11–

13]. Its been noted that human body carries messages in the form of motions of the body joints [1–3, 11]. Researchers have considered stylized dance movements to recognize human emotion, where anger has been identified as an easily distinguished emotion [12, 13]. Very recently, the first embedded hardware device that monitors the number of people inside the confined space using Kinect sensor was developed in [4].

This paper is one of the first studies on the potential use of Kinect in situation awareness systems to provide additional cues in response to an emergency situation. First of all, we develop a new framework for using Kinect gait sensor to identify a person in distress, through observation of a person's upper body arm movement. The classification relies on five supervised and unsupervised machine-learning methods, including KNN, SVM, Decision tree, Neural Network and Naive Bayes methods. Secondly, we propose to use evolutionary computing to evolve an optimal weight for each of the classifiers, which allows to achieve a high recognition of a person's state from the arm movement. The experiments conducted on the data set of eight subjects demonstrate a high recognition performance of the method. The methodology is presented in the next section.

3   Methodology

Figure 1 depicts the flowchart of the proposed method for human arm movement recognition based on Kinect sensor captured upper body motion.

3.1   Feature Extraction

The Kinect version 2 with the Microsoft SDK can keep track of 25 body joints, labeled in Fig. 2. From these joints, features for arm movement recognition are extracted.

In the developed system, emotions are recognized based on the right arm movement. With only the top body joints, extracted features are sufficient to identify the emotion. Extracted features includes raw joints positions in X and Y coordinates, distances

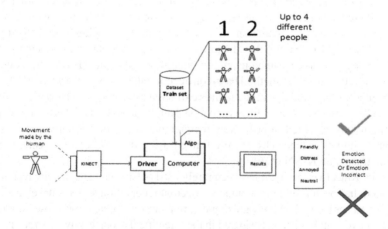

Fig. 1. Overall diagram

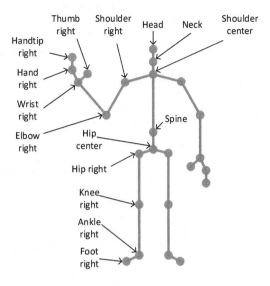

Fig. 2. Kinect skeleton joints

between two joints, angles calculated among three joints, velocity of joints during movement and their acceleration. These features are shown in Table 1.

Table 1. Extracted features

Features	Type	IDe
[RIGHT] shoulder (x, y), elbow (x, y) and hand (x, y)	Position	0, 1, 2, 3, 4, 5
Hand_shoulder, hand_head	Distance	6, 7
Head-shoulder-elbow, head-shoulder-hand, head-elbow-hand, shoulder-elbow-hand	Angle	8, 9, 10, 11
Position features	Velocity	12, 13, 14, 15, 16, 17
Angle features	Velocity	18, 19, 20, 21
Position features	Acceleration	22, 23, 24, 25, 26, 27
Angle features	Acceleration	28, 29, 30, 31

The feature computation is performed similarly to [6], that introduces new Kinect gait features: JRD–Joint Relative Distance and JRA–Joint Relative Angle.

In our paper, distance features are extracted using the Eq. (1):

$$f = \sqrt{\left(hand_x - shoulder_x\right)^2 + \left(hand_y - shoulder_y\right)^2} \tag{1}$$

Angle features are computed as:

$$\alpha = \arccos(\frac{(x_b - x_a)(x_c - x_a) + (y_b - y_a)(y_c - y_a)}{\sqrt{(x_b - x_a)^2 + (y_b - y_a)^2}\sqrt{(x_c - x_a)^2 + (y_c - y_a)^2}}) \tag{2}$$

Joint and angle velocities are computed using equation below:

$$v = \frac{x_1 - x_2}{\left(\dfrac{1}{30}\right)} = (x_1 - x_2)30 \tag{3}$$

Finally, joint and angles accelerations, which are the second derivatives of the position, are computed by:

$$a = \frac{v_1 - v_2}{\left(\dfrac{1}{30}\right)} = (v_1 - v_2)30 \tag{4}$$

Equations (3) and (4) are discrete derivatives, based on the frame per second rate obtained from the Kinect. The Kinect version 2 records 30FPS a new frame. Thus, dt in both (3) and (4) computes the velocity and acceleration with coefficient 30.

Figure 3 shows a space of four emotions: friendly wave (labelled as wave), annoyed, in distress and neutral, for the two extracted features which are angle *head-shoulder-hand* and angle *head-shoulder-elbow*. As it can be observed from Fig. 3 the distress emotion is well separated from others, while the three remaining emotions overlap.

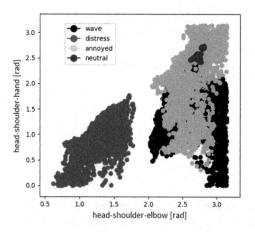

Fig. 3. Separation of four emotions based on angle features

Figure 4 shows the covariance features graph for the first 21 extracted features. The blue colors demonstrates a very low correlation among features, while the dark red color found on the diagonal and in a few other areas shows a high correlation among features. This figure indicates that quite a few of the extracted features are independent, including majority of position and angles features, which is an indication that a classifier algorithm trained on them can perform very well.

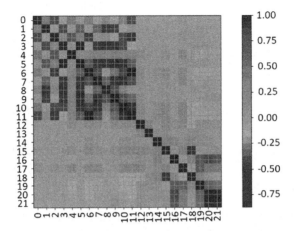

Fig. 4. Covariance features graph (Color figure online)

3.2 Machine Learning Classification Methods

K nearest neighbors, support machine vector, decision tree, Neural Network and Naive Bayes are classifiers implemented for body emotion recognition [5].

K Nearest Neighbors computes the distance between the test sample and the train set for the K nearest neighbors to the sample. Support Vector Machine classifier fits a separating hyperplane between support vectors to separate classes in multiple area (as shown in Fig. 5). Decision Tree classifier builds a conditional tree based on assumption from the training set. Neural Network constructs a perceptron network, shown in Fig. 6 (left). Figure 6 (right) illustrates the back propagation of the error, which adapts the perceptron weight at each level of the network.

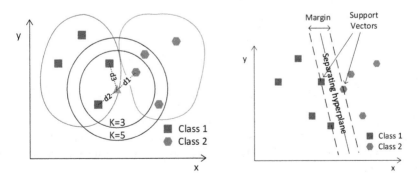

Fig. 5. KNN classifier (left) and SVM classifier (right)

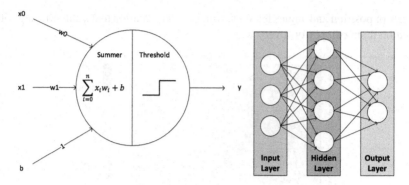

Fig. 6. Neural network perceptron (left) and layers (right)

3.3 Genetic Algorithm

An implementation of a genetic algorithm with a selection and cross mutation is used to improve the feature selection for some classifiers. The genetic algorithm, shown in Fig. 7, includes the evaluation and the evolution stages. During the evaluation step, multiple classifiers of a specific type are trained and tested. The genetic algorithm will look at the best configuration to use among the classifiers. The population evolution process includes selection, combination, mutation and completion. Each step has a chance to succeed based on a random number for the specific species. The selection has a 5% chance, the combination has a 5% chance, and the mutation has a 10% chance. The completion stabilizes the population size.

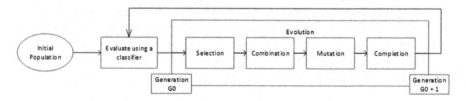

Fig. 7. Genetic algorithm

This genetic algorithm was implemented for KNN, SVM and NB classifiers with an initial population of 40 species, which were generated randomly. Each of the species holds a list of features, up to 21 features can be included and only one instance of the same feature is stored. Figure 8 shows the evolution of a population with the KNN classifier up to 200 generations. The population mean value tends to increase to remove less fitted species inside the population. Note that species can achieve up to 98.96% of recognition rate with the KNN classifier. These species are obtained from the combination and mutation.

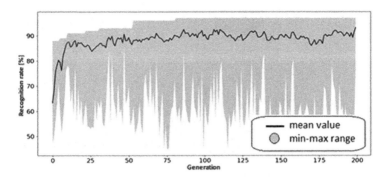

Fig. 8. KNN genetic algorithm evolution

The genetic algorithm provides the best extracted feature set to be used with the individual classclassifiers. Figure 9 shows the best fit convergence for genetic algorithm.

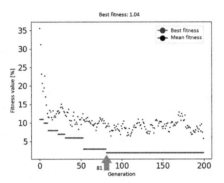

Fig. 9. Genetic algorithm best fitness convergence

4 Experimental Results

This section describes the data collection and outcomes of gesture recognition from all of the classifier methods, trained using genetic algorithms, and the resulting recognition rate.

4.1 Database Description

The database contains Kinect recordings of eight subjects with four emotions expressed through the upper body movement: a friendly wave, a distress, an annoyance and neutral. To obtain the data, each person was asked to stand still at two different distances in front of the Kinect (see Fig. 10). Then a subject is asked to move one hand to express one of the four movements. Each video was collected in the same way, which makes a total of 192 sets of recordings, each movement repeated 3 times.

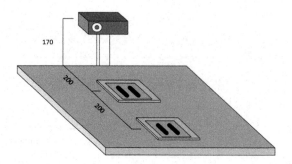

Fig. 10. Dataset collection setup

4.2 Classifiers Parameter Tuning

Classifiers can be configured using different parameters. Table 2 lists parameters tuned to get the best recognition rate for each of the KNN, SVM and DT classifiers.

Table 2. Classifier tuned parameters

Classifier	Parameters tuned
KNN	The number of neighbors and the distance algorithm (Euclidean)
SVM	The kernel type among 'linear', 'poly' and 'rbf'
DT	The tree depth among 20, 50, 100, 200, 300, 500, 1000, 2000 and 5000

Deep Neural Network was used with a single hidden layer which contains 1000 perceptrons. The number of perceptrons was chosen from the total amount of features possible at the input layer. Naïve Bayes algorithm was used without changing parameters.

4.3 Classification Results

Before using binary classifiers, namely SVM, KNN and DTree, we tune them to ensure maximum performance.

Table 3. Classifier parameters and recognition rates

Classifier	Features	Frame Nb.	Recognition rate [%]
KNN, neighbors = 1 distance = 2	0, 8, 9, 10, 13	46	98.96
SVM, kernel = poly	4, 6, 8, 10, 11	77	95.83
DT, depth = 200	8, 9, 10, 11, 13	70	94.79
NN, hidden layers = 1	8, 9, 10, 11, 13	131	93.75
NB	9, 10, 11, 13	51	95.83

Table 3 shows the recognition rates for all classifiers tested. An interesting observation is that joint angles features were utilized by all classifiers. The total number of features to provide for testing the classifier depended on the number of frames.

Figure 11 shows all 5 classifier emotion recognition rate using cross-validation at a fixed number of frames. The best recognition is achieved by KNN classifier, that at 46 frames resulted in 98.96% emotion recognition rate.

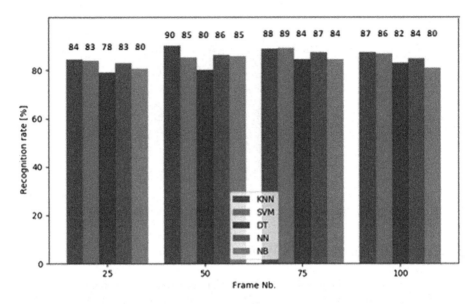

Fig. 11. Emotion recognition rates at different frames

Figure 12 shows the best reachable cross validation recognition rate for all 5 classifiers after cross-validation across all frames. We observe that KNN and SVM are two of the best performing classifiers.

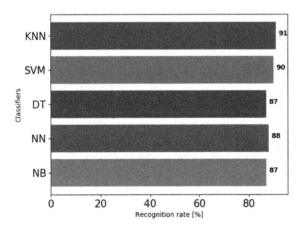

Fig. 12. Overall emotion recognition rates

Figure 13 shows the probability of the emotion to be in the N first emotions recognized by the classifier. While both SVM and KNN are two of the best performing individual classifiers, SVM reaches 100% recognition at rank 3 (which means that the given emotion was one identified among top three emotions), while KNN at rank 4.

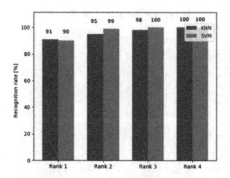

Fig. 13. Identifying correct emotion among top ranked emotions

5   Conclusions and Future Work

The fact that body movement can express emotions and intentions is being explored in this research. Several classifiers tested show a high recognition rate after cross validation. KNN and SVM classifiers outperformed DT, NB and NN classifiers in the current study. Emotions recognition based on arm movement using Kinect sensor achieved 98% recognition rate on a small dataset of eight subjects. As a future work, the methods will be tested on a larger data set collected under varied conditions, ensemble of classifiers will be implemented, and an embedded system will be implemented in real-time to build a prototype ready for use in situation awareness systems.

Acknowledgements. Authors would like to acknowledge NSERC, MITACS funding and Switzerland international exchange program. We also grateful to all members of the Biometric Technologies Laboratory, Department of Computer Science, University of Calgary.

References

1. Kozlowski, L.T., Cutting, J.E.: Recognizing the sex of a walker from a dynamic point-light display. Atten. Percept. Psychophy. **21**(6), 575–580 (1977)
2. Cutting, J.E.: Generation of synthetic male and female walkers through manipulation of a biomechanical invariant. Perception **7**(4), 393–405 (1978)
3. Hill, H., Pollick, F.E.: Exaggerating temporal differences enhances recognition of individuals from point light displays. Psychol. Sci. **11**(3), 223–228 (2000)
4. Munir, S., Arora, R., Hesling, C., Li, J., Francis, J., Shelton, C., Martin, C., Rowe, A., Berges, M.: Real-time fine grained occupancy estimation using depth sensors on ARM embedded platforms. In: 23rd IEEE RTETA Symposium (2017)

5. Gavrilova, M.L., Monwar, M.: Multimodal Biometrics and Intelligent Image Processing for Security Systems. IGI Global, Hershey (2013)
6. Ahmed, F., Paul, P.P., Gavrilova, M.L.: DTW-based kernel and rank-level fusion for 3D gait recognition using Kinect. Vis. Comput. **31**(6–8), 915–924 (2015)
7. Pollick, F.E., Paterson, H.M., Bruderlin, A., Sanford, A.J.: Perceiving affect from arm movement. Cognition **82**(2), B51–B61 (2001)
8. Xia, L., Chen, C., Aggarwal, J.: View invariant human action recognition using histograms of 3D joints. In: IEEE Conference CVPR, pp. 20–27 (2012)
9. Yang, X., Tian, Y.: Effective 3D action recognition using eigenjoints. J. Vis. Commun. Image Represent. **25**(1), 2–11 (2014)
10. Theodorakopoulos, I., Kastaniotis, D., Economou, G., Fotopoulos, S.: Pose-based human action recognition via sparse representation in dissimilarity space. J. Vis. Commun. Image Represent. **25**(1), 12–23 (2014)
11. Roether, C.L., Omlor, L., Christensen, A., Giese, M.A.: Critical features for the perception of emotion from gait. J. Vis. **9**(6), 1–15 (2009)
12. Saha, S., Datta, S., Konar, A., Janarthanan, R.: A study on emotion recognition from body gestures using Kinect sensor. In: Proceedings of Communications and Signal Processing (ICCSP), pp. 056–060 (2014)
13. Senecal, S., Cuel, L., Aristidou, A., Magnenat-Thalmann, N.: Continuous body emotion recognition system during theater performances. Comput. Animat. Virtual Worlds **27**(3–4), 311–320 (2016)
14. Wang, Y., Howard, N., Karcprzyk, J., Frieder, P., Sheu, P., Fiorini, R., Gavrilova, M.L., Patel, S., Peng, J., Widrow, B.: Cognitive informatics: towards cognitive machine learning and autonomous knowledge manipulation. Int. J. Cogn. Inf. Nat. Intell. (IJCINI) **12**(1), 1–18 (2017)
15. Gavrilova, M., Wang, Y., Ahmed, F., Paul, P.P.: KINECT sensor gesture and activity recognition for consumer cognitive systems. IEEE Consum. Electron. Mag. Spec. Issue. Consum. Electron. **4**, 88–96 (2017)
16. Melin, P., Castillo, O., Kacprzyk, J. (eds.): Nature-Inspired Design of Hybrid Intelligent Systems. SCI, vol. 667. Springer, Cham (2017). https://doi.org/10.1007/978-3-319-47054-2
17. Ball, A., Rye, D., Ramos, F., Velonaki, M.: Unsupervised clustering of people from 'skeleton' data. In: 2012 Conference on Human Robot Interaction, pp. 225–226 (2012)
18. Preis, J., Kessel, M., Linnhoff-Popien, C., Werner, M.: Gait recognition with Kinect. In: Workshop on Kinect in Pervasive Computing (2012)
19. Gabel, M., Gilad-Bachrach, R., Renshaw, E., Schuster, A.: Full body gait analysis with kinect. In: Annual International Conference of the IEEE Engineering in Medicine and Biology Society, pp. 1964–1967 (2012)
20. Sas, D., Saeed, K.: Comprehensive performance evaluation of various feature extraction methods for OCR purposes. In: Saeed, K., Homenda, W. (eds.) CISIM 2015. LNCS, vol. 9339, pp. 411–422. Springer, Cham (2015). https://doi.org/10.1007/978-3-319-24369-6_34
21. Markowska-Kaczmar, U., Kwasnicka, H., Szczepkowski, M.: Genetic algorithm as a tool for stock market modelling. In: Rutkowski, L., Tadeusiewicz, R., Zadeh, L.A., Zurada, J.M. (eds.) ICAISC 2008. LNCS (LNAI), vol. 5097, pp. 450–459. Springer, Heidelberg (2008). https://doi.org/10.1007/978-3-540-69731-2_44

Particle Swarm Optimization with Single Particle Repulsivity for Multi-modal Optimization

Michal Pluhacek[✉]ⓘ, Roman Senkerikⓘ, Adam Viktorinⓘ,
and Tomas Kadavyⓘ

Faculty of Applied Informatics, Tomas Bata University in Zlin,
T. G. Masaryka 5555, 760 01 Zlin, Czech Republic
{pluhacek,senkerik,aviktorin,kadavy}@utb.cz

Abstract. This work presents a simple but effective modification of the velocity updating formula in the Particle Swarm Optimization algorithm to improve the performance of the algorithm on multi-modal problems. The well-known issue of premature swarm convergence is addressed by a repulsive mechanism implemented on a single-particle level where each particle in the population is partially repulsed from a different particle. This mechanism manages to prolong the exploration phase and helps to avoid many local optima. The method is tested on well-known and typically used benchmark functions, and the results are further tested for statistical significance.

Keywords: Particle Swarm Optimization · PSO · Convergence Repulsivity

1 Introduction

The Particle Swarm Optimization (PSO) [1–3] remains one of the most used and highly popular metaheuristic optimizers. Due to its simplicity, the fast and robust performance it has served as inspiration for many new swarm intelligence methods in the past decade.

M. Pluhacek—This work was supported by the Ministry of Education, Youth and Sports of the Czech Republic within the National Sustainability Programme Project no. LO1303 (MSMT-7778/2014), further by the European Regional Development Fund under the Project CEBIA-Tech no. CZ.1.05/2.1.00/03.0089 and by Internal Grant Agency of Tomas Bata University under the Projects no. IGA/CebiaTech/2018/003. This work is also based upon support by COST (European Cooperation in Science & Technology) under Action CA15140, Improving Applicability of Nature-Inspired Optimisation by Joining Theory and Practice (ImAppNIO), and Action IC1406, High-Performance Modelling and Simulation for Big Data Applications (cHiPSet). The work was further supported by resources of A.I. Lab at the Faculty of Applied Informatics, Tomas Bata University in Zlin (ailab.fai.utb.cz).

© Springer International Publishing AG, part of Springer Nature 2018
L. Rutkowski et al. (Eds.): ICAISC 2018, LNAI 10841, pp. 486–494, 2018.
https://doi.org/10.1007/978-3-319-91253-0_45

The PSO is still subject to intensive research [4–7] that aims to understand its inner dynamic and improve its performance. One of the well-known issues of original PSO design is the premature convergence into local minima on complex multi-modal functions. In this work, a single-particle repulsive strategy is proposed to address this problem.

The repulsive mechanism was previously successfully implemented into PSO in [7] and more recently, it was utilized in [8]. However, the above-mentioned mechanism is a reactive method; it requires computing the diversity of the swarm and only reacts when the swarm has converged in the local extreme. The method proposed in this work is a proactive approach to prevent premature convergence of the swarm.

The rest of the paper is organized as follow: in Sect. 2, the original PSO algorithm is described. Following is the description of the proposed modification. The experiment is designed afterward, and the results are presented in Sect. 5. The results are discussed in Sect. 6, followed by the conclusion.

2   Particle Swarm Optimization Algorithm

The PSO algorithm is based on the behavior of fish and birds. The particles simulate the movement of an artificial swarm over the multi-dimensional search space. Each particle is trying to find the best possible solution for the problem.

The knowledge of global best solution, (typically noted $gBest$) is shared among the individuals (particles) in the swarm. Furthermore, each particle has the knowledge of its own (personal) best solution (noted $pBest$). Last important part of the algorithm is the velocity of each particle that is taken into account during the calculation of the particle movement. The new position of each particle is then given by (1), where x_i^{t+1} is the new particle position; x_i^t refers to current particle position and v_i^{t+1} is the new velocity of the particle.

$$x_i^{t+1} = x_i^t + v_i^{t+1} \tag{1}$$

To calculate the new velocity, the distances from $pBest$ and $gBest$ are taken into account, alongside with current velocity that is multiplied by inertia weigh value (2)

$$v_{ij}^{t+1} = w \cdot v_{ij}^t + c_1 \cdot Rand_1 \cdot (pBest_{ij} - x_{ij}^t) + c_2 \cdot Rand_2 \cdot (gBest_j - x_{ij}^t) \tag{2}$$

Where:

v_{ij}^{t+1} - New velocity of the ith particle in iteration $t+1$. (component j of the dimension D).

w - Inertia weight.

v_{ij}^t - Current velocity of the ith particle in iteration t. (component j of the dimension D).

$c_1, c_2 = 2$ - Acceleration constants.

$pBest_{ij}$ - Local (personal) best solution found by the ith particle. (component j of the dimension D).

$gBest_j$ - Best solution found in a population. (component j of the dimension D).
x_{ij}^t - Current position of the ith particle (component j of the dimension D) in iteration t.
$Rand_{1,2}$ - Pseudo random number, interval $(0, 1)$.

There are two different approaches for position updating [9]. In the asynchronous method, each particle updates its new position and $pBest$ (and potentially $gBest$) instantly. In the synchronous strategy, the new positions and $pBests$ ($gBest$) are updated at the end of the iteration simultaneously for all particles. Each of this methods offer different advantages, and their performance is slightly different as is discussed in [9]. In this work, we choose to compare the proposed modification to both strategies for position updating. However, the proposed modification itself is using the asynchronous updating model.

3   Proposed Modification

In the original PSO, the exploration phase often ends prematurely, and the algorithm starts exploiting the current sub-optima almost instantly. This leads to poor performance when facing complex multi-modal problems. To prevent the swarm from instant convergence, a partial repulsivity is proposed in this work. The swarm is indexed as a logical ring topology; therefore, it is easy to assign each particle a single neighbor with indexes higher by one. Given the ring topology, the last particle in the population is neighboring the first particle in the population.

The velocity calculation formula is altered to the form given by (3) and (4).

$$v_{ij}^{t+1} = w \cdot v_{ij}^t + c_1 \cdot Rand_1 \cdot (pBest_{ij} - x_{ij}^t) + c_2 \cdot Rand_2 \cdot (gBest_j - x_{ij}^t)$$
$$-c_3 \cdot Rand_3 \cdot (x_{kj}^t - x_{ij}^t) \tag{3}$$

where

c_3 - repulsive constant
$Rand_3$ - Pseudo-random number, uniform distribution, interval $(0, 1)$.
k - index of neighboring particle in the population (4)

$$k = (i \bmod NP) + 1 \tag{4}$$

where

i - index of the active particle, $i \in (1; NP)$
NP - population size

Each particle is partially repulsed from its right-hand side neighbor in the ring topology. This unique repulsivity on single-particle level helps slow down the convergence of the swarm but does not prevent the convergence completely. As no two particles share the same point of repulsivity, the natural movement of the swarm is maintained. Using this method, the exploration phase is prolonged, while the exploitation phase is not abandoned. After extensive testing, the value of c_3 has been set to 0.15. For this value, the algorithm seems to achieve the best performance. The algorithm is not oversensitive for the value of the repulsive constant, and a fine-tuning is not necessary.

4   Experiment

In the experimental part, the performance of the newly proposed modification was tested using well-known unimodal and multimodal functions in several different dimensional settings. To provide a fair comparison, the newly proposed modification is compared to both asynchronous and synchronous updating PSOs. The control parameters were set as follows (in accordance with common values and recommendations in [6]):

$NP = 30$;
$max.\ CFE = 30\ 000$;
$dim = 10,\ 30,\ 50$;
$w = 0.7298$;
$c_1,\ c_2 = 1.49618$;
$c_3 = 0.15$;

The following set of four common test functions was used:
The Sphere function is given by (5).

$$f(x) = \sum_{i=1}^{\dim} x_i^2 \tag{5}$$

Function minimum:
Position for E_n: $(x_1, x_2 \ldots x_n) = (0,\ 0, \ldots,\ 0)$
Value for E_n: $f(x) = 0$

The Rosenbrock's function is given by (6).

$$f(x) = \sum_{i=1}^{\dim-1} 100(x_i^2 - x_{i+1})^2 + (1 - x_i)^2 \tag{6}$$

Function minimum:
Position for E_n: $(x_1, x_2 \ldots x_n) = (1,\ 1, \ldots,\ 1)$
Value for E_n: $f(x) = 0$

The Rastrigin's function is given by (7).

$$f(x) = 10\dim + \sum_{i=1}^{\dim} x_i^2 - 10\cos(2\pi x_i) \tag{7}$$

Function minimum:
Position for E_n: $(x_1, x_2 \ldots x_n) = (0,\ 0, \ldots,\ 0)$
Value for E_n: $f(x) = 0$

Schwefel's function is given by (8).

$$f(x) = \sum_{i=1}^{\dim} -x_i \sin(\sqrt{|x|}) \tag{8}$$

Optimum position for E_n: $(x_1, x_2 \ldots x_n) = (420.969, 420.969, \ldots, 420.969)$
Optimum value for E_n: $f(x) = -418.983 \cdot dimension$

5   Results

In Table 1, the mean results of original PSO with asynchronous and synchronous updating are compared with the proposed PSO with single particle repulsivity (noted PSO Repuls.). Alongside the mean results comparison, statistical evaluation was performed using the Wilcoxon rank sum test with a level of significance alpha 0.05. All pairs of algorithms were tested, and the winner is identified if it managed to out-perform both remaining algorithms with statistical significance. Otherwise, the winner is not designated. Further, the mean history of best results during the optimization (convergence lines) is compared in Figs. 1, 2, 3 and 4 for $dim = 10$;

Table 1. Mean results overview.

Function	PSO Async.	PSO Sync.	PSO Repuls.	Winner
dim = 10				
Sphere	0.00E+00	0.00E+00	2.94E−28	-
Rosenbrock	2.83E+01	1.92E+01	2.24E+01	-
Rastrigin	7.23E+00	7.96E+00	4.59E+00	PSO Repuls.
Schwefel	9.44E+02	9.68E+02	9.72E+01	PSO Repuls.
dim = 30				
Sphere	3.42E−15	1.01E−12	3.57E−09	PSO Async.
Rosenbrock	1.07E+02	7.49E+01	1.46E+02	-
Rastrigin	1.19E+02	1.28E+02	6.84E+01	PSO Repuls.
Schwefel	4.48E+03	4.76E+03	2.03E+03	PSO Repuls.
dim = 50				
Sphere	3.62E−05	1.32E−02	2.37E−03	PSO Async.
Rosenbrock	1.84E+02	3.38E+02	5.47E+02	PSO Async.
Rastrigin	4.05E+02	4.93E+02	2.35E+02	PSO Repuls.
Schwefel	8.62E+03	8.61E+03	4.53E+03	PSO Repuls.

5.1   Convergence in Higher Dimensions

In the higher dimensions, the convergence on Sphere and Rosenbrock function are too similar for meaningful graphical depiction. Therefore, only the convergence on Rastrigin and Schwefel function is depicted. Figures 5 and 6 show the comparison of convergence on Rastrigin function in 30 and 50 dimensions. Figures 7 and 8 depict similar comparison on Schwefel function.

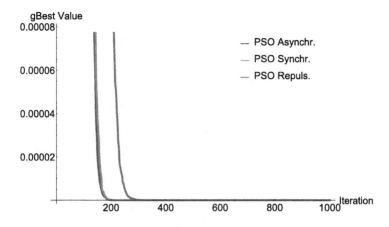

Fig. 1. Mean best value history comparison - Sphere function; $dim = 10$

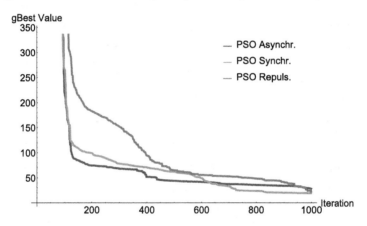

Fig. 2. Mean best value history comparison - Rosenbrock function; $dim = 10$

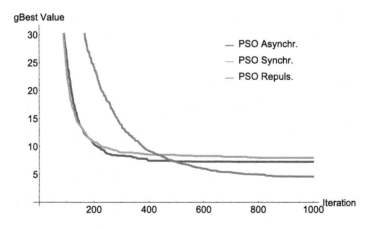

Fig. 3. Mean best value history comparison - Rastrigin function; $dim = 10$

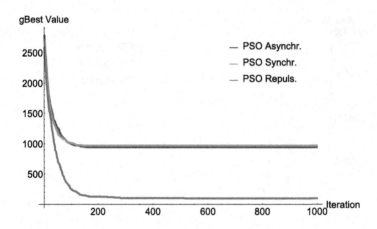

Fig. 4. Mean best value history comparison - Schwefel function; $dim = 10$

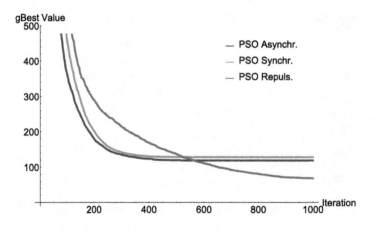

Fig. 5. Mean best value history comparison - Rastrigin function; $dim = 30$

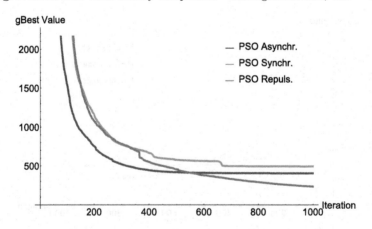

Fig. 6. Mean best value history comparison - Rastrigin function; $dim = 50$

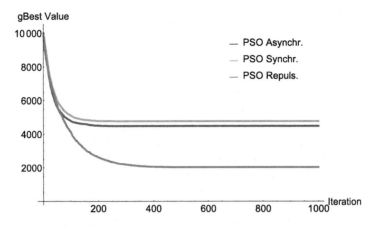

Fig. 7. Mean best value history comparison - Schwefel function; $dim = 30$

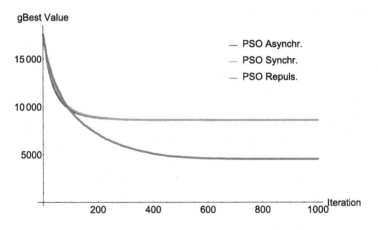

Fig. 8. Mean best value history comparison - Schwefel function; $dim = 50$

6   Results Discussion

According to Table 1, the proposed modification of PSO managed to obtain best results with statistical significance according to Wilcoxon rank sum test for Rastrigin and Schwefel function in all dimensional settings. This result is very encouraging in regards to the scalability of the method and its performance on multi-modal functions.

In the case of $dim = 10$, the performance of all three compared methods is similar. In higher dimensions, the asynchronous PSO seems to perform better on simple uni-modal functions.

According to the presented convergence plots (Figs. 1, 2, 3, 4, 5, 6, 7 and 8), the proposed modification achieved the anticipated convergence behavior on multi-modal functions, prolonging the exploration phase and managing to avoid local minima for a longer period. Therefore, the proposed modification manages to obtain significantly better results than both PSO with asynchronous and synchronous updating.

7 Conclusion

The aim of this work was to present a simple modification to the original PSO formula that improves its performance of multi-modal optimization problems. The repulsive behavior was implemented on the single-particle level, and therefore the swarm as a whole manages the capability to converge into a promising region, but the exploration phase is prolonged helping to avoid some local extremes. The results were tested for statistical significance and proved very encouraging for the future research in this direction.

Following this study, the method will be tested on an extensive collection of benchmark functions and developed further with the aim to improve its robustness. The modification proposed in this paper might be easily implemented into advanced PSO modifications as well as other swarm-based algorithms.

References

1. Kennedy, J., Eberhart, R.: Particle swarm optimization. In: Proceedings of the IEEE International Conference on Neural Networks, pp. 1942–1948 (1995)
2. Shi, Y., Eberhart, R.: A modified particle swarm optimizer. In: Proceedings of the IEEE International Conference on Evolutionary Computation. IEEE World Congress on Computational Intelligence, pp. 69–73 (1998)
3. Kennedy, J.: The particle swarm: social adaptation of knowledge. In: Proceedings of the IEEE International Conference on Evolutionary Computation, pp. 303–308 (1997)
4. Nickabadi, A., Ebadzadeh, M.M., Safabakhsh, R.: A novel particle swarm optimization algorithm with adaptive inertia weight. Appl. Soft Comput. **11**(4), 3658–3670 (2011). ISSN 1568-4946
5. Eberhart, R.C., Shi, Y.: Comparing inertia weights and constriction factors in particle swarm optimization. In: Proceedings of the IEEE Congress on Evolutionary Computation, San Diego, USA, pp. 84–88 (2000)
6. Van Den Bergh, F., Engelbrecht, A.P.: A study of particle swarm optimization particle trajectories. Inf. Sci. **176**(8), 937–971 (2006)
7. Riget, J., Vesterstrøm, J.S.: A diversity-guided particle swarm optimizer-the ARPSO. Technical Report, vol. 2, p. 2002. Department of Computer Science, University of Aarhus, Aarhus, Denmark (2002)
8. Han, F., Liu, Q.: An improved hybrid PSO based on ARPSO and the Quasi-Newton method. In: Tan, Y., Shi, Y., Buarque, F., Gelbukh, A., Das, S., Engelbrecht, A. (eds.) ICSI 2015. LNCS, vol. 9140, pp. 460–467. Springer, Cham (2015). https://doi.org/10.1007/978-3-319-20466-6_48
9. Engelbrecht, A.P.: Particle swarm optimization: iteration strategies revisited. In: 2013 BRICS Congress on Computational Intelligence and 11th Brazilian Congress on Computational Intelligence, Ipojuca, pp. 119–123 (2013)

Hybrid Evolutionary System to Solve Optimization Problems

Krzysztof Pytel[(✉)]

Faculty of Physics and Applied Informatics, University of Lodz, Łódź, Poland
kpytel@uni.lodz.pl

Abstract. The article presents an Evolutionary System designed to solve optimization problems. The system consists of Genetic Algorithm and Evolutionary Strategy, working together to improve the efficiency of optimization and increase the resistance to stuck to suboptimal solutions. In the system, we combined the ability of the Genetic Algorithm to explore the search space and the ability of the Evolutionary Strategy to exploit the search space. The system maintains the right balance between the ability to explore and exploit the search space. Genetic Algorithm and Evolutionary Strategy can exchange information about the solutions found till now and periodically migrate the best individuals between populations. The efficiency of the system has been investigated by an example of function optimization. The results of the experiments suggest that the proposed system can be an effective tool in solving complex optimization problems.

Keywords: Genetic Algorithms · Evolutionary Strategies
Artificial intelligence · Function optimization

1 Introduction

Evolutionary Algorithms (EA) is a group of methods inspired by observation of nature. The principle of these methods is based on the theory of evolution developed by Charles Darwin for living organisms. The natural processes of evolution in the world of living organisms - i.e. mutation, genetic recombination and natural selection, especially the rule of the "survival of the fittest" can be simulated using computer programs. The Evolutionary Algorithm processes the population of individuals. An individual in the form of a chromosome represents a potential solution to the problem. EA works in certain environments, which can be defined on the basis of the problem solved by the algorithm. Fitness function (a numeric value that determines the adaptation of an individual to the environment) is assigned to all individuals. This function describes the ability of an individual to act as a parent for the next generation of a population and to transfer his genetic material to descendants. A characteristic feature of Evolutionary Algorithms is that in the process of evolution they do not use the knowledge specific for a given problem, except for the fitness function assigned

© Springer International Publishing AG, part of Springer Nature 2018
L. Rutkowski et al. (Eds.): ICAISC 2018, LNAI 10841, pp. 495–504, 2018.
https://doi.org/10.1007/978-3-319-91253-0_46

to all individuals. The optimization by EA uses the process of exploration and exploitation. Exploration is the process of searching for a new region of a search space where an optimum can exist. It consists of probing a much larger portion of the search space with the hope of finding other promising solutions that can be refined. Exploitation is the process of searching for regions within the neighborhood of previously visited points. It consists of probing a limited region of the search space with the hope of improving a solution found till now. In this operation, algorithm tends to intensify a local search. The EA must keep the right balance between exploration and exploitation of the search space. Examples of EA are Genetic Algorithms (GA), Evolutionary Strategies (ES), Genetic Programming (GP).

The Genetic Algorithm (GA) is the most popular type of EA. It seeks the solution of an optimization problem, by applying the mechanism of natural evolution (such as selection, mutation, cross-over of individuals and reproduction). Individuals of the GA could be coded by binary strings (binary representation), real numbers (a real number representation) or composite structures of genes. The probability of selection and probability of mutation are main parameters of the GA. They describe the ability to explore and to exploit the search space. In Genetic Algorithms the exploitation is done through the selection process. Cross-over and mutation are methods of exploring the search space. Reproduction in GAs is closely related to maintaining the diversity of the population, the selection pressure and avoiding premature convergence to the local optima. Finding the right balance between exploration and exploitation is a very important problem. If the exploitation ability is too large, the algorithm can get stuck in the local optima. If the exploration ability is too large, the algorithm will waste time on poor solutions and cannot focus on solutions found till now. GA can solve sophisticated optimization problems, so they are usually used for seeking approximate results of the NP-complete problems. More information on Genetic Algorithms can be found in publications [3,5,7].

Evolutionary Strategies (ES) are a variant of AE that uses primarily mutation and selection as search operators. Solutions in ES are usually coded using real-valued vectors. ES uses the principle that small changes have small effects. The mutation is usually performed by adding a normally distributed random value to each individual's genes. The size of mutation (i.e. the standard deviation of the normal distribution) is often governed by a self-adaptation process. If the ratio of successful mutations is less than the specified threshold (it is common to use the 1/5th success rule), the mutation parameters are increased to obtain a greater diversity of individuals. If the ratio of successful mutations is greater than the specified threshold, the mutation parameters are decreased to increase the accuracy of the search and accelerate the convergence of the algorithm. The simplest Evolutionary Strategy $(1 + 1) - ES$ operates on a population of two individuals: the current point (a parent) and the result of its mutation (a child). If the child's fitness is equal to or better than the parent's fitness, the child becomes the parent in the next generation. Otherwise, the new child is created in the next loop. In strategy $(1 + \lambda) - ES$, λ children can be generated and compete with the parent. In $(1, \lambda) - ES$ the best child becomes the parent in the next generation while the current parent is always disregarded. Evolutionary Strategies $(\mu/\rho+, \lambda) - ES$ can use the population of ρ parents and also recombination as an additional operator. This makes them less prone to get stuck in the local optima. More information on Evolutionary Strategies can be found in publications [1, 2].

A hybrid intelligent system is a system which employs, a combination of methods and techniques from artificial intelligence subfields, for example, Genetic Algorithms, the Fuzzy Logic, Artificial Neural Networks etc. Such systems are able to use the advantage of the methods and techniques of artificial intelligence while avoiding their disadvantages. They can be used to improve the effectiveness of the methods or where simple methods do not produce the expected results in satisfactory time. More information on intelligent systems can be found in publications [15, 16]. The system proposed in this article will use a Genetic Algorithm co-operating with Evolutionary Strategy. Hybrid intelligent systems using artificial intelligence methods can be used in different optimization problems, for example in multiobjective optimization [9–11], Connected Facility Location Problem [12], Clustering Problem [13] or Function Optimization Problem [14].

2   The Proposed Hybrid Evolutionary System

Genetic Algorithms can solve sophisticated optimization problems by use operators of cross-over and mutation to search the space for possible solutions. They are usually used for seeking approximate results of the NP-complete problems. Its convergence is very much dependent on initial solution. A disadvantage of GA is low efficiency in the final search stage. A binary representation of individuals is usually used for GA, but there are many examples of using real numbers or complex data structures as a representation of individuals. Recombination in Evolutionary Strategy introduces competition between parents and descendants for the selection of a new individual for next generation. The ES uses primarily mutation and selection operators but they are at risk of getting stuck in suboptimal solutions. The advantage of ES is the rapid convergence to the optimum value. ESs should be applied to continuous problems whereas GAs are most useful to discrete problems. Hybridization of both algorithms is a promising method for solving many problems of continuous and discrete optimization.

The proposed hybrid system (GA-ES) combine advantages and avoid disadvantages of a Genetic Algorithm and an Evolutionary Strategy. The proposed hybrid system block diagram is shown in Fig. 1. It combines the ability of a GAs to find the areas of possible optima and the ability of ESs to quickly converge to the optima. Both of them are different types of Evolutionary Algorithms but can use the same individuals' representation, operators of selection and mutation. In the system, both algorithms work in parallel. Both algorithms start with their own, randomly initial population. After a predetermined number of generations, the best individuals from both algorithms will be compared. Depending on the result of this comparison, transposition of individuals between the algorithms may be performed:

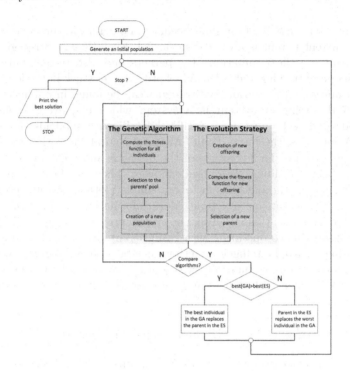

Fig. 1. The proposed hybrid system block diagram

- if the best individual in the GA is better than the best individual in the ES, it means that a new area of the optima is found. The best individual in the GA replaces the parent in the ES.
- if the best individual in the ES is better than the best individual in the GA, it means that a new optimal solution is found as a result of the ES. The best individual in the ES replaces the worst individual in the GA population.

3   Problem Formulation

Optimization is the process of selecting the best element (with regard to some criterion) from all feasible solutions. Optimization problems can be divided into two categories: continuous or discrete. The Function Optimization Problem (FOP) is a continuous problem in which certain parameters (variables) need to be determined to achieve the best measurable performance (objective function) under given constraints. For function $f(x)$, called the objective function, that has a domain of real numbers of set S, the maximum optimal solution occurs where $f(x_0) \geq f(x)$ over set S and the minimum optimal solution occurs where $f(x_0) \leq f(x)$ over set S.

Formally, optimization is the minimization or maximization of a function subject to constraints on its variables. Let's denote:

- x is the vector of variables (parameters);
- $f(x)$ is the objective function that we want to maximize or minimize;
- c is the vector of constraints that the variables must satisfy. It may consist of several restrictions that we place on the variables.

The function optimization problem (e.g. a function maximizing problem) can be stated as follows:

$$\begin{cases} max \quad f(x) \\ \text{subject to: } c_i(x) \leq 0 \text{ for } i = 1, 2, ..., k \\ x \in S \end{cases} \tag{1}$$

where:

- $x = [x_1, x_2, x_3, ..., x_n] \in \Re; n \in N$ - is an n-dimensional vector of decision variables,
- $f(x)$ - is the objective function of variables x,
- $c_i(x)$ - are constrains,
- S - the search area.

The FOP problem can be used as a benchmark for testing optimization methods. There is a lot of functions with varying complexity and difficulty, presented in the literature [17,18]. These functions could be used to test the proposed hybrid system and compare results to other optimization methods. Various methods of solving the FOP are discussed in a literature, for example [4,6,8].

4   Computational Experiment

The goal of our experiments is verification of the idea of the hybrid Genetic Algorithm and Evolutionary Strategy system in solving the function optimization problem. We built the system and used functions of a wide range of complexity and difficulty in a diverse environment for tests. For tests we used a set of functions presented in literature:

- $f_1(x_1, x_2)$ - an easy function of two variables (similar to cosinemixture [13] function). The function has many local optima and was used for testing the algorithm's ability to find the global optimum. The function is given by the formula:

$$f_1(x_1, x_2) = (sin(x_1) + 0.6 * sin(20 * x_1)) * sin(x_2) \tag{2}$$

where:

$$x_1, x_2 \in (0, \pi) \tag{3}$$

The value of maximum 1.6 at point $(\pi/2, \pi/2)$.

– $f_2(x_1, x_2, ..., x_{10})$ - the function of multiple variables (similar to alpine2 [13] function). A function with a low inclination angle, used for testing the ability of the algorithm to determine the exact solution. The function is given by the formula:

$$f_2(x_1, x_2, ..., x_{10}) = \prod_{i=1}^{10} sin(x_i) \tag{4}$$

where:

$$x_1, x_2, ..., x_{10} \in (0, \pi) \tag{5}$$

The value of maximum 1 at point $(\pi/2, \pi/2, \pi/2, \pi/2, \pi/2, \pi/2, \pi/2, \pi/2, \pi/2, \pi/2)$.

– f_3 - the function proposed in [8]. It is a sophisticated function with many local optima with different values, which permits to estimate the ability of the algorithm to solve difficult optimization problems. The first generation of individuals was placed in the local optimum (point [5, 5]). The algorithm should find the total optimum (point [50, 50]), avoiding the local optima. The function is given by the formula:

$$f_3(x_1, x_2) = \sum_{1}^{7} h_i * e^{-\mu_i * ((x_1 - x_{i1})^2 + (x_2 - x_{i2})^2)} \tag{6}$$

where: $h_1 = 1.5, h_2 = 1, h_3 = 1, h_4 = 1, h_5 = 2, h_6 = 2, h_7 = 2.5$

$\mu_1 = \mu_2 = \mu_3 = \mu_4 = \mu_5 = \mu_6 = \mu_7 = 0.01$

$(x_{11}, x_{12}) = (5, 5), (x_{21}, x_{22}) = (5, 30), (x_{31}, x_{32}) = (25, 25), (x_{41}, x_{42}) = (30, 5), (x_{51}, x_{52}) = (50, 20), (x_{61}, x_{62}) = (20, 50), (x_{71}, x_{72}) = (50, 50)$

The value of maximum 2.5 at point (50, 50).

– the Rastrigin function - is a non-convex, non-linear multimodal function. It was first proposed as a 2-dimensional function and has been generalized to an n-dimensional domain (in experiments we used 2, 5 and 10 dimensions). Finding the minimum of this function is a fairly difficult problem due to its large search space and its large number of local minima. The function is usually evaluated on the hypercube $x_i \in [-5.12, 5.12]$, for all $i = 1, ..., d$ and the minimum $f(x^*) = 0$ is at point $x^* = (0, ..., 0)$. The function is given by the formula:

$$f_{Ra}(x) = 10d + \sum_{i=1}^{d} [x_i^2 - 10cos(2\pi x_i)] \tag{7}$$

– the Styblinski-Tang function - is a continuous, multimodal, non-convex function, generalized to an n-dimensional domain (in experiments we used 2, 5 and 10 dimensions). Finding the minimum of this function is a fairly difficult problem due to its large search space and its shape. The function is usually evaluated on the hypercube $x_i \in [-5, 5]$, for all $i = 1, ..., d$. The minimum is at point $x^* = (-2.903534, ..., -2.903534)$, and his value is $f(x^*) = -39.16599d$. The function is given by the formula:

$$f_{ST}(x) = \frac{1}{2} + \sum_{i=1}^{d} [x_i^4 - 16x_i^2 + 5x_i] \tag{8}$$

– the Rosenbrock function (also referred to as the Valley or Banana function) - is a popular test problem for gradient-based optimization algorithms. The function is generalized to an n-dimensional domain (in experiments we used 2 dimensions). The function is unimodal, and the global minimum lies in a narrow, parabolic valley. However, even though this valley is easy to find, convergence to the minimum is difficult. The function is usually evaluated on the hypercube $x_i \in [-5, 10]$, for all $i = 1, ..., d$ and the minimum $f(x*) = 0$ is at point $x^* = (1, ..., 1)$. The function is given by the formula:

$$f_{Ro}(x) = \sum_{i=1}^{d-1} [100(x_{i+1} - x_i^2)^2 + (x_i - 1)^2] \tag{9}$$

– the Shubert function - 2-dimensional function with several local minima and many global minima. The function is usually evaluated on the square $x_i \in [-10, 10]$, for all $i = 1, 2$. The minimum value is $f(x) = -186.7309$. The function is given by the formula:

$$f_{Sh}(x) = (\sum_{i=1}^{5} icos((i+1)x_1 + i))(\sum_{i=1}^{5} icos((i+1)x_2 + i)) \tag{10}$$

Some test functions are a minimization problem. In this case, they were converted into the problem of maximization by negation and adding an additional constant C. The constant C was determined for each function during the initial experiments.

The values of parameters of the Genetic Algorithm and the Evolutionary Strategy was fixed during the initial experiments. In the experiments we accepted the following values of parameters of the Genetic Algorithm:

– the genes of individuals are represented by real numbers,
– the probability of cross-over = 0.8,
– the probability of mutation = 0.15,
– the number of individuals in the population = 25.

For the Evolutionary Strategy, model $(1+1) - ES$ was chosen and the mutation performed by adding a number generated randomly according to a normal distribution. To maintain the rule of 1/5 successes, if for the next 10 generations there was no improvement, the size of the mutation was enlarged (increased σ parameter).

The best individuals from both algorithms were compared after every 50 generations and, depending on the result of this comparison, transposition of individuals was performed between the algorithms. The number of generations between

Table 1. The average running time of algoritms

Function	Stop criterion	Constant C	SGA		ES		GA-ES	
			Time [s]	σ (Time)	Time [s]	σ (Time)	Time [s]	σ (Time)
f_1	1,596	-	0.0732	0.0282	0.0309	0.0074	0.0365	0.0069
f_2	0,999	-	0.0379	0.0059	14.403*	44.833*	0.0614	0.0082
f_3	2,50048	-	0.1506	0.0778	0.0683	0.0399	0.0415	0.0057
$f_{Ra}2d$	74.999	75	0.0794	0.3128	0.7091	0.5797	0.0342	0.0064
$f_{Ra}5d$	149.999	150	0.7763	0.3344	-*	-*	0.0997	0.0352
$f_{Ra}10d$	289.5	290	1.8962	0.7065	-*	-*	1.0523	0.2440
$f_{ST}2d$	656.663	500	0.0725	0.0195	0.0578	0.0770	0.0398	0.0056
$f_{ST}5d$	1641.6	1250	0.2098	0.0807	-*	-*	0.0504	0.0077
$f_{ST}10d$	2383.0	2500	0.9966	0.3044	-*	-*	0.0982	0.0281
f_{Ro}	1210120.99	1210121	0.3772	0.3469	0.0324	0.0038	0.0641	0.0184
f_{Sh}	271,795	187	0.1182	0.0623	0.1492	0.1939	0.0448	0.0070

* - at least once, the algorithm got stuck in the local optima.

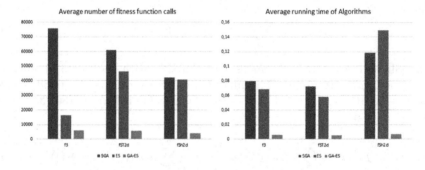

Fig. 2. The average running time and number of fitness function calls in the genetic algorithm (SGA), the evolutionary strategy and the proposed system (GA-ES)

comparisons was determined experimentally during the preliminary tests. The system was stopped when the best individual reached the predetermined value of the optimized function.

In the experiment, we compared the results of the proposed GA-ES system, the Evolutionary Strategy (ES), and the Standard Genetic Algorithm (SGA) described in [5] and adapted it to optimize the test functions. Each algorithm was executed 10 times on a standard PC computer (CPU: Intel i3, RAM: 8GB, Windows 10 operating system).

Table 1 shows the average time needed to reach the predetermined value of the optimized function. Table 2 shows the average number of fitness function calls needed to reach the predetermined value of the optimized function. The graph in Fig. 2 shows the average running time and the number of fitness function calls in the Genetic Algorithm (SGA), the Evolutionary Strategy and the proposed (GA-ES) system.

Table 2. The average number of fitness function calls needed to reach the predetermined value of the optimized function

Function	SGA		ES		GA-ES	
	The number of fitness function calls	σ (number)	The number of fitness function calls	σ (number)	The number of fitness function calls	σ (number)
f_1	43317	35232	3459	7272	4427	2535
f_2	31380	3622	3136*	1268*	8320	1096
f_3	75927	49888	16300	15662	5990	1520
$f_{Ra}2d$	53547	28302	750909	671132	7092	2587
$f_{Ra}5d$	609717	298351	-*	-*	41730	20003
$f_{Ra}10d$	967040	306805	-*	-*	459019	116278
$f_{ST}2d$	60942	28981	46345	137035	5720	1397
$f_{ST}5d$	193842	90410	-*	-*	11310	2297
$f_{ST}10d$	857025	281590	-*	-*	28990	7606
f_{Ro}	651025	680334	3868	2522	43295	27897
f_{Sh}	42222	36062	40722	70610	4030	737

* - at least once, the algorithm got stuck in the local optima.

5   Final Remarks

The proposed hybrid Genetic Algorithm-Evolutionary Strategy system was able to find a solution near the optimum for all tested functions.

In all tasks, the GA-ES system needed less fitness function calls, than SGA. Depending on the task, the system required from 3.3% to 47.4% of the fitness function calls, compared to the SGA. The standard deviation was lower, which may show a greater stability of the system.

In 4 tasks, the GA-ES system needed less fitness function calls, than ES. ES was unable to solve 4 tasks from the tests, and in addition, in one of the tasks, the algorithm gets stuck to the local optima at least one time.

In one task, a running time of GA-ES was longer than the SGA (62% worse). In all other tasks, the running time of GA-ES was 9% to 275% shorter compared to SGA.

In 2 tasks, a running time of GA-ES was longer than ES (worse than 8% to 97%). In all other tasks, the running time of GA-ES was 0.4% to 68% shorter.

Optimization of functions has shown, that the system has greater convergence and accuracy in comparison to the SGA and the system is more resistant to premature convergence to the local optimum compared to the ES.

In the proposed system it is possible to perform parallel calculations by the Genetic Algorithm and the Evolutionary Strategy, eg. by using multiple processors or processor cores.

The proposed system is an efficient tool for solving function optimization problems. It could be used for solving a very wide range of optimization problems.

References

1. Bäck, T., Hoffmeister, F., Schwefel, H.-P.: A survey of evolution strategies. In: Proceedings of the Fourth International Conference on Genetic Algorithms, vol. 2, no. 9. Morgan Kaufmann (1991)
2. Beyer, H.-G., Schwefel, H.-P.: Evolution strategies: a comprehensive introduction. J. Nat. Comput. 1(1), 3–52 (2002)
3. Goldberg, D.E.: Genetic Algorithms in Search, Optimization, and Machine Learning Reading. Addison-Wesley, Boston (1989)
4. Jensi, R., Jiji, G.W.: An improved krill herd algorithm with global exploration capability for solving numerical function optimization problems and its application to data clustering. Appl. Soft Comput. 46, 230–245 (2016)
5. Michalewicz, Z.: Genetic Algorithms + Data Structures = Evolution Programs. Springer, Berlin (1992). https://doi.org/10.1007/978-3-662-07418-3
6. Karaboga, D., Basturk, B.: A powerful and efficient algorithm for numerical function optimization: artificial bee colony (ABC) algorithm. J. Glob. Optim. 39(3), 459–471 (2007)
7. Kwasnicka, H.: Evolutionary Computation in Artificial Intelligence. Publishing House of the Wroclaw University of Technology, Wroclaw (1999). (in Polish)
8. Potter, M.A., De Jong, K.A.: A cooperative coevolutionary approach to function optimization. In: Davidor, Y., Schwefel, H.-P., Männer, R. (eds.) PPSN 1994. LNCS, vol. 866, pp. 249–257. Springer, Heidelberg (1994). https://doi.org/10.1007/3-540-58484-6_269
9. Pytel, K., Nawarycz, T.: Analysis of the distribution of individuals in modified genetic algorithms. In: Rutkowski, L., Scherer, R., Tadeusiewicz, R., Zadeh, L.A., Zurada, J.M. (eds.) ICAISC 2010. LNCS (LNAI), vol. 6114, pp. 197–204. Springer, Heidelberg (2010). https://doi.org/10.1007/978-3-642-13232-2_24
10. Pytel, K.: The fuzzy genetic strategy for multiobjective optimization. In: Proceedings of the Federated Conference on Computer Science and Information Systems, Szczecin (2011)
11. Pytel, K., Nawarycz, T.: The fuzzy-genetic system for multiobjective optimization. In: Rutkowski, L., Korytkowski, M., Scherer, R., Tadeusiewicz, R., Zadeh, L.A., Zurada, J.M. (eds.) EC/SIDE -2012. LNCS, vol. 7269, pp. 325–332. Springer, Heidelberg (2012). https://doi.org/10.1007/978-3-642-29353-5_38
12. Pytel, K., Nawarycz, T.: A fuzzy-genetic system for ConFLP problem. In: Advances in Decision Sciences and Future Studies, vol. 2. Progress & Business Publishers, Krakow (2013)
13. Pytel, K.: Hybrid fuzzy-genetic algorithm applied to clustering problem. In: Proceedings of the 2016 Federated Conference on Computer Science and Information Systems, Gdańsk (2016) https://doi.org/10.15439/2016F232
14. Pytel, K.: Hybrid multievolutionary system to solve function optimization problems. In: Proceedings of the 2017 Federated Conference on Computer Science and Information Systems, Prague, Czech Republik (201). https://doi.org/10.15439/2017F85
15. Rutkowska, D.: Intelligent Computational Systems. Academic Publishing House PLJ, Warsaw (1997)
16. Rutkowska, D., Pilinski, M., Rutkowski, L.: Neural Networks, Genetic Algorithms and Fuzzy Systems. PWN Scientific Publisher, Warsaw (1997)
17. Test Functions Index. http://infinity77.net/global_optimization/test_functions.html
18. Virtual Library of Simulation Experiments: Test Functions and Datasets. http://www.sfu.ca/~ssurjano/optimization.html

Horizontal Gene Transfer as a Method of Increasing Variability in Genetic Algorithms

Wojciech Rafajłowicz[✉]

Department of Computer Engineering,
Faculty of Electronics, Wrocław University of Science and Technology,
Wybrzeże Wyspiańskiego 27, 50-370 Wrocław, Poland
wojciech.rafajlowicz@pwr.edu.pl

Abstract. A horizontal (or lateral) gene transfer, well known in biology is used as an additional mutation factor in genetic algorithms used for optimization. Numerical results indicate the usefulness of this concept for problems of moderate size.

1 Introduction

The main source of evolution in living organisms is vertical transfer of genes. By this, we understand transfer from parent to the offspring regardless of reproduction type (with or without crossing). Any other method of transfer is called lateral gene transfer.

Examples of this kind of transfer can be found both in eukaryotic and prokaryotic organisms. A typical biological organism has much more non-coding DNA – sometimes as much as 97%.

In most cases structure of DNA has the result that the whole codon (part coding one aminoacid) is transferred. We would simulate this feature (see the next section) only partially.

The simplified diagram showing horizontal gene transfer (HGT) is shown in Fig. 1. Parts of DNA from a dead, disintegrated cell can, in some rare cases, penetrate the cellular wall and be incorporated into living cell DNA. Details can be found in the biological literature e.g., [3,5,13].

The probability of these complex changes is relatively small see [7] or [11].

Genetic algorithms are a group of methods that attempt simulation of evolution of living organisms. This process is used to find the best possible solution as an analogy of survival of the fittest. It was tried for the first time by Rechenberg [10] to solve real value problems and later developed further by Schwefel [12]. The history of genetic and evolutionary computation is described in [2].

Genetic algorithms are still developed further and used in many applications [4,9]. Many modifications are proposed like [8].

L. Rutkowski et al. (Eds.): ICAISC 2018, LNAI 10841, pp. 505–513, 2018.
https://doi.org/10.1007/978-3-319-91253-0_47

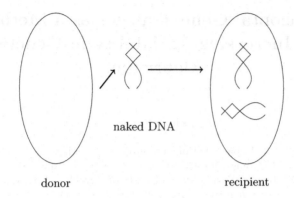

naked DNA

donor   recipient

Fig. 1. A simplified diagram of horizontal DNA transfer in biology.

2   Proposed Modification

The proposed method extends a traditional genetic algorithm with an additional mutation step. For obvious reasons this additional mutation has to happen with a different probability than a typical mutation.

An additional step works as follows. We assume that the whole previous population is now dead (this is not an exact statement because we draw parents of the new population from old individuals). The DNA from dead cells is floating around and can be horizontally transferred. One of those genotypes is chosen at random (as they are dead, their fitness cannot be measured). From this genotype, we draw a random part of it, by choosing a position where a fragment starts and its length. The distribution of random numbers is a normal distribution with μ and σ as algorithm parameters. The selected part is inserted in exactly the same position in the mutated genotype as it was in the "old" one. This process is shown in Fig. 3. The positioning of the proposed mutation is shown in Fig. 2 as a block with bold text.

3   Test Problems

As benchmarks three typical functions were used. These functions are used in many papers like [1] or [6] and it is also present as an element of many benchmark packages, sometimes in some modified form. Plots shown in subsequent subsections are for $d = 2$ for obvious reasons.

3.1   The Ackley Function

$$20 + e - 20e^{-0.2\sqrt{\frac{1}{d}\sum_{i=1}^{d} x_i^2}} - e^{\frac{1}{d}\sum_{i=1}^{n} cos(2\pi x_i)} \tag{1}$$

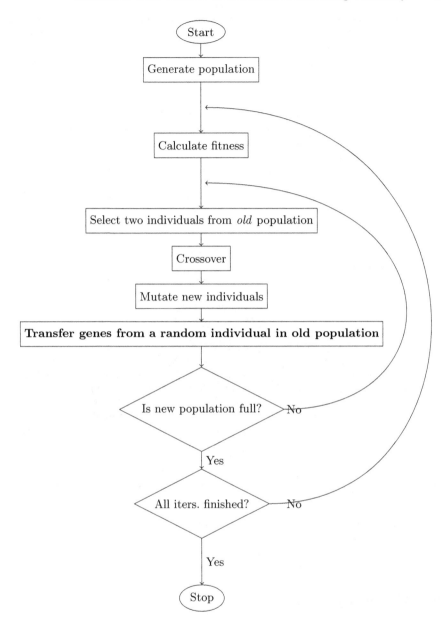

Fig. 2. The positioning of the proposed mutation in a genetic algorithm.

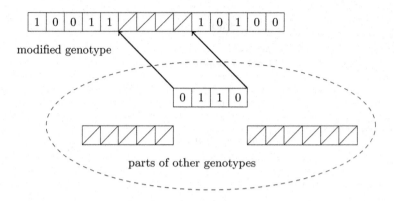

modified genotype

parts of other genotypes

Fig. 3. Genotype transfer simulated in GA.

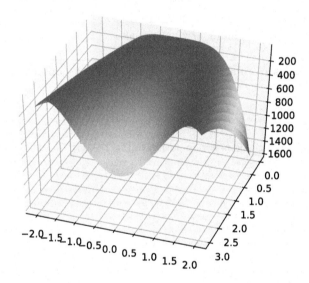

Fig. 4. Rosenbrock function.

3.2   The Rosenbrock Function

This is a typical test function for unimodal problems.

$$\sum_{i=1}^{d-1} \left(100 \cdot (x_{i+1} - x_i^2)^2 + (x_i - 1)^2\right) \tag{2}$$

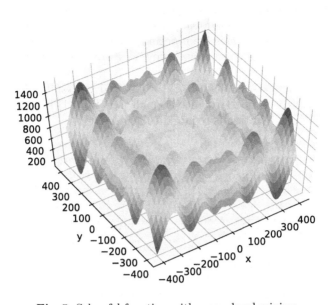

Fig. 5. Schwefel function with many local minima.

Table 1. Results of numerical trials – 50 trials for each combination of problem and algorithm.

Algorithm	Problem	n	m	Ptr	μ	σ	Iterations	Mean f	Std. dev.	Median
GA-HGT	Ackley	200	200	0.005	25.0	12.5	1000	86.047	0.741	86.026
GA	Ackley	200	200	X	25.0	12.5	1000	85.876	0.705	85.766
GA	Rosenbrock	200	200	X	25.0	12.5	1000	279776.160	7223.173	278333.511
GA-HGT	Rosenbrock	200	200	0.1	25.0	12.5	1000	279969.052	7008.858	280386.168
GA-HGT	Schwefel	200	200	X	25.0	12.5	1000	4203.501	0.633	4203.445
GA-HGT	Schwefel	200	200	0.1	25.0	12.5	1000	4203.488	0.517	4203.357

3.3 The Schwefel Function

This is one more difficult function with a large number of local minimas. The plot is visible in Fig. 5.

$$418.9829 \cdot d + \sum_{i=1}^{d} x_i sin\sqrt{|x_i|} \tag{3}$$

4 Numerical Results

Numerical experiments with 10-dimensional problems were carried out in order to assess how transferring parts of the genotype affects results. These results are shown in Table 1 and in the Figures.

In case of the Ackley function, we should start with assessing how the probability of gene transfer affects the average solution. In Fig. 7 we can see how the average result depends on this probability. It is clearly visible that in the case of the Ackley function, the biggest gain occurs when this probability is small. We can clearly see that HGT can be used only as a supplement to other mutation methods.

The best result used, namely $p_{HGT} = 0.005$ gives results visible in Fig. 6. The resulting increase in performance is small (see Table 1) but we can see that a smaller number of iteration is required to achieve it.

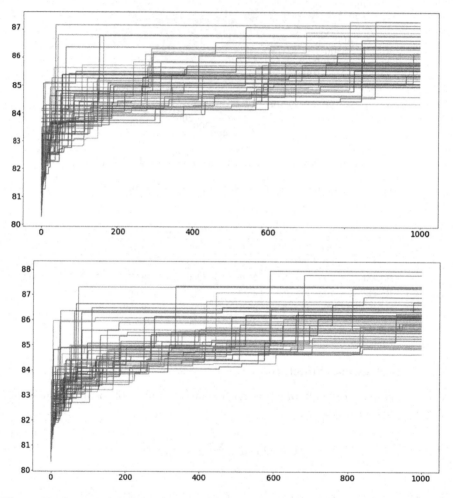

Fig. 6. Traditional GA (upper) and GA-HGT (lower) compared for the Ackley test problem. Results from 50 trial runs.

In case of the Schwefel, function the results were similar and the average result is better. Simulation results can be seen in Fig. 8.

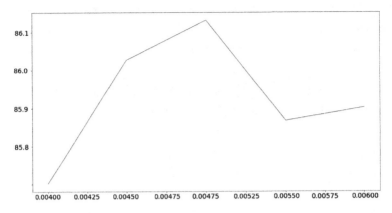

Fig. 7. The average result of the Ackley test function for different gene transfer probability.

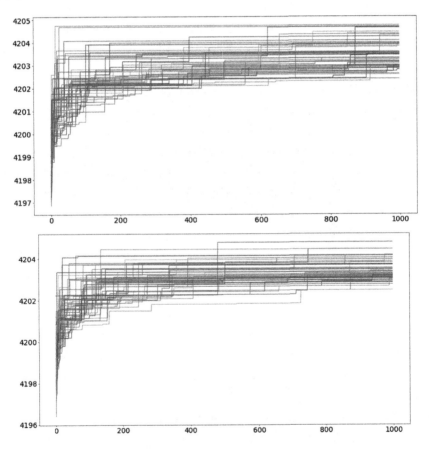

Fig. 8. Traditional GA (upper) and GA-HGT (lower) compared for the Schwefel test problem. Results from 50 trial runs.

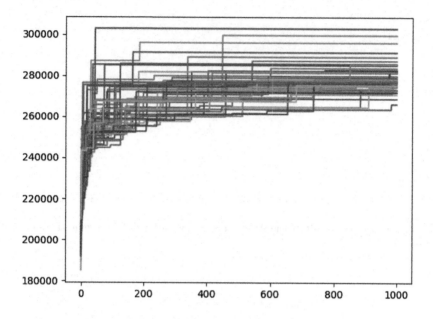

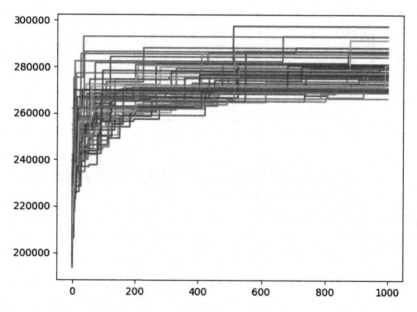

Fig. 9. Traditional GA (upper) and GA-HGT (lower) compared for the Rosenbrock test problem. Results from 50 trial runs.

The results are similar for the Rosenbrock test function, too (Fig. 9). Here we can also see that the best result is achieved in a smaller number of generations (see Fig. 4).

5 Summary

A horizontal gene transfer occurred as a viable alternative method of increasing performance of genetic algorithms.

One of the possible avenues of using this method, which was not tested in this paper, are genetic algorithms for discrete problems like job scheduling or the traveling salesman problem since the order in the cut-and-transferred part is kept.

References

1. Eiben, A.E., Bäck, T.: Empirical investigation of multi-parent recombination operators in evolution strategies. Evol. Comput. **5**(3), 347–365 (1997)
2. Fogel, D.B.: Evolutionary Computation: The Fossil Record. Wiley-IEEE Press, Hoboken (1998)
3. Keeling, P., Palmer, J.: Horizontal gene transfer in eukaryotic evolution. Nat. Rev. Genet. **9**(8), 605–618 (2008)
4. Li, M., et al.: Accurate determination of geographical origin of tea based on tera-hertz spectroscopy. Appl. Sci. **7**(2), 172 (2017)
5. Kishnapillai, V.: Horizontal gene transfer. J. Genet. **75**(2), 219–232 (1996)
6. Ortiz-Boyer, D., Hervás-Martínez, C., García-Pedrajas, N.: CIXL2: a crossover operator for evolutionary algorithms based on population features. J. Artif. Intell. Res. (JAIR) **24**, 1–48 (2005)
7. Prise, K.M., et al.: A review of studies of ionizing radiation-induced double-strand break clustering. Radiat. Res. **156**(5), 572–576 (2001)
8. Rafajłowicz, W.: Cosmic rays inspired mutation in genetic algorithms. In: Rutkowski, L., Korytkowski, M., Scherer, R., Tadeusiewicz, R., Zadeh, L.A., Zurada, J.M. (eds.) ICAISC 2017. LNCS (LNAI), vol. 10245, pp. 418–426. Springer, Cham (2017). https://doi.org/10.1007/978-3-319-59063-9_37
9. Ramadan, B.M.S.M., et al.: Hybridization of genetic algorithm and priority list to solve economic dispatch problems. In: Region 10 Conference (TENCON). IEEE (2016)
10. Rechenberg, I.: Evolution Strategy: Optimization of Technical Systems by Means of Biological Evolution, vol. 104. Fromman-Holzboog, Stuttgart (1973)
11. Scally, A.: The mutation rate in human evolution and demographic inference. Curr. Opin. Genet. Dev. **41**, 36–43 (2016)
12. Schwefel, H.P.: Evolution strategy and numerical optimization. Technical University of Berlin (1975)
13. Thomas, C., Nielsen, K.: Mechanisms of, and barriers to, horizontal gene transfer between bacteria. Nat. Rev. Microbiol. **3**(9), 711–721 (2005)

Evolutionary Induction of Classification Trees on Spark

Daniel Reska, Krzysztof Jurczuk$^{(\boxtimes)}$, and Marek Kretowski

Faculty of Computer Science, Bialystok University of Technology,
Wiejska 45a, 15-351 Bialystok, Poland
{d.reska,k.jurczuk,m.kretowski}@pb.edu.pl

Abstract. Evolutionary-based approaches have recently been increasingly proposed for data mining tasks, but their real applicability depends on efficiency and scalability for large-scale data. It is clear that parallel and distributed processing support is indispensable herein. Apache Spark is one of the most promising cluster-computing engines for Big Data. In this paper, we investigate the application of Spark to speed up an evolutionary induction of classification trees in the Global Decision Tree (GDT) system. The system simultaneously searches for the tree structure and tests in non-terminal nodes due to specialized genetic operators. As the original GDT system is implemented in C++, the Java-based module is developed for Spark-based acceleration of the most computationally demanding fitness evaluation. The training dataset is transformed to Resilient Distributed Dataset, which enables in-memory processing of dataset's parts on workers. Preliminary experimental validation on large-scale artificial and real-life datasets shows that the proposed solution is efficient and scales well.

Keywords: Decision tree · Evolutionary algorithms · Spark
Distributed computing · Data mining · Large-scale data

1 Introduction

In the last decade, the availability of large data volumes in business, industry and research increased tremendously. This gives huge opportunities as well as challenges for knowledge discovery from data [20]. Typical data mining approaches, originating from classical statistical pattern recognition or machine learning algorithms, are usually computationally complex. They become often useless when confronted with contemporary Big Data warehouses. One of the possible solutions is adaptation and extension of existing methods by using parallel and/or distributed processing [12]. Nowadays a lot of researchers commitment is directed at developing flexible frameworks for migrating calculation to computing clusters or graphical processing units (GPU)s, as former low-level approaches like e.g. MPI were relatively demanding to apply. Spark [21] is a novel alternative for cluster-computing, especially well suited for in-memory computing. It offers

© Springer International Publishing AG, part of Springer Nature 2018
L. Rutkowski et al. (Eds.): ICAISC 2018, LNAI 10841, pp. 514–523, 2018.
https://doi.org/10.1007/978-3-319-91253-0_48

fault-tolerance, robust horizontal scalability as well as rich monitoring and diagnostic tools.

Evolutionary algorithms [16] are currently one of the most popular population-based meta-heuristics for solving optimization and search problems. These nature-inspired techniques try to mimic biological evolution, where more fitted individuals have a better chance to survive and reproduce. Evolutionary algorithms are known for their robustness and are successfully applied to a wide range of problems. On the other hand, these are not the fastest techniques when run on a single processor, so especially for the big data mining, efficient parallel and/or distributed implementations and accelerations are indispensable [2,11].

In this paper, we investigate speeding up an evolutionary induction of decision trees. Global Decision Tree (GDT) is a data mining system that enables induction of various variants of decision trees: univariate, oblique and mixed ones. In a single run of an evolutionary algorithm, the tree structure, tests and predictions in the leaves are searched. The resulting decision trees are generally simple and accurate, but for large-scale data computation times are long [3]. To overcome this limitation, we tried to develop more time efficient implementations on computing clusters using MPI/OpenMP [5,6] and on GPUs [13] using CUDA. In this paper, we explore the benefits of Spark as an acceleration engine for the GDT system.

The rest of the paper is organized as follow. Firstly, Spark is briefly introduced and its applications in an evolutionary search and especially evolutionary data mining are listed. In the next section, the most important features of global induction of decision trees are shortly recalled. Then Spark-based acceleration of the GDT system is presented and vital implementation issues are also shortly discussed. In Sect. 4 experimental validation of the proposed solution on large-scale artificial and real datasets is described. The paper is concluded in the last section and possible directions of future works are sketched.

1.1 Spark

Apache Spark [21] is an open-source distributed computing engine for large-scale data processing. Spark provides high-level APIs in Java, Scala, Python and R and offers tools for structured data processing, machine learning, graph processing and data streaming.

Apache Spark architecture is based on a concept of Resilient Distributed Dataset (RDD) - an immutable distributed data structure that provides fault tolerance and can be processed in parallel.

In contrast to Hadoop MapReduce, where intermediate results are stored on disk, Spark processes data in distributed shared memory model, preferably in the RAM of the cluster nodes, which makes it much more suited for iterative algorithms and interactive data exploration. Furthermore, Spark offers a much broader set of high-level functional-style data operators that simplify the implementation of distributed applications.

1.2 Related Works

One of the first attempts to apply Spark to evolutionary algorithms was proposed by Deng et al. [8]. The authors implemented a parallel version of differential evolution where the population is treated as an RDD and only the fitness evaluation is distributed to workers. Teijero et al. [19] also tried to parallelize differential evolution, focusing on individual's mutation. The proposed three master-slave implementations were not efficient, so the authors decided to switch to the island model with local-range mutations and rare migrations. In [18] a parallel genetic algorithm for pairwise test suite generation was proposed. A population was stored as an RDD and the fitness was evaluated on workers, whereas genetic operators were applied in subpopulation corresponding to partitions.

As for evolutionary data mining approaches using Spark, in [9] fuzzy rule-based classifiers were generated. The fitness function of multi-objective algorithm scanned for entire training datasets and this computationally expensive operation was divided among cluster nodes. Another multi-objective fuzzy approach for subgroup discovery was presented in [17]. Evolutionary search for a set of rules was executed in separate dataset partitions and repeated for each value of the target variable. Then reduction of the rules obtained in each partition based on the global measures was carried out. In [10] the authors tried to scale a genetic programming-based solution for symbolic regression and proposed a fitness evaluation service based on Spark.

2 Global Decision Tree Induction Framework

The GDT system [7,14] enables induction of several types of decision trees, depending among others on the type of a predictive task to be solved, the permitted test types in nodes and models in leaves, etc. All variants of the algorithm share the same typical evolutionary process [16] with an unstructured, fixed size population and a generational selection.

Every individual in a population corresponds to a single decision tree, whose size and structure is dynamically changed during the evolution. There is no special encoding and the trees are represented in the actual form. In the simplest case of binary, univariate, classification trees, only inequality tests on continuous-valued features with two outcomes are placed in the internal nodes, whereas class labels are associated with leaves. More complicated variants can, for example, rely on oblique tests in non-terminal nodes or multiple-linear models in leaves.

Individuals in an initial population are created with a simplified top-down induction, which is applied to randomly selected small sub-samples of the learning data.

As decision trees are "peculiar", hierarchical structures, the variation operators should be appropriately designed and applied so that the evolution is efficient and robust. In the GDT system, two groups of specialized genetic operators were developed: mutation-like operators which are applied to single individuals and cross-over ones which operate on pairs of trees. In each group, several variants of operators were proposed depending, among others, on the tree type and node type. Each time, the choices of the variant and affected nodes (or node) are random, but probability distributions are not uniform across both nodes and variants. For nodes the location in the tree (modifying nodes from upper levels result in more global changes) and quality of the subtree (less accurate nodes should be modified more often) are taken into account. For a drawn node only matching variants are considered, but the user preferences can be also included.

When a non-terminal node is to be mutated, it can be pruned into a leaf or its test can be modified, recreated or copied from another node. These changes can be purely random or can be based on a local search (so-called memetic extensions). As for a leaf node, it can be transformed into a stump (one internal node and leaves) and a new test needs to be created. Concerning simpler symmetric cross-over variants, only tests or whole sub-tress can be exchanged between two individuals. In more complicated asymmetric scenarios, a subtree from a donor position in one tree is duplicated and implanted to a receiver position in the second tree. In certain variants of both mutation and cross-over operators, randomly chosen dipoles[1] can be used to guide the decision tree modifications.

The fitness function, which drives the evolutionary search, should as much as possible close reflect the goal of the algorithm. In many of data mining tasks, one tries to find the best predictor, but simultaneously simplicity of such a predictor is often desired. It is well known that a classifier which perfectly classifies the training data is usually much worse, when tested on unseen data, due to the over-fitting problem. In case of decision trees, the predictor complexity can be reduced just to the number of nodes or can rely on the complexity of tests in internal nodes and/or models in the leaves. In the GDT system many forms of the fitness function, both single-objective or multi-objective, can be applied. As, in this paper, only univariate decision trees applied to classification problem are considered, the simple weighted form of the fitness function can be used:

$$Fitness(T) = Accuracy(T) - \alpha * Size(T), \qquad (1)$$

where $Accuracy(T)$ represents the reclassification quality of the tree T, $Size(T)$ stands for the number of nodes in T and α is the user-supplied parameter, which can be possibly tuned up for the given dataset (default value $\alpha = 0.001$).

3 Spark-Accelerated Evolutionary Induction

For large-scale data, the most time-consuming operation in evolutionary induction is clearly the fitness evaluation, as it requires re-classifying of the whole training dataset for every tree in each iteration. Thus, in the proposed Spark-based acceleration, we decided to concentrate only on distributing the training dataset, as it enables the most productive parallelization. The rest of the evolution is unaffected in principle and is realized sequentially.

Firstly, the training dataset is loaded line-by-line and transformed into an RDD of elements representing observation groups of the same size. The number

---

[1] A dipole is a pair of observations; if observations come from different classes, a dipole is called mixed.

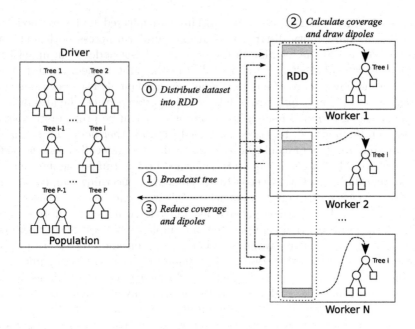

Fig. 1. Distributed fitness evaluation and dipoles searching.

of groups should correspond to the number of available computing nodes so the dataset chunks can be effectively cached in their memory. During the induction, all observations are passed through every transferred decision tree and arrangement of observations in its leaves is obtained (Fig. 1). It is realized by typical pair of **map-reduce** operations evoked on the grouped RDD: each group emits a locally processed copy of the tree (**map(group)** → **tree**) and the local trees are then reduced into a final result. Moreover, for each processed group a set of dipoles, which can be potentially used in genetic operators, is randomly chosen and implicitly reduced. Finally, the class distributions in each leaf can be calculated and the overall accuracy is estimated.

The proposed method is highly iterative, therefore an effective distribution of the dataset in the cluster memory is one of the crucial implementation aspects. The dataset can be loaded from a local file and distributed on the cluster (using **SparkContext::parallelize()** method from Spark API) or loaded from HDFS with **SparkContext::textFile()** in case of larger files. Each file line is parsed into an observation object which is randomly assigned to a group with a numeric ID. This creates a key-value RDD of **<groupID, observation>** elements that are grouped using **RDD::groupByKey()**. As the number of groups equals the number of computing nodes, this operation triggers a global dataset repartition that results in a uniform distribution of the data over the cluster. As this is a one-time operation, the cost of repartition is negligible and the lack of data skew is highly beneficial.

The Spark-based solution uses a multi-process architecture (Fig. 2). The original GDT system, implemented in C++, was modified to communicate with the

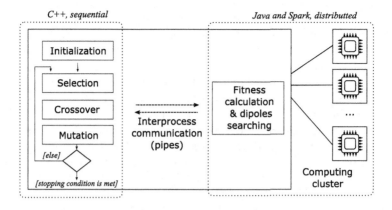

Fig. 2. Implementation concerns.

main Spark process (Driver) that reimplements the operations, which require access to the dataset, in Java. The Driver dispatches the work to multiple Spark Worker processes that are distributed over the cluster nodes. Both the Spark Driver and GDT applications are running on the same machine and use named pipes mechanism for inter-process request/response communication. As a result, the core evolution is performed in one process (C++ part), whereas the distributed fitness evaluation and dipoles searching are realized in Spark (Java part). Such a heterogeneous implementation provided a convenient way to independently modify and test both components.

4   Experimental Validation

The proposed solution is validated on both artificial and real-life large-scale datasets. The first group comprises 4 variants of *Chess* dataset (inspired by 3×3 chessboard) with increasing number of instances/observations. Moreover, two real-life datasets (*Suzy* and *Higgs*), which are the biggest classification problems available in UCI Repository [4], are analysed. The characteristics of the considered datasets are given in Table 1.

Table 1. Characteristics of the datasets: name, number of instances, number of attributes, and number of classes.

Dataset	Instances	Attributes	Classes
Chess1M	1 000 000	2	2
Chess5M	5 000 000	2	2
Chess10M	10 000 000	2	2
Chess20M	20 000 000	2	2
Suzy	5 000 000	18	2
Higgs	11 000 000	28	2

All presented results were obtained with a default set of parameter's values from the sequential version of the GDT system. In this paper, we are focused only on time performance thus results for the classification accuracy are not included. Moreover, the proposed solution only accelerates evolution so it does not really affect the resulting classifiers. For detailed information about accuracy performance, we refer a reader to our previous papers [7,14].

The experiments were performed on a cluster of 18 SMP workstations connected by a Gigabit Ethernet network. Each cluster node was equipped with a quad-core Intel Xeon E3-1270 3.4 GHz CPU, 16 GB RAM and was running Ubuntu 16.04 (Linux 4.4). 16 worker nodes were used by Spark executors, one node was dedicated to Spark Master and HDFS NameNode and the last node was running Spark Driver and GDT C++ processes. The cluster used Apache Hadoop 2.7.3 and Apache Spark 2.2.0 [1] deployed in a standalone mode with a single executor on each worker node.

The experiments were run on a different number of total CPU cores, distributed uniformly over the executor worker nodes. The speedups calculated for all *ChessXM* variants and the different number of cores (from 2 up to 64) are presented in Fig. 3. The speedup is defined as the ratio of the baseline execution time to the time of a given Spark run. The baseline time was obtained by running the solution on a single-threaded version of the Java module with the Spark integration completely disabled.

It should be noticed that for the smallest dataset (1 million of instances), the proposed Spark-based acceleration is completely useless. For larger variants, the observed speedups are much better, but not very impressive (e.g. around 10 for 32 cores). It could be also noted that moving from 32 to 64 cores, results in the visibly slower increase of speedup. Such behavior can be easily

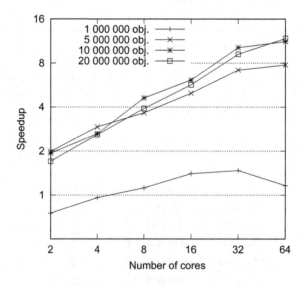

Fig. 3. Performance evaluation: speedup obtained on a different number of cores for the *Chess* dataset variants with increasing number of observations.

explained by an inevitable executor overhead (e.g. scheduler delay, task serialization/deserialization, result read/write and shuffle time) introduced by Spark. Figure 4 illustrates this situation for the *Higgs* dataset. Each computational task contains a short (about 8 ms) overhead that becomes more significant in the overall performance as the actual computing time decreases with the increase of a number of cores.

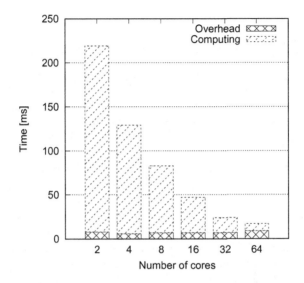

Fig. 4. Performance evaluation: detailed time-sharing information of induction task with a different number of cores. Evaluation performed on the *Higgs* dataset.

The results obtained on the real-life data are presented in Table 2. Although datasets contain a larger number of attributes, the observed patterns are rather expected and analogous to previously seen on 3 larger artificial datasets. It can be also noticed that for the smaller dataset (*Suzy*), increasing the number of cores from 32 to 64, does not give any significant acceleration as a slightly increased overhead cancels out the speedup gain.

In this paper, Spark is applied to speed up our global decision tree induction, but in the well-know Spark's *MLlib* library [15], distributed versions of classical top-down decision tree induction algorithms are available. In order to have at least an approximate reference, the basic algorithm (with default settings) was launched on *Suzy* and *Higgs* datasets. The estimated accuracies are practically the same as obtained by the GDT system, but the decision trees generated by *MLlib* are clearly overgrown (63 nodes for both datasets as opposed to 5 to 9 and 5 to 11 nodes, correspondingly). However, the induction time on 64 cores is only 1–2 min, which is substantially shorter than the evolutionary approach (around 30 min). Such a difference in tree complexity can be clearly explained by the fact that the *MLlib* algorithm is based on the top-down approach whereas the GDT system uses global induction. Global evolutionary inducers are known to provide smaller (less complicated) models [3].

Table 2. Obtained speedup for real-life datasets for a different number of CPU cores.

Dataset	Speedup on different number of cores					
	2	4	8	16	32	64
Suzy	1.74	2.64	3.79	5.80	8.16	8.03
Higgs	1.72	2.49	3.65	6.15	10.22	14.23

5   Conclusion

In this paper, Apache Spark is applied to speed up an evolutionary induction of classification trees in the Global Decision Tree (GDT) system. The most computationally demanding operations are fitness evaluations as they require reclassifying of the whole training dataset and for large-scale data it results in too long induction time. Hence, the dataset is transformed into RDD and it enables distributed and in-memory calculations on workers. Even preliminary experimental results show that the time of evolutionary induction can be significantly reduced for the largest datasets. It should be also noticed that the Spark-based solution can be easily scaled up just by connecting new computing stations.

In the future works, we plan to perform more extensive testing of the proposed approach. We are especially interested in revealing what are the limits of Spark-based induction: how big datasets could be processed on a given hardware in a fixed time. Our future investigations will also deal with a hybrid, e.g. Spark+CUDA, approaches as well as parallelization of other more elaborated decision trees, like model or oblique trees.

Acknowledgments. This work was supported by the grant S/WI/2/18 from Bialystok University of Technology founded by Polish Ministry of Science and Higher Education.

References

1. The Apache Software Foundation. Apache Spark - Lightning-Fast Cluster Computing (2018). https://spark.apache.org/
2. Alba, E., Tomassini, M.: Parallelism and evolutionary algorithms. IEEE Trans. Evol. Comput. **6**(5), 443–462 (2002)
3. Barros, R.C., Basgalupp, M.P., Carvalho, A.C., Freitas, A.A.: A survey of evolutionary algorithms for decision-tree induction. IEEE Trans. SMC, Part C **42**(3), 291–312 (2012)
4. Blake, C., Keogh, E., Merz, C.: UCI repository of machine learning databases (1998). http://www.ics.uci.edu/~mlearn/MLRepository.html
5. Czajkowski, M., Jurczuk, K., Kretowski, M.: A parallel approach for evolutionary induced decision trees. MPI+OpenMP implementation. In: Rutkowski, L., Korytkowski, M., Scherer, R., Tadeusiewicz, R., Zadeh, L.A., Zurada, J.M. (eds.) ICAISC 2015. LNCS (LNAI), vol. 9119, pp. 340–349. Springer, Cham (2015). https://doi.org/10.1007/978-3-319-19324-3_31

6. Czajkowski, M., Jurczuk, K., Kretowski, M.: Hybrid parallelization of evolutionary model tree induction. In: Rutkowski, L., Korytkowski, M., Scherer, R., Tadeusiewicz, R., Zadeh, L.A., Zurada, J.M. (eds.) ICAISC 2016. LNCS (LNAI), vol. 9692, pp. 370–379. Springer, Cham (2016). https://doi.org/10.1007/978-3-319-39378-0_32

7. Czajkowski, M., Kretowski, M.: Evolutionary induction of global model trees with specialized operators and memetic extensions. Inf. Sci. **288**, 153–173 (2014)

8. Deng, C., Tan, X., Dong, X., Tan, Y.: A parallel version of differential evolution based on resilient distributed datasets model. In: Gong, M., Pan, L., Song, T., Tang, K., Zhang, X. (eds.) BIC-TA 2015. CCIS, vol. 562, pp. 84–93. Springer, Heidelberg (2015). https://doi.org/10.1007/978-3-662-49014-3_8

9. Ferranti, A., Marcelloni, F., Segatori, A., Antonelli, M., Ducange, P.: A distributed approach to multi-objective evolutionary generation of fuzzy rule-based classifiers from big data. Inf. Sci. **415–416**, 319–340 (2017)

10. Funika, W., Koperek, P.: Towards a scalable distributed fitness evaluation service. In: Wyrzykowski, R., Deelman, E., Dongarra, J., Karczewski, K., Kitowski, J., Wiatr, K. (eds.) PPAM 2015. LNCS, vol. 9573, pp. 493–502. Springer, Cham (2016). https://doi.org/10.1007/978-3-319-32149-3_46

11. Gong, Y.J., Chen, W.N., Zhan, Z.H., Zhang, J., Li, Y., Zhang, Q., Li, J.J.: Distributed evolutionary algorithms and their models: a survey of the state-of-the-art. Appl. Soft Comput. **34**, 286–300 (2015)

12. Grama, A., Karypis, G., Kumar, V., Gupta, A.: Introduction to Parallel Computing. Addison-Wesley, Boston (2003)

13. Jurczuk, K., Czajkowski, M., Kretowski, M.: Evolutionary induction of a decision tree for large-scale data: a GPU-based approach. Soft Comput. **21**(24), 7363–7379 (2017)

14. Kretowski, M., Grzes, M.: Evolutionary induction of mixed decision trees. Int. J. Data Warehous. Min. (IJDWM) **3**(4), 68–82 (2007)

15. Meng, X., Bradley, J., Yavuz, B., Sparks, E., Venkataraman, S., Liu, D., Freeman, J., Tsai, D., Amde, M., Owen, S., et al.: MLlib: machine learning in apache spark. J. Mach. Learn. Res. **17**(1), 1235–1241 (2016)

16. Michalewicz, Z.: Genetic Algorithms + Data Structures = Evolution Programs. Springer Science & Business Media, Heidelberg (2013). https://doi.org/10.1007/978-3-662-03315-9

17. Pulgar-Rubior, F., Rivera-Rivas, A., Perez-Godoy, M., Gonzalez, P., Carmona, C., del Jesus, M.: MEFASD-BD: multi-objective evolutionary fuzzy algorithm for subgroup discovery in big data environments - a MapReduce solution. Knowl.-Based Syst. **117**, 70–78 (2017)

18. Qi, R., Wang, Z., Li, S.: A parallel genetic algorithm based on Spark for pairwise test suite generation. J. Comput. Sci. Technol. **31**(2), 417–427 (2016)

19. Teijeiro, D., Pardo, X.C., González, P., Banga, J.R., Doallo, R.: Implementing parallel differential evolution on spark. In: Squillero, G., Burelli, P. (eds.) EvoApplications 2016. LNCS, vol. 9598, pp. 75–90. Springer, Cham (2016). https://doi.org/10.1007/978-3-319-31153-1_6

20. Wu, X., Zhu, X., Wu, G.Q., Ding, W.: Data mining with big data. IEEE Trans. Knowl. Data Eng. **26**(1), 97–107 (2014)

21. Zaharia, M., et al.: Apache Spark: a unified engine for big data processing. Commun. ACM **59**(11), 56–65 (2016)

How Unconventional Chaotic Pseudo-Random Generators Influence Population Diversity in Differential Evolution

Roman Senkerik[1]([✉]) [ID], Adam Viktorin[1] [ID], Michal Pluhacek[1] [ID],
Tomas Kadavy[1] [ID], and Ivan Zelinka[2] [ID]

[1] Faculty of Applied Informatics, Tomas Bata University in Zlin,
T. G. Masaryka 5555, 760 01 Zlin, Czech Republic
{senkerik,aviktorin,pluhacek,kadavy}@utb.cz
[2] Faculty of Electrical Engineering and Computer Science,
Technical University of Ostrava, 17. listopadu 15,
708 33 Ostrava-Poruba, Czech Republic
ivan.zelinka@vsb.cz

Abstract. This research focuses on the modern hybridization of the discrete chaotic dynamics and the evolutionary computation. It is aimed at the influence of chaotic sequences on the population diversity as well as at the algorithm performance of the simple parameter adaptive Differential Evolution (DE) strategy: jDE. Experiments are focused on the extensive investigation of totally ten different randomization schemes for the selection of individuals in DE algorithm driven by the default pseudo random generator of Java environment and nine different two-dimensional discrete chaotic systems, as the chaotic pseudo-random number generators. The population diversity and jDE convergence are recorded for 15 test functions from the CEC 2015 benchmark set in $30D$.

Keywords: Differential Evolution · Complex dynamics
Deterministic chaos · Population diversity · Chaotic map

1 Introduction

This research deals with the mutual intersection of the two different soft-computing fields, which are the complex sequencing and inner dynamics given by the selected chaotic systems, and evolutionary computation techniques (ECT's).

Together with this persistent development in above-mentioned mainstream research topics, the popularity of hybridizing of chaos and metaheuristic algorithms is growing every year. Recent research in chaotic approach for metaheuristics uses various chaotic maps in the place of pseudo-random number generators (PRNG). Also, vice-versa, the metaheuristic approach in chaos control/synchronization is more popular in recent years.

© Springer International Publishing AG, part of Springer Nature 2018
L. Rutkowski et al. (Eds.): ICAISC 2018, LNAI 10841, pp. 524–535, 2018.
https://doi.org/10.1007/978-3-319-91253-0_49

The original concept of embedding chaotic dynamics into the evolutionary/swarm algorithms as chaotic pseudo-random number generator (CPRNG) is given in [1]. Firstly, the Particle Swarm Optimization (PSO) algorithm with elements of chaos was introduced as CPSO [2], followed by the initial testing of chaos embedded Differential evolution (DE) [3], PSO with an ensemble of chaotic systems [4]. Recently the chaos driven heuristic concept has been utilized in several swarm-based algorithms like ABC algorithm [5], Firefly [6] and other metaheuristic algorithms [7–10], as well as many applications with DE [11].

The original chaos-based approach is tightly connected with the importance of randomization within heuristics as compensation of a limited amount of search moves. This idea has been carried out in several papers describing different techniques to modify the randomization process [12]. The importance and influence of randomization operations were also profoundly experimentally tested in simple control parameter adjustment DE strategy jDE [13].

The focus of this research is the deeper insight into the population dynamics of the selected DE strategy (jDE) [14] when the directly embedded CPRNG is driving the indices selection. Many DE variants have been recently developed with the emphasis on control parameters self-adaptivity, learning techniques and better understandability of population dynamics. DE has been modified and extended several times using new proposals of versions, and the performances of different DE variants have been widely studied and compared with other ECTs.

Currently, DE [15] is a well-known evolutionary algorithm complex optimization problems in the continuous domain. Over recent decades, DE has won most of the evolutionary algorithm competitions in the leading scientific conferences [16], as well as being applied to several applications.

The organization of this paper is following: Firstly, the motivation for this research is proposed. The next sections are focused on the description of the concept of chaos driven jDE, the experiment background, and results discussions.

2 Motivation and Related Research

Recently, chaos with its properties like ergodicity, stochasticity, self-similarity, and density of periodic orbits became very popular and modern tool for improving the performance of various ECTs. Nevertheless, the questions remain unanswered, as to why it works, why it may be beneficial to use the chaotic sequences for pseudo-random numbers driving the selection, mutation, crossover or other processes in particular heuristics. This research represents a follow up to previous initial experiments with different sampling rates applied to the chaotic sequences resulting in keeping, partially/fully removing of traces of chaos [17,18]. These works show the possible hidden features of chaotic maps (discretized flows), that can be beneficial for the heuristic. Not the distribution of the CPRNG used, but the unique sequencing given by the chaotic attractor hidden dynamics seems to be the key feature that may improve the heuristic performance. In some instances, the population dynamics (which can be also considered as a chaotic or at the edge of chaos) seems to self-synchronize with the chaotic attractor dynamics and sequencing.

This research is also an extension and continuation of the previous successful experiment with the single/multi-chaos driven PSO [4] and jDE [19], where the positive influence of hidden complex dynamics for the heuristic performance has been experimentally shown.

The motivation and the novelty of the research are given by the investigating the influence of the original chaotic sequences to the population diversity, tightly connected with the algorithm performance of the basic control parameter adjustment DE strategy: jDE. This strategy was selected as a compromise between original simple DE and the most recent Success-History based Adaptive Differential Evolution (SHADE) variants [16], where the influence of chaotic dynamics may be suppressed by the complex adaptive process and operations with the archive.

3   Chaotic Systems for CPRNGs

Following nine well known and frequently studied discrete dissipative chaotic maps were used as the CPRNGs for jDE. Systems of the interest were: *Arnold Cat Map*, *Burgers Map* (1), *Delayed Logistic Map*, *Dissipative Standard Map*, *Henon Map*, *Ikeda Map*, *Lozi Map* (2), *Sinai Map* and *Tinkerbell Map*. With the settings and definitions as in [20], systems exhibit typical chaotic behavior. Please refer to the (1) and (2) for the examples of maps definition. Also, Fig. 1 shows the short chaotic sequences for Burgers and Lozi maps. These plots support the claims that due to the presence of self-similar chaotic sequences the heuristic is forced to neighborhood-based selection (or alternative communication in swarms).

$$X_{n+1} = aX_n - Y_n^2$$
$$Y_{n+1} = bY_n + X_nY_n$$
$$(1)$$

$$X_{n+1} = 1 - a|X_n| + bY_n$$
$$Y_{n+1} = X_n$$
$$(2)$$

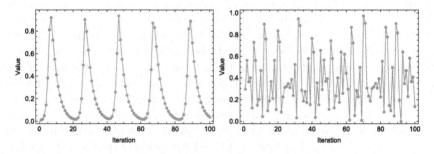

Fig. 1. Chaotic sequences normalized to the typical range of 0−1 for CPRNG; Burgers Map (left) with significant sequencing and periodicity; Lozi Map (right) with patterns of self-similarity.

4   The Concept of ChaosDE with Discrete Chaotic System as Driving CPRNG

The general idea of CPRNG is to replace the default PRNG with the chaotic system (either discrete map or discretized time-continuous flow). As the chaotic system is a set of equations with a static start position, we created random start positions of the chaotic systems, to have different start position for different experiments. Thus we are utilizing the typical feature of chaotic systems, which is extreme sensitivity to the initial conditions, popularly known as "butterfly effect", as the random seed. This random position is initialized with the default PRNG. Once the start position of the chaotic system has been obtained, the system generates the next sequence using its current position. Used approach is based on the following definition (3):

$$rndreal = \quad \text{mod (abs}\,(rndChaos)\,, 1.0) \tag{3}$$

5   Differential Evolution

This section describes the basics of original DE and jDE strategies. The original DE [15] has four static control parameters – a number of generations G, population size NP, scaling factor F and crossover rate CR. In the evolutionary process of DE, these four parameters remain unchanged and depend on the initial user setting. jDE algorithm, on the other hand, adapts the F and CR parameters during the evolution. The mutation strategy for jDE is adapted from the original DE. The concept of essential operations in jDE algorithm is shown in following sections, for a detailed description of either original DE or for jDE see [15].

5.1   The jDE Algorithm

In this research, we have used jDE and chaotic C-jDE with original DE "rand/1/bin" (4) mutation strategy and binomial crossover (5). The generated ensemble of two control parameters F_i and CR_i is assigned to each i-th individual of the population and survives with the solution if an individual is transferred to the new generation. The initialization of values of F and CR is designed to be either fully random with uniform distribution for each individual in the population or can be set according to the recommended values in the literature. If the newly generated solution is not successful, i.e., the trial vector has worse fitness than the compared original active individual; the new (possibly) reinitialized control parameters values disappear together with not successful solution. The both aforementioned DE control parameters may be randomly mutated with predefined probabilities τ_1 and τ_2. If the mutation condition happens, a new random value of $CR \in [0, 1]$ is generated, possibly also a new value of F which is mutated in $[F_l, F_u]$. These new control parameters are after that stored in the new population. Input parameters are typically set to $F_l = 0.1$, $F_u = 0.9$, $\tau_1 = 0.1$, and $\tau_2 = 0.1$ as originally given in [14].

Mutation Strategies and Parent Selection. The parent indices (vectors) are selected either by standard PRNG with uniform distribution or by CPRNG in case of chaotic versions. Mutation strategy "rand/1/bin" uses three random parent vectors with indexes $r1$, $r2$ and $r3$, where $r1 = U[1, NP]$, $r2 = U[1, NP]$, $r3 = U[1, NP]$ and $r1 \neq r2 \neq r3$. Mutated vector $v_{i,G}$ is obtained from three different vectors x_{r1}, x_{r2}, x_{r3} from current generation G with the help of scaling factor F_i as follows:

$$v_{i,G} = x_{r1,G} + F_i \left(x_{r2,G} - x_{r3,G} \right) \tag{4}$$

Crossover and Selection. The trial vector $u_{i,G}$ which is compared with original vector $x_{i,G}$ is completed by crossover operation (5). CR_i value in jDE algorithm is not static.

$$u_{j,i,G} = \begin{cases} v_{j,i,G} \text{ if } U[0,1] \leq CR_i \text{ or } j = j_{rand} \\ x_{j,i,G} \qquad \text{otherwise} \end{cases} \tag{5}$$

Where j_{rand} is a randomly selected index of a feature, which has to be updated ($j_{rand} = U[1, D]$), D is the dimensionality of the problem.

The vector which will be placed into the next generation $G + 1$ is selected by elitism. When the objective function value of the trial vector $u_{i,G}$ is better than that of the original vector $x_{i,G}$, the trial vector will be selected for the next population. Otherwise, the original will survive (6).

$$x_{i,G+1} = \begin{cases} u_{i,G} \text{ if } f(u_{i,G}) < f(x_{i,G}) \\ x_{i,G} \qquad \text{otherwise} \end{cases} \tag{6}$$

6 Experiment Design

For the population diversity analysis and performance comparisons in this research, the CEC 15 benchmark was selected. The dimension D was set to 30. Every instance was repeated 51 times with the maximum number of objective function evaluations set to 300 000 ($10,000 \times D$). The convergence and population diversity were recorded for all tested algorithm – original jDE and nine versions of C_jDE with different CPRNGs. All algorithms used the same set of control parameters: population size $NP = 50$ and initial settings $F = 0.5$, $CR = 0.8$. Experiments were performed in the environment of *Java*; jDE, therefore, has used the built-in *Java linear congruential pseudorandom number generator* representing traditional pseudorandom number generator in comparisons.

The Population Diversity (PD) measure used in this paper was described in [21] and is based on the sum of deviations (8) of individual's components from their corresponding means (7).

$$\overline{x_j} = \frac{1}{NP} \sum_{i=1}^{NP} x_{ij} \tag{7}$$

$$PD = \sqrt{\frac{1}{NP} \sum_{i=1}^{NP} \sum_{j=1}^{D} (x_{ij} - \overline{x_j})^2} \tag{8}$$

Where i is the population member iterator and j is the vector component iterator.

7 Results

Statistical results for the comparisons are shown in comprehensive Tables 1 and 2. Table 1 shows the mean results, with the following highlighting: the bold values depict the best-obtained results (based on the mean values); italic values are considered to be significantly different (according to the *Wilcoxon sum-rank test* with the significance level of 0.05; performed for each pair of original jDE and C_jDE; † - performance of C_JDE was significantly worse, ‡ - significantly better). Table 2 depicts the minimum values, also here the bold values represent the best results (the minimum found). Ranking of the algorithms given in Fig. 2 was evaluated based on the *Friedman test with Nemenyi post hoc test*. Figures 3, 4, 5 and 6 depict the graphical comparisons of the convergence plots and corresponding population diversity plots provided for the selected four benchmark functions. The Fig. 7 shows the detailed comparisons of population diversity plots (with confidence intervals) for the selected pair of jDE and C_jDE where the performance is different. Since the C_jDE driven by Sinai map performed the best algorithm according to the rank (see Fig. 2), it was selected for graphical pair comparisons. The results discussion is in the next section.

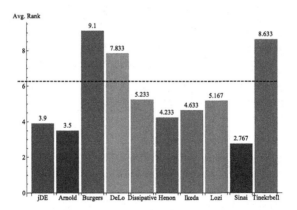

Fig. 2. Ranking of all algorithms based on the 51 runs and 15 functions of CEC2015 benchmark in 30D. The dashed line represents the Nemenyi Critical Distance.

Table 1. Results comparisons for the mean results of jDE and C-jDE; CEC 2015 Benchmark set, 30D, 51 runs

system\f	1	2	3	4	5	6	7	8	9	10	11	12	13	14	15
jDE	21401.4	0.	20.286	32.256	2086.7	4135.8	7.2354	741.24	102.636	1433.4	400.25	105.76	97.63	32509.6	**100.**
Arnold C-jDE	**16646.9**	0.	20.293	32.590	2099.1	4409.8	7.2613	583.23	102.624	1369.4	399.62	**105.71**	97.28	**32418.8**	**100.**
Burgers C-jDE	*69321.1*†	*5.97E-03*†	20.292	*36.088*†	2194.5	*11138.2*†	*8.1019*†	*5193.7*†	*102.868*†	*4492.0*†	*481.07*†	*106.23*†	98.71	32792.4	**100.**
DeLo C-jDE	*40161.4*†	0.	*20.302*†	33.127	2114.9	*7465.7*†	*7.8623*†	*2816.9*†	*102.743*†	*3243.2*†	*438.05*†	*105.93*†	98.02	32641.8	**100.**
Dissipative C-jDE	22160.6	0.	*20.300*†	33.355	2151.1	4063.8	7.4814	*2003.9*†	102.641	*2319.0*†	**394.88**	105.75	96.96	32471.7	**100.**
Henon C-jDE	21485.9	0.	**20.282**	**32.116**	2157.4	3919.5	7.3704	889.64	**102.622**	1755.1	409.00	105.79	96.69	32580.4	**100.**
Ikeda C-jDE	23716.4	0.	20.288	32.353	2072.9	3849.8	7.5059	*1096.4*†	102.66	1708.9	396.50	105.78	97.44	32460.7	**100.**
Lozi C-jDE	23822.9	0.	20.286	34.024	2047.4	4518.0	7.4047	*933.81*†	102.644	1589.1	411.97	105.79	97.11	32498.1	**100.**
Sinai C-jDE	18312.1	0.	20.288	32.689	**2059.2**	**3645.0**	**7.2131**	**544.72**	102.636	**1342.4**	395.34	105.81	**96.65**	32430.5	**100.**
Tinkerbell C-jDE	*71880.6*†	*5.59E-04*†	20.298	33.003	2108.5	*9197.2*†	*8.5355*†	*5506.2*†	*102.793*†	*4572.4*†	*465.64*†	*106.03*†	98.19	*32970.9*†	**100.**

The bold values in Table 1 depict the best-obtained results (based on the mean values); italic values are considered to be significantly different (according to the Wilcoxon sum-rank test with the significance level of 0.05; performed for each pair of original jDE and C-jDE; † - performance of C-jDE was significantly worse; ‡ - significantly better).

Table 2. The best (minimum found) results for jDE and C-jDE; CEC 2015 Benchmark set, 30D, 51 runs

system\f	1	2	3	4	5	6	7	8	9	10	11	12	13	14	15	Total
jDE	21401.4	0.	20.2857	32.2562	2086.74	4135.78	7.2354	741.242	102.636	1433.36	400.253	105.755	97.626	32509.6	100.	2
Arnold C-jDE	1377.1	0.	**20.2001**	18.8103	1648.51	546.402	5.9085	21.008	102.249	700.564	300.851	**104.359**	89.781	31135.3	100.	4
Burgers C-jDE	8982.6	0.	20.2125	19.8992	1309.47	1615.65	**3.2634**	393.240	102.424	908.528	301.05	105.058	87.478	**31092.3**	100.	4
DeJo C-jDE	**1131.8**	0.	20.2016	**18.5366**	1492.13	**204.353**	6.2112	192.475	102.452	867.952	300.979	104.538	91.976	31284.6	100.	5
Dissipative C-jDE	1961.2	0.	20.1868	23.7039	1494.52	464.55	5.9809	42.766	102.437	**608.195**	300.774	104.776	86.767	31153.1	100.	3
Henon C-jDE	1160.1	0.	20.2089	23.418	1625.39	487.359	5.8133	62.843	102.248	625.237	300.776	104.661	87.742	31135.3	100.	2
Ikeda C-jDE	2033.8	0.	20.1451	20.6607	1395.12	821.604	5.9344	13.695	102.425	799.635	**300.746**	104.598	87.335	31156.0	100.	3
Lozi C-jDE	2678.8	0.	20.1898	19.5244	1437.67	750.209	6.3772	28.547	**102.224**	721.005	300.78	104.585	89.382	31228.7	100.	3
Sinai C-jDE	1334.1	0.	20.1894	22.1101	**1201.76**	575.933	4.5881	**11.374**	102.302	611.908	300.791	104.429	89.401	31109.5	100.	4
Tinkerbell C-jDE	8152.0	0.	20.2243	22.0193	1566.62	1238.78	3.3666	143.234	102.282	904.377	300.928	104.73	**86.013**	31309.9	100.	3

The bold values in Table 2 depict the best-obtained results (based on the min. values).

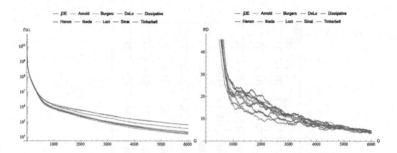

Fig. 3. Convergence (left) and population diversity (right) of CEC2015 *f1* in 30*D*.

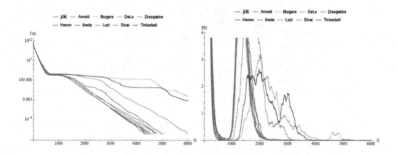

Fig. 4. Convergence (left) and population diversity (right) of CEC2015 *f2* in 30*D*.

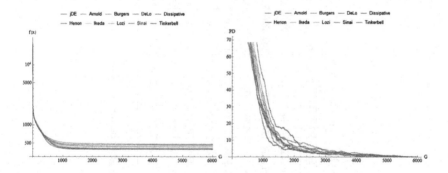

Fig. 5. Convergence (left) and population diversity (right) of CEC2015 *f11* in 30*D*.

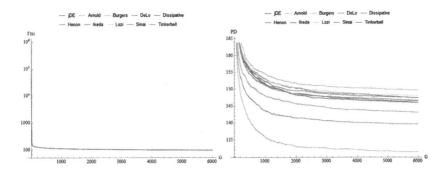

Fig. 6. Convergence (left) and population diversity (right) of CEC2015 *f13* in 30*D*.

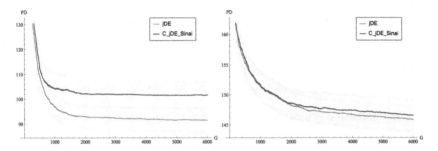

Fig. 7. Detailed population diversity plots for the selected pair of jDE and C_jDE driven by Sinai chaotic system (left *f7*) and (right *f13*), CEC2015 in 30*D*.

8   Conclusions

The research of randomization issues and insights into the inner dynamic of metaheuristic algorithms was many times addressed as essential and beneficial. The results presented here support the approach for multi-chaotic generators [22] or ensemble systems, where we can profit from the combined/selective population diversity (i.e. exploration/exploitation) tendencies, sequencing-based either stronger or moderate progress towards the function extreme, all given by the smart combination of multi-randomization schemes. The findings can be summarized as:

Obtained graphical comparisons and data in Tables 1 and 2 support the claim that jDE is sensitive to the chaotic dynamics driving the selection (mutation) process through CPRNG. At the same time, it is clear that (selection of) the best CPRNGs are problem-dependent. By using the CPRNG inside the heuristic, its performance is (significantly) different: either better or worse against other compared versions.

Mostly the performance of compared pairs of jDE and C_jDE is similar, or in some cases, the chaotic versions performed significantly worse. Such a worse performance was repeatedly observed for three chaotic maps: Delayed logistic, Burgers, and Tinkerbell. On the other hand, these maps usually secured

robust progress towards function extreme (local) followed by premature population stagnation phase, thus repeatedly secured finding of minimum values.

Overall, C_jDE versions seem to be very effective regarding finding the min.
values of the objective function (See Table 2).

It is possible to identify 3 groups of population diversity behavior in comparison with original j_DE: less decreasing (Sinai, Henon, Ikeda maps), more
decreasing (Lozi, Arnold, Dissipative maps) and significantly more decreasing
(Delayed Logistic, Tinkerbell, Burgers maps).

The selected paired diversity plots in Fig. 7 show that the diversity of the
population is maintained higher for a longer period. Therefore the exploration
phase supported by Sinai map based CPRNG is longer. This in return is beneficial for the result of the optimization.

The population diversity analysis supports the theory, that unique features
of the chaos transformed into the sequencing of CPRNG values may create
the subpopulations (or inner neighborhood selection schemes, i.e. lower population diversity based on the particular chaotic system and its CPRNG distribution). Thus the metaheuristic can benefit from the searching within those sub-
populations and quasi-periodic exchanges of information between individuals.

Acknowledgements. This work was supported by the Ministry of Education, Youth
and Sports of the Czech Republic within the National Sustainability Programme
Project no. LO1303 (MSMT-7778/2014), further by the European Regional Development Fund under the Project CEBIA-Tech no. CZ.1.05/2.1.00/03.0089 and by Internal
Grant Agency of Tomas Bata University. IGA/CebiaTech/2018/003. This work is also
based upon support by COST Action CA15140 (ImAppNIO), and COST Action IC406
(cHiPSet). Prof. Zelinka acknowledges following grants/projects: SGS No. 2018/177,
VSB-TUO and by the EU's Horizon 2020 research and innovation programme under
grant agreement No. 710577.

References

1. Caponetto, R., Fortuna, L., Fazzino, S., Xibilia, M.G.: Chaotic sequences to
   improve the performance of evolutionary algorithms. IEEE Trans. Evol. Comput.
   7(3), 289–304 (2003)
2. dos Santos Coelho, L., Mariani, V.C.: A novel chaotic particle swarm optimization
   approach using Hénon map and implicit filtering local search for economic load
   dispatch. Chaos Solitons Fractals **39**(2), 510–518 (2009)
3. Davendra, D., Zelinka, I., Senkerik, R.: Chaos driven evolutionary algorithms for
   the task of PID control. Comput. Math. Appl. **60**(4), 1088–1104 (2010)
4. Pluhacek, M., Senkerik, R., Davendra, D., Oplatkova, Z.K., Zelinka, I.: On the
   behavior and performance of chaos driven PSO algorithm with inertia weight.
   Comput. Math. Appl. **66**(2), 122–134 (2013)
5. Pluhacek, M., Senkerik, R., Davendra, D.: Chaos particle swarm optimization with
   eensemble of chaotic systems. Swarm Evol. Comput. **25**, 29–35 (2015)
6. Metlicka, M., Davendra, D.: Chaos driven discrete artificial bee algorithm for
   location and assignment optimisation problems. Swarm Evol. Comput. **25**, 15–28
   (2015)

7. Gandomi, A.H., Yang, X.S., Talatahari, S., Alavi, A.H.: Firefly algorithm with chaos. Commun. Nonlinear Sci. Numer. Simul. **18**(1), 89–98 (2013)
8. Wang, G.G., Guo, L., Gandomi, A.H., Hao, G.S., Wang, H.: Chaotic Krill Herd algorithm. Inf. Sci. **274**, 17–34 (2014)
9. Zhang, C., Cui, G., Peng, F.: A novel hybrid chaotic ant swarm algorithm for heat exchanger networks synthesis. Appl. Therm. Eng. **104**, 707–719 (2016)
10. Jordehi, A.R.: Chaotic bat swarm optimisation (CBSO). Appl. Soft Comput. **26**, 523–530 (2015)
11. Wang, G.G., Deb, S., Gandomi, A.H., Zhang, Z., Alavi, A.H.: Chaotic cuckoo search. Soft. Comput. **20**(9), 3349–3362 (2016)
12. dos Santos Coelho, L., Ayala, H.V.H., Mariani, V.C.: A self-adaptive chaotic differential evolution algorithm using gamma distribution for unconstrained global optimization. Appl. Math. Comput. **234**, 452–459 (2014)
13. Zamuda, A., Brest, J.: Self-adaptive control parameters' randomization frequency and propagations in differential evolution. Swarm Evol. Comput. **25**, 72–99 (2015)
14. Brest, J., Greiner, S., Boskovic, B., Mernik, M., Zumer, V.: Self-adapting control parameters in differential evolution: a comparative study on numerical benchmark problems. IEEE Trans. Evol. Comput. **10**(6), 646–657 (2006)
15. Das, S., Mullick, S.S., Suganthan, P.N.: Recent advances in differential evolution-an updated survey. Swarm Evol. Comput. **27**, 1–30 (2016)
16. Tanabe, R., Fukunaga, A.S.: Improving the search performance of shade using linear population size reduction. In: 2014 IEEE Congress on Evolutionary Computation (CEC), pp. 1658–1665. IEEE (2014)
17. Senkerik, R., Pluhacek, M., Zelinka, I., Davendra, D., Janostik, J.: Preliminary study on the randomization and sequencing for the chaos embedded heuristic. In: Abraham, A., Wegrzyn-Wolska, K., Hassanien, A.E., Snasel, V., Alimi, A.M. (eds.) Proceedings of the Second International Afro-European Conference for Industrial Advancement AECIA 2015. AISC, vol. 427, pp. 591–601. Springer, Cham (2016). https://doi.org/10.1007/978-3-319-29504-6_55
18. Senkerik, R., Pluhacek, M., Viktorin, A., Kadavy, T.: On the randomization of indices selection for differential evolution. In: Silhavy, R., Senkerik, R., Kominkova Oplatkova, Z., Prokopova, Z., Silhavy, P. (eds.) CSOC 2017. AISC, vol. 573, pp. 537–547. Springer, Cham (2017). https://doi.org/10.1007/978-3-319-57261-1_53
19. Senkerik, R., Pluhacek, M., Zelinka, I., Viktorin, A., Kominkova Oplatkova, Z.: Hybridization of multi-chaotic dynamics and adaptive control parameter adjusting jDE strategy. In: Matoušek, R. (ed.) ICSC-MENDEL 2016. AISC, vol. 576, pp. 77–87. Springer, Cham (2017). https://doi.org/10.1007/978-3-319-58088-3_8
20. Sprott, J.C., Sprott, J.C.: Chaos and time-series analysis, vol. 69. Citeseer (2003)
21. Poláková, R., Tvrdík, J., Bujok, P., Matoušek, R.: Population-size adaptation through diversity-control mechanism for differential evolution. In: MENDEL, 22th International Conference on Soft Computing, pp. 49–56 (2016)
22. Viktorin, A., Pluhacek, M., Senkerik, R.: Success-history based adaptive differential evolution algorithm with multi-chaotic framework for parent selection performance on CEC2014 benchmark set. In: 2016 IEEE Congress on Evolutionary Computation (CEC), pp. 4797–4803. IEEE (2016)

An Adaptive Individual Inertia Weight Based on Best, Worst and Individual Particle Performances for the PSO Algorithm

G. Spavieri[1], D. L. Cavalca[2], R. A. S. Fernandes[1,2(✉)], and G. G. Lage[1]

[1] Department of Electrical Engineering, Federal University of Sao Carlos,
Sao Carlos, Brazil
`ricardo.asf@ufscar.br`
[2] Graduate Program in Computer Science, Federal University of Sao Carlos,
Sao Carlos, Brazil

Abstract. Due to the growing need for metaheuristics with features that allow their implementation for real-time problems, this paper proposes an adaptive individual inertia weight in each iteration considering global and individual analysis, i.e., the best, worst and individual particles' performance. As a result, the proposed adaptive individual inertia weight presents faster convergence for the Particle Swarm Optimization (PSO) algorithm when compared to other inertia mechanisms. The proposed algorithm is also suitable for real-time problems when the actual optimum is difficult to be attained, since a feasible and optimized solution is found in comparison to an initial solution. In this sense, the PSO with the proposed adaptive individual inertia weight was tested using eight benchmark functions in the continuous domain. The proposed PSO was compared to other three algorithms, reaching better optimized results in six benchmark functions at the end of 2000 iterations. Moreover, it is noteworthy to mention that the proposed adaptive individual inertia weight features rapid convergence for the PSO algorithm in the first 1000 iterations.

Keywords: Adaptive inertia weight · Benchmark functions
Particle Swarm Optimization

1 Introduction

The optimization of processes and systems consists in a useful task for enterprises due to the possibility of minimizing operational costs and/or maximizing the use of available resources, and the application of optimization methods to solve such problems may be applied to distinct fields of knowledge. Thus, according to their mathematical modeling, different problems can be solved by a variety of approaches, such as metaheuristics.

© Springer International Publishing AG, part of Springer Nature 2018
L. Rutkowski et al. (Eds.): ICAISC 2018, LNAI 10841, pp. 536–547, 2018.
https://doi.org/10.1007/978-3-319-91253-0_50

In the above-mentioned category, some algorithms that have been widely used for many applications are Genetic Algorithms (GA) [1], Particle Swarm Optimization (PSO) [2], and Ant Colony Optimization (ACO) [3], among others. However, in this paper, it will be highlighted the advances on PSO. In this sense, the idealization of a PSO algorithm featuring faster and more efficient convergence has been a challenge for the scientific community. Some advances related to the PSO are: (i) the use and modifications of the inertia parameter; (ii) hybridizations; new equations and definitions to update a particle's velocity and position; development of multi-objective and bi-level algorithms.

In this context, this paper proposes an adaptive individual inertia weight which is calculated in each iteration considering global analysis (the best and worst particles' performance) and individual analysis (individual particle's performance). As a result, the proposed adaptive individual inertia weight presents faster convergence for the PSO algorithm when compared to other inertia mechanisms. For this reason, it is suitable for real-time applications. In this way, the PSO with the proposed adaptive individual inertia weight was tested and compared with three other PSO algorithms, namely: constant inertia weight PSO, linearly decreasing inertia weight PSO and dynamic adaptive inertia weight PSO. In order to guarantee a good comparison, it was used eight benchmark functions in the continuous domain, namely: Sphere, Rosenbrock, Rastrigin, Griewank, Alpine, Schaffer, Schwefel and Ackley.

2   Inertia Weight Mechanisms

The first inertia weight mechanism related to the PSO algorithm was proposed in [4], and it consists in a positive constant or a positive linear or nonlinear function of time that multiplies the current velocity of the particles in each iteration of the algorithm. The objective of this classic mechanism is to play the role of balancing the global and local search characteristics of convergence, preventing the search for local optima only. The authors attested that when the inertia weight is small, the PSO works more like a local search algorithm. Thus, if there is an acceptable solution within the initial search space, the PSO will find the global optimum quickly; otherwise it will not find the global optimum. On the other hand, if the inertia weight is large, the PSO works more like a global search method and will take more iterations to find the global optimum, being more likely to fail. The authors also observed that, by using a linearly decreasing function of time instead of a fixed constant, the algorithm tends to initially exploit the search space and, then, perform the exploration of a local area to find the global optimum. The authors in [5] proposed the use of a random value of inertia weight to enable the PSO to track the optimum in a dynamic environment, in which it is difficult to predict whether in a given time exploration or exploitation would be desired.

In accordance with the idea of a decreasing inertia weight over iterations, the work developed in [6] proposes a nonlinear decreasing variant of inertia weight mechanism in which a new parameter, the nonlinear modulation exponent, controls the overall variation of the inertia along iterations. This mechanism allows the swarm to employ an aggressive tuning during initial iterations to better search the solution space and quickly arrives near the optimum and, then, gradually employs fine tuning during later iterations so that the optimum is approached with better accuracy. Unlike the widespread use of decreasing inertia weight, in [7], the authors propose a linearly increasing inertia weight mechanism. This mechanism was tested for minimizing four different nonlinear functions and the results indicated that the linearly increasing inertia weight, compared to the linearly decreasing inertia weight, greatly improves the accuracy and convergence speed of global search, with almost no additional computational burden.

In addition to the constant, random, linear or nonlinear decreasing or increasing inertia weight mechanism, another category that has been researched is the adaptive inertia weight strategies. In these mechanisms, the inertia weight is adapted based on one or more feedback parameters that monitor the search process of the optimal solution [8]. In [9], the authors propose the use of a Fuzzy System (FS) with nine rules to dynamically adapt the population inertia weight. The proposed FS uses the normalized current best performance evaluation and the current inertia weight as input variables and the output variable is the change of the inertia weight. Experiments with three benchmark functions showed that PSO with a FS for tuning its inertia weight can improve its performance compared to the PSO with a linearly decreasing inertia weight for, at least, these three benchmark functions. In [10], the authors propose an adaptive inertia weight mechanism in which each particle of the swarm has its own inertia weight based on its rank according to the following equation:

$$\omega_i = \omega_{min} + (\omega_{max} - \omega_{min})\frac{rank_i}{n_p}, \tag{1}$$

where $rank_i$ is the position of the ith particle when the swarm is ordered based on the particles' best fitness and n_p is the population size of the swarm. In each iteration, the particle that presents the best performance is assigned as the highest particle of the rank and receives the maximum inertia weight of the swarm in order to keep moving at the current direction. On the other hand, the particle that presents the worst performance receives the minimum inertia weight, making it easier to this particle change its direction. This approach provided results comparable to or better than those obtained by the PSO with constant inertia weight when applied to the economic load dispatch problem.

In [11], the authors propose an adaptive inertia weight mechanism based on two characteristic parameters that describe the evolving state of the algorithm: the evolution speed factor and the aggregation degree factor. The speed factor is defined as the following ratio:

$$h_i^k = \left| \frac{min[f(pbest_i^{k-1}), f(pbest_i^k)]}{max[f(pbest_i^{k-1}), f(pbest_i^k)]} \right|, \tag{2}$$

where $f(pbest_i^k)$ is the best previous fitness of the ith particle in the kth iteration. Under the above definition, it can be noticed that $0 < h \leq 1$. This parameter reflects the evolutionary speed of each particle, that is, the smaller the value of h, the faster the speed. The aggregation degree factor is defined as:

$$s = \left| \frac{min[fbest^k, mean(f^k)]}{max[fbest^k, mean(f^k)]} \right|, \qquad (3)$$

where $f(best^k)$ is the best fitness achieved out of all the particles in the kth iteration and is the arithmetic mean of the fitness of all the particles in the kth iteration. In order to give to the algorithm a better ability to rapidly search and move out of the local optima, the dynamic inertia weight is defined directly proportional to the aggregation degree factor and to the speed factor:

$$\omega_i^k = \omega_{in} - \alpha(1 - h_i^k) + \beta, \qquad (4)$$

where ω_{in} is the initial value of inertia weight; α and β are two constants typically within the range $[0, 1]$. The experiments performed with this adaptive inertia weight mechanism produced results that shows a remarkably improvement in the ability of PSO to jump out of the local optima and a significant enhancement on the convergence precision.

In order to develop an inertia weight mechanism that takes into account the global and the individual particles' fitness, this work proposes an adaptive individual inertia weight mechanism based on the relative performance of each particle compared to the best and the worst current performances, which will be detailed in the following section.

3   Proposed Adaptive Inertia Weight Mechanism

The basic idea of the adaptive individual inertia weight mechanism proposed in this paper is to set the inertia weight of each particle based on the relative fitness compared to the best and the worst global fitness. For the purpose of implementation, for each iteration of the PSO, the inertia weight of each particle is calculated by:

$$\omega_i^{k+1} = \omega_{min} + (\omega_{max} - \omega_{min})\frac{f_i^k - fworst^k}{fbest^k - fworst^k}, \qquad (5)$$

where the subscript i denotes the ith particle; the superscript k denotes the kth iteration; ω_{max} and ω_{min} are constants that represent, respectively, the maximum and the minimum possible inertia weight; $fbest^k$ and $fworst^k$ are, respectively, the best and the worst fitness values out of all the swarm in the kth iteration; and f_i^k is the fitness of the ith particle in the kth iteration. Since in the proposed mechanism each particle has its own inertia weight, the velocity update equation is slightly changed to:

$$v_i^{k+1} = \omega_i^k v_i^k + r_1\varphi_1(pbest_i^k - x_i^k) + r_2\varphi_2(gbest^k - x_i^k), \qquad (6)$$

where v_i^k and x_i^k are the velocity and the position of particle i in the kth iteration, respectively; $pbest_i^k$ and $gbest^k$ are, respectively, particle's best position and the global best position up to the kth iteration; φ_1 and φ_2 are positive constants, called the cognitive and social parameter respectively; and, finally, r_1 and r_2 are random numbers uniformly distributed between 0 and 1.

As it can be noticed from Eq. 5, the swarm inertia weight distribution is linearly disposed in the range of the swarm worst and best fitness values in the current iteration and, differently from the proposed mechanisms presented in [4–7], it is not a constant value, a random value, or a value directly dependent on the iteration number. In this mechanism, the particles that have the best fitness in the current iteration will receive the highest inertia weights in the next iteration, so they will tend to keep moving at the current best search direction. On the other hand, the particles that have the worst fitness in the current iteration will be the ones that receive the smallest inertia weight values in the next iteration, encouraging them to change their direction. Thus, the preferable search directions will be those followed by the particles with the best fitness values. Despite the fact that the adaptive inertia weight mechanism proposed in this paper is similar to the mechanism proposed in [10], the difference between these approaches is the method used to rank the particles' performances. While in [10] each particle of the swarm has an inertia weight different from all of the others, even those that showed the same fitness, in the approach presented in this paper, if two particles have the same fitness, then the same inertia weight value will be assigned to both of them.

The main differences between the dynamic adaptive inertia weight mechanism presented in [11] and the mechanism proposed in this paper are related to the feedback parameters that are used to monitor and adjust the search process of the PSO. The dynamic adaptive inertia weight considers a more complex set of feedback parameters that requires the fitness values of both the current and the previous iterations. In this sense, the main advantage of the proposed mechanism is the simplicity of the feedback parameters (that involves only the fitness of the current iteration) and of the basic principle used to define the inertia weights of each particle.

In the Sect. 5, the proposed PSO with adaptive inertia weight will be properly tested and compared to the most popular mechanisms of inertia: the constant inertia weight (CWPSO) and the linearly decreasing inertia weight (LDWPSO). In addition, the proposed approach will be also compared with other adaptive inertia weight, the dynamic adaptive inertia (DAWPSO) presented in [11]. From this point, the proposed adaptive individual inertia weight PSO will be referred to as PAWPSO.

4   Experimental Setup

To validate and verify the performance of the proposed inertia weight mechanism, it was tested using four nonlinear functions considered in [12], and four additional benchmark functions.

The first one is the Sphere function, and it consists in a nonlinear function $f_1 : \mathbb{R}^n \to \mathbb{R}$ such that:

$$f_1(x) = \sum_{i=1}^{n} x_i^2, \tag{7}$$

where $x \in \mathbb{R}^n | -100 < x_i < 100, i = 1, ..., n$. The second one is the Rosenbrock function, and it consists in a nonlinear, non-convex and non-separable function $f_2 : \mathbb{R}^n \to \mathbb{R}$ such that:

$$f_2(x) = \sum_{i=1}^{n-1} [100 \left(x_{i+1} - x_i^2 \right)^2 + (x_i - 1)^2], \tag{8}$$

where $x \in \mathbb{R}^n | -2.048 < x_i < 2.048, i = 1, ..., n$.

The third function is the generalized Rastrigin function, and it consists in a nonlinear and non-convex function $f_3 : \mathbb{R}^n \to \mathbb{R}$ such that:

$$f_3(x) = \sum_{i=1}^{n} (x_i^2 - 10cos(2\pi x_i) + 10), \tag{9}$$

where $x \in \mathbb{R}^n | -5.12 < x_i < 5.12, i = 1, ..., n$. The fourth function is the generalized Griewank function, and it consists in a nonlinear and non-convex function $f_4 : \mathbb{R}^n \to \mathbb{R}$ such that:

$$f_4(x) = \frac{1}{4000} \sum_{i=1}^{n} x_i^2 - \prod_{i=1}^{n} cos\left(\frac{x_i}{\sqrt{i}} \right) + 1, \tag{10}$$

where $x \in \mathbb{R}^n | -600 < x_i < 600, i = 1, ..., n$.

The fifth function is the Alpine function, and it consists in a nonlinear and non-convex function $f_5 : \mathbb{R}^n \to \mathbb{R}$ such that:

$$f_5(x) = \sum_{i=1}^{n} |x_i sin(x_i) + 0.1x_i|, \tag{11}$$

where $x \in \mathbb{R}^n | -5.12 < x_i < 5.12, i = 1, ..., n$.

The sixth function is the Schaffer function, and it consists in a nonlinear, non-convex and non-separable function $f_6 : \mathbb{R}^2 \to \mathbb{R}$ such that:

$$f_6(x_1, x_2) = 0.5 + \frac{sin^2 \left(\sqrt{x_1^2 + x_2^2} \right) - 0.5}{[1 + 0.001(x_1^2 + x_2^2)]^2}, \tag{12}$$

where $x \in \mathbb{R}^n | -100 < x_i < 100, i = 1, ..., n$.

The seventh function is the Schwefel function, and it consists in a nonlinear, non-convex, non-separable, asymmetrical function with many local optima and the second better local optimum is geometrically far from the global optimum. The function is defined as $f_7 : \mathbb{R}^n \to \mathbb{R}$ such that:

$$f_7(x) = 418.9829n - \sum_{i=1}^{n} \left[x_i sin \left(\sqrt{|x_i|} \right) \right], \tag{13}$$

where $x \in \mathbb{R}^n | -500 < x_i < 500, i = 1, ..., n$.

The eighth and last function is the Ackley function, and it consists in a nonlinear, non-convex function that has many local minima. This function is defined as $f_8 : \mathbb{R}^n \to \mathbb{R}$ such that:

$$f_8(x) = -20exp\left(-0.2\sqrt{\frac{1}{n}\sum_{i=1}^{n}x_i^2}\right) - exp\left(\frac{1}{n}\sum_{i=1}^{n}cos(2\pi x_i)\right) + 20 + exp(1),$$

(14)

where $x \in \mathbb{R}^n| - 32 < x_i < 32, i = 1, ..., n$.

Observe that all functions are nonlinear, and that functions f_2 to f_8 are non-convex, which characterize possible multiple minima depending on their domain.

In this paper, the dimensions of the domains of the considered functions are: $n = 5$ for f_1; $n = 10$ for f_2; $n = 30$ for f_3; $n = 50$ for f_4; $n = 10$ for f_5; $n = 5$ for f_7, and, at last, $n = 50$ for f_8. Notice that the Schaffer function, f_6, is the only function that is exclusively defined in the $\mathbb{R}^2$.

The global minima solutions for all benchmark functions is 0 (zero). In all experiments, the swarm size was 50 and it was initialized in exactly the same way. The cognitive and social factors were both set to 1.25, the maximum velocity was set to the right bound of the hypercube space for in each function, and every function was tested 20 times independently, each time with 2000 iterations. As follows, Table 1 lists the inertia weight parameters considered for the four mechanisms for each benchmark function.

Table 1. Inertia parameters setup.

Function	CWPSO	LDWPSO		DAWPSO			PAWPSO	
	ω	ω_{min}	ω_{max}	ω_{in}	α	β	ω_{min}	ω_{max}
Sphere	0.9	0.4	0.9	0.7	0.3	0.2	0.4	0.9
Rosenbrock	0.9	0.4	0.9	0.7	0.3	0.2	0.4	0.9
Rastrigin	0.9	0.7	0.9	0.8	0.1	0.1	0.7	0.9
Griewank	0.9	0.4	0.9	0.7	0.3	0.2	0.4	0.9
Alpine	0.9	0.4	0.9	0.7	0.3	0.2	0.4	0.9
Schaffer	0.9	0.7	0.9	0.8	0.1	0.1	0.7	0.9
Schwefel	0.9	0.7	0.9	0.8	0.1	0.1	0.7	0.9
Ackley	0.9	0.7	0.9	0.8	0.1	0.1	0.7	0.9

5   Experimental Results and Discussions

The metric used to evaluate the accuracy that each algorithm can reach up to a given number of iterations was the Mean Optimum Fitness (MOF), i.e., the mean of the global best fitness obtained in the 20 runs of the algorithm. Thus, Table 2 presents the obtained MOF for the eight benchmark functions. The MOF indicates that the adaptive inertia weight mechanism proposed in this paper can

Table 2. Mean Optimum Fitness.

Function	CWPSO	LDWPSO	DAWPSO	PAWPSO
Sphere	$2.1e^{-12}$	$2.6e^{-264}$	0.0	$4.9e^{-30}$
Rosenbrock	71.8	25.2	6.9	2.4
Rastrigin	217.9	122.5	107.2	103.7
Griewank	112.0	31.8	27.3	4.6
Alpine	$1.7e^{-3}$	$1.3e^{-15}$	$1.4e^{-15}$	$1.3e^{-15}$
Schaffer	$2.8e^{-18}$	0.0	$9.7e^{-4}$	0.0
Schwefel	248.8	242.8	231.0	213.2
Ackley	19.3	19.0	19.2	19.0

attain better optimum fitness results for six out of the eight benchmark functions. These results indicate that the proposed mechanism has better global search performance when compared to the other three inertia weight PSO variants. The results obtained for the Rosenbrock and the Griewank functions highlight the performance increase accomplished by the proposed adaptive individual inertia weight PSO. The only exceptions were for the Sphere function, in which all four algorithms reached an excellent MOF, but the best value was reached by the DAWPSO, and for the Ackley function, in which the best MOF was achieved by the LDWPSO.

To analyze and compare the convergence speed of the proposed inertia weight mechanism, Figs. 1 to 8 indicate the average fitness of the particles along the iterations.

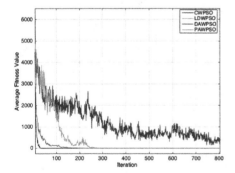

Fig. 1. Sphere function minimization convergence characteristic.

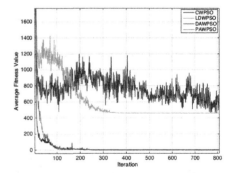

Fig. 2. Rosenbrock function minimization convergence characteristic.

Figures 1 to 8 clearly illustrate that the proposed PAWPSO convergence is faster than two of the other tested algorithms for six out of the eight analyzed benchmark functions, being outperformed by the DAWPSO only for the Sphere

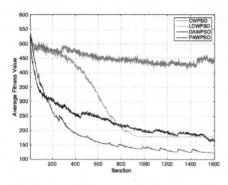

Fig. 3. Rastrigin function minimization convergence characteristic.

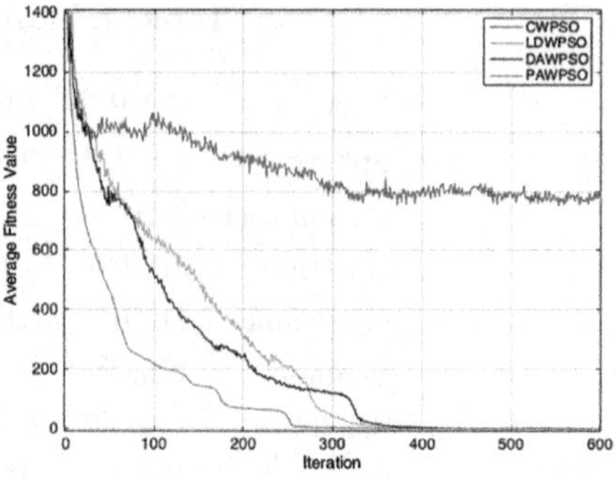

Fig. 4. Griewank function minimization convergence characteristic.

and Rosenbrock functions. It can be argued, nevertheless, that the fast convergence characteristic of the PAWPSO at the early iterations indicates that the proposed inertia weight mechanism is more efficient than the constant and the linearly decreasing inertia weight mechanism for applications that, for some reason, the maximum number of function evaluations is restricted.

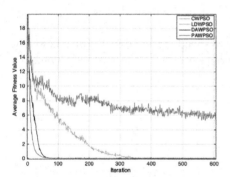
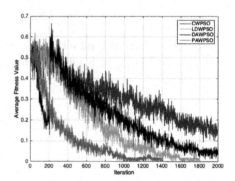

Fig. 5. Alpine function minimization convergence characteristic.

Fig. 6. Schaffer function minimization convergence characteristic.

In order to better understand the adaptive behavior of the inertia weight in PAWPSO and compare it to the DAWPSO (second better algorithm), Fig. 9 presents the typical inertia weight trajectories for both algorithms. These trajectories were obtained for the Alpine function and are the inertia weight trajectories of the particle that presented the best global fitness value at the final iteration.

Figure 9 shows that the adaptive behavior of the inertia weight in PAWPSO is very different than in DAWPSO. In the DAWPSO, the inertia weight varies in early iterations, when the evolution speed and aggregation degree factors

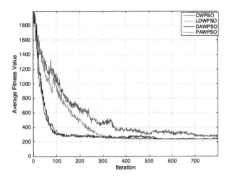

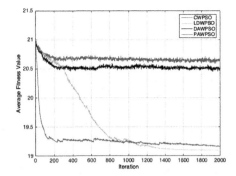

Fig. 7. Schwefel function minimization convergence characteristic.

Fig. 8. Ackley function minimization convergence characteristic.

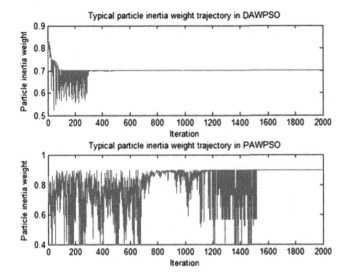

Fig. 9. Typical inertia weight trajectories in DAWPSO and in PAWPSO.

both keep changing according to the evolving state of the algorithm, tending to keep at the initial value ω_{in} in the later iterations. In the PAWPSO, the inertia weight keep ranging within its minimum and maximum values in most part of its trajectory, tending to keep at the maximum value in the later iterations. This difference can be understood by the fact that, in the DAWPSO, the particle inertia weight is defined by parameters which only refers to the particle: its evolution speed and aggregation degree. In the PAWPSO, the particle's inertia weight depends on the evolving state of the whole swarm, frequently changing according to the particle's relative fitness when compared to rest of the particles.

6   Conclusions

This paper proposed an adaptive individual inertia weight for the PSO algorithm, where the inertia parameter of each particle is updated as a function of the best and worst particle's fitness and the maximum and minimum values of inertia obtained for the swarm in each iteration. The proposed PSO algorithm was compared with three other algorithms with inertia mechanisms, being one of them with constant inertia weight, one with linearly decreasing inertia weight and one with a dynamic adaptive inertia weight. The obtained results allowed the conclusion that the proposed adaptive inertia weight provides a faster convergence in the first 1000 iterations for all of the considered benchmark functions. This characteristic makes the proposed PSO algorithm suitable for real-time applications. After 2000 iterations, the proposed PSO algorithm achieved better feasible results than those obtained by the other three PSO algorithms for six out of eight nonlinear benchmark functions. However, it is noteworthy to mention that the proposed algorithm only presented a slightly worse outcome for the Sphere and Ackley functions. Regardless, in both cases, the presented results were very close to zero, that is, the optimal solution.

Acknowledgements. This paper was supported by FAPESP (grant number 2015/12599-0), CNPq (grant number 420298/2016-9) and CAPES.

References

1. Goldberg, D.: Genetic Algorithms in Search, Optimization and Machine Learning. Wesley, Hoboken (1989)
2. Kennedy, J., Eberhart, R.: Particle swarm optimization. In: Proceedings of the IEEE International Conference on Neural Network, vol. 4, pp. 1942–1948 (1995)
3. Dorigo, M., Gambardella, L.M.: Ant colony system: a cooperative learning approach to the traveling salesman problem. IEEE Trans. Evol. Comput. **1**(1), 53–66 (1997)
4. Shi, Y., Eberhart, R.C.: A modified particle swarm optimizer. In: Proceedings of the IEEE International Conference on Evolutionary Computation, pp. 69–73 (1998)
5. Eberhart, R.C., Shi, Y.: Tracking and optimizing dynamic systems with particle swarms. In: Proceedings of the Congress on Evolutionary Computation, vol. 1, pp. 94–100 (2001)
6. Chatterjee, A., Siarry, P.: Nonlinear intertia weight variation for dynamic adaptation in particle swarm optimization. Comput. Oper. Res. **33**(3), 859–871 (2006)
7. Zheng, Y., Ma, L., Zhang, L., Qian, J.: Empirical study of particle swarm optimizer with an increasing inertia weight. IEEE Congr. Evol. Comput. **1**, 221–226 (2003)
8. Nickabadi, A., Ebadzadeh, M.M., Safabakhsh, R.: A novel particle swarm optimization algorithm with adaptive inertia weight. Appl. Soft Comput. **11**(4), 3658–3670 (2011)
9. Shi, Y., Eberhart, R.C.: Fuzzy adaptive particle swarm optimization. In: Proceedings of the Congress on Evolutionary Computation, vol. 1 pp. 101–106 (2001)

10. Panigrahi, B.K., Pandi, V.R., Das, S.: Adaptive particle swarm optimization approach for static and dynamic economic load dispatch. Energy Convers. Manag. **49**(6), 1407–1415 (2008)
11. Xueming, Y., Jinsha, Y., Jiangye, Y., Huina, M.: A modified particle swarm optimizer with dynamic adaptation. Appl. Math. Comput. **189**(2), 1205–1213 (2007)
12. Shi, Y., Eberhart, R.C.: Empirical study of particle swarm optimization. In: Proceedings of the Congress on Evolutionary Computation, vol. 3, pp. 1945–1950 (1999)

A Mathematical Model and a Firefly Algorithm for an Extended Flexible Job Shop Problem with Availability Constraints

Willian Tessaro Lunardi[1]([✉])(iD), Luiz Henrique Cherri[2], and Holger Voos[1]

[1] Interdisciplinary Centre for Security, Reliability and Trust (SnT),
University of Luxembourg, 6 rue Coudenhove-Kalergi,
1359 Luxembourg City, Luxembourg
{willian.tessarolunardi,holger.voos}@uni.lu
[2] Institute of Mathematics and Computer Sciences (ICMC), University of São Paulo,
400 Avenida Trabalhador São-Carlense, São Paulo 13566-590, Brazil
lhcherri@icmc.usp.br

Abstract. Manufacturing scheduling strategies have historically ignored the availability of the machines. The more realistic the schedule, more accurate the calculations and predictions. Availability of machines will play a crucial role in the Industry 4.0 smart factories. In this paper, a mixed integer linear programming model (MILP) and a discrete firefly algorithm (DFA) are proposed for an extended multi-objective FJSP with availability constraints (FJSP-FCR). Several standard instances of FJSP have been used to evaluate the performance of the model and the algorithm. New FJSP-FCR instances are provided. Comparisons among the proposed methods and other state-of-the-art reported algorithms are also presented. Alongside the proposed MILP model, a Genetic Algorithm is implemented for the experiments with the DFA. Extensive investigations are conducted to test the performance of the proposed model and the DFA. The comparisons between DFA and other recently published algorithms shows that it is a feasible approach for the stated problem.

Keywords: Firefly algorithm · Flexible job-shop scheduling
Metaheuristics · Mixed integer linear programming
Availability constraints

1 Introduction

The flexible job shop problem (FJSP) is an extension of the job shop problem (JSP) where is assumed that there is often more than one machine that is able to process a particular manufacturing task. The FJSP can be decomposed into two sub-problems: the machine selection problem (MS) and the operations sequencing problem (OS). Most of the FJSP studies have purely focused on assumptions that machines are continuously available. Nevertheless, in a real-world situation,

© Springer International Publishing AG, part of Springer Nature 2018
L. Rutkowski et al. (Eds.): ICAISC 2018, LNAI 10841, pp. 548–560, 2018.
https://doi.org/10.1007/978-3-319-91253-0_51

continuous availability of machines is not normally feasible. Machine unavailable periods might be consequent of pre-scheduling, preventive maintenance, shift pattern, or the overlap of two consecutive time horizons in the rolling time horizon planning algorithm.

There are various types of availability constraints in production systems. They can be categorized as *fixed* and *non-fixed*. The unavailable period of a fixed availability constraint starts at a fixed time point. Unavailable periods can also be categorized as *crossable* when it allows an operation to be interrupted and resumed, and *non-crossable* when it prevents the interruption of any operation. *Resumable* means that an operation can continue the processing when it is released from an interruption resultant of an unavailable period and *non-resumable* means an operation must be reprocessed fully after interrupted by an unavailable period [9].

Most existing literature focuses on the problem of integrating production scheduling with unavailable periods in the context of a single machine, parallel machine and flow shop (especially two-machine problems). The FJSP with non-resumable operations was addressed in [3]. The periods of unavailability are non-crossable, non-fixed and flexible within an end-time window and have to be determined during the scheduling procedure. In [14], a Genetic Algorithm (GA) was proposed to solve the multi-purpose machine (MPM) scheduling problem with fixed non-crossable unavailable periods in a job shop environment with non-resumable operations. A filtered beam search (FBS) [9], was proposed to solve the FJSP with non-fixed and fixed non-crossable unavailable periods and non-resumable operations.

In this paper, we put forward a mixed integer linear model (MILP) and a discrete firefly algorithm (DFA) for solving the FJSP with fixed crossable unavailable periods and resumable operations. In order to evaluate the performance of our methods, as well to be close to situations that may happen in industrial reality, we propose a new set of instances with fixed availability data. In addition, as the FJSP-FCR is an extension of the traditional FJSP. We also used traditional FJSP instances for the computational experiments. These instances include 35 open problems for FJSP. Through experimental studies, the merits of this work are clearly demonstrated.

The remainder of this paper is structured as follows. The problem formulation and the model are presented in Sect. 2. The discrete firefly algorithm and solution representation are discussed in Sect. 3. Numerical results are reported in Sect. 4. Finally, conclusions are presented at the end of this work.

2 MILP Model

The formulation of the FJSP-FCR can be given as follows. There is a set of n jobs and a set of m machines. Each job i consists of a sequence of J_i operations. M denotes the set of all machines. Each operation $O_{ij}(i = 1, \ldots, n; j = 1, \ldots, J_i)$ has to be processed on a machine k out of a set of given compatible machines M_{ij} ($k \in M_{ij}, M_{ij} \subseteq M$). In this work, we extend the classical FJSP formulation and we consider that operations are resumable and machines are not continuously available. Each machine k has M_k crossable unavailable periods. We denote U_{kr} as the rth crossable unavailable period on machine k, with su_{kr} and cu_{kr} being respectively the unavailable period starting and completion time.

The notations used in this paper are summarized below.

Indices

$$k : \text{ index of machines, } k = 1, \ldots, m;$$
$$r : \text{ index of unavailabilities, } r = 1, \ldots, M_k;$$
$$i, h : \text{ index of jobs, } i, h = 1, \ldots, n;$$
$$j, g : \text{ index of operation sequences, } j, g = 1, \ldots, J_i;$$

Parameters

$$J_i : \text{ total number of operation of job i;}$$
$$M_k : \text{ total number of unavailable periods at machine } k;$$
$$O_{ij} : \text{ the } j\text{th operation of job } i;$$
$$M_{ij} : \text{ machines able to perform operation } O_{ij};$$
$$p_{ijk} : \text{ processing time of } O_{ij} \text{ on machine } k;$$
$$U_{kr} : \text{ the } r\text{th unavailable period of machine } k;$$
$$su_{kr} : \text{ starting time of unavailable period } U_{kr};$$
$$cu_{kr} : \text{ completion time of unavailable period } U_{kr};$$
$$\lambda : \text{ an weight coefficient;}$$
$$L : \text{ an arbitrary large positive number;}$$

Decision variables

$$C_{max} : \text{ maximal completion time of the machines;}$$
$$W_{max} : \text{ maximal workload of the machines } (\max_{k}\{W_k\});$$
$$s_{ijk} : \text{ starting time of operation } O_{ij} \text{ on machine } k;$$
$$c_{ijk} : \text{ completion time of the operation } O_{ij};$$

$$v_{ijk} : \begin{cases} 1 & \text{if } O_{ij} \text{ is performed on machine } k \\ 0 & \text{otherwise;} \end{cases}$$

$$u_{ijkr} : \begin{cases} 1 & \text{if } s_{ijk} < su_{kr} < c_{ijk} \ \ \forall i, j, r \ \forall k \in M_{ij} \\ 0 & \text{otherwise;} \end{cases}$$

$$y_{ijkr} : \begin{cases} 1 & \text{if } U_{kr} \text{ precedes operation } O_{ij} \text{ on machine } k \\ 0 & \text{otherwise;} \end{cases}$$

$$z_{ijhgk} : \begin{cases} 1 & \text{if } O_{ij} \text{ precedes operation } O_{hg} \text{ on machine } k \\ 0 & \text{otherwise.} \end{cases}$$

The mixed integer programming model for the FJSP-FCR can be given as follows:

$$\text{minimize } \lambda_1 C_{max} + \lambda_2 W_{max} + \lambda_3 \sum_{k=1}^{m} \sum_{i=1}^{n} \sum_{j=1}^{J_i} p_{ijk} v_{ijk} \qquad (1)$$

$$\text{s.t. } C_{max} \geqslant \sum_{k \in M_{ij}} c_{ijk}, \qquad \forall i, j = J_i \qquad (2)$$

$$W_{max} \geqslant \sum_{i=1}^{n} \sum_{j=1}^{J_i} p_{ijk} v_{ijk}, \qquad \forall k \qquad (3)$$

$$
\begin{aligned}
c_{ijk} \geqslant\ & s_{ijk} + p_{ijk} \\
& + \sum_{\forall r} (cu_{rk} - su_{rk}) u_{ijkr} \\
& - (1 - v_{ijk})\, L, \qquad\qquad \forall i, j, k \in M_{ij}
\end{aligned}
\qquad (4)
$$

$$s_{ijk} \geqslant c_{hgk} - z_{ijhgk}\, L, \qquad \forall i < h, j, g, k \in M_{ij} \cap M_{hg} \qquad (5)$$

$$s_{hgk} \geqslant c_{ijk} - (1 - z_{ijhgk})\, L, \qquad \forall i < h, j, g, k \in M_{ij} \cap M_{hg} \qquad (6)$$

$$c_{ijk} \leqslant su_{kr} + u_{ijkr}\, L + y_{ijkr} L \qquad \forall i, j, r, k \in M_{ij} \qquad (7)$$

$$s_{ijk} \geqslant cu_{kr} - (1 - y_{ijkr})\, L, \qquad \forall i, j, r, k \in M_{ij} \qquad (8)$$

$$\sum_{k \in M_{ij}} s_{ijk} \geqslant \sum_{k \in M_{ij}} c_{ij-1k}, \qquad \forall i, j = 2, \dots, J_i \qquad (9)$$

$$\sum_{k \in M_{ij}} v_{ijk} = 1, \qquad \forall i, j \qquad (10)$$

$$s_{ijk} \leqslant v_{ijk}\, L, \qquad \forall i, j, k \in M_{ij} \qquad (11)$$

$$c_{ijk} \leqslant v_{ijk}\, L, \qquad \forall i, j, k \in M_{ij} \qquad (12)$$

Objective function (1) ensures the minimization of maximal completion time, maximal workload, and total workload of the machines and is supported by constraints (2) and (3). Constraints (11) and (12) ensures that the start and the completion time of operation on a specific machine is zero if it is not performed on this machine. The duration of the operation, considering its processing time and all the unavailabilities it passes through, is ensured by Constraints (4). Constraints (5) and (6) guarantee that two operations do not overlap on the same machine. Constraints (7) and (8) certify that the operations do not overlap the unavailabilities and, if it occurs, it is accounted to increase the operation time (performed by Constraints (4)). The precedence of each job operations is established by Constraints (9). Constraints (10) states that one machine can be selected from the set of available machines for each operation. The parameter L is an upper bound to the maximum processing time and unavailable time and is calculated as $\sum_{i}^{n} \sum_{j}^{J_i} \max_{\forall k \in M_{ij}} p_{ijk} + \max_{k=1,\dots,m} \left(\sum_{r=1}^{M_k} cu_{rk} - su_{rk} \right)$.

3   Firefly Algorithm

The firefly algorithm is a nature-inspired meta-heuristic for solving continuous problems and has been motivated by the simulation of the social behavior of

---

Algorithm 1. Firefly Algorithm

1: Objective function $f(x)$, $x = (x_1, \ldots, x_d)^T$
2: Generate initial pop. P of fireflies $x_i (i = 1, 2, \ldots, c)$
3: Light intensity $I_i = f(x_i)$
4: Define light absorption coefficient γ
5: **while** $(t < MaxGeneration)$ **do**
6:       **for each** $x_i \in P$ **do**
7:             **for each** $x_j \in P$ **do**
8:                   **if** $(I_i < I_j)$ **then** Move x_i towards x_j **end if**
9:                   Vary β with distance r via $exp[-\gamma r]$
10:                  Evaluate solutions and update light intensity
11:             **end for** j
12:       **end for** i
13:       Rank fireflies and find the current global best
14: **end while**

---

fireflies. The two fundamental functions of its flashing lights are to attract mating partners (communication), and to attract potential prey.

In essence, FA uses the three following idealized rules: all fireflies are unisex; attractiveness β is proportional to their brightness, in this way for any two flashing fireflies, the less bright one will move towards the brighter one; the brightness of a firefly is affected or determined by the landscape of the objective function. The pseudo code shown in Algorithm 1 summarizes the basic steps of the FA.

3.1   Variations of Light Intensity and Attractiveness

The variation of light intensity and formulation of the attractiveness are two important issues. For simplicity, we can always assume the attractiveness of a firefly is determined by its brightness, which in turn is associated with the encoded objective function.

The attractiveness function $\beta(r)$ can be any monotonically decreasing functions such as the following generalized form

$$\beta(r) = \beta_0 e^{-\gamma r^m}, \ m \geqslant 1, \tag{13}$$

where β_0 is the attractiveness at $r = 0$, and r is the distance between two fireflies. The Eq. (13) can be approximated as

$$\beta(r) = \frac{\beta_0}{1 + \gamma r^2}. \tag{14}$$

The distance between any two fireflies i and j, at position x_i and x_j, can be defined as a Cartesian distance:

$$r_{ij} = \|x_i - x_j\| = \sqrt{\sum_{k=1}^{d} (x_{ik} - x_{jk})^2}, \tag{15}$$

where x_{ik} is the kth component of the spatial cordinate x_i of ith firefly.

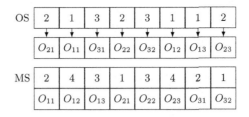

Fig. 1. Example of OS string and MS string of a firefly.

The random movement of a firefly i towards another more brighter firefly j is determined by

$$x_i = x_i + \beta_0 e^{-\gamma r_{ij}^2}(x_i - x_j) + \alpha \, \epsilon_i, \tag{16}$$

where the second term considers a firefly's attractiveness, the third term is randomization with α being the randomization parameter, and ϵ_i is a vector of random numbers drawn from a Gaussian distribution or uniform distribution. For most applications we can take $\beta_0 = 1$, $\alpha \in [0, 1]$. The parameter γ is crucially important in determining the speed of the convergence and how the FA algorithm behaves. For most applications, it typically varies from 0.001 to 1000. In this implementation of the algorithm, we used $\beta_0 = 1.0$, $\alpha \in [0, 1]$ and $\gamma = 0.1$.

3.2 Firefly Representation for the FJSP

In our proposed algorithm, each firefly represents an FJSP solution, i.e. operation sequence and machine assignment. The algorithm starts with an initial population of fireflies. Each firefly is attracted by other fireflies to varying degrees, on the basis of the objective value of those solutions and the distance between them, i.e. how different they are. The population of fireflies evolves by each firefly randomly (not directly) moving toward the most attractive solution.

The FJSP contains two sub-problems, in this way, our representation contains two strings. The MS string denotes the selected machine for the corresponding operations of each job. The hth part of the MS string can assume any value $k \in M_v$ and represents the assigned machine for operation v.

The OS string represents the order in which the operations will be processed in their respective machines. This representation uses an unpartitioned permutation with J_i repetitions of the job numbers, i.e. each job number appears J_i times in the OS string. By scanning the OS string from left to right, the fth appearance of a job number refers to the fth operation of this job. In this way, any permutation of the OS string can be decoded into a feasible solution and avoid the use of a repair mechanism. When a firefly is decoded, the OS string is translated into a sequence of operation at first. Figure 1 presents an example of OS and the MS strings.

The computation of the makespan can be obtained using graph traversal algorithms, commonly used in temporal planning. During the computation of the makespan for the FJSP-FCR, due to the advent of the unavailable periods, before

Table 1. Update of the movement of firefly i towards a brighter firefly j.

	MS string	OS string
Firefly j	1 4 1 3 2 2 3 4	1 2 3 2 1 1 2 3
Firefly i	2 4 3 1 3 4 2 1	2 1 3 2 3 1 1 2
H_{ij} and S_{ij}	$\{(1, 1), (3, 1), (4, 3), (5, 2), (6, 2), (7, 3), (8, 4)\}$	$\{(1, 2), (5, 6), (6, 7), (7, 8)\}$
$\|H_{ij}\|$ and $\|S_{ij}\|$	7	4
Attractiveness $\beta(r)$	0.17	0.38
$rand \in [0, 1]$	$\{0.35, 0.1, 0.09, 0.14, 0.33, 0.49, 0.32\}$	$\{0.52, 0.05, 0.12, 0.69\}$
Movement β-step	$\{(3, 1), (4, 3), (5, 2)\}$	$\{(5, 6), (6, 7)\}$
Position after β-step	2 4 1 3 2 4 2 1	2 1 3 2 1 1 3 2
Position after α-step	2 4 1 3 2 2 4 1	2 1 3 2 1 1 2 3

updating the outcome edges and vertices, is necessary to check whether there is an overlap of the operation with an unavailable period in the machine route. Thus, if the starting of the operation is overlapping the unavailable interval, the starting of the operation must be delayed to the end of the unavailable period; if the starting of the unavailable period is overlapping the operation, the processing time of the operation must be increased by the extension of the unavailable period.

3.3 Discrete Firefly Algorithm for the FJSP

The FA has been originally developed for solving continuous optimization problems and cannot be directly applied to solve discrete optimization problems. The main challenges for using the FA to solve FJSP are computing the discrete distance between two fireflies, and how they move in the coordination. In this work, the discretization is done for the following issues.

3.4 Distance

The discrete distance between two fireflies is defined by the distance between the permutation of its strings. There are two possible ways to measure the distance between two permutations: (a) Swapping distance (S_{ij}), i.e. the number of minimal required swaps in a permutation i in order to obtain j; and (b) Hamming distance (H_{ij}), i.e. the number of non-corresponding elements in the sequence of i compared with sequence j.

The distance between two MS strings can be measured by using Hamming distance. The minimal number of swaps cannot be used for the MS string since two different strings can contain different elements. Given two MS strings, $MS_i = \{2\ 4\ 3\ 1\ 3\ 4\ 2\ 1\}$ and $MS_j = \{1\ 4\ 1\ 3\ 2\ 2\ 3\ 4\}$, every bit is compared and

Table 2. The experimental results on Fattahi instances.

Instance	n	o	m	OOY		WLH		DFA	
				C_{max}	CPU	C_{max}	CPU	C_{max}	CPU
MFJS01	5	3	6	468	0.20	468	0.21	468	0.11
MFJS02	5	3	7	446	0.32	446	0.32	446	0.18
MFJS03	6	3	7	466	0.90	466	0.91	466	0.36
MFJS04	7	3	7	554	2.54	554	2.56	554	1.99
MFJS05	7	3	7	514	1.64	514	1.78	514	1.28
MFJS06	8	3	7	634	3.80	634	3.88	634	4.46
MFJS07	8	4	7	879	43.33	879	44.54	879	9.34
MFJS08	9	4	8	884	977	884	1050.55	884	15.23
MFJS09	11	4	8	[877.9; 1111] 20.98%	3600	[861; 116] 22.85%	3600	1055	31.22
MFJS10	12	4	8	[1012; 1208] 16.23%	3600	[1008.2; 1220] 17.36%	3600	1196	39.32

the number of bits whose are not equal are recorded, the Hamming distance is $H_{ij} = 7$. The distance between two OS strings of two fireflies can be measured with the so-called swapping distance. Given two OS strings, $OS_i = \{2\ 1\ 3\ 2\ 3\ 1\ 1\ 2\}$ and $OS_j = \{1\ 2\ 3\ 2\ 1\ 1\ 2\ 3\}$, the swapping distance is $S_{ij} = 4$.

3.5   Attraction and Movement

In this study we break up the movement given in Eq. (16) into two sub-steps: β-step and α-step. The attraction steps β and α are not interchangeable, thereby, β-step must be computed before α-step while finding the new position. Both steps are illustrated in details on Table 1, where the firefly i updates its position towards the a best firefly j. The parameters used in this illustration are as follows: $\beta_0 = 1, \gamma = 0.1, \alpha = 1$.

Moving Towards Another Firefly: β-Step. The β-step brings the iterated firefly closer to another firefly. An insertion mechanism and a pair-wise exchange mechanism are used to advance the MS string and OS string of a firefly towards the brighter firefly position. At first, all necessary insertions in the MS string and all pair-wise exchanges in the OS string, to make the elements of the current firefly equal to the best firefly, are computed and store in H_{ij} and S_{ij}. The Hamming distance and swap distance are respectively defined by $|H_{is}|$ and $|S_{ij}|$. The β probability is computed using Eq. (14). Secondly, it is defined which elements of H_{ij} and S_{ij} will be used to change the current solution. A random number $rand \in [0, 1]$ is generated for each element, and if $rand \leqslant \beta$, then the corresponding insertion/pair-wise exchange is performed on the elements of the current firefly.

Table 3. The experimental results (computational time in terms of seconds) on the proposed instances with fixed available periods.

Instance	n	o	m	u	WLH		GA		DFA	
					C_{max}	CPU	C_{max}	CPU	C_{max}	CPU
FCR01	5	3	6	6	513	0.20	513	3.57	513	0.13
FCR02	5	3	7	9	548	0.56	552	7.18	548	0.16
FCR03	6	3	7	14	620	2.50	620	5.80	620	0.44
FCR04	7	3	7	17	746	27.46	748	5.86	746	2.33
FCR05	7	3	7	20	693	20.94	709	11.23	693	4.28
FCR06	8	3	7	20	774	4.83	777	11.31	774	6.17
FCR07	8	4	7	12	[1000; 1024] 2.34%	3600	1044	28.46	1024	10.34
FCR08	9	4	8	35	[1414; 1467] 3.61%	3600	1478	35.22	1418	16.78
FCR09	11	4	8	49	[1410.96; 2051] 31.21%	3600	1976	65.65	1944	34.70
FCR10	12	4	8	52	[1815.26; 2631] 31.00%	3600	2337	70.64	2320	43.01

Random Movement: α-Step. The α-step is much simpler than the β-step. The random movement of firefly $\alpha(rand - 1/2)$ is approximated as $\alpha(rand_{int})$ given Eq. 17.

$$x_i = x_i + \alpha(rand_{int}). \tag{17}$$

It allows us to shift the permutation into one of the neighbouring permutations, by choosing an element position using $\alpha(rand_{int})$ and swap with another position in the string which also chosen at random, where $rand_{int}$ is a positive integer generated between the minimum and maximum number of elements in the string.

4   Numerical Results

To solve the MILP models, we used the IBM ILOG CPLEX 12.7 solver with default parameters and a time limit of 3600 seconds. The DFA proposed in this work, and the Genetic Algorithm (GA) proposed in a previous work [7], were coded in C++. The MILP models, the DFA and the GA were run on an Intel Core i7 2.70 GHz, with 8 GB of RAM memory. The best and average results from 50 different runs were collected for performance comparison. Observations among the MILP model, the proposed DFA and GA, and other state-of-the-art reported algorithms are also provided to determine their performance. To demonstrate the efficiency of the proposed methods, the computational time is further compared.

The instances used in the experiments can be characterized by number of jobs n, number of machines m, number of operations o, and number of unavailable periods u. The DFA parameters consist of the population size P, a maximum number of generations G, firefly's attractiveness β_0, light absorption γ, and randomization α. We kept fixed the following parameters: $\beta_0 = 1.0$, $\alpha \in [0, 1]$, and

Table 4. Comparison of the DFA with other algorithms on Brandimarte instances.

Instance	n	m	o	TABC		MA		DFA	
				C_{max}	CPU	C_{max}	CPU	C_{max}	CPU
Mk01	10	6	7	40	3	40	20	40	5
Mk02	10	6	7	26	3	26	28	26	16
Mk03	15	8	10	204	1	204	53	204	3
Mk04	15	8	10	60	66	60	30	60	11
Mk05	15	4	10	173	78	172	36	172	19
Mk06	10	15	15	60	173	59	80	59	63
Mk07	20	5	5	139	66	139	37	139	43
Mk08	20	10	15	523	2	523	77	523	4
Mk09	20	10	15	307	304	307	75	307	34
Mk10	20	15	15	202	418	202	90	202	94

$\gamma = 0.1$. The variation of P and G was based on the size of each instance. We used $P = 125$ and $G = 100$ for small instances, i.e. less or equal to 6 jobs and 5 machines; $P = 250$ and $G = 200$ for medium instances i.e. less or equal to 10 jobs and 8 machines; $P = 500$ and $G = 300$ for instances that does not belong to another group.

4.1 Fattahi Instances

We compare the proposed model and the DFA experimentally with [8] (OOY), a concise MILP model for the FJSP and has proven to be effective when compared to other state-of-the-art MILP models, as shown in [2]. Both models were implemented in the same platform and experiments were conducted in the same computer, mentioned in Sect. 4. The weight coefficients employed in this experiment are: $\lambda_1 = 1.0, \lambda_2 = 0.0$, and $\lambda_3 = 0.0$. Table 2 shows the numerical results of the experiments involving the Fattahi instances.

The proposed model contains additional constraints (compared to OOY model) to address the FJSP-FCR. Even with the additional constraints to address the available periods, our model can achieve similar results solving the standard FJSP. CPLEX found the optimal solution for the instances MFJS01-08. The DFA obtained the best solutions for all instances.

4.2 FJSP-FCR Instances

Due to the lack of literature, in order to compare the DFA with other algorithms using the FJSP-FCR instances, we implemented a GA for this experiment. The results of this experiment are presented in the Table 3. This set of instances can be obtained in JSON format through the following URL: https://github.com/snt-robotics/fjsp_fcr.

Table 5. The experimental results on Kacem instances of a multi-objective optimization experiment.

Algorithm	4 × 5				8 × 8				10 × 7				10 × 10				15 × 10			
	f_1	f_2	f_3	CPU	f_1	f_2	f_3	CPU	f_1	f_2	f_3	CPU	f_1	f_2	f_3	CPU	f_1	f_2	f_3	CPU
AL + CGA	16	10	34	–	15	13	79	–	–			–	7	5	45	–	23	11	93	–
PSO + SA		–		–	15	12	75	–		–		–	7	6	44	–	12	11	91	–
AIA		–		–	14	12	77	0.76		–		–	7	5	43	8.97	11	11	93	109.22
P-DABC	11	10	32	–	14	12	77	–	12	11	61	–	8	7	41	–	12	11	91	–
SMF	12	8	32	2.6	14	12	77	39.5	11	10	62	109.5	7	6	42	39.1	11	10	93	864.6
PSO + TS	12	8	32	0.34	14	12	77	1.67	–			–	7	6	43	2.05	11	11	93	10.88
WLH	12	8	32	0.06	14	12	77	0.54	11	10	62	0.36	7	5	43	1.07	11	11	93	211.74
GA	11	10	32	0.29	14	12	77	0.90	11	10	62	1.57	7	6	42	2.02	12	12	93	18.76
DFA	12	8	32	0.11	14	12	77	0.64	11	10	62	0.84	7	5	43	1.27	11	11	93	6.86

$n \times m$ total number of jobs and machines. − equals not available. f_1, f_2 and f_3 are respectively the C_{max}, W_{max}, and total workload of the machines.

In this experiment, we can see that the DFA achieve better results, and is more efficient and effective than the GA. The MILP model has found best solutions for the FCR01,..., FCR06, and for the other instances, bounds were provided.

4.3   Brandimarte Instances

To better demonstrate the effectiveness of the DFA we compare results with other state-of-the-art algorithms for the FJSP using the Brandimate instances. We compare the DFA with an artificial bee colony algorithm (TABC) [4] and a memetic algorithm (MA) [12]. The TABC was implemented on an Intel 2.4 GHz Core 2 Duo processor with 4.0 GB of RAM memory in C++. The MA was implemented on an Intel Core i7-3520M 2.9 GHz processor with 8.0 GB of RAM memory in Java. The weight coefficients employed in this experiment are: $\lambda_1 = 1.0$, $\lambda_2 = 0.0$, and $\lambda_3 = 0.0$. Table 4 shows the comparison on the 10 Brandimarte instances. In this experiment, we can see that the DFA can achieve similar results to state-of-the-art algorithms.

4.4   Kacem Instances

Kacem et al. proposed five multi-objective FJSP instances. Using these instances our MILP model (WLH), and the DFA are compared with the hybrid particle swarm optimization and tabu search (PSO + TS) [13], implemented on a Pentium IV 1.8 GHz in C++; the discrete artificial bee colony (DABC) [6], implemented on a Pentium IV 1.8 GHz with MB of RAM memory in C++; the artificial immune algorithm (AIA) [1] implemented on a 2.0 GHz processor with 256 MB of RAM memory in C++; the simulation modeling (SMF) [11], implemented on a Pentium IV 2.4 GHz personal with 512 MB RAM memory in Matlab; the hybrid evolutionary and fuzzy logic (AL + CGA) [5]; the GA proposed in [7], and the hybrid particle swarm optimization and simulating annealing (PSO + SA) [10]. The weight coefficients used in this experiment are: $\lambda_1 = 0.5, \lambda_2 = 0.3$, and $\lambda_3 = 0.2$. Table 5 shows the comparison of the results on the five Kacem instances.

5   Conclusion

Planning and scheduling with machine availability constraint become increasingly more important as a better understanding of their importance in various applications. We put forward a new MILP model and an FA for the FJSP-FCR. New instances are provided. We further presented computational experiments on classical instances in order to provide comparisons with other state-of-the-art algorithms. The numerical results make clear that the MILP model is important for comparisons with non-exact methods, providing good bounds to many small and medium size instances. The experiments among the DFA and others recently published algorithms shows that it is a feasible approach for the considered problem.

References

1. Bagheri, A., Zandieh, M., Mahdavi, I., Yazdani, M.: An artificial immune algorithm for the flexible job-shop scheduling problem. Future Gener. Comput. Syst. **26**(4), 533–541 (2010)
2. Demir, Y., İşleyen, S.K.: Evaluation of mathematical models for flexible job-shop scheduling problems. Appl. Math. Model. **37**(3), 977–988 (2013)
3. Gao, J., Gen, M., Sun, L.: Scheduling jobs and maintenances in flexible job shop with a hybrid genetic algorithm. J. Intell. Manuf. **17**(4), 493–507 (2006)
4. Gao, K.Z., Suganthan, P.N., Chua, T.J., Chong, C.S., Cai, T.X., Pan, Q.K.: A two-stage artificial bee colony algorithm scheduling flexible job-shop scheduling problem with new job insertion. Expert Syst. Appl. **42**(21), 7652–7663 (2015)
5. Kacem, I., Hammadi, S., Borne, P.: Pareto-optimality approach for flexible job-shop scheduling problems: hybridization of evolutionary algorithms and fuzzy logic. Math. Comput. Simul. **60**(3), 245–276 (2002)
6. Li, J.Q., Pan, Q.K., Gao, K.Z.: Pareto-based discrete artificial bee colony algorithm for multi-objective flexible job shop scheduling problems. Int. J. Adv. Manuf. Technol. **55**(9), 1159–1169 (2011)
7. Lunardi, W.T., Voos, H.: Comparative study of genetic and discrete firefly algorithm for combinatorial optimization. In: 33rd ACM/SIGAPP Symposium on Applied Computing, Pau, France, 9–13 April 2018 (2018)
8. Özgüven, C., Özbakır, L., Yavuz, Y.: Mathematical models for job-shop scheduling problems with routing and process plan flexibility. Appl. Math. Model. **34**(6), 1539–1548 (2010)
9. Wang, S., Yu, J.: An effective heuristic for flexible job-shop scheduling problem with maintenance activities. Comput. Ind. Eng. **59**(3), 436–447 (2010)
10. Xia, W., Wu, Z.: An effective hybrid optimization approach for multi-objective flexible job-shop scheduling problems. Comput. Ind. Eng. **48**(2), 409–425 (2005)

11. Xing, L.N., Chen, Y.W., Yang, K.W.: Multi-objective flexible job shop schedule: design and evaluation by simulation modeling. Appl. Soft Comput. **9**(1), 362–376 (2009)
12. Yuan, Y., Xu, H.: Multiobjective flexible job shop scheduling using memetic algorithms. IEEE Trans. Autom. Sci. Eng. **12**(1), 336–353 (2015)
13. Zhang, G., Shao, X., Li, P., Gao, L.: An effective hybrid particle swarm optimization algorithm for multi-objective flexible job-shop scheduling problem. Comput. Ind. Eng. **56**(4), 1309–1318 (2009)
14. Zribi, N., El Kamel, A., Borne, P.: Minimizing the makespan for the MPM job-shop with availability constraints. Int. J. Prod. Econ. **112**(1), 151–160 (2008)

On the Prolonged Exploration of Distance Based Parameter Adaptation in SHADE

Adam Viktorin$^{(\boxtimes)}$ ⓘ, Roman Senkerik ⓘ, Michal Pluhacek ⓘ,
and Tomas Kadavy ⓘ

Faculty of Applied Informatics, Tomas Bata University in Zlin, T. G. Masaryka 5555,
760 01 Zlin, Czech Republic
{aviktorin,senkerik,pluhacek,kadavy}@utb.cz

Abstract. In this paper, a prolonged exploration ability of distance based parameter adaptation is subject to a test via clustering analysis of the population in Success-History based Adaptive Differential Evolution (SHADE). The comparative study is done on the CEC 2015 benchmark set in two dimensional settings – 10D and 30D. It is shown, that the exploration phase of distance based adaptation in SHADE (Db_SHADE) lasts for more generations and therefore avoids the premature convergence into local optima.

Keywords: Distance based parameter adaptation · SHADE
Db_SHADE · Clustering analysis · DBSCAN · Exploration
Exploitation

1 Introduction

The algorithm of Differential Evolution (DE) has been proposed in 1995 by Storn and Price [1] and since then, it has been one of the best performing algorithms for heuristic numerical optimization. Over the years, researchers had found out, that performance of the canonical version is very sensitive to the setting of control parameters – scaling factor F, crossover rate CR and population size NP [2,3]. This fact led to a need for adaptation of these control parameters, which is a

A. Viktorin—This work was supported by the Ministry of Education, Youth and Sports of the Czech Republic within the National Sustainability Programme Project no. LO1303 (MSMT-7778/2014), further by the European Regional Development Fund under the Project CEBIA-Tech no. CZ.1.05/2.1.00/03.0089 and by Internal Grant Agency of Tomas Bata University under the Projects no. IGA/CebiaTech/2018/003. This work is also based upon support by COST (European Cooperation in Science & Technology) under Action CA15140, Improving Applicability of Nature-Inspired Optimisation by Joining Theory and Practice (ImAppNIO), and Action IC1406, High-Performance Modelling and Simulation for Big Data Applications (cHiPSet). The work was further supported by resources of A.I.Lab at the Faculty of Applied Informatics, Tomas Bata University in Zlin (ailab.fai.utb.cz).

© Springer International Publishing AG, part of Springer Nature 2018
L. Rutkowski et al. (Eds.): ICAISC 2018, LNAI 10841, pp. 561–571, 2018.
https://doi.org/10.1007/978-3-319-91253-0_52

current topic in the DE field with many researchers working on new adaptation schemes. The whole path to self-adaptive DE variants has been neatly surveyed in following publications [4–6]. And the fact, that self-adaptation is very useful for the optimization has been proven in numerous competitions. For example, one of the most successful DE variants titled Success-History based Adaptive Differential Evolution (SHADE) by Tanabe and Fukunaga [7] has formed a basis for the last four CEC competition winners (CEC 2014 – L-SHADE [8], CEC 2015 – SPS-L-SHADE-EIG [9], CEC 2016 – LSHADE_EpSin [10] and CEC 2017 – jSO [11]) and placed 3rd in the year of its proposal (CEC 2013).

The main benefit of the self-adaptation is in balancing the exploration and exploitation abilities of the algorithm in order to optimize a given problem. Therefore, it is important to study the effect of adaptation on these two key features. In this paper, a novel distance based parameter adaptation [12] is a subject to the population clustering analysis, which should provide an insight into its prolonged exploration. The prolonged exploration should be beneficial in an optimization of problems of larger scale and therefore, the comparative analysis is provided on the CEC 2015 benchmark set in $10D$ and $30D$.

The remaining content of the paper is structured as follows: The next section describes SHADE algorithm with distance based parameter adaptation (Db_SHADE), Sect. 3 follows with experimental design, Sect. 4 depicts results of the comparative analysis of the performance of SHADE and Db_SHADE algorithms along with clustering analysis and Sect. 5 concludes the paper.

2   From DE to Db_SHADE

In order to describe the Db_SHADE algorithm, it is important to start from the canonical Differential Evolution (DE) by Storn and Price [1].

The canonical 1995 DE is based on the idea of evolution from a randomly generated set of solutions of the optimization task called population $\boldsymbol{P}$, which has a preset size of NP. Each individual (solution) in the population consists of a vector $\boldsymbol{x}$ of length D (each vector component corresponds to one attribute of the optimized task) and objective function value $f(\boldsymbol{x})$, which mirrors the quality of the solution. The number of optimized attributes D is often referred to as the dimensionality of the problem and such generated population $\boldsymbol{P}$, represent the first generation of solutions.

The individuals in the population are combined in an evolutionary manner in order to create improved offspring for the next generation. This process is repeated until the stopping criterion is met (either the maximum number of generations, or the maximum number of objective function evaluations, or the population diversity lower limit, or overall computational time), creating a chain of subsequent generations, where each following generation consists of better solutions than those in previous generations – a phenomenon called elitism.

The combination of individuals in the population consists of three main steps: Mutation, crossover and selection.

In the mutation, attribute vectors of selected individuals are combined in simple vector operations to produce a mutated vector $\boldsymbol{v}$. This operation uses a control parameter – scaling factor F. In the crossover step, a trial vector $\boldsymbol{u}$ is created by selection of attributes either from mutated vector $\boldsymbol{v}$ or the original vector $\boldsymbol{x}$ based on the crossover probability given by a control parameter – crossover rate CR. And finally, in the selection, the quality $f(\boldsymbol{u})$ of a trial vector is evaluated by an objective function and compared to the quality $f(\boldsymbol{x})$ of the original vector and the better one is placed into the next generation.

From the basic description of the DE algorithm, it can be seen, that there are three control parameters, which have to be set by the user – population size NP, scaling factor F and crossover rate CR. It was shown in [2,3], that the setting of these parameters is crucial for the performance of DE. Fine-tuning of the control parameter values is a time-consuming task and therefore, many state-of-the-art DE variants use self-adaptation in order to avoid this cumbersome task. Which is also a case of SHADE algorithm proposed by Tanabe and Fukunaga in 2013 [7] and since it is used in this paper, the algorithm is described in more detail in the next section along with the novel distance based parameter adaptation.

2.1 SHADE

As aforementioned, SHADE algorithm was proposed with a self-adaptive mechanism of some of its control parameters in order to avoid their fine-tuning. Control parameters in question are scaling factor F and crossover rate CR. It is fair to mention, that SHADE algorithm is based on Zhang and Sanderson's JADE [13] and shares a lot of its mechanisms. The main difference is in the historical memories $\boldsymbol{M}_F$ and $\boldsymbol{M}_{CR}$ for successful scaling factor and crossover rate values with their update mechanism.

Following subsections describe individual steps of the SHADE algorithm: Initialization, mutation, crossover, selection and historical memory update.

Initialization. The initial population $\boldsymbol{P}$ is generated randomly and for that matter, a Pseudo-Random Number Generator (PRNG) with uniform distribution is used. Solution vectors $\boldsymbol{x}$ are generated according to the limits of solution space – *lower* and *upper* bounds (1).

$$\boldsymbol{x}_{j,i} = U\left[lower_j,\ upper_j\right] \text{ for } j = 1,\ \ldots,\ D; i = 1,\ \ldots,\ NP, \qquad (1)$$

where i is the individual index and j is the attribute index. The dimensionality of the problem is represented by D, and NP stands for the population size.

Historical memories are preset to contain only 0.5 values for both, scaling factor and crossover rate parameters (2).

$$M_{CR,i} = M_{F,i} = 0.5 \text{ for } i = 1,\ldots,H, \qquad (2)$$

where H is a user-defined size of historical memories.

Also, the external archive of inferior solutions $\boldsymbol{A}$ has to be initialized. Because of no previous inferior solutions, it is initialized empty, $\boldsymbol{A} = \varnothing$. And index k for historical memory updates is initialized to 1.

The following steps are repeated over the generations until the stopping criterion is met.

Mutation. Mutation strategy "current-to-$p$best/1" was introduced in [13] and it combines four mutually different vectors in a creation of the mutated vector v. Therefore, $x_{pbest} \neq x_{r1} \neq x_{r2} \neq x_i$ (3).

$$v_i = x_i + F_i (x_{pbest} - x_i) + F_i (x_{r1} - x_{r2}), \tag{3}$$

where x_{pbest} is randomly selected individual from the best $NP \times p$ individuals in the current population. The p value is randomly generated for each mutation by PRNG with uniform distribution from the range $[p_{min}, 0.2]$ and $p_{min} = 2/NP$. Vector x_{r1} is randomly selected from the current population P. Vector x_{r2} is randomly selected from the union of the current population P and external archive A. The scaling factor value F_i is given by (4).

$$F_i = C[M_{F,r}, 0.1], \tag{4}$$

where $M_{F,r}$ is a randomly selected value (index r is generated by PRNG from the range 1 to H) from M_F memory and C stands for Cauchy distribution. Therefore the F_i value is generated from the Cauchy distribution with location parameter value $M_{F,r}$ and scale parameter value of 0.1. If the generated value F_i higher than 1, it is truncated to 1 and if it is F_i less or equal to 0, it is generated again by (4).

Crossover. In the crossover step, trial vector u is created from the mutated v and original x vectors. For each vector component, a PRNG with uniform distribution is used to generate a random value. If this random value is less or equal to given crossover rate value CR_i, current vector component will be taken from a trial vector, otherwise, it will be taken from the original vector (5). There is also a safety measure, which ensures, that at least one vector component will be taken from the trial vector. This is given by a randomly generated component index j_{rand}.

$$u_{j,i} = \begin{cases} v_{j,i} & \text{if } U[0,1] \leq CR_i \text{ or } j = j_{rand} \\ x_{j,i} & \text{otherwise} \end{cases}. \tag{5}$$

The crossover rate value CR_i is generated from a Gaussian distribution with a mean parameter value $M_{CR,r}$ selected from the crossover rate historical memory M_{CR} by the same index r as in the scaling factor case and standard deviation value of 0.1 (6).

$$CR_i = N[M_{CR,r}, 0.1]. \tag{6}$$

When the generated CR_i value is less than 0, it is replaced by 0 and when it is greater than 1, it is replaced by 1.

Selection. The selection step ensures, that the optimization will progress towards better solutions because it allows only individuals of better or at least equal objective function value to proceed into the next generation $G+1$ (7).

$$x_{i,G+1} = \begin{cases} u_{i,G} & \text{if } f(u_{i,G}) \leq f(x_{i,G}) \\ x_{i,G} & \text{otherwise} \end{cases}, \tag{7}$$

where G is the index of the current generation.

Historical Memory Updates. Historical memories M_F and M_{CR} are initialized according to (2), but their components change during the evolution. These memories serve to hold successful values of F and CR used in mutation and crossover steps. Successful in terms of producing trial individual better than the original individual. During every single generation, these successful values are stored in their corresponding arrays S_F and S_{CR}. After each generation, one cell of M_F and M_{CR} memories is updated. This cell is given by the index k, which starts at 1 and increases by 1 after each generation. When it overflows the memory size H, it is reset to 1. The new value of k-th cell for M_F is calculated by (8) and for M_{CR} by (9).

$$M_{F,k} = \begin{cases} \text{mean}_{WL}\left(S_F\right) & \text{if } S_F \neq \emptyset \\ M_{F,k} & \text{otherwise} \end{cases}, \tag{8}$$

$$M_{CR,k} = \begin{cases} \text{mean}_{WL}\left(S_{CR}\right) & \text{if } S_{CR} \neq \emptyset \\ M_{CR,k} & \text{otherwise} \end{cases}, \tag{9}$$

where $\text{mean}_{WL}()$ stands for weighted Lehmer (10) mean.

$$\text{mean}_{WL}\left(S\right) = \frac{\sum_{k=1}^{|S|} w_k \bullet S_k^2}{\sum_{k=1}^{|S|} w_k \bullet S_k}, \tag{10}$$

where the weight vector w is given by (11) and is based on the improvement in objective function value between trial and original individuals in current generation G.

$$w_k = \frac{\text{abs}\left(f\left(u_{k,G}\right) - f\left(x_{k,G}\right)\right)}{\sum_{m=1}^{|S_{CR}|} \text{abs}\left(f\left(u_{m,G}\right) - f\left(x_{m,G}\right)\right)}. \tag{11}$$

And since both arrays S_F and S_{CR} have the same size, it is arbitrary which size will be used for the upper boundary for m in (11).

The last Eq. (11) is the subject of change in the novel Db_SHADE algorithm, which is described in the next section.

2.2 Db_SHADE

The original adaptation mechanism for scaling factor and crossover rate values uses weighted forms of means (10), where weights are based on the improvement in objective function value (11). This approach promotes exploitation over exploration and therefore might lead to premature convergence, which could be a problem especially in higher dimensions.

Distance approach is based on the Euclidean distance between the trial and the original individual, which slightly increases the complexity of the algorithm by exchanging simple difference for Euclidean distance computation for the price of stronger exploration. In this case, scaling factor and crossover rate values connected with the individual that moved the furthest will have the highest weight (12).

$$w_k = \frac{\sqrt{\sum_{j=1}^{D}\left(u_{k,j,G} - x_{k,j,G}\right)^2}}{\sum_{m=1}^{|S_{CR}|} \sqrt{\sum_{j=1}^{D}\left(u_{m,j,G} - x_{m,j,G}\right)^2}}. \tag{12}$$

Therefore, the exploration ability is rewarded and this should lead to avoidance of the premature convergence in higher dimensional objective spaces. Such approach might be also useful for constrained problems, where constrained areas could be overcome by increased changes of individual's components.

Below is the pseudo-code of the Db_SHADE algorithm for a clear overview.

Algorithm 1. Db_SHADE

1: Set NP, H and stopping criterion;
2: $G = 0$, $\boldsymbol{x}_{best} = \{\}$, $k = 1$, $p_{min} = 2/NP$, $\boldsymbol{A} = \emptyset$;
3: Randomly initialize (1) population $\boldsymbol{P} = (\boldsymbol{x}_{1,G}, \ldots, \boldsymbol{x}_{NP,G})$;
4: Set $\boldsymbol{M}_F$ and $\boldsymbol{M}_{CR}$ according to (2);
5: $\boldsymbol{P}_{new} = \{\}$, $\boldsymbol{x}_{best} =$ best from population $\boldsymbol{P}$;
6: **while** stopping criterion not met **do**
7:   $\boldsymbol{S}_F = \emptyset$, $\boldsymbol{S}_{CR} = \emptyset$;
8:   **for** $i = 1$ to NP **do**
9:     $\boldsymbol{x}_{i,G} = \boldsymbol{P}[i]$;
10:    $r = U[1, H]$, $p_i = U[p_{min}, 0.2]$;
11:    Set F_i by (4) and CR_i by (6);
12:    $\boldsymbol{v}_{i,G}$ by mutation (3);
13:    $\boldsymbol{u}_{i,G}$ by crossover (5);
14:    **if** $f(\boldsymbol{u}_{i,G}) < f(\boldsymbol{x}_{i,G})$ **then**
15:      $\boldsymbol{x}_{i,G+1} = \boldsymbol{u}_{i,G}$;
16:      $\boldsymbol{x}_{i,G} \to \boldsymbol{A}$;
17:      $F_i \to \boldsymbol{S}_F$, $CR_i \to \boldsymbol{S}_{CR}$;
18:    **else**
19:      $\boldsymbol{x}_{i,G+1} = \boldsymbol{x}_{i,G}$;
20:    **end if**
21:    **if** $|\boldsymbol{A}| > NP$ **then**
22:      Randomly delete $|\boldsymbol{A}| - NP$ individuals from $\boldsymbol{A}$;
23:    **end if**
24:    $\boldsymbol{x}_{i,G+1} \to \boldsymbol{P}_{new}$;
25: **end for**
26: **if** $\boldsymbol{S}_F \neq \emptyset$ and $\boldsymbol{S}_{CR} \neq \emptyset$ **then**
27:    Update $\boldsymbol{M}_{F,k}$ (8) and $\boldsymbol{M}_{CR,k}$ (9) with distance based weights from (12), $k{+}{+}$;

28:    **if** $k > H$ **then**
29:      $k = 1$;
30:    **end if**
31: **end if**
32: $\boldsymbol{P} = \boldsymbol{P}_{new}$, $\boldsymbol{P}_{new} = \{\}$, $\boldsymbol{x}_{best} =$ best from population $\boldsymbol{P}$;
33: **end while**
34: **return** $\boldsymbol{x}_{best}$ as the best found solution

---

3   Experimental Settings

The clustering analysis was performed for each generation of the optimization process for both SHADE and Db_SHADE in order to provide a clear comparison. The clustering and convergence histories were recorded for 51 independent runs on each of the 15 test functions in the CEC 2015 benchmark set in $10D$ and $30D$. Following is the description of setting of both SHADE variants and the clustering algorithm.

3.1   SHADE and Db_SHADE Settings

In order to provide the most comparable results, both algorithms had the same setting of control and other parameters:

1. Population size $NP = 100$,
2. Historical memory size $H = 10$,
3. External archive size $|A| = NP$,
4. Dimensionality of problems $D = \{10, 30\}$,
5. Stopping criterion – maximum number of objective function evaluations $MAXFES = 10,000 \times D$,
6. Number of runs $runs = 51$.

3.2   Cluster Analysis

The clustering algorithm selected for this experiment is Density Based Spatial Clustering of Applications with Noise (DBSCAN) [14], which conveniently works on the basis of cluster density, and therefore is able to discover clusters of arbitrary shapes.

The DBSCAN algorithm requires setting of two control parameters and a distance measure. The settings with an explanation are as follows:

1. Core point distance $Eps = 1\%$ of the decision space – for CEC2015 benchmark set $Eps = 2$,
2. Minimal number of points to form a cluster $MinPts = 4$ (minimal number of individuals for mutation),
3. Distance measure = Chebyshev distance [15] – if the distance between any corresponding attributes of two individuals is higher than 1% of the decision space, they are not considered directly density-reachable.

4   Results and Discussion

This section provides results of the performance comparative analysis (Table 1 for $10D$ and Table 2 for $30D$) and results of the clustering analysis (Table 3 and Fig. 1). The comparison between SHADE and Db_SHADE algorithms is depicted for median and mean values over the 51 runs and Wilcoxon rank-sum test with

Table 1. SHADE vs. Db_SHADE on CEC2015 in 10D.

f	SHADE		Db_SHADE		Result
	Median	Mean	Median	Mean	
1	0.00E+00	0.00E+00	0.00E+00	0.00E+00	=
2	0.00E+00	0.00E+00	0.00E+00	0.00E+00	=
3	2.00E+01	1.89E+01	2.00E+01	1.92E+01	=
4	3.07E+00	2.97E+00	3.06E+00	2.98E+00	=
5	2.21E+01	3.42E+01	2.98E+01	4.52E+01	=
6	2.20E−01	2.97E+00	4.16E−01	8.08E−01	=
7	1.67E−01	1.88E−01	1.73E−01	1.91E−01	=
8	8.15E−02	2.69E−01	4.28E−02	2.06E−01	=
9	1.00E+02	1.00E+02	1.00E+02	1.00E+02	=
10	2.17E+02	2.17E+02	2.17E+02	2.17E+02	=
11	3.00E+02	1.66E+02	3.00E+02	2.01E+02	=
12	1.01E+02	1.01E+02	1.01E+02	1.01E+02	=
13	2.78E+01	2.78E+01	2.79E+01	2.76E+01	=
14	2.94E+03	4.28E+03	2.98E+03	4.66E+03	=
15	1.00E+02	1.00E+02	1.00E+02	1.00E+02	=

Table 2. SHADE vs. Db_SHADE on CEC2015 in 30D.

f	SHADE		Db_SHADE		Result
	Median	Mean	Median	Mean	
1	3.73E+01	2.62E+02	2.12E+01	2.42E+02	=
2	0.00E+00	0.00E+00	0.00E+00	0.00E+00	=
3	2.01E+01	2.01E+01	2.01E+01	2.01E+01	=
4	1.41E+01	1.41E+01	1.32E+01	1.31E+01	=
5	1.55E+03	1.50E+03	1.54E+03	1.52E+03	=
6	5.36E+02	5.73E+02	3.37E+02	3.48E+02	+
7	7.17E+00	7.26E+00	6.81E+00	6.74E+00	+
8	1.26E+02	1.21E+02	5.27E+01	7.38E+01	+
9	1.03E+02	1.03E+02	1.03E+02	1.03E+02	=
10	6.27E+02	6.22E+02	5.29E+02	5.32E+02	+
11	4.53E+02	4.50E+02	4.10E+02	4.16E+02	+
12	1.05E+02	1.05E+02	1.05E+02	1.05E+02	=
13	9.52E+01	9.50E+01	9.47E+01	9.50E+01	=
14	3.21E+04	3.24E+04	3.22E+04	3.24E+04	=
15	1.00E+02	1.00E+02	1.00E+02	1.00E+02	=

0.05 significance level results are provided in the last column. When there is no difference, the "=" sign is used, when the Db_SHADE algorithm performs better than the SHADE algorithm, "+" sign is used and in the last case, "–" is used.

As it can be seen in Table 1, there is no significant difference in performance between SHADE and Db_SHADE algorithms in $10D$ decision space, however, the situation in $30D$ decision space is different. There are five significant wins for the Db_SHADE algorithm on multimodal ($f6$, $f7$, $f8$) and hybrid ($f10$, $f11$) functions. Average clustering of the population on these functions is depicted in Fig. 1 and numerical results are provided in Table 3. Table 3 provides a number of runs, where clustering occurred (# runs) and average generation, where the first cluster appeared (Avg. CO).

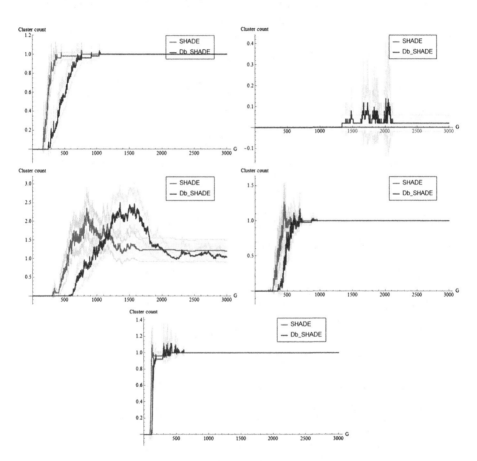

Fig. 1. Average clustering of the population comparison between SHADE and Db_SHADE on CEC2015 test functions $f6$ (top left), $f7$ (top right), $f8$ (middle left), $f10$ (middle right) and $f11$ (bottom) in $30D$.

The average generation in which clusters appear is a good indicator of the end of exploration phase of the algorithm. This happens because the population clusters in one part of the decision space, and it is improbable for it to escape this area if the scaling factor is lower than 1, which is a case of the SHADE algorithm. As can be seen in Fig. 1 and Table 3, first clusters appear later in the population of a Db_SHADE algorithm in 4 out of 5 cases, the only exception is function $f7$, where the population clusters only twice over the 51 runs in the case of Db_SHADE and never in the case of a SHADE algorithm. These results, therefore, support the claim, that the distance based parameter adaptation is beneficial for the prolonging of the exploration phase and might help in avoiding the premature convergence into local optima. However, there are other possibilities for use of the clustering analysis in balancing exploration and exploitation and these will be the focus of future studies – partial population restart based on the clustering of the population, population size management, scaling factor historical memory management.

Table 3. SHADE vs. Db_SHADE clustering analysis on selected functions from CEC2015 in 30D.

	SHADE		Db_SHADE	
f	# runs	Avg. CO	# runs	Avg. CO
6	51	2.62E+02	51	4.90E+02
7	0	–	2	1.40E+03
8	51	5.45E+02	51	8.91E+02
10	51	3.65E+02	51	5.01E+02
11	51	1.21E+02	51	1.57E+02

5   Conclusion

This paper studied the difference between exploration phase length in SHADE algorithm and SHADE algorithm with distance based parameter adaptation (Db_SHADE). The results support the claim, that the exploration phase is maintained for a longer period in the case of distance based adaptation and that this is beneficial for optimization of problems of a larger scale, where SHADE algorithm suffers from premature convergence. However, there is still a lot of work to do because the premature convergence is not avoided, only postponed for a longer period. This problem should be addressed in future studies based on the results provided in this paper. Possible future research directions, therefore, contain intelligent population size management, scaling factor memory control and partial population restart based on the insight to population dynamic.

References

1. Storn, R., Price, K.: Differential Evolution-A Simple and Efficient Adaptive Scheme for Global Optimization over Continuous Spaces, vol. 3. ICSI, Berkeley (1995)
2. Gämperle, R., Müller, S.D., Koumoutsakos, P.: A parameter study for differential evolution. Adv. Intell. Syst. Fuzzy Syst. Evol. Comput. **10**, 293–298 (2002)
3. Liu, J., Lampinen, J.: On setting the control parameter of the differential evolution method. In: Proceedings of the 8th International Conference on Soft Computing (MENDEL 2002), pp. 11–18 (2002)
4. Neri, F., Tirronen, V.: Recent advances in differential evolution: a survey and experimental analysis. Artif. Intell. Rev. **33**(1–2), 61–106 (2010)
5. Das, S., Suganthan, P.N.: Differential evolution: a survey of the state-of-the-art. IEEE Trans. Evol. Comput. **15**(1), 4–31 (2011)
6. Das, S., Mullick, S.S., Suganthan, P.N.: Recent advances in differential evolution-an updated survey. Swarm Evol. Comput. **27**, 1–30 (2016)
7. Tanabe, R., Fukunaga, A.: Success-history based parameter adaptation for differential evolution. In: 2013 IEEE Congress on Evolutionary Computation (CEC), pp. 71–78. IEEE, June 2013
8. Tanabe, R., Fukunaga, A.S.:. Improving the search performance of SHADE using linear population size reduction. In: 2014 IEEE Congress on Evolutionary Computation (CEC), pp. 1658–1665. IEEE, July 2014
9. Guo, S.M., Tsai, J.S.H., Yang, C.C., Hsu, P.H.: A self-optimization approach for L-SHADE incorporated with eigenvector-based crossover and successful-parent-selecting framework on CEC 2015 benchmark set. In: 2015 IEEE Congress on Evolutionary Computation (CEC), pp. 1003–1010. IEEE, May 2015
10. Awad, N.H., Ali, M.Z., Suganthan, P.N., Reynolds, R.G.: An ensemble sinusoidal parameter adaptation incorporated with L-SHADE for solving CEC2014 benchmark problems. In: 2016 IEEE Congress on Evolutionary Computation (CEC), pp. 2958–2965. IEEE, July 2016
11. Brest, J., Maučec, M.S., Bošković, B.: Single objective real-parameter optimization: algorithm jSO. In: 2017 IEEE Congress on Evolutionary Computation (CEC), pp. 1311–1318. IEEE, June 2017
12. Viktorin, A., Senkerik, R., Pluhacek, M., Kadavy, T., Zamuda, A.: Distance based parameter adaptation for differential evolution. In: IEEE Symposium Series on Computational Intelligence (SSCI), pp. 2612–2618. IEEE (2017, in press)
13. Zhang, J., Sanderson, A.C.: JADE: adaptive differential evolution with optional external archive. IEEE Trans. Evol. Comput. **13**(5), 945–958 (2009)
14. Ester, M., Kriegel, H.P., Sander, J., Xu, X.: A density-based algorithm for discovering clusters in large spatial databases with noise. In: KDD, vol. 96, no. 34, pp. 226–231, August 1996
15. Deza, M.M., Deza, E.: Encyclopedia of distances. In: Deza, M.M., Deza, E. (eds.) Encyclopedia of Distances, pp. 1–583. Springer, Berlin (2009). https://doi.org/10.1007/978-3-642-00234-2_1

Investigating the Impact of Road Roughness on Routing Performance: An Evolutionary Algorithm Approach

Hulda Viljoen[✉] and Jacomine Grobler

Department of Industrial and Systems Engineering,
University of Pretoria, Pretoria, South Africa
hulda.viljoen@up.ac.za

Abstract. This paper investigates the use of evolutionary and other meta-heuristic algorithms for routing problems where vehicle operating cost (VOC) and specifically, road roughness, has a significant impact. Three algorithms were implemented, namely a greedy heuristic, simulated annealing and CMA-ES. Simulated annealing delivered statistically significant results and was used to evaluate routes with and without VOC.

1 Introduction

Road conditions have a significant influence on freight and passenger vehicles [1]. Some of the potential effects of deteriorating riding quality are: damage to the cargo, increased fuel consumption, and accelerated vehicle wear. All of these aspects have consequences and financial implications for the transportation company. Roads deteriorate at different rates which are influenced by the initial quality of construction, the usage, the maintenance, and rehabilitation techniques applied [2].

Various Vehicle Operating Cost (VOC) models exist which provide an indication of the effects of road roughness on tyre wear, additional damage to the vehicle structure and fuel consumption. There is, however, no defining model for routing which includes VOC in the consideration of optimal routes. By minimising the total VOC of a route versus just the total distance travelled, significant savings can be attained as the VOC is a more accurate representation of actual costs incurred.

The aim of this paper is to determine whether the consideration of VOC in routing will have a significant influence on the results of said routing. A decision has to be made for each road section on a given road network. If the road condition is not ideal and the calculated cost of using the road will have a significant impact, a longer route with better road quality might be chosen instead.

In this paper, three algorithms were used as basis for the investigation, namely a greedy heuristic, simulated annealing [3] and CMA-ES [4]. Next, the

© Springer International Publishing AG, part of Springer Nature 2018
L. Rutkowski et al. (Eds.): ICAISC 2018, LNAI 10841, pp. 572–582, 2018.
https://doi.org/10.1007/978-3-319-91253-0_53

simulated annealing algorithm was used to solve two instances of a Vehicle Routing Problem (VRP). The first considered the condition of the roads travelled, whereas the second only considered the total distance travelled. The results showed that the route considering the road condition is longer than the second route, but the total road roughness is lower which translates to a lower cost incurred due to the road condition.

The significance of this paper is the novel inclusion of road roughness when considering route optimisation as the road roughness has a direct correlation to the VOC as well as the first use of CMA-ES in routing with road roughness.

The rest of the paper is organised as follows: Sect. 2 provides an overview of the existing literature, explains the problem, and provides solution strategies. Section 3 discusses the greedy heuristic, simulated annealing and CMA-ES approach. The algorithm evaluation is described in Sect. 4 and the paper is concluded in Sect. 5.

2 Literature Review

The literature review first discusses routing and the challenges faced in that field. The next subsection focuses on VOC and the factors contributing to VOC. The influence of road conditions on the VOC is also reviewed. Finally, a number of solution strategies to the problem are discussed.

2.1 Vehicle Routing

The *Vehicle Routing Problem* (VRP) has its origin in the Travelling Salesperson (TSP) whereby, in its simplest form, a single entity (in this case referred to as a vehicle) is tasked with having to visit a set number of locations with the added constraint of having to start and finish at the same location, which is a depot of sorts. The objective of such a problem becomes more complex as various factors can be considered: the minimisation of the total travelling distance, the minimisation of travelling time, etc.

Adding factors to make it more applicable to real-life scenarios only adds to the complexity and the intractability of the problem. Some of these factors include: time-frame window restrictions, fixed sequence of delivery, traffic, varying capacity, and green routing.

Since this paper aims to serve as a proof-of-concept, the VRP was simplified by using only one vehicle with unlimited capacity and unlimited time to complete the deliveries. The mathematical formulation of the problem is as follows: [5]

$$\sum_{i=0}^{n} \sum_{j=1, i \neq j}^{n} d_{ij} x_{ij} \tag{1}$$

subject to:

$$\sum_{i=0, i \neq j}^{n} x_{ij} = 1 \qquad \forall j \in \boldsymbol{N} \tag{2}$$

$$\sum_{j=0, j \neq i}^{n} x_{ij} = 1 \qquad \forall i \in \boldsymbol{N} \tag{3}$$

$$u_i - u_j + p x_{ij} \leq p - 1 \qquad (1 \leq i \neq j \leq n) \tag{4}$$

The objective function (1) is, therefore, the sum of the distances between the customers multiplied by the effect of International Roughness Index (IRI). The values used here are for a light truck, as derived in the linear regression models of Chatti et al. [6]. The correlation used in this model is:

$$VOC = (0.00162 \times IRI + 0.1925) \times distance \; (in \; mile) \tag{5}$$

2.2   Vehicle Operating Costs

VOC refers to the combined monetary value of operating a vehicle. This cost is different for each type of vehicle and is heavily dependent on the type of usage. The VOC includes fuel consumption cost, tyre wear cost, repair and maintenance cost. These factors are mainly dependent on the vehicle type, pavement surface characteristics, roadway geometry, and environmental conditions [2].

Various models have been developed for calculating VOC. The Highway Development and Management Tool (HDM-4) [7] is regarded as the most comprehensive model. This model quantifies the effect of pavement roughness on VOC. Chatti and Zaabar [8] utilised this model and applied it to United States of America conditions to estimate the increase in fuel consumption for different types of vehicles as an effect of pavement roughness.

A pilot study conducted by Steyn et al. [9] developed simplified equations to calculate fuel consumption and tyre wear. These equations are functions of the travelling speed and the riding quality which can be measured by the IRI. The models and data of this study can be used to evaluate the potential VOCs on specific travelling routes, which in turn can be used to make a choice between a route that may be longer in distance, but perhaps more cost effective. Since road roughness also has an impact on the freight, the potential damage to said freight can be quantified which may lead to choosing an alternative route with a smoother road.

Another study by Wang et al. [2] explored the effects of deteriorating riding quality on VOC as a function of the rehabilitation treatment applied to an ageing road. To understand the deterioration of the road, Fig. 1 illustrates the exponential increase in IRI over time. The application of treatments to improve the road surface causes a step-wise decrease in the IRI, which systematically starts increasing again. Treatment options include a thin overlay, chip seal and crack seal. The size of the step depends on the treatment applied. As can be seen from the graph, the IRI does not return to its initial value; treatment merely slows the increase in IRI. Terminal IRI refers to a predetermined point at which the road is deemed irreparable and not fit for use. The significance of this terminal IRI point to the project is to take into account that such a road will be unavailable for routing and alternative roads will have to be used until this road is rebuilt. The primary significance of this graph is that the condition of a road is not constant. As the IRI increases, the vehicle operating cost also increases. This relationship is not necessarily linear, but a clear correlation exists between the two. Both of these studies, however, do not include the weight of the loaded truck when calculating the VOC.

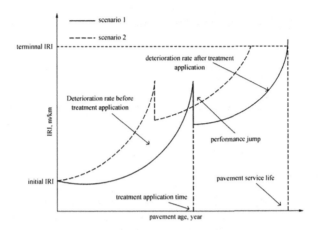

Fig. 1. Illustration of IRI development before and after treatment [2]

There is no defining model in literature for routing which includes VOC in the consideration of optimal routes.

2.3   Solution Strategies

There are three main strategies by which VRPs are solved: exact, heuristic and meta-heuristic.

Exact techniques are mathematically based and solves the optimisation problem optimally. Since a VRP is NP-hard, only problems of small size can be solved in this manner. Heuristics are applied to problems for which the solution space is too large to search exhaustively. Toro et al. [10] identifies three classes:

1. Two phases - The problem is divided into two phases: firstly, customers are allocated to a route, and secondly, the order of the visit is determined. The route can also be assigned first and the cluster allocation second.
2. Construction - An empty solution is filled by considering feasible alternatives.
3. Iterative improvement - A feasible solution is iteratively changed until a local optimum is reached.

Heuristics face the problem of falling into a local optimum and, therefore, a meta-heuristic approach is followed in this paper. Some meta-heuristics that have successfully been applied to VRPs are Tabu Search [11–13], Genetic Algorithms [14, 15], Ant Colony Optimisation [16], Simulated Annealing [17], Iterative Location Search [18], and Variable Neighbourhood Search [19].

From the literature review it is clear that a VRP with VOC is worth solving and that there are numerous ways of approaching the problem. For this paper the focus lies on a proof-of-concept for the inclusion of IRI in the objective function. After this concept has been established, the solution algorithm design can be improved.

3   Selected Algorithms

Two algorithms were selected namely simulated annealing and CMA-ES which were benchmarked against a greedy heuristic. These are discussed in more detail here.

3.1   Greedy Heuristic

A greedy heuristic (Algorithm 1) considers the current position and searches for the shortest distance, or in this case the lowest distance and IRI value combination to a next customer, until all customers have been visited, and then returns to the depot.

D represents the distance matrix and T is the customer visitation sequence.

Algorithm 1. greedyHeuristic

Input: D
Output: T
1 T = [depot]
2 **while** *notInRoute* **do**
3 │   (closestCity,closestDistance) = nextCustomer(D,notInRoute,T[-1])
4 │   T += T[closestCity]
5 │   notInRoute -= notInRoute[closestCity]
6 **end**
7 T + T[depot]

3.2   Simulated Annealing

The simulated annealing algorithm mimics the natural process of annealing in metals. At a high temperature particles in a metal move around rapidly and have a high amount of energy. However, as the temperature reduces during the annealing process particles lose energy and their movement becomes restricted. Initially, the simulated annealing algorithm is allowed to make numerous non-improving moves to explore the solution space. As time passes the algorithm gradually limits the number of non-improving moves until eventually none are allowed. By following this process, the algorithm is less likely to fall into a local optimum.

The cooling schedule is achieved through the Bolzmann equation (6), which is the difference in two objective values divided by the current temperature. The Bolzmann equation gradually reduces the acceptance probability. The temperature is diminished by 0.1% every 10 iterations.

I represents the IRI matrix, the objective function is given as z, and P is a probability.

Algorithm 2. simulatedAnnealing

Input: D,I,T,numberOfWalks,P,temperature
Output: Tbest, zBest
1 Tcurrent = T
2 zCurrent = z
3 Tbest = T
4 zBest = zCurrent
5 **for** $i \in numberOfWalks$ **do**
6   **if** *(i+1)%10 == 0* **then**
7 | reduce temperature
8   **end**
9   (Tcandidate, zCandidate) = randomSwap(D,I,Tcurrent)
10   P = exp(-(Zcandidate - zCurrent)/temperature)
11   **if** *(Zcandidate $\leq$ zCurrent) or random() $\leq$ P* **then**
12 | Tcurrent = TCandidate
13 | zCurrent = zCandidate
14   **end**
15   newBest,tBest,zBest = checkBest(Tcurrent,zCurrent,Tbest,zBest)
16 **end**

$$P = \exp(\Delta z / temperature) \tag{6}$$

Simulated annealing was chosen for the simplicity of the algorithm and the implementation thereof. The use of simulated annealing in routing is also supported by its successful track record as shown by Yu et al. [17].

3.3 CMA-ES

CMA-ES is a stochastic, non-linear optimisation algorithm consisting of four main phases which are solution generation, selection and recombination, covariance matrix update, and step size update. During the first generation phase, a population of solutions is generated at each iteration according to a multivariate normal distribution such that

$$\boldsymbol{x}_i(t+1)N(\boldsymbol{m}(t), \sigma^2_{CMA}(t))\boldsymbol{C}(t) \tag{7}$$

where $N(\boldsymbol{m}(t), \sigma_{CMA}^2(t))$ denotes a normal distribution with mean $\boldsymbol{m}(t)$ at time t and standard deviation $\sigma_{CMA}(t)$ which denotes the step size of the algorithm at time t. $\boldsymbol{C}(t)$ is the covariance matrix at time t. The solutions are evaluated and sorted whereafter the mean of the population is adjusted to facilitate selection and recombination as follows:

$$\boldsymbol{m}(t+1) = \sum_{k=1}^{n_s} w_k \boldsymbol{x}_k \tag{8}$$

where w_k is the kth recombination weight. The covariance matrix is updated as:

$$\boldsymbol{C}(t+1) = (1 - c_{cov})\boldsymbol{C}(t) + \frac{c_{cov}}{\mu_{cov}} p_{cCMA} p_{cCMA}^T + c_{cov}\left(1 - \frac{1}{\mu_{cov}}\right)$$
$$\times \sum_{k=1}^{n_s} w_k \left(\frac{x_k(t+1) - \boldsymbol{m}(t)}{\sigma_{CMA}(t)}\right)\left(\frac{x_k(t+1) - \boldsymbol{m}(t)}{\sigma_{CMA}(t)}\right)^T \tag{9}$$

where

$$\mu_{cov} \geq 1, \tag{10}$$
$$\mu_{cov} = \mu_{eff}, \text{ and} \tag{11}$$
$$c_{cov} \approx \min(\mu_{cov}, \mu_{eff}, n_x^2)/n_x^2 \tag{12}$$

The learning rate for the covariance matrix update is denoted by c_{cov}, μ_{eff} is the variance effective selection mass and μ_{cov} denotes a parameter which weighs between the rank-one update (estimates the covariance matrix using only the previous iteration) and rank-μ update (uses all previous iterations).

The CMA-ES algorithm makes use of cumulative step-size adaptation. The anisotropic evolution path, p_{cCMA}, associated with the covariance matrix and the isotropic evolution path, p_σ, associated with the step size are two evolution paths used. p_{cCMA} is calculated as follows:

$$p_{cCMA} = (1 - c_{cCMA})p_{cCMA} + \sqrt{c_{cCMA}(2 - c_{cCMA})\mu_{eff}}\left(\frac{\boldsymbol{m}(t+1) - \boldsymbol{m}(t)}{\sigma_{CMA}(t)}\right) \tag{13}$$

where μ_{eff} is given by

$$\mu_{eff} = \left(\sum_{k=1}^{n_s} w_k^2\right)^{-1} \tag{14}$$

and c_{cCMA} is the backward time horizon of the anisotropic evolution path.

Finally, the step size, $\sigma_{CMA}(t+1)$, is updated as follows:

$$\sigma_{CMA}(t+1) = \sigma_{CMA}(t)\exp\left(\frac{c_\sigma}{d_\sigma}\left(\frac{\|p_\sigma(t+1)\|}{E\|N(\boldsymbol{0}, \boldsymbol{I})\|} - 1\right)\right) \tag{15}$$

where d_σ is the damping parameter in the CMA-ES algorithm, $\frac{1}{c_\sigma}$ is the backward time horizon of the isotropic evolution path, p_σ:

$$p_\sigma = (1 - c_\sigma)p_\sigma + \sqrt{c_\sigma(2 - c_\sigma)\mu_{eff}}\,\boldsymbol{C}(t)^{-0.5}\left(\frac{\boldsymbol{m}(t+1) - \boldsymbol{m}(t)}{\sigma_{CMA}(t)}\right). \tag{16}$$

CMA-ES was chosen based on its parallel search capability. Fewer individuals are needed compared to a genetic algorithm and fewer parameters have to be tuned.

4   Algorithm Evaluation

Two random sets of data were generated whereby one represents the distances between customers and the other the IRI values associated with each segment. Ten datasets were generated in total, two of each size. Symmetrical datasets were used, implying that the distance from customer i to j is the same as from j to i. The distances were varied between 0 and 100 miles and the IRI between 0 and 2.6 m/km since 2.6 is when a road is considered to be in a terminal condition [2].

In the simulated annealing, a number of operators can be used to change the current solution to find a better solution. A swap move operator was chosen to randomly swap the sequence of two customers. The objective function is then recalculated and if it is an improvement on the previous solution, this new solution is accepted. However, since a meta-heuristic is used, even if the new solution is not an improvement, it is accepted with an ever-decreasing probability, P.

CMA-ES operates in a continuous space and an individual is interpreted as a list of customer priorities. The initial solution is mapped to this continuous space by arranging the greedy solution evenly between −100 and 100.

To find good values for the two parameters, namely the starting temperature and the initial acceptance probability, numerous runs were conducted. Each dataset was considered separately. The probability was set and the temperature adjusted between 20 and 50 with 30 runs at each parameter combination. Next, the probability was adjusted and the temperature set at a constant value. This process continued for probabilities ranging from 20% to 90% at 10% increments. For the CMA-ES, σ was varied between 0.2 and 0.5. The average objective function value and standard deviation for each set-up was recorded and the best parameters are shown in Table 1.

The two smallest datasets were solved in an exact manner using the branch-and-bound method. The best result of the simulated annealing matched this global optimum.

For benchmarking purposes the results were compared to that of a simple greedy heuristic. The CMA-ES delivered no improvement over the greedy heuristic and therefore the standard deviation is zero. The dashes in the sigma column indicate that the same results were obtained regardless of the sigma value. The three algorithms were compared in a pairwise manner. On every dataset-algorithm combination, a Mann-Whitney U test was performed at 95% significance. For the biggest eight datasets, the simulated annealing performed better than the CMA-ES and the greedy heuristic, for the dataset 25_2 the CMA-ES was better and for the 25_1 the algorithms performed the same. The reason for the simulated annealing outperforming the CMA-ES is due to the mapping mechanism used in the CMA-ES which operates in the continuous space. A possible reason for the simulated annealing performing so well is due to the fact

Table 1. Parameter testing for simulated annealing and CMA-ES

Dataset	Simulated annealing				CMA-ES			Greedy
	P	Temp	Mean	Stdev	Sigma	Mean	Stdev	
25_1	20	70	233.256	8.855	0.2	236.859	6.202	241.838
25_2	30	30	205.933	5.987	0.2	**197.619**	5.138	210.031
50_1	40	90	**296.423**	0.524	-	296.524	0	296.524
50_2	60	20	**309.532**	0.592	-	309.719	0	309.719
100_1	50	20	**549.584**	0.361	-	549.723	0	549.723
100_2	80	80	**451.139**	0.539	-	451.279	0	451.279
500_1	30	80	**1118.022**	1.034	-	1118.211	0	1118.211
500_2	80	30	**994.240**	0.272	-	994.289	0	994.289
1000_1	20	20	**1358.225**	0.046	-	1358.233	0	1358.233
1000_2	50	80	**1427.691**	0.267	-	1427.739	0	1427.739

that it is more intelligent than the greedy heuristic and that it utilises a better method of searching through the solution space than the CMA-ES which has to work through a mapping mechanism.

To show the value of considering IRI when making a route decision, the simulated annealing algorithm was used as it outperformed the other algorithms. For each dataset, a route was calculated which considers the IRI of the roads. The corresponding total distance travelled and total IRI was calculated based on this developed route. Next, the algorithm was run with just the distance matrix and a route was optimised. Again, the corresponding total distance travelled and total IRI was calculated. These results can be seen in Table 2. For every dataset, the total IRI is smaller in the first instance than in the second, whereas the total distance travelled is larger for the first instance than the second.

Table 2. Distances and associated IRI values for two instances

Dataset	Minimise distance and IRI		Minimise distance	
	Distance	IRI	**Distance**	IRI
25_1	1296.11	25.65	1064.09	31.96
25_2	1070.36	30.59	975.00	33.65
50_1	1525.19	58.67	1519.14	60.54
50_2	1591.27	61.25	1578.64	60.42
100_1	2768.88	132.59	2790.49	132.98
100_2	2318.98	125.89	2318.98	125.89
500_1	5747.05	644.74	5717.52	669.09
500_2	5114.96	612.75	5012.31	640.84
1000_1	7578.27	1289.72	6979.53	1290.21
1000_2	7337.89	1291.46	7222.49	1282.91

5   Conclusion

Three algorithms were used as basis for the investigation into the inclusion of the effect of road roughness on routing. A greedy heuristic, simulated annealing and CMA-ES algorithms were used to optimise ten datasets. Simulated annealing showed statistically significant results when compared to the other two algorithms and was therefore used to further investigate the effect of road roughness on routing decisions. It was confirmed that the consideration of the road condition in the form of IRI will have an influence on the route choices. The algorithm will choose a longer path to avoid road sections with high IRI values due to the fact that it adds to the objective function. When only the total distance was used as an objective function, the resulting IRI was higher.

The equations used in this paper to calculate the effect of IRI on VOC are very simplistic. Further research is to be conducted on other factors contributing to vehicle operating cost and how it can be related to the roads travelled. The next step would be to implement more constraints on the model and focus on improving the design of the optimisation algorithms.

Acknowledgements. This work is based on the research supported wholly/in part by the National Research Foundation of South Africa (Grant Numbers 109273).

References

1. Steyn, W.J., Monismith, C.L., Nokes, W.A., Harvey, J.T., Holland, T.J., Burmas, N.: Challenges confronting road freight transport and the use of vehicle-pavement interaction analysis in addressing these challenges. J. South Afr. Inst. Civil Eng. **54**(1), 14–21 (2012)
2. Wang, Z., Wang, H.: Life-cycle cost analysis of optimal timing of pavement preservation. Front. Struct. Civil Eng. **11**(1), 17–26 (2016)
3. Kirkpatrick, S., Gelatt, C.D., Vecchi, M.P.: Optimization by simulated annealing. Science **220**(4598), 671–680 (1983)
4. Hansen, N., Ch, P.K.P.E.: Reducing the time complexity of the derandomized evolution strategy with covariance matrix adaptation (CMA-ES). In: Evolutionary Computation. Citeseer (2003)
5. Miller, C.E., Tucker, A.W., Zemlin, R.A.: Integer programming formulation of traveling salesman problems. J. ACM **7**(4), 326–329 (1960)
6. Chatti, K., Zaabar, I.: Estimating the effects of pavement condition on vehicle operating costs. NCHRP 720, Transportation Research Board, Washington, D.C. (2012)
7. Kerali, H.R., Robinson, R., Paterson, W.D.O.: Role of the new HDM-4 in highway management. In: Proceedings Fourth International Conference on Managing Pavements, Durban, South Africa, pp. 17–21, May 1998
8. Zaabar, I., Chatti, K.: Identification of localized roughness features and their impact on vehicle durability. In: HVTT, vol. 11, pp. 1–13 (2010)
9. Steyn, W.J.vdM., Nokes, W., Du Plessis, L., Agacer, R., Burmas, N., Holland, T.J., Popescu, L.: Selected road condition, vehicle and freight considerations in pavement life cycle assessment. In: International Symposium on Pavement Life Cycle Assessment, Davis, CA, USA, 14–16 October 2014

10. Toro, O., Eliana, M., Escobar, Z., Antonio, H., Granada, E.: Literature review of vehicle routing problem in the green transportation context. Revista Luna Azul **42**, 362–387 (2016)
11. Li, X., Leung, S.C.H., Tian, P.: A multistart adaptative memory based tabu search algorithm for the heterogeneous fixed fleet open vehicle routing problem. Experts Syst. Appl. **39**, 365–374 (2012)
12. Renaud, J., Laporte, G., Boctor, F.: A tabu search heuristic for the multi depot vehicle routing problem. COR **23**, 229–235 (1996)
13. Prins, C., Prodhon, C., Ruiz, A., Soriano, P., Wolfler, R.: Solving the CLRP by a coop. Lagrangean relaxation-granular heuristic. Transp. Sci. **41**(4), 470–483 (2007)
14. Lacomme, P., Prins, C., Ramdane-Chérif, W.: A genetic algorithm for the capacitated arc routing problem and its extensions. In: Boers, E.J.W. (ed.) EvoWorkshops 2001. LNCS, vol. 2037, pp. 473–483. Springer, Heidelberg (2001). https://doi.org/10.1007/3-540-45365-2_49
15. Jorgensen, M., Larsen, J.: Solving dial-a ride problem using genetic algorithm. JOR Soc. **58**, 1321–1331 (2006)
16. Donati, A., Montemanni, R., Casagrande, N., Rizzoli, A., Gambardella, L.: Time dependent VRP with a multi ant colony system. EJOR **185**(3), 1174–1191 (2008)
17. Yu, V.F., Redi, A.P., Hidayat, Y.A., Wibowo, O.J.: A simulated annealing heuristic for the hybrid vehicle routing problem. Appl. Soft Comput. **53**, 119–132 (2017)
18. Azi, N., Gendreau, M., Potvin, J.: An adaptive large neighborhood search for a vehicle routing problem with multiple trips. CIRRELT-2010-08 (2010)
19. Hemmelmayr, V., Doerner, K.: A variable neighboorhood search heuristic for periodical vehicle routing problems. EJOR **195**(3), 791–802 (2009)

Pattern Classification

Integration Base Classifiers in Geometry Space by Harmonic Mean

Robert Burduk$^{(\boxtimes)}$

Department of Systems and Computer Networks,
Wroclaw University of Science and Technology,
Wybrzeze Wyspianskiego 27, 50-370 Wroclaw, Poland
robert.burduk@pwr.edu.pl

Abstract. One of the most important steps in the formation of multiple classifier systems is the fusion process. The fusion process may be applied either to class labels or confidence levels (discriminant functions). In this paper, we propose an integration process which takes place in the geometry space. It means that the fusion of base classifiers is done using decision boundaries. In our approach, the final decision boundary is calculated by using the harmonic mean. The algorithm presented in the paper concerns the case of 3 basic classifiers and two-dimensional features space. The results of the experiment based on several data sets show that the proposed integration algorithm is a promising method for the development of multiple classifiers systems.

Keywords: Ensemble selection · Multiple classifier system
Decision boundary

1 Introduction

The problem of using simultaneously multiple base classifiers for making decisions as to the membership of an object in a class label has been discussed in the works associated with the classification systems for about twenty years [6,18]. The systems consisting of more than one base classifier are called ensembles of classifiers (EoC) or multiple classifiers systems (MCSs) [4,7,9,17]. The reasons for the use of a classifier ensemble include, for example, the fact that single classifiers are often unstable (small changes in input data may result in creation of very different decision boundaries).

The task of constructing MCSs can be generally divided into three steps: generation, selection and integration [1]. In the first step a set of base classifiers is trained. There are two ways, in which base classifiers can be trained. The classifiers, which are called homogeneous are of the same type. However, randomness is introduced to the learning algorithms by initializing training objects with different weights, manipulating the training objects or using different features subspaces. The classifiers, which are called heterogeneous, belong to different machine learning algorithms, but they are trained on the same data set. In this

© Springer International Publishing AG, part of Springer Nature 2018
L. Rutkowski et al. (Eds.): ICAISC 2018, LNAI 10841, pp. 585–592, 2018.
https://doi.org/10.1007/978-3-319-91253-0_54

paper, we will focus on homogeneous classifiers which are obtained by applying the same classification algorithm to different learning sets.

The second phase of building MCSs is related to the choice of a set of classifiers or one classifier from the whole available pool of base classifiers. If we choose one classifier, this process will be called the classifier selection. But if we choose a subset of base classifiers from the pool, it will be called the ensemble selection. Generally, in the ensemble selection, there are two approaches: the static ensemble selection and the dynamic ensemble selection [1]. In the static classifier selection one set of classifiers is selected to create EoC during the training phase. This EoC is used in the classification of all the objects from the test set. The main problem in this case is to find a pertinent objective function for selecting the classifiers. Usually, the feature space in this selection method is divided into different disjunctive regions of competence and for each of them a different classifier selected from the pool is determined. In the dynamic classifier selection, also called instance-based, a specific subset of classifiers is selected for each unknown sample [3]. It means that we are selecting different EoCs for different objects from the testing set. In this type of the classifier selection, the classifier is chosen and assigned to the sample based on different features or different decision regions [5]. The existing methods of the ensemble selection use the validation data set to create the so-called competence region or level of competence. These competencies can be computed by K nearest neighbours from the validation data set. In this paper, we will use the static classifier selection and regions of competence will be designated by the decision boundary of the base classifiers.

The integration process is widely discussed in the pattern recognition literature [12,16]. One of the existing ways to categorize the integration process is using the outputs of the base classifiers selected in the previous step. Generally, the output of a base classifier can be divided into three types [10].

- The abstract level – the classifier ψ assigns the unique label j to a given input x.
- The rank level – in this case for each input (object) x, each classifier produces an integer rank array. Each element within this array corresponds to one of the defined class labels. The array is usually sorted and the label at the top is the first choice.
- The measurement level – the output of a classifier is represented by a confidence value (CV) that addresses the degree of assigning the class label to the given input x. An example of such a representation of the output is a posteriori probability returned by Bayes classifier. Generally, this level can provide richer information than the abstract and rank levels.

For example, when considering the abstract level, voting techniques [15] are most popular. As majority voting usually works well for classifiers with a similar accuracy, we will use this method as a baseline.

In this paper we propose the concept of the classifier integration process which takes place in the geometry space. It means that we use the decision boundary in the integration process. In other words, the fusion of base classifiers is done using decision boundaries. In our approach, the final decision boundary is calculated by using the harmonic mean. The algorithm presented in the paper concerns the case of 3 basic classifiers and a two-dimensional features space.

The geometric approach discussed in [11] is applied to find characteristic points in the geometric space. These points are then used to determine the decision boundaries. Thus, the results presented in [13] do not concern the process of integration of base classifiers, but a method for creating decision boundaries.

The remainder of this paper is organized as follows. Section 2 presents the basic concept of the classification problem and EoC. Section 3 describes the proposed method for the integration base classifiers in the geometry space in which the harmonic mean is used to determine the decision boundary. The experimental evaluation is presented in Sect. 4. The discussion and conclusions from the experiments are presented in Sect. 5.

2   Basic Concept

Let us consider the binary classification task. It means that we have two class labels $\Omega = \{0, 1\}$. Each pattern is characterized by the feature vector x. The recognition algorithm Ψ maps the feature space x to the set of class labels Ω according to the general formula:

$$\Psi(x) \in \Omega. \tag{1}$$

Let us assume that $k \in \{1, 2, ..., K\}$ different classifiers $\Psi_1, \Psi_2, \ldots, \Psi_K$ are available to solve the classification task. In MCSs these classifiers are called base classifiers. In the binary classification task, K is assumed to be an odd number. As a result of all the classifiers' actions, their K responses are obtained. Usually all K base classifiers are applied to make the final decision of MCSs. Some methods select just one base classifier from the ensemble. The output of only this base classifier is used in the class label prediction for all objects. Another option is to select a subset of the base classifiers. Then, the combining method is needed to make the final decision of EoC.

The majority vote is a combining method that works at the abstract level. This voting method allows counting the base classifiers outputs as a vote for a class and assigns the input pattern to the class with the majority vote. The majority voting algorithm is as follows:

$$\Psi_{MV}(x) = \arg\max_{\omega} \sum_{k=1}^{K} I(\Psi_k(x), \omega), \tag{2}$$

where $I(\cdot)$ is the indicator function with the value 1 in the case of the correct classification of the object described by the feature vector x, i.e. when $\Psi_k(x) = \omega$. In the majority vote method each of the individual classifiers takes an equal part in building EoC.

3   Proposed Method

The author's earlier work [2] presents results of the integration base classifier in the geometric space in which the base classifiers use Fisher's classification rule, while the process of the base classifier selection is performed in the regions of competence defined by the intersection points of decision boundary functions. In this paper we use different linear classifiers, the selection process is not performed and in addition, the harmonic mean is used to determine the decision boundary of EoC. The calculation of one decision boundary for several base classifiers is a fusion method which is carried out in a geometric space.

The calculation of one decision boundary is performed as follows.

Algorithm 1. Algorithm for finding decision boundary of combining classifier Ψ_{HM}

Input   : Decision boundaries for the 3 basic classifiers Ψ_1, Ψ_2, Ψ_3
Output: Decision boundary of combining classifier Ψ_{HM}

1 **for** $x \in \mathbb{R}^1$ **do**
2   $\quad g_{min}(x) = min(g_{\Psi_1}(x), g_{\Psi_2}(x), g_{\Psi_3}(x));$
3   $\quad g_{max}(x) = max(g_{\Psi_1}(x), g_{\Psi_2}(x), g_{\Psi_3}(x));$
4   $\quad g_{med}(x) = median(g_{\Psi_1}(x), g_{\Psi_2}(x), g_{\Psi_3}(x));$
5   $\quad h_1 = g_{max}(x) - g_{med}(x);$
6   $\quad h_2 = g_{med}(x) - g_{min}(x);$
7   $\quad hm = $ harmonic mean$(h_1, h_2);$
8   $\quad$**if** $g_{med}(x) > (g_{max}(x) + g_{min}(x))/2$ **then**
9   $\quad\quad\mid\ g_{\Psi_{HM}}(x) = g_{max}(x) - hm;$
10 $\quad$**else**
11 $\quad\quad\mid\ g_{\Psi_{HM}}(x) = g_{min}(x) + hm;$
12 $\quad$**end**
13 **end**

Graphical interpretation of the proposed method for two-dimensional data set and three base classifiers is shown in Figs. 1 and 2. Decision boundaries defined by 3 linear classifiers are presented in Fig. 1. In addition, there are also marked values of h_1, h_2 which are distances between the decision boundaries $g_{\Psi_1}(x), g_{\Psi_2}(x), g_{\Psi_3}(x))$ in point x. From these distances the harmonic mean is calculated. The decision boundaries defined by the majority vote method (for the base classifier presented in Fig. 1) and the method of integration base classifiers in the geometry space by using the harmonic mean are presented in Fig. 2. The decision boundary defined by the majority vote method is piecewise linear function (green line) when the decision boundary defined by the proposed method is a polynomial function (red line).

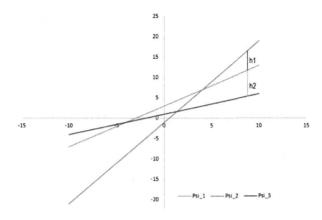

Fig. 1. Decision boundaries defined by 3 linear classifiers (Color figure online)

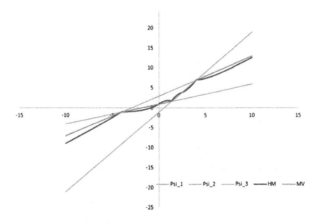

Fig. 2. Decision boundaries defined by the majority vote method (MV) and the method of integration base classifiers in the geometry space by using the harmonic mean (HM) (Color figure online)

4 Experimental Studies

The main aim of the experiments was to compare the quality of classifications of the proposed base classifiers integration algorithm Ψ_{HM} with the base classifiers $\Psi_1, ..., \Psi_3$ and their ensemble based on majority voting rule (MV) Ψ_{MV}.

In the experiment 3 base linear classifiers were used. One of them works according to Fisher linear discriminant rule, the next base classifier uses the Logistic regression model and the third uses the nearest mean rule. In the experimental research we use 10 publicly available data sets. The used data sets come from UCI machine learning repository and the KEEL Project. All of them have binary class labels. For all data sets the feature selection process [8,14] was performed to indicate two most informative features. The numbers of attributes and available examples are introduced in Table 1. All the experiments are performed following the standard 10-fold-cross-validation method.

Table 1. Description of data sets selected for the experiments

Data set	Examples	Attributes
Blood transfusion	748	5
Bupa	345	6
Haberman's survival	306	3
Magic gamma telescope	19020	11
Phoneme	5404	5
Pima Indians diabetes	768	8
Sonar	208	60
Spambase	4597	57
Twonorm	7400	20
Wdbc	569	30

Table 2. Classification accuracy and mean rank positions for the base classifiers Ψ_1, Ψ_2, Ψ_3, the majority voting method Ψ_{MV} and the proposed method Ψ_{HM} produced by the Friedman test

Data set	Ψ_1	Ψ_2	Ψ_3	Ψ_{MV}	Ψ_{HM}
Blood	0.767	0.767	0.675	0.766	0.775
Bupa	0.583	0.583	0.568	0.583	0.583
Haberman	0.745	0.742	0.719	0.745	0.745
Magic	0.780	0.790	0.749	0.790	0.791
Phoneme	0.748	0.742	0.711	0.748	0.748
Pima	0.767	0.764	0.727	0.767	0.770
Sonar	0.755	0.760	0.702	0.755	0.760
Spambase	0.748	0.768	0.757	0.755	0.763
Twonorm	0.721	0.721	0.721	0.721	0.721
Wdbc	0.851	0.896	0.840	0.882	0.859
Mean rank	3.10	2.45	4.60	2.75	2.10

Table 2 shows the results of the classification accuracy (ACC) and the mean ranks obtained by the Friedman test. Table 3 shows the results of the Matthews correlation coefficient (MCC) and the mean ranks obtained by the Friedman test. The MCC is generally regarded as a balanced measure which can be used even if the classes are of very different sizes and can be calculated using the formula: $MMC = \dfrac{TP*TN-FP*FN}{\sqrt{(TP+FP)(TP+FN)(TN+FP)(TN+FN)}}$.

The average ranks are the lowest for the proposed method for both ACC and MCC measures. A much larger difference occurs for MCC measure. In the case of ACC measure, the proposed method proved to be more effective than the MV method in case of 5 data sets. In the case of MCC measure better results were

Table 3. Matthews correlation coefficient and mean rank positions for the base classifiers Ψ_1, Ψ_2, Ψ_3, the majority voting method Ψ_{MV} and the proposed method Ψ_{HM} produced by the Friedman test

Data set	Ψ_1	Ψ_2	Ψ_3	Ψ_{MV}	Ψ_{HM}
Blood	0.154	0.154	0.165	0.148	0.220
Bupa	0.046	0.046	0.149	0.046	0.046
Haberman	0.194	0.165	0.306	0.194	0.194
Magic	0.498	0.523	0.438	0.523	0.526
Phoneme	0.345	0.319	0.420	0.346	0.351
Pima	0.464	0.457	0.411	0.464	0.471
Sonar	0.511	0.518	0.407	0.509	0.518
Spambase	0.461	0.507	0.483	0.477	0.493
Twonorm	0.442	0.442	0.441	0.442	0.442
Wdbc	0.682	0.776	0.664	0.747	0.701
Mean rank	3.50	2.90	3.30	3.25	2.05

obtained for 6 data sets. This observation confirms that the proposed method of integration base classifiers in the geometry space Ψ_{HM} proposed in the paper can improve the quality of the classification as compared to the MV method. However, the post-hoc Nemenyi test at $p = 0.05$, $p = 0.1$, is not powerful enough to detect any significant differences between the proposed algorithm and the MV method.

5   Conclusion

In this paper we have proposed a concept of a classifier integration process taking place in the geometry space. It means that we use the decision boundary in the integration process but we do not consider information produced by the base classifiers such as class labels, a ranking list of possible classes or confidence levels. In the presented approach, the final decision boundary is calculated by using the harmonic mean. The algorithm presented in the paper concerns the case of 3 basic classifiers and two-dimensional features space.

The experiments have been carried out on ten benchmark data sets. The aim of the experiments was to compare the proposed algorithm Ψ_{HM} and the majority voting method Ψ_{MV}. Two measures were used to determine the quality of the classification: accuracy and Matthews correlation coefficient. The results of the experiment show that the proposed integration algorithm is a promising method for the development of multiple classifiers systems.

Acknowledgments. This work was supported in part by the National Science Centre, Poland under the grant no. 2017/25/B/ST6/01750.

References

1. Britto, A.S., Sabourin, R., Oliveira, L.E.: Dynamic selection of classifiers-a comprehensive review. Pattern Recogn. **47**(11), 3665–3680 (2014)
2. Burduk, R.: Integration base classifiers based on their decision boundary. In: Rutkowski, L., Korytkowski, M., Scherer, R., Tadeusiewicz, R., Zadeh, L.A., Zurada, J.M. (eds.) ICAISC 2017. LNCS (LNAI), vol. 10246, pp. 13–20. Springer, Cham (2017). https://doi.org/10.1007/978-3-319-59060-8_2
3. Cavalin, P.R., Sabourin, R., Suen, C.Y.: Dynamic selection approaches for multiple classifier systems. Neural Comput. Appl. **22**(3–4), 673–688 (2013)
4. Cyganek, B.: One-class support vector ensembles for image segmentation and classification. J. Math. Imaging Vis. **42**(2–3), 103–117 (2012)
5. Didaci, L., Giacinto, G., Roli, F., Marcialis, G.L.: A study on the performances of dynamic classifier selection based on local accuracy estimation. Pattern Recogn. **38**, 2188–2191 (2005)
6. Drucker, H., Cortes, C., Jackel, L.D., LeCun, Y., Vapnik, V.: Boosting and other ensemble methods. Neural Comput. **6**(6), 1289–1301 (1994)
7. Giacinto, G., Roli, F.: An approach to the automatic design of multiple classifier systems. Pattern Recogn. Lett. **22**, 25–33 (2001)
8. Guyon, I., Elisseeff, A.: An introduction to variable and feature selection. J. Mach. Learn. Res. **3**, 1157–1182 (2003)
9. Korytkowski, M., Rutkowski, L., Scherer, R.: From ensemble of fuzzy classifiers to single fuzzy rule base classifier. In: Rutkowski, L., Tadeusiewicz, R., Zadeh, L.A., Zurada, J.M. (eds.) ICAISC 2008. LNCS (LNAI), vol. 5097, pp. 265–272. Springer, Heidelberg (2008). https://doi.org/10.1007/978-3-540-69731-2_26
10. Kuncheva, L.I.: Combining Pattern Classifiers: Methods and Algorithms. Wiley, Hoboken (2004)
11. Li, Y., Meng, D., Gui, Z.: Random optimized geometric ensembles. Neurocomputing **94**, 159–163 (2012)
12. Ponti, Jr., M.P.: Combining classifiers: from the creation of ensembles to the decision fusion. In: 2011 24th SIBGRAPI Conference on Graphics, Patterns and Images Tutorials (SIBGRAPI-T), pp. 1–10. IEEE (2011)
13. Pujol, O., Masip, D.: Geometry-based ensembles: toward a structural characterization of the classification boundary. IEEE Trans. Pattern Anal. Mach. Intell. **31**(6), 1140–1146 (2009)
14. Rejer, I.: Genetic algorithms for feature selection for brain computer interface. Int. J. Pattern Recogn. Artif. Intell. **29**(5), 1559008 (2015)
15. Ruta, D., Gabrys, B.: Classifier selection for majority voting. Inf. Fusion **6**(1), 63–81 (2005)
16. Tulyakov, S., Jaeger, S., Govindaraju, V., Doermann, D.: Review of classifier combination methods. In: Marinai, S., Fujisawa, H. (eds.) Machine Learning in Document Analysis and Recognition, pp. 361–386. Springer, Heidelberg (2008). https://doi.org/10.1007/978-3-540-76280-5_14
17. Woźniak, M., Graña, M., Corchado, E.: A survey of multiple classifier systems as hybrid systems. Inf. Fusion **16**, 3–17 (2014)
18. Xu, L., Krzyzak, A., Suen, C.Y.: Methods of combining multiple classifiers and their applications to handwriting recognition. IEEE Trans. Syst. Man Cybern. **22**(3), 418–435 (1992)

Similarity of Mobile Users Based on Sparse Location History

Pasi Fränti[(✉)], Radu Mariescu-Istodor, and Karol Waga

School of Computing, University of Eastern Finland, Joensuu, Finland
`pasi.franti@uef.fi`

Abstract. We propose a method to measure similarity of users based on their sparse location history such as geo-tagged photos or check-in activity of user. The method is useful when complete movement trajectories are not available. We map each activity point into the nearest location in a predefined set of fixed places. The problem is then formulated as histogram comparison. We compare the performance of similarity measures such as L_1, L_2, L_∞, ChiSquared, Bhatta-charyya and Kullback and Leibler divergence using both crisp and fuzzy histograms. Results show that user can be recognized with fair accuracy, and that all similarity measures are suitable except L_2 and L_∞, which perform poorly.

Keywords: User similarity · Mobile activity · GPS data analysis
Histogram matching · Fuzzy pattern recognition · Location-based services

1 Introduction

Similarity of users has been widely used in recommender systems based on the assumption that similar users are interested in similar things. Collaborative recommender systems [1] estimate relevance of an item to a given user based on ratings given by similar users. To find similar users most of the recommendation systems searches for the common items that the users have rated [1]. This approach is used in online shops, movie databases and similar recommendation systems.

The knowledge of similar users has been applied to improve retail experience by finding correlation between buying and browsing behavior. Similarity of users has also been used for recommending events and friends in [9], and provide good initial guess for personalizing the recommendations for new users [21]. Similarity of users have also been measured based on how they tag for bookmarking purpose [13].

In [11], we studied whether social network can be used for improving location-aware recommendations. The results showed that user's own understanding of the similarity correlates more with the similarity of their page likings and less with the similarity of their location histories, for which only minor correlation was detected. The same order of importance was observed in [12]: people value most the things they have common, then the places where they are active, and least important was to know the same people.

© Springer International Publishing AG, part of Springer Nature 2018
L. Rutkowski et al. (Eds.): ICAISC 2018, LNAI 10841, pp. 593–603, 2018.
https://doi.org/10.1007/978-3-319-91253-0_55

In location-aware recommendations, however, opinions of local experts in the given area can be more valuable than just the similarity of the user [2]. This can be useful for improving rating of the services by utilizing users whose opinions matter most. In general, knowing the similarity of location history can provide additional information for improving recommendation. In this paper, we study how the similarity of users can be measured from a limited amount of location data. Sample location histories of three users are shown in Fig. 1.

One approach is to analyze complete trajectories of the user movements. Several similarity measures were compared in [16] based on the complete trajectories. In [21], potential friends are recommended based on users' movement trajectories. So-called *stay cells* are also created based on detected stops, which are considered important places because user stayed there a longer time. Similarity of trajectories are then measured based on their longest common subsequence giving higher weight to longer patterns. In [15], a revised version of the longest common subsequence is applied by partitioning the trajectories based on speed and detected turn points. The similarity score is based on both geographic similarity and the semantic similarity. In [8], similarity of the location and their temporal semantics were also taken into account.

In [4], similarity of a person's days is assessed based on the trajectory by discovering their semantic meaning. The data is collected from tracking users' cars and pre-processed by detecting stop points. Most common pairs of stops are assumed to be user's home and work locations. Dynamic time warping of the raw trajectories using geographic distances of the points is reported to work best. In [20], personalized search for similar trajectories is performed by taking into account user preferences of which parts of the query trajectory is more important.

Complete trajectories are not always available and the similarity must then be measured based on sparse location data such as visits, favorite places or check-ins. In [14], user data is hierarchically clustered into geographic regions. A graph is constructed from the clustered locations so that a node is a region user has visited, and an edge between two nodes represents the order of the visits to these regions. This method still relies on the visiting order.

In this paper, we study how the similarity of users can be measured based on their location history when the entire GPS trajectories are not available. We use single location points originating from geo-tagged photos, and the start and end points of movements. Other activity points that could be used are stay points, i.e. the places where the user stayed longer than 30 min, as in [23].

We propose to measure the similarity between users by taking each user activity as an observation into a histogram that represents places in the region. The problem then reduces to comparison of the two discrete distributions. We consider several measures based on normalized frequency vectors: L_1, L_2, L_∞, ChiSquared, Bhattacharyya and Kullback and Leibler divergence. Fuzzy histograms are also considered.

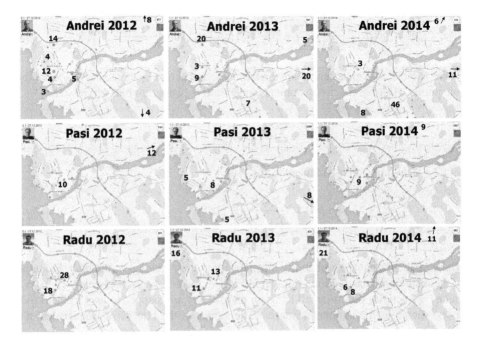

Fig. 1. Activity data of three users during 2012–2014.

2 Sparse Location Histograms

Mopsi (http://cs.uef.fi/mopsi) is a prototype media sharing platform in which users can collect geo-tagged photos, track their routes, and recommend places of interest [19] to other users by upgrading them to *services*. These services include hotels, restaurants, cafeterias and many others that the users consider relevant to themselves and to others. The locations of these services serve as the histogram bins, which we denote as *places*. User similarities are calculated based on the locations where the users have collected data. We call these as *activity points*.

2.1 Locations, Distance and Places

Calculating similarity of two users starts by constructing histograms of the users. We process the activity points of a user by mapping them to their nearest place. Every activity point adds the count of the corresponding histogram bin by one. Distance calculation is based on *haversine* distance of the two locations given as latitude (φ) and longitude (λ) coordinates. The formula to calculate the haversine distance (in kilometers) is defined as:

$$hav = 2 \cdot R \cdot \arcsin(\psi) \tag{1}$$

$$\psi = \sqrt{\sin^2\left(\frac{\varphi_2 - \varphi_1}{2}\right) + \cos\left(\varphi_1\right)\cos\left(\varphi_2\right)\sin^2\left(\frac{\lambda_2 - \lambda_1}{2}\right)} \qquad (2)$$

where $R = 6372.8$ km, φ_1 and φ_2 are the latitudes, and λ_1 and λ_2 are the longitudes of the two points. User has n activity points mapped into m histogram bins $h(i)$ so that:

$$\sum_{i=1}^{m} h(i) = n \qquad (3)$$

The histograms are normalized as follows:

$$p(i) = \frac{h(i)}{\sum\limits_{j=1}^{m} h(j)} \forall i \in [1, m] \qquad (4)$$

where $p(i)$ represent the probability that an activity point belongs to the bin i. Example of the histogram construction is shown in Fig. 2, where three users have $n_1 = 9$, $n_2 = 7$ and $n_3 = 7$ activity points (small icons on map). They are mapped to $m = 8$ places (hot spots) shown as the thumbnail images. The values of the bins (visit frequencies) are shown below each place.

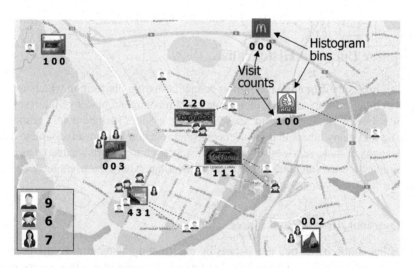

Fig. 2. Converting location history of three users to histogram of $m = 8$ bins.

The choice of using existing places is just one possibility. Another alternative might be to cluster all activity data, and in this way, automatically determine the hot spot places. Semantics of the places are also not considered. For example, McDondald's and its local competitor, Hesburger, are considered different places. We are only interested whether the user can be distinguished based on the location only without considering what is there.

2.2   Histogram Matching

The problem is to calculate similarity (or distance) between two distributions, represented as histograms. The histogram is usually a one-dimensional array consisting of numerical values, for example, pixel intensities of an image. In this case, there is an explicit ordering of the bins, and the values of the neighboring bins highly correlate with each other. Mapping an observation into the histogram is straightforward.

The bin values can also be nominal, or multivariate as in our case, so that there does not exists any natural ordering of the bins. However, since the observations appear in a metric space, the observations (activity points) can still be mapped to the histogram by simple distance calculations. The problem therefore reduces to histogram comparison [5, 6].

Figure 3 demonstrates the process using *Bhattacharyya coefficient* originally proposed as a similarity measure between statistical populations. First, product $p_i \cdot q_i$ of two frequencies are calculated, and their square roots are then summed over the all histogram bins. The higher the frequencies, the higher the product. The result is converted to a distance in range [0, 1] by applying logarithmic scaling. This provides natural bounds on the Bayes misclassification probability.

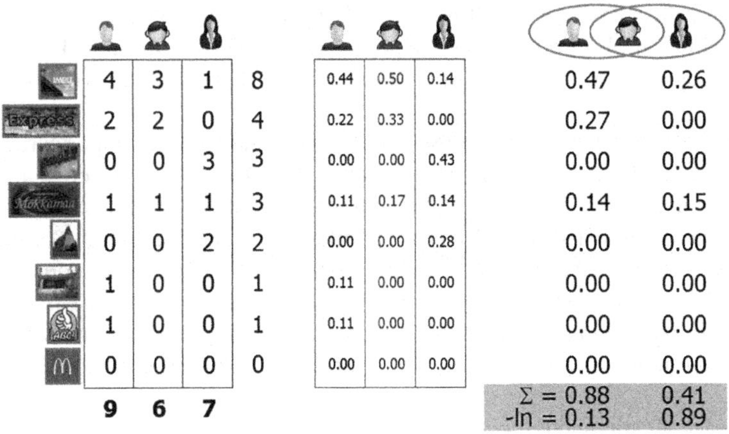

Fig. 3. Distance calculations of two histograms using Bhattacharyya distance.

We also consider various L-norms such as L_1, L_2 and L_∞, and classical Chi Squared distance. Kullback and Leibler distance [22] generalizes Shannon's concept of probabilistic uncertainty called entropy by calculating minimum cross entropy of two probability distributions [15]. These and the Bhattacharyya coefficient are defined below.

$$L_1 = \sum_i |p_i - q_i| \tag{5}$$

$$L_2 = \sum_i (p_i - q_i)^2 \tag{6}$$

$$L_\infty = \max_{1 \le i \le n} |p_i - q_i| \tag{7}$$

$$d_{ChiSq} = \sum_i \frac{(p_i - q_i)^2}{p_i + q_i} \tag{8}$$

$$d_{KLD} = \sum_i \left(p_i \cdot \log \frac{p_i}{q_i} + q_i \cdot \log \frac{q_i}{p_i} \right) \tag{9}$$

$$S_{BC} = \sum_i \sqrt{p_i \cdot q_i} \tag{10}$$

where p_i and q_i are the relative frequencies of the histogram bins i, and the summation is done over all m places.

All the above techniques calculate the distance of each bin independently. In case of sparse observations, it may happen that strongly peaked histograms would become mismatched due to slight translation. So-called *earth mover distance* (EMD) [17] aims at solving this by transforming surplus from one bin to the bins that have deficit. In case of one-dimensional numeric data this is straightforward to calculate by processing the histogram sequentially from left to right. However, in multivariate case the optimal moving of the surplus becomes more complicated problem. It was noted in [7] that the problem could be solved as transportation problem but faster algorithms would be needed.

In [18], the peaks of the histograms were considered as more important. Improved performance of L_1, L_2 and EMD was demonstrated in case of time-series analysis by calculating the sum of the peak weights multiplied by their closeness factors.

Sparseness of the data may also cause problems when there are too few observations compared to the number of available bins. Fuzzy histograms were proposed in [10] motivated by its successful application in image processing field. In case of one-dimensional histograms, observations are divided into several neighboring bins. We generalize the idea into multivariate case by utilizing the *k-nearest neighbors* (kNN) concept as follows.

For each location activity, we find its k nearest places and calculate fuzzy counts similarly as done in the well-known fuzzy C-means algorithm [3]:

$$w_i = \left(\sum_{j=1, j \ne i}^{k} \frac{d(x - h_i)}{d(x - h_j)} \right)^{-1} \tag{11}$$

Otherwise, the histogram comparison can be done in the same way as with the crisp variant.

3   Experiments

We used data from Mopsi that has been collected using native mobile application available in all platforms with the following user distribution: WindowsPhone (55%), Android (28%), iOS (14%) and Symbian (3%). There are 36243 photos and 8963 trajectories, and by 9.5.2015, there have been 909 registered mobile users. Of these, 94 users have collected more than 5 photos and 5 routes. In the following, we focus on a small subset of users whose location activity we are familiar with.

3.1   Data Extraction

For creating the histogram bins, we selected 293 services from Mopsi (http://cs.uef.fi/ mopsi). These include cafes, restaurants, holiday resorts, shops, parks and other services within the bounding box that covers the Joensuu sub-region, see Fig. 4. This region covers Joensuu downtown, its suburban, neighboring municipalities and large sparsely inhabited rural areas. We use here only the location of the services. The coverage of the bins is dense in the downtown area but sparse in rural area. For example, one service is the only one within 20 km radius whereas there are about 75 services within the 3×3 blocks (300 m $\times$ 500 m area) around the market place.

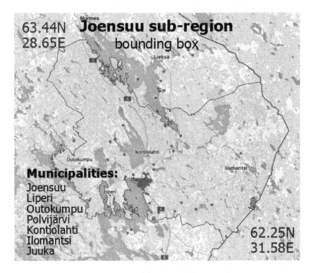

Fig. 4.   Distribution of the places (histogram bins).

We then recorded activities of the users from the years 2011–2014 as follows: (1) places where they took photos, (2) places where tracking a route was started or ended. Each activity is counted to the frequency of its nearest service (histogram bin). Our first test consists of three active users (A = Andrei, P = Pasi, R = Radu) called A-P-R trio in the following. These users have 5831 location points in total. The most popular places with the corresponding visit frequencies are listed in Table 2.

The data is divided per year resulting into four subsets for each user, see Table 1. In total, we will have 12 subsets (pseudo users) denoted as: A11, A12, A13, A14, P11, P12, P13, P14, R11, R12, R13, R14. By default, the subsets A11–A14 should be similar to each other, and dissimilar to the other subsets. In practice, the situation is more complicated. The biggest frequency is in the bin corresponding to the location of user's home. However, both Andrei and Radu moved in 2014 causing significant changes in their histograms. All users have the same work place (#8). This and the homes are emphasized in Table 2.

Table 1. Three test users and the summary of their location data.

	2011	2012	2013	2014
Andrei	206	757	432	329
Pasi	1263	545	636	751
Radu	37	292	324	259

Table 2. Ten most popular locations (histogram bins) and the corresponding frequencies.

	Andrei				Pasi				Radu				Total
	2011	2012	2013	2014	2011	2012	2013	2014	2011	2012	2013	2014	
1	20	0	29	150	47	7	6	8	1	2	2	3	275
2	13	11	87	36	64	17	16	7	0	1	12	2	266
3	0	0	0	0	51	54	54	69	0	1	0	0	229
4	12	107	87	0	0	0	1	2	0	0	2	0	211
5	1	0	0	1	34	9	20	11	0	0	52	54	182
6	6	29	10	3	6	2	1	3	7	54	35	16	172
7	7	92	6	0	18	4	7	15	0	13	4	2	168
8	22	6	5	4	36	9	6	18	1	12	11	21	151
9	0	3	4	0	0	0	0	1	12	82	41	7	150
10	0	0	0	0	73	6	48	13	0	0	0	0	140

3.2   Test Setup and Results

We calculate the similarity (or distance) value between all pairs of the 12 subsets. The task is then to decide which of the subsets belong to the same user by thresholding. The expected result is that $3 \cdot 4 \cdot 4 = 48$ pairs (33%) should be recognized as the same user, and $3 \cdot 4 \cdot 8 = 96$ pairs (67%) should fail the test.

For thresholding, we study the effect of different alternatives. First choice (average) is to use the average of all similarity values. This is the simplest non-parametric threshold that attempts to adapt the method to the data. Second choice (apriori) is obtained by selecting the threshold value (for the particular method) that passes 48 pairs (or as close to this as possible) based on a priori information that 33% values should pass the similarity test, and 96 should fail. The last choice (oracle) is the threshold that provides best accuracy for the particular method.

The results in Table 3 show that L_1, Chi^2, BHA and KLD provide good results (8%, 8%, 10%, 11%) using the average as threshold, and only slightly better if the optimal threshold (Oracle) was known (7%, 7%, 8%, 10%). The two other methods, L_2 and L_∞, perform poorly (35%, 43%). The a priori information does not help, and even if the optimal threshold is known their results are inferior (15%, 18%).

Table 3. Classification accuracy for the APR trio.

	Threshold (crisp)			Accuracy (crisp)			Accuracy (fuzzy)		
	Mean	Apr.	Oracle	Mean	Apr.	Oracle	Mean	Apr.	Oracle
L_1	0.31	0.27	0.28	8%	8%	7%	10%	10%	10%
Chi2	1.22	1.24	1.18	8%	7%	7%	17%	11%	10%
BHA	0.46	0.46	0.48	10%	10%	8%	15%	14%	11%
KLD	0.82	0.89	0.88	11%	11%	10%	36%	21%	15%
L_2	0.84	0.80	0.88	35%	47%	15%	35%	49%	14%
L_∞	0.79	0.72	0.87	43%	43%	18%	38%	47%	21%

Fuzzy histograms were also considered with neighborhood of fixed size $k = 3$. The classification errors systematically increased without clear reason. Especially KLD works significantly worse when using the fuzzy counts than with the crisp counts. Another observation is that the choice of the threshold becomes now critical. Using average as the threshold no long works with most measures.

The BHA and Chi^2 measures made classification error with users A13 and R13. In this case, the users reported to have recorded lots of joint bicycle trips at that year. In the data, this shows as increased counts in the bins representing rural areas.

Analysis of the other classification errors reveals the following details. All methods recognize A14 as different user than A11, A12 and A13. The same happens also between R11 and R14 with all methods except KLD. Both users changed their homes in 2014 causing different histogram bins to become dominant within the user. We therefore hypothesized that the methods might be affected too much by the dominant values. To test this, we performed additional tests by removing the top-10 location values. The changes are summarized in Table 4. The revised experiments show even weaker performance. The changes did not make A14 match with A11–A13. On the contrary, it caused more mismatch results and showed, that the methods somewhat rely on the detection of user's most popular places like work and home.

Table 4. Classification accuracy for the APR trio when 10 most popular bins have been eliminated from the calculations.

Method	All data	Excluding Top-10	Observation
L_1	8%	24%	Loses its ability
Chi^2	8%	13%	A11 becomes similar with P11, P13, P14
BHA	10%	13%	A11 becomes similar with P11, P13, P14 P11–R14 no longer matches
KLD	11%	13%	A11 and R14 become similar, no other effects
L_2	35%	40%	Works slightly worse
L_∞	42%	43%	Works slightly worse

4 Conclusion

Locations of people show their preferences and interests. In this paper, we have performed detailed study how histogram matching can be used to measure similarity of users based on their location history. We have shown that L_1, ChiSquared, Bhattacharyya and Kullback and Leibler divergence all are applicable to the problem. However, L_2 and L_∞ work poorly. Fuzzy histograms also worked worse than the corresponding crisp variant.

As future work, we consider the following open questions. Automatic method for selecting the threshold could be constructed based on expected distribution, as well as optimizing the number of bins. Possible use of double normalization, log-scaling of frequencies, cosine distance, and fuzzy modeling for sparse data can also provide further improvement. The size of test data vs. recognition accuracy should be studied further. Generalization of earth mover distance for the multivariate case is also worth to consider. Finally, histogram-based comparison might be better to replace prototype based comparison using clustering. These are all points of future studies.

References

1. Adomavicius, G., Tuzhilin, A.: Toward the next generation of recommender systems: a survey of the state-of-the-art and possible extensions. IEEE Trans. Knowl. Data Eng. **17**(6), 734–749 (2005)
2. Bao, J., Zheng, Y.H., Mokbel, M.F.: Location-based and preference-aware recommendation using sparse geo-social networking data. In: International Conference on Advances in Geographic Information Systems (SIGSPATIAL), Redondo Beach, CA, pp. 199–208 (2013)
3. Bezdek, J.C., Ehrlich, R., Full, W.: FCM: the fuzzy c-means clustering algorithm. Comput. Geosci. **10**(2–3), 191–203 (1984)
4. Biagioni, J., Krumm, J.: Days of our lives: assessing day similarity from location traces. In: Carberry, S., Weibelzahl, S., Micarelli, A., Semeraro, G. (eds.) UMAP 2013. LNCS, vol. 7899, pp. 89–101. Springer, Heidelberg (2013). https://doi.org/10.1007/978-3-642-38844-6_8
5. Cha, S.-H.: Comprehensive survey on distance/similarity measures between probability density functions. Int. J. Math. Models Methods Appl. Sci. **4**(1), 300–307 (2007)
6. Cha, S.-H.: Taxonomy of nominal type histogram distance measures. In: American Conference on Applied Mathematics, Harvard, MA, USA, pp. 325–330 (2008)
7. Cha, S.-H., Srihari, S.N.: On measuring the distance between histograms. Pattern Recogn. **29**(13), 1768–1774 (2008)
8. Chen, X., Pang, J., Xue, R.: Constructing and comparing user mobility profiles for location-based services. In: ACM Symposium on Applied Computing, pp. 261–266 (2013)
9. De Pessemier, T., Minnaert, J., Vanhecke, K., Dooms, S., Martens, L.: Social recommendations for events. In: ACM Conference on Recommender Systems, Hong Kong, China (2013)
10. Fober, T., Hullermeier, E.: Similarity measures for protein structures based on fuzzy histogram comparison. In: IEEE International Conference on Fuzzy Systems, Barcelona, pp. 1–7 (2010)
11. Fränti, P., Waga, K., Khurana, C.: Can social network be used for location-aware recommendation? In: International Conference on Web Information Systems & Technologies (WEBIST 2015), Lisbon, Portugal (2015)

12. Guy, I., Jacovi, M., Perer, A., Ronen, I., Uziel, E.: Same places, same things, same people?: Mining user similarity on social media. In: ACM Conference on Computer Supported Cooperative Work, Savannah, GA, USA, pp. 41–50 (2010)
13. Li, X., Guo, L., Zhao, Y.: Tag-based social interest discovery. In: Conference on World Wide Web, Beijing, China, pp. 675–684 (2008a)
14. Li, Q., Zheng, Y., Xie, X., Chen, Y., Liu, W., Ma, W.-Y.: Mining user similarity based on location history. In: ACM SIGSPATIAL International Conference on Advances in Geographic Information Systems, Paper #34, Irvine, CA, USA (2008b)
15. Liu, H., Schneider, M.: Similarity measurement of moving object trajectories. In: International Workshop on GeoStreaming, pp. 19–22 (2012)
16. Mariescu-Istodor, R., Fränti, P.: Grid-based method for GPS route analysis for retrieval. ACM Trans. Spat. Algorithms Syst. **3**(3), 8:1–8:28 (2017)
17. Rubner, Y., Tomasi, C., Guibas, L.J.: A metric for distributions with applications to image databases. In: IEEE International Conference Computer Vision, pp. 59–66 (1992)
18. Strelkov, V.V.: A new similarity measure for histogram comparison and its application in time series analysis. Pattern Recogn. Lett. **29**(13), 1768–1774 (2008)
19. Waga, K., Tabarcea, A., Fränti, P.: Recommendation of points of interest from user generated data collection. In: IEEE International Conference on Collaborative Computing: Networking, Applications and Worksharing (CollaborateCom), Pittsburgh, USA (2012)
20. Wang, H., Liu, K.: User oriented trajectory similarity search. In: International Workshop on Urban Computing, pp. 103–110 (2012)
21. Yang, X., Steck, H., Guo, Y., Liu, Y.: On Top-k recommendation using social networks. In: ACM Conference on Recommender Systems, Dublin, Ireland, pp. 67–74 (2012)
22. Ying, J.J.C., Lu, E.H.C., Lee, W.C., Wen, T.C.M., Tseng, V.S.: Mining user similarity from semantic trajectories. In: International Workshop on Location Based Social Networks, San Jose, CA, USA (2010)
23. Zheng, V.W., Zheng, Y., Xie, X., Yang, Q.: Collaborative location and activity recommendations with GPS history data. In: ACM International Conference on World Wide Web, Raleigh, NC, USA, pp. 1029–1038 (2010)

Medoid-Shift for Noise Removal to Improve Clustering

Pasi Fränti[(⊠)] and Jiawei Yang

School of Computing, University of Eastern Finland, Joensuu, Finland
pasi.franti@uef.fi

Abstract. We propose to use medoid-shift to reduce the noise in data prior to clustering. The method processes every point by calculating its *k-nearest neighbors* (k-NN), and then replacing the point by the medoid of its neighborhood. The process can be iterated. After the data cleaning process, any clustering algorithm can be applied that is suitable for the data.

Keywords: Clustering · Noise removal · Outlier detection

1 Introduction

Outliers are points that deviate from the typical data. They can be anomalies that represent significant information that are wanted to be detected such as credit card frauds [1], and they can affect statistical conclusions based on significance tests [14]. In *clustering*, outliers are usually considered harmful as they can affect the clustering quality. It is therefore desired that the outliers can be detected and removed.

Detection of outliers is typically considered as a separate pre-processing step prior to clustering. Another approach is to perform the clustering first, and then label those points as outliers which did not fit into any cluster, see Fig. 1. DBSCAN is an example of this kind of approach [5]. However, it is a kind of *chicken-or-egg problem*: removing outliers first would improve the clustering, but on the other hand, performing clustering first would make the outlier detection easier.

Outlier removal can also be integrated in the clustering directly by modifying the cost function [13]. The benefit is that the clustering can provide more information to make better decision whether a point is an outlier, and yet, better clustering can be obtained as the outliers can be removed before the final clustering. In [9], outlier removal was integrated within the k-means algorithms so that the farthest points from their centroid were removed, and k-means is then repeated for the modified dataset. This iterative process removes not only outliers but also border points with the hope of obtaining more stable cluster centroids.

In this paper, we propose a new approach where we do not remove the outliers at all. Instead, we pre-process the data so that the effect of potential outliers will be reduced. The key idea is similar to smoothing in image processing, where the pixel values are altered based on the local window. We do the same in the feature space by

© Springer International Publishing AG, part of Springer Nature 2018
L. Rutkowski et al. (Eds.): ICAISC 2018, LNAI 10841, pp. 604–614, 2018.
https://doi.org/10.1007/978-3-319-91253-0_56

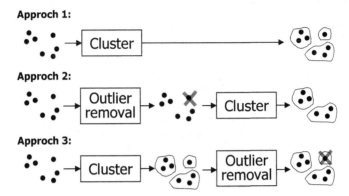

Fig. 1. Three approaches how to deal with outliers: (1) ignore; (2) remove outliers by pre-processing; (3) detect outliers from the clustering result.

finding *k-nearest neighbors* (k-NN) for every pixel, and then processing the pixel value in this neighborhood. We apply replacing the original value by the medoid value of its neighbors.

2 Existing Work and Their Limitations

Density-based approaches analyze the neighborhood of the points, and consider low density points as outliers. For example, a data point is marked as an outlier if there are at most k points are within a given distance d [11]. Another method calculates the k-nearest neighbors (k-NN), and use the distance to the kth neighbor [15], or the average distance to all k neighbors [10] as the outlier indicator. Points with the largest distance are considered as outliers. However, these approaches do not take into account that clusters may have significantly different densities.

Topology of the neighborhood has also been considered. In [3], undirected graph is constructed from the k-NN pointers. A link between two points is created only if they are mutual neighbors of each others. Connected components are considered as data points, and isolated points as outliers. In ODIN [10], the *indegree* of the nodes in the k-NN graph is used as the indicator of outliers because isolated points appear less frequently in the k-neighborhood of others.

However, if the number of outliers is large, traditional outlier removal techniques might not be efficient. In case when the number of outliers is of the same order of magnitude than the size of data, or even larger, it becomes significantly more challenging to differentiate outliers from normal points, see Fig. 2. Removing low density points might work into certain extent but setting the correct threshold is not trivial, especially if the data points have varying densities.

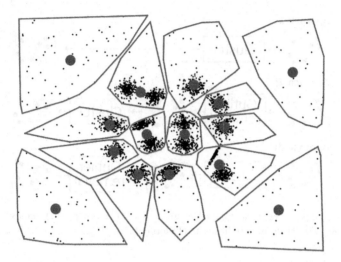

Fig. 2. Example of noisy data and how it affects clustering.

3   Medoid-Shift Noise Removal

In this work, we consider an alternative approach by formulating the outlier detection as noise removal problem. Instead of trying to detect whether a point is outlier, we process the data points locally based on the neighborhood. We propose a method called *medoid-shift noise removal*. The idea is to find k-nearest neighbors, and then replace the point by the medoid of its *k* neighbors. This will effectively remove the effect of isolated points and reduce the effect of abnormal points in the border areas of the clusters.

The idea is closely related to mean-shift filtering used in image processing [4], and mean-shift clustering algorithm [19]. The first one takes pixels value and its coordinates as the feature vector, and transforms each feature towards the mean of its neighbors. It has been used for detecting fingerprint and contamination defects in multicrystalline solar wafers [17]. The idea resembles also low-pass and median filtering used for image denoising.

The difference to mean-shift clustering is that we apply the shift merely for noise removal, and that we do not iterate the process until convergence. In other words, we do not force to use mean-shift clustering algorithm but the mean-shift process is merely a pre-processing stage. The choice of the algorithm is left as a separate question. This is because the mean-shift process itself can be good for removing noisy point, but its clustering performance is far from perfect.

3.1   Method Description

The core algorithm applies the following three steps for every point x:

1. Calculate the k-nearest neighbors $kNN(x)$ of the point x.
2. Calculate the mean of the neighbors.
3. Replace point x by the medoid M.

The algorithm modifies the data so that the effect of the noisy points is minimized. It is implemented as separate pre-processing step so that it is independent on the clustering method applied. The process can be iterated several times to have stronger noise removal effect. The number of iterations is the parameter.

The iterative variant is similar to the CERES algorithm in [9] where most remote points are iteratively removed in each cluster. The difference is that the outliers are chosen based on the distance to their cluster centroid in the intermediate clustering solution. It is therefore possible that points can be falsely removed if the cluster is not correctly determined. In our method, we remove the points based on their local neighborhood regardless of the clustering process. Figure 3 demonstrates the process.

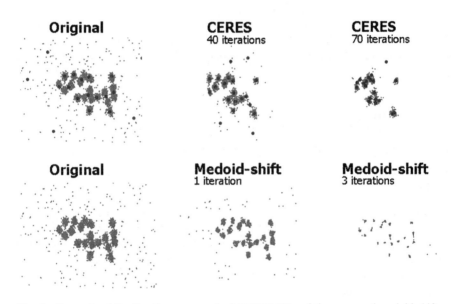

Fig. 3. Example of the iterative processed of CERES [9] and the proposed medoid-shift.

3.2   Mean or Median?

Both *mean* and *median* have been used in the mean-shift clustering concept [16, 19]. The benefit of using mean is that it is trivial to calculate for numeric data. Its main disadvantage is that it can cause blurring effect where very noisy points bias the calculation of the mean of clean points as well. Median is less sensitive to this because single noisy point in the neighborhood may not affect the calculation of the median at all, and it has therefore less effect on the clean points. Median can also reach its *root signal* [18] when repeated until converge.

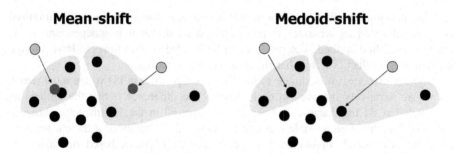

Fig. 4. Effect of mean (left) and median (right) of two sample points in a noisy neighborhood.

However, with multivariate data it is not obvious how median is defined. We use the *medoid*, which is the point in the set that has minimal total distance to all other points in the set. A disadvantage of medoid is that its calculation can be more time-consuming. A toy example of using mean and median is shown in Fig. 4.

4 Experiments

We test the proposed method with two clustering algorithms:

- K-means [6]
- Random swap [7].

Both algorithms aim at minimizing sum-of-squared errors. The first one is commonly used but does not always find the correct clustering solution even with the clean data. Random swap, however, finds the correct clustering result with all our datasets when no noise added. Therefore, after noise added, any errors the clustering algorithm makes can be addressed to the noise. For the effect of the noise removal, we compare the following methods:

- LOF [2]
- ODIN [10]
- Mean-shift (proposed)
- Medoid-shift (proposed).

Table 1. Datasets used in the experiments. (http://cs.uef.fi/sipu/datasets/)

Dataset	Size	Clusters
S1	5000	15
S2	5000	15
S3	5000	15
S4	5000	15
A1	3000	20
A2	5250	35
A3	7500	50
Unbalance	6500	8

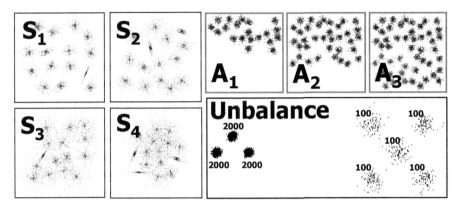

Fig. 5. Datasets used in the experiments.

We evaluate the clustering quality by the following two measures:

- CI = Centroid index [8]
- NMI = Normalized mutual information [12].

Centroid index is cluster level measure, which counts how many cluster centroids are wrongly located. Value CI = 0 indicates that all cluster centroids are correct with

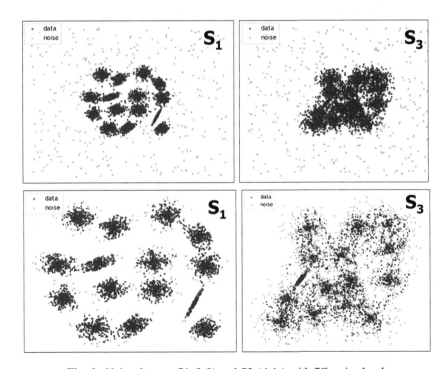

Fig. 6. Noisy datasets S1 (left) and S3 (right) with 7% noise level.

respect to the ground truth. *Normalized mutual information* is a point-level measure, which calculates the amount of information the clustering shares with the ground truth. Value 1 indicates perfect match to that of the ground truth.

4.1   Datasets

We use the eight benchmark datasets summarized in Table 1. The datasets are also shown in Fig. 5. The S sets have varying level of cluster overlap, A sets have varying number of clusters, and unbalance set have clusters with different densities.

Table 2. Summary of the point-level clustering results (NMI).

Noise type 1

Prepro-cess:	Combination	S1	S2	S3	S4	A1	A2	A3	Un	Av.
none	RS	0.87	0.82	0.78	0.67	0.84	0.87	0.88	0.61	0.75
	KM	0.85	0.82	0.78	0.67	0.83	0.87	0.89	0.90	0.82
noise removal	LOF+RS	0.85	0.84	0.76	0.69	0.88	0.90	0.91	0.70	0.81
	LOF+KM	0.90	0.84	0.81	0.67	0.89	0.90	0.91	**0.91**	0.85
	ODIN+RS	0.85	0.80	0.76	0.66	0.81	0.85	0.86	0.70	0.78
	ODIN+KM	0.85	0.79	0.76	0.62	0.80	0.85	0.86	0.84	0.80
medoid-shift	medoid+RS	**0.93**	**0.89**	0.83	0.68	**0.91**	**0.92**	**0.92**	0.90	**0.87**
	medoid+KM	0.91	0.88	**0.84**	**0.69**	0.90	**0.92**	**0.92**	0.89	**0.87**
	mean+RS	0.87	0.83	0.77	0.67	0.87	0.89	0.90	0.76	0.82
	mean+KM	0.88	0.83	0.78	0.66	0.87	0.89	0.89	0.90	0.84

Noise type 2

Prepro-cess:	Combination	S1	S2	S3	S4	A1	A2	A3	Un	Av.
none	RS	0.80	0.76	0.70	0.70	0.77	0.82	0.84	**0.81**	0.77
	KM	0.80	0.76	0.70	0.66	0.78	0.81	0.83	0.79	0.77
noise removal	LOF+RS	0.79	0.74	0.68	0.68	0.75	0.82	0.83	0.64	0.74
	LOF+KM	0.79	0.74	0.67	0.65	0.77	0.82	0.83	0.82	0.76
	ODIN+RS	0.78	0.74	0.68	0.67	0.68	0.67	0.63	0.77	0.70
	ODIN+KM	0.78	0.71	0.68	0.64	0.75	0.79	0.81	0.75	0.74
medoid-shift	medoid+RS	**0.86**	**0.84**	**0.77**	0.72	**0.85**	**0.86**	**0.87**	0.80	**0.82**
	medoid+KM	**0.86**	**0.84**	**0.77**	**0.73**	**0.85**	**0.86**	**0.87**	0.80	**0.82**
	mean+RS	0.80	0.76	0.68	0.69	0.81	0.83	0.86	0.63	0.76
	mean+KM	0.83	0.76	0.68	0.66	0.82	0.85	0.85	0.78	0.78

4.2 Noise Models

We consider two types of noise:

1. Random noise
2. Data-dependent noise.

In the first case, uniformly distributed random noise is added to the data. Random values are generated in each dimension between $[x_{mean} - 2 \cdot range, x_{mean} + 2 \cdot range]$, where x_{mean} is the mean of all data points, and *range* is the maximum distance of any point from the mean: $range = \max(|x_{max} - x_{mean}|, |x_{mean} - x_{min}|)$. The amount of noise is 7% of data size. In the second case, 7% of the original points are copied and moved to random direction. Noisy datasets are shown in Fig. 6.

Table 3. Summary of the cluster-level results (CI).

Noise type 1

Prepro-cess:	Combination	S1	S2	S3	S4	A1	A2	A3	Un	Av.
none	RS	4	4	4	3	6	7	14	4	5.8
	KM	4	3	3	3	5	8	13	2	5.1
noise removal	LOF+RS	3	4	3	2	5	5	9	2	4.1
	LOF+KM	2	4	3	3	4	6	10	3	4.4
	ODIN+RS	4	4	4	3	5	7	14	3	5.5
	ODIN+KM	4	4	4	4	6	8	11	2	3.5
medoid-shift	medoid+RS	0	0	2	2	0	2	4	2	1.5
	medoid+KM	1	1	1	1	0	2	3	2	**1.4**
	mean+RS	4	3	6	3	3	7	13	3	5.3
	mean+KM	4	4	4	2	4	7	14	2	5.1

Noise type 2

Prepro-cess:	Combination	S1	S2	S3	S4	A1	A2	A3	Un	Av.
none	RS	4	4	4	3	5	7	13	3	5.4
	KM	4	4	4	3	4	6	12	3	5.0
noise removal	LOF+RS	4	4	4	3	5	7	13	2	5.3
	LOF+KM	4	4	4	3	5	7	11	2	5.0
	ODIN+RS	4	4	4	4	5	8	13	3	5.6
	ODIN+KM	4	4	4	3	4	7	11	2	4.8
medoid-shift	medoid+RS	0	0	1	2	0	1	5	3	1.5
	medoid+KM	0	0	1	1	0	1	4	2	**1.1**
	mean+RS	4	4	4	3	4	7	11	3	5.0
	mean+KM	2	4	4	3	3	5	12	3	4.5

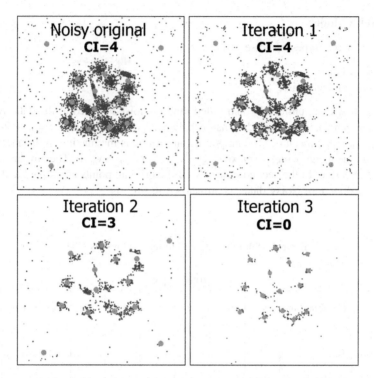

Fig. 7. Visualization of the medoid-shift noise removal on the noisy S1 dataset.

4.3   Results

The clustering results are summarized in Tables 2 and 3, both at the point level (NMI) and at the cluster level (CI). The point level results show that the medoid-shift improves both k-means and random swap clustering result from about NMI = 0.82 to 0.87. The medoid-shift is significantly better than the mean-shift, which is not really effective with this amount of noise. Visual examples are shown in Fig. 7.

The cluster-level results reveal that the task is challenging. The medoid-shift finds the correct clustering (CI = 0) only with the three easier datasets: S1, S2 and A1. The competitive outlier removal methods fail almost in all cases. The density-based method (LOF+KM) manages to improve the k-means clustering result in case of A2 and A3 datasets.

The choice of the clustering algorithm also matters but less with the noisy data. Random swap optimizes the cluster centroids better whereas k-means makes fails to put centroids in the low density areas. However, since noise has lower density, it mostly cancels the inferior optimization capability of k-means, and overall, both algorithms perform equally in practice. To sum up, the noise reduction is more significant than the optimization capability of the clustering algorithm.

5   Conclusions

Mean-shift and medoid-shift were proposed as a separate noise removal process before clustering. The results show that medoid-shift is more effective than mean-shift, and that they both improves k-means clustering. Best results were achieved using three iterations of the process. The proposed approach outperforms two existing outlier removal methods: LOF and ODIN.

All methods have parameters to tune. The parameter k is needed in all methods. In addition, mean-shift and medoid-shift have the number of iterations (our recommended value is 3). LOF has the top-N parameter, which we set to 3% of the data size. ODIN has the threshold parameter, which we set to 1. Our future work includes studying the effect of parameters, and how their effect could be reduced.

References

1. Ali, A.M., Angelov, P.: Anomalous behaviour detection based on heterogeneous data and data fusion. Soft Comput. 1–15 (2018). https://doi.org/10.1007/s00500-017-2989-5
2. Breunig, M.M., Kriegel, H., Ng, R.T., Sander, J.: LOF: identifying density-based local outliers. In: ACM SIGMOD International Conference on Management of Data, vol. 29, no. 2, pp. 93–104, May 2000
3. Brito, M.R., Chavez, E.L., Quiroz, A.J., Yukich, J.E.: Connectivity of the mutual k-nearest-neighbor graph in clustering and outlier detection. Stat. Prob. Lett. **35**(1), 33–42 (1997)
4. Comaniciu, D., Meer, P.: Mean shift: a robust approach toward feature space analysis. IEEE Trans. Pattern Anal. Mach. Intell. **24**(5), 603–619 (2002)
5. Ester, M., Kriegel, H.P., Sander, J., Xu, X.: A density-based algorithm for discovering clusters in large spatial databases with noise. In: International Conference on Knowledge Discovery and Data Mining, KDD, pp. 226–231 (1996)
6. Forgy, E.: Cluster analysis of multivariate data: efficiency vs. interpretability of classification. Biometrics **21**, 768–780 (1965)
7. Fränti, P.: Efficiency of random swap clustering. J. Big Data **5**(13), 1–29 (2018)
8. Fränti, P., Rezaei, M., Zhao, Q.: Centroid index: cluster level similarity measure. Pattern Recognit. **47**(9), 3034–3045 (2014)
9. Hautamäki, V., Cherednichenko, S., Kärkkäinen, I., Kinnunen, T., Fränti, P.: Improving k-means by outlier removal. In: Kalviainen, H., Parkkinen, J., Kaarna, A. (eds.) SCIA 2005. LNCS, vol. 3540, pp. 978–987. Springer, Heidelberg (2005). https://doi.org/10.1007/11499145_99
10. Hautamäki, V., Kärkkäinen, I., Fränti, P.: Outlier detection using k-nearest neighbour graph. In: International Conference on Pattern Recognition, ICPR 2004, Cambridge, UK, pp. 430–433, August, 2004
11. Knorr, E.M., Ng, R.T.: Algorithms for mining distance-based outliers in large datasets. In: International Conference on Very Large Data Bases, New York, USA, pp. 392–403 (1998)
12. Kvålseth, T.O.: Entropy and correlation: some comments. IEEE Trans. Syst. Man Cybern. **17**(3), 517–519 (1987)
13. Ott, L., Pang, L., Ramos, F., Chawla, S.: On integrated clustering and outlier detection. In: Advances in Neural Information Processing Systems, NIPS, pp. 1359–1367 (2014)

14. Pollet, T.V., van der Meij, L.: To remove or not to remove: the impact of outlier handling on significance testing in testosterone data. Adapt. Hum. Behav. Physiol. **3**(1), 43–60 (2017)
15. Ramaswamy, S., Rastogi, R., Shim, K.: Efficient algorithms for mining outliers from large data sets. In: ACM SIGMOD Record, vol. 29, no. 2, pp. 427–438, June 2000
16. Sheikh, Y.A., Khan, E.A., Kanade, T.: Mode-seeking by medoidshifts. In: IEEE International Conference on Computer Vision, ICCV, Rio de Janeiro, Brazil, October 2007
17. Tsai, D.-M., Luo, J.-Y.: Mean shift-based defect detection in multicrystalline solar wafer surfaces. IEEE Trans. Ind. Inf. **7**(1), 125–135 (2011)
18. Yin, L., Yang, R., Gabbouj, M., Neuvo, Y.: Weighted median filters: a tutorial. IEEE Trans. Circ. Syst. II: Analog Digit. Signal Process. **43**(3), 157–192 (1996)
19. Cheng, Y.: Mean shift, mode seeking, and clustering. IEEE Trans. Pattern Anal. Mach. Intell. **17**(8), 790–799 (1995)

Application of the Bag-of-Words Algorithm in Classification the Quality of Sales Leads

Marcin Gabryel[1(✉)], Robertas Damaševičius[2], and Krzysztof Przybyszewski[3,4]

[1] Institute of Computational Intelligence, Czestochowa University of Technology,
Al. Armii Krajowej 36, 42-200 Częstochowa, Poland
marcin.gabryel@iisi.pcz.pl
[2] Software Engineering Department, Kaunas University of Technology,
Studentu 50, Kaunas, Lithuania
[3] Information Technology Institute, University of Social Sciences, 90-113 Łódź, Poland
[4] Clark University, Worcester, MA 01610, USA

Abstract. The article presents a sales lead classification method using an adapted version of the Bag-of-Words algorithm. The data collected on the website of a financial institution and evaluated by that institution undergo a classification process. It is expected that the customer submitting data through a web form should be a person interested in a particular financial product. It often happens that instead of a person, i.e. a human user, it is a bot – a computer program that simulates human behavior. However, bots deliver lower quality sales leads. The way in which a web form is handled by a bot differs from the way in which it is completed by a human user. It is therefore possible to analyze the behavior on the website and to link it with the evaluation of the submitted data. The Bag-of-Words algorithm has been adapted to deal with this particular task. Experimental research based on the real-life data obtained from a bank shows how effective this algorithm is in the sales leads quality classification.

Keywords: Bot detection · Online Ad-fraud · Security

1 Introduction

The article presents a method for determining the value of a sales lead on the basis of data collected on a website and evaluation prepared by a financial institution. A sales lead is a person or company that is potentially interested in a particular product or service offered. This term is usually associated with digital marketing, where the main purpose is to persuade a person or company to leave personal data or other valuable information about themselves on various websites. Institutions' websites feature web forms which customers fill in and submit. Then, those customers are contacted and offered a given service. The customer who uses a website and submits their data via a web form is usually a person, i.e. a human user, however, this is not always the case. This kind of websites is promoted by online digital advertising, which is provided by affiliate networks, from where publishers choose advertisements and post them on their private websites.

© Springer International Publishing AG, part of Springer Nature 2018
L. Rutkowski et al. (Eds.): ICAISC 2018, LNAI 10841, pp. 615–622, 2018.
https://doi.org/10.1007/978-3-319-91253-0_57

However, there also quite frequently operate fraudulent publishers, who – through such advertising – send their own databases. For this purpose, they use special computer programs – so-called – bots that simulate human behavior. The data entered by a bot are usually data obtained from other sources. In many cases, these are the data of people who have previously applied for some other financial service (loan, deposit, bank account, cable TV, etc.). Dishonest publishers expect to earn a profit for delivering a lead through their advertising as well as to earn a commission on a possible sale of another product. That is why banking institutions depend primarily on data acquisition directly from a human.

The problem of online digital advertising fraud is becoming increasingly commonplace. Very little research work exists to systematically summarize fraud and their major characteristics [1]. Researchers mostly focus on searching for botnets or DDoS attacks by following log files on www servers [5]. Botnets are networks formed by malware-compromised machines which have become a serious threat to the Internet [3]. Experts believe that about 16–25% of computers connected to the Internet are members of botnets [4]. Such networks are designed to operate illegally on a scale large enough to threaten the functioning of private and public services in several countries around the world. One of the biggest botnets detected so far is MethBot. This "bot farm" generates $3 to $5 million in fraudulent revenue per day by targeting the premium video advertising ecosystem [6]. Literature offers a number of papers on techniques for detecting botnets. In [8] the authors propose BotGrab, a general botnet detection system that considers both malicious activities and the history of coordinated group activities in the network to identify bot-infected hosts. Paper [9] presents a bot detection system that aims to detect randomized bot command and control traffic and also aims at early bot detection. In [10], anomaly score based botnet detection is proposed to identify botnet activities by using the similarity measurement and the periodic characteristics of botnets. However, there are rather few papers devoted to protection against fraudulent pay-per-click or bot-delivered sales leads.

The data used in this article comes from the real website of a large Polish bank. Activities within this particular institution do not reach as high a scale as botnets do. The most common problem is single publishers, appearing from time to time, that try to provide, through their advertising, the databases with personal information which they have acquired. The data obtained from the financial institution were referenced by a call center and individual sales leads were qualified as positive or negative. The sales lead, against which the decision to grant a loan was considered, is treated as positive data. Negative data means the data of a customer who was not interested, who possesses incorrect data, who resigned from the service, whose data was a duplicate, or with whom no contact was established. The method presented in this article will allow for automatic assignment of the appropriate class to the new personal data being received and thus their appropriate queuing during the call center operations.

This paper is divided into several sections. The next section presents the main idea of the classical Bag-of-Words algorithm. Section 3 provides the description of the presented classification algorithm. Section 4 presents the research results and effectiveness of the new algorithm. The paper ends with the conclusions.

2 Bag-of-Words Algorithm

Applications, as well as descriptions of the Bag-of-Words algorithm, can be found in numerous papers [2]. The classical Bag-of-Words algorithm is based on the concept of text search methods within collections of documents. Terms are stored in dictionaries with an emphasis on appearing in various documents. A term can be represented by simple words (1-gram) or composed words (2, 3,..., n-gram) that occur in the document. Each term is used as an attribute of the data set represented in the attribute-value form. The set of documents, after preprocessing, can be represented as shown in Table 1.

Table 1. Representation of documents.

	t_1	t_2	$\ldots$	t_m	C
d_1	a_{11}	a_{12}		a_{1m}	c_1
d_2	a_{21}	a_{22}		a_{2m}	c_2
$\ldots$			$\ldots$		
d_n	a_{n1}	a_{n2}		a_{nm}	c_n

In Table 1 n documents are represented and each document consists of m terms. Each document d_i can be understood as a histogram $d_i = [a_{i1}, a_{i2}, \ldots, a_{im}]$. Value a_{ij} refers to the value of the jth term of the ith document. In the presented Bag-of-Words algorithm a_{ij} is a count of the number of occurrences of term t_j in document d_i.

Documents are compared by comparing the distances between two histograms d_{i1} and d_{i2}, which can be measured with different distance metrics, e.g. Manhattan (L1), Euclidean distance or Earth Movers Distance [7], or others. Classification, i.e. determining to which class c_i a particular document belongs, uses all documents from a given database along with class labels assigned to them.

The Bag-of-Words algorithm is also applied in computer vision – it is successfully used in image classification and image retrieval. Then characteristic features recurring in pictures are treated as terms.

3 Description of the Presented Method

The presented Bag-of-Words algorithm has been modified for the purposes of classifying the data collected from the website.

Let us consider herein a set of given data of sales leads $\mathbf{X} = \{\mathbf{x}_1, \ldots, \mathbf{x}_N\}$, where N – a number of all data, $\mathbf{x}_i = [\mathbf{x}_{i,1}, \ldots, \mathbf{x}_{i,n}]$ is a vector of values of features, $i = 1, \ldots N$, $x_{i,j}$ is a value of single feature j of lead i, $j = 1, \ldots n$ and n is the number of all types of features of all leads. Each lead $\mathbf{x}_i$ has class $c(\mathbf{x}_i)$, where $c(\mathbf{x}_i) \in \Omega$, $\Omega = \{\omega_1, \ldots, \omega_C\}$ is a set of all classes and C is the number of all classes.

The next steps of the algorithm are as follows:

1. Creating a dictionary:
   a. For each class c and for each type of feature i create dictionaries $D_i^c = \{d_{i1}^c, \ldots, d_{iN_i^c}^c\}$, where each one comprises N_i^c of unique words d_{ij}^c, $j = 1, \ldots, N_c, i = 1, \ldots, n$ on the basis of $\mathbf{X}$.
   b. Calculate the number of occurrences designated as $h_i^c(d_{ij}^c)$ of each of the words d_{ij}^c for each class c and each type i, according to the following formula:

$$h_i^c\left(d_{ij}^c\right) = \sum_{k=1}^{N} \delta_{ij}^c(k), \tag{1}$$

where

$$\delta_{ij}^c(k) = \begin{cases} 1 & \text{if } d_{ij}^c = x_{k,i} \text{ and } c\left(\mathbf{x}_i\right) = \omega_c \\ 0 & \text{otherwise} \end{cases} \tag{2}$$

and $j = 1, \ldots, N_c, k = 1, \ldots, N$.

2. Classification process:
   a. There are given data of the query of sales lead $\mathbf{x}_q = [x_1^q, \ldots, x_n^q]$.
   b. Searching the dictionary for words – for each class c and each feature type i find those values d_{iq}^c which have the same values as features x_i^q

$$\wedge_{c \in \{1,\ldots,C\}} \vee_{p \in \{1,\ldots,N_c\}} x_i^q = d_{ip}^c, \tag{3}$$

where $i = 1, \ldots, n$.

   c. Calculate the number of occurrences of d_{ip}^c counted for each one from class c. Having the maximum value, class c_{win} is the class to which query sales lead $\mathbf{x}_q$ belongs:

$$c\left(\mathbf{x}_q\right) = \omega_{c_{win}} \Leftrightarrow \sum_{i=1}^{n} \left[\frac{h_i^{c_{win}}\left(d_{ip}^{c_{win}}\right)}{N_{c_{win}}}\right] = \max_{c=1,\ldots,C} \sum_{i=1}^{n} \left[\frac{h_i^c\left(d_{ip}^c\right)}{N_c}\right]. \tag{4}$$

The algorithm is straightforward enough to be implemented directly in a database. Similar solutions are presented in papers [11, 12].

4   Experimental Research

The research was conducted using authentic data collected from the website of a large Polish bank. The website contains a contact form inviting customers to provide basic customer data (name, surname, telephone number, e-mail address and consent to the processing of personal data) so that the bank can make a loan offer. After the data is submitted, the bank attempts to contact the customer via a call center. As a result of the call, a positive rating is assigned (when the data are correct, it is possible to call the customer, or a loan has been granted) or a negative rating (when the data are incorrect or duplicated, or the customer cannot be contacted). When you visit the site, about 60 different data are collected to identify the client's computer, browser and to keep track of the customer's behavior on the website. The data collected include:

- information about the source – tracking the origin of leads. This information allows you to find out whether the client has reached the site as a result of advertising and who the publisher of a particular advertisement is,
- time spent filling in the application form – time counted from the moment the website opens until the moment of sending the form data,
- information whether a mobile device was used – it is checked whether the client uses the mobile version of the website or a smartphone,
- behavior of the mouse pointer – moves made by the mouse pointer is monitored along with subsequent clicks and others,
- presence of cookies – it is checked whether the customer has visited the website before,
- the number of keys pressed – it is checked whether the client has been using the keyboard,
- the number of copy-paste sequences usage – whether the data is copied from the system clipboard,
- client's fingerprint – makes it possible to identify client's computer and browser,
- client's IP number and Internet service provider,
- no data is retrieved about characters typed from the keyboard, so there is no information about client's personal data, phone and email address.

In the time period of two months of 2017, approximately 47,000 data were collected. The bank tried to contact all the persons whose details were submitted through their web form. As a result, each sales lead was given a label assessing the customer as positive or negative. These data made it possible to examine the effectiveness of the presented algorithm.

The data collected from the bank's website were divided into two parts. The dictionary was created using 90% of the data. The remaining 10% were used to test the algorithm. The algorithm was implemented in the Java language. Its operation is fast enough so as not to require extensive computing power. The algorithm operation was evaluated with the use of four indexes of performance metrics in classification problems [13]: accuracy, precision, recall and F1 score.

Accuracy is the ratio between the data that are correctly classified and the total number of samples, as calculated using the equation:

$$Accuracy = \frac{TP + TN}{TP + TN + FN + FP} \tag{5}$$

where: TP, TN, FN and FP can be derived from a confusion matrix which is shown in Table 2.

Table 2. A confusion matrix.

		Classified	
		Positive	Negative
Data	Positive	TP: True Positive	FN: False Negative
	Negative	FP: False Positive	TN: True Negative

Precision is the ratio of the number of correctly classified data to the total number of irrelevant and relevant data classified:

$$Precision = \frac{TP}{TP + FP}. \tag{6}$$

Recall is the ratio between the number of data that are correctly classified to the total number of positive data.

$$Recall = \frac{TP}{TP + FN}. \tag{7}$$

F1 score is the harmonic mean of Precision and Recall:

$$F1\ score = \frac{2 \cdot Precision \cdot Recall}{Precision + Recall} \tag{8}$$

The results of the experimental study are presented in Table 3. For the sake of comparison the same experiments were conducted using the Naïve Bayesian Classifier. Their results show that for the learning data the algorithm presented in this paper works much better than the Naïve Bayesian Classifier. The accuracy for the testing data is 76%. In the case of the testing data the Bayesian Classifier obtains slightly better results for the recall and F1 score.

Table 3. The results of the experimental study.

	Proposed BoW algorithm		Naïve Bayesian Classifier	
	Training	Testing	Training	Testing
Accuracy	0.76	0.76	0.65	0.53
Precision	0.80	0.55	0.63	0.52
Recall	0.89	0.39	0.71	0.59
F1 score	0.85	0.46	0.67	0.56

5   Conclusions

The BoW algorithm presented in this paper is an effective algorithm that is not only effective in text or image classification. As shown here, it can also be successfully used to solve other types of problems. The experimental research has shown the effectiveness of this tool for sales leads classification in terms of the rating provided by banking institutions. The presented method allows for preliminary sorting of personal data submitted online to the database of a given institution and allows for quick contact with those persons who have received a positive rating. Sales leads with negative rating are moved to the end of the lead queue. This gives a financial institution an opportunity to promptly reach those customers who really expect this contact.

Sales leads, which are submitted through web forms located on the websites of financial institutions, are often provided by bots operated by dishonest publishers. Those publishers are counting to make a profit from any commission fees for e.g. granting a loan or credit. The research conducted under this paper shows that the data provided by bots is given a negative rating.

The classification made on the presented learning data can probably be done much more effectively by other methods of computational intelligence [14, 15]: artificial neural networks [16–18], fuzzy systems [19, 20], intelligent algorithms supported by, for example, evolutionary algorithms [21, 22] and parallel computations [23, 24]. However, the cost of these calculations is much greater than simple calculations performed by the Bag-of-Words algorithm. Its big advantage also includes the possibility of a straightforward extension of the knowledge base with new sales leads while the system is being operated.

References

1. Zhu, X., Tao, H., Wu, Z., Cao, J., Kalish, K., Kayne, J.: Fraud Prevention in Online Digital Advertising. Springer, Heidelberg (2017). https://doi.org/10.1007/978-3-319-56793-8
2. Martins, C.A., Monard, M.C., Matsubara, E.T.: Reducing the dimensionality of bag-of-words text representation used by learning algorithms. In: Proceedings of 3rd IASTED International Conference on Artificial Intelligence and Applications, pp. 228–233 (2003). Author, F., Author, S.: Title of a proceedings paper. In: Editor, F., Editor, S. (eds.) Conference 2016, LNCS, vol. 9999, pp. 1–13. Springer, Heidelberg (2016)
3. Silva, S.S.C., Silva, R.M.P., Pinto, R.C.G., Salles, R.M.: Botnets: a survey. Comput. Netw. 57(2), 378–403 (2013)
4. AsSadhan, B., Moura, J., Lapsley, D., Jones, C., Strayer, W.: Detecting botnets using command and control traffic. In: Eighth IEEE International Symposium on Network Computing and Applications, NCA 2009, pp. 156–162 (2009)
5. Seyyar, M.B., Çatak, F.Ö., Gül, E.: Detection of attack-targeted scans from the apache HTTP server access logs. Appl. Comput. Inf. 14(1), 28–36 (2018)
6. WhiteOps: The Methbot operation. https://www.whiteops.com/methbot. Accessed 01 Feb 2018
7. Rubner, Y., Tomasi, C., Guibas, L.J.: The earth mover's distance as a metric for image retrieval. Int. J. Comput. Vis. 40(2), 99–121 (2000)
8. Yahyazadeh, M., Abadi, M.: BotGrab: a negative reputation system for botnet detection. Comput. Electr. Eng. 41, 68–85 (2015)
9. Soniya, B., Wilscy, M.: Detection of randomized bot command and control traffic on an endpoint host. Alex. Eng. J. 55(3), 2771–2781 (2016)
10. Chen, C.-M., Lin, H.-C.: Detecting botnet by anomalous traffic. J. Inf. Secur. Appl. 21, 42–51 (2015)

11. Gabryel, M.: A bag-of-features algorithm for applications using a NoSQL database. In: Dregvaite, G., Damasevicius, R. (eds.) ICIST 2016. CCIS, vol. 639, pp. 332–343. Springer, Cham (2016). https://doi.org/10.1007/978-3-319-46254-7_26

12. Gabryel, M.: The bag-of-features algorithm for practical applications using the MySQL database. In: Rutkowski, L., Korytkowski, M., Scherer, R., Tadeusiewicz, R., Zadeh, L.A., Zurada, J.M. (eds.) ICAISC 2016. LNCS (LNAI), vol. 9693, pp. 635–646. Springer, Cham (2016). https://doi.org/10.1007/978-3-319-39384-1_56

13. Olson, D.L., Delen, D.: Advanced Data Mining Techniques, 1st edn. Springer, Heidelberg (2008). https://doi.org/10.1007/978-3-540-76917-0

14. Woźniak, M., Połap, D.: Adaptive neuro-heuristic hybrid model for fruit peel defects detection. Neural Netw. **98**, 16–33 (2018). https://doi.org/10.1016/j.neunet.2017.10.009

15. Starczewski, A., Krzyżak, A.: A modification of the Silhouette index for the improvement of cluster validity assessment. In: Rutkowski, L., Korytkowski, M., Scherer, R., Tadeusiewicz, R., Zadeh, L.A., Zurada, J.M. (eds.) ICAISC 2016. LNCS (LNAI), vol. 9693, pp. 114–124. Springer, Cham (2016). https://doi.org/10.1007/978-3-319-39384-1_10

16. Korytkowski, M.: A novel convolutional neural network with Glial cells. In: Rutkowski, L., Korytkowski, M., Scherer, R., Tadeusiewicz, R., Zadeh, L.A., Zurada, J.M. (eds.) ICAISC 2016. LNCS (LNAI), vol. 9693, pp. 670–679. Springer, Cham (2016). https://doi.org/10.1007/978-3-319-39384-1_59

17. Bologna, G., Hayashi, Y.: Characterization of symbolic rules embedded in deep DIMLP networks: a challenge to transparency of deep learning. J. Artif. Intell. Soft Comput. Res. **7**(4), 265–286. https://doi.org/10.1515/jaiscr-2017-0019

18. Villmann, T., Bohnsack, A., Kaden, M.: Can learning vector quantization be an alternative to SVM and deep learning? - Recent trends and advanced variants of learning vector quantization for classification learning. J. Artif. Intell. Soft Comput. Res. **7**(1), 65–81. https://doi.org/10.1515/jaiscr-2017-0005

19. Nowicki, R.K., Starczewski, J.T.: A new method for classification of imprecise data using fuzzy rough fuzzification. Inf. Sci. **414**, 33–52 (2017)

20. Riid, A., Preden, J.-S.: Design of fuzzy rule-based classifiers through granulation and consolidation. J. Artif. Intell. Soft Comput. Res. **7**(2), 137–147 (2017). https://doi.org/10.1515/jaiscr-2017-0010

21. Łapa, K., Cpałka, K.: Evolutionary approach for automatic design of PID controllers. In: Gawęda, A.E., Kacprzyk, J., Rutkowski, L., Yen, G.G. (eds.) Advances in Data Analysis with Computational Intelligence Methods. SCI, vol. 738, pp. 353–373. Springer, Cham (2018). https://doi.org/10.1007/978-3-319-67946-4_16

22. Rotar, C., Iantovics, L.B.: Directed evolution - a new metaheuristc for optimization. J. Artif. Intell. Soft Comput. Res. **7**(3), 183–200. https://doi.org/10.1515/jaiscr-2017-0013

23. Marszałek, Z.: Parallelization of modified merge sort algorithm. Symmetry **9**(9), 176 (2017)

24. Bilski, J., Smoląg, J.: Parallel architectures for learning the RTRN and elman dynamic neural networks. IEEE Trans. Parallel Distrib. Syst. **26**(9), 2561–2570 (2015)

Probabilistic Feature Selection in Machine Learning

Indrajit Ghosh[✉]

Agro-Computing Research Laboratory, Department of Computer Science,
Ananda Chandra College, Jalpaiguri 735101, India
ighosh2002@gmail.com

Abstract. In machine learning, Case Based Reasoning is a prominent technique for harvesting knowledge from past experiences. The past experiences are represented in the form of a repository of cases having a set of features. But each feature may not have the equal relevancy in describing a case. Measuring the relevancy of each feature is always a prime issue. A subset of relevant features describes a case with adequate accuracy. An appropriate subset of relevant features should be selected for improving the performance of the system and to reduce dimensionality. In case based domain, feature selection is a process of selecting an appropriate subset of relevant features. There are various real domains which are inherently case based and features are expressed in terms of linguistic variables. To assign a numerical weight to each linguistic feature, a lot of feature subset selection algorithms have been proposed. But the weighting values are usually determined using subjective judgement or a trial and error basis.

This work presents an alternative concept in this direction. It can be efficiently applied to select the relevant linguistic features by measuring the probability in term of numerical values. It can also rule out irrelevant and noisy features. Applications of this approach in various real world domain show an excellent performance.

Keywords: Probabilistic feature selection · Machine learning
Case Based Reasoning

1 Introduction

In Machine Learning, Case Based Reasoning (CBR) is a prominent technique for harvesting knowledge from past experiences. The past experiences are represented in the form of a repository of cases, well known as case base. The CBR reasons using analogy concepts based on remembering from the case base.

There are different domains like agriculture, health care, law, engineering, weather forecasting etc. which are inherently case based and real field cases are available. Most of the CBR approach make use of general domain knowledge in addition to knowledge represented by cases. The domain knowledge are represented as a set of cases having a set of features and a set of outcomes. A feature

© Springer International Publishing AG, part of Springer Nature 2018
L. Rutkowski et al. (Eds.): ICAISC 2018, LNAI 10841, pp. 623–632, 2018.
https://doi.org/10.1007/978-3-319-91253-0_58

is an individual parameter of an event or object to be stored in case base. But each feature may not have the equal contribution in describing a case. So the measurement of relevancy is one of the major issues. Moreover, the exponential growth of cases in various domains may include many challenges such as noisy data, irrelevant and redundant features and high dimensionality in term of features.

The classification is a prime task in CBR domain. It has been enjoying increasing interest in the machine learning community. The classifiers use a multiple of techniques to decide whether an unknown instance described by a vector of features belongs to a certain class. In this context, several well known approaches like Artificial Neural Networks, Probabilistic Neural Networks, k-Nearest Neighbour approach, Bayesian classifiers, Radial Basis Function networks etc. have been reported. But the task of determining relevant subset of features is one of the central problems [1]. The computational complexity and limitations of training algorithms are also vital problems [2–4].

Feature selection is a process of selecting a subset of relevant features that maximize the relevance to the goal and minimize redundancy and dimensionality. It is an effective technique for data preprocessing leading to robustness, quickness and efficiency of a system. There are various real domains such as insect pest identification, irrigation, crop yield prediction, medical diagnosis etc., where most of the features are expressed in terms of linguistic variables. So measurement criterion of relevance of linguistic features is also a prime issue. For appropriate modelling, the relevancy of each linguistic feature should be expressed in term of some numerical scale.

There are many feature subset selection methods based on assigning a numerical weight to each feature. It has long been the focus of researchers in various fields and many alternative algorithms have been proposed, which include [5–26]. But the weighting values are usually determined using subjective judgement or a trial and error basis.

Case Based Reasoning is both a hybrid paradigm for computational techniques and a model of human cognition. Human cognition develops from the heuristics and experience on a domain. Heuristics is basically a probabilistic abstraction of knowledge from the past experience without having any measurable parameters. Probabilistic approach is inherent in human cognition.

From probabilistic learning point of view, some feature based models have been proposed, which include [27–36]. But most of them are not applicable for feature based supervised learning and suffer from computational complexity. Till today, no such simple models of probabilistic feature selection for domains having linguistic features have been reported.

This work proposes a very simple probabilistic model for feature subset selection in a domain where the features are expressed in terms of linguistic variables and real field cases are available. It is assumed that a feature set is associated with a probability space and if a particular feature in the feature set has more contribution in determining a class then it must have a greater probability corresponding to that class. A feature may have different probabilities for different classes. Probability value defines the relevancy of each feature. It also suggests a way to limit the dimensionality and storage of a case base.

2 Methodology

2.1 Architecture

The system consists of a case base consisting of case tables for each type of cases and a case indexing mechanism. Each case table contains the definite labelled cases obtained from the domain. Each case is represented with a set of features accepted as case descriptors. Features are useful to match among previous cases and a new case. Cases are tuples {F, P, C}, where F is the set of mutually independent features $f_1, f_2, f_3, \dots f_m$, P is the set of corresponding probability values $p_1, p_2, p_3, \dots p_m$, C is the output case description and m is the total number of features.

The initial feature set to cover all possible cases of the domain are considered by consultation with the field experts, field observers and previous case histories. Cases are collected from the real world domain along with some benchmark datasets of UCI repository and sequentially inserted through input interface in the corresponding case table using case indexing mechanism. To avoid biasness, cases are inserted as first come first serve basis.

When a new observed case is inserted, the proposed probabilistic evaluation process scans the corresponding class table and measures the probability of each feature corresponding to that case table. This process continues till saturation.

Saturation is defined by a learning rate threshold δ in the order of 10^{-3}, as the probability values are considered upto three digits after decimal. When saturation is attained, the probability distribution against the number of cases retained, are shown in the Figs. (2, 3, 4, 5 and 6). The relevant features are selected based on a threshold value θ of probability at saturation. The value of θ is determined by consulting with the scientists of Tea Research Association and various tea garden experts. The architecture of the system is presented in Fig. 1.

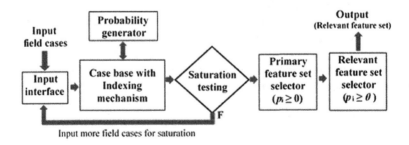

Fig. 1. System architecture

2.2   Proposed Mathematical Foundation

When a training case appears, the initial probability of the features are not known. So an unbiased probability should be assigned to each feature. This unbiased probability can be estimated as $\beta = (1/k)$, where $k \leq m$, is the number of features observed for the case. The case is inserted in the corresponding case table and the probabilistic evaluation function (1) measures the probability of each feature and updates the probability space.

For a class having n cases, the proposed probability function $p_{n,i}$ of ith feature (f_i) is given by:

$$p_{n,i} = \frac{\gamma_{n,i}}{n.m} \times \frac{\sum\limits_{j=1}^{n} \beta_{j,i}}{\sum\limits_{i=1}^{m} \beta_{n,i}} \tag{1}$$

where $\gamma_{n,i}$ is the frequency of the ith feature for n cases.

For i^{th} feature, the learning rate $\Delta(p_i)$ for each iteration is given as:

$$\Delta(p_i) = p_{i(new)} - p_{i(old)}$$
$$or \quad \Delta(p_i) = p_{n,i} - p_{(n-1),i} \tag{2}$$

When the learning rate $\Delta(p_i)$ approaches to zero or a small value (at the flat part of the curve), the system is saturated and no more cases are required to train the system. The probability space at this stage is stable enough and can be used to select relevant features. The features having $p_i \geq \theta$ are considered as a subset of relevant features.

2.3   Algorithm

Name: **Relevant feature subset selector**.
Version: FCBR 1.0.
INPUT:
    F_{in} ; Observed feature set
    θ ; Probability threshold
    δ ; Learning rate threshold
    m   ; No of total features
OUTPUT:
    F'_{out} ; Primary feature subset
    F_{out} ; Relevant feature subset
INITIALIZE:
    F_{out}
    F'_{out}
    $i = 1$

```
do (i ≤ m):
    Select case tables using F_in
    Append observed feature set
    Calculate probabilities p_i of each feature
    if (Δ(p_i) ≥ δ)
        Incorporate f_i, p_i in F'_out
    endif
    Increment i
enddo
Select F'_out
Get j   ; j is the number of features in primary subset
i = 1
do (i ≤ j)   ; Selects relevant features from primary subset
    if (p_i ≥ θ)
        Incorporate f_i, p_i in F_out
    endif
    increment i
enddo
Return F_out , F'_out
end
```

3   Validation

For validation, the model has been tested using UCI benchmark datasets along with the dataset obtained from various real field domains. As an example, the application of the model in tea cultivation is presented for better understanding.

Tea is one of the major crop in Darjeeling and Jalpaiguri districts of India. Tea gardens are organized agricultural sector where real field cases are available. In tea gardens, tea bushes are often subject to the attack of various insect pests throughout the year. When a bush is attacked by a particular type of insect pest, some set of signs and symptoms should be observed on a specific part of the bush. These set of signs and symptoms (features) lead to proper identification of the insect pests active in the field in general. But tea industry is an wide spread one where the agro-climatic conditions differ from garden to garden even from section to section and thus the set of signs and symptoms become region specific. The complete set of features is very rare to be observed. Moreover, among all the features, some may have greater significance than the other less important one. These region specific characteristics demand region specific training.

There are 8 major insect pests of tea plants which make considerable damage to the crop. So the case base has been designed to contain 8 case tables. Each case table is allotted to contain the cases of attack due to a particular insect pest. The labeled case tables against each insect pest is presented in Table 1.

Total 31 possible sign and symptoms of attack leading to cover the identification of these 8 insect pests are identified by using domain knowledge as shown

Table 1. Labeled case tables against 8 pests

Insect pests	Case tables
Red spider	Case table 1
Helopeltis	Case table 2
Scarlet mites	Case table 3
Thrips	Case table 4
Aphid	Case table 5
Jussid	Case table 6
Purple mite	Case table 7
Pink mite	Case table 8

in Table 2. This general feature set contains all the possible subsets of features due to the attack of these 8 insect pests.

Table 2. General feature set of the domain

Feature codes	Feature descriptions	Feature codes	Feature descriptions
f_1	Site of damage: *Young leaf*	f_{16}	Leaf colour: *Yellow*
f_2	Site of damage: *Matured leaf*	f_{17}	Leaf colour: *Pale*
f_3	Site of damage: *Bud & young leaf*	f_{18}	Leaf colour: *Purple bronze*
f_4	Site of damage: *Tender stem*	f_{19}	Leaf colour: *Brown*
f_5	Leaf surface: *Upper*	f_{20}	Leaf spot: *Brown ring*
f_6	Leaf surface: *Lower*	f_{21}	Leaf spot: *Red brown patch*
f_7	Leaf appearance: *Dry up*	f_{22}	Leaf spot: *Sand papery lines*
f_8	Leaf appearance: *Leathery*	f_{23}	Mid rib colour: *Brownish*
f_9	Leaf appearance: *Dusted white*	f_{24}	Mid rib colour: *Upper brown*
f_{10}	Leaf appearance: *Curved downward*	f_{25}	Edge colour: *Brownish*
f_{11}	Leaf appearance: *Curved upwards*	f_{26}	Edge colour: *Pinkish*
f_{12}	Leaf appearance: *Crinkled curled*	f_{27}	Tip colour: *Brownish*
f_{13}	Leaf appearance: *Deformed*	f_{28}	Vein colour: *Brownish*
f_{14}	Leaf appearance: *Margin brown*	f_{29}	Finger tip test: *Red smear*
f_{15}	Leaf colour: *Copper bronze*	f_{30}	Bush appearance: *Flush stunted*
		f_{31}	Bush appearance: *Defoliation*

Observed set of features leading to the definite known cases of attack of these 8 insect pests have been supplied to the system sequentially as first cum first server basis to train the system. After each entry, an iterative probabilistic function (1) measures the probability of each feature corresponding to each case table and saturation is tested with δ. At saturation, the relevant subset of features corresponding to each major pest are identified.

For better understanding, a case study for Red spider attack is presented as a representative one. The observed set of features leading to the case of red spider attack have been inserted in the case table of the system sequentially as first come first serve basis to avoid biasness. The proposed probability function measures the probabilities of all 31 features corresponding to Red spider (Case table-1), in each iteration. The probabilities of features having non zero probabilities (Primary subset) are plotted against the number of cases retained as shown in the graphs (Figs. 2, 3, 4, 5 and 6).

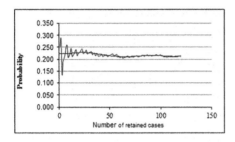

Fig. 2. Probability curve of feature (f_2).

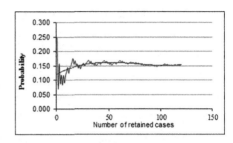

Fig. 3. Probability curve of feature (f_{19}).

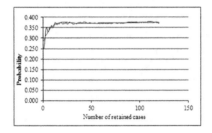

Fig. 4. Probability curve of feature (f_{21}).

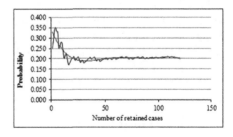

Fig. 5. Probability curve of feature (f_{29}).

It is observed that, after insertion of a reasonable number of cases, the probabilities become saturated. The probability values at saturation (at flat part of

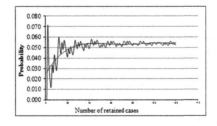

Fig. 6. Probability curve of feature (f_{30}).

Table 3. Primary feature subset for Red spider attack

Feature codes	Feature descriptions	Probability
f_2	Site of damage IS Matured leaf	$p_{1,2} = 0.213$
f_{19}	Leaf colour IS Brown	$p_{1,19} = 0.152$
f_{21}	Leaf spots IS Reddish brown patch	$p_{1,21} = 0.374$
f_{29}	Finger tip test IS Red smear	$p_{1,29} = 0.208$
f_{30}	Bush appearance IS Flush stunted	$p_{1,30} = 0.053$

Table 4. Relevant feature subset for Red spider attack

Feature codes	Feature descriptions	Probability
f_2	Site of damage IS Matured leaf	$p_{1,2} = 0.213$
f_{21}	Leaf spots IS Reddish brown patch	$p_{1,21} = 0.374$
f_{29}	Finger tip test IS Red smear	$p_{1,29} = 0.208$

the curve) are considered as final values. The features with nonzero probability form the primary feature subset for Red spider attack, as shown in the Table 3.

But all of them may not be relevant enough. For better performance, the relevant features are selected based on the threshold value θ of probability. The value of θ is domain and case specific and can be set by the consultation with the domain experts. For the present case of Red spider, considering $\theta = 0.200$, the finally selected relevant feature subset is presented in Table 4 along with their probabilities. Other features are irrelevant and ruled out.

Thus the selected relevant feature subset for Red spider attack is $F_{out} = \{f_2, f_{21}, f_{29}\}$.

4   Discussion

From the above example, it is evident that the proposed approach might be considered as a potential approach for feature selection using CBR. In feature selection task, the consistency of training data is a vital problem which is managed easily in this system. The system deals with data obtained from UCI repository along with definite real field cases. As the system gets trained with standard and real field cases, no heuristics or approximate estimation is concerned and thus reducing the uncertainty to a great extent. The only uncertainty concerned in the system is the uncertainty related to the real field observations for the new cases. This model sorts out the irrelevant features and the noisy features are ruled out. It is computationally simple and memory efficient.

The performance analysis of this model with other feature selection methods using various benchmark dataset of UCI machine learning repository is under study and the results obtained so far are very interesting. But it is not presented in this paper due to space limit. It will be published in the next paper.

My future direction of work will be to upgrade this prototype to a mobile App based user friendly intelligent agricultural Toolkit for rural farmers to identify insect pests, diseasis, soil fertility deficiency etc. I hope that it will contribute a lot both to rural as well as national economy.

References

1. Payne, T.R., Edwards, P.: Implicit feature selection with the value difference metric. In: Prade, H. (ed.) 13th European Conference on Artificial Intelligence (ECAI-1998). Wiley, Hoboken (1998)
2. Minsky, M., Papert, S.: Perceptrons. The MIT Press, Cambridge (1988)
3. Roy, A.: Summery of panel discussion at ICNN97 on connectionist learning. Connectionist learning: is it time to reconsider the foundations. JNNS Newsl. Neural Netw. **11**(2) (1998)
4. Roy, A.: Artificial neural networks - a science in trouble. Vivek Q. Artif. Intell. **13**(2), 17–24 (2000)
5. Aha, D.W., Bankart, R.L.: Feature selection for case-based classification of cloud types: an empirical comparison. In: Workshop on Case-Based Reasoning, Technical Report WS-94-01. AAAI Press (1994)
6. Nguyen, H.V., Gopalkrishnan, V.: Feature extraction for outlier detection in high-dimensional spaces. In: The 4th Workshop on Feature Selection in Data Mining (2010)
7. Das, M., Liu, H.: Feature selection for classification. Intell. Data Anal.: Int. J. **1**(3), 131–156 (1997)
8. Blum, A., Langley, P.: Selection of relevant features and examples in machine learning. Artif. Intell. **97**(1–2), 245–271 (1997)
9. Guyon, I., Elisseeff, A.: An introduction to variable and feature selection. J. Mach. Learn. Res. **3**, 1157–1182 (2003)
10. Liu, H., Motoda, H.: A selective sampling approach to active feature selection. Artif. Intell. **159**, 49–74 (2004)
11. John, G., Kohavi, R., Pfleger, K.: Irrelevant feature and subset selection problem. In: Proceedings of the Eleventh International Machine Learning Conference, pp. 121–129 (1994)
12. Vafaie, H., De Jong, K.: Robust feature selection algorithms. In: Proceeding of the Fifth Conference on Tools for Artificial Intelligence, pp. 356–363 (1993)
13. Song, L., Smola, A., Gretoon, A., Borgwardt, K., Bedo, J.: Supervised feature selection via dependence estimation. In: International Conference on Machine Learning (2007)
14. Xu, Z., Jin, R., Ye, J., Lyu, M.R., King, I.: Discriminative semi-supervised feature selection via manifold regularization. In: Proceedings of the 21th International Joint Conference on Artificial Intelligence, IJCAI 2009 (2009)
15. Zhao, Z., Liu, H.: Spectral feature selection for supervised and unsupervised learning. In: International Conference on Machine Learning, ICML-2007 (2007)
16. Zhao, Z., Wang, L., Liu, H.: Efficient spectral feature selection with minimum redundancy. In: Proceedings of the 24th AAAI Conference on Artificial Intelligence, AAAI-2010 (2010)
17. Yu, L., Liu, H.: Redundancy based feature selection for microarray data. In: Proceedings of the Tenth ACM SIGKDD Conference on Knowledge Discovery and Data Mining, ACM SIGKDD-2004 (2004)
18. Queiros, C.E., Gelsema, E.: On feature selection. In: Proceedings of the Seventh International Conference on Pattern Recognition, pp. 128–130 (1984)

19. Chang, S., Dasgupta, N., Carin, L.: A Bayesian approach to unsupervised feature selection and density estimation using expectation propagation. In: IEEE Conference on Computer Vision and Pattern Recognition, vol. 2, pp. 1043–1050 (2005)
20. Dy, J.G., Brodley, C.E.: Feature selection for unsupervised learning. J. Mach. Learn. Res. **5**, 845–889 (2004)
21. He, X., Cai, D., Niyogi, P.: Laplacian score for feature selection. In: Advances in Neural Information Processing Systems 18, pp. 507–514. MIT Press (2006)
22. Piramuthu, S.: Evaluating feature selection methods for learning in data mining application. Eur. J. Oper. Res. **156**(2), 483–494 (2004)
23. Das, S.: Filters, wrappers and a boosting-based hybrid for feature selection. In: Proceedings of the Eighteenth International Conference on Machine Learning, pp. 74–81 (2001)
24. Kumar, V., Minz, S.: Feature selection: a literature review. Smart Comput. Rev. **4**(3), 211–229 (2014)
25. Tang, J., Alelyani, S., Liu, H.: Feature selection for classification: a review. In: Data Classification: Algorithms and Applications. CRC Press (2013)
26. Molina, L.C., Belanche, L., Nebot, A.: Feature selection algorithms: a survey and experimental evaluation. In: Proceedings of ICDM, pp. 306–313 (2002)
27. Dadaneh, B.Z., Markid, H.Y., Zakerolhosseini, A.: Unsupervised probabilistic feature selection using ant colony optimization. Expert Syst. Appl. **53**, 27–42 (2016)
28. Uysal, A.K., Gunal, S.: A novel probabilistic feature selection method for text classification. Knowl.-Based Syst. **36**, 226–235 (2012)
29. Alibeigi, M., Hashemi, S., Hamzeh, A.: Unsupervised feature selection using feature density functions. Int. J. Comput. Electr. Autom. Control Inf. Eng. **3**(3), 847–852 (2009)
30. Kohavi, R.: Feature subset selection as search with probabilistic estimates. In: AAAI Fall Symposium on Relevance, pp. 122–126 (1994)
31. Salehi, E., Nayachavadi, J., Gras, R.: A statistical implicative analysis based algorithm and MMPC algorithm for detecting multiple dependencies. In: The 4th Workshop on Feature Selection in Data Mining (2010)
32. Liu, H., Setiono, R.: Feature selection and classification - a probabilistic wrapper approach. In: Proceedings of the Ninth International Conference on Industrial and Engineering Application of AI and ES, pp. 419–424 (1996)
33. Pearl, J.: Probabilistic Reasoning in Intelligent Systems. Morgan Kaufmann, San Mateo (1988)
34. Agresti, A.: Categorical Data Analysis: Probability and Mathematical Statistics. Wiley, Hoboken (1990)
35. Kherfi, M.L., Ziou, D.: Relevance feedback for CBIR: a new approach based on probabilistic feature weighting with positive and negative examples. IEEE Trans. Image Process. **15**(4), 1017–1030 (2006)
36. Bolon-Canedo, V., Sanchez-Marono, N., Alonso-Betanzos, A.: A review of feature selection methods on synthetic data. Knowl. Inf. Syst. **34**(3), 483–519 (2013)

Boost Multi-class sLDA Model for Text Classification

Maciej Jankowski$^{(\boxtimes)}$

Faculty of Cybernetics, Military University of Technology in Warsaw,
Warsaw, Poland
`maciej.jankowski@wat.edu.pl`

Abstract. Text classification is an important problem in Natural Language Processing. It differs from many other classification tasks by the large number of features that have to be used during training. One of the solution for reducing dimensionality of feature space, is the usage of Latent Dirichlet Allocation. After this step, the smaller problem can be solved using standard classifiers. In [11], authors propose combination of LDA and Softmax classifier called Multi-class sLDA, that does both tasks simultaneously. However, to use the method, we have to choose a number of topics - hyperparameter of the model. This step requires analysis and human supervision. In this paper, we propose Boost Multi-class sLDA model, based on ensemble of many Multi-class sLDA models, that does not require the choice of topic number. Moreover, our model achieves significantly better classification accuracy, than Multi-class sLDA for any number of topics.

1 Introduction

Topic models are very popular methods of text analysis. The most popular algorithm for topic modeling is Latent Dirichlet Allocation (LDA) [5]. Recently, many new methods were proposed, that enable the usage of this model in large scale processing. One of the extension to the LDA is Supervised Latent Dirichlet Allocation (sLDA) [9], which adds to LDA a response variable Y, connected to each document. This response variable is real valued and drawn from a linear regression. For classification purposes, continuous output is not appropriate, therefore a new model called Multi-class sLDA was proposed in [11]. This model can be successfully used for classification, competing with state of the art classifiers. Its advantage lies in the intermediate step that reduces dimension and can possibly find useful features like synonyms and polysemes.

One of the problem is, that choice of the number of topics K, is not happening automatically and requires some previous analysis. A few methods were proposed so far to automatize this step [6,10,12,15], but none of them works very well in supervised tasks. In this paper, we develop an ensemble algorithm that consists of many Multi-class sLDA models with different numbers of topics. We show, that this ensemble works better than any single model. In addition to

© Springer International Publishing AG, part of Springer Nature 2018
L. Rutkowski et al. (Eds.): ICAISC 2018, LNAI 10841, pp. 633–644, 2018.
https://doi.org/10.1007/978-3-319-91253-0_59

improving accuracy, the usage of this ensemble allows us to avoid the manual step of choosing the number of topics.

The rest of this paper is structured as follows: in Sect. 2, we briefly discuss the problem of dimensionality reduction in text analysis and the role LDA-based methods to achieve this goal. In Sect. 3, we provide definition of Multi-class sLDA model. We also describe in detail one of the approach to carry out inference, estimation and prediction in this model. In Sect. 4 we introduce our approach called Boost Multi-class sLDA. In Sect. 5, we study the performance of our model. Finally in Sect. 6 we summarize our findings.

2   Dimensionality Reduction for Text Classification

Classification of texts is a challenging subject. Treating words as individual features, leads to a very large set. This may pose problems for classifiers. Some features may be correlated (synonyms), others may have multiple meanings. Computational cost for large number of features may be prohibitive for large corpora. Because of those reasons, we are interested in preprocessing raw text in a way that reduces its dimension. One way to achieve that is to use LDA [5] in an unsupervised manner. However, in the context of classification, choosing the best number of topics, may not be possible using only unsupervised methods. If for example, our task is to classify movies as good or bad, we want to find topics that are somehow related to sentiments. However, if the dominant structure in reviews is genre, this is something that may be found. It is possible, that best value of K for LDA, is not the best value of K for Multi-class sLDA.

3   Multi-class sLDA Model

In this section, we describe Multi-class Supervised Latent Dirichlet Allocation (Multi-class sLDA) [11], a supervised method for classification, that builds on previous models for Topic Modeling [5] and [9]. Notation used in this paper is summarized in Table 2. We use $Dir(\alpha)$ for Dirichlet distribution with parameter α, and $Mult(1, \tau)$ for multinomial distribution with single trial and probability of success of each outcome described by vector τ.

Let K be a fixed number of topics. For a given text corpus T, we define V to be a number of words in dictionary, M be a number of documents in corpus and C be a number of classes, to which each document can belong. We further assume, that there is a corpus dependent parameter $\alpha \in \mathbb{R}^K$, parameter $\beta \in \mathbb{R}^{K \times V}$ and parameter $\eta \in \mathbb{R}^{C \times K}$. Each document $d \in \{1, \ldots, M\}$, is assumed to be generated from the following process:

1. Draw topic proportions $\theta_d | \alpha \sim Dir(\alpha)$
2. For each word $w_{d,n}, n \in \{1, 2, \ldots, N_d\}$:
   (a) Draw topic assignment $z_{d,n} | \theta \sim Mult(1, \theta_d)$
   (b) Draw word $w_{d,n} | z_{d,n}, \beta_{z_{d,n}} \sim Mult(1, \beta_{z_{d,n}})$

3. Draw class label $c_d|z_d, \eta \sim softmax(\bar{z}_d, \eta)$, where

$$\bar{z}_d = \frac{1}{N_d} \sum_{n=1}^{N_d} z_{d,n}$$

is the empirical topic frequencies $(\bar{z}_d \in \mathbb{R}^K)$, and the softmax distribution is given by

$$p(c_d|\bar{z}_d, \eta) = \frac{\exp(\eta_{c_d}^T \bar{z}_d)}{\sum_{l=1}^{C} \exp(\eta_l^T \bar{z}_d)}, \qquad c_d = 1, \ldots, C$$

In the above definition, $z_{d,n} \in \mathbb{R}^K$, is an indicator vector with all elements equal to zero except one that equals 1, for example $(0, \ldots, 0, 1, 0, \ldots, 0)$. Each of those vectors, denote a single integer between 1 and K. When we write $\beta_{z_{d,n}}$, then subscript means the integer denoted by indicator vector (e.g. $\beta_{(0,1,0)}$ means β_2 and $\beta_{(0,0,1)}$ means β_3). Similarly, we often write $\beta_{i,w_{d,n}}$. Since $w_{d,n}$ is also indicator variable, when we write $\beta_{i,(0,0,1)}$ we mean $\beta_{i,3}$.

The classification problem is as follows: we have input of M vectors, each constitutes a topic proportion θ_d. For each θ_d, we want to estimate the probability of the class label taking on each of the C different possible values.

3.1   Multi-class sLDA Computation

Standard approach to finding parameters of mixture models is to use Expectation Maximization algorithm [1]. In this algorithm, we alternate between two steps: E - where we find posterior distribution of mixture component given parameters, and M - in which we estimate parameters using distribution from E step. In this form, the algorithm cannot be used for Multi-class sLDA, because E step is intractable [5]. The main challenge is therefore, to approximate the distribution $p(z|w_d)$, in an efficient way. One of the solution, called Variational Inference, is based on approximating the distribution with a family of simpler distributions. Standard approach in LDA based models is to use fully factorized distribution. The method is known as mean field approximation. We will not present details of this algorithm, for references see [5,9,11,17]. Instead, we will focus on those details, that are needed to develop Boost Multi-class sLDA algorithm.

3.2   Approximate Inference

In inference part (E-step in variational EM), we approximate values of parameters $\gamma_d \in \mathbb{R}^{N_d \times K}$ (parameters of Dirichlet prior for θ) and $\phi_d \in \mathbb{R}^K$ (Dirichlet prior for z). Inference is carried out for a single document, therefore index d responsible for a document does not vary, but is fixed. Let $\pi = \{\alpha, \beta_d, \eta\}$ denote the set of model parameters and $q(\theta_d, z_d|\gamma, \phi)$ is join variational distribution of θ_d and z_d which factorizes as follows

$$q(\theta_d, z_d|\gamma, \phi) = q(\theta|\gamma) \prod_{n=1}^{N_d} q(z_n|\phi_n) \tag{1}$$

The variational objective function $\mathcal{L}$ also known as the evidence lower bound (ELBO), is an expectation with respect to latent variables z that follow an approximating distribution q. For Multi-class sLDA, ELBO is

$$\mathcal{L}_d(\gamma_d, \phi_d; \pi) = \mathbb{E}_q[\log p(\theta_d|\alpha)] + \sum_{n=1}^{N_d} \mathbb{E}_q[\log p(Z_{d,n}|\theta_d)]$$

$$+ \sum_{n=1}^{N_d} \mathbb{E}_q[\log p(w_{d,n}|Z_{d,n}, \beta_d)] + \mathbb{E}_q[\log p(c|Z_d, \eta)] + \mathbf{H}(q) \quad (2)$$

where $\mathbf{H}(q)$, is the entropy of the variational distribution. Maximizing this lower bound with respect to γ and ϕ leads to the following pair of updates of an iterative fixed-point method

$$\gamma_{d,i} = \alpha_i + \sum_{n=1}^{N_d} \phi_{d,n,i}, \qquad i \in \{1, \dots, K\} \quad (3)$$

$$\phi_{d,n,i} \propto \beta_{i,w_n} \exp\left(\Psi(\gamma_{d,i}) + \frac{1}{N}\eta_{c,i} - (h^T \phi_n^{(old)})^{-1} h_i \right) \quad (4)$$

where

$$h^T \phi_n = \sum_{c=1}^{C} \prod_{n=1}^{N_d} \left(\sum_{k=1}^{K} \phi_{dnk} \exp\left(\frac{1}{N_d}\eta_{ck}\right) \right)$$

3.3   Estimation

Main goal of estimation (M-step in variational EM), is to find maximum likelihood estimates of topics β_d, $d \in \{1, \dots, M\}$, and class coefficients η_c, $c \in \{1, \dots, C\}$. Corpus log-likelihood is

$$\mathcal{L}(\mathcal{T}) = \sum_{d=1}^{M} \log p(w_d, c_d|\alpha, \eta, \beta) \quad (5)$$

Optimizing with respect to $\beta_{k,i}$, $k \in \{1, \dots, K\}, i \in \{1, \dots, V\}$, leads to the following update rule for β

$$\beta_{k,i} \propto \sum_{d=1}^{M} \sum_{n=1}^{N_d} w_{d,n}^i \phi_{d,n,i} \quad (6)$$

where

$$w_{d,n}^i = \begin{cases} 1, & \text{if } w_{d,n} = (0, \dots, 0, \underset{i}{1}, 0, \dots, 0) \\ 0, & \text{if } w_{d,n} = (0, \dots, 0, \underset{j \neq i}{1}, 0, \dots, 0) \end{cases}$$

Terms in ELBO containing η are

$$\mathcal{L}_{[\eta]}(\mathcal{T}) = \sum_{d=1}^{D} \left(\eta_{c_d}^T \bar{\phi}_d - \log \left(\underbrace{\sum_{e=1}^{C} \prod_{n=1}^{N_d} \left(\sum_{k=1}^{K} \phi_{dnk} \exp\left(\frac{1}{N_d}\eta_{ek}\right) \right)}_{u} \right) \right)$$

Optimization for η is done using conjugate gradient. This method requires first derivative

$$\frac{\partial \mathcal{L}_{[\eta]}(\mathcal{T})}{\partial \eta_{ci}} = \sum_{d=1}^{M} I[c_d = c]\bar{\phi}_{di} - \sum_{d=1}^{M} \frac{1}{u}\frac{\partial u}{\partial \eta_{ci}} \tag{7}$$

3.4   Prediction

To classify new unseen document, we calculate approximate probability of each class c given document w_d, and choose a class with the highest probability

$$c^* = \operatorname*{argmax}_{c \in \{1,...,C\}} \mathbb{E}_q[\eta_c^T \bar{z}] = \operatorname*{argmax}_{c \in \{1,...,C\}} \eta_c^T \bar{\phi}$$

The denominator is constant, therefore we do not need to calculate it. However, it is possible to approximate probability of each class given document w_d

$$p(c|w_d) = \exp\left(\eta_c^T \bar{z} - \log \left(\sum_{l=1}^{C} \exp(\eta_l^T \bar{z}) \right) \right).$$

Because we do not know real value of $\bar{z}$, we approximate the probability as described in [11], using expectation with respect to variational distribution $\mathbb{E}_q$

$$p(c|w_d) \approx \mathbb{E}_q\left[\exp\left(\eta_c^T \bar{z} - \log \left(\sum_{l=1}^{C} \exp(\eta_l^T \bar{z}) \right) \right) \right]$$

$$\geq \exp\left(\mathbb{E}_q[\eta_c^T \bar{z}] - \mathbb{E}_q\left[\log \left(\sum_{l=1}^{C} \exp(\eta_l^T \bar{z}) \right) \right] \right)$$

$$\geq \exp\left(\eta_c^T \bar{\phi} - \log \left(\sum_{l=1}^{C} \prod_{n=1}^{N} \left(\sum_{k=1}^{K} \phi_{n,k} \exp\left(\frac{1}{N}\eta_{l,k}\right) \right) \right) \right) \tag{8}$$

4   Boost Multi-class sLDA

One of the problem with topic models is, that we need to know the number of topics, before we start to train the model. This is a standard example of model selection. The simplest approach is to run estimation for different values of K, and test each model on a validation set. Finally choosing the number that gives the best performance. In the context of LDA model, many approaches to this problem have been proposed in the literature [6,10,12,15]. Instead of trying to find the best number of topics, we can train many models and validate them, finally rejecting those that do not pass the acceptance criteria. A very popular approach for criticizing Bayesian models is called Posterior Predictive Checking (PPC) [3]. This idea has been applied to LDA model in [13]. In [8], authors propose a model known as Hierarchical Dirichlet Process (HDP), which chooses the best number of topic automatically and separately for each document. All those methods (with the exception of HDP), were designed for LDA, which is unsupervised model.

4.1   Ensemble

In this section, we develop Boost Multi-class sLDA, an ensemble algorithm for Multi-class sLDA, based on AdaBoost algorithm. We will show, that ensemble of many different models, each with different number of topics, outperforms the best single model. These results, allow us to avoid the manual step of choosing single value of K.

The following description addresses binary classification problem, but it can be easily extended to multiclass case. We therefore assume, that response variable Y takes values in set $\{0, 1\}$. We denote each Multi-class sLDA classifier as $\mathcal{M}_l$: $\{0, 1\}^{N_d \times V} \rightarrow \{-1, 1\}$. This function is defined as

$$\mathcal{M}_l(w_d) = \underset{c \in \{-1, 1\}}{\mathrm{argmax}} \, p(Y = c | w_d),$$

and can be calculated as described in Sect. 3.4. Ensemble of L Multi-class sLDA classifiers is a function $\mathcal{M} : \{0, 1\}^{N_d \times V} \rightarrow \{-1, 1\}$, that depends on the choice of the ensemble algorithm

$$\mathcal{M}(w_d) = f(\mathcal{M}_1(w_d), \ldots, \mathcal{M}_L(w_d))$$

4.2   AdaBoost for Multi-class sLDA

AdaBoost is very popular learning algorithm introduced in [2]. The purpose of the algorithm is to apply classification algorithm to repeatedly modified versions of the data. This way, we obtain many instances of the classifier $\mathcal{M}_1, \ldots, \mathcal{M}_L$, each with different characteristics. To predict outcome for a new unseen document, we apply weighted majority voting

$$\mathcal{M}(w_d) = \mathrm{sign} \left(\sum_{l=1}^{L} \alpha_l \mathcal{M}_l(w_d) \right),$$

where $\alpha_1, \ldots, \alpha_L$ are computed by AdaBoost. Algorithm works as follows, in first iteration $l = 1$, all observations are given weights $r_d^{(l)} \in \mathbb{R}$, $d = 1, \ldots, M$, that control, how much attention is classification algorithm giving to those observations. Then, classifier is trained on those weighted observations. In next step, weights are recalculated in such a way, that misclassified observations have their weights increased. New classifier is trained on those new weights $r_d^{(l+1)} \in \mathbb{R}$, $d = 1, \ldots, M$. Described steps are repeated many times. For further details and analysis, see for example [4] or [7].

In this paper we present a combination of Multi-class sLDA and AdaBoost algorithms. For those two algorithms to work together, we have to modify corpus level ELBO in such a way, that it take into account observation weights. No changes to inference step are required, because it is done separately for each document.

4.3   Changes to Estimation of Parameters

Corpus level ELBO is a sum of document level ELBOs. Therefore, we can apply observation weights $r_d^{(l)}$, $d = 1, \ldots, M$, to elements of this sum. Equation 5 changes to

$$\mathcal{L}(\mathcal{T}|\mathcal{M}_l) = \sum_{d=1}^{M} r_d^{(l)} \log p(w_d, c_d|\mathcal{M}_l).$$

Update rules for β were obtained by differentiating 5 with respect to $\beta_{k,i}$. Simple calculation shows, that Eq. 6 changes to

$$\beta_{k,i}^{(l)} \propto \sum_{d=1}^{M} \sum_{n=1}^{N_d} r_d^{(l)} w_{d,n}^i \phi_{d,n,i}^{(l)} \tag{9}$$

Optimization for η is based on conjugate gradient that requires first derivative of optimized function. Equation 7 changes to

$$\frac{\partial \mathcal{L}_{[\eta]}(\mathcal{T}|\mathcal{M}_l)}{\partial \eta_{ci}^{(l)}} = \sum_{d=1}^{M} r_d^{(l)} \mathbb{1}[c_d = c] \bar{\phi}_{di}^{(l)} - \sum_{d=1}^{M} r_d^{(l)} \frac{1}{u} \frac{\partial u}{\partial \eta_{ci}^{(l)}}. \tag{10}$$

Final version is presented in Algorithm 1.

5   Empirical Study

In this section, we provide an example of the use of a Ensemble Multi-class sLDA model on real data. For this purpose, we use two datasets: first consists of 942 (742 train, 200 test), documents containing a set of SMS labeled messages that have been extracted from SMS Spam Collection Data Set available on UCI Machine Learning Repository site and first introduced in [14]. The second dataset, is a collection of 773 (573 train, 200 test) political blogs available as part of the LDA R package [16]. Both datasets were split to train and test datasets. Train datasets were used for training models. Final accuracy was measured on test datasets.

Algorithm 1. Boost Multi-class sLDA

---

1. Initialize weights for first model $r^{(1)} \in \mathbb{R}^M$, $r_d^{(1)} = 1/M$, $d = 1, 2, \ldots, M$
2. For $l = 1$ to L:
   (a) Fit Multi-class sLDA model $\mathcal{M}_l$ to the training data using weights $r^{(l)}$, using VEM algorithm with the following modifications
      i. Inference part is done exactly as described in Sect. 3.2
      ii. Estimate parameters as described in Sect. 4.3
   (b) Compute

$$\mathrm{err}_l = \frac{\sum_{d=1}^{M} r_d^{(l)} I[c_d \neq \mathcal{M}_l(w_d)]}{\sum_{d=1}^{M} r_d}.$$

   (c) Compute $\alpha_l = \log((1 - \mathrm{err}_l)/\mathrm{err}_l)$.
   (d) Set $r_d^{(l+1)} \leftarrow r_d^{(l)} \cdot \exp[\alpha_l \cdot I[c_d \neq \mathcal{M}_l(w_d)])]$, $d = 1, 2, \ldots, M$.
3. Output

$$\mathcal{M}(w_d) = \mathrm{sign}\left(\sum_{l=1}^{L} \alpha_l \mathcal{M}_l(w_d) \right),$$

5.1 Multi-class sLDA for Various Number of Topics

In this experiment, we have compared Multi-class sLDA models with different number of topics. We have chosen $K = 5, 10, 15, 20, 30, 40, 50, 75, 100$, and for each value we have trained the model and reported its error rate on test as well as on train datasets. Results are presented in Fig. 1. Left picture (A) contains error rates for SMSSpam dataset. As shown, the model overfitted a little bit for 15 topics. Overall, error on test dataset oscillates somewhere between 18% and 29%. This discrepancy is a clear indication, that the choice of K, has great influence on the accuracy of the algorithm. Right picture (B), contains results for Poliblog dataset. As we can see, this model suffers from severe overfitting. While error on train dataset is around 22%, error on test dataset equals 29% at best.

Both pictures, also contain results on test set for ensemble. As we will see in the next section, boosting improves the accuracy and does not require the choice of K.

5.2 Boost Multi-class sLDA for Varying Number of Topics

In the second experiment, we have researched the ability of Boost Multi-class sLDA to improve classification accuracy, when the number of topics is unknown. We have chosen 5 models with K equal to 5, 10, 15, 20 and 30. First model was trained using $K = 5$, second model used $K = 10$ and so on, up to $K = 30$. After training 5 models, we started next iteration for $K = 5$ again, then $K = 10$ and so forth. We stopped, when the accuracy on a test was not improving. Results for SmsSpam and Poliblog datasets, were presented in Fig. 2. Left picture (A), contains error rate for successive boosting iterations for SmsSpam test dataset. As we can see, Boost Multi-class sLDA kept improving up to 38th iteration,

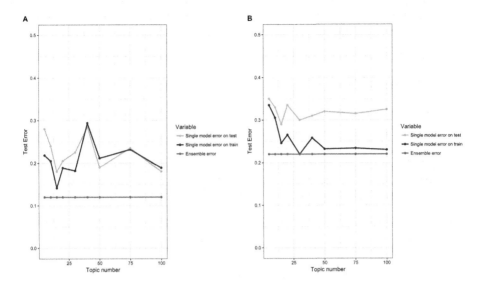

Fig. 1. Multi-class sLDA error rate for different number of topics compared with ensemble error rate, applied to SmsSpam (A) and Poliblog (B) dataset

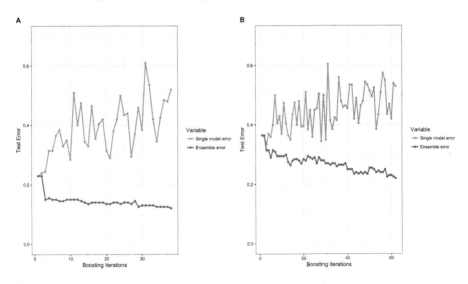

Fig. 2. Boost Multi-class sLDA for 5 LDA models with 5, 10, 15, 20 and 30 number of topics applied to SmsSpam (A) and Poliblog (B) datasets

finally reaching error level equal 12%. It is improvement in comparison to the best Multi-class sLDA model with $K = 15$, that achieved 14.1% on test dataset (see Table 1). Right picture (B), contains results of similar experiment for Poliblog dataset. This time, boosting achieved the best result equal to 22% error rate, after 63 iterations. In comparison, best Multi-class sLDA model for $K = 30$ reached only 29% error rate.

Table 1. Comparison of best Multi-class sLDA and Boost Multi-class sLDA

Model	SmsSpam	Poliblog
Multi-class sLDA on train dataset (best)	0.141 (15 topics)	**0.22** (30 topics)
Multi-class sLDA on test dataset (best)	0.18 (15 topics)	0.29 (15 topics)
Boost Multi-class sLDA on test dataset (5, 10, 15, 20, 30 topics)	**0.12** (38 iterations)	**0.22** (63 iterations)
Boost Multi-class sLDA on test dataset (15 topics - best)	**0.12** (35 iterations)	0.23 (92 iterations)

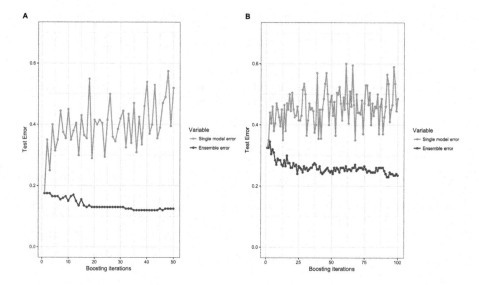

Fig. 3. Boost Multi-class sLDA applied to SmsSpam 15 topics (A) and Poliblog 15 topics (B)

5.3   Boost Multi-class sLDA for Best Number of Topics

Knowing the best number of topics for Multi-class sLDA, we have checked how Boost Multi-class sLDA can improve classification accuracy, if applied to a single model with optimal value of K. In Fig. 3, we present results for SmsSpam and Poliblog datasets. Left picture (A) contains 50 iterations of boosting applied to Multi-class sLDA model with $K = 15$, which is the optimal choice according to previous results. The lowest error equal to 12%, was achieved after 35 iterations. Recall, that this is exactly the same error as for Boost Multi-class sLDA with varying number of topics. This shows, that for SmsSpam dataset, the choice of K is not that important. We can choose many values, run Boost Multi-class sLDA and expect optimal result. Right picture (B), contains summary of similar

Table 2. Notation used in paper

$K \in \mathbb{N}$	Number of topics
$V \in \mathbb{N}$	Number of words in dictionary
$M \in \mathbb{N}$	Number of documents
$C \in \mathbb{N}$	Number of classes
$c_d \in \{1, \dots, C\}$	Class label for document d
$N_d \in \mathbb{N}, \quad d = 1, \dots, M$	Number of words in document d
$w_d \in \{0,1\}^{N_d \times V}$	Single document
$w_{d,n} \in \{0,1\}^V$	Indicator vector that denotes a single word
$w_{d,n}^i \in \{0,1\}$	i-th coordinate of indicator vector
$\mathcal{T} = \{w_d, c_d\}_{d=1}^M$	Corpus
$\alpha \in \mathbb{R}^K$	Dirichlet prior for per document topic proportions
$\theta_{d,k} \in \mathbb{R}$	Proportion of topic k in document d
$\theta_d \in \mathbb{R}^K$	Topic proportions in document d
$\theta \in \mathbb{R}^{M \times K}$	Topic proportions in corpus
$z_{d,n} \in \{0,1\}^K$	Topic assignment for single word (one-hot)
$\bar{z}_d = \frac{1}{N_d} \sum_{n=1}^{N_d} z_{d,n} \in \mathbb{R}^K$	Empirical topic frequencies for single document d
$z_d \in \{0,1\}^{N_d \times K}$	Topic assignments for single document d
$\beta \in \mathbb{R}^{K \times V}$	Topics - distributions over vocabulary
$\beta_k \in \mathbb{R}^V$	Single topic - distribution over vocabulary
$\eta = (\eta_1, \dots, \eta_C) \in \mathbb{R}^{C \times K}$	Parameters of multinomial logistic regression model
$\eta_c \in \mathbb{R}^K$	c-th parameter of multinomial logistic regression model
$\phi_{d,n} \in \mathbb{R}^K$	Free variational parameters of multinomial distribution of topic assignments for $z_{d,n}$
$\phi_d \in \mathbb{R}^{N_d \times K}$	Free variational parameters of multinomial distribution of topic assignments for d-th document
$\phi \in \mathbb{R}^{M \times N_d \times K}$	Free variational parameters of multinomial distribution of topic assignments
$\gamma_d \in \mathbb{R}^K$	Free variational parameters of Dirichlet prior for d-th document
$\gamma \in \mathbb{R}^{M \times K}$	Free variational parameters of Dirichlet prior

experiment carried out for Poliblog dataset. Boost Multi-class sLDA reached 23% error rate after 92 iterations. It is slightly worse than ensemble for varying number of topics which equals 22%.

6   Summary

We have developed Boost Multi-class sLDA model, for text classification that is an ensemble of Multi-class sLDA models. We have demonstrated, that our model outperforms Multi-class sLDA, reducing error rate by 6% (SmsSpam) and 7% (Poliblog). We have also shown, that the model does not require a previous choice of hyperparameter K, that was required in Multi-class sLDA.

References

1. Dempster, A.P., Laird, N.M., Rubin, D.B.: Maximum likelihood from incomplete data via the EM algorithm. J. R. Stat. Soc. Ser. B (Methodol.) **39**(1), 1–38 (1977)
2. Freund, Y., Schapire, R.: A decision-theoretic generalization of online learning and an application to boosting. J. Comput. Syst. Sci. **55**, 119–139 (1997)
3. Rubin, D.: Bayesianly justifiable and relevant frequency calculations for the applied statistician. Ann. Stat. **12**(4), 1151–1172 (1984)
4. Hastie, T., Tibshirani, R., Friedman, J.: The Elements of Statistical Learning. Springer, New York (2001). https://doi.org/10.1007/978-0-387-84858-7
5. Blei, D., Ng, A., Jordan, M.: Latent Dirichlet allocation. J. Mach. Learn. Res. **3**, 993–1022 (2003)
6. Griffiths, T., Steyvers, M.: Finding scientific topics. Proc. Natl. Acad. Sci. **101**, 5228–5235 (2004). https://doi.org/10.1073/pnas.0307752101
7. Bishop, C.M.: Pattern Recognition and Machine Learning (Information Science and Statistics). Springer, New York (2006)
8. Teh, Y.W., Jordan, M.I., Beal, M.J., Blei, D.M.: Hierarchical Dirichlet processes. J. Am. Stat. Assoc. **101**(476), 1566–1581 (2006)
9. Mcauliffe, J.D., Blei, D.M.: Supervised topic models. In: Advances in Neural Information Processing Systems (2008)
10. Cao, J., Xia, T., Li, J., Zhang, Y., Tang, S.: A density-based method for adaptive LDA model selection. Neurocomputing **72**(7–9), 1775–1781 (2008). 16th European Symposium on Artificial Neural Networks
11. Wang, C., Blei, D., Fei-Fei, L.: Simultaneous image classification and annotation. In: Computer Vision and Pattern Recognition (2009)
12. Arun, R., Suresh, V., Veni Madhavan, C.E., Narasimha Murthy, M.N.: On finding the natural number of topics with latent Dirichlet allocation: some observations. In: Zaki, M.J., Yu, J.X., Ravindran, B., Pudi, V. (eds.) PAKDD 2010. LNCS (LNAI), vol. 6118, pp. 391–402. Springer, Heidelberg (2010). https://doi.org/10.1007/978-3-642-13657-3_43
13. Mimno, D., Blei, D.: Bayesian checking for topic models. In: Proceedings of the Conference on Empirical Methods in Natural Language Processing. Association for Computational Linguistics (2011)
14. Almeida, T.A., Gomez Hidalgo, J.M., Yamakami, A.: Contributions to the study of SMS spam filtering: new collection and results. In: Proceedings of the 2011 ACM Symposium on Document Engineering (DOCENG 2011), Mountain View, CA, USA (2011)
15. Deveaud, R., Sanjuan, E., Bellot, P.: Accurate and effective latent concept modeling for ad hoc information retrieval. Revue des Sciences et Technologies de l'Information - Série Document Numérique, Lavoisier, 61–84 (2014)
16. Chang, J.: LDA: Collapsed Gibbs Sampling Methods for Topic Models. R package version 1.4.2 (2015). https://CRAN.R-project.org/package=lda
17. Blei, D., Kucukelbir, A., McAuliffe, J.: Variational inference: a review for statisticians. J. Am. Stat. Assoc. **112**(518), 859–877 (2017)

Multi-level Aggregation
in Face Recognition

Adam Kiersztyn[1]([⊠]), Paweł Karczmarek[1], and Witold Pedrycz[2,3,4]

[1] Institute of Mathematics and Computer Science,
The John Paul II Catholic University of Lublin, ul. Konstantynów 1H,
20-708 Lublin, Poland
{adam.kiersztyn,pawelk}@kul.pl
[2] Department of Electrical and Computer Engineering, University of Alberta,
Edmonton T6R 2V4 AB, Canada
wpedrycz@ualberta.ca
[3] Department of Electrical and Computer Engineering, Faculty of Engineering,
King Abdulaziz University, Jeddah 21589, Saudi Arabia
[4] Systems Research Institute, Polish Academy of Sciences, Warsaw, Poland

Abstract. This paper presents the results of an in-depth analysis of the impact of aggregation of different parts of the face to its recognition process. A novel approach is based on the aggregation of distances determined between histograms, which describe different parts of the face as well as various color channels. In addition, we propose to include thresholding to local descriptors and demonstrate that this type of image processing highly improves the accuracy of classification process. This paper also describes a new approach to converting color images to grayscale images using the variation of each channel in the neighborhood of a given pixel.

Keywords: Aggregation · Facial features · Face recognition
Grayscale

1 Introduction

Human face recognition (or more generally, shape recognition) has been one of the most analyzed biometrics problems in recent years. The main factor that makes this research centre is its comprehensive applications, which includes verification of documents, border inspection, control of access to protected facilities, or even control of access to devices. Over many years of studies, various main trends have been developed, including geometric methods, statistical methods, local descriptors [1,3,10,15] Gabor wavelets, Granular Computing, deep learning, sparse representation, fuzzy fusion and aggregation [7,16], etc. Most of these methods are negatively impacted by unfavorable factors such as changes in lighting, pose, facial expression, etc. One of the main directions of research is to use a wide class of algorithms referred to as local descriptors. LBP (local binary pattern) is one of

© Springer International Publishing AG, part of Springer Nature 2018
L. Rutkowski et al. (Eds.): ICAISC 2018, LNAI 10841, pp. 645–656, 2018.
https://doi.org/10.1007/978-3-319-91253-0_60

the most important local descriptors [1]. This algorithm comes with a number of modifications, whose broad and comprehensive review can be found in [3,4]. A very interesting extension is a Multi-scale Block LBP (MBLBP) [17,21], where the whole blocks of pixels are considered instead of individual pixels. Particular generalizations of the LBP method are efficient Local Ternary Pattern (LTP) and Differential Local Ternary Pattern (DLTP). Another class of local descriptors concerns the methods using words or, more generally, dictionaries of words describing pixels. One of the representatives of this class is Full Ranking descriptor [5]. In this method, a dictionary of a predetermined length and fixed length of the "words" is generated. A slightly different approach is used by Chain Code-Based Local Descriptor (CCBLD) [8], where the length of words is not predetermined. A thorough analysis of this method and its extensions can be found in [10].

In addition, the issue of data aggregation used in the process of face recognition has been widely considered. Various methods of data aggregation at the level of the classification results are described in many studies, for instance, [6,9,11–13]. The most often-considered factor is an aggregation of classification results obtained for particular parts of the face.

In this study, our objective is a comprehensive analysis of data aggregation. The aggregation is completed at a different level than the one over which classification is realized. Our goal is to complete a thorough analysis of the impact of aggregation of different parts of the face and the different color channels on the performance of the classifier. We complete an essential consideration of the impact of color selection and the method of conversion into grayscale on the quality of recognition. A kind of a fuzzy set-based method of conversion of RGB channels into greyscale is implemented and experimented with. In addition, on a basis of LBP, a method of a fuzzy aggregation at the level of construction of histograms containing the facial descriptions being the results of LBP transformation will be presented. The research of threshold histograms brings very interesting results and allows connecting local descriptors with a fuzzy variability of pixel neighborhood. A considerable part of the paper will be devoted to the analysis of aggregation of different color channels (red, green, blue, grey, and weighted grey) for various combinations of face parts. The originality of the presented research lies in the aggregation of individual parts of the face as well as the individual color channels. Particularly, noteworthy is the innovative way of converting the RGB channels to the weighted gray. Another key and novel point of the study is the introduction of weighted LBP, as an example of local descriptors modification, based on the introduction of thresholds of acceptability for pixels. A selection of local descriptors as the main tool is motivated by their robustness to changing lighting condition and small geometrical displacements of a face. Local descriptors are furthermore easy to implement and provide many opportunities for further improvement and modifications.

The paper is structured as follows. Section 2 presents the general concept of the local file descriptor at the example of LBP. This section describes the proposed modifications and methods of data aggregation. The results of various experiments are described in Sect. 3. The last section covers conclusions and future works.

2   Description of LBP Algorithm and Possible Stages of Data Aggregation

The operation of many local descriptors used at face recognition coincides with the course of classic LBP algorithm, thus our deliberations concerning multi-level aggregation in face recognition will be based on possible modifications of this popular and intuitive method. For this purpose, in the beginning, we will briefly present this approach and point out some aspects where it is justified to apply a certain type of data aggregation. The LBP algorithm is usually used for images in grayscale, however, currently most devices used in the monitoring or supervision systems have the possibility of recording color images. Therefore, it seems reasonably practicable to complete a certain type of aggregation at the level at which images are converted into that in grayscale. In the proposed solution we consider the degree of variability of each of the base RGB channels. A certain modification is suggested to the classic conversion of a color image into grayscale. Recall that in a classic Luma approach, the intensity of grey level is determined, among others, by means of the following formula

$$Gray = 0.21 Red + 0.72 Green + 0.07 Blue.$$

Of course, other weights of individual channels are also used, e.g.

$$Gray = 0.299 Red + 0.587 Green + 0.114 Blue.$$

In addition to RGB space, image analysis also considers other color spaces, such as $c_1 c_2 c_3$ color space, $I_1 I_2 I_3$ color space, or $X_1 X_2 X_3$ color space. The intention of the authors was to present potential levels of analysis on which data aggregation is possible, rather than a thorough analysis of specific conversion methods or concise local descriptors.

Here, the degree of variations of each channel in the 3×3 square surrounding of a considered pixel is additionally taken into account. For each of the channels, a variation in a given neighborhood is determined as

$$v[x, y] = \max_{-1 \leq i,j \leq 1} p\left[[x + i, y + j]\right] - \min_{-1 \leq i,j \leq 1} p\left[[x + i, y + j]\right] \tag{1}$$

where $p[x, y]$ denotes the value of the pixel at coordinates (x, y). For a color image three variations are specified independently, namely vR, vG, vB which are then used to determine weighted grayscale (WG) in accordance with the novel formula

$$WG = (0.21 + vR/v) \cdot Red/2 + (0.72 + vG/v) \cdot Green/2 + (0.07 + vB/v) \cdot Blue/2 \tag{2}$$

where $v = vR + vG + vB$. The main idea behind the proposed method consists of assigning higher importance to the channel, which is characterized by the greater diversity of values in relation to the total variability. The proposed method of converting color images to grayscale enhances the edges. Edge enhancement helps improve quality of recognition, since greater importance is associated with the

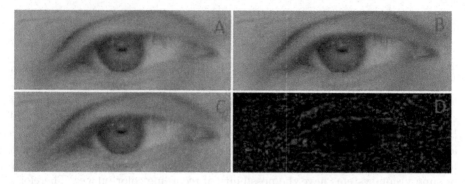

Fig. 1. Comparison of the LUMA algorithm (B) and its modification (C) on the example image of the eye (A), with the marked differences between the analyzed scaling methods (D).

characteristic points. A comparison of the conversion algorithm using the LUMA and the proposed modifications is visualized in Fig. 1. We will now describe the functioning of the LBP algorithm with the proposed modifications. Let us consider a grayscale image. The surrounding of each point of the image is analyzed and then it is re-coded according to the following rule: if the value of a given pixel differs from the pixel surroundings is greater than the one currently examined it is re-coded to 1, otherwise, to 0. Then, from the values of pixels from the surrounding, the binary number is formed which is later re-coded as a decimal number and, instead of the input value, the resulting value becomes recorded. In the classic LBP algorithm and many related methods, histograms are created on a basis of the data obtained from a transformation of a given image. The modification proposed here, which can be applied to each local descriptor, is carried out as follows, when constructing the histograms, the degree of variability of the central pixel's surrounding is taken into consideration. The essence of this process is described by using the example of LBP. During the execution of the algorithm, the value of a pixel, on the basis of its surrounding, is re-coded in accordance with the diagram presented in Fig. 2. Additionally, a variation of surrounding is calculated simultaneously, based on (2).

In case of surrounding presented in Fig. 2, the value of variations is 120. This value is recorded and used at a stage of forming the histograms. Namely, the difference is that in a classic LBP, the re-coded value of each pixel is taken into account in a histogram being developed for an image (or its part). In the introduced modification, we propose histograms created only on a basis of those values for which a variation of the examined pixel exceeds a given value. It is worth noting that it is possible to use several thresholds and to concatenate histograms created in this way. More specifically, in the novel modification proposed here a weighting of the classical method LBP (or similar local descriptors) is completed. For each pixel of the analyzed image, the hesitation of its surroundings is determined according to the formula (1). The hesitation is expressed for images in a grayscale. It is possible, however, to take into account variations of each of RGB

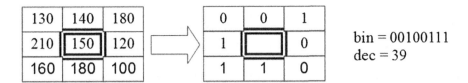

Fig. 2. Realization of the LBP algorithm

channels. In addition, the whole image is converted according to the LBP algorithm (or other) as shown in Fig. 2. At the next step, a histogram of occurrences of each value is formed. In the proposed modification N histograms are built. For each of them a different threshold of acceptability is selected. If the value of the pixel after the conversion by LBP exceeds the threshold of acceptability, then the value is included in the histogram for a given threshold. At the next step histograms for the different thresholds of acceptability are concatenated into a single histogram.

In many cases, in order to improve the specificity of description, the picture data is divided into parts and histograms are determined separately for each the regions, the outputs are then concatenated into a single histogram describing the overall image. In this study, we consider a slightly different approach, namely, we suggest that the course of histogram construction takes into account the degree of variability of a given pixel's surrounding. Two operations are conducted simultaneously in the image analysis. The first is an above-mentioned transformation, and the second one constitutes a calculation of variations of a given surrounding in accordance with the use of (1). In the case of dividing an image into several parts, the corresponding histograms should be concatenated. Having descriptions of each images comprising one sequence of numbers, the distance between these routes (histograms) can be examined. The distance can be calculated, for instance, according to the histogram intersection described in the form

$$d(X, Y) = 1 - \frac{\sum_{i=1}^{n} \min(x_i, y_i)}{\sum_{i=1}^{n} \max(x_i, y_i)} \tag{3}$$

The reason for choosing this method is the speed and insensitivity to a large number of zeros in the histograms. Other ways of determining similarities are the measure of Hellinger, correlation or the distance χ^2. The results obtained by the measure may generally give better recognition than the others, which are comparable only to the measure of Hellinger. However, this method of determining the distance is faster than Hellinger, correlation and χ^2.

2.1   Aggregation of Face Features and Color Channels

Having distances between the examined images (corresponding histograms) for particular parts of the face and particular channels, a number of aggregation of these distances can be realized using various operations. More precisely, in the case of using any of the local descriptors (in particular LBP) for different parts of the face, and for different channels (red, green, blue, gray, weighted gray (WG)) the result is a distance between images. The classification is made based on these distances. The aggregation of obtained distances within the individual parts of the face is considered in many works. We propose the aggregation of the various parts of the face as well as the individual channels. Let $a(f;c)$ be the aggregation of the distances between the histograms for the set of parts of the face f and the color channels c. In the case where there are N parts of the face, and M different channels (R,G,B, grayscale, weighted gray, and other cases) many aggregations can be made for the individual components. It is easy to see that for all combinations of at least two-elements, we have $2^{NM} - NM - 1$. This value is multiplied by the number of aggregation functions under consideration (in the case of absence of combining various functions) and we arrive at a large broad number of possibilities. It is also possible to combine different aggregation functions, different parts of the face and the different color channels. For example, for the two parts of the face (f_1, f_2) and the three color channels (c_1, c_2, c_3) for each individual aggregation function should be considered 21 cases. In the experimental section only a selected set of features aggregations and color channels are described. The scheme of the general process of operation of a local descriptor, for example, LBP with the marked locations of the optional modifications (italics) is presented in Fig. 3.

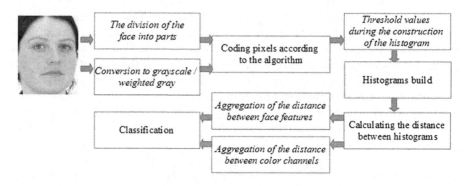

Fig. 3. Scheme of local descriptor processing with potential modifications.

3   Experimental Studies

The PUT (Poznan University of Technology) database [14] was used in the experiments. The choice of this database is motivated by owing to the fact that developers of this dataset, apart from color images, provided also characteristic points coordinates and coordinates of key areas. In the case of other databases, it would be necessary to compute coordinates of those points. The analysis covered a subset of PUT database consisting of 1,100 photographs of 100 persons,

11 photographs per person. Moreover, the choice of this database was implied by the fact that it contains color photos with precisely defined characteristic points. In the further part of the study, the following marking for particular parts of the face will be used: Left Eyebrow (LEB), Left Eye (LEyes), Left Eyeball (LEL), Right Eyebrow (REB), Right Eye (REyes), Right Eyeball (REL), Mouth (Lips), Tip of Nose (No), Nose (Noses).

3.1   Modification of LBP Algorithm Based on Aggregation of Data Describing Variations

In the above-mentioned modification, we propose histograms created only on a basis of those values for which a variation of the examined pixel exceeds a given threshold value. It is worth noting that it is acceptable to use several thresholds and to concatenate histograms created in this way. Sample values of subsequent ranks at various sets of thresholds for images in grayscale, are presented in Fig. 4.

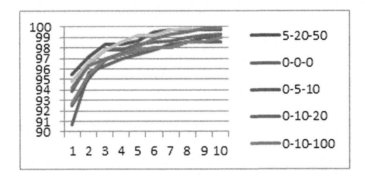

Fig. 4. Rank 1–10 comparison of the effectiveness of the modified LBP algorithm with different threshold values. Notation of the form x-y-z means that the acceptance thresholds are equal to x, y, z, respectively. More precisely, expression 5-20-50 denotes, that the pixel values are taken into account in the construction of the histogram then if the variance of the neighborhood exceeds 5, 20, and 50, respectively.

From an analysis of the last chart Fig. 4 it may be observed that consideration of the degree of variability of surrounding allows improving the quality of recognition. This observation is correct also for the remaining channels being analyzed. We can note that the nearest neighbor classifier with respect to each channel gives better results than when using the traditional method. The use of threshold values can be regarded as insertion of a certain degree of fuzzification at the significance level of pixels. The effectiveness of the suggested modification is best evidenced by the difference between recognition rate expressed in percentage points for the modified and traditional version of the algorithm. The values of differences are presented in Table 1.

Table 1. Differences between the weighted LBP and classical LBP (in percentage)

Color	Gray	Red	Green	Blue	WG
Difference	4.83	3.00	3.66	3.83	3.33

The proposed method is also effective for other commonly used databases. Comparison of the effectiveness of the classic LBP and proposed modifications for selected databases include Table 2.

Table 2. Comparison of the effectiveness of new methods with classic ones. The Py descriptor denotes a division of an image into $y \times y$ parts. Analysis of the ORL [18] and Yale [2, 20] databases

Database	Method	P1	P2	P3	P4	P5	P6	P7	P8	P9	P10
ORL	Classic LBP	83.7	93.5	95.7	95.7	**95.5**	94.7	94.2	93.0	**93.0**	92.2
ORL	Classic MBLBP3	94.2	96.2	96.7	97.2	95.5	95.0	95.2	94.0	93.0	92.7
ORL	Weighted LBP	**89.5**	94.2	**96.2**	95.7	95.2	94.7	94.2	**93.5**	92.2	**92.5**
ORL	Weighted MBLBP3	**97.0**	**99.0**	98.5	**99.5**	**98.0**	**96.0**	**96.0**	**95.0**	**95.0**	**94.0**
Yale	Classic LBP	69.7	71.4	72.9	**73.5**	73.3	73.8	73.3	74.4	74.6	75.1
Yale	Classic MBLBP3	**71.8**	73.6	**73.3**	75.0	75.5	76.8	77.0	78.5	79.3	78.2
Yale	Weighted LBP	69.7	**73.6**	**73.0**	72.3	72.3	73.0	**74.3**	**75.6**	**76.3**	**76.9**
Yale	Weighted MBLBP3	71.0	**75.6**	73.0	**75.0**	**78.9**	**80.9**	**81.5**	**82.2**	**83.5**	**84.2**

As one can see, the proposed modification significantly improves the recognition rate. Therefore, further study of this problem is justified. Analogous results are obtained for other databases, in the case of Gray Feret [19], recognition rate difference between the classical and the modification reaches nearly 20%.

The purpose of this work is to present the effect of selecting a color channel and parts of the face in the recognition process. The analysis is based on the classic LBP method, as the main idea is to present the effect of aggregation of the distance between picture representations on the quality of recognition, rather than to present a new method. For most local descriptors, the analysis uses images in grayscale. We will show that it seems reasonably practical to apply these tools to particular channels separately, and then aggregate the results received. Table 3 presents a summary of results obtained by means of the classical LBP method for three basic R, G, and B channels, classic grayscale, and weighed grayscale. The Py descriptor denotes, as before, a division of an image into $y \times y$ parts. The analysis of the above results confirms that at the lack of division of image into parts, red channel and weighed grayscale dominate. The difference between these values and classically used grayscale is 1.8%, which is a significant figure.

It is also interesting to compare different methods for particular channels being analyzed. The comparison of classic LBP and popular MB LBP (see [17, 21] for details) modification with blocks of the size 3, 5, and 7, respectively, are presented in Table 4. We may observe that in every case the application of traditional grayscale is worse than any other channel.

Table 3. Values of recognition rate

	P1	P2	P3	P4	P5	P6	P7	P8	P9	P10
Blue	90.50	98.83	99.33	99.50	99.33	99.05	99.66	99.50	99.50	99.30
Gray	91.40	98.25	99.16	99.58	99.58	99.66	99.58	99.58	99.58	99.50
Green	91.50	99.00	99.50	99.50	99.33	99.50	99.66	99.50	99.50	99.33
Red	93.20	98.25	99.33	99.41	99.41	99.66	99.50	99.66	99.41	99.50
Weighted Gray	93.20	98.83	99.16	99.33	99.33	99.33	99.50	99.50	99.50	99.33

Table 4. Summary recognition rate for the various methods, the whole face

	LBP	MBLBP3	MBLBP5	MBLBP7
Blue	90.50	95.33	98.16	99.33
Gray	91.41	97.25	99.08	99.41
Green	91.50	96.33	99.33	99.50
Red	93.16	97.92	99.16	99.25
Weighted Gray	93.16	97.00	98.50	98.83

We may observe that the best recognition is achieved using Red and Green channel. However, in the case of classic LBP the best recognition is the one achieved using Red and WG channels. An obvious question arises at this point. What effect can be obtained using an aggregation of particular channels at the level of determination of distances between particular representations of images? The results of the comparison of all possible two-color combinations of analyzed colors by using typical aggregation function maximum are presented in Table 5.

Table 5. Recognition rates with an aggregation using maximum function

	Blue	Gray	Green	Red	Weighted Gray
Blue	90.50	91.33	91.83	93.33	92.83
Gray	91.33	91.41	91.33	93.16	92.50
Green	91.83	91.33	91.50	93.50	93.16
Red	93.33	93.16	93.50	93.16	94.00
Weighted Gray	92.83	92.50	93.16	94.00	93.16

3.2 Aggregation of Different Parts of the Face and Different Channels

The recognition rate value for particular parts of the face, with breakdown into particular channels received by means of the classical BP method, at the lack of division into parts, is presented in Table 6.

Table 6. Recognition rate value for different face features

	Blue	Green	Red	Gray	WG
Face	90.5	91.50	93.17	91.42	93.17
Left Eyebrow	80.67	81.00	86.08	82.58	82.50
Left Eyeball	51.83	62.17	60.47	60.00	60.33
Left Eye	79.50	84.83	82.58	82.75	78.83
Mouth	83.33	88.17	82.83	85.75	84.33
Tip of Nose	81.67	87.17	86.00	85.25	86.17
Nose	83.83	84.83	83.58	83.33	84.00
Right Eyebrow	86.83	87.83	90.42	87.83	87.33
Right Eyeball	58.83	60.50	64.25	60.67	58.50
Right Eye	78.50	84.33	83.00	82.33	82.17

A key at this point is to check what effect the aggregation of particular parts of the face and available channels have on recognition. Distances between different images will be subjected to aggregation; the sum and maximum functions will be used as aggregation functions. When analyzing the data we can see that aggregation by means of the sum function produces better results and, apart from this, in line with presumptions, aggregation of more parts of the face produces better results. In the case of aggregation of distances for Lips, No, Reb for all channels, apart from Blue, recognition rate reaches or exceeds the value of 99%. At this point it is worth mentioning that in the case of aggregation of all analyzed parts of the face within each color we obtain the rate at the level presented in Table 7.

Table 7. Aggregation of all considered part of the face of each color separately

	Blue	Green	Red	Gray	WG
Max(Leb, Lel, LE, Lips, No, Noses, Reb, Rel, RE)	92.50	94.00	95.83	95.00	93.83
Sum(Leb, Lel, LE, Lips, No, Noses, Reb, Rel, RE)	99.83	100	99.83	100	99.83

4   Conclusions and Future Studies

The paper describes a number of original techniques of aggregation of photos of a human face. It should be noted that the methods proposed here exhibit a significant degree of generality and may be used in a wide class of algorithms. Special attention should be paid to the ways of conversion of pictures to grayscale images using the variability of particular channels. The results obtained for so-called "weighted grey" concept present in one of the key algorithms, which is

LBP, indicate a potential of this method. In addition, interesting results have been obtained in the case of aggregation of data at the level of building histograms.

It is advisable to conduct further, in-depth analysis of the mechanisms of the weighted conversion to grayscale. In particular, one should verify that for other databases remain valid conclusions received for PUT database. Moreover in the course of further research one should carefully examine the impact of thresholds of acceptability for recognition, and explore other models of grayscale weighting.

Acknowledgements. The authors are supported by National Science Centre, Poland [grant no. 2014/13/D/ST6/03244]. Support from the Canada Research Chair (CRC) program and Natural Sciences and Engineering Research Council is gratefully acknowledged (W. Pedrycz).

References

1. Ahonen, T., Hadid, A., Pietikäinen, M.: Face recognition with local binary patterns. In: Pajdla, T., Matas, J. (eds.) ECCV 2004. LNCS, vol. 3021, pp. 469–481. Springer, Heidelberg (2004). https://doi.org/10.1007/978-3-540-24670-1_36
2. Belhumeur, P.N., Hespanha, J.P., Kriegman, D.J.: Eigenfaces vs. fisherfaces: recognition using class specific linear projection. IEEE Trans. Pattern Anal. Mach. Intell. **19**, 711–720 (1997)
3. Bereta, M., Karczmarek, P., Pedrycz, W., Reformat, M.: Local descriptors in application to the aging problem in face recognition. Pattern Recogn. **46**, 2634–2646 (2013)
4. Bereta, M., Pedrycz, W., Reformat, M.: Local descriptors and similarity measures for frontal face recognition: a comparative analysis. J. Vis. Commun. Image Represent. **24**, 1213–1231 (2013)
5. Chan, C.H., Yan, F., Kittler, J., Mikolajczyk, K.: Full ranking as local descriptor for visual recognition: a comparison of distance metrics on S_n. Pattern Recognit. **48**(4), 1328–1336 (2015)
6. Dolecki, M., Karczmarek, P., Kiersztyn, A., Pedrycz, W.: Utility functions as aggregation functions in face recognition. In: 2016 IEEE Symposium Series on Computational Intelligence (SSCI), Athens, pp. 1–6 (2016)
7. Karczmarek, P., Pedrycz, W., Reformat, M., Akhoundi, E.: A study in facial regions saliency: a fuzzy measure approach. Soft. Comput. **18**, 379–391 (2014)
8. Karczmarek, P., Kiersztyn, A., Pedrycz, W., Rutka, P.: Chain code-based local descriptor for face recognition. In: Burduk, R., Jackowski, K., Kurzyński, M., Woźniak, M., Żołnierek, A. (eds.) Proceedings of the 9th International Conference on Computer Recognition Systems CORES 2015. AISC, vol. 403, pp. 307–316. Springer, Cham (2016). https://doi.org/10.1007/978-3-319-26227-7_29
9. Karczmarek, P., Pedrycz, W., Kiersztyn, A., Rutka, P.: A study in facial features saliency in face recognition: an analytic hierarchy process approach. Soft. Comput. **21**(24), 7503–7517 (2017)
10. Karczmarek, P., Kiersztyn, A., Pedrycz, W., Dolecki, M.: An application of chain code-based local descriptor and its extension to face recognition. Pattern Recogn. **65**, 26–34 (2017). https://doi.org/10.1016/j.patcog.2016.12.008

11. Karczmarek, P., Kiersztyn, A., Pedrycz, W.: An evaluation of fuzzy measure for face recognition. In: Rutkowski, L., Korytkowski, M., Scherer, R., Tadeusiewicz, R., Zadeh, L.A., Zurada, J.M. (eds.) ICAISC 2017. LNCS (LNAI), vol. 10245, pp. 668–676. Springer, Cham (2017). https://doi.org/10.1007/978-3-319-59063-9_60
12. Karczmarek, P., Kiersztyn, A., Pedrycz, W.: On developing Sugeno fuzzy measure densities in problems of face recognition. Int. J. Mach. Intell. Sens. Sig. Process. **2**, 80–96 (2017)
13. Karczmarek, P., Kiersztyn, A., Pedrycz, W.: Generalized Choquet integral for face recognition. Int. J. Fuzzy Syst. **20**(3), 1047–1055 (2018). https://doi.org/10.1007/s40815-017-0355-5
14. Kasinski, A., Florek, A., Schmidt, A.: The PUT face database. Image Process. Comm. **13**(3–4), 59–64 (2008)
15. Kurach, D., Rutkowska, D., Rakus-Andersson, E.: Face classification based on linguistic description of facial features. In: Rutkowski, L., Korytkowski, M., Scherer, R., Tadeusiewicz, R., Zadeh, L.A., Zurada, J.M. (eds.) ICAISC 2014. LNCS (LNAI), vol. 8468, pp. 155–166. Springer, Cham (2014). https://doi.org/10.1007/978-3-319-07176-3_14
16. Kwak, K.C., Pedrycz, W.: Face recognition: a study in information fusion using fuzzy integral. Pattern Recogn. Lett. **26**, 719–733 (2005)
17. Liao, S., Zhu, X., Lei, Z., Zhang, L., Li, S.Z.: Learning multi-scale block local binary patterns for face recognition. In: Lee, S.-W., Li, S.Z. (eds.) ICB 2007. LNCS, vol. 4642, pp. 828–837. Springer, Heidelberg (2007). https://doi.org/10.1007/978-3-540-74549-5_87
18. AT&T Laboratories Cambridge. The Database of Faces. www.cl.cam.ac.uk/research/dtg/attarchive/facedatabase.html. Accessed 1 Aug 2016
19. Phillips, P.J., Wechsler, J., Huang, J., Rauss, P.: The FERET database and evaluation procedure for face recognition algorithms. Image Vis. Comput. **16**(5), 295–306 (1998)
20. [dataset] Yale database. vision.ucsd.edu/datasets/yale_face_dataset_original
21. Zhang, L., Chu, R., Xiang, S., Liao, S., Li, S.Z.: Face detection based on multi-block LBP representation. In: Lee, S.-W., Li, S.Z. (eds.) ICB 2007. LNCS, vol. 4642, pp. 11–18. Springer, Heidelberg (2007). https://doi.org/10.1007/978-3-540-74549-5_2

Direct Incorporation of L_1-Regularization into Generalized Matrix Learning Vector Quantization

Falko Lischke[1(✉)], Thomas Neumann[1], Sven Hellbach[1], Thomas Villmann[2],
and Hans-Joachim Böhme[1]

[1] HTW Dresden, Friedrich-List-Platz 1, 01069 Dresden, Germany
{lischke,neumann,hellbach,boehme}@htw-dresden.de
[2] Saxony Institute for Computational Intelligence and Machine Learning,
University of Applied Sciences Mittweida, 09648 Mittweida, Germany
thomas.villmann@hs-mittweida.de

Abstract. Frequently, high-dimensional features are used to represent data to be classified. This paper proposes a new approach to learn interpretable classification models from such high-dimensional data representation. To this end, we extend a popular prototype-based classification algorithm, the matrix learning vector quantization, to incorporate an enhanced feature selection objective via L_1-regularization. In contrast to previous work, we propose a framework that directly optimizes this objective using the alternating direction method of multipliers (ADMM) and manifold optimization. We evaluate our method on synthetic data and on real data for speech-based emotion recognition. Particularly, we show that our method achieves state-of-the-art results on the Berlin Database of Emotional speech and show its abilities to select relevant dimensions from the eGeMAPS set of audio features.

1 Introduction

Thanks to the growing availability of sensors and increasing capacity of storage devices, high-dimensional data becomes more and more ubiquitous. For example, the area of human machine interaction is increasingly relying on data-based classification, for example to recognize speech or emotional state of the human user. Such recognition may be performed based on different modalities, e. g. facial expression [2,28,44] and voice [22,29,38]. To recognize emotions from the sound of the voice, one common approach is to compute a potentially large number of audio features (such as eGeMAPS, Interspeech2011, etc.) [14,36], which are statistical parameters like mean of characteristic frequencies or intensities to describe the voice.

F. Lischke—This work was supported in part by SAB grant number 100231931.

L. Rutkowski et al. (Eds.): ICAISC 2018, LNAI 10841, pp. 657–667, 2018.
https://doi.org/10.1007/978-3-319-91253-0_61

In this line, a classifier model should learn to categorize sound regarding their audio features from a set of training samples. Different classifiers can be used to assign audio features to an emotion, multi-layer perceptrons networks (MLP) [23] and support vector machines (SVM) for example [7,14]. MLPs have the disadvantage that they are difficult to analyze because of their complex structure and distributed information representation in the network for high-dimensional features, whereas SVMs depend heavily on the kernel and its parameters [17,41]. In contrast, prototype-based learning methods are more intuitive and comprehensible. Training and test results are easily to interpret [5]. For this reason, this paper is primarily concerned with the generalized matrix learning vector quantization (GMLVQ) as a classifier [33]. This model takes as the cost function to be minimized an approximation of the overall classification error. Further, the GMLVQ cost function is equipped with a scheme to weight the importance of the data features regarding the classification task.

As mentioned above, high-dimensional features are a difficulty in terms of computational complexity and class discrimination. A reduction of the data features to the relevant ones for a given classification task would reduce the complexity of the classifier model. One possibility to enforce this feature sparsity is to supplement the original cost function with a penalty term for regularization yielding a modified cost function. Frequently, the L_0-norm in the features space as a regularization is best suited for selecting features because nonzero features are penalized directly. The disadvantage is that the resulting modified cost function becomes non-convex, such that the respective optimization problem is a NP-hard [25]. The thightest convex approximation to the L_0-norm is the L_1-norm [13]. However, a cost function equipped with L_1-norm regularization cannot be optimized directly by stochastic gradient descent learning (SGDL) because the L_1-norm is not differentiable due to the absolute value function inside. Riedel et al. [30] solve this problem by instead using a differentiable approximation of the L_1-norm. However, this might result in suboptimal solutions and potentially slow training speed. Here, we show how we can use the L_1-norm directly. In particular, the contributions of this paper are:

- we present of a framework for the consistent regularization of a GMLVQ model using manifolg optimization and ADMM
- we show how the framework can be used to directly incorporate the L_1-regularization into the GMLVQ objective
- we demonstrate the approach on synthetic data and on emotion recognition tasks from the commonly used eGeMAPS audio classification feature set

2   Related Work

The development of the feature sets used for speech-based emotion recognition can be seen monitoring the respective changes in approaches presented for the Interspeech challenges since 2009. In the first Interspeech Emotion Challenge 2009, 384 features were specified, including F_0 frequency and MFCC [35].

The classifier was freely choosable. For the Interspeech Speaker Challenge 2011, an SVM with linear kernel and a feature set of 4368 features was predefined [36]. Since 2012 the feature set contains 6373 features and linear SVM are used as classifiers [37]. SVM are still frequently used for classification, as in [19,38,43]. Instead of the predefined feature sets, more and more attempts are being made to have them learned automatically from artificial neural networks (MLP) [20,43]. In order to learn optimal features for MLP, it is assumed that the available data contains a large amount of variations provided by a huge database. In contrast, prototype-based methods frequently can work successfully with fewer data [5].

A popular prototype-based classification method is learning vector quantization (LVQ), introduced by Kohonen [21]. The generalized LVQ (GLVQ) [32] of Sato and Yamada was proposed as a modification of the LVQ, replacing the learning heuristic by SGDL derived from a cost function approximating the classification error. Yet, the accuracy of the GLVQ is highly dependent on an approximately equal scaling and importance of the input features regarding the classification task. Since these two aspects are not always fulfilled, the additional introduction of scaling factors allows different weighting of the input features. This modified GLVQ is called generalized relevance LVQ (GRLVQ) [16]. The relevance profile calculated in GRLVQ provides information on the importance of each data dimension. For high-dimensional data in particular, a large part of the relevance profile is characterized by small, non-vanishing relevance values. This property is kept if the relevance profile is extended to a relevance matrix. In addition to the relevance of features, a relevance matrix also contains relevances of feature correlations. This method is termed general matrix LVQ (GMLVQ) [33].

Feature selection methods are used to reduce the risk of classifier model overfitting and increase the generalizability of classifiers [11]. As mentioned above, Riedel et al. [30] use an L_1-norm for regularization in GRLVQ. Their method is based on the Least Absolute Selection and Shrinkage Operator method (LASSO), whereas the original LASSO [39] uses convex optimization. An alternative to LASSO for feature selection is the L_1-norm SVM [4,45], which has an advantage over the standard SVM if there are redundant noise features. In Saeys et al. [31], multiple feature selection methods are combined to yield a robust feature set and to compensate for weak points of individual methods.

3   Method

This paper extend upon the GMLVQ classification method. The GMLVQ training the cost function E based on a finite training set $X = \{(x_i, y_i) \subset \mathbb{R}^n \times \{1, \ldots, C\} | i = 1, \ldots, m\}$ with n features for each of the m training samples belonging to one of the C predefined classes. Training data X is used to adapt a set of M prototypes $W = \{w_k \in \mathbb{R}^n, k = 1, \ldots, M\}$ and a positive semidefinite, symmetric classification correlation matrix Λ describing the feature relevances and correlations regarding the classification task [18]. A class label is assigned to each prototype indicating its assignment to a certain class. The matrix Λ can also be decomposed as $\Lambda = \Omega^T \Omega$ an arbitrary matrix $\Omega \in \mathbb{R}^{n \times n}$. The GMLVQ cost function is

$$E(\Omega, W, X) = \sum_{x \in X} \frac{\|\Omega(x - w^+)\|_2^2 - \|\Omega(x - w^-)\|_2^2}{\|\Omega(x - w^+)\|_2^2 + \|\Omega(x - w^-)\|_2^2}, \tag{1}$$

where w^+ is the closest prototype to the training sample x with the same class label as x and w^- is the closest prototype for data x whose label is not the same. The matrix Ω projects the data and prototypes onto a latent space where the classification takes place. Regularization of the Ω-matrix regarding numerical instabilities during SGDL is studied in [34]. However, their approach does not deal with feature sparsity.

If a complete row is zero in Ω, the corresponding feature does not have any influence on classification allowing a feature selection scheme. The feature sparseness can be measured directly with the L_0-norm or alternatively approximated using the convex relaxation of the L_0-norm, namely the L_1-norm. This methodology is known for its good feature selection capabilities [26,27,45].

To incorporate this idea of feature selection into GMLVQ, we add a regularization term $R(\Omega) = \|\Omega\|_1$ to the optimization objective

$$\underset{\Omega, W}{\text{minimize}}\ E(\Omega, W, X) + \xi R(\Omega), \tag{2}$$

where $\xi > 0$ is a regularization parameter to control sparsity of Ω. We can see that the feature selecting property of the L_1-norm acts on the feature correlation matrix Ω, thus shrinking its entries. When solving the minimization problem in Eq. 2, Ω entries can become very small if no constraints are applied. This was actually observed in [6] even without the L_1 regularization. To avoid this effect, entries in Ω are constrained to lie on the n^2 dimensional hypersphere, $\|\Omega\|_F^2 = 1$.

We now show how Eq. (2) can be optimized directly without approximation of the L_1-norm. We suggest for optimization the Alternating Direction Method of Multipliers (ADMM) as a proximal algorithm [9]. Other possibilities would be the stochastic approach [9] or the FISTA algorithm [3], for example, which are not considered here. The ADMM was chosen because it can deal with many loss functions, converges quickly to an adequately good approximate solution and, particularly, allows the optimization of non-convex functions. To bring Eq. (2) into a form that can be optimized using ADMM requires incorporation of a second variable $\phi \in \mathbb{R}^{n \times n}$, so that the data and regularization term are decoupled:

$$\underset{\Omega, W, \phi}{\text{minimize}}\ E(\Omega, W) + \xi R(\phi)$$
$$\text{subject to } \Omega = \phi,\ \|\Omega\|_2^2 = 1, \tag{3}$$

with the corresponding ADMM update steps for iteration k with penalty parameter ρ according to

$$(\Omega, W)^{k+1} = \underset{\Omega, W}{\text{argmin}} \left(E(\Omega, W) + \frac{\rho}{2} \|\Omega - \phi^k + u^k\|_2^2 \right),\ \text{s.t. } \|\Omega\|_F^2 = 1 \tag{4}$$

$$\phi^{k+1} = \underset{\phi}{\text{argmin}} \left(R(\phi) + \frac{\rho}{2} \|\Omega^{k+1} - \phi + u^k\|_2^2 \right) \tag{5}$$

$$u^{k+1} = u^k + \Omega^{k+1} - \phi^{k+1} \tag{6}$$

where u^k is a dual variable that keeps track of the constraint $\Omega = \phi$ from (3).

It is numerically difficult to solve the minimization problem in Eq. (4) because of the constraint and both, W and Ω, have to be optimized at the same time. To solve this problem, a manifold optimization [1] is used for this update step. For this, we use the Trust region optimizer over the hypersphere manifold from the PyManopt toolbox [40]. It is a second order method that approximates the cost function by a quadratic surface and then updates the current estimate based on that.

The minimization problem in Eq. (5) corresponds to the proximal operator of the L_1-norm. In this case, the solution of the proximal operator can be found in closed form as the soft-thresholding operator S [12]:

$$\phi_{i,j}^{k+1} = S_{\xi/\rho}(\Omega^{k+1} + u^k) = (1 - \frac{\xi}{\rho|\Omega_{i,j}^{k+1} + u_{i,j}^k|})_+(\Omega_{i,j}^{k+1} + u_{i,j}^k) \tag{7}$$

The notation $(\cdot)_+$ is equivalent to $max(0, \cdot)$. The soft-thresholding operator S is simultaneously a shrinkage operator, which means that relevance values are moved toward zero.

4   Results

4.1   Artificial Data

We first test the presented method on an artificial data set from Bojer et al. [8]. The data set consists of 3 classes each with 2 clusters with 15 data points in $\mathbb{R}^{10}$. The first two dimensions (x_0, x_1) are shown in Fig. 1a. The dimensions x_2 to x_5 are copies of dimension x_0 affected by Gaussian noise with variances 0.05, 0.1, 0.2 and 0.5. The following two dimensions (x_6, x_7) contain uniform noise in the intervals $[-0.5, 0.5]$ and $[-0.2, 0.2]$, respectively. Furthermore, dimensions (x_8, x_9) consist of Gaussian noise with variances 0.5 and 0.2. Based on the generated data dimensions, it can be seen that only the first two are relevant for classification. Dimensions x_2 to x_5 are copies of a relevant dimension with additional noise and the last four contains pure noise, so they are not relevant for classification at all. In experiments, the data are randomly divided into a training and test set of the same size.

For classification, 2 prototypes are used per class. Initially, each center of gravity of the clusters is assigned a prototype of the respective class. Relevances on the diagonal of matrix Ω are set to $1/\sqrt{n}$ at the beginning. We also compare against a decision tree and SVM with linear kernel. Regularization parameter ξ in Eq. 3 is tested for the values $\xi = 0, 0.5, 1, 5$ to proof an increasing sparseness of features. For a value of $\xi = 0$ the presented method corresponds to a GMLVQ solved with manifold optimization.

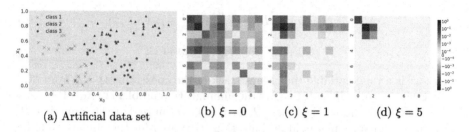

(a) Artificial data set   (b) $\xi = 0$ (c) $\xi = 1$ (d) $\xi = 5$

Fig. 1. Illustration of the first two dimensions of the artificial dataset (a) and comparison of relevance matrices with increasing regularization factor ξ (b–d)

Table 1. Accuracies of different classifiers for the two different data sets

Classifier	Accuracy	
	Artificial dataset	EmoDB
Decision tree	0.867	0.466
SVM	0.756	0.696
manifold-GMLVQ	0.711	0.636
L1 manifold-GMLVQ ($\xi = 0.5$)	0.822	0.652
L1 manifold-GMLVQ ($\xi = 1$)	0.644	0.652
L1 manifold-GMLVQ ($\xi = 5$)	0.844	0.525

For the artificial data set, we expect increasing sparsity with larger regularization parameter ξ. Figure 1b to d depicts relevance matrices Λ for $\xi = 0, 1, 5$. It is obvious how ξ increases the sparseness of relevance values. Influence of dimensions that contain less information is vanishing. For $\xi = 1$, x_3 no longer has any influence on classification because all related row and column entries are zero. A comparison of accuracies in Table 1 for the artificial dataset shows that the results of our method are comparable to selected Decision Tree and SVM classifiers. Regularization with $\xi = 5$ outperforms both SVM and GMLVQ without regularization. The accuracy is close to that of the decision tree, although only 3 features are still used for classification. The number of used features can be significantly reduced from 10 to 3 without great loss of accuracy. As expected, relevant features are selected and irrelevant are vanished.

4.2   Audio Emotion Recognition

In further tests, our method will be applied to a real data set with high-dimensional audio features. Berlin Database of Emotional Speech (EmoDB) [10] contains 535 emotional utterances. It includes 10 emotionally neutral German short sentences spoken by 10 professional actors. Each actor spoke every phrase with each of the 7 basic emotions: anger, boredom, disgust, fear, happiness, sadness and neutral. Out of the 700 utterances, only those are still included in the dataset that have been labelled as natural sounding and identifiable with respect

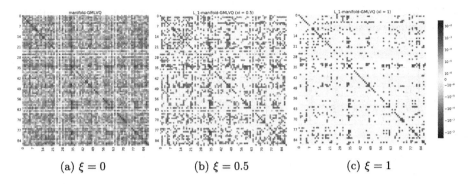

(a) $\xi = 0$ (b) $\xi = 0.5$ (c) $\xi = 1$

Fig. 2. eGeMAPS feature matrices Λ of person ID 3 for different values of ξ

Table 2. eGeMAPS features that no longer apply for classification after regularization with $\xi = 1$

F0semitoneFrom27.5 Hz_sma3nz_mean FallingSlope (8)	loudness_sma3_stddev RisingSlope (17)	mfcc1_sma3_stddevNorm (23)
mfcc2_sma3_stddev Norm (25)	mfcc3_sma3_stddev Norm (27)	shimmerLocaldB_sma3nz_ amean (32)
F3bandwidth_sma3nz_ stddevNorm (55)	alphaRatioV_sma3nz_ stddevNorm (59)	slopeV0-500_sma3nz_stddev Norm (63)
slopeV500-1500_sma3nz_ stddevNorm (65)	mfcc3V_sma3nz_stddev Norm (73)	mfcc4V_sma3nz_stddev Norm (75)

to the emotion of 20 people. EmoDB is a common dataset to examine and test newly introduced features (e.g. [14,24]) or methods for the automatic learning of features using convolutional networks (e.g. [20,42]). For each EmoDB audio sample, the 88-dimensional eGeMAPS [14] feature vector is extracted by the OpenS-MILE toolkit [15]. This feature set was proposed for automatic speech analysis using a minimalistic set of voice parameters. These were hand-selected based on their potential to indicate affective changes in voice and theoretical significance. The eGeMAPS features are initially scaled to zero mean and unit variance. The initialization of Ω and prototypes is equivalent to the previous experiments. To simulate how the system would generalize for a new speaker, we perform splits according to the speaker ID, so effectively we perform Leave-One-Speaker-Out cross-validation. Figure 2 shows the determined test accuracies. Contrary to the attempt with artificial data, the decision tree is not suitable for this problem because it achieves only a low accuracy. The best accuracy is achieved with a SVM classifier. Our L_1-regularized GMLVQ is slightly worse, but with the added benefit of selecting meaningful features. Also, our L_1 GMLVQ improves upon standard GMLVQ. The learned matrices Λ for person ID 3 are depicted in Fig. 2.

Table 3. Confusion matrices summarized over all persons for EmoDB data with different regularization factors

	anger	boredom	disgust	fear	happiness	sadness	neutral
anger	98	1	2	6	19	1	0
boredom	1	63	6	0	0	4	7
disgust	8	2	25	5	2	2	2
fear	12	1	2	37	6	7	4
happiness	26	2	5	6	29	0	3
sadness	0	5	8	1	0	42	6
neutral	3	7	5	7	4	7	46

(a) Confusion matrix for $\xi = 0$

	anger	boredom	disgust	fear	happiness	sadness	neutral
anger	98	1	2	7	19	0	0
boredom	1	63	10	0	0	2	5
disgust	7	1	28	4	1	3	2
fear	17	0	3	35	4	7	3
happiness	30	2	2	6	27	0	4
sadness	0	4	10	1	0	42	5
neutral	0	9	4	4	4	2	56

(b) Confusion matrix for $\xi = 1$

The sparseness increases the larger ξ becomes, so that more and more features for the classification are not available because almost all corresponding Λ entries are zero. For $\xi = 1$, rows and columns of 12 audio features are set to zero, see Table 2, which means that they only play a minor role in determining emotions. These features mainly refer to the standard deviation of individual voice characteristics. This demonstrates that our method can be used to automatically detect irrelevant features, which could in the future be used to improve upon hand-selected feature sets such as eGeMAPS and Interspeech. Despite regularization, the accuracy for $\xi = 0.5$ and 1 increases. Regularization apparently removes correlations from the matrix that have a negative influence on accuracy. The confusion matrices in Table 3 show slight differences in the recognition. In both cases, however, the main diagonal is stronger than the false predictions. The occupancy shows that happiness is often false classified as anger. This indicates that the relevance of the features for distinguishing these two emotions scarcely differs.

5   Conclusion

In our tests, we have shown that the GMLVQ with L_1-regularization leads to sparseness in the feature matrix Λ. The optimization used for this purpose is solved without an approximation of the L_1-norm. Instead, the L_1-regularization was directly incorporated in the GMLVQ objective. In both experiments, GMLVQ with L_1-regularization based on our framework achieves similar accuracies as standard classifiers. In addition, the accuracy could be increased by an additional regularization of the manually selected eGeMAPS features. This shows that an objective feature selection can result in higher accuracy than subjectively selected features. Our optimization framework offers an opportunity to select features from more extensive feature sets based on objective criteria.

We presented an optimization framework which also provides a general tool for regularization of GMLVQ that can be extended in multiple interesting ways in the future. In fact, any regularizer that has a proximal operator can be integrated, which is not possible with the classical gradient-based GMLVQ framework. For example, group sparsity or hierarchical schemes could be readily integrated, enabling new and powerful feature selection approaches.

References

1. Absil, P.A., Mahony, R., Sepulchre, R.: Optimization Algorithms on Matrix Manifolds. Princeton University Press, Princeton (2009)
2. Ali, H., Hariharan, M., Yaacob, S., Adom, A.H.: Facial emotion recognition using empirical mode decomposition. Expert Syst. Appl. **42**(3), 1261–1277 (2015)
3. Beck, A., Teboulle, M.: A fast iterative shrinkage-thresholding algorithm for linear inverse problems. SIAM J. Imaging Sci. **2**(1), 183–202 (2009)
4. Bi, J., Bennett, K., Embrechts, M., Breneman, C., Song, M.: Dimensionality reduction via sparse support vector machines. JMLR **3**(Mar), 1229–1243 (2003)
5. Biehl, M., Hammer, B., Villmann, T.: Prototype-based models in machine learning. Wiley Interdisc. Rev.: Cogn. Sci. **7**(2), 92–111 (2016)
6. Biehl, M., Hammer, B., Schleif, F.M., Schneider, P., Villmann, T.: Stationarity of matrix relevance learning vector quantization. Mach. Learn. Rep. **3**, 1–17 (2009)
7. Bishop, C.M.: Pattern Recognition and Machine Learning (Information Science and Statistics). Springer, New York (2006)
8. Bojer, T., Hammer, B., Schunk, D., Von Toschanowitz, K.: Relevance determination in learning vector quantization. In: Proceedings of ESANN (2001)
9. Boyd, S., Parikh, N., Chu, E., Peleato, B., Eckstein, J.: Distributed optimization and statistical learning via the alternating direction method of multipliers. Found. Trends Mach. Learn. **3**(1), 1–122 (2011)
10. Burkhardt, F., Paeschke, A., Rolfes, M., Sendlmeier, W., Weiss, B.: A database of German emotional speech. In: Interspeech, vol. 5, pp. 1517–1520 (2005)
11. Chandrashekar, G., Sahin, F.: A survey on feature selection methods. Comput. Electr. Eng. **40**(1), 16–28 (2014)
12. Donoho, D.L.: De-noising by soft-thresholding. IEEE TIT **41**(3), 613–627 (1995)
13. Donoho, D.L.: For most large underdetermined systems of linear equations the minimal ℓ1-norm solution is also the sparsest solution. CPAMA **59**(6), 797–829 (2006)
14. Eyben, F., Scherer, K.R., Schuller, B.W., Sundberg, J., André, E., Busso, C., Devillers, L.Y., Epps, J., Laukka, P., Narayanan, S.S., Truong, K.P.: The Geneva minimalistic acoustic parameter set (GeMAPS) for voice research and affective computing. IEEE TAC **7**(2), 190–202 (2016)
15. Eyben, F., Weninger, F., Gross, F., Schuller, B.: Recent developments in openSMILE, the munich open-source multimedia feature extractor. In: Proceedings of the 21st ACM, pp. 835–838. ACM (2013)
16. Hammer, B., Villmann, T.: Generalized relevance learning vector quantization. Neural Netw. **15**(8), 1059–1068 (2002)
17. Hsu, C.W., Lin, C.J.: A comparison of methods for multiclass support vector machines. IEEE TNN **13**(2), 415–425 (2002)
18. Kaden, M., Lange, M., Nebel, D., Riedel, M., Geweniger, T., Villmann, T.: Aspects in classification learning - review of recent developments in learning vector quantization. Found. Comput. Decis. Sci. **39**(2), 79–105 (2014)
19. Kanth, N.R., Saraswathi, S.: Efficient speech emotion recognition using binary support vector machines multiclass SVM. In: 2015 IEEE ICCIC, December 2015
20. Kim, J., Truong, K.P., Englebienne, G., Evers, V.: Learning spectro-temporal features with 3D CNNs for speech emotion recognition. arXiv preprint arXiv:1708.05071 (2017)
21. Kohonen, T.: Learning vector quantization. In: Kohonen, T. (ed.) Self-Organizing Maps. SSINF, vol. 30, pp. 175–189. Springer, Heidelberg (1995). https://doi.org/10.1007/978-3-642-97610-0_6

22. Korkmaz, O.E., Atasoy, A.: Emotion recognition from speech signal using mel-frequency cepstral coefficients. In: 2015 9th ELECO, pp. 1254–1257, November 2015

23. Lee, J., Tashev, I.: High-level feature representation using recurrent neural network for speech emotion recognition. In: Interspeech 2015. ISCA, September 2015

24. Mao, Q., Dong, M., Huang, Z., Zhan, Y.: Learning salient features for speech emotion recognition using convolutional neural networks. IEEE Trans. Multimedia **16**(8), 2203–2213 (2014)

25. Murty, K.G., Kabadi, S.N.: Some NP-complete problems in quadratic and nonlinear programming. Math. Program. **39**(2), 117–129 (1987)

26. Ng, A.Y.: Feature selection, $L1$ vs. $L2$ regularization, and rotational invariance. In: Proceedings of the 21th ICML, ICML 2004, p. 78. ACM, New York (2004)

27. Obozinski, G., Taskar, B., Jordan, M.: Multi-task feature selection. Statistics Department, UC Berkeley, Technical report 2 (2006)

28. Ofodile, I., Kulkarni, K., Corneanu, C.A., Escalera, S., Baro, X., Hyniewska, S., Allik, J., Anbarjafari, G.: Automatic recognition of deceptive facial expressions of emotion (2017)

29. Palo, H., Mohanty, M., Chandra, M.: Efficient feature combination techniques for emotional speech classification. IJST **19**(1), 135–150 (2016)

30. Riedel, M., Rossi, F., Kästner, M., Villmann, T.: Regularization in relevance learning vector quantization using l_1-norms. In: Verleysen, M. (ed.) Proceedings of ESANN 2013, pp. 17–22 (2013)

31. Saeys, Y., Abeel, T., Van de Peer, Y.: Robust feature selection using ensemble feature selection techniques. In: Daelemans, W., Goethals, B., Morik, K. (eds.) ECML PKDD 2008. LNCS (LNAI), vol. 5212, pp. 313–325. Springer, Heidelberg (2008). https://doi.org/10.1007/978-3-540-87481-2_21

32. Sato, A., Yamada, K.: Generalized learning vector quantization. In: Advances in Neural Information Processing Systems, pp. 423–429 (1996)

33. Schneider, P., Biehl, M., Hammer, B.: Adaptive relevance matrices in learning vector quantization. Neural Comput. **21**(12), 3532–3561 (2009)

34. Schneider, P., Bunte, K., Stiekema, H., Hammer, B., Villmann, T., Biehl, M.: Regularization in matrix relevance learning. IEEE Trans. Neural Netw. **21**(5), 831–840 (2010)

35. Schuller, B., Steidl, S., Batliner, A.: The INTERSPEECH 2009 emotion challenge. In: 10th Annual Conference of the ISCA (2009)

36. Schuller, B., Steidl, S., Batliner, A., Schiel, F., Krajewski, J.: The INTERSPEECH 2011 speaker state challenge. In: 12th Annual Conference of the ISCA (2011)

37. Schuller, B., Steidl, S., Batliner, A., et al.: The INTERSPEECH 2017 computational paralinguistics challenge: addressee, cold and snoring. In: ComParE, Interspeech 2017, pp. 3442–3446 (2017)

38. Sinith, M.S., Aswathi, E., Deepa, T.M., Shameema, C.P., Rajan, S.: Emotion recognition from audio signals using support vector machine. In: 2015 IEEE RAICS, pp. 139–144, December 2015

39. Tibshirani, R.: Regression shrinkage and selection via the lasso. J. Roy. Stat. Soc. Ser. B (Methodol.) 267–288 (1996)

40. Townsend, J., Koep, N., Weichwald, S.: Pymanopt: a python toolbox for optimization on manifolds using automatic differentiation. J. Mach. Learn. Res. **17**(137), 1–5 (2016)

41. Villmann, T., Bohnsack, A., Kaden, M.: Can learning vector quantization be an alternative to SVM and deep learning? JAISCR **7**(1), 65–81 (2017)

42. Wang, K., An, N., Li, B.N., Zhang, Y., Li, L.: Speech emotion recognition using fourier parameters. IEEE TAC **6**(1), 69–75 (2015)
43. Wen, G., Li, H., Huang, J., Li, D., Xun, E.: Random deep belief networks for recognizing emotions from speech signals. Comput. Intell. Neurosci. **2017** (2017)
44. Zhang, Y., Zhang, L., Hossain, M.A.: Adaptive 3D facial action intensity estimation and emotion recognition. Expert Syst. Appl. **42**(3), 1446–1464 (2015)
45. Zhu, J., Rosset, S., Tibshirani, R., Hastie, T.J.: 1-norm support vector machines. In: Advances in Neural Information Processing Systems, pp. 49–56 (2004)

Classifiers for Matrix Normal Images: Derivation and Testing

Ewaryst Rafajłowicz[(✉)]

Faculty of Electronics, Wrocław University of Science and Technology,
Wrocław, Poland
ewaryst.rafajlowicz@pwr.edu.pl

Abstract. We propose a modified classifier that is based on the maximum a posteriori probability principle that is applied to images having the matrix normal distributions. These distributions have a special covariance structure, which is interpretable and easier to estimate than general covariance matrices. The modification is applicable when the estimated covariance matrices are still not well-conditioned. The proposed classifier is tested on synthetic images and on images of gas burner flames. The results of comparisons with other classifiers are also provided.

Keywords: Matrix normal distribution · Bayesian classifier
Classification of flames

1 Introduction

Our aim is to discuss classifiers dedicated to image recognition. We consider images as matrices of their gray levels. In opposite to the mainstream literature in the field, we avoid a feature selection as the first step before image recognition. Conversely, we consider images to be recognized as whole entities. Additionally, we avoid the vectorization of matrices because for large images (10 MPix, say) it is not only inconvenient but also destroys their covariance and neighborhood structures.

We adopt random matrices as models of images. We confine our attention to random matrices having Gaussian distributions since their description requires the knowledge of the mean and covariance matrices. However, we have to confine even more, because the covariance matrix of a 100×100 pixels image has about 10^8 elements and their estimation would require a huge amount of data. Thus, we need a class of random matrices with a more specialized covariance structure. The so-called matrix normal distribution (see [7]) seems to be a good choice, because their overall covariance matrices have a special structure that can be fully described by two matrices of covariances between rows and columns, respectively. These matrices are much smaller and easier to estimate (see Section... for details). Furthermore, it seems that inter-row and inter-column covariances are dominating for a sufficiently broad class of real-life images.

© Springer International Publishing AG, part of Springer Nature 2018
L. Rutkowski et al. (Eds.): ICAISC 2018, LNAI 10841, pp. 668–679, 2018.
https://doi.org/10.1007/978-3-319-91253-0_62

Clearly, images as inputs of classifiers re-appeared in the literature starting from the perceptron for classifying hand-written digits ([3] and the bibliography therein). Also later on random matrices with spatially Markov structure, generated by cliques, were extensively studied (see [5]). However, according to our knowledge, random normal matrices as models of images to be recognized are not considered. The research that is closest to ours was done by Krzysko [4] who considered the problem of recognizing random normal matrices that arise by stacking vectors, having their own interpretation, into a matrix. This idea was extended in [8] to recognizing sequences of images by stacking them into longer matrices.

The paper is organized as follows.

- In Sect. 2 we state the problem and describe basic properties of random normal matrices for the reader's convenience.
- The derivation of the Bayes recognizer is presented in Sect. 2 and then its empirical version is proposed, which is based on the plug-in principle, using the matrix learning sequence for the maximum likelihood estimation of inter-row and inter-column covariances.
- The quality of the proposed algorithm is compared with the well known methods: the nearest neighbor (NN) and the support vector machine (SVM). The comparison is done by extensive simulations of the Monte-Carlo (MC) type, assuming that the assumed covariance structure is an adequate one.
- Finally, the proposed algorithm is compared with NN classifying a sequence of real-life images of gas burner flames. This time the method of cross-validation is used for comparisons and we are not sure whether the assumed covariance structure is an adequate one.

We notice that a similar sequence of gas burner flames was used for comparisons in [10,11]. In the latter paper a more traditional classifier – based on features extraction – was used. In the former, the classifier is based on whole images (without features extraction), but a special structure of the covariance matrix is not assumed. Instead, firstly medoid-based clustering is done. Thus, the results are incomparable with presented here, because of different assumptions.

2   Problem Statement

MND as Class Densities and Their Estimation. We assume that probability distributions of gray-level images from class $j = 1, 2, \ldots, J$ have MND with probability density functions (p.d.f.'s) of the form:

$$f_j(\mathbf{X}) = \frac{1}{c_j} \exp\left[-\frac{1}{2} \operatorname{tr}[U_j^{-1}(\mathbf{X} - \mathbf{M}_j) V_j^{-1} (\mathbf{X} - \mathbf{M}_j)^T]\right], \tag{1}$$

where ^{T} stands for the transposition and $\det[.]$ denotes the determinant of a matrix in the brackets, while F the normalization constants are given by:

$$c_j \stackrel{def}{=} (2\pi)^{0.5\,n\,m} \det[U_j]^{0.5\,n} \det[V_j]^{0.5\,m}, \tag{2}$$

where $n \times m$ matrices M_j's denote the class means matrices. Concerning the covariance structure of MND class densities:

1. $n \times n$ matrix U_j denotes the covariance matrix between rows of an image from j-th class,
2. $m \times m$ matrix V_j stands for the covariance matrix between columns of an image from j-th class.

The above definitions are meaningful only when $\det[U_j] > 0$, $\det[V_j] > 0$. Later on, we shall write shortly,

$$\mathbf{X} \sim \mathcal{N}_{n,m}(\mathbf{M}_j, U_j, V_j), \text{ for } j = 1, 2, \ldots, J \tag{3}$$

Remark 1. *In order to reveal to what extent MND is a special case of a general class of Gaussian p.d.f.'s let us notice that the formally equivalent description of MND is the following:*

$$vec(\mathbf{X}) \sim \mathcal{N}_{nm}(vec(\mathbf{M}_j), \Sigma_j), \text{ for } j = 1, 2, \ldots, J, \tag{4}$$

where $vec(\mathbf{X})$ is the operation of stacking columns of matrix $\mathbf{X}$, while Σ_j is a $nm \times nm$ covariance matrix of j-th class, which is the Kronecker product (denoted as $\otimes$) of U_j and V_j, i.e.,

$$\Sigma_j \overset{def}{=} U_j \otimes V_j, \quad j = 1, 2, \ldots, J. \tag{5}$$

Thus, MND densities have special covariance structure in comparison to a general multivariate Gaussian densities. Namely, their covariance matrices do not have inter rows-columns covariances, which makes them much easier to estimate.

Now, let us consider the case when $\mathbf{M}_j$, U_j, V_j,s are unknown and they have to be estimated from learning sequences.

Further on, we assume that we have J learning sequences of the following form: $\mathbf{X}_i^{(j)}$, $i = 1, 2, \ldots N_j$, $j = 1, 2, \ldots, J$. They are used for calculating estimates, further denoted by the same letter with the hat $\hat{\,}$.

We estimate the class mean matrices and a priori class probabilities in a classical way, i.e.,

$$\hat{M}_j = N_j^{-1} \sum_{i=1}^{N_j} \mathbf{X}_i^{(j)}, \quad \hat{p}_j = N_j/N, \quad j = 1, 2, \ldots, J. \tag{6}$$

The necessary condition for estimating the covariance matrices is to have a sufficient number of observations. In [6] it was proved that the following requirements should be satisfied:

$$N_j \geq \max\left\{\frac{n}{m}, \frac{m}{n}\right\} + 1, \quad j = 1, 2, \ldots, J. \tag{7}$$

These conditions are less demanding than the classic one: $N_j > mn$. The reason is that the overall covariance matrix has a special structure.

The maximum likelihood estimates (MLE) of the inter-row and inter-column covariance matrices for classes are coupled by the following set of equations (see [6,12]):

$$\hat{U}_j = \frac{1}{N_j\, m} \sum_{i=1}^{N_j} (\mathbf{X}_i - \hat{\mathbf{M}}_j)\, \hat{V}_j^{-1}\, (\mathbf{X}_i - \hat{\mathbf{M}}_j)^T, \tag{8}$$

$$\hat{V}_j = \frac{1}{N_j\, n} \sum_{i=1}^{N_j} (\mathbf{X}_i - \hat{\mathbf{M}}_j)^T\, \hat{U}_j^{-1}\, (\mathbf{X}_i - \hat{\mathbf{M}}_j) \tag{9}$$

for $j = 1, 2, \ldots, J$. Equations (8) and (9) can be solved by the freezing method. In more detail:

Step 1. $\hat{U}_j$ in (8) is calculated replacing $\hat{V}_j^{-1}$ by the identity matrix,

Step 2. $\hat{U}_j$ is substituted into (9) and $\hat{V}_j$ is calculated,

Step 3. $\hat{V}_j$ is substituted into (8) and return to Step 2 (until convergence).

It was proved in [12] that one can perform only one iteration between Step 3 and Step 2 to obtain the efficient estimates of U_j and V_j.

When the estimates $\hat{U}_j$ and $\hat{V}_j$ are obtained, then one can easily define and calculate estimates $\hat{c}_j$ for c_j's. Indeed, the empirical, plug-in type, version of (2) is the following:

$$\hat{c}_j = (2\,\pi)^{0.5\, n\, m}\, \det[\hat{U}_j]^{0.5\, n}\, \det[\hat{V}_j]^{0.5\, m}. \tag{10}$$

Thus, we have estimators for all the unknown parameters of the matrix normal class densities.

A Modified MAP Classifier for Matrices. In this section we firstly derive the maximum a priori probability (MAP) classifier, assuming for a while that f_j's are known. Then we consider its empirical version of the plug-in type. Finally, we add an important modification, namely occasionally switching to the nearest mean classifier, when the estimates of the covariance matrices are ill-conditioned. When new images appear and improve the estimated covariance matrices, then we return to the proposed classifier, which is later called the matrix classifier (MCL).

Denote by $p_j > 0$, $j = 1, 2, \ldots, J$, $\sum p_j = 1$, a priori class probabilities. It is well known that in a general case the MAP classifier assigns $\mathbf{X}$ to class j^* such that

$$j^* = \arg\max_j[p_j\, f_j(\mathbf{X})], \tag{11}$$

where $\arg\max_j[\ ]$ stands for the argument for which the maximum is attained. It is also well known that this rule is the optimal one when the 0–1 loss function is used (see, e.g., [2]). Applying this rule to (1) and using the fact that the logarithm is a strictly increasing function we obtain the following decision rule:

When $\mathbf{M}_j$ and U_j, V_j, p_j, $j = 1, 2, \ldots, J$ are known, then from (11) and (1) we obtain: classify matrix (image) $\mathbf{X}$ to Class j^* if

$$j^* = \arg\min_j \left[\frac{1}{2} \operatorname{tr}[U_j^{-1}(\mathbf{X} - \mathbf{M}_j) V_j^{-1} (\mathbf{X} - \mathbf{M}_j)^T] \right] - \log(p_j/c_j) \qquad (12)$$

The expressions in the brackets play the role of the Mahalanobis distances. The matrices U_j^{-1} and V_j^{-1} de-correlate rows and columns of an image, respectively. Thus, in a general case, the optimal classifier is quadratic in $\mathbf{X}$ and we have to know (or to estimate) all parameters: $\mathbf{M}_j$ and U_j, V_j, p_j, $j = 1, 2, \ldots J$. Their estimators were proposed in the previous subsection. Substituting them into (12), we obtain the following empirical MAP classifier: classify image (matrix) $\mathbf{X}$ to the class $\hat{j}$, where

$$\hat{j} = \arg\min_{1 \leq j \leq J} \left[\frac{1}{2} \operatorname{tr}[\hat{U}_j^{-1}(\mathbf{X} - \hat{\mathbf{M}}_j) V_j^{-1} (\mathbf{X} - \hat{\mathbf{M}}_j)^T] \right] - \log(\hat{p}_j/\hat{c}_j). \qquad (13)$$

Although this classifier is the empirical version of the optimal one, it does not give a full guarantee that it works well in practice. The reasons are (at least) the following:

1. Condition (7) is the minimal one in the sense that it yields the existence of the inverse matrices $\hat{U}_j^{-1}$ and $\hat{V}_j^{-1}$. However, we even do not know whether $\hat{U}_j$ and $\hat{V}_j$ are sufficiently well-conditioned, which is a prerequisite for obtaining their inverses in a numerically stable way.
2. It is not clear to what extent (13) retains its good properties in the case when underlying class distributions are not matrix normal.

We tackle problem 1 in this section, while problem 2 is discussed in the last section.

Recall that for symmetric and positive definite matrix A, its (Euclidean) condition number κ is defined as follows:

$$\kappa = \frac{\lambda_{max}(A)}{\lambda_{min}(A)}, \qquad (14)$$

where $\lambda_{max}(A)$ and $\lambda_{min}(A)$ denote the largest and the smallest eigenvalue of A, that are real and positive. It is also known that κ acts as an amplifier of errors that occur during matrix inversion. When κ is in the range between 1 and about 100, one can expect low numerical errors in matrix inversion algorithms. In practice, especially for large matrices, easily computable bounds for κ are used. However, in our computational experiments, reported in the section before last, we used directly (14) for matrices of a moderate size.

Taking the above discussion into account, we propose the following algorithm for classifying images (matrices) having the matrix normal distribution with class densities given by (1).

Matrix Classifier (MCL)

Step 0. Select $0 < \kappa_{max} < 100$.
Step 1. Initial learning phase: calculate $\hat{p}_j$, $\hat{U}_j$, $\hat{V}_j$, $\hat{c}_j$, $\hat{M}_j$, $j = 1, 2, \ldots, J$ for the learning sequence.
Step 2. Check whether all the inequalities:

$$\hat{\kappa}_j^{(1)} \stackrel{def}{=} \frac{\lambda_{max}(U_j)}{\lambda_{min}(U_j)} < \kappa_{max}, \; j = 1, 2, \ldots, J \tag{15}$$

as well as

$$\hat{\kappa}_j^{(2)} \stackrel{def}{=} \frac{\lambda_{max}(V_j)}{\lambda_{min}(V_j)} < \kappa_{max}, \; j = 1, 2, \ldots, J \tag{16}$$

are fulfilled. If so, go to Step 3, otherwise, go to Step 4.
Step 3. Classify new image (matrix) $\mathbf{X}$ according to the rule described as (13), acquire the next image (matrix) for classification and repeat Step 3.
Step 4. Classify new image (matrix) $\mathbf{X}$ according to the nearest mean rule, i.e., classify it to the class

$$\tilde{j} = \arg\min_j ||\mathbf{X} - \hat{M}_j||^2, \tag{17}$$

where the squared distance $||\mathbf{X} - \hat{M}_j||^2$ is defined as follows:

$$||\mathbf{X} - \hat{M}_j||^2 = \text{tr}[(\mathbf{X} - \hat{\mathbf{M}}_j)(\mathbf{X} - \hat{\mathbf{M}}_j)^T]. \tag{18}$$

Update the estimates of $\hat{U}_{\tilde{j}}$ and $\hat{V}_{\tilde{j}}$ by adding current $\mathbf{X}$ to the learning sequence as $(\mathbf{X}, \tilde{j})$ and go to Step 2.

Notice that the update of $\hat{U}_{\tilde{j}}$ and $\hat{V}_{\tilde{j}}$ can be done recursively. A distinguishing feature of the MCL algorithm is that it has not only the learning phase (as usual), but also it learns at the recognition phase when the result of classification may be unreliable. Conditions (15) and (16) serve as indicators of this. The additional learning is supported by a surrogate classifier, which can be arbitrary. Here, we use the nearest mean classifier for this purpose as it is closest to the MAP classifier and does not need to store the whole learning sequence.

3 Performance Evaluation by Monte Carlo Experiments

In this section, we provide the results of extensive Monte Carlo (MC) simulations for evaluating the performance of MCL. We take advantage that in MC simulations we can repeat 1000 times the learning and the testing phase, having proper classifications for comparisons with MCL outputs. The results of these simulations are also used for comparisons of MCL performance with the classification errors committed by the nearest neighbor (NN) and the support vector machine (SVM) classifiers, implemented in the Wolfram's Mathematica ver. 11.

Fig. 1. Excerpt of samples from Class 1.

Fig. 2. Excerpt of samples from Class 2.

The following parameters were used during simulations. Classified 6×6 matrices (images) were drawn at random from two classes, each having the MND with

- mean matrices M_1 being a 6×6 matrix of zeros, while M_2 is a 6×6 constant matrix with 0.05 as elements.
- covariance matrices: $U_1 = 0.1\, I_6$, V_1 is a 6×6 tridiagonal matrix with 2 at the main diagonal and 1 along the bands running above and below the main diagonal, $U_2 = V_1$, $V_2 = I_6$, where I_6 is a 6×6 identity matrix,
- $p_1 = p_2 = 0.5$.

Samples of the simulated learning sequences are shown in Figs. 1 and 2 for classes 1 and 2, respectively. As one can notice, class densities were selected in such a way that the samples are not easily distinguishable. The results of testing are shown in Figs. 3 and 4. Each point at these figures was obtained from 1000 MC simulation runs. They present statistics of the percentage of errors (% Err) committed by MCL, SVM and NN classifiers as a function of the learning sequence length N_d, i.e., $N_1 = N_2 = N_d$. From Fig. 3 it is clear that MCL provides a much smaller mean and median of errors than those of NN and SVM classifiers. Furthermore, also maximal classification errors, calculated as the maximal % of errors from 1000, is essentially smaller for MCL (see the left panel in Fig. 4). The variability of % Err, measured by the dispersion, is also smaller for MCL than for NN and SVM classifiers (see the right panel in Fig. 4). Thus, one can expect a more reliable classification for different learning sequences.

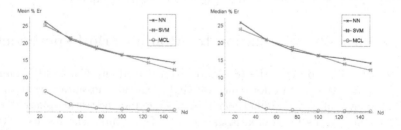

Fig. 3. The mean and the median of % of classification errors.

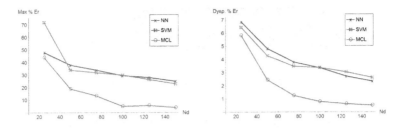

Fig. 4. The maximal % Err (Left), the dispersion of % Err (right).

As the conclusion from the above MC simulations, one can say that the MCL classifier outperforms the NN and SVM classifiers when MND assumptions hold.

4   Testing MCL on Images of Flames of a Gas Burner

Our aim in this section is to test the MCL on real data, namely on images of flames of an industrial gas burner. We also compare its performance with the 10 nearest neighbors (10–NN) classifier, where the number of neighbors was selected as the best after a series of preliminary experiments. The images of flames are classified to three classes (see Fig. 6). Namely,

Class 1 – improper combustion – too high air supply rate,
Class 2 – proper combustion – proper air supply rate,
Class 3 – improper combustion – too low air supply rate.

The result of the classification of a current flame image can be used to decide whether to decrease, leave unchanged or increase the air supply rate, respectively. This classification is rather rough and serves mainly for illustration purposes. We refer the reader to [9,13] for more detailed explanations of relationships between flames images and the combustion process.

Preparation of Flames Images. The original images of flames were stored as 170×352 RGB images. Before using them for classification, they were converted to $[0, 1]$ gray-levels scale and their size is reduced to 17×36 by 10 times down-sampling. An example of an original image and its downsampled version is shown in Fig. 5.

Fig. 5. An original image and its down-sampled version.

Fig. 6. Examples of images: Classes 1–3 (from the left).

Learning, Testing and Comparing Classifiers. The sequence of flames images that was at our disposal consists of 145 items. They were classified to the above-mentioned classes by the author. The resulting labeled sequence will be further denoted as H_{145}.

The Methodology of Cross-Validation Testing

In opposite to the MC testing, here we do not have the freedom of generating learning and testing examples freely. Instead, we used the well-known cross-validation (CV) methodology of testing:

Step 1. Select from H_{145} (at random with the same probabilities) a learning sequence of the length 45 and denote it as L_{45}. The rest of H_{145} denote as T_{100} and use it for testing.
Step 2. Learn MCL and 10-NN classifiers, using the L_{45} sequence.
Step 3. Test the classifiers from Step 2, applying them to T_{100}, calculate and store the percentages of errors committed by them.
Step 4. Repeat Steps 1–3 1000 times.

Notice that this is an intensive testing procedure because we have to estimate three matrices of the means and six covariance matrices 1000 times when learning MCL. Also, 10-NN method is time-consuming at the testing phase.

In order to illustrate the testing procedure, we provide the results of learning from only one of its runs.

A priori probabilities, estimated from the frequencies of occurrence in this run, are the following $\hat{p}_1 = 0.37$, $\hat{p}_2 = 0.34$, $\hat{p}_3 = 0.29$. In other CV runs this pattern is similar.

Mean images for each class in this particular CV run are shown in Fig. 7 and again, in other runs these images are similar. In Fig. 8 the inter-column 36×36 covariance matrices for the same CV run and for all three classes are shown as images (black pixels correspond to larger (co-)variance). In the case of covariance matrices, the variability between CV runs is larger than for the mean, but the general pattern remains similar. Namely, Class 1 images have almost uncorrelated columns and the correlation grows for Classes 2 and 3. In Fig. 9 the inter-column 17×17 covariance matrices for the same CV run and for all three classes are shown as images (black pixels correspond to larger (co-)variance). As above, the variability between CV runs is larger than for the mean, but the general pattern, although different than for columns, remains similar. Namely, for Class 1 almost all rows are similarly correlated, while for Class 2 mainly the central rows have higher correlations. In opposite, for Class 3 almost all rows are correlated. This justifies the hope that MCL will have a low classification error also in this case.

Fig. 7. Mean images for classes in one of the CV runs.

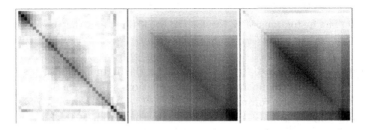

Fig. 8. Inter-column 36×36 covariance matrices in one CV run.

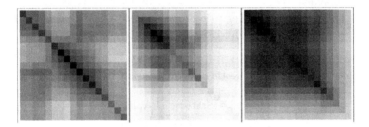

Fig. 9. Inter-rows 17×17 covariance matrices in one CV run.

The Results of Cross-Validation Testing. The results of testing of the MCL (with $\kappa_{max} = 50$) and 10-NN algorithm, using the above described CV methodology, are shown in Table 1. The analysis of Table 1 is equivocal. The mean % Err of the 10-NN method is smaller than that of the MCL, but the comparison of the medians yields the opposite results. A partial explanation of this behaviour is shown in Fig. 10. As one can notice, % Err of the MCL is much more concentrated around the mean than that of 10-NN (although the estimated dispersions suggest the opposite). Another partial explanation of this behavior is presented in the next subsection.

Table 1. The CV comparison of MCL and 10-NN % errors.

% Err.	Mean	Med.
10-NN	12.5	12.2
MCL	15	10

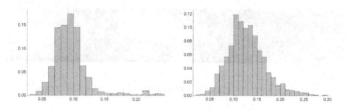

Fig. 10. The histograms of % Err for the MCL (left) and the 10-NN (right).

Baringhaus-Henze Test for Classes. The Baringhaus-Henze test (see, e.g., [1]) for the multivariate normality of observations from Classes 1–3 was performed. The results are summarized in Table 2. As one can observe, for observations from Classes 2 and 3 the test rejects the hypothesis that observations are multivariate normal. In fact, this test should be performed before MCL was used. We provide its results here, as one more partial explanation that the 10-NN classifier behaves comparably well or slightly better than the MCL, which is optimal, but only when observations have multivariate normal distributions with the special structure of the covariance matrix.

Table 2. The results of testing the multivariate normality of classes.

	Statistic	P-value
Class 1	0.999997	0.379991
Class 2	1.00041	0
Class 3	1.00006	0

5   Recommendations and Conclusions

A modified MAP classifier for images (matrices) having matrix normal distribution was derived and extensively tested both on synthetic MND data as well on down-sampled images of gas burner flames. Conclusions and recommendations from testing MCL are the following.

- When images (matrices) come from classes having matrix normal distributions, then the MCL essentially outperforms the SVM and the NN classifiers and can safely be recommended. The Baringhaus-Henze test can be used in order to test MND for classes.
- When for the observations the Baringhaus-Henze test fails, then the classification errors of the MCL and the 10-NN are comparable and it can be more reliable to apply the xx-NN or another nonparametric classifier (see [2]).

It can be of interest to develop nonparametric classifiers dedicated to classifying images (matrices).

References

1. Baringhaus, L., Henze, N.: A consistent test for multivariate normality based on the empirical characteristic function. Metrika **35**(1), 339–348 (1988)
2. Devroye, L., Gyorfi, L., Lugosi, G.: A Probabilistic Theory of Pattern Recognition. Springer, Berlin (2013)
3. Haykin, S.S., et al.: Neural Networks and Learning Machines. Pearson, Upper Saddle River (2009)
4. Krzysko, M., Skorzybut, M., Wolynski ,W.: Classifiers for doubly multivariate data. Discussiones Mathematicae: Probability & Statistics, p. 31 (2011)
5. Li, S.Z.: Markov Random Field Modeling in Image Analysis. Springer, Berlin (2009). https://doi.org/10.1007/978-1-84800-279-1
6. Manceur, A.M., Dutilleul, P.: Maximum likelihood estimation for the tensor normal distribution: algorithm, minimum sample size, and empirical bias and dispersion. J. Comput. Appl. Math. **239**, 37–49 (2013)
7. Ohlson, M., Ahmad, M.R., Von Rosen, D.: The multilinear normal distribution: Introduction and some basic properties. J. Multivar. Anal. **113**, 37–47 (2013)
8. Rafajłowicz, E.: Data structures for pattern and image recognition with application to quality control Acta Polytechnica Hungarica, Informatics (under review)
9. Rafajłowicz, E., Pawlak-Kruczek, H., Rafajłowicz, W.: Statistical classifier with ordered decisions as an image based controller with application to gas burners. In: Rutkowski, L., Korytkowski, M., Scherer, R., Tadeusiewicz, R., Zadeh, L.A., Zurada, J.M. (eds.) ICAISC 2014. LNCS (LNAI), vol. 8467, pp. 586–597. Springer, Cham (2014). https://doi.org/10.1007/978-3-319-07173-2_50
10. Rafajłowicz, E., Rafajłowicz, W.: Image-driven decision making with application to control gas burners. In: Saeed, K., Homenda, W., Chaki, R. (eds.) CISIM 2017. LNCS, vol. 10244, pp. 436–446. Springer, Cham (2017). https://doi.org/10.1007/978-3-319-59105-6_37
11. Skubalska-Rafajłowicz, E.: Sparse random projections of camera images for monitoring of a combustion process in a gas burner. In: Saeed, K., Homenda, W., Chaki, R. (eds.) CISIM 2017. LNCS, vol. 10244, pp. 447–456. Springer, Cham (2017). https://doi.org/10.1007/978-3-319-59105-6_38
12. Werner, K., Jansson, M., Stoica, P.: On estimation of covariance matrices with Kronecker product structure. IEEE Trans. Sig. Process. **56**(2), 478–491 (2008)
13. Wójcik, W., Kotyra, A.: Combustion diagnosis by image processing. Photonics Lett. Pol. **1**(1), 40–42 (2009)

Random Projection for k-means Clustering

Sami Sieranoja and Pasi Fränti[(✉)]

School of Computing, University of Eastern Finland, Joensuu, Finland
{sami.sieranoja,pasi.franti}@uef.fi

Abstract. We study how much the k-means can be improved if initialized by random projections. The first variant takes two random data points and projects the points to the axis defined by these two points. The second one uses furthest point heuristic for the second point. When repeated 100 times, cluster level errors of a single run of k-means reduces from CI = 4.5 to 0.8, on average. We also propose simple projective indicator that predicts when the projection-heuristic is expected to work well.

Keywords: Clustering · Random projection · K-means

1   Introduction

K-means groups N data points into k clusters by minimizing the sum of squared distances between the data points and their nearest cluster centers (*centroid*). It takes any initial solution as an input and then improves it iteratively. A random set of points is the most common choice for the initialization. K-means is very popular because of its simple implementation. It has also been used as a part of other clustering algorithms such as genetic algorithms [14, 25], random swap [16], spectral clustering [34] and density clustering [3].

The best property of k-means is its excellent fine-tuning capability. Given a rough location of initial cluster centers, it can optimize their locations locally. The main limitation is that k-means cannot optimize the locations globally. Problems appear especially when clusters are well separated, or when stable clusters block the movements of the centroids, see Fig. 1. A poor initialization can therefore lead to an inferior local minimum.

To compensate the problem, alternative initialization heuristics have been considered. These include *Maxmin* heuristic, *sorting* heuristic, *density* and *projection* heuristics. Comparative studies [8, 22, 27, 30] have found that no single technique would outperform the others in all cases. A clear state-of-the-art is missing.

Another way to improve k-means is to repeat it several times [11]. The idea is simply to restart k-means from different initial solutions, and then keeping the best result. This requires that the initialization technique include some randomness in the process to produce different initial solutions. We refer this as *repeated k-means* (RKM).

For the initialization, projection-based heuristics aim at finding one-dimensional projection that would allow the data to be divided to roughly equal size clusters [1, 31, 35]. Most obvious choices are *diagonal* and the *principal* axis. Diagonal axis works well

© Springer International Publishing AG, part of Springer Nature 2018
L. Rutkowski et al. (Eds.): ICAISC 2018, LNAI 10841, pp. 680–689, 2018.
https://doi.org/10.1007/978-3-319-91253-0_63

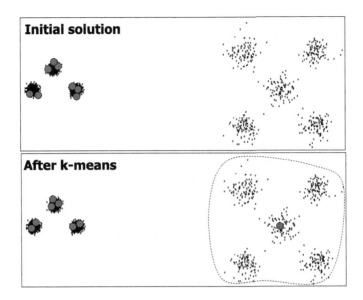

Fig. 1. Example of sub-optimal convergence of k-means.

with image data when the points are strongly correlated. Principal axis finds the direction with maximum variance. It has been widely used also in split-based algorithms that iteratively split one cluster at a time [4, 15, 23, 35].

In general, projection heuristic transforms the clustering to an easier segmentation problem, which can be optimally solved for a given objective function using dynamic programming [32]. However, it is unlikely that the data points would nicely fit along a linear axis in the space. Even a more complex non-linear principal curve was also considered in [9] but the final result still depended on the fine-tuning by k-means. Open questions are when the projection-based technique is expected to work, and how to choose the best projection axis.

In this paper, we study two simple variants to be used with repeated k-means: *random* and *furthest point* projections. The first takes two random data points without any sanity checks. The key idea is the randomness; single selection may provide poor initialization but when repeating several times, the chances to find one good initialization increases significantly.

The second heuristic is slightly more deterministic but still random. We start by selecting a random point, which will be the first reference point. We then calculate its furthest point and select it as the second reference point. The projection axis is the line passing through the two reference points. We again rely on randomness, but now the choices are expected to be more sensible, potentially providing better results using fewer trials.

The projection technique works when the data has one-dimensional structure, see Fig. 2. Otherwise, it may not provide additional benefit compared to the simple random initialization. We therefore introduce a simple *projective indicator* that predicts how well the data can be represented by a linear projection. This indicator can also be used to evaluate the goodness of the random projection before performing the actual clustering, which can speed-up the process.

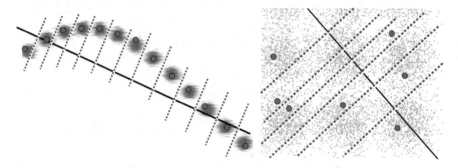

Fig. 2. The principle of the projection heuristic: when it works (left) and when not (right).

2   Projection-Based Initialization

The goal is to find a simple one-dimensional projection that can be used to create trivial equal size clusters along the axis, and yet, provide meaningful initial clustering. We merely seek to find a better initialization to k-means than the random heuristic, and generating it as simply as possible.

2.1   One-Dimensional Projections

The original idea was to project the data points on a single direction and then solving the initial partition along this direction [2]. Simplest way to do it is to sort the data according to the chosen axis, and then select every $(N/k)^{th}$ point. In [33], the points are sorted according to their distance to origin. If the attributes are non-negative, then this is essentially the same as projecting the data to the diagonal axis. Such projection is trivial to implement by calculating the average of the attribute values. The diagonal axis has also been used for speeding-up nearest neighbor searches in clustering [28].

In [1], the points are sorted according to the dimension with the largest variance. This adapts to the data slightly better than just using the diagonal. A more general approach would be to use *principal axis*, which is the direction along which the variance is maximal. It has also been used in divisive clustering algorithms [4, 15, 23, 35]. However, calculation of the principal axis takes $O(DN)$-$O(D^2N)$ depending on the variant [15]. To speed-up the process, only diagonal covariance matrix was computed in [31]. A more complex *principal curve* has also been used for clustering [9].

2.2   Random Projection in Higher Dimensions

Higher dimensional projections have also been considered. In [12], the data was projected in 2–dimensional subspace by selecting the highest variance attribute as the first dimension, and as second the one that has minimum absolute correlation with the first. According to [10], Gaussian clusters can be projected into $O(\log k)$ dimensions while still preserving the approximate level of separation between the clusters. It was shown that random projection can change the shape of the clusters to be more spherical. The following methodology was also suggested: perform random projection into a lower dimensional space, perform clustering, and use the result for obtaining clustering in the high dimensional space. This is exactly what we do here: perform one-dimensional projection followed by k-means in the full space.

Random projections have also been used in clustering ensemble [13] where individual clustering result is aggregated into a combined similarity matrix. Final clustering is then obtained by agglomerative clustering using the similarity matrix instead of the original data. This was shown to provide better and more robust clustering than PCA + EM (k-means variant for Gaussian clusters).

The diversity of the individual solution was also found to be an important element of the algorithm in [13]. We had similar observation in [18], where having some level of randomness in the algorithm was essential in swap-based clustering. Both these results suggest that random projections might be suitable for k-means initialization better than some fixed deterministic projection.

In [7], random projections were motivated by the need for speed in clustering of data streams. Quasilinear time complexity, $O(N \cdot \log N)$, was set as a requirement. The speed was also the main motivation in [5] where the problem was considered as feature extraction, which aims at selecting such (linear) combination of the original features that the clustering quality would not be compromised.

In [6], k-means is applied several times, each time increasing the dimensionality after the convergence of k-means. In other words, least accurate version of the data is clustered first, and at each of the following steps, the accuracy is gradually increased. This approach is analogous to cooling in simulated annealing.

2.3   Random Projection

Although the above-mentioned solutions are also based on random projections, they do it in higher-dimensions. Their main goal was not to loose the clustering accuracy. However, we merely seek a simple initialization to k-means. We perform the random projection in one-dimensional space for two particular reasons:

- Simplicity
- Speed-up benefits

The random projection works as follows:

1. Select random data point r_1
2. Select random data point r_2

Unlike higher dimensional projections, one-dimensional projection induces unique sorting of the data. This could also be used to speed-up the nearest centroid searches without any further loss in the accuracy [28]. Significant reduction of processing time was reported when used jointly with the *centroid activity detection* [24]: speed-up factor of 27:1 (Bridge dataset) and 62:1 (Miss America dataset); both 16-dimensional image data. Although speed-up is not our goal, it is worth to taken into account.

2.4 Furthest Point Projection

Our second heuristic is based on the classical furthest point heuristic where the next centroid is always selected as the one that is furthest to all other centroids [21]. However, instead of selecting every centroid as the next furthest, we only need to find two reference points. To involve some level of randomness in the process, we select the first reference point randomly:

1. Select random data point r_1
2. Find the furthest point r_2 of r_1

The projection axis is then defined by the reference points r_1 and r_2. The process is illustrated in Fig. 3. Since the first point is chosen randomly, the heuristic may create different solution in different runs, which is the purpose.

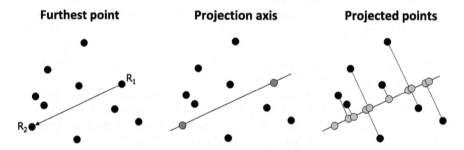

Fig. 3. Sample data and the projections.

2.5 Repeated K-means

Both heuristics are used with the k-means algorithm as follows. We repeat the algorithm $R = 100$ times, and select the best final result. The pseudo code of this overall algorithm is shown in Fig. 4. For a single projection axis, we predict its goodness as follows:

$$\Delta R = \sum_{a,b} (dist(a,b))/N^2, \forall a, b \tag{1}$$

$$\Delta L = \sum_{x} dist(x, p_x)/N \tag{2}$$

$$Projective = \frac{\Delta R}{\Delta L} \tag{3}$$

```
Repeated K-means(x):
f_best = ∞;
FOR r=1 to 100 DO
        y ← Projection(x);
        y ← Sort(y);
        FOR i=1 to k DO
            mid = (int) i*(N/k) + 0.5;
            c[i] = y[mid];

        f ← Evaluate(x,c);
        IF f < f_best THEN   f ← f_best

RandomProjection(x)
r1 ← RandomNumber(1,N);
r2 ← RandomNumber(1,N);
FOR i=1 to N DO
        p[i] ← Project(x[i], r1, r2);

FurthestPointProjection(x):
r1 ← Random(1,N);
r2 ← FindFurthest(X, x[r0]);
FOR i=1 to N DO
        p[i] ← Project(x[i], r1, r2);
```

Fig. 4. Pseudo code of the considered projection-based techniques.

The divider ΔR is the average pairwise distance of all data points in the original space, and the nominator ΔL is the average distance of the points (x) to the projection axis (p_x).

3   Experiments

We compare the two projection-based heuristics against the two most common heuristics: *random* initialization, *Maxmin* heuristic [21]. We use the datasets shown in Table 1. These include several artificial datasets, and real image datasets.

In case of artificial data with known ground truth, the success of the clustering is evaluated by *Centroid Index* (CI) [17], which counts how many cluster centroids are wrongly located. Value CI = 0 indicates correct clustering. Reference results are also given for single run of k-means, and when k-means is re-started 100 times (repeated). In case of image data, we calculate sum of squared errors (Fig. 5).

Table 1. Datasets used in the experiments.

Data set	Ref.	Type of data	Points (N)	Clusters (k)	Dimension (d)
Bridge	[16]	Image blocks	4096	256	16
House	[16]	*RGB* image	34112	256	3
Miss America	[16]	Residual blocks	6480	256	16
Europe		Differential values	169673	256	2
Birch, Bich2	[36]	Artificial	100000	100	2
$S_1 - S_4$	[19]	Artificial	5000	15	2
$A_1 - A_3$	[26]	Artificial	3000-7500	20-50	2
Unbalance	[29]	Artificial	6500	8	2
DIM-32	[20]	Artificial	1024	16	32

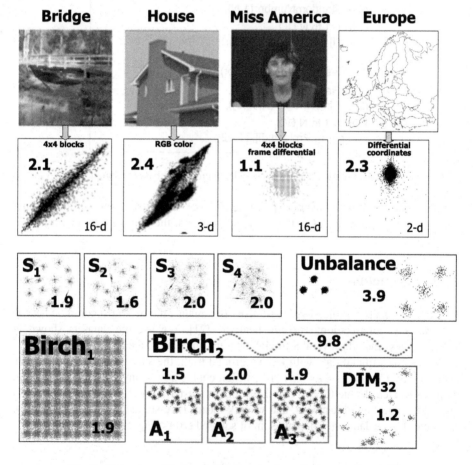

Fig. 5. Datasets and their calculated projective values according to (3).

The results are summarized in Table 2 where the correct clustering results are emphasized by green color. As expected, single run of k-means rarely finds the correct clustering regardless of the initialization. On average, CI = 4.5 centroids are wrongly located with random heuristic, and about CI = 2 with the other heuristics.

Table 2. Performance comparison of the various k-means initialization heuristics: random, furthest point, and the two projection-based: random (Proj-Rand) and the furthest point (Proj-FP).

Artificial data (Single run):

Method	S1	S2	S3	S4	A1	A2	A3	Unb	B1	B2	D32	Av.
Random	1.9	1.4	1.3	0.9	2.5	4.6	6.6	3.9	6.7	16.6	3.6	4.5
Furthest point	0.7	1.0	0.7	1.0	1.0	2.6	2.9	0.9	5.5	7.3	0.0	2.1
Proj-Rand	1.2	0.9	0.8	0.6	1.2	3.3	5.2	4.0	5.3	0.2	0.9	2.2
Proj-FP	1.2	0.9	0.8	0.6	1.1	3.3	5.0	4.0	5.4	0.0	1.0	2.1

Artificial data (100 repeats):

Method	S1	S2	S3	S4	A1	A2	A3	Unb	B1	B2	D32	Av.
Random	0.0	0.0	0.0	0.0	0.3	1.8	2.9	2.9	2.8	11	1.1	2.0
Furthest point	0.0	0.0	0.0	0.0	0.0	0.5	0.6	0.0	2.8	3.9	0.0	0.7
Proj-Rand	0.0	0.0	0.0	0.0	0.0	0.9	2.0	3.9	1.9	0.0	0.0	0.8
Proj-FP	0.0	0.0	0.0	0.0	0.0	0.8	1.9	4.0	1.8	0.0	0.0	0.8

Image data (100 repeats):

Method	Bridge	House	Miss America	Europe
Random	176.71	6.43	5.83	3.66
Furthest point	172.09	6.93	5.68	3.56
Proj-Rand	176.66	6.44	5.83	3.65
Proj-FP	176.83	6.46	5.83	3.65

However, when the k-means is repeated 100 times, the results are significantly better. The projection-based heuristics solve now 7 of the 11 artificial datasets, having CI = 0.8 centroids incorrectly located, on average. We also observe that the different heuristics work better with different datasets. In specific, Maxmin heuristic works best when the clusters have different densities and the projection-based better with data having high projective values (Birch2).

None of the image sets have high projective values, and therefore, the projection heuristic works basically as the random heuristic. There is also no significant difference between the two projection-techniques. The furthest point projection works slightly better but when repeated, the minor difference disappear. Figure 6 demonstrates what kinds of projections are generated with the A2 dataset.

Random projection   **Furthest points projection**

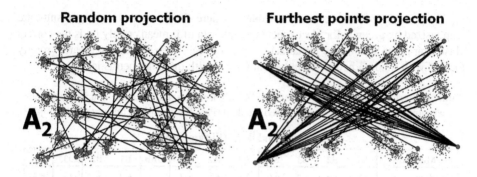

Fig. 6. Examples of the random projections axes (left), and the furthest point projection axes (right).

4   Conclusions

Both the random projection and furthest points heuristics outperform random initialization, and provide competitive performance to the Maxmin heuristic. The performance of the heuristics can be predicted by the projective indicator, which helps to recognize when the data is suitable to be projected. This could also be useful in split-based algorithms where the data is hierarchically split into smaller clusters.

References

1. Al-Daoud, M.B., Roberts, S.A.: New methods for the initialisation of clusters. Pattern Recogn. Lett. **17**(5), 451–455 (1996)
2. Anderberg, M.R.: Cluster Analysis for Applications. Academic Press, NewYork (1973)
3. Bai, L., Cheng, X., Liang, J., Shen, H., Guo, Y.: Fast density clustering strategies based on the k-means algorithm. Pattern Recogn. **71**, 375–386 (2017)
4. Boley, D.: Principal direction divisive partitioning. Data Min. Knowl. Disc. **2**(4), 325–344 (1998)
5. Boutsidis, C., Zouzias, A., Mahoney, M.W., Drineas, P.: Randomized dimensionality reduction for k-means clustering. IEEE Trans. Inf. Theory **61**(2), 1045–1062 (2015)
6. Cardoso, A., Wichert, A.: Iterative random projections for high-dimensional data clustering. Pattern Recogn. Lett. **33**, 1749–1755 (2012)
7. Carraher, L.A., Wilsey, P.A., Moitra, A., Dey, S.: Random projection clustering on streaming data. In: IEEE International Conference on Data Mining Workshops, pp. 708–715 (2016)
8. Celebi, M.E., Kingravi, H.A., Vela, P.A.: A comparative study of efficient initialization methods for the k-means clustering algorithm. Expert Syst. Appl. **40**, 200–210 (2013)
9. Cleju, I., Fränti, P., Wu, X.: Clustering based on principal curve. In: Kalviainen, H., Parkkinen, J., Kaarna, A. (eds.) SCIA 2005. LNCS, vol. 3540, pp. 872–881. Springer, Heidelberg (2005). https://doi.org/10.1007/11499145_88
10. Dasgupta, S.: Experiments with random projection. In: Uncertainty in Artificial Intelligence, pp. 143–151 (2000)
11. Duda, R.O., Hart, P.E.: Pattern Classification and Scene Analysis. Wiley, New York (1973)
12. Erisoglu, M., Calis, N., Sakallioglu, S.: A new algorithm for initial cluster centers in k-means algorithm. Pattern Recogn. Lett. **32**(14), 1701–1705 (2011)

13. Fern, X.Z., Brodley, C.E.: Random projection for high dimensional data clustering: a cluster ensemble approach. In: International Conference on Machine Learning (ICMC), Washington, DC (2003)
14. Fränti, P.: Genetic algorithm with deterministic crossover for vector quantization. Pattern Recogn. Lett. **21**(1), 61–68 (2000)
15. Fränti, P., Kaukoranta, T., Nevalainen, O.: On the splitting method for VQ codebook generation. Opt. Eng. **36**(11), 3043–3051 (1997)
16. Fränti, P.: Efficiency of random swap clustering. J. Big Data **5**(13), 1–29 (2018)
17. Fränti, P., Rezaei, M., Zhao, Q.: Centroid index: cluster level similarity measure. Pattern Recogn. **47**(9), 3034–3045 (2014)
18. Fränti, P., Tuononen, M., Virmajoki, O.: Deterministic and randomized local search algorithms for clustering. In: IEEE International Conference on Multimedia and Expo, Hannover, Germany, pp. 837–840, June 2008
19. Fränti, P., Virmajoki, O.: Iterative shrinking method for clustering problems. Pattern Recogn. **39**(5), 761–765 (2006)
20. Fränti, P., Virmajoki, O., Hautamäki, V.: Fast agglomerative clustering using a k-nearest neighbor graph. IEEE Trans. Pattern Anal. Mach. Intell. **28**(11), 1875–1881 (2006)
21. González, R., Tou, J.: Pattern Recognition Principles. Addison-Wesley, Boston (1974)
22. He, J., Lan, M., Tan, C.-L., Sung, S.-Y., Low, H.-B.: Initialization of cluster refinement algorithms: a review and comparative study. In: IEEE International Joint Conference on Neural Networks (2004)
23. Huang, C.-M., Harris, R.W.: A comparison of several vector quantization codebook generation approaches. IEEE Trans. Image Process. **2**(1), 108–112 (1993)
24. Kaukoranta, T., Fränti, P., Nevalainen, O.: A fast exact GLA based on code vector activity detection. IEEE Trans. Image Process. **9**(8), 1337–1342 (2000)
25. Krishna, K., Murty, M.N.: Genetic k-means algorithm. IEEE Trans. Syst. Man Cybern. Part B **29**(3), 433–439 (1999)
26. Kärkkäinen, I., Fränti, P.: Dynamic local search algorithm for the clustering problem. Research Report A-2002-6 (2002)
27. Peña, J.M., Lozano, J.A., Larrañaga, P.: An empirical comparison of four initialization methods for the k-means algorithm. Pattern Recogn. Lett. **20**(10), 1027–1040 (1999)
28. Ra, S.-W., Kim, J.-K.: A fast mean-distance-ordered partial codebook search algorithm for image vector quantization. IEEE Trans. Circ. Syst. **40**, 576–579 (1993)
29. Rezaei, M., Fränti, P.: Set-matching methods for external cluster validity. IEEE Trans. Knowl. Data Eng. **28**(8), 2173–2186 (2016)
30. Steinley, D., Brusco, M.J.: Initializing k-means batch clustering: a critical evaluation of several techniques. J. Classif. **24**, 99–121 (2007)
31. Su, T., Dy, J.G.: In search of deterministic methods for initializing k-means and Gaussian mixture clustering. Intell. Data Anal. **11**(4), 319–338 (2007)
32. Wu, X.: Optimal quantization by matrix searching. J. Algorithms **12**(4), 663–673 (1991)
33. Wu, X., Zhang, K.: A better tree-structured vector quantizer. In: IEEE Data Compression Conference, Snowbird, UT, pp. 392–401 (1991)
34. Yan, D., Huang, L., Jordan, M.I.: Fast approximate spectral clustering. In: ACM SIGKDD International Conference on Knowledge Discovery and Data Mining, 907–916, 2009
35. Yedla, M., Pathakota, S.R., Srinivasa, T.M.: Enhancing k-means clustering algorithm with improved initial center. Int. J. Comput. Sci. Inf. Technol. **1**(2), 121–125 (2010)
36. Zhang, T., Ramakrishnan, R., Livny, M.: BIRCH: a new data clustering algorithm and its applications. Data Min. Knowl. Disc. **1**(2), 141–182 (1997)

Modified Relational Mountain Clustering Method

Kristina P. Sinaga, June-Nan Hsieh, Josephine B. M. Benjamin,
and Miin-Shen Yang[⊠]

Department of Applied Mathematics, Chung Yuan Christian University,
Chung-Li 32023, Taiwan
msyang@math.cycu.edu.tw

Abstract. The relational mountain clustering method (RMCM) is a simple and effective algorithm that can be used to obtain cluster centers and partitions for a relational data set. However, the performance of RMCM heavily depends on the choice of parameters of relational mountain function. In order to solve this problem, we propose a modified RMCM (M-RMCM) by using the correlation self-comparison method to estimate the parameters of the modified relational mountain function, and then applied a validity index to estimate the number of clusters. The proposed M-RMCM can provide good cluster centers, partitions and the number of clusters for most relational data sets in which the results will not be sensitive to parameters. The simulations and comparisons show the superiority and effectiveness of the proposed M-RMCM.

Keywords: Clustering algorithms · Mountain method · Relational data
Relational mountain method

1 Introduction

Cluster analysis is a tool for clustering data with similar characteristics into groups that can discover the structure of the data set. In most clustering algorithms, such as k-means [7] and fuzzy c-means (FCM) [1, 16], they need to have some prior information that includes initial guesses of centers, partitions and the number of clusters. In general, clustering performances for these clustering algorithms are always affected by these prior assumptions. It is known that the mountain method, a simple and effective algorithm proposed by Yager and Filev [12], can be used to obtain prior information for most clustering algorithms.

However, the mountain method [12] is computed in the amount of computations growing exponentially with the increase in the dimensionality of the data. Chiu [2] proposed the subtractive clustering method by considering the mountain function on the data points instead of the grid nodes to reduce the computational time. Pal and Chakraborty [8] proposed a scheme to improve the accuracy of the prototypes obtained by Yager's mountain method and Chiu's modified type. Yager and Filev [13] applied the mountain method to the generation of fuzzy rules and Velthuizen et al. [10] applied it for clustering large data sets, such as the segmentation of magnetic resonance images of the human brain. Although these methods can provide the prior information for most

© Springer International Publishing AG, part of Springer Nature 2018
L. Rutkowski et al. (Eds.): ICAISC 2018, LNAI 10841, pp. 690–701, 2018.
https://doi.org/10.1007/978-3-319-91253-0_64

clustering algorithms, their performance heavily depends on parameters of mountain function and stopping condition. To resolve these problems, Yang and Wu [15] proposed the modified mountain clustering method that uses a correlation comparison technique to estimate the parameters and also proposed a validity index to explore the cluster numbers. Wu et al. [11] used the concept of data transformation to apply the modified mountain method for clustering switching regression data set.

Relational data are different from feature data. Each feature data point in a data set represents the individual and unique role. For an example with 2-dimension feature data, scatter plot can be used to compare clustering results for their performances. However, for a relational data set, we only have information about the relation between two objects so that relational data present only similarity or dissimilarity (distance) measures. Some methods were derived from FCM for relational data, such as relational fuzzy c-means (RFCM) [5], non-Euclidean RFCM [4] and robust FRC [3]. These algorithms still require good prior information for clustering. Pal et al. [9] first proposed the relational mountain clustering method (RMCM) to deal with relational data. Although the RMCM can obtain prior information for relational data sets, the performance still heavily depends on the choice of parameters.

In this paper, we will propose a modified RMCM (M-RMCM) to solve these parameter selection problems in RMCM. The remainder of this paper is organized as follows. In Sect. 2, we briefly review the mountain and modified mountain methods. In Sect. 3, we first introduce the RMCM and then propose the M-RMCM. In M-RMCM for relational data, we adopt the correlation comparison technique to estimate the parameters and then use a validity index to explore the number of clusters. We then provide some examples to show the performance of the M-RMCM in Sect. 4. We finally make conclusions in Sect. 5.

2 Mountain Clustering Methods

2.1 Mountain Method

The mountain method was first proposed by Yager and Filev [12] as an approximate clustering method which can provide prior information for clustering algorithms and also obtain cluster centers and partitions. Let $X = \{x_1, x_2, \cdots, x_n\}$ be a set of n data points in a q-dimensional Euclidean space $\Re^q$. We make grids to the data space. Let $I = I_1 \times \cdots \times I_q$ be a q-dimensional hypercube where the interval I_p, $p = 1, \cdots, q$ is set to be the range of the pth coordinate of the data set, and so $X \subset I$. Each interval I_p is subdivided into r_p equidistant points. This discretization forms a grid on the hypercube, which has $N = \prod_{p=1}^{q} r_p$ grid points. Let $\{N_i\}$ denote the set of all grid points, $i = 1, \cdots, N$. The grid points are candidate cluster centers for the given data set in the mountain method. Let $d(x_j, N_i)$ denote the distance from x_j to the grid point N_i and the mountain function for each N_i is calculated with

$$M_1(N_i) = \sum_{j=1}^{n} e^{-\alpha d(x_j, N_i)}, \quad i = 1, \cdots, N \tag{1}$$

The mountain function of the grid points can be seen as a measure of the density of the data points in the neighborhood. A grid point surrounded by many data points will get a large mountain function value and whose neighborhood is sparsely populated will get a small mountain function value.

The parameter α in the mountain function (1) decides the neighborhood radius and the data points outside the radius will have a small influence on the mountain function. Once the mountain function for each grid point is calculated, the first cluster center estimate is the grid point with a maximal mountain function value. Thus, N_1^* is selected as the first cluster center estimate among all grid points with

$$M_1(N_1^*) = \max_i\{M_1(N_i)\}.$$

To find the rest cluster center N_k^*, Yager and Filev [12] defined the revised mountain function as $M_k(N_i) = M_{k-1}(N_i) - M_{k-1}(N_{k-1}^*) \cdot e^{-\gamma d(N_{k-1}^*, N_i)}$, $k = 2, 3, \cdots$ so that the kth cluster center N_k^* is found by $M_k(N_k^*) = \max_i\{M_k(N_i)\}, k = 2, 3, \cdots$. The parameter γ determines the neighborhood radius that will have measurable reductions in the mountain function. Each updating of the revised mountain function $M_k(N_i)$ leads to one new cluster center. This process will get c cluster centers if the stopping condition satisfies $M_{c+1}(N_{c+1}^*) < \delta$.

2.2 Modified Mountain Clustering Algorithm

Since the mountain method is computed in the amount of computations that will be growing exponentially with the increase in the dimensionality of the data, Chiu [2] proposed the subtractive clustering method by considering the mountain function on the data points instead of the grid nodes to decreasing the amount of computations. However, Chiu's [2] subtractive method needs to have appropriate values of parameters to be selected, and obtaining these parameters that work well across various data sets is difficult. Therefore, Yang and Wu [15] proposed a new way for improving the mountain method, called modified mountain method, by measuring the cluster validity as a stopping condition.

Yang and Wu [15] proposed a method improving the dependent of the mountain method with respect to the parameters and then establishing the index of choosing a cluster number. They defined the modified mountain function for each data point as

$$M_1(x_i) = \sum_{j=1}^{n} e^{-m\beta d^2(x_j, x_i)}, \quad i = 1, \cdots, n$$

$$\beta = \left(\sum_{j=1}^{n} \|x_j - \bar{x}\|^2 / n\right)^{-1} \quad \text{with} \quad \bar{x} = \sum_{j=1}^{n} x_j / n$$

(2)

In fact, the parameter $m\beta$ in the modified mountain function (2) and the parameter α in the mountain function (1) are both used to determine the approximate density shape of the data set. Since the parameter β in (2) is set to be a normalization term of the dissimilarity measure $d(x_j, x_i)$, the m in (2) is the only parameter need to be estimated.

Difference to the method proposed by Chiu [2], Yang and Wu [15] proposed the correlation self-comparison (CSC) procedure to get the value of the parameter m. We will also apply this technique in our new proposed method which will be illustrated in next section.

After the correlation self-comparison algorithm is implemented, the parameter m in the modified mountain function (2) is acquired. Since negative revised mountain function values may occur, Yang and Wu [15] also modified the revised mountain function to solve this unreasonable result and the modified revised mountain function after extracting kth cluster center is

$$M_k(x_i) = M_{k-1}(x_i) - M_{k-1}(x_i) \cdot e^{-\beta d^2\left(x_{k-1}^*, x_i\right)}, \quad k = 2, 3, \cdots \tag{3}$$

where x_i is the feature vector and x_k^* is the kth cluster center which satisfies

$$M_k\left(x_k^*\right) = \max_i\{M_k(x_i)\}, \quad k = 2, 3, \cdots. \tag{4}$$

We can obtain newly identified cluster centers by the iterations of Eqs. (3) and (4). Since the stopping condition in both mountain method and subtractive clustering method are sensitive to the parameter δ, Yang and Wu [15] proposed a validity index to replace the stopping rule. A suitable cluster number estimate will optimize the validity index which will also be used in relational data environment in next section.

3   Modified Relational Mountain Clustering Method

Clustering methods are tools for clustering an object data set $X = \{x_1, \cdots, x_n\}$ as a partition of X into c subgroups. Each subgroup represents a homogeneous subset in X and data points in each cluster are more similar than in other clusters. In general, each object can be represented by a set of q measurements, that is, x_i can be represented by $x_i \in \Re^q$. For a relational data set, we only have the values of (dis)similarity between objects x_i and x_j, denoted by R_{ij} with $0 \leq R_{ij} \leq 1$. The data set X will now reforms a relational matrix $R = \left[R_{ij}\right]_{n \times n}$ and R_{ij} represents the relation strength of the objects x_i and x_j. Sometimes we may not have the q measurements of an object data set $X = \{x_1, \cdots, x_n\}$, but only have its relational data R. Many methods are derived from FCM to cluster the relational data, such as the relational fuzzy c-means (RFCM) [5], non-Euclidean RFCM [4], and robust FRC [3]. These algorithms still require good prior information for clustering. Pal et al. [9] first proposed the relational mountain clustering method (RMCM) algorithm to deal with the relational data. Similar to the original mountain method [12], the performance of RMCM still depends on the choice of parameters. We first give a brief review of the RMCM and then propose the modified RMCM (M-RMCM) to solve the problems of parameter selections in RMCM.

3.1   Relational Mountain Clustering Method

For relational data, Pal et al. [9] modified Chiu's [2] mountain method and then created the relational mountain clustering method (RMCM). To make the mountain method work for relational data, Pal et al. [9] used the following two assumptions:

(i) $R = [R_{ij}]_{n \times n}$ is a dissimilarity relation, and
(ii) $0 \le R_{ij} = R(x_i, x_j) \le 1$.

The relational mountain function at each object is then defined by

$$RM_1(x_i) = \sum_{j=1}^{n} e^{-R^2(x_i, x_j)/\sigma_a^2}, \quad i = 1, \cdots, n. \tag{5}$$

Thus, the object x_1^* is selected as the first cluster center estimate among all objects with

$$RM_1(x_1^*) = \max_i \{RM_1(x_i)\}.$$

and the revised mountain function is defined as

$$RM_k(x_i) = RM_{k-1}(x_i) - RM_{k-1}(x_{k-1}^*) \cdot e^{-R^2(x_i, x_{k-1}^*/\sigma_b^2)}, \quad k = 2, 3, \cdots \tag{6}$$

and x_k^* is the kth cluster exemplars (cluster center estimators) if $RM_k(x_k^*) = \max_i \{RM_k(x_i)\}$. The process will get c cluster exemplars if the stopping condition $RM_{c+1}(x_{c+1}^*)/RM_1(x_1^*) < \delta$ is satisfied where δ is a small positive quantity. The parameters σ_a, σ_b in Eqs. (5) and (6) are similar to the parameters α, γ in the mountain method. The results still depend on these parameters.

3.2   The Proposed Modified Relational Mountain Clustering Method

Following the modification technique proposed by Yang and Wu [15], we define the modified relational mountain function for each object x_i as

$$MRM_1(x_i) = \sum_{j=1}^{n} e^{-m\beta R^2(x_i, x_j)}, \quad i = 1, \cdots, n \tag{7}$$

The role of the parameter β is a normalization term which normalizes the dissimilarity measure R_{ij} and defined by

$$\beta = \left(\sum_{i<j} (R_{ij} - \bar{R})^2 / C_2^n \right)^{-1} \quad \text{and} \quad \bar{R} = \sum_{i<j} R_{ij} / C_2^n. \tag{8}$$

The parameter $m\beta$ in (7) and σ_a in (5) are both to determine an approximate density shape for the relational data set. Since the parameter β is estimated by (8), and so m is the only parameter that needs to be estimated. We apply the correlation self-comparison technique to estimate m. We now rewrite the modified relational mountain function as $MRM_1^p(x_i) = \sum_{j=1}^n e^{-p\beta R^2(x_i, x_j)}$ with $p = 1, 2, \cdots$. The correlation self-comparison procedure for M-RMCM can be summarized as follows.

Correlation Self-comparison Algorithm

1. Set $p = 1$ and $w = 0 \cdot 99$.
2. Calculate the correlation between $\{MRM_1^p(x_1), \cdots, MRM_1^p(x_n)\}$ and $\{MRM_1^{p+1}(x_1), \cdots, MRM_1^{p+1}(x_n)\}$.
3. **IF** the correlation is greater than or equal to the specified number w,
   THEN choose $MRM_1^p(x_i)$ (i.e. $m = p$) to be the modified relational mountain function;
   ELSE $p = p + 1$ and **GOTO** step 2.

Since the dissimilarity relation measure R must satisfy the assumption of $0 \le R_{ij} \le 1$, the modified relational mountain function (7) is more sensitive to the parameter m than it in the modified mountain function (2). Hence, the increasing quantity of the correlation self-comparison procedure in M-RMCM is set to be 1 (i.e. $p = p + 1$) which is less than that in M-MCM. After the parameter m is estimated by the correlation self-comparison, the modified relational mountain function is acquired and the revised mountain function is then defined by

$$MRM_k(x_i) = MRM_{k-1}(x_i) - MRM_{k-1}(x_i) \cdot e^{-\beta R^2(x_i, x_{k-1}^*)}, \ k = 2, 3, \cdots \quad (9)$$

where x_k^* is the kth cluster exemplars (cluster center estimators) if $MRM_k(x_k^*) = \max_i \{MRM_k(x_i)\}$.

In RMCM, the c cluster exemplars are extracted if the stopping condition $RM_{c+1}(x_{c+1}^*)/RM_1(x_1^*) < \delta$ is satisfied. Although this condition can be used to estimate the cluster number, the results are sensitive to the parameter δ. If the parameter δ is too large, we may miss some important clusters and we may get too many clusters when the parameter δ is too small. In order to get a suitable cluster number estimate, we propose a validity measure for the proposed relational mountain method by modifying the concept of Yang and Wu [15]. The validity function is defined as

$$MV(c) = \sum_{k=2}^{c} pot(k), \ c = 2, 3, \cdots, n - 1 \quad (10)$$

where c denotes the number of clusters. The function $pot(k)$ measures the potential of kth extracted exemplar x_k^* being a suitable cluster exemplar and is defined as

$$pot(k) = MRM_1(x_k^*) \frac{MRM_1(x_k^*)}{MRM_1(x_1^*)} - n \cdot \exp(-m\beta r_k^2), \ k = 2, 3, \cdots \quad (11)$$

where $r_k = \min\{R(x_k^*, x_{k-1}^*), R(x_k^*, x_{k-2}^*), \cdots, R(x_k^*, x_1^*)\}$ is the minimum relation measure among x_k^* and all $(k-1)$ previous extracted cluster exemplars. For the kth extracted cluster, we use the term $MRM_1\left(x_k^*\right)/MRM_1\left(x_1^*\right)$ to measure the compactness and βd_k^2 to measure the separation. We will find the optimal cluster number c^* if $MV(c^*) = \max_c\{MV(c)\}$.

4   Numerical Examples and Comparisons

In this section, we present several examples to compare the propose M-RMCM to RMCM. The first two data sets (Norm-mix and IRIS) had been discussed by Pal et al. [9] and we will show that our method can easily solve the parameter selection problems in the two known-cluster-number data sets. The last two examples will demonstrate the ability of the M-RMCM in detecting suitable cluster number estimate.

Example 1. The data set Norm-mix, proposed by Pal et al. [9], consists of 800 points (200 points from each of $c = 4$ classes) in $\Re^2$ that are randomly drawn from a mixture bivariate normal distributions, as shown in Fig. 1. Each distribution has the same covariance matrix $\sum_i = \begin{Bmatrix} 0.5 & 0.0 \\ 0.0 & 0.5 \end{Bmatrix}$. Population means of four components are (3, 3), (8, 8), (13, 13) and (18, 18). We then use this data set to construct a relational data set. The result of RMCM for this relational data have been shown in Pal et al. [9] which is also summarize in Table 1. The columns 2 and 3 of Table 1 list the four exemplars extracted by the M-RMCM and RMCM algorithms. In order to compare the accuracy, we show mean squared error (MSE) between the extracted exemplars and true centers. We find that the proposed M-RMCM gives more accurate results than the RMCM.

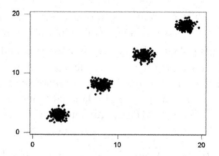

Fig. 1. Scatter plot of Norm-mix data.

Table 1. Exemplars extracted by RMCM and M-RMCM for Norm-mix data

Cluster no.	M-RMCM	RMCM $(\sigma_a = 0.25, \sigma_b = 0.3)$	True centers
1	(7.99, 8.07)	(8.00, 7.95)	(8.0, 8.0)
2	(18.26, 18.24)	(17.23, 16.98)	(18.0, 18.0)
3	(3.02, 3.05)	(2.36, 2.24)	(3.0, 3.0)
4	(13.03, 13.04)	(12.76, 13.18)	(13.0, 13.0)
MSE	0.1321	0.655396	

Example 2. We compare the RMCM and M-RMCM methods for the IRIS data set in this example. The IRIS data set has n = 150 points in a four-dimensional space with sepal length, sepal width, petal length, petal width from each of three species, Iris Setosa, Iris Versicolor, and Iris Virginica. The cluster 1 in the IRIS data is well separated from those of clusters 2 and 3. Similar to Example 1, we use the raw data to construct a relational data set for implementing M-RMCM and RMCM. The classification errors are shown in the second row of Table 2 and the error percentages are in the third row. Note that, according to studies in the literature, the error counts obtained by k-means or fuzzy c-means (FCM) for this data set are approximately 16. In Pal et al. [9], eight combinations of σ_a and σ_b are selected for RMCM. The best result is $Error = 11$ with $\sigma_a = 0.075$, $\sigma_b = 0.075$ and the worst result is $Error = 72$ with $\sigma_a = 0.2$, $\sigma_b = 0.4$. The M-RMCM produces an error of 16 and the error percentage is 10.67%. We find that parameter selection plays an important role in RMCM. For better values of σ_a and σ_b, the exemplars extracted by the RMCM are good, but for worse values of σ_a and σ_b, it produces very worse results. The FCM algorithm and the M-RMCM both give the same results in this data set. Although the results of the RMCM (with $\sigma_a = 0.075$, $\sigma_b = 0.075$) is better than the results of FCM and M-RMCM, the FCM and the M-RMCM are more stable than the RMCM. The RMCM actually depends on parameter selections, and the FCM depends on initial value assignments. In this example, the M-RMCM shows the superiority in free of parameter selection and initialization.

Table 2. Classification errors for IRIS data using M-RMCM, RMCM, and FCM

M-RMCM	RMCM $(\sigma_a = 0.2, \sigma_b = 0.4)$	RMCM $(\sigma_a = 0.075, \sigma_b = 0.075)$	FCM $(m = 2.0)$
Error = 16	Error = 72	Error = 11	Error = 16
10.67%	48%	7.33%	10.67%

Example 3. The data set of this example is from Yang and Shih [14] which collected the portraits of 15 members from three family A, B and C. Numbers 1–15 are marked corresponding to the portraits as follows:

	Dad	Mom	Children
Family A	4	12	2, 7
Family B	8	11	1, 5, 9, 14
Family C	3	15	6, 10, 13

According to Yang and Shih [14], pairwise similarities are assigned as: very similar = 0.8, similar = 0.6, not so similar = 0.4, different = 0.2, quite different = 0. Since the relational measure R in Yang and Shih [14] is a similarity relation, we need to convert the similarity to be the dissimilarity and obtain the relational matrix R as shown in Table 3. Since the results of RMCM are influenced by the parameters, we try to estimate σ_a and σ_b using M-RMCM. Comparing Eqs. (5) and (6) to Eqs. (7) and (9), we can find that $\sigma_a = 1/\sqrt{m\beta}$ and $\sigma_b = 1/\sqrt{\beta}$. In this example, the parameters m and β in M-RMCM are 2 and 1.7736, respectively. According to above equations, we have parameters $\sigma_a = 0.53$ and $\sigma_b = 0.75$ for RMCM. On the other hand, we also choose another parameters $\sigma_a = 0.2$ and $\sigma_b = 0.3$ for the RMCM. These results of M-RMCM and RMCM are shown in Table 4. This example shows that the parameters estimated by M-RMCM are also suitable for RMCM. We know that the results of RMCM heavily depend on the parameters, but Pal et al. [9], the authors of RMCM, did not offer a method to estimate them. In this paper, we provide the proposed M-RMCM for estimating the parameters in RMCM.

Table 3. Relational matrix of 15 members comes from 3 families.

$$R = \begin{bmatrix} 0.0 & 0.8 & 1.0 & 0.6 & 0.2 & 0.6 & 0.8 & 0.2 & 0.2 & 0.8 & 0.2 & 0.8 & 0.6 & 0.2 & 1.0 \\ 0.8 & 0.0 & 0.8 & 0.4 & 0.8 & 0.8 & 0.2 & 0.2 & 0.6 & 1.0 & 1.0 & 0.2 & 0.8 & 0.8 & 1.0 \\ 1.0 & 0.8 & 0.0 & 1.0 & 1.0 & 0.4 & 0.8 & 1.0 & 0.8 & 0.4 & 1.0 & 1.0 & 0.4 & 1.0 & 1.0 \\ 0.6 & 0.4 & 1.0 & 0.0 & 1.0 & 0.6 & 0.2 & 1.0 & 1.0 & 0.8 & 1.0 & 0.8 & 1.0 & 1.0 & 1.0 \\ 0.2 & 0.8 & 1.0 & 1.0 & 0.0 & 0.8 & 0.6 & 0.4 & 0.4 & 0.8 & 0.4 & 0.8 & 0.8 & 0.2 & 0.8 \\ 0.6 & 0.8 & 0.4 & 0.6 & 0.8 & 0.0 & 0.8 & 0.6 & 0.8 & 0.2 & 0.8 & 0.8 & 0.2 & 0.6 & 0.6 \\ 0.8 & 0.2 & 0.8 & 0.2 & 0.6 & 0.8 & 0.0 & 0.6 & 0.8 & 1.0 & 0.6 & 0.2 & 0.8 & 0.8 & 0.6 \\ 0.2 & 0.2 & 1.0 & 1.0 & 0.4 & 0.6 & 0.6 & 0.0 & 0.6 & 1.0 & 1.0 & 1.0 & 1.0 & 0.4 & 1.0 \\ 0.2 & 0.6 & 0.8 & 1.0 & 0.4 & 0.8 & 0.8 & 0.6 & 0.0 & 0.6 & 0.2 & 1.0 & 0.8 & 0.2 & 0.8 \\ 0.8 & 1.0 & 0.4 & 0.8 & 0.8 & 0.2 & 1.0 & 1.0 & 0.6 & 0.0 & 0.8 & 1.0 & 0.4 & 0.6 & 0.2 \\ 0.2 & 1.0 & 1.0 & 1.0 & 0.4 & 0.8 & 0.6 & 1.0 & 0.2 & 0.8 & 0.0 & 1.0 & 1.0 & 0.2 & 1.0 \\ 0.8 & 0.2 & 1.0 & 0.8 & 0.8 & 0.8 & 0.2 & 1.0 & 1.0 & 1.0 & 1.0 & 0.0 & 1.0 & 1.0 & 1.0 \\ 0.6 & 0.8 & 0.4 & 1.0 & 0.8 & 0.2 & 0.8 & 1.0 & 0.8 & 0.4 & 1.0 & 1.0 & 0.0 & 0.6 & 0.2 \\ 0.2 & 0.8 & 1.0 & 1.0 & 0.2 & 0.6 & 0.8 & 0.4 & 0.2 & 0.6 & 0.2 & 1.0 & 0.6 & 0.0 & 0.6 \\ 1.0 & 1.0 & 1.0 & 1.0 & 0.8 & 0.6 & 0.6 & 1.0 & 0.8 & 0.2 & 1.0 & 1.0 & 0.2 & 0.6 & 0.0 \end{bmatrix}$$

Table 4. The results for the three different families data set

Cluster	TRUE	RMCM $(\sigma_a = 0.53, \sigma_b = 0.75)$	RMCM $(\sigma_a = 0.2, \sigma_b = 0.3)$	M-RMCM
Family-1	2, 4, 7, 12	2, 4, 7, 12	2, 3, 4, 7, 12	2, 4, 7, 12
Family-2	1, 5, 8, 9, 11, 14	1, 5, 8, 9, 11, 14	1, 5, 6, 8, 9, 11, 14	1, 5, 8, 9, 11, 14
Family-3	3, 6, 10, 13, 15	3, 6, 10, 13, 15	10, 13, 15	3, 6, 10, 13, 15

Example 4. In this example, we discuss the influence of stopping conditions in both M-RMCM and RMCM. The number of extracted exemplars can be specified by the users (if the cluster number is known) or decided by the stopping conditions. In RMCM, the parameter δ will control the number of clusters. However, in M-RMCM, the cluster number is estimated by a proposed validity index. Here, we use

the data set from Hwang et al. [6] which contains the 15 intuitionistic fuzzy set (IFS) patterns $\{x_1, \cdots, x_{15}\}$. In Hwang et al. [6], they used their proposed similarity measure to calculate the similarity matrix $S = [S_{ij}]$ between these 15 IFS patterns. We obtain the (dissimilarity) relational matrix $R = [R_{ij}]$ with $R_{ij} = 1 - S_{ij}$ as shown in Table 5. For this relational matrix, there are three well-separated clusters $(C_1 = \{1, 2, 3, 4, 5\},\ C_2 = \{6, 7, 8, 9, 10\},\ C_3 = \{11, 12, 13, 14, 15\})$ can be found. In RMCM, we test a lot of parameter combinations and only two kinds of combinations can work well. We can find the appropriate parameters σ_a and σ_b are between 0.2 and 0.3. The stopping parameter δ for acquiring cluster number is very important in RMCM. Even though we choose appropriate parameters σ_a and σ_b, RMCM still performs inappropriately when the parameter δ is chosen inappropriately. In M-RMCM, the validity function $MV(c)$ for this data set is shown in Fig. 2 which indicates that $c = 3$ is a good cluster number estimate for this data set. The extracted clusters listed in Table 6 show that M-RMCM works well with the cluster number $c = 3$ for this relational data set. However, the results from RMCM heavily depend on parameter selection. Overall, the M-RMCM method works well for most relational data sets without parameter selection.

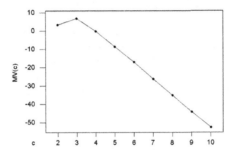

Fig. 2. The validity function $MV(c)$ for the IFS data set

Table 5. Relational (dissimilarity) matrix for 15 intuitionistic fuzzy set (IFS) patterns

0.000	0.015	0.039	0.058	0.113	0.513	0.539	0.566	0.555	0.525	0.340	0.372	0.372	0.359	0.340
0.015	0.000	0.054	0.072	0.126	0.521	0.546	0.573	0.562	0.532	0.350	0.381	0.381	0.369	0.350
0.039	0.054	0.000	0.020	0.077	0.493	0.521	0.548	0.537	0.506	0.313	0.346	0.346	0.333	0.313
0.058	0.072	0.020	0.000	0.058	0.483	0.511	0.539	0.528	0.496	0.299	0.333	0.333	0.320	0.299
0.113	0.126	0.077	0.058	0.000	0.451	0.481	0.511	0.498	0.465	0.256	0.292	0.292	0.278	0.256
0.513	0.521	0.493	0.483	0.451	0.000	0.054	0.109	0.086	0.091	0.263	0.225	0.225	0.240	0.263
0.539	0.546	0.521	0.511	0.481	0.054	0.000	0.058	0.034	0.058	0.302	0.267	0.267	0.281	0.302
0.566	0.573	0.548	0.539	0.511	0.109	0.058	0.000	0.054	0.086	0.343	0.309	0.309	0.323	0.343
0.555	0.562	0.537	0.528	0.498	0.086	0.034	0.054	0.000	0.063	0.326	0.292	0.292	0.306	0.326
0.525	0.532	0.506	0.496	0.465	0.091	0.058	0.086	0.063	0.000	0.281	0.244	0.244	0.259	0.281
0.340	0.350	0.313	0.299	0.256	0.263	0.302	0.343	0.326	0.281	0.000	0.049	0.049	0.049	0.039
0.372	0.381	0.346	0.333	0.292	0.225	0.267	0.309	0.292	0.244	0.049	0.000	0.049	0.049	0.049
0.372	0.381	0.346	0.333	0.292	0.225	0.267	0.309	0.292	0.244	0.049	0.049	0.000	0.020	0.049
0.359	0.369	0.333	0.320	0.278	0.240	0.281	0.323	0.306	0.259	0.049	0.049	0.020	0.000	0.030
0.340	0.350	0.313	0.299	0.256	0.263	0.302	0.343	0.326	0.281	0.039	0.049	0.049	0.030	0.000

Table 6. The results for 15 intuitionistic fuzzy set (IFS) patterns

Method	Parameters			Cluster number	Clusters
	σ_a	σ_b	δ		
RMCM	0.1	0.1	0.80	12	$\{x_1,x_2\}, \{x_3,x_4\}, \{x_5\}, \{x_6\}, \{x_7\}, \{x_8\},$ $\{x_9\}, \{x_{10}\}, \{x_{11}\}, \{x_{12}\}, \{x_{13},x_{14}\}, \{x_{15}\}$
			0.85	4	$\{x_1,x_2\}, \{x_3,x_4,x_5\}, \{x_6,x_7,x_8,x_9,x_{10}\},$ $\{x_{11},x_{12},x_{13},x_{14},x_{15}\}$
			0.90	4	$\{x_1,x_2\}, \{x_3,x_4,x_5\}, \{x_6,x_7,x_8,x_9,x_{10}\},$ $\{x_{11},x_{12},x_{13},x_{14},x_{15}\}$
	0.2	0.2	0.80	4	$\{x_1,x_2\}, \{x_3,x_4,x_5\}, \{x_6,x_7,x_8,x_9,x_{10}\},$ $\{x_{11},x_{12},x_{13},x_{14},x_{15}\}$
			0.85	3	$\{x_1,x_2,x_3,x_4,x_5\}, \{x_6,x_7,x_8,x_9,x_{10}\},$ $\{x_{11},x_{12},x_{13},x_{14},x_{15}\}$
			0.90	2	$\{x_1,x_2,x_3,x_4,x_5\},$ $\{x_6,x_7,x_8,x_9,x_{10},x_{11},x_{12},x_{13},x_{14},x_{15}\}$
	0.3	0.3	0.80	4	$\{x_1,x_2\}, \{x_3,x_4,x_5\}, \{x_6,x_7,x_8,x_9,x_{10}\},$ $\{x_{11},x_{12},x_{13},x_{14},x_{15}\}$
			0.85	3	$\{x_1,x_2,x_3,x_4,x_5\}, \{x_6,x_7,x_8,x_9,x_{10}\},$ $\{x_{11},x_{12},x_{13},x_{14},x_{15}\}$
			0.90	2	$\{x_1,x_2,x_3,x_4,x_5\},$ $\{x_6,x_7,x_8,x_9,x_{10},x_{11},x_{12},x_{13},x_{14},x_{15}\}$
	0.4	0.4	0.80	2	$\{x_1,x_2,x_3,x_4,x_5\},$ $\{x_6,x_7,x_8,x_9,x_{10},x_{11},x_{12},x_{13},x_{14},x_{15}\}$
			0.85	2	$\{x_1,x_2,x_3,x_4,x_5\},$ $\{x_6,x_7,x_8,x_9,x_{10},x_{11},x_{12},x_{13},x_{14},x_{15}\}$
			0.90	2	$\{x_1,x_2,x_3,x_4,x_5\},$ $\{x_6,x_7,x_8,x_9,x_{10},x_{11},x_{12},x_{13},x_{14},x_{15}\}$
	0.5	0.5	0.80	2	$\{x_1,x_2,x_3,x_4,x_5\},$ $\{x_6,x_7,x_8,x_9,x_{10},x_{11},x_{12},x_{13},x_{14},x_{15}\}$
			0.85	2	$\{x_1,x_2,x_3,x_4,x_5\},$ $\{x_6,x_7,x_8,x_9,x_{10},x_{11},x_{12},x_{13},x_{14},x_{15}\}$
			0.90	2	$\{x_1,x_2,x_3,x_4,x_5\},$ $\{x_6,x_7,x_8,x_9,x_{10},x_{11},x_{12},x_{13},x_{14},x_{15}\}$
M-RMCM	No parameter			3	$\{x_1,x_2,x_3,x_4,x_5\}, \{x_6,x_7,x_8,x_9,x_{10}\},$ $\{x_{11},x_{12},x_{13},x_{14},x_{15}\}$

5 Conclusions

In real applications, most data sets can be illustrated in relational types which have the information only with the relationship between two pair of data points. The RMCM proposed by Pal et al. [9] can be used as an approximately clustering method to obtain the prior information for a relational data set. However, the results of RMCM always depend on the parameters σ_a, σ_b and δ. In this paper, we propose a modified RMCM (M-RMCM) which used the correlation self-comparison to estimate the parameters and used a validity index to choose a suitable number of clusters. The simulation and comparison results show the superiority and robustness of the proposed M-RMCM.

References

1. Bezdek, J.C.: Pattern Recognition With Fuzzy Objective Function Algorithms. Plenum, New York (1981)
2. Chiu, S.L.: Fuzzy model identification based on cluster estimation. J. Intel. Fuzzy Syst. **2**, 267–278 (1994)
3. Dave, R.N., Sen, S.: Robust fuzzy clustering of relational data. IEEE Trans. Fuzzy Syst. **10**, 713–727 (2002)
4. Hathaway, R.J., Bezdek, J.C.: NERF c-means: non-euclidean relational fuzzy clustering. Pattern Recogn. **27**, 429–437 (1994)
5. Hathaway, R.J., Davenport, J.W., Bezdek, J.C.: Relational duals of the c-means clustering. Pattern Recogn. **22**, 205–212 (1989)
6. Hwang, C.M., Yang, M.S., Hung, W.L., Lee, M.G.: A similarity measure of intuitionistic fuzzy sets based on Sugeno integral with its application to pattern recognition. Inf. Sci. **189**, 93–109 (2012)
7. MacQueen, J.: Some methods for classification and analysis of multivariate observations. In: Proceedings of 5th Berkeley Symposium, vol. 1, pp. 281–297 (1967)
8. Pal, N.R., Chakraborty, D.: Mountain and subtractive clustering method: improvements and generalizations. Int. J. Intell. Syst. **15**, 329–341 (2000)
9. Pal, K., Pal, N.R., Keller, J.M., Bezdek, J.C.: Relational mountain (density) clustering method and web log analysis. Int. J. Intell. Syst. **20**, 375–392 (2005)
10. Velthuizen, B.P., Hall, L.O., Clarke, L.P., Silbiger, M.L.: An investigation of mountain method clustering for large data sets. Pattern Recogn. **30**, 1121–1135 (1997)
11. Wu, K.L., Yang, M.S., Hsieh, J.N.: Mountain c-regressions method. Pattern Recogn. **43**, 86–98 (2010)
12. Yager, R.R., Filev, D.P.: Approximate clustering via the mountain method. IEEE Trans Syst. Man Cybern. **24**, 1279–1284 (1994)
13. Yager, R.R., Filev, D.P.: Generation of fuzzy rules by mountain clustering. J. Intell. Fuzzy Syst. **2**, 209–219 (1994)
14. Yang, M.S., Shih, H.M.: Cluster analysis based on fuzzy relations. Fuzzy Sets Syst. **120**, 197–212 (2001)
15. Yang, M.S., Wu, K.L.: A modified mountain clustering algorithm. Pattern Anal. Appl. **8**, 125–138 (2005)
16. Yang, M.S., Tian, Y.C.: Bias-correction fuzzy clustering algorithms. Inf. Sci. **309**, 138–162 (2015)

Relative Stability of Random Projection-Based Image Classification

Ewa Skubalska-Rafajłowicz[✉]

Department of Computer Engineering, Faculty of Electronics,
Wrocław University of Science and Technology, Wrocław, Poland
ewa.rafajlowicz@pwr.edu.pl

Abstract. Our aim is to show that randomly generated transformation of high-dimensional data vectors, for example, images, could provide low dimensional features which are stable and suitable for classification tasks. We examine two types of projections: (a) global random projections, i.e., projections of the whole images, and (b) concatenated local projections of spatially-organized parts of an image (for example rectangular image blocks). In both cases, the transformed images provide good features for correct classification. The computational complexity of designing the transformation is linear with respect to the size of images and in case (b) it does not depend on the form of image partition. We have analyzed the stability of classification results with respect to random projection and to different randomly generated training sets. Experiments on the images of ten persons taken from the Extended Yale Database B demonstrate that the methods of classification based on Gaussian random projection are effective and positively comparable with PCA-based methods, both from the point of view of stability and classification accuracy.

1 Introduction

It is well known that many classification methods are unstable. This means that small changes in the training set or in other classifier design factors may result in drastic changes in classifier decisions and consequently, in their accuracies [5,7].

In this paper, we explore properties of random projection used as a method of large digital images dimension reduction from the point of view of image object recognition.

Random projections are one of the methods used for simple image data dimensionality reduction [3,8,12,16,27], among others. In these approaches, images are treated as large vectors in the Euclidean space of very high dimensionality.

Random projections (RP) [1,2,20,29] are used in signal processing (see for example [21,23]). RP are closely connected to the Johnson-Lindenstrauss embedding [18], popular in many computer science applications.

Random projection-based methods of image dimensionality reduction are independent of the data. Thus, adding or removing images from the classification

© Springer International Publishing AG, part of Springer Nature 2018
L. Rutkowski et al. (Eds.): ICAISC 2018, LNAI 10841, pp. 702–713, 2018.
https://doi.org/10.1007/978-3-319-91253-0_65

system does not require any changes of the transformation. Furthermore, the computational complexity of designing the transformation is linear with respect to the size of an image.

The main message of this paper is that Gaussian random projections allow us to handle a very large number of input variables without over-fitting and can produce highly accurate classification predictions. Furthermore, variability (instability) of such predictions measured by standard deviation is relatively small in comparison to the variability (instability) introduced by randomly selected training sets. Five popular methods of classification such as random forest (RF), naive Bayes (NB), nearest neighbors (NN), support vector machine (SVM), and logistic regression (LR) are examined. Experiments show that these methods can provide results with very different accuracy but their variability with respect to the RP is very similar and relatively small.

The paper is organized as follows. Section 2 provides a general idea of random projection dimensionality reduction for classification purposes. The details of the image classification method are shortly presented in Sect. 3. The next Sections contain results of experiments which concentrates on the accuracy and stability of face classification obtained using the RP-based image transformations. Experiments were performed on a set of face images taken from the Extended Yale Database B. It consists of 640 images of ten persons. Some comparisons with $1D$-PCA and a block–based $2D$-PCA method of image dimensionality reduction in the context of face classification are also presented. Section 4 provides results for random projections performed on the whole image treated as a vector. The next Section shows similar experiments for projections of images partitioned into blocks. Finally, some comments and conclusions are summarized.

2   Random Projections

We can project vector $x \in R^d$ onto k dimensional subspace using simple transformation $Sx = \sum_{i=1}^{k} s_i^T x$. Matrix $S = [s_1, s_2, \ldots, s_k]^T \in R^{k \times d}$ defines random projection of dimensionality k. Each entry of S is generated independently from a normal distribution with mean zero and variance 1. Random vector $V = [s_1^T x, \ldots, s_k^T x]^T$ is a linear combination of independent normal vectors with zero means. So, it is also a zero mean normal random vector and its covariance matrix is a diagonal matrix of the form $||x||^2 I$. It is well known (see for example [29]) that $||V||^2/||x||^2$ follows χ_k^2 distribution and

$$E_S\left\{\frac{1}{k}||Sx||^2\right\} = ||x||^2,$$

where $||.||$ is the Euclidean norm of an adequate dimensionality and E_S denotes the expectation with respect to random matrix S. Notice that $\frac{1}{\sqrt{k}}Sx$ is a non-orthogonal transformation. Orthogonalization of S is sometimes used, but it is known that in very high dimensions S is almost orthogonal [20]. Random projections are closely related to the Johnson-Lindenstrauss lemma

[11,18,20,29], which states that any set A, say, of N points in an Euclidean space can be embedded in a Euclidean space of lower dimension ($\sim O(\log N)$) with relatively small distortion $0 < \varepsilon < 1$ of the distances between any pair of points from A. This property is a basis of usage of random projections as an effective method of dimensionality reduction.

Generally, if $x, y \in \mathcal{R}^d$ and $< \frac{x}{||x||}, \frac{y}{||y||} > = r$ then one can obtain (see [24] for the proof):

$$(1 - \varepsilon)kr \leq < \frac{Sx}{||x||}, \frac{Sy}{||y||} > \leq kr(1 + \varepsilon)$$

with similar probability bounds as for the Euclidean distances [20,29]. If $x = y$ then $r = 1$ and we arrive at the formula (see for example [29]):

$$\mathcal{P}\left\{ | \frac{||Sx||^2}{k||x||^2} - 1| \geq \varepsilon \right\} \leq 2 \exp\left(-k(\varepsilon^2/4 - \varepsilon^3/6)\right). \tag{1}$$

Furthermore, it is known that the mentioned probability bounds are in many cases rather loose and combined with Bonferroni union bounds (for many point configurations) lead to very large lower bounds on projection dimension k [29].

In general, we do not need an exact representation of the data structure in the classification. It is enough for points close to each other in the original space to be close to each other after the projection. Similarly, the distant points should remain distant after a transformation reducing the dimension. Such properties translate directly into the efficiency of distance-based classifiers, such as the NN method, for example. There are also classifiers that additionally separate objects belonging to different classes, and group objects belonging to the same class, as for example the LR method. Thus, in practice one may use much smaller values of k, which has been shown in many experiments and which is also confirmed by the results obtained in this work.

3   An Image and Its Classification

An image can be classified in a number of ways, depending on our a priori knowledge and many other factors. In this paper, we will concentrate on the usage of random projection as a method of feature selection. In the first approach, images under consideration are treated as vectors and these vectors are randomly projected onto lower-dimensional, let's say k-dimensional, space. In the second approach, we have to establish an image partitioning pattern. Each image is divided into many blocks of the same size. In this case, every image block is considered as a separate image and it is projected independently from other image blocks, forming only a part of a new feature vector. Thus, the dimension of the feature vector is kM, where M stands for a number of blocks. Thus, every image is split up into blocks of $m_b \times n_b$ pixels (32×32 pixels in our experiments). For the sake of simplicity, we have assumed that $m = 0 \ mod \ m_b$ and $n = 0 \ mod \ n_b$.

In both cases, the feature vectors are normalized since we want to accept images taken under different illumination conditions. Thus, endpoints of normalized feature vectors lie on a k-dimensional, or an $M \times k$-dimensional, sphere.

A set of common classification methods have been used for classification, such as random forest (RF), naive Bayes (NB), nearest neighbors (NN), support vector machine (SVM) with Gaussian kernels, and logistic regression (LR) [6,15, 22,26]. These methods were applied using the procedure "classify" (and adequate options), as provided by Mathematica ver. 11. The 1D-PCA [28] dimensionality reduction method was used for comparison purposes.

It should be noted that RPs require only $O(kD)$ operations for providing a transformation matrix whereas PCA needs $0(kD^2)$ or $O(D^2N) + O(D^3)$. If the rank of the covariance matrix is r then computational complexity for SVD is O(DrN) [8,17]. Hotelling's power method [10,17] an iterative procedure that finds only the largest eigenvalues and their corresponding eigenvectors, is more reasonable since its complexity is kD^2.

3.1   Data Used for Experiments

The methods of classification based on random projections have been evaluated on the data taken from the Extended Yale Face Database B. The original database contains 16128 images of 28 human subjects under 9 poses and 64 illumination conditions [14]. In the experiments we have used images of the first 10 subjects taken from the Cropped Yale Face Database B. The images of the subjects are previously manually aligned, cropped, and then re-sized to 168×192 images by Lee and [19]. In our experiments, the images are additionally cropped to the size of 160×192 pixels. Thus, the dimension in our experiments is equal to 30720.

3.2   Stability of the Classifier with Respect to RP

Let us assume that the efficiency of the selected classifier built on the basis of $X \in R^D$ data is measured by the accuracy of the classifier F_X. The accuracy of the classifier depends on how the data set is divided into the training set L and the testing set T. We can estimate the expected value of $F_X(L)$ randomly dividing X onto to subsets, repeat the process many times and calculate the average value. Let $|X| = n$ and $|L| = N$. The number of possible partitions of X is $\frac{n!}{N!(n-N)!}$ so usually only a small number of possibilities could be checked. The standard deviation of $F_X(L)$ can be used as a measure of the instability of the classifier [5–7]. RPs introduce another type of randomness into the process of classification. The accuracy of the classifier now depends not only on the training/testing sets but also on projection matrix S. The dimension of projection k is a parameter which is assumed to be known. We propose to define the stability of the classifier with respect to RP in a similar way as previously, i.e., as the standard deviation of $F_X(S, L = l)$ averaged over possible L. Thus, it can be estimated as

$$\hat{\sigma}_{RP} = \frac{1}{\sqrt{p}}(\sum_{i=1}^{p} \hat{\sigma}_{li}^2)^{0.5},$$

where

$$\hat{\sigma}_l^2 = \frac{1}{p-1} \sum_{j=1}^{p} (F_X(S_j, l) - \bar{F}_X(S, l))^2,$$

and $\bar{F}_X(S, l)$ is a mean with respect to S estimated for a fixed training set $L = l$.

4   Random Projections of the Whole Image Treated as a Vector

Random projection of the whole image treated as a vector is the very simple method of obtaining low dimensional feature vectors representing large images.

Let X be an image consisting of m rows and n columns ($X \in R^{m \times n}$). Furthermore, let $y = Svec(X) \in R^k$ be the projection of X into k dimensions, where S is a randomly generated projection matrix. Notice, that for the sake of simplicity, we have skipped factor $\frac{1}{\sqrt{k}}$. It should be emphasized that projection matrix S is randomly chosen only once, at the beginning of each experiment and after that, the same transformation S is applied for every data point (a vectorized image).

We have used normalized vectors $y_{norm} = \frac{y}{||y||}$ as vector features since it allows us to have similar features for images of the same object (in the similar pose) which are taken in different illumination conditions.

Table 1 shows mean classification accuracy for 10 different learning sets averaged over 10 random projections of the whole images. Each learning set consists of 320 examples, i.e., 32 learning examples are randomly chosen for each class. Dimension of the random projections was set to $k = 10, k = 20, k = 50, k = 100, k = 200$ and $k = 300$. Standard deviations, which represent the variability of the classification results with respect to RP, are given in brackets. The last column provides standard deviations between means from columns (2)–(11). These values show the influence of the training sequence on the results of the classification. It differs from the standard deviation computed for all 100 experiments (10 different RP × 10 different training sets). The last value contains the influence of both random factors, i.e., RP and random training sets. The adequate values are shown in the last but one column (in brackets). It is clearly visible that starting from $k = 100$ RPs stabilize the classification results. The best results were given by the LR method, and for smaller values of k, the SVM method and these results are presented in Table 1. Other methods such as RF, SVM, NB, NN have been found to be similarly stable, but less accurate (no results are shown for lack of the space). Table 2 presents the results for the same learning series as in the previous Table, obtained with the reduction of the dimension using the PCA method. The corresponding mean values and standard deviations are shown for 10, 20, 50, 100 and 300 principal components. In this case, also the LR method proved to be the best due to both accuracy and stability. When using 300 PC, the NN and SVM methods exhibited signs of instability. Moreover, the LR method turned out to be slightly more accurate than in the case of the use of the RP dimensionality reduction method (see Table 1).

Table 1. Mean classification accuracy for 10 learning sets (based on 10×32 learning examples) averaged over 10 random projections of the whole images. Dimension of the random projections was: $k = 10, k = 20, k = 50, k = 100, k = 200$ and $k = 300$. Standard deviations are given in brackets.

k method	1	2	3	4	5	6	7	8	9	10	Mean $\hat{\sigma}_{RP}$	Std
10 SVM	73.0	69.4	69.5	68.0	69.9	72.5	71.5	74.7	71.2	70.5	70.95	
	(3.13)	(4.92)	(2.69)	(2.81)	(3.01)	(1.94)	(3.26)	(2.53)	(4.21)	(2.73)	(3.08)	(1.99)
10 LR	46.04	41.05	42.93	42.53	43.29	41.88	42.76	43.13	40.92	41.22	42.51	
	(6.95)	(5.74)	(4.78)	(5.30)	(5.50)	(5.57)	(8.21)	(6.19)	(7.50)	(7.80)	(6.15)	(1.59)
20 SVM	69.97	69.46	73.75	75.72	73.34	72.09	71.91	65.91	70.56	71.53	71.42	
	(5.05)	(2.43)	(2.38)	(3.15)	(3.97)	(3.71)	(3.25)	(3.87)	(3.05)	(4.32)	(3.44)	(2.69)
20 LR	66.75	64.01	67.34	66.88	66.19	65.38	63.22	64.75	63.06	64.88	65.25	
	(4.46)	(3.76)	(3.13)	(3.37)	(7.72)	(7.22)	(5.03)	(2.62)	(4.54)	(5.19)	(4.74)	(1.53)
50 LR	85.53	83.84	87.66	88.31	86.72	86.13	85.84	82.5	84.53	84.78	85.68	
	(2.18)	(2.15)	(1.99)	(1.85)	(1.98)	(1.05)	(2.38)	(1.49)	(2.45)	(1.92)	(1.89)	(1.80)
100 LR	91.22	88.41	90.88	91.88	90.44	91.44	91.19	87.05	88.53	89.16	90.02	
	(1.09)	(1.58)	(0.94)	(0.95)	(1.40)	(1.23)	(1.24)	(1.74)	(1.63)	(1.00)	(1.25)	(1.62)
200 LR	93.44	91.25	94.28	95.25	93.12	93.84	93.31	90.09	90.81	92.47	92.78	
	(1.50)	(1.17)	(1.15)	(1.08)	(1.37)	(0.94)	(1.17)	(1.59)	(1.57)	(1.88)	(1.31)	(1.63)
300 LR	94.94	92.03	94.88	95.53	93.94	94.34	94.56	91.97	90.43	93.28	93.59	
	(0.79)	(0.97)	(0.51)	(0.83)	(0.71)	(0.85)	(1.02)	(1.09)	(1.40)	(1.05)	(0.91)	(1.63)

Table 2. Mean classification accuracy obtained for PCA projections of dimensionality $k = 10, 20, 50, 100, 300$ and for 10 different training sets

k	10	20	50	100	200	300
RF	68.31	82.06	88.34	88.7	87.64	84.89
	(2.47)	(2.48)	(1.29)	(2.18)	(1.87)	(1.47)
NB	54.22	71.19	76.53	76.31	68.81	60.66
	(2.84)	(1.60)	(1.45)	(2.67)	(1.86)	(2.57)
SVM	79.97	87.03	94.03	94.97	91.11	68.19
	(1.00)	(1.27)	(0.85)	(1.08)	(2.33)	(13.64)
NN	60.5	80.93	88.56	88.1	86.22	73.56
	(1.39)	(0.95)	(0.75)	(1.60)	(1.56)	(18.91)
LR	81.03	91.37	95.19	96.01	96.60	96.31
	(1.14)	(1.62)	(0.98)	(1.10)	(0.95)	(1.04)

5   Projections of Images Partitioned into Blocks

It is well-known that image pixel values are spatially correlated. Motivated by this elementary image property we propose to use randomly-generated image transformation in order to obtain image features of low dimensionality in such a way that the image transformation consists of local projections of the spatially-organized parts of an image (rectangular image blocks). In contrast to the 2D-PCA method where rows (or columns) of the image under consideration form separate blocks [30] or modular PCA [13], we reduce the dimension of vectorized rectangular image blocks using randomly-generated projection matrices [25]. We will show that such transformed images provide good features for sufficiently accurate and stable classification.

Let $S \in R^{k \times D}$ stand for projection matrix, where k denotes dimension of each image and each image block after projection. Linear transformation $S\,vec(X)$ is, in general, a non-orthogonal projection of X into k dimensional space. Symbol vec stands for vectorization. Denote by M the number of image blocks of the same size, i.e., consisting of $m_b \times n_b$ pixels ($m_b n_b = d = D/M$, where $D = mn$. Denote by X_i, $i = 1, \ldots, M$ the subsequent image blocks. Let $S = [S_1, S_2, \ldots, S_M]$ consists of M sub-matrices of the same size: $S_i \in R^{k \times d}$, $i = 1, \ldots, M$. Each entry of projection matrix S and as a consequence, each entry of every sub-matrix S_i is generated independently from standard normal distribution $\mathcal{N}(0, 1)$. It should be stressed that the overall number of random numbers that we have to generate is equal to the number of image pixels $D = m\,n$, independently of dimensions of blocks (sub-images).

Let $y_i = S_i vec(X_i) \in R^k$ be the projection of block X_i into k dimensions. If $vec(X)$ is an adequate concatenation of $vec(X_i)$, $i = 1, \ldots, M$, then

$$S\,vec(X) = \sum_{i=1}^{M} S_i vec(X_i).$$

Thus, y consists of partial projections of the transformed image and as a result, it contains more spatially restricted image information (M vectors of k dimension) in comparison to the global projection $S\,vec(X)$ which results in only one k-dimensional vector. Observe that each coordinate of $S\,vec(X)$ is a sum of M corresponding coordinates of block projections. As previously, after generation of S, all images are transformed using the same transformation matrix.

Table 3 provides the classification accuracy for the different number of random projections k. In the experiments, only original images (i.e., without performing a histogram equalization) were examined. 320 randomly selected images (32 for each class) were used as a training set. The rest of the images (i.e., 320) form a testing sequence. The process of a learning set selection was repeated ten times. Results for each set are shown in a separate column in Table 3. Dimension of the random projections was set to $k = 10, k = 20, k = 50$ and $k = 70$ with respect to partition of each image into $M = 30$ blocks. Thus, the final feature dimensions were 300, 600, 1500 and 2100. The last column provides standard deviations between means from columns (2)–(11). As previously, five methods of

Table 3. Mean classification accuracy for 10 learning sets averaged over 10 random projections of dimensionality $k = 10, k = 20, k = 50$ and $k = 70$ with partition into $M = 30$ image blocks. Standard deviations are given in brackets.

k	1	2	3	4	5	6	7	8	9	10	Avg mean $\hat{\sigma}_{RP}$	Std
10 RF	87.09	85.56	89.13	89.22	89.22	87.66	86.84	85.0	86.03	87.31	87.31	
	(1.68)	(1.87)	(0.95)	(1.73)	(1.19)	(1.16)	(1.44)	(2.55)	(2.34)	(1.48)	(1.63)	(1.62)
10 NB	63.56	66.16	66.97	63.16	61.47	66.09	68.75	55.41	67.22	63.41	64.22	
	(2.79)	(0.39)	(2.24)	(2.13)	(3.33)	(2.44)	(1.32)	(1.79)	(4.39)	(2.10)	(2.40)	(3.83)
10 SVM	88.35	85.13	89.97	89.44	89.22	90.06	87.25	82.78	85.53	86.66	87.44	
	(0.75)	(0.63)	(1.23)	(1.45)	(1.37)	(1.56)	(1.93)	(1.69)	(2.07)	(1.83)	(1.45)	(2.42)
10 NN	77.69	78.97	83.03	81.59	83.25	83.84	81.63	76.41	79.38	83.81	80.96	
	(0.86)	(1.30)	(0.81)	(0.66)	(0.81)	(0.87)	(0.74)	(0.85)	(0.91)	(1.10)	(0.87)	(2.68)
10 LR	96.59	94.19	96.59	97.81	95.78	95.53	96.69	93.59	95.0	96.14	95.78	
	(0.71)	(1.06)	(0.81)	(0.81)	(0.69)	(0.90)	(0.74)	(1.09)	(0.85)	(1.09)	(0.85)	(1.26)
20 RF	88.81	88.03	91.22	90.72	90.47	90.28	88.53	86.16	87.75	90.22	89.21	
	(1.18)	(1.81)	(1.39)	(1.83)	(2.23)	(1.03)	(1.81)	(1.63)	(1.47)	(2.06)	(1.60)	(1.62)
20 NB	65.50	70.16	70.06	65.44	64.16	68.56	69.34	57.22	69.78	63.13	66.34	
	(1.71)	(1.44)	(2.71)	(2.13)	(2.52)	(2.41)	(1.48)	(1.59)	(2.17)	(1.57)	(1.93)	(4.14)
20 SVM	81.78	80.75	87.09	83.69	82.47	85.22	82.03	76.59	80.28	80.75	82.07	
	(0.86)	(0.90)	(0.81)	(0.67)	(0.58)	(1.10)	(0.90)	(0.83)	(0.84)	(0.98)	(0.82)	(2.88)
20 NN	77.47	79.66	84.59	83.59	84.47	84.72	81.22	76.78	81.0	84.75	81.83	
	(1.11)	(1.26)	(1.37)	(0.90)	(1.39)	(1.53)	(1.60)	(0.98)	(1.07)	(1.02)	(1.19)	(3.07)
20 LR	98.03	95.53	98.16	98.91	96.75	97.25	97.5	95.06	95.97	97.34	97.05	
	(0.96)	(0.83)	(0.63)	(0.34)	(0.68)	(0.55)	(0.68)	(0.73)	(0.81)	(0.61)	(0.67)	(1.23)
50 RF	90.91	88.94	92.09	92.09	91.56	90.28	89.56	86.88	88.75	91.16	90.22	
	(1.96)	(1.38)	(0.85)	(1.21)	(1.55)	(1.25)	(1.99)	(2.40)	(1.03)	(0.91)	(1.46)	(1.68)
50 NB	66.25	71.86	71.09	65.09	64.53	69.53	71.34	56.88	72.59	65.16	67.43	
	(1.23)	(1.99)	(1.05)	(1.18)	(0.90)	(1.32)	(2.01)	(0.95)	(1.50)	(1.22)	(1.32)	(4.84)
50 SVM	62.0	69.06	76.69	67.63	69.13	66.66	69.13	63.84	66.88	62.81	67.38	
	(0.74)	(0.97)	(1.11)	(0.71)	(0.67)	(0.99)	(0.72)	(0.92)	(0.53)	(0.86)	(0.80)	(17.65)
50 NN	78.0	79.78	84.72	83.59	85.56	85.34	81.88	75.81	81.34	85.09	82.11	
	(0.84)	(0.61)	(0.86)	(0.78)	(0.99)	(0.71)	(1.13)	(1.01)	(1.22)	(0.49)	(0.85)	(11.43)
50 LR	98.38	96.09	98.84	92.28	97.34	97.56	98.09	95.47	97.09	98.41	96.96	
	(0.41)	(0.75)	(0.59)	(0.21)	(0.53)	(0.51)	(0.80)	(0.80)	(0.49)	(0.40)	(0.55)	(3.81)
70 RF	89.31	90.16	92.66	92.41	91.09	90.31	90.13	86.81	89.16	90.01	90.21	
	(1.93)	(1.60)	(1.50)	(1.71)	(1.09)	(0.99)	(1.93)	(1.95)	(1.76)	(1.31)	(1.54)	(2.79)
70 NB	66.41	71.22	71.59	65.63	64.31	69.38	71.44	57.84	72.56	64.59	67.50	
	(2.03)	(1.40)	(1.61)	(1.12)	(1.46)	(1.39)	(0.86)	(1.41)	(0.93)	(0.88)	(1.29)	(21.32)
70 SVM	52.28	59.34	69.25	54.41	57.72	51.56	60.66	53.84	57.59	50.81	56.75	
	(0.91)	(1.64)	(1.12)	(0.91)	(0.74)	(0.85)	(1.14)	(1.14)	(0.79)	(0.90)	(1.00)	(5.54)
70 NN	78.22	79.78	85.06	83.97	85.44	86.41	82.03	76.72	81.34	85.25	82.42	
	(0.96)	(0.74)	(0.67)	(1.09)	(1.32)	(1.05)	(0.89)	(0.71)	(0.72)	(0.95)	(0.89)	(3.34)
70 LR	98.72	96.47	99.0	99.09	97.44	97.69	98.19	96.25	96.84	98.28	97.80	
	(0.50)	(0.69)	(0.29)	(0.31)	(0.32)	(0.54)	(0.44)	(0.57)	(0.40)	(0.59)	(0.46)	(1.03)

classification were applied: RF, NB, SVM, NNs, and LR and as in the previous experiments, the LR behaved much better than the other methods. The level of instability (standard deviation value) introduced by random projections is smaller (or in a few cases comparable) than the level of instability introduced by the randomness of training data. In addition, for the projection size of 10 and 50, the average value and standard deviation for fixed projection matrixes and random training sequences were calculated. The experiments were repeated 10 times for 10 different RP matrices (see Table 4). The values of standard deviations obtained remain in general agreement with the corresponding results in the previous table (Table 3).

Table 4. Mean classification accuracy for 10 random projections of dimensionality $k = 10$ and $k = 50$ averaged over 10 randomly selected learning sets. Standard deviations are given in brackets.

1	2	3	4	5	6	7	8	9	10	Avg mean
95.85	96.69	96.44	96.19	96.22	95.66	96.35	96.85	96.84	96.13	96.32
(1.54)	(1.11)	(0.92)	(1.09)	(1.09)	(1.43)	(1.09)	(0.94)	(0.94)	(0.89)	(1.14)
98.03	97.75	98.03	97.44	97.75	97.72	97.72	98.55	97.72	97.13	97.79
(0.90)	(1.11)	(0.61)	(1.22)	(1.06)	(1.63)	(1.53)	(0.96)	(1.46)	(1.15)	(1.21)

5.1 Block-Based PCA

For comparison, similar calculations were carried out using the PCA method to reduce the dimension of each of the image blocks. Table 5 provides the results of classification and their variability when using the LR method, while Table 6 contains the results obtained for the same problems while using the NB method. Both methods are similarly stable, while the NB method definitely exceeds the accuracy of other methods, including the LR method (due to the lack of space, we do not provide results for the other methods studied). Recall that when using PCA reduction in relation to the whole image, the NB method was one of those giving worse results. It should be mentioned that the probability bounds for inadequacy of the projected sets of points are based on Bonferroni union bounds and they lead to significantly overestimated approximations of these probabilities. Thus, however the known bounds suggest very large values of k, the usage of even only ten projection dimensions occurred to be sufficient for classification purposes.

The performed experiments were restricted to a relatively small number of image classes (ten subjects). A higher number of classes would probably require increasing the number of projections.

Table 5. Mean classification accuracy (LR method) for PCA of dimensionality $k = 10$ (first row), $k = 20$, $k = 50$ and $k = 70$ (last row) in 30 image blocks. Standard deviations are given in brackets.

k	1	2	3	4	5	6	7	8	9	10	Mean
10	94.06	94.07	96.25	97.19	94.69	94.69	94.69	91.25	92.5	92.19	94.19 (1.81)
20	96.56	96.56	96.25	97.19	95.0	95.63	95.31	93.13	95.31	93.13	95.41 (1.38)
50	97.81	98.12	97.5	96.25	94.38	95.94	95.94	92.81	97.5	95.31	96.16 (1.68)
70	98.44	97.5	95.63	96.56	95.0	95.93	94.06	92.19	96.88	96.56	95.88 (1.79)

Table 6. Classification accuracy and its average value (NB method) for modular PCA of dimensionality $k = 10, k = 20, k = 50$ and $k = 70$ with $M = 30$ image blocks - 10 training samples. Standard deviations are given in brackets.

k	1	2	3	4	5	6	7	8	9	10	Mean
10	97.81	96.56	97.81	98.125	96.88	98.13	98.62	94.06	96.88	97.81	96.97 (1.30)
20	99.69	98.43	99.38	98.75	98.44	99.38	97.81	97.19	99.06	99.06	98.72 (0.77)
50	99.38	97.5	98.75	99.06	98.44	98.44	96.25	95.63	98.44	96.56	97.85 (1.29)
70	98.44	95.63	97.81	98.75	97.19	97.19	93.43	93.13	97.5	95.31	96.44 (1.98)

6   Conclusions

In this paper, we have examined the stability of classification methods in combination with dimensionality reduction of images based on Gaussian RP. This method of dimension reduction is very easy to implement - the computational complexity of designing the transformation is linear with respect to the size of the images and does not depend on a form of image partition (block-based approach). It should be emphasized that the RP-based approach is independent of the learning data. Thus, adding or removing images from the classification system does not require a change of transformation. Furthermore, we have shown experimentally that RP-based classifiers can provide both efficient and stable classification results.

Acknowledgments. This research was supported by grant 041/0145/17 at the Faculty of Electronics, Wrocław University of Science and Technology.

References

1. Achlioptas, D.: Database-friendly random projections: Johnson-Lindenstrauss with binary coins. J. Comput. Syst. Sci. **66**, 671–687 (2003)
2. Ailon, N., Chazelle, B.: The fast Johnson-Lindenstrauss transform and approximate nearest neighbors. SIAM J. Comput. **39**(1), 302–322 (2009)
3. Amador, J.J.: Random projection and orthonormality for lossy image compression. Image Vis. Comput. **25**, 754–766 (2007)
4. Baraniuk, R., Davenport, M., DeVore, R., Wakin, M.: A simple proof of the restricted isometry property for random matrices. Constr. Approx. **28**(3), 253–263 (2008)
5. Breiman, L.: Arcing classifiers. Ann. Stat. **26**(3), 801–849 (1998)
6. Breiman, L.: Random forests. Mach. Learn. **45**(1), 5–32 (2001)
7. Briand, B., Ducharme, G.R., Parache, V., Mercat-Rommens, C.: A similarity measure to assess the stability of classification trees. Comput. Stat. Data Anal. **53**(4), 1208–1217 (2009)
8. Brigham, E., Mannila, H.: Random projection in dimensionality reduction: applications to image and text data. In: Proceedings of the Conference on Knowledge Discovery and Data Mining, vol. 16, pp. 245–250 (2001)
9. Fodor, I.K.: A survey of dimension reduction techniques. Technical report, Lawrence Livermore National Lab., CA (US) (2002)
10. Du, Q., Fowler, J.E.: Low-complexity principal component analysis for hyperspectral image compression. Int. J. High Perform. Comput. Appl. **22**, 438–448 (2008)
11. Frankl, P., Maehara, H.: Some geometric applications of the beta distribution. Ann. Inst. Stat. Math. **42**(3), 463–474 (1990)
12. Fowler, J.E., Du, Q.: Anomaly detection and reconstruction from random projections. IEEE Trans. Image Process. **21**(1), 184–195 (2012)
13. Gottmukkal, R., Asari, V.K.: An improved face recognition technique based on modular PCA approach. Pattern Recogn. Lett. **24**(4), 429–436 (2004)
14. Georghiades, A.S., Belhumeur, P.N., Kriegman, D.J.: From few to many: illumination cone models for face recognition under variable lighting and pose. IEEE Trans. Pattern Anal. Mach. Intell. **21**(6), 643–660 (2001)
15. James, G., Witten, D., Hastie, T., Tibshirani, R.: An Introduction to Statistical Learning. Springer, New York (2013). https://doi.org/10.1007/978-1-4614-7138-7
16. Jeong, K., Principe, J.C.: Enhancing the correntropy MACE filter with random projections. Neurocomputing **72**(1–2), 102–111 (2008)
17. Jolliffe, I.: Principal Component Analysis, 2nd edn. Springer, NewYork (2002). https://doi.org/10.1007/b98835
18. Johnson, W.B., Lindenstrauss, J.: Extensions of Lipshitz mapping into Hilbert space. Contemp. Math. **26**, 189–206 (1984)
19. Lee, K.-C., Ho, J., Driegman, D.: Acquiring linear subspaces for face recognition under variable lighting. IEEE Trans. Pattern Anal. Mach. Intell. **27**(5), 684–698 (2005)
20. Matoušek, J.: On variants of the Johnson-Lindenstrauss lemma. Random Struct. Algorithms **33**(2), 142–156 (2008)
21. Marzetta, T.L., Tucci, G.H., Simon, S.H.: A random matrix-theoretic approach to handling singular covariance estimates. IEEE Trans. Inf. Theory **57**, 6256–6271 (2011)
22. Ng, A.Y., Jordan, M.I.: On discriminative vs. generative classifiers: a comparison of logistic regression and Naive Bayes. In: Advances in Neural Information Processing Systems, vol. 14, pp. 841–848 (2002)

23. Skubalska-Rafajłowicz, E.: Random projections and Hotelling's T 2 statistics for change detection in high-dimensional data streams. Int. J. Appl. Math. Comput. Sci. **23**(2), 447–461 (2013)
24. Skubalska-Rafajłowicz, E.: Neural networks with sigmoidal activation functions – dimension reduction using normal random projection. Nonlinear Anal.: Theory Methods Appl. **71**(12), e1255–e1263 (2009)
25. Skubalska-Rafajłowicz, E.: Spatially-organized random projections of images for dimensionality reduction and privacy-preserving classification. In: Proceedings of 10th International Workshop on Multidimensional (nD) Systems (nDS), pp. 1–5 (2017)
26. Steinwart, I., Christmann, A.: Support Vector Machines. Springer, New York (2008). https://doi.org/10.1007/978-0-387-77242-4
27. Tsagkatakis, G., Savakis, A.: A random projections model for object tracking under variable pose and multi-camera views. In: Proceedings of the Third ACM/IEEE International Conference on Distributed Smart Cameras, ICDSC, pp. 1–7 (2009)
28. Turk, M., Pentland, A.: Eigenfaces for recognition. J. Cogn. Neurosci. **3**(1), 71–86 (1991)
29. Vempala, S.: The Random Projection Method. American Mathematical Society, Providence (2004)
30. Yang, J., Zhang, D., Frangi, A.F., Yang, J.: Two-dimensional PCA: a new approach to appearance-based face representation and recognition. IEEE Trans. Pattern Anal. Mach. Intell. **26**(1), 131–137 (2004)

Cost Reduction in Mutation Testing with Bytecode-Level Mutants Classification

Joanna Strug[1] and Barbara Strug[2(✉)]

[1] Faculty of Electrical and Computer Engineering,
Cracow University of Technology, ul. Warszawska 24, 31-155 Krakow, Poland
joanna.strug@pk.edu.pl
[2] Department of Physics, Astronomy and Applied Computer Science,
Jagiellonian University, Lojasiewicza 11, 30-059 Krakow, Poland
barbara.strug@uj.edu.pl

Abstract. The paper presents the application of classification based approach to software quality domain. In particular it deals with the issue of reducing the cost of mutation testing. The presented approach is based on the similarity of mutants represented at the bytecode level. The distance matrix for mutants is used in kNN algorithm to predict if a given test set detects a mutant or not. Experimental results are also presented in this paper on the basis of two systems. The obtained results show the usefulness of the proposed method.

Keywords: Machine learning · Mutation testing
Bytecode distance · Classification · Test evaluation

1 Introduction

Testing is an important step in developing a software system. If it is carried out with a help of tests being able to detect faults with a high accuracy, its results will provide valuable information that allows to determine the degree to which the system satisfies certain requirements and to decide if it is ready for deployment or needs to be further improved [13,14].

Nowadays mutation testing [3] is considered to be the most accurate and dependable technique for assessing and measuring fault detection ability of a test set [2,7]. The idea behind mutation testing is to assess the tests ability to detect real faults by assessing it ability to detect small, artificially inserted into the system, modifications. It is done by creating a number of modified version of the system (so called mutants), each having only one change introduced accordingly to a specific rule (called mutation operator [3]), and running them against the tests from the set under assessment. When at least one of the tests is able to detect the modification given by a mutant, the mutant is considered to be detected (killed) by the set. The ratio of the number of mutants killed by the tests

© Springer International Publishing AG, part of Springer Nature 2018
L. Rutkowski et al. (Eds.): ICAISC 2018, LNAI 10841, pp. 714–723, 2018.
https://doi.org/10.1007/978-3-319-91253-0_66

over the total number of non-equivalent [3, 7] mutants determines the accuracy of the test set. This ratio, called mutation score, is a very dependable measurement of the tests ability to detect real faults, even if it calculated basing on the tests ability to detect artificially made changes. Effectiveness of mutation testing has been shown by several studies (detailed references can be found in [7]).

However, the effectiveness of mutation testing has a price - its application can be very time consuming, because it typically produces a large number of mutants that needs to be run against the assessed tests. Several researchers has already taken on the problem by proposing various methods aiming at limiting either the number of generated mutants (e.g. selective mutation proposed by Offut at. [11]), the number of mutants that needs to be run (e.g. mutant sampling proposed by Acree [1], mutant clustering proposed by Hussain [5]) or the time required for running the mutants (e.g. parallel execution proposed first by Mathur and Krauser [12]). A survey of costs reduction approaches can be found for example in [7].

The approach presented in this paper shares some concepts with mutants sampling and mutants clustering, as it also required a subset of all mutants be run against tests. In mutants sampling such a subset was selected randomly and the mutation score has only been calculated for the sample. The approaches proposed by Hussain [5] and Ji et al. [6], both used clustering algorithms to divide the mutants into groups. Ji et al. used a domain specific analysis to weight the mutants, so only mutant being representative for some group were later run. Hussain grouped all mutants accordingly to their detectability and used the results to reduce a size of a test set.

Our approach to reducing the number of mutants to be run uses machine learning methods. The subset of mutants that are run is considered to be a training set and the fault detection ability of the remaining mutants is predicted basing on their similarity to the training set. The paper follows out previous work in this area [15–17]. These works were based on the conversion of mutants to a control flow graph representation. While allowing us to obtain satisfactory results, it was a time consuming due to the conversion to graph form, especially for larger programs. In the current work the comparison of mutants is done at bytecode level, instead on their more complex graph representation. This approach does not require any additional preprocessing step as the bytecode is generated automatically by the compiler. As the result this approach scales very well and can be used even for large programs.

Using bytecode in mutation testing is not a novelty [10], but approaches using bytecode mutations are not directly relevant to our work, as they focus mostly on problems related to translating higher level mutation operators into a bytecode level ones and do not try to reduce the number of mutants to be run. However, introducing changes into a byte code directly reduces the compilation time, thus the approaches may also be seen as a way to reduce the costs of applying mutation testing.

2   Classification of Mutants

To be able to run a classification algorithm a way of representing programs is needed. In this research Java programs and its bytecode representation were used. Java bytecode is a intermediate, machine independent, representation of a Java program. It results from compiling a Java program [8]. The bytecode representation of a program was used in this research, because of two main reasons:

1. it is straightforward attainable, thus no additional effort needs to be put into getting it,
2. its structure is simple and rather regular (see [8]), especially if compared with a program source code or graph representation, thus its analyzing is less expensive in term of computational costs.

In addition to it using bytecode allows to avoid any problems with scaling from small benchmark examples to real-life problems as it does not generate any additional costs.

```
public int search(int v){          public int search(int v){          public int search(int v){
    int i;                             int i;                             int i;
    for(i=0;i<size;i=i+1)              for(i=0;++i<size;i=i+1)            for(i=0;i<size;i=i+1)
        if(values[i]==v) return i;         if(values[i]==v) return i;         if(values[i]≥v) return i;
    return -1;                         return -1;                         return -1;
}                                  }                                  }

        a) original                       b)  AOIS mutant                    c) ROR mutant
```

Fig. 1. A search method (a) and two of its mutants: (b) AOIS and (c) ROR [9] mutants

2.1   Mutants Generation

Mutated program represented at the code level differs from the original level by one small change defined by the type of the mutation operator applied to it. Typical mutation operators result in insertion, deletion or replacement of operators or operand in statements of the original program. In case of programs implemented in Java the most popular generator of mutants is muJava [9]. For example, the mutant depicted in Fig. 1(b) was generated by applying an AOIS operator [9] that inserted the ++ operator before the variable i in the condition of the for loop. At the bytecode level the mutation resulted in an additional instruction (Fig. 2(b)). The second mutant shown in Fig. 1(c) was obtained by using a ROR operator [9] that replaced a binary relational operator == with > operator in the condition of the if statement. At the bytecode level the change was reflected in a similar way, by replacing if_icmpne with if_icmple instruction (Fig. 2(c)). All the above described changes are underlined in the Fig. 1(b) and (c) and Fig. 2(b) and (c).

```
public int search(int);          public int search(int);          public int search(int);
  Code:                            Code:                            Code:
    0: iconst_0                      0: iconst_0                      0: iconst_0
    1: istore_2                      1: istore_2                      1: istore_2
    2: iload_2                       2: iinc       2, 1               2: iload_2
    3: aload_0                       5: iload_2                       3: aload_0
    4: getfield    #2               6: aload_0                       4: getfield    #2
    7: if_icmpge   29               7: getfield    #2                7: if_icmpge   29
   10: aload_0                      10: if_icmpge   32               10: aload_0
   11: getfield    #3               13: aload_0                      11: getfield    #3
   14: iload_2                      14: getfield    #3               14: iload_2
   15: iaload                       17: iload_2                      15: iaload
   16: iload_1                      18: iaload                       16: iload_1
   17: if_icmpne   22              19: iload_1                       17: if_icmple   22
   20: iload_2                      20: if_icmpne   25               20: iload_2
   21: ireturn                      23: iload_2                      21: ireturn
   22: iload_2                      24: ireturn                      22: iload_2
   23: iconst_1                     25: iload_2                      23: iconst_1
   24: iadd                         26: iconst_1                     24: iadd
   25: istore_2                     27: iadd                         25: istore_2
   26: goto      2                  28: istore_2                     26: goto      2
   29: iconst_m1                    29: goto      2                  29: iconst_m1
   30: ireturn                      32: iconst_m1                    30: ireturn
                                    33: ireturn
     a) original                      b) AOIS mutant                   c) ROR mutant
```

Fig. 2. A bytecode representation of Java programs from Fig. 1

2.2   Classification Process

The classification process is carried out in several steps. The first is the preparation of input data. In case of mutation testing for each program (classification example) a set of mutants is generated and the set of tests is also provided. The aim of the prediction is to classify each mutant as detectable or not detectable by a given test t. For each test t a vector of the size equal to the number of mutants contains 1 if test t detects a given mutant and 0 otherwise. This data constitutes the first set of inputs for the classification process.

As the programs are represented by their bytecode we do not have actual features for each mutant to classify. So in order to be able to use a kNN classifier a distance matrix has to be calculated. To calculate the distance between two mutants the number of differences are taken into account. As only the first order mutants are used there is only one place in which any given mutant differs from the original program. Thus any two mutants can differ in at most two places. At the bytecode level this changes may transfer to change in an instruction and/or addition or deletion of one or more instructions. Thus the distance is calculated by comparing bytecode of two different mutants and counting the number of changes in the bytecode. The distance matrix is the second input element to the classification process.

After preparing the input data the actual classification process is started. In the first step a training set is selected as a predefined percentage of the whole set of mutants. The selection is performed by a random sampling. The chosen set is checked to exclude the situation in which selected mutants are either all detected or all undetected by any test. If this is a case such a training set is discarded and another one selected. All the remaining mutants are considered to be the test set for which the prediction is performed. Then the kNN algorithm is run for each item of the test set and the number of correctly predicted outputs as well as the number of false positives and false negatives is counted. The whole process is repeated a number of times and the obtained values are averaged to yield the final result.

The predicted mutation score is calculated over the whole set of mutants that is for the mutants belonging to the test set a predicted value is used while for the mutants from the training set the actual value is used. Although it is possible to calculate the mutation score only for the mutants belonging to the test set such a value would be of little use as it would be impossible to compare it with the actual mutation score.

In the next section two examples for which this process has been applied are presented.

Table 1. Actual values for mutation score

	Example 1	Example 2
TS 1	0.421	0.448
TS 2	0.711	0.621
TS 3	0.474	0.253
TS 4	—	0.356
TS 5	—	0.483

3   Experimental Results

The experiment were carried out for two different programs. The mutants for them were generated using Mujava tool [9]. The remaining part of the experiments were supported tool implemented in C++ and Python. One of the examples was a simple search, for which Mujava generated 38 mutants. The second one was binary search, presented in Fig. 1a, for the second examples there were 87 mutants. The mutants were then compiled into bytecode form by a standard Java compiler.

For each set of mutants a k-NN classification algorithm was run using the precalculated distance matrix computed for the bytecode. For the first example three test sets were used and for the second one - five test sets. The set of mutants was randomly divided into training set and test set. The size of the training set was set at 25%, 40%, 50% and 90% of the whole set of mutants. In case of the first example the percentage translated to using training sets of 10,

15, 20 and 34 mutants, respectively. In the second example the training set sizes were equal to 22, 33, 44 and 78 mutants, respectively. In case of mutation testing the quality of classification is measured not only by standard values of correctly classified elements, false positives and false negatives but also by the value of predicted mutation score. Ideally the predicted mutation score should be as close to the actual one as possible. The actual mutation scores for all tests and both examples are presented in Table 1. As the mutation score gives us an insight into the quality of the test set used it plays here an important role. Obtaining a much higher mutation score the it actually should be may prevent the tester from working on the test set improvements. On the other hand predicting a much lower mutation score may involve, probably unnecessary, work on improving the tests. Still, the later case is less dangerous, as improving a good test set is not as bad as failing to improve a poor one.

In all examples the experiment has been run 100 times and the results were averaged. It has to be noticed that as the result of rounding errors the averages obtained over all 100 experiments do not always add up to 100%. Tables 2 and 3 present results obtained for the first example with values of k set to 3 and 5, respectively. Table 4 presents results obtained for the second example with the value of k set to 5. In all tables the results for mutants classified incorrectly are presented separately for those classified as detectable, while actually they are not (column labelled false positive) and for those classified as not detected, while they actually are detected by a given test set (column labelled false negative) because the meaning of these misclassifications in context of testing is different. Classifying a mutant as not detected leads to overtesting, while the misclassification of the second type can result in missing real errors in code and thus is more dangerous as it may lead to undetected errors in code.

Table 2. The classification of mutants for the example 1 with $k = 3$

	Training set size	Correct	False positive	False negative	Mutation score
TS 1	25%	55.47%	17.38%	26.21%	0.355
TS 1	40%	60.29%	12.15%	26.44%	0.333
TS 1	50%	65.36%	10.3%	23.29%	0.358
TS 1	90%	65.01%	6.51%	27.21%	0.350
TS 2	25%	66.41%	26.72%	5.84%	0.866
TS 2	40%	67.55%	26.6%	4.7%	0.844
TS 2	50%	67.35%	26.93%	4.61%	0.817
TS 2	90%	71.0%	19.6%	9.4%	0.723
TS 3	25%	53.27%	20.57%	25.0%	0.478
TS 3	40%	54.69%	30.08%	13.97%	0.583
TS 3	50%	55.55%	32.95%	10.24%	0.576
TS 3	90%	55.0%	4.2%	40.8%	0.372

Table 3. The classification of mutants for the example 1 with k = 5

	Training set size	Correct	False positive	False negative	Mutation score
TS 1	25%	53.46%	19.67%	25.99%	0.373
TS 1	40%	56.26%	9.96%	32.71%	0.281
TS 1	50%	58.88%	9.43%	30.69%	0.319
TS 1	90%	65.24%	3.97%	30.16%	0.365
TS 2	25%	63.5%	25.38%	10.28%	0.823
TS 2	40%	70.82%	21.52%	8.6%	0.875
TS 2	50%	71.74%	21.96%	6.26%	0.829
TS 2	90%	75.2%	11.10%	13.70%	0.736
TS 3	25%	54.73%	14.6%	29.79%	0.307
TS 3	40%	56.61%	10.29%	32.06%	0.287
TS 3	50%	62.03%	3.41%	33.39%	0.317
TS 3	90%	67.26%	5.28%	26.73%	0.403

Table 4. The classification of mutants for the example 2 with k = 5

	Training set size	Correct	False positive	False negative	Mutation score
TS 1	25%	54.59%	13.44%	30.82%	0.317
TS 1	40%	65.68%	19.81%	13.56%	0.457
TS 1	50%	60.39%	14.04%	24.29%	0.392
TS 1	90%	64.35%	22.33%	12.32%	0.458
TS 2	25%	56.67%	28.46%	13.84%	0.730
TS 2	40%	57.46%	29.64%	11.34%	0.730
TS 2	50%	58.67%	28.01%	11.79%	0.701
TS 2	90%	66.28%	21.12%	20.70%	0.630
TS 3	25%	65.1%	15.26%	18.5%	0.227
TS 3	40%	61.29%	24.74%	12.42%	0.313
TS 3	50%	64.24%	19.25%	15.22%	0.277
TS 3	90%	67.22%	14.15%	17.33%	0.231
TS 4	25%	57.39%	14.7%	26.78%	0.264
TS 4	40%	59.25%	9.62%	29.42%	0.331
TS 4	50%	61.92%	11.81%	24.62%	0.385
TS 4	90%	63.14%	23.98%	11.88%	0.460
TS 5	25%	50.47%	30.62%	17.66%	0.579
TS 5	40%	51.88%	32.52%	14.28%	0.602
TS 5	50%	52.35%	32.69%	13.41%	0.578
TS 5	90%	54.23%	24.42%	20.35%	0.487

The results depicted in tables show that the classification worked quite well for all test sets, especially taking into account the fact that the training set has been selected randomly and the distance matrix has been calculated only on the basis of the number of changes. It must be mentioned here that although in some domains the classification rate of 60% would be considered very poor, in the domain of mutation testing it is sufficient to be used for the prediction of the quality of tests. As expected the larger the size of the training set is the more accurate the classification is. The classification results for the training set size set to 25% of the mutants are usually too low to be useful, but the results obtained for the size set to 40% and 50% are good enough to be useful in practice. As the authors in [4] claim that the best results in mutation testing are obtained with the sample size set to 90% of all mutants the test were also carried out with this value. Although our results confirm this claim, the improvement in the quality of classification in comparison to lower sample sizes is not high enough to justify using such a large sample size.

It has to be mentioned here that the test used in our experiments were not very good, that is their mutation score is not very high. In the first example tests T1 and T3 have the mutations core below 0.5, what means that they are able to detect less then a half of all mutants. Also in the second example tests T1, T2, T4 and T5 have the mutation score below 0.5.

This fact results in another important issue that can be observed in all experiments, which is related to the difference of the predicted mutation score against the actual one. For the test with the actual mutation score above 0.5 the predicted mutation score tends to be slightly higher then the actual one, while for those tests with mutation score below 0.5 the predicted mutation score is usually lower then the actual one. In the context of the problem it means that the classification process tends to overrate the good test and underrate the weak ones. The only exception from this observation is test T5 in the second example which has the actual mutation score of 0.48, but in all experiments it is consistently overrated. It may be due to the structure of this test, but this issue needs further examination. Such results are very useful in practical application as assuming that a poor test set is even poorer would make programmers and testers work on improving it. On the other hand slightly overestimating the quality of good tests would allow us to avoid unnecessary work on good tests.

4   Conclusions

The paper deals with a dynamic approach to the problem of the reduction of costs of mutation testing. A classification approach was proposed in this paper allowing to reduce the number of mutants to be executed. It reduces the number of executed mutants depending on the program for which they are generated rather than using some statical method based on the operators or programming language. The approach needs still more experiments to fully confirm its validity, but the results obtained so far are encouraging. However the fact that for 40 and 50% of mutants used as a training set the obtained results are satisfactory means that we can reduce the number of executed mutants by half and thus the cost of mutation testing can also be reduced by half.

Although not all languages compile to a bytecode or other intermediary format, some of the most often used, general purpose languages (such as $C\#$ or Java) usually have some form of it. Hence the approach presented in this paper has a potential to be applied in other contexts too.

Two main directions for future work include devising more elaborated methods for calculating the distance between mutants and defining different methods of selecting the training set. Currently the distance is calculated basing on the number of changes, but the classification may work better if it will considered for example the type or the placement of the change.

As for the training set selection currently only the random sampling has been used, with just one small adjustment described in the paper, which allows us to avoid selecting a training set containing elements of only one class. Yet there is huge potential for different selection methods. The first idea of how to deal with it will be based on selecting proportional number of mutants of each type (generated by a given type of mutation operators) but this idea needs further development.

One more aspect that is also planned to be taken into account is the choice of the k nearest neighbours. As it has already been mentioned due to the nature of the problem, i.e., using the first order mutants, the number of differences between mutants is limited and it in turn results with the distance matrix containing identical values for many pairs of mutants. Thus, in case when for a given item there are more then k neighbours within identical distance from it, the selection of neighbours should involve checking the type of the neighbour. The modification could potentially improve the quality of the classification, probably combined with more sophisticated distance calculations mentioned earlier.

References

1. Acree, A.T.: On Mutation, Ph.D. Thesis, Georgia Institute of Technology, Atlanta, Georgia (1980)
2. Andrews, J.H., Briand, L.C., Labiche, Y.: Is mutation an appropriate tool for testing experiments? In: Proceedings of ICSE, pp. 402–411 (2005)
3. DeMillo, R.A., Lipton, R.J., Sayward, F.G.: Hints on test data selection: help for the practicing programmer. Computer 11(4), 34–41 (1978)
4. Derezinska, A., Rudnik, M.: Evaluation of mutant sampling criteria in object-oriented mutation testing. In: Proceedings of the 2017 Federated Conference on Computer Science and Information Systems, FedCSIS 2017, pp. 1315–1324 (2017)
5. Hussain, S.: Mutation Clustering, Masters Thesis, Kings College London, Strand, London (2008)
6. Ji, C., Chen, Z., Xu, B., Zhao, Z.: A novel method of mutation clustering based on domain analysis. In: Proceedings of the 21st International Conference on Software Engineering and Knowledge Engineering. Knowledge Systems Institute Graduate School (2009)
7. Jia, Y., Harman, M.: An analysis and survey of the development of mutation testing. IEEE Trans. Softw. Eng. 37, 649–678 (2011)

8. Lindholm, T., Yellin, F., Bracha, G., Buckley, A.: The Java Virtual Machine Specification, Java SE 8 Edition, 1st edn. Addison-Wesley Professional, Boston (2015)
9. Ma, Y., Offutt, J., Kwon, Y.R.: MuJava: a mutation system for java. In: Proceedings of ICSE 2006, pp. 827–830 (2006)
10. Ma, Y., Offutt, J., Kwon, Y.R.: MuJava: an automated class mutation system. Softw. Test. Verif. Reliab. **15**(2), 97–133 (2005)
11. Mathur, A.P.: Performance, effectiveness, and reliability issues in software testing. In: Proceedings of the 5th International Computer Software and Applications Conference, pp. 604–605 (1991)
12. Mathur, A.P., Krauser, E.W.: Mutant unification for improved vectorization. Purdue University, West Lafayette, Indiana, Technique report SERC-TR-14-P (1988)
13. Myers, G., Sandler, C., Badgett, T.: The Art of Software Testing. Wiley, London (2011)
14. Roman, A.: Testing and Software Quality. PWN, Warsaw (2015). (in Polish)
15. Strug, J., Strug, B.: Machine learning approach in mutation testing. In: Nielsen, B., Weise, C. (eds.) ICTSS 2012. LNCS, vol. 7641, pp. 200–214. Springer, Heidelberg (2012). https://doi.org/10.1007/978-3-642-34691-0_15
16. Strug, J., Strug, B.: Classifying mutants with decomposition kernel. In: Rutkowski, L., Korytkowski, M., Scherer, R., Tadeusiewicz, R., Zadeh, L.A., Zurada, J.M. (eds.) ICAISC 2016. LNCS (LNAI), vol. 9692, pp. 644–654. Springer, Cham (2016). https://doi.org/10.1007/978-3-319-39378-0_55
17. Strug, J., Strug, B.: Using classification for cost reduction of applying mutation testing. In: Proceedings of FedCSIS 2017, pp. 99–108 (2017)

Probabilistic Learning Vector Quantization with Cross-Entropy for Probabilistic Class Assignments in Classification Learning

Andrea Villmann[1,2], Marika Kaden[1], Sascha Saralajew[1,3],
and Thomas Villmann[1(✉)]

[1] Saxony Institute for Computational Intelligence and Machine Learning,
University of Applied Sciences Mittweida, Mittweida, Germany
[2] Schulzentrum Döbeln-Mittweida, Mittweida, Germany
thomas.villmann@hs-mittweida.de
[3] Dr. Ing. h.c. F. Porsche AG Weissach, Weissach, Germany

Abstract. Classification learning by prototype based approaches is an attractive strategy to achieve interpretable classification models. Frequently, those models optimize the classification error or an approximation thereof. Current deep network approaches use the cross entropy maximization instead. Therefore, we propose a prototype based classifier based on cross-entropy as a probabilistic classifier. As we deduce, the proposed probabilistic classifier is a generalization of the robust soft-learning vector quantizer and allows to handle label noise in training data, i.e. the classifier is able to take into account probabilistic class assignments during learning.

1 Introduction

Learning of complex classification task by deep learning approaches is a standard way to tackle those tasks nowadays. One major advantage of this strategy is the use of pre-trained modules for the deep architecture, which become more and more available for different application ranges far behind the standard image classification problems. However, full training of deep architectures or unsupervised pre-training of task specific modules usually require a huge amount of training data, which maybe are not available for a given problem. For example, patient databases in medicine frequently contain only a few hundreds of entries. Prototype based approaches like nearest neighbor classifiers are able to deal also with only a few data available for training. Additionally, those classifiers are also attractive because of their interpretability of the resulting classification model. Further, probabilistic networks are frequently favored over crisp classification schemes [1, p. 121ff]. Yet, dealing with probabilistic class assignments as well as label noise is crucial and requires sophisticated methods [2].

© Springer International Publishing AG, part of Springer Nature 2018
L. Rutkowski et al. (Eds.): ICAISC 2018, LNAI 10841, pp. 724–735, 2018.
https://doi.org/10.1007/978-3-319-91253-0_67

One of the most prominent prototype based classifiers is learning vector quantization (LVQ) as introduced by Kohonen [3]. Here, each prototype is predetermined to be responsible for a certain class. During the learning phase, the prototype vectors are distributed in the data space by a simple but intuitive attraction and repelling scheme to minimize the classification error based on heuristics. After training, data are classified according to the nearest prototype principle [4]. The prototypes serve as references for the local class distributions. Hence, the interpretability of LVQ networks is obvious.

Stochastic gradient descent learning (SGDL) for LVQ was proposed in [5]. The underlying cost function is a smooth approximation of the classification error. This respective approximation may violate the interpretability of the model due to non-intuitive prototype localizations obtained by this model [6,7]. A probabilistic variant of LVQ is soft LVQ (SLVQ) and its modification robust SLVQ (RSLVQ) [8]. The cost function to be minimized in SLVQ is the negative log-likelihood ratio between the probabilities of correct and incorrect classification based on the estimation of the class distribution using Gaussian mixtures. SLVQ as well as its robust counterpart suffer from the well-known difficulties regarding the estimation of Gaussian mixtures in high-dimensional data spaces [9]. Further, if several prototypes are used for each class, the prototype distribution learning shows a strong tendency for sticking together. Last but not least, ratios are frequently numerically instable in optimization.

In this paper we propose to use cross-entropy learning for LVQ motivated by the information theoretic background of deep learning and multi-perceptron networks [10,11]. Thus the log-likelihood ratio in RSLVQ is replaced by the the cross-entropy obtaining a probabilistic LVQ (PLVQ) model based on information processing principles [12], but still based on mixture of Gaussians to estimate the class distributions. The resulting new PLVQ is able to deal with probabilistic class information and label noise during learning and, hence, extends the range of LVQ classifiers. Additionally to this new feature, we reduce the difficulties regarding Gaussian mixture learning for high-dimensional data in PLVQ applying a adaptive projection learning as discussed in [13] for LVQ. Finally, we incorporate neighborhood cooperativeness strategies in prototype learning to avoid the stickiness of the prototypes.

2   Cross-Entropy in Learning in LVQ

Learning vector quantization (LVQ) assumes a set $W = \{\mathbf{w}_1, \ldots, \mathbf{w}_N\}$ of prototypes $\mathbf{w}_k \in \mathbb{R}^n$ to represent and to classify data $\mathbf{x} \in \mathbb{R}^n$. For this purpose, each prototype is assigned to be responsible for a certain class by the class label $c(\mathbf{w}_k) \in \mathscr{C} = \{1, \ldots, C\}$. Each class is represented by at least one prototype. For network learning a training data set of observed pairs $(\mathbf{X}, \mathbf{T}) = \{\mathbf{x}_i, \mathbf{t}_i\}_{i=1}^{N_D}$ is supposed where $\mathbf{t}_i \in [0,1]^C$ provides the probabilistic *target* class information with $t_{ij} \in [0,1]$ and $\sum_j t_{ij} = 1$. For unique mutually exclusive classification training data $t_{ij} \in \{0,1\}$ is required. For this latter case the respective target class assignment is denoted by $\tau(\mathbf{x})$.

If for mutual exclusive class information, however, class label noise is supposed, this can be modeled by $\tau_c(\mathbf{x}) = 1 - \zeta$ with $\zeta \in (0, 0.5)$ determining a noise level for this label c. We assume for this case that for all components $\tau_k(\mathbf{x})$ of the target vector with $k \neq c$ the class assignment probabilities are given as $\tau_k(\mathbf{x}) = \frac{\zeta}{C-1}$.

2.1   Standard Soft Learning Vector Quantization (SLVQ)

Soft learning vector quantization is designed to find an estimator for the posterior probabilities $p(c|\mathbf{x})$ for a given class $c \in \mathscr{C}$ for unique classification training data. SLVQ considers

$$P_W(\mathbf{x}) = \sum_{j=1}^{N} p(\mathbf{x}|\mathbf{w}_j) p(\mathbf{w}_j) \tag{1}$$

as the probability density for $\mathbf{x}$ generated by the model where the model parameters are the prototypes $W = \{\mathbf{w}_1, \ldots, \mathbf{w}_M\}$ and $p(\mathbf{x}|\mathbf{w}_j)$ is the probability that $\mathbf{x}$ is generated by the jth model component determined by the prototype vector $\mathbf{w}_j$. The probabilities $p(\mathbf{w}_j)$ are the priors for the model components.

The model based estimator of the joint probability for $\mathbf{x}$ and an arbitrarily given but fixed class $c \in \mathscr{C}$ is

$$P_W(\mathbf{x}, c) = \sum_{j: c(\mathbf{w}_j) = c} p(\mathbf{x}|\mathbf{w}_j) p(\mathbf{w}_j) \tag{2}$$

and, analogously, we have

$$P_W(\mathbf{x}, \neg c) = \sum_{j: c(\mathbf{w}_j) \neq c} p(\mathbf{x}|\mathbf{w}_j) p(\mathbf{w}_j) \tag{3}$$

as model based probability of the complement. Further in [8] the probabilities

$$p_c(\mathbf{w}_j|\mathbf{x}) = \frac{p(\mathbf{x}|\mathbf{w}_j) p(\mathbf{w}_j)}{P_W(\mathbf{x}, c)} \tag{4}$$

and

$$p_{\neg c}(\mathbf{w}_j|\mathbf{x}) = \frac{p(\mathbf{x}|\mathbf{w}_j) p(\mathbf{w}_j)}{P_W(\mathbf{x}, \neg c)} \tag{5}$$

are introduced describing the (posterior) probabilities that a data point $\mathbf{x}$ is assigned to the prototype $\mathbf{w}_j$ given that this data point was generated by class c and under the opposite assumption that this data point was not generated any other class than c.

The class prediction probability $p(c|\mathbf{x})$ of SLVQ for each class $c \in \mathscr{C}$ is

$$p(c|\mathbf{x}) = \sum_{j: c(\mathbf{w}_j) = c} p(\mathbf{w}_j|\mathbf{x}) \tag{6}$$

whereby we calculate

$$p\left(\mathbf{w}_j|\mathbf{x}\right) = \frac{p\left(\mathbf{x}|\mathbf{w}_j\right)p\left(\mathbf{w}_j\right)}{\sum_l p\left(\mathbf{x}|\mathbf{w}_l\right)p\left(\mathbf{w}_l\right)} \tag{7}$$

according to the Bayes rule. Here, $p\left(\mathbf{w}_j|\mathbf{x}\right)$ describes the prototype winning probability. This leads to

$$p_W\left(c|\mathbf{x}\right) = \frac{P_W\left(\mathbf{x},c\right)}{P_W\left(\mathbf{x}\right)} \tag{8}$$

as the predicted class probability of the model with $\sum_{c=1}^{C} p\left(c|\mathbf{x}\right) = 1$ is valid.

SLVQ uses the negative log-likelihood ratio

$$L_{SLVQ}\left(X,W\right) = -\sum_k \ln\left(\frac{P_W\left(\mathbf{x}_k,\tau_k\right)}{P_W\left(\mathbf{x}_k,\neg\tau_k\right)}\right) \tag{9}$$

as cost function to be minimized by SGDL for given training data $X = (\mathbf{X}, t\left(\mathbf{x}_k\right))$, whereas the robust variant RSLVQ takes

$$L_{RSLVQ}\left(X,W\right) = -\sum_k \ln\left(\frac{P_W\left(\mathbf{x}_k,\tau_k\right)}{P_W\left(\mathbf{x}_k\right)}\right) \tag{10}$$

into account using the abbreviation $\tau_k = \tau\left(\mathbf{x}_k\right)$ for the target class.

Following Seo and Obermayer in [8] the probability $p\left(\mathbf{x}|\mathbf{w}_j\right)$ can be specified by a Gaussian ansatz

$$p\left(\mathbf{x}|\mathbf{w}_j\right) = K_j \exp\left(-f\left(\mathbf{x},\mathbf{w}_j,\boldsymbol{\Omega}\right)\right) \tag{11}$$

where $f\left(\mathbf{x},\mathbf{w}_j,\boldsymbol{\Omega}\right)$ is a parametrized dissimilarity measure for the data point $\mathbf{x}$ and component vector $\mathbf{w}_j$ and K_j is a normalization constant to ensure $p\left(\mathbf{x}|\mathbf{w}_j\right)$ being a probability.[1] According to this Gaussian assumption, the component vectors $\mathbf{w}_j$ are interpreted as prototype vectors representing the center of a Gaussian probability distribution. The priors $p\left(\mathbf{w}_j\right)$ are chosen to be

$$p\left(\mathbf{w}_j\right) = \frac{1}{N} \tag{12}$$

giving no bias to any prototype.

Usually, the dissimilarity measure f is chosen as the squared Euclidean distance $d_E\left(\mathbf{x},\mathbf{w}_j\right) = \left(\mathbf{x} - \mathbf{w}_j\right)^2$ and $\boldsymbol{\Omega}$ is the identity matrix. Hence, the adaptation of the Gaussian mixture for modeling the probability distributions is,

---

[1] In [8] it is assumed without any explanation that the normalization constants K_j are set to be $K_j = 1 \; \forall j$. Strictly speaking, this may violate the probability property for $p\left(\mathbf{x}|\mathbf{w}_j\right)$, i.e. $\int p\left(\mathbf{x}|\mathbf{w}_j\right) d\mathbf{x} = 1$. However, it does not influence the gradients needed for the online learning algorithm despite a constant factor. Therefore, we also make use of this convention in this paper.

in fact, a Parzen windows estimation problem, which may become crucial for high-dimensional tasks [14,15].

Therefore, we recommend to use the quadratic form

$$d_\Omega\left(\mathbf{x}, \mathbf{w}_j\right) = \left(\boldsymbol{\Omega}\left(\mathbf{x} - \mathbf{w}_j\right)\right)^2 \tag{13}$$

instead of d_E because it offers a nice opportunity to avoid this problem: If $\boldsymbol{\Omega} \in \mathbb{R}^{m \times n}$ with $m \ll n$ we have a low-dimensional projection of the data, i.e. we consider a limited rank matrix $\boldsymbol{\Omega}$ as known from *Limited Rank GMLVQ* [13]. Hence, the Parzen estimation takes place in the low-dimensional space $\mathbb{R}^m$.

Yet, another difficulty arises if more than one prototype per class are used in the model. The prototypes belonging to the same class may stick together because they are attracted by the same data and, hence, they become more and more similar. This observation is well-known in unsupervised vector quantization [16]. One possibility to diminish this difficulty is a careful choice or adaptation of *local* projection matrices $\boldsymbol{\Omega}_j$ for each prototype $\mathbf{w}_j$. In consequence, each prototype follows a local dissimilarity measure, which, however, reduces the interpretability of the model. Another option to prevent the clustering tendency is to focus the prototypes locally by application of a neighborhood cooperativeness based on dissimilarity ranks as applied in neural gas quantizer [17]. The winning rank of a prototype $r\left(\mathbf{w}_k, \mathbf{x}\right)$ is defined as

$$r\left(\mathbf{w}_j, \mathbf{x}\right) = \sum_{k=1}^{N} H\left(d_\Omega\left(\mathbf{x}, \mathbf{w}_j\right) - d_\Omega\left(\mathbf{x}, \mathbf{w}_k\right)\right)$$

with

$$H\left(z\right) = \begin{cases} 1 & z > 0 \\ \\ 0 & z \leq 0 \end{cases} \tag{14}$$

is the Heaviside function. Therefore, we suggest to scale the dissimilarity measure $d_\Omega\left(\mathbf{x}, \mathbf{w}_j\right)$ with these ranks obtaining

$$f\left(\mathbf{x}, \mathbf{w}_j, \boldsymbol{\Omega}\right) = \exp\left(-\frac{r\left(\mathbf{w}_j, \mathbf{x}\right)}{\lambda}\right) \cdot d_\Omega\left(\mathbf{x}, \mathbf{w}_j\right) \tag{15}$$

for use in SLVQ/RSLVQ, where the parameter λ determines the range of neighborhood cooperativeness. We denote these scaled quantities as *winning rank scaled dissimilarites*. In the beginning of the learning process the λ-value is initialized to be large and then slowly decreased to a small but non-vanishing final value.

2.2   Cross-Entropy as an Information Theoretic Objective for the Probabilistic LVQ

We introduce in the following a probabilistic variant of LVQ (PLVQ). This PLVQ combines RSLVQ with cross-entropy optimization learning as known from deep learning. To do so, we first recapitulate the information theoretic concepts for classification learning and incorporate these issues afterwards into the desired PLVQ.

Cross-Entropy as an Information Theoretic Objective for Classification Learning: Information theoretic concepts provide a natural way for classification learning [18]. We suppose a class probability vector $\mathbf{p}(\mathbf{x}) = (p_1(\mathbf{x}), \ldots, p_C(\mathbf{x}))$ and $\mathbf{p}_W(\mathbf{x}) = (p_W(1|\mathbf{x}), \ldots, p_W(C|\mathbf{x}))$ as the respective predicted class probability vector by a probabilistic classifier model depending on the parameter set W. The mutual information between them is maximized if the corresponding Kullback-Leibler-divergence

$$D_{KL}(\mathbf{p}(\mathbf{x}) \| \mathbf{p}_W(\mathbf{x})) = \sum_k p_k(\mathbf{x}) \cdot \log\left(\frac{p_k(\mathbf{x})}{p_W(k|\mathbf{x})}\right) \qquad (16)$$

is minimized. Further we have

$$\begin{aligned} D_{KL}(\mathbf{p}(\mathbf{x}) \| \mathbf{p}_W(\mathbf{x})) &= \sum_k p_k(\mathbf{x}) \cdot \log(p_k(\mathbf{x})) - \sum_k p_k(\mathbf{x}) \cdot \log(p_W(k|\mathbf{x})) \\ &= H_S(\mathbf{p}(\mathbf{x})) - Cr(\mathbf{p}(\mathbf{x}), \mathbf{p}_W(\mathbf{x})) \end{aligned}$$

where $H_S(\mathbf{p}(\mathbf{x}))$ is the Shannon entropy and

$$Cr(\mathbf{p}(\mathbf{x}), \mathbf{p}_W(\mathbf{x})) = \sum_k p_k(\mathbf{x}) \cdot \log(p_W(k, \mathbf{x})) \qquad (17)$$

is the cross-entropy [12, p. 221ff]. Divergence based classification learning was already studied in [9]. In this paper, however, a Rényi-divergence

$$D_\alpha(\mathbf{p}(\mathbf{x}) \| \mathbf{p}_W(\mathbf{x})) = \frac{1}{1-\alpha} \log\left(\sum_k (p_k(\mathbf{x}))^\alpha \cdot (p_W(k|\mathbf{x}))^{1-\alpha}\right) \qquad (18)$$

is used instead of the Kullback-Leibler-divergence [19]. For $\alpha = 2$ this divergence becomes numerically easy [12]. In the limit $\alpha \searrow 1$ the Rényi-divergence becomes the Kullback-Leibler-divergence [20]. Further, cross-entropy is the standard cost function for deep learning multi-layer networks [21]. This decision is based on the observation that gradient of the divergence $D_{KL}(\mathbf{p}(\mathbf{x}) \| \mathbf{p}_W(\mathbf{x}))$ with respect to an arbitrary network parameter β is calculated as

$$\frac{\partial D_{KL}(\mathbf{p}(\mathbf{x}) \| \mathbf{p}_W(\mathbf{x}))}{\partial \beta} = \underbrace{\frac{\partial H_S(\mathbf{p}(\mathbf{x}))}{\partial \beta}}_{=0} - \frac{\partial Cr(\mathbf{p}(\mathbf{x}), \mathbf{p}_W(\mathbf{x}))}{\partial \beta} \qquad (19)$$

because the Shannon entropy of the data is independent from any network parameter.

Otherwise, the sum $\sum_i D_{KL}(\mathbf{p}(\mathbf{x}_i) \| \mathbf{p}_W(\mathbf{x}_i))$ of divergences $D_{KL}(\mathbf{p}(\mathbf{x}_i) \| \mathbf{p}_W(\mathbf{x}_i))$ with $\mathbf{t}_i = \mathbf{p}(\mathbf{x}_i)$ can be related to the *log-likelihood ratio* $\log\frac{p(\mathbf{T}|\mathbf{X})}{p_W(\mathbf{C}|\mathbf{X})}$

in multivariate statistics for mutually exclusive classes in the training set $(\mathbf{X}, \mathbf{T}) = \{\mathbf{x}_i, \mathbf{t}_i\}_{i=1}^{N_D}$ [22]. To see this, we consider *true conditional probability* $p(\mathbf{t}_i|\mathbf{x}_i)$ for a given target class membership vector $\mathbf{t}_i$ as a multinomial distribution

$$p(\mathbf{t}_i|\mathbf{x}_i) = \prod_{j=1}^{C} p_j(\mathbf{x}_i)^{t_{ij}}$$

whereas for the estimated (by the network) conditionals $p_W(\mathbf{c}_i|\mathbf{x}_i)$ based on the training data information $\mathbf{t}_i$ we get

$$p_W(\mathbf{c}_i|\mathbf{x}_i) = \prod_{j=1}^{C} (p_W(j, \mathbf{x}_i))^{t_{ij}}$$

with $t_{ij} \in \{0, 1\}$ and $\sum_j t_{ij} = 1$ according to the mutual exclusiveness in the training data. Further, we have the probabilities $p(\mathbf{T}|\mathbf{X}) = \prod_{i=1}^{N_D} p(\mathbf{t}_i|\mathbf{x}_i)$ and $p_W(\mathbf{C}|\mathbf{X}) = \prod_{i=1}^{N_D} p_W(\mathbf{c}_i|\mathbf{x}_i)$. Thus we obtain

$$\log \frac{p(\mathbf{T}|\mathbf{X})}{p_W(\mathbf{C}|\mathbf{X})} = \log \prod_{i=1}^{N_D} \frac{p(\mathbf{t}_i|\mathbf{x}_i)}{p_W(\mathbf{c}_i|\mathbf{x}_i)}$$

$$= \sum_{i=1}^{N_D} \log \frac{p(\mathbf{t}_i|\mathbf{x}_i)}{p_W(\mathbf{c}_i|\mathbf{x}_i)}$$

$$= \sum_{i=1}^{N_D} \log \left(\prod_{j=1}^{C} p_j(\mathbf{x}_i)^{t_{i,j}} \right) - \sum_{i=1}^{N_D} \log \left(\prod_{j=1}^{C} (p_W(k|\mathbf{x}))^{t_{i,j}} \right)$$

$$= \sum_{i=1}^{N_D} \sum_{j=1}^{C} t_{i,j} \log(p_j(\mathbf{x}_i)) - \sum_{i=1}^{N_D} \sum_{j=1}^{C} t_{i,j} \log(p_W(k|\mathbf{x}))$$

$$= \sum_{i=1}^{N_D} H_S(\mathbf{t}_i) - Cr(\mathbf{t}_i, \mathbf{p}_W(\mathbf{x}_i))$$

which is the desired result.

The PLVQ Based on Cross-Entropy. Now, we are ready to consider PLVQ taking the previously explained information into account. We consider the cost function

$$E_{PLVQ}(X, W) = \sum_{\mathbf{x}} D_{KL}(\mathbf{t}(\mathbf{x}) || \mathbf{p}_W(\mathbf{x})) \tag{20}$$

for PLVQ with

$$\frac{\partial_s E_{PLVQ}(X, W)}{\partial \mathbf{w}_l} = -\frac{\partial Cr(\mathbf{t}(\mathbf{x}), \mathbf{p}_W(\mathbf{x}))}{\partial \mathbf{w}_l} \tag{21}$$

as the stochastic gradient. Now we take the SLVQ model class prediction probability $p_W(c|\mathbf{x})$ from (8) as the probabilistic output of the network, which can be plugged into the cross-entropy (17) obtaining

$$Cr_W(\mathbf{x}) = \sum_{c=1}^{C} t_c(\mathbf{x}) \cdot \log\left(\frac{P_W(\mathbf{x},c)}{P_W(\mathbf{x})}\right) \tag{22}$$

$$= \sum_{c=1}^{C} t_c(\mathbf{x}) \ln\left(\frac{\sum_{j:c(\mathbf{w}_j)=c} \exp\left(-f(\mathbf{x},\mathbf{w}_j,\boldsymbol{\Omega})\right)}{\sum_l \exp\left(-f(\mathbf{x},\mathbf{w}_l,\boldsymbol{\Omega})\right)}\right)$$

$$= \sum_{c=1}^{C} t_c(\mathbf{x}) \ln\left(S_c(\mathbf{x})\right) \tag{23}$$

whereby we specified the log to be the natural logarithm and the sum

$$S_c(\mathbf{x}) = \sum_{j:c(\mathbf{w}_j)=c} S_W(j,\mathbf{x}) \tag{24}$$

estimates the model based class prediction probability $p_W(c|\mathbf{x})$ from (8).

We remark at this point that the model based cross-entropy $Cr_W(\mathbf{x})$ in (22) contains the logarithm of the probability ratios $\frac{P_W(\mathbf{x},c)}{P_W(\mathbf{x})}$ known from the cost function (10) of RSLVQ. Particularly, $Cr_W(\mathbf{x})$ in (22) reduces to the RSLVQ cost function in case of mutually exclusive training data. Thus, the obtained PLVQ model is more general than the original RSLVQ approach.

Further, we mention that the addends

$$S_W(j,\mathbf{x}) = \frac{\exp\left(-f(\mathbf{x},\mathbf{w}_j,\boldsymbol{\Omega})\right)}{\sum_l \exp\left(-f(\mathbf{x},\mathbf{w}_l,\boldsymbol{\Omega})\right)} = \frac{\exp(-f_j)}{\sum_l \exp(-f_l)}$$

are the estimators for prototype winning probabilities from (7) but now explained as a softmax function. This observation allows to calculate the derivative $\frac{\partial S_W(j)}{\partial f_k}$ easily as

$$\frac{\partial S_W(j,\mathbf{x})}{\partial f_k} = S_W(j,\mathbf{x})\left(\delta_{jk} - S_W(k,\mathbf{x})\right) \tag{25}$$

according to the properties of the softmax function [21,23,24]. Furthermore, we have

$$\frac{\partial \ln\left(S_c(\mathbf{x})\right)}{\partial S_c(\mathbf{x})} = \frac{1}{S_c(\mathbf{x})} \tag{26}$$

and

$$\frac{\partial S_c(\mathbf{x})}{\partial S_W(j,\mathbf{x})} = \delta_{c(\mathbf{w}_j),c} = \begin{cases} 1 & c(\mathbf{w}_j) = c \\ \\ 0 & c(\mathbf{w}_j) \neq c \end{cases} \tag{27}$$

Thus, we get for the derivative of the cross entropy (22)

$$\frac{\partial Cr_W(\mathbf{x})}{\partial \mathbf{w}_l} = -\sum_{c=1}^{C} t_c(\mathbf{x}) \frac{\partial \ln\left(S_c(\mathbf{x})\right)}{\partial \mathbf{w}_l} \tag{28}$$

which can be calculated as

$$\frac{\partial Cr_W(\mathbf{x})}{\partial \mathbf{w}_l} = \sum_{c=1}^{C} t_c(\mathbf{x}) \frac{\partial \ln(S_c(\mathbf{x}))}{\partial S_c(\mathbf{x})} \frac{\partial S_c(\mathbf{x})}{\partial S_W(l,\mathbf{x})} \frac{\partial S_W(l,\mathbf{x})}{\partial f_l} \frac{\partial f_l}{\partial \mathbf{w}_l}$$

$$= \sum_{c=1}^{C} t_c(\mathbf{x}) \frac{1}{S_c(\mathbf{x})} \delta_{c(\mathbf{w}_l),c} \cdot S_W(l,\mathbf{x})(1 - S_W(l,\mathbf{x})) \frac{\partial f_l}{\partial \mathbf{w}_l}$$

$$= \sum_{c=1}^{C} \frac{t_c(\mathbf{x})}{S_c(\mathbf{x})} \delta_{c(\mathbf{w}_l),c} \cdot S_W(l,\mathbf{x})(1 - S_W(l,\mathbf{x})) \frac{\partial f_l}{\partial \mathbf{w}_l}$$

Because each prototype is uniquely responsible for a particular class $c_l = c(\mathbf{w}_l)$, i.e. $\delta_{c_l,c} = 1$ is valid only for $c = c_l$, the last term simplifies to

$$\frac{\partial Cr_W(\mathbf{x})}{\partial \mathbf{w}_l} = \begin{cases} \frac{t_{c_l}(\mathbf{x})}{S_{c_l}(\mathbf{x})} S_W(l,\mathbf{x})(1 - S_W(l,\mathbf{x})) \frac{\partial f_l}{\partial \mathbf{w}_l} & c_l = c \\ 0 & c_l \neq c \end{cases} \tag{29}$$

as the respective gradient. Again, we observe the similarity to RSLVQ updates. As we already mentioned, we favor the winning rank scaled dissimilarities f_k from (15) for use in RSLVQ and, hence, also in PLVQ.

3 Experiments

The illustrating *triangle dataset* consists of 2D-data belonging to three overlapping classes, see Fig. 1. The label certainty ζ (compare to Sect. 2) was chosen in dependence on the localization of the data point. Thereby, an unique class assignment to class c with a respective target value $t_{ic} = 1$ for a data vector $\mathbf{x}_i$ corresponds to a ζ-value one. The target value in regions with increased uncertainty was set randomly as $t_c(\mathbf{x}) \in [0.5, 1]$ such that $\zeta = 1 - t_{ic}$ whereas for the respective overlapping class k we have $t_{ik} = \zeta$. We trained a RSLVQ as well as a PLVQ with one prototype for each class. However, only PLVQ uses this additional information of uncertainty. The resulting prototype configurations are visualized in Fig. 2. We observe that the prototypes of PLVQ try to be responsible for areas with high label certainty whereas the RSLVQ-prototypes are located more in the class centers despite the weak certainty information in this region. To reflect this behavior precisely, we introduce the quantity

$$\zeta_\mathbf{w}(X) = \frac{\#\{\mathbf{x} \in X | \mathbf{w} = \mathbf{w}_{s(\mathbf{x})} \wedge c(\mathbf{x}) = c(\mathbf{w}_{s(\mathbf{x})})\}}{\#\{\mathbf{x} \in X | \mathbf{w} = \mathbf{w}_{s(\mathbf{x})}\}}$$

denoted as the *classification certainty* of the prototype $\mathbf{w}$ regarding the prototype class $c(\mathbf{w})$. Hence, the classification certainty of $\mathbf{w}$ is the fraction of data points in the receptive field $R_\mathbf{w}$ belonging to the same class like $\mathbf{w}$. The overall model classification certainty is the average $\zeta(X,W) = \frac{1}{|W|} \sum_{\mathbf{w} \in W} (\zeta_\mathbf{w}(X))$. For the RSLVQ we obtain $\zeta(X,W) = 0.919$ whereas the PLVQ yields $\zeta(X,W) = 1$ indicating an improvement of the classification certainty.

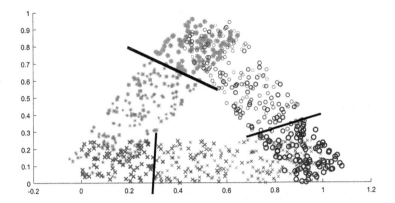

Fig. 1. Triangle dataset - three overlapping classes. Bold symbols of a class (color) c indicate that for the respective target value $t_{ic} = 1$ corresponding to a certainty value $\zeta = 1$. Otherwise, $t_{ic} < 1$ with the certainty value $\zeta = 1 - t_{ic}$ is valid and for the overlapping class c the target $t_{ik} = \zeta$ holds. The different certainty areas within the classes are separated by black lines. (Color figure online)

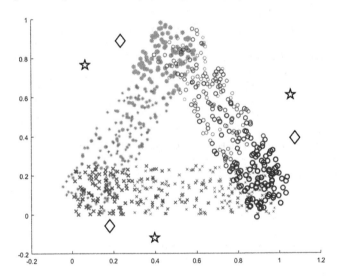

Fig. 2. Learned prototypes for standard RSLVQ (★) and PLVQ (◇). The prototypes for the PLVQ moved to the class areas with high certainty value to ensure a secure classification whereas standard RSLVQ-prototypes do not benefit from this information and, therefore, are located in regions with high uncertainty.

4   Conclusions

In this contribution we investigate a probabilistic variant of LVQ (PLVQ) based on information theoretic quantities. Particularly, the Kullback-Leibler-divergence is considered leading to a cross-entropy cost function. The resulting model is equivalent to RSLVQ, if unique class assignments for training data are supposed. However, the new model is able to handle probabilistic class assignments for training data, too, such that PLVQ can be seen as a generalization of RSLVQ.

References

1. Hertz, J.A., Krogh, A., Palmer, R.G.: Introduction to the Theory of Neural Computation. Santa Fe Institute Studies in the Sciences of Complexity: Lecture Notes, vol. 1. Addison-Wesley, Redwood City (1991)
2. Frénay, B., Verleysen, M.: Classification in the presence of label noise: a survey. IEEE Trans. Neural Netw. Learn. Syst. **25**(5), 845–869 (2014)
3. Kohonen, T.: Learning vector quantization. Neural Netw. **1**(Suppl. 1), 303 (1988)
4. Biehl, M., Hammer, B., Villmann, T.: Prototype-based models in machine learning. Wiley Interdiscip. Rev.: Cogn. Sci. **7**(2), 92–111 (2016)
5. Sato, A., Yamada, K.: Generalized learning vector quantization. In: Touretzky, D.S., Mozer, M.C., Hasselmo, M.E. (eds.) Proceedings of the 1995 Conference on Advances in Neural Information Processing Systems, vol. 8, pp. 423–429. MIT Press, Cambridge (1996)
6. Kaden, M., Lange, M., Nebel, D., Riedel, M., Geweniger, T., Villmann, T.: Aspects in classification learning - review of recent developments in Learning Vector Quantization. Found. Comput. Decis. Sci. **39**(2), 79–105 (2014)
7. Villmann, T., Bohnsack, A., Kaden, M.: Can learning vector quantization be an alternative to SVM and deep learning? J. Artif. Intell. Soft Comput. Res. **7**(1), 65–81 (2017)
8. Seo, S., Obermayer, K.: Soft learning vector quantization. Neural Comput. **15**, 1589–1604 (2003)
9. Torkkola, K.: Feature extraction by non-parametric mutual information maximization. J. Mach. Learn. Res. **3**, 1415–1438 (2003)
10. LeCun, Y., Bengio, Y., Hinton, G.: Deep learning. Nature **521**, 436–444 (2015)
11. Xu, D., Principe, J.: Training MLPs layer-by-layer with the information potential. In: Proceedings of the International Joint Conference on Neural Networks, IJCNN 1999, Los Alamitos, pp. 1045–1048. IEEE Press (1999)
12. Principe, J.C.: Information Theoretic Learning. Springer, Heidelberg (2010). https://doi.org/10.1007/978-1-4419-1570-2
13. Bunte, K., Schneider, P., Hammer, B., Schleif, F.-M., Villmann, T., Biehl, M.: Limited rank matrix learning, discriminative dimension reduction and visualization. Neural Netw. **26**(1), 159–173 (2012)
14. Principe, J.C., Fischer III, J.W., Xu, D.: Information theoretic learning. In: Haykin, S. (ed.) Unsupervised Adaptive Filtering. Wiley, New York (2000)
15. Hild, K.E., Erdogmus, D., Principe, J.: Blind source separation using Rényi's mutual information. IEEE Signal Process. Lett. **8**(6), 174–176 (2001)
16. Martinetz, T.: Selbstorganisierende neuronale Netzwerkmodelle zur Bewegungssteuerung. Ph.D.-thesis, Technische Universität München, München, Germany (1992)
17. Martinetz, T.M., Berkovich, S.G., Schulten, K.J.: 'Neural-gas' network for vector quantization and its application to time-series prediction. IEEE Trans. Neural Netw. **4**(4), 558–569 (1993)

18. Deco, G., Obradovic, D.: An Information-Theoretic Approach to Neural Comput-
    ing. Springer, Heidelberg, New York, Berlin (1997). https://doi.org/10.1007/978-
    1-4612-4016-7
19. Rényi, A.: On measures of entropy and information. In: Proceedings of the Fourth
    Berkeley Symposium on Mathematical Statistics and Probability, Berkeley. Uni-
    versity of California Press (1961)
20. Rényi, A.: Probability Theory. North-Holland Publishing Company, Amsterdam
    (1970)
21. Goodfellow, I., Bengio, Y., Courville, A.: Deep Learning. MIT Press, Cambridge
    (2016)
22. Wittner, B.S., Denker, J.S.: Strategies for teaching layered networks classification
    tasks. In: Anderson, D.Z. (ed.) Neural Information Processing Systems, pp. 850–
    859. American Institute of Physics (1988)
23. Bengio, Y.: Learning deep architectures for AI. Found. Trends Mach. Learn. **2**(1),
    1–127 (2009)
24. Bengio, Y.: Practical recommendations for gradient-based training of deep archi-
    tectures. In: Montavon, G., Orr, G.B., Müller, K.-R. (eds.) Neural Networks: Tricks
    of the Trade. LNCS, vol. 7700, pp. 437–478. Springer, Heidelberg (2012). https://
    doi.org/10.1007/978-3-642-35289-8_26

Multi-class and Cluster Evaluation Measures Based on Rényi and Tsallis Entropies and Mutual Information

Thomas Villmann[✉] and Tina Geweniger

Saxony Institute for Computational Intelligence and Machine Learning,
University of Applied Sciences Mittweida, Mittweida, Germany
thomas.villmann@hs-mittweida.de

Abstract. The evaluation of cluster and classification models in comparison to ground truth information or other models is still an objective for many applications. Frequently, this leads to controversy debates regarding the informative content. This particularly holds for cluster evaluations. Yet, for imbalanced class cardinalities, similar problems occur. One possibility to handle evaluation tasks in a more natural way is to consider comparisons in terms of shared or non-shared information. Information theoretic quantities like mutual information and divergence are designed to answer respective questions. Besides formulations based on the most prominent Shannon-entropy, alternative definitions based on relaxed entropy definitions are known. Examples are Rényi- and Tsallis-entropies. Obviously, the use of those entropy concepts result in an readjustment of mutual information etc. and respective evaluation measures thereof. In the present paper we consider several information theoretic evaluation measures based on different entropy concepts and compare them theoretically as well as regarding their performance in applications.

1 Introduction

Information theoretic measures for cluster/classification evaluation are a popular alternative to standard quantities like classification error or cluster separability/compactness [1–4]. Examples are the partition entropy for evaluation of fuzzy clustering solutions [5,6] or the conn-index for prototype-based clustering [7]. The comparison of cluster/classification models with ground truth data or other models usually is done by the statistical evaluation of respective confusion matrices [8].

Recently, an information theoretic approach for the latter problem was proposed [9]. In this work, the Shannon mutual information between the ground truth (true information) and the cluster/classification assignment obtained from the model (system output) is investigated for artificial and real data examples. It was shown in this paper that the Shannon mutual information is suitable to discriminate classification/cluster solutions regarding their performance in comparison to ground truth. Thereby, the Shannon mutual information is based on the Shannon entropy and, hence, also the numerical calculation of the respective

© Springer International Publishing AG, part of Springer Nature 2018
L. Rutkowski et al. (Eds.): ICAISC 2018, LNAI 10841, pp. 736–749, 2018.
https://doi.org/10.1007/978-3-319-91253-0_68

quantities are based on the corresponding Shannon entropy. Yet, it is well known that the calculation of the Shannon entropy suffers from numerical instabilities in the presence of noisy estimated probability values due to the logarithm magnifying these effects. More robust alternatives are information theoretic quantities based on Rényi or Tsallis entropies [10,11]. For the Rényi entropy the logarithm is applied to the sum of the estimated probabilities, which diminishes this sensitivity. The Tsallis entropy goes completely without the logarithm.

In the present work we examine, how cluster/classification evaluation measures based on information theoretic quantities behave, if the underlying Shannon entropy is replaced by a Rényi- or Tsallis-entropy. We work out the details of the different theoretical properties of the corresponding quantities. Additionally, we compare their behavior, if they are applied to illustrative and instructive artificial examples as well as for real world problems. For this purpose, first we briefly review the information theoretic concepts based on the Shannon entropy. Thereafter, we present the corresponding quantities in terms of Rényi- and Tsallis-entropy paying attention to their mathematical properties. Respective numerical studies for artificial as well as real word data complete the considerations.

2 Information Theoretic Cluster/Classification Evaluation Based on Shannon Entropy

2.1 General Description of the Evaluation Approach

In [12] the authors introduced an information theoretic approach to evaluate multi-class/cluster assignment models. It is based on the Shannon entropy and mutual information yielding a measure for the overall system performance. Figure 1 depicts a Venn diagram of the joint entropy [13] representing the relation between entropy and mutual information.

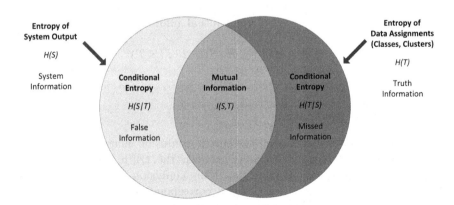

Fig. 1. Venn diagram for the visualization of the relations between the information theoretic quantities

We consider a system output S and the true information T regarding a classification or clustering problem, whereby the system delivers an estimation for the cluster solution or the classification assignment. The following five information theoretic measures can be derived for this scenario [12]:

- the total amount of truth information determined as the expected information amount known as the Shannon entropy $H(T)$ [14]
- the total amount of system information induced by the assignments provided by the model is the Shannon entropy $H(S)$
- the mutual information $I(T, S)$ representing the amount of matching information of the true classes/clusters and the model classes/clusters
- the amount of missed given by the conditional entropy $H(T|S)$
- the amount of false information $H(S|T)$

The conditional entropies $H(T|S)$ and $H(S|T)$ are determined according to the so-called *chain rule of the conditional entropy*

$$H(S|T) = H(S,T) - H(T) \text{ and } H(T|S) = H(S,T) - H(S) \qquad (1)$$

where $H(S,T)$ is the joint entropy. Hence, the mutual information can be calculated as

$$I(S,T) = H(S) - H(S|T) = H(T) - H(T|S) \qquad (2)$$

such that

$$I(S,T) = H(S) + H(T) - H(S,T) \qquad (3)$$

is valid [15].

For an optimal classification/cluster model the missed as well as the false information should vanish maximizing the mutual information $I(S,T)$ at the same time. This idea is picked up in [12] to motivate the score

$$\varsigma_S = H(T|S) + H(S|T) \qquad (4)$$

for cluster/classification evaluation based on Shannon entropy. Obviously, we can express the score in terms of the mutual information $I(S,T)$, such that $\varsigma_S = \varsigma_{MI}$ with

$$\varsigma_{MI} = H(T) + H(S) - 2 \cdot I(S,T) \qquad (5)$$

according to relation (2).

2.2   Numerical Realization

In multi-class/cluster assignment systems a confusion matrix $\mathbf{C} \in \mathbb{R}^{N_T \times N_S}$ is used to represent the relationship between the truth classes/cluster and the system classes/cluster assignments. If the confusion matrix is normalized, i.e. $\sum_j c_{ij} = 1$ is valid, this relationship corresponds to the joint probability function $p_{t,s}(i,j) = c_{ij}$. The respective probability functions $p_t(i)$ and $p_s(j)$ can be calculated by summing up the row respectively column values of $\mathbf{C}$. The conditional probabilities are obtained as $p_{s|t}(i,j) = p_{t,s}(i,j)/p_t(i)$ and

$p_{t|s}(i,j) = p_{t,s}(i,j)/p_s(j)$. These probability functions can now be used to determine the above mentioned information theoretic measures:

$$H(T) = -\sum_{i=1}^{N_T} p_t(i) \log(p_t(i)) \tag{6}$$

$$H(S) = -\sum_{j=1}^{N_S} p_s(j) \log(p_s(j)) \tag{7}$$

$$H(T|S) = -\sum_{i=1}^{N_T} \sum_{j=1}^{N_S} p_{t,s}(i,j) \log(p_{t|s}(i,j)) \tag{8}$$

$$H(S|T) = -\sum_{i=1}^{N_T} \sum_{j=1}^{N_S} p_{t,s}(i,j) \log(p_{s|t}(i,j)) \tag{9}$$

The Shannon mutual information can be expressed in terms of the Kullback-Leibler-divergence

$$I(T,S) = D_{KL}(p_{t,s} \| p_t \cdot p_s) \tag{10}$$
$$= \sum_{i=1}^{N_T} \sum_{j=1}^{N_S} p_{t,s}(i,j) \log\left(\frac{p_{t,s}(i,j)}{p_t(i) \cdot p_s(j)}\right)$$

such that we can calculate $I(T,S)$ without explicit determination of $H(T|S)$ and $H(S|T)$.

It is known that the numerical computation of these entropies/divergences is unstable for very small probability values regarding small variations due to the logarithmic function inside the above quantities based on Shannon entropy [16]. A more robust alternative compared to Shannon entropy are the Rényi and the Tsallis entropy, which are considered in the next two sections.

3   Information Theoretic Cluster/Classification Evaluation Based on Rényi Entropy

To avoid the above mentioned numerical instabilities regarding the Shannon entropy we now investigate its replacement by the Rényi-entropy [10]. The total amount of truth and system information now read as

$$H_\alpha(T) = \frac{1}{1-\alpha} \log\left(\sum_{i=1}^{N_T} (p_t(i))^\alpha\right) \tag{11}$$

$$H_\alpha(S) = \frac{1}{1-\alpha} \log\left(\sum_{j=1}^{N_S} (p_s(j))^\alpha\right) \tag{12}$$

respectively, where $\alpha > 0$ and $\alpha \neq 1$ is a parameter. In the limit $\alpha \to 1$ the Shannon entropy is obtained. The Rényi-mutual-information (RMI) is defined by

$$I_\alpha(T, S) = \frac{1}{\alpha - 1} \log \left(\sum_{i=1}^{N_T} \sum_{j=1}^{N_S} \frac{(p_{t,s}(i,j))^\alpha}{(p_t(i))^{\alpha-1} \cdot (p_s(j))^{\alpha-1}} \right) \tag{13}$$

as outlined in [17]. In analogy to ς_{MI} from (5), we define the Rényi evaluation score

$$\varsigma_{RMI}(\alpha) = H_\alpha(T) + H_\alpha(S) - 2 \cdot I_\alpha(S, T) \tag{14}$$

based on the RMI. However, in contrast to Eq. (3) being valid in the Shannon case, the inequality

$$I_\alpha(T, S) \neq H_\alpha(T) + H_\alpha(S) - H_\alpha(T, S) \tag{15}$$

holds for RMI. Thus, the RMI is not based on the Rényi-entropy in a straightforward way [18] and, hence, we can not find an expression for $\varsigma_{RMI}(\alpha)$ in terms of conditional Rényi-entropies in an easy manner like in (4) for the Shannon case [19]. Moreover, the RMI can not be easily related to the Rényi-divergence of Kullback-Leibler-type

$$D_\alpha^R(p_{t,s} \| q_{t,s}) = \frac{1}{\alpha - 1} \log \left(\sum_{i=1}^{N_T} \sum_{j=1}^{N_S} (p_{t,s}(i,j))^\alpha (q_{t,s}(i,j))^{1-\alpha} \right)$$

as mentioned in [20]. Yet, the limit $\lim_{\alpha \to 1} D_\alpha^R(p_{t,s} \| q_{t,s}) = D_{KL}(p_{t,s} \| q_{t,s})$ is valid.

These problems arise due to the difficulties to define an appropriate conditional Rényi entropy [21]. Several attempts were proposed to obtain respective quantities consistent with desired properties [22,23]. In this paper we consider the following variants studied in [24,25]:

– Jizba-Arimitsu conditional Rényi-entropy

$$H_\alpha^{JA}(S|T) = H_\alpha(S, T) - H_\alpha(T) \tag{16}$$

– Arimoto conditional Rényi-entropy

$$H_\alpha^A(S|T) = \frac{\alpha}{1-\alpha} \log \left(\sum_{i=1}^{N_T} p_t(i) \left(\sum_{j=1}^{N_S} (p_{s|t}(i,j))^\alpha \right)^{1/\alpha} \right) \tag{17}$$

– Hayashi conditional Rényi-entropy

$$H_\alpha^H(S|T) = \frac{1}{1-\alpha} \log \left(\sum_{i=1}^{N_T} p_t(i) \left(\sum_{j=1}^{N_S} (p_{s|t}(i,j))^\alpha \right) \right) \tag{18}$$

It can be shown that $H_\alpha^H(S|T) \le H_\alpha^A(S|T)$ is valid using the Jensen's inequality [24]. Obviously, $H_\alpha^{JA}(S|T)$ can be interpreted as an extension of the conditional Shannon entropy $H(S|T)$ because (16) corresponds to Shannon chain rule (1). Analogously, we find $H_\alpha^{JA}(T|S)$, $H_\alpha^A(T|S)$, and $H_\alpha^H(T|S)$.

Based on these quantities we define the scores

$$\varsigma_\alpha^K = H_\alpha^K(S|T) + H_\alpha^K(T|S) \tag{19}$$

with $K \in \{JA, A, H\}$ in agreement with the Shannon entropy based score ς_S from (4)

Further, we can define the following mutual information measures

$$I_\alpha^A(S,T) = H_\alpha(S) - H_\alpha^A(S|T) \tag{20}$$

and

$$I_\alpha^H(S,T) = H_\alpha(S) - H_\alpha^H(S|T) \tag{21}$$

which both are not symmetric, non-negative and equal to zero if S and T are statistically independent with $I_\alpha^A(S,T) \le I_\alpha^H(S,T)$ [24]. The symmetric variants read as

$$\tilde{I}_\alpha^A(S,T) = \frac{H_\alpha(S) + H_\alpha(T)}{2} - \frac{H_\alpha^A(T|S) + H_\alpha^A(S|T)}{2} \tag{22}$$

and

$$\tilde{I}_\alpha^H(S,T) = \frac{H_\alpha(S) + H_\alpha(T)}{2} - \frac{H_\alpha^H(T|S) + H_\alpha^H(S|T)}{2} \tag{23}$$

which can be used to define respective scores $\varsigma_{RMI}^A(\alpha)$ and $\varsigma_{RMI}^H(\alpha)$ in analogy to $\varsigma_{RMI}(\alpha)$ from (14). However, we calculate

$$\varsigma_{RMI}^A(\alpha) = H_\alpha(S) + H_\alpha(T) - \underbrace{\left(H_\alpha(S) + H_\alpha(T) - H_\alpha^A(T|S) - H_\alpha^A(S|T)\right)}_{=2\cdot\tilde{I}_\alpha^A(S,T)}$$

$$= \varsigma_\alpha^A$$

and

$$\varsigma_{RMI}^H(\alpha) = H_\alpha(S) + H_\alpha(T) - \underbrace{\left(H_\alpha(S) + H_\alpha(T) - H_\alpha^H(T|S) - H_\alpha^H(S|T)\right)}_{=2\cdot\tilde{I}_\alpha^H(S,T)}$$

$$= \varsigma_\alpha^H$$

such that the score based on mutual information for Rényi entropies does not lead to new quantitative measures.

4 Information Theoretic Cluster/Classification Evaluation Based on Tsallis Entropy

Another robust information measure is the Tsallis entropy

$$T_\alpha(S) = \frac{1}{\alpha - 1}\left(1 - \sum_{j=1}^{N_S}(p_s(j))^\alpha\right) \tag{24}$$

for S [11] (analogously we have $T_\alpha\,(T)$). It is closely related to the Rényi-entropy via the equation

$$H_\alpha\,(S) = \frac{1}{1-\alpha} \log\left(1 + (1-\alpha)\,T_\alpha\,(S)\right)$$

but it does not share the property of additivity. The joint entropy reads as

$$T_\alpha\,(T,S) = \frac{1}{\alpha-1}\left(1 - \sum_{i=1}^{N_T}\sum_{j=1}^{N_S}\left(p_{,ts}\,(i,j)\right)^\alpha\right)$$

whereas the conditional Tsallis entropy is consistently defined by

$$T_\alpha\,(T|S) = \frac{1}{\alpha-1}\left(1 - \frac{\sum_{i=1}^{N_T}\sum_{j=1}^{N_S}\left(p_{,ts}\,(i,j)\right)^\alpha}{\sum_{j=1}^{N_S}\left(p_s\,(j)\right)^\alpha}\right) \tag{25}$$

as pointed out in [26]. Accordingly, the relation

$$T_\alpha\,(T|S) = \frac{T_\alpha\,(T,S) - T_\alpha\,(S)}{1 + (1-\alpha)\cdot T_\alpha\,(S)} \tag{26}$$

holds (and analogously for $T_\alpha\,(S|T)$) [27], where

$$T_\alpha\,(T,S) = T_\alpha\,(S) + T_\alpha\,(T|S) + (1-\alpha)\cdot T_\alpha\,(T|S)\cdot T_\alpha\,(S) \tag{27}$$

is the joint entropy [26], which simplifies to

$$T_\alpha\,(T,S) = T_\alpha\,(S) + T_\alpha\,(T) + (1-\alpha)\cdot T_\alpha\,(T)\cdot T_\alpha\,(S) \tag{28}$$

for independent quantities S and T determining the subadditivity [27,28].

We can use the conditional entropies $T_\alpha\,(T|S)$ and $T_\alpha\,(S|T)$ for the score

$$\varsigma_\alpha^{\mathrm{T}} = T_\alpha\,(T|S) + T_\alpha\,(S|T) \tag{29}$$

in analogy to ς_S from (4). Further, the Tsallis mutual information

$$I_\alpha^{\mathrm{T}}\,(S,T) = D_\alpha^{\mathrm{T}}\,(p_{t,s} \,\|\, p_t\cdot p_s) \tag{30}$$

is of Kullback-Leibler-type, i.e.

$$D_\alpha^{\mathrm{T}}\,(p_{t,s} \,\|\, p_t\cdot p_s) = \frac{1}{1-\alpha}\left(1 - \sum_{i=1}^{N_T}\sum_{j=1}^{N_S} \frac{\left(p_{t,s}\,(i,j)\right)^\alpha}{\left(p_t\,(i)\right)^{\alpha-1}\cdot\left(p_s\,(j)\right)^{\alpha-1}}\right)$$

as explained in [29]. A symmetry relation is given by $\frac{D_\alpha^{\mathrm{T}}(f\|g)}{\alpha} = \frac{D_{1-\alpha}^{\mathrm{T}}(g\|f)}{1-\alpha}$ [29], which here leads to the equivalence $\frac{I_\alpha^{\mathrm{T}}(S,T)}{\alpha} = \frac{I_{1-\alpha}^{\mathrm{T}}(T,S)}{1-\alpha}$ generating an α-symmetry with respect to $\alpha = 0.5$ and $\alpha \in (0,1)$. Now

$$\varsigma_{TMI}\,(\alpha) = I_\alpha^{\mathrm{T}}\,(S,T) \tag{31}$$

can serve as a score. Unfortunately, this kind of mutual information cannot be directly related to conditional entropies $T_\alpha(S|T)$ and $T_\alpha(T|S)$ from (26) due to the subadditivity property of the joint entropy [30]. Therefore, another quantity came into play for a generalized mutual information, which adopts the chain rule (1) and the definition (3) of the joint entropy for Shannon entropy: Tsallis proposed

$$\tilde{M}_\alpha^T(S,T) = T_\alpha(S) - T_\alpha(S|T)$$

as Tsallis mutual entropic information [28]. Unfortunately, the inequality $\tilde{M}_\alpha^T(S,T) \neq \tilde{M}_\alpha^T(T,S)$ holds, violating the symmetry property of a mutual information. Taking further into account the case of completely dependent variables finally the Tsallis mutual entropy is obtained as

$$M_\alpha^T(S,T) = \frac{T_\alpha(S) + T_\alpha(T) - T_\alpha(S,T) + (1-\alpha)\cdot T_\alpha(S)\cdot T_\alpha(T)}{1 + (1-\alpha)\cdot[\max\{T_\alpha(S), T_\alpha(T)\}]} \qquad (32)$$

using the subadditivity relation (28) as well as the expression (26) for the conditional Tsallis entropy [30]. It is consistent with all required properties for a mutual information, i.e. it vanishes for independent variables, it is symmetric and if S is coincident with T then $M_\alpha^T(S,T) = T_\alpha(S)$. Thus, the last score considered in this paper is

$$\varsigma_{TME}(\alpha) = M_\alpha^T(S,T) \qquad (33)$$

denoted as Tsallis mutual entropy score.

5   Numerical Experiments

We reconsider for the new evaluation measures the same experiments, an artificial and a real world example, as provided in [12] for evaluation of measures based on Shannon entropy.

5.1   Artificial Illustrative Example

In the first illustrative but artificial experiment four confusion matrices are given representing four hypothetical classifiers A–D, see Fig. 2. Thereby, the classifiers A–C yield the same accuracy as well as have equal statistical κ-coefficients [31]. They differ regarding their off-diagonal variations. The classifier D has reduced accuracy and κ-value.

The results for the different scores and measures are depicted in Tables 1, 2 and 3. As reference always serve the scores/measures based on Shannon entropy. All Shannon-scores favor the sequence C→D→B→A with decreasing performance in line. The parameterized measures based on Rényi entropy with $\alpha = 0.5$ are in nice agreement with the Shannon-based quantities. For Tsallis-based measures, at a first glance, their compliance holds only for ζ_α^T being based on the conditional entropy. For the two scores $\zeta_{TMI}(\alpha)$ and $\zeta_{TME}(\alpha)$ the order seems

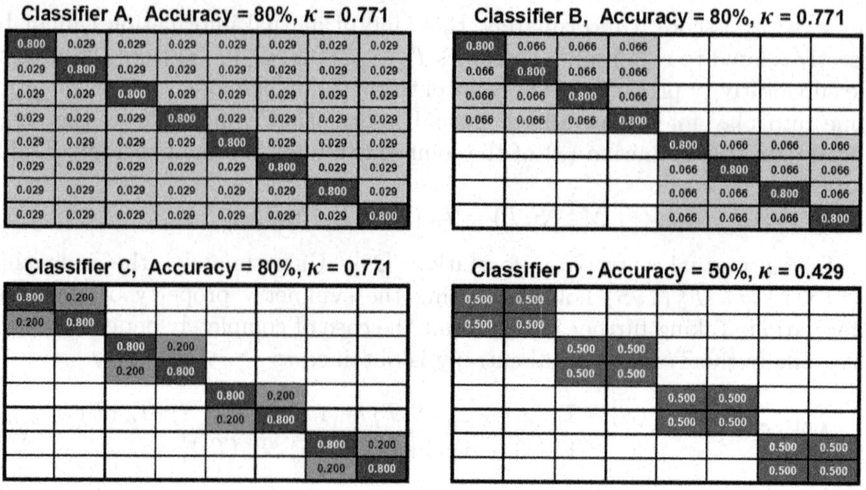

Fig. 2. Confusion matrices obtained from the hypothetical classifiers A–D. Example taken from [12].

to be reverse. Yet, taking into account the definitions of the scores according to (31) and (33), it becomes obvious that they have to be interpreted differently: Since both are identical either to the mutual information or the mutual entropy, higher values indicate better performance. Therefore, the same ordering for the classifiers A–D is obtained.

Yet, if the parameter α is set to $\alpha = 2$, the ranking C→B→A→D results for all non-Shannon-measures. Thus, this is a systematic deviation. Taking a closer

Table 1. Scores based on conditional entropies for artificial classifiers A–D. Lower values indicate better classifier performance.

Data set	Shannon	Rényi			Tsallis
	ς_S	ς_α^{JA}	ς_α^A	ς_α^H	ς_α^T
	α not avail.	$\alpha = 0.5$			
Class.A	1.8005	2.9359	2.9419	2.9479	4.3334
Class.B	1.4294	2.0436	2.0396	2.0356	2.6672
Class.C	1.0008	1.1756	1.1756	1.1756	1.3666
Class.D	1.3863	1.3863	1.3863	1.3863	1.6569
–		$\alpha = 2.0$			
Class.A	–	0.8862	0.8743	0.8803	0.7159
Class.B	–	0.8441	0.8521	0.8481	0.6886
Class.C	–	0.7713	0.7713	0.7713	0.6400
Class.D	–	1.3863	1.3863	1.3863	1.0000

Table 2. Mutual information values for artificial classifiers A–D. Higher values indicate better classifier performance.

Data set	Shannon	Rényi			Tsallis
	I	I_α	$\tilde{I}_\alpha^A$	$\tilde{I}_\alpha^H$	I_α^T
	α not avail.	$\alpha = 0.5$			
Class.A	1.1824	0.6025	0.6115	0.6085	1.4986
Class.B	1.3626	1.0636	1.0576	1.0596	2.3176
Class.C	1.5790	1.4917	1.4917	1.4917	2.9736
Class.D	1.3863	1.3863	1.3863	1.3863	2.8284
	–	$\alpha = 2.0$			
Class.A	–	1.6363	1.6363	1.6333	0.5163
Class.B	–	1.6574	1.6574	1.6594	0.5312
Class.C	–	1.6938	1.6938	1.6938	0.5550
Class.D	–	1.3863	1.3863	1.3863	0.3750

Table 3. Scores based on mutual information/entropy for artificial classifiers A–D. Lower values indicate better classifier performance - except for $\varsigma_{TME}(\alpha)$ and $\varsigma_{TME}(\alpha)$, where higher values are desirable.

Data set	Shannon	Renyi			Tsallis	
	ς_{MI}	$\varsigma_{RMI}(\alpha)$	$\varsigma_{RMI}^A(\alpha)$	$\varsigma_{RMI}^H(\alpha)$	$\varsigma_{TMI}(\alpha)$	$\varsigma_{TME}(\alpha)$
	α undef	$\alpha = 0.5$				
Class.A	1.8005	2.9599	2.9419	2.9479	**0.5202**	**1.4986**
Class.B	1.4294	2.0276	2.0396	2.0356	**0.8249**	**2.3176**
Class.C	1.0008	1.1756	1.1756	1.1756	**1.0513**	**2.9736**
Class.D	1.3863	1.3863	1.3863	1.3863	**1.0000**	**2.8284**
	–	$\alpha = 2.0$				
Class.A	–	0.8743	0.8743	0.8803	**4.1362**	**0.5163**
Class.B	–	0.8521	0.8521	0.8481	**4.2455**	**0.5312**
Class.C	–	0.7713	0.7713	0.7713	**4.4400**	**0.5550**
Class.D	–	1.3863	1.3863	1.3863	**3.0000**	**0.3750**

look to classifier D, this result is not surprising. Yet, here we have to carry out more detailed investigations to study the influence of the α-parameter.

5.2   Real World Example

The second experiment reconsiders the evaluation of three classifier approaches C1–C3 regarding the same real world data set. Again, this evaluation was performed before in [12] using evaluation measures based on Shannon entropy. The evaluated classifiers are used in subsequent publications which report an

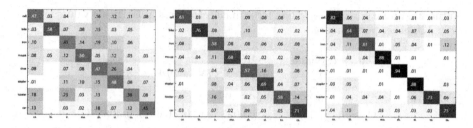

Fig. 3. Confusion matrices obtained from classifiers C1–C3 for a real world data set as reported in [12].

Table 4. Scores based on conditional entropies for the real world example. Lower values indicate better classifier performance.

Data set	Shannon	Rényi			Tsallis
	ς_S	ς_α^{JA}	ς_α^{A}	ς_α^{H}	ς_α^{T}
	α not avail.	$\alpha = 0.5$			
C1	3.0373	3.4381	3.4990	3.4220	5.4482
C2	2.4464	3.1353	3.1946	3.1527	4.7594
C3	1.6692	2.6786	2.7057	2.6368	3.8143
	–	$\alpha = 2.0$			
C1	–	2.5516	2.4930	2.4096	1.4414
C2	–	1.6299	1.5995	1.5698	1.1146
C3	–	0.8971	0.9458	0.9046	0.7229

Table 5. Mutual information values for the real world example. Higher values indicate better classifier performance.

Data set	Shannon	Rényi			Tsallis
	I	I_α	$\tilde{I}_\alpha^{A}$	$\tilde{I}_\alpha^{H}$	I_α^{T}
	α not avail.	$\alpha = 0.5$			
C1	0.5800	0.3726	0.3261	0.3646	0.3400
C2	0.8685	0.4980	0.4857	0.5067	0.4409
C3	1.2410	0.7741	0.7187	0.7532	0.6419
	–	$\alpha = 2.0$			
C1	–	0.9092	0.7959	0.8376	1.4824
C2	–	1.3012	1.2565	1.2714	2.6738
C3	–	1.6303	1.6132	1.6338	4.1053

increasing performance from C1 to C3. In particular, we have the accuracies $a_1 = 46.80\%, a_2 = 64.78\%$, and $a_3 = 78.16\%$. The corresponding confusion matrices are depicted in Fig. 3.

The resulting scores and measures for evaluation are depicted in Tables 4, 5 and 6. For this real world example all measure show complete agreement with the Shannon measures., taking into account the reverse interpretation of the Tsallis scores based on mutual informatio/entropy as explained in the previous section.

Table 6. Scores based on mutual information/entropy for the real world example. Lower values indicate better classifier performance - except for $\varsigma_{TMI}(\alpha)$ and $\varsigma_{TME}(\alpha)$, where higher values are desirable.

Data set	Shannon	Renyi			Tsallis	
	ς_{MI}	$\varsigma_{RMI}(\alpha)$	$\varsigma_{RMI}^{A}(\alpha)$	$\varsigma_{RMI}^{H}(\alpha)$	$\varsigma_{TMI}(\alpha)$	$\varsigma_{TME}(\alpha)$
	α not avail.	$\alpha = 0.5$				
C1	2.9767	3.4059	3.4990	3.4220	**0.3400**	**0.9184**
C2	2.4247	3.1700	3.1946	3.1527	**0.4409**	**1.2856**
C3	1.6573	2.5950	2.7057	2.6368	**0.6419**	**1.7262**
	–	$\alpha = 2.0$				
C1	–	2.2663	2.4930	2.4090	**1.4824**	**0.1539**
C2	–	1.5102	1.5995	1.5698	**2.6738**	**0.3181**
C3	–	0.9116	0.9458	0.9046	**4.1053**	**0.5174**

6   Conclusion

In the present work we investigate information theoretic evaluation measures for cluster/classification assessments. Usually, those quantities are based on the Shannon entropy, which causes frequently numerical instabilities during the calculation. Therefore, we consider more robust alternatives based on Rényi and Tsallis entropy. It turns out that these approaches deliver similar assessment results and, hence, can serve as robust numerical alternatives. Future work should include a deeper investigation of noise effects regarding the numerical quantization of the stability for the considered evaluation measures and the dependence on the parameter α.

References

1. Bishop, C.M.: Pattern Recognition and Machine Learning. Information Science and Statistics. Springer, New York (2006)
2. Geweniger, T., Fischer, L., Kaden, M., Lange, M., Villmann, T.: Clustering by fuzzy neural gas and evaluation of fuzzy clusters. Comput. Intell. Neurosci. Article ID 165248 (2013). https://doi.org/10.1155/2013/165248

3. Geweniger, T., Schleif, F.-M., Villmann, T.: Probabilistic prototype classification using t-norms. In: Villmann, T., Schleif, F.-M., Kaden, M., Lange, M. (eds.) Advances in Self-organizing Maps and Learning Vector Quantization. AISC, vol. 295, pp. 99–108. Springer, Cham (2014). https://doi.org/10.1007/978-3-319-07695-9_9

4. Kaden, M., Lange, M., Nebel, D., Riedel, M., Geweniger, T., Villmann, T.: Aspects in classification learning - review of recent developments in learning vector quantization. Found. Comput. Decis. Sci. **39**(2), 79–105 (2014)

5. Bezdek, J.C.: Cluster validity with fuzzy sets. J. Cybern. **3**, 58–73 (1974)

6. Pal, N., Bezdek, J.: On the cluster validity for the fuzzy c-means modell. IEEE Trans. Inf. Theory **3**(3), 370–379 (1995)

7. Tasdemir, K., Merényi, E.: A validity index for prototype-based clustering of data sets with complex cluster structures. IEEE Trans. Syst. Man Cybern. **41**(4), 1039–1053 (2011)

8. Duda, R.O., Hart, P.E.: Pattern Classification and Scene Analysis. Wiley, New York (1973)

9. Vinh, N.X., Epps, J., Bailey, J.: Information theoretic measures for clusterings comparison: variants, properties, normalization and correction of chance. J. Mach. Learn. Res. **11**, 2837–2854 (2010)

10. Rényi, A.: On measures of entropy and information. In: Proceedings of the Fourth Berkeley Symposium on Mathematical Statistics and Probability. University of California Press, Berkeley (1961)

11. Tsallis, C.: Possible generalization of Bolzmann-Gibbs statistics. J. Math. Phys. **52**, 479–487 (1988)

12. Holt, R.S., Mastromarino, P.A., Kao, E.K., Hurley, M.B.: Information theoretic approach for performance evaluation of multi-class assignment systems. In: Proceedings SPIE Defense, Security, and Sensing (Orlando). SPIE, vol. 7697, pp. 1–12. SPIE The International Society for Optical Engineering, MIT Press (2010)

13. Kapur, J.N.: Measures of Information and Their Application. Wiley, New Delhi (1994)

14. Shannon, C.E.: A mathematical theory of communication. Bell Syst. Tech. J. **27**, 379–432 (1948)

15. Mackay, D.J.C.: Information Theory, Inference and Learning Algorithms. Cambridge University Press, Cambridge (2003)

16. Onicescu, O.: Theorie de l'Information Energie informationelle. C. R. Acad. Sci. Ser. A-B Tome **263**, 841–842 (1966)

17. Rényi, A.: Probability Theory. North-Holland Publishing Company, Amsterdam (1970)

18. Hild, K.E., Erdogmus, D., Principe, J.: Blind source separation using Rényi's mutual information. IEEE Sig. Process. Lett. **8**(6), 174–176 (2001)

19. Csiszár, I.: Information-type measures of differences of probability distributions and indirect observations. Studia Sci. Math. Hungaria **2**, 299–318 (1967)

20. Verdú, S.: α-mutual information. In: Information Theory and Applications Workshop (ITA), San Diego, pp. 1–6. IEEE Press (2015)

21. Csiszár, I.: Axiomatic characterization of information measures. Entropy **10**, 261–273 (2008)

22. Fehr, S., Berens, S.: On the conditional Rényi entropy. IEEE Trans. Inf. Theory **60**(11), 6801–6810 (2014)

23. Teixeira, A., Matos, A., Antunes, L.: Conditional Rényi entropies. IEEE Trans. Inf. Theory **58**(7), 4273–4277 (2012)

24. Iwamoto, M., Shikata, J.: Revisiting conditional Rényi entropies and generalizing Shannons bounds in information theoretically secure encryption. Cryptology ePrint Archive 440/2013 (2013)
25. Ilić, V.M., Djordjević, I.B., Stanković, M.: On a general definition of conditional Rényi entropies. In: Proceedings of the 4th International Electronic Conference on Entropy and Its Application (ECEA 2017), vol. 2, pp. 1–6. MDPI Open Access (2018)
26. Manije, S.T., Gholamreza, M.B., Mohammad, A.: Conditional Tsallis entropy. Cybern. Inf. Technol. 13(2), 37–42 (2013)
27. Abe, S., Rajagopal, A.K.: Nonadditive conditional entropy and its signicance for local realism. Physica A 289, 157–164 (2001)
28. Furuichi, S.: Information theoretical properties of Tsallis entropies. J. Math. Phys. 47, 023302 (2006)
29. Tsallis, C.: Generalized entropy-based criterion for consistent testing. Phys. Rev. E 58, 1442–1445 (1998)
30. Sparavigna, A.C.: Mutual information and nonadditive entropies: the case of Tsallis entropy. Int. J. Sci. 4(10), 1–4 (2015)
31. Mould, R.F.: Introductory Medical Statistics, 3rd edn. Institute of Physics Publishing, London (1998)

Verification of Results in the Acquiring Knowledge Process Based on IBL Methodology

Lukasz Was[1], Piotr Milczarski[2(✉)] [iD], Zofia Stawska[2] [iD], Slawomir Wiak[1], Pawel Maslanka[2] [iD], and Marek Kot[3]

[1] Institute of Mechatronics and Information Systems, Technical University of Lodz, Stefanowski Str. 18/22, 90-924 Lodz, Poland
{lukasz.was,slawomir.wiak}@p.lodz.pl
[2] Faculty of Physics and Applied Informatics, University of Lodz, Pomorska Str. 149/153, 90-236 Lodz, Poland
{piotr.milczarski,zofia.stawska,pmaslan}@uni.lodz.pl
[3] Department of Dermatology and Venereology, Medical University of Lodz, Haller Square 1, 90-647 Lodz, Poland
marek.kot@umed.lodz.pl
http://www.imsi.pl, http://www.wfis.uni.lodz.pl, http://umed.pl

Abstract. In the paper, we discuss IBL - Instance-Based Learning - as a method of acquiring knowledge, and apply it to the verification of the shape symmetry/asymmetry of the skin lesions. The test verifying whether the asymmetry of the lesion presented in PH2 dataset is conducted using IB3 algorithms. We also verify the construction of the DASMShape asymmetry measure and its results. We achieved classification ratio on DAS values from PH2 around 59% in comparison to 84% achieved on the defined DASMShape measure. It implies that the data verification using IBL algorithms is very vital in order to design reliable dermatological diagnosis supporting systems.

Keywords: Knowledge acquisition · Advisory systems
Expert systems · Artificial intelligence · Inference · Dermatology
Instance-Base Learning

1 Introduction

The effective knowledge acquisition schemes are necessary for constructing large knowledge bases, where problems appear in a clear way, unknown in systems with several dozen or at most several hundred rules. It is required to develop methods for acquiring knowledge, enabling the reduction of the time for building an advisory systems and methods of verification of the obtained knowledge base in terms of its consistency, completeness, as well as the elimination of information redundancy. The purpose of acquiring knowledge for advisory systems is to increase the efficiency of these systems. Instance-based learning (IBL) methods

© Springer International Publishing AG, part of Springer Nature 2018
L. Rutkowski et al. (Eds.): ICAISC 2018, LNAI 10841, pp. 750–760, 2018.
https://doi.org/10.1007/978-3-319-91253-0_69

are extension of k nearest neighbor classification [11]. The drawback is that IBL algorithms do not maintain a set of abstractions of models created from the instances. On the other hand, k-NN algorithms have large space requirement [6,14].

Aha et al. [1,2] discussed how storage requirement can be reduced significantly, with only minor loss in learning rate and accuracy. They also focused and expanded their research towards noisy instances, since a lot of real-life datasets having training instances and k-NN algorithms do not work well with noise [23].

The paper presents research that is an integral part of the creation of an automatic system to support the diagnosis of skin dermatological changes. We present the problem of obtaining knowledge from an expert and we question the reliability of data using specific methods of acquiring knowledge. We work on data describing skin lesions. Skin lesions may be evaluated using i.e. the total dermatology score (TDS) methodology, 3 point check list of dermatology or 7 point rules as well as the ABCD rule etc. The authors want to verify the knowledge acquired from PH2 [18] database [12,13] using the mentioned rules and methodologies. Then, we confront the results with knowledge obtained from experts using our own modified authoring rules and scales (3PCLD and 7PCL [22]) as well as using the reasoning process and methods.

In our research, we use PH2 as reference dataset. Calculation of the asymmetry of shape, structure and color is one of the most important factors improving the diagnosis of skin lesions e.g. according to 3 point check list. In the paper, we focus on estimating using IBL whether the proposed shape asymmetry measure DASMShape [21] is defined properly.

The paper is organized as follows. In Sect. 2 we present and discuss methods of knowledge acquisition. Instance-Based Learning - Learning from examples is thoroughly described. The stages and methods of acquiring the knowledge form the experts are discussed in Sect. 3. Dermatological asymmetry measure is presented in Sect. 4. In addition, results and observations of different models of DASMShape are shown in Sect. 4. The final results are given in Sect. 5. Finally, Sect. 6 presents the conclusions.

2   Methods Used to Acquire Knowledge

Acquiring knowledge based on examples is divided into two methods:

- IBL - Instance-Based Learning - Learning from examples [2];
- CBR - Case-Based Reasoning - Case-based inference.

IBL algorithms are supervised learning algorithms. They learn from labeled examples. IBL algorithms [7–9] can be used incrementally, where the input is a sequence of instances. Each instance is described by n attribute-value pairs. One attribute is a category attribute. We assume that there is exactly one category attribute for each instance, and the categories are disjoint, although IBL algorithms can be extended to learn multiple, possibly overlapping categories simultaneously.

We distinguish the following IBL methods:

- The k-nearest neighbors algorithm (k-NN)
- The algorithm of locally weighted linear regression
- RBF (Radial Basis Function)

Properties of IBL:

- They are sometimes called 'lazy' learning methods, because they delay the processing until the new instance has to be classified.
- When a new example appears, a set of similar examples is created in the memory and the new example is graded.
- One of the features of Instance-Based Learning is that they can create different approximations of a purpose function for an example that we need to classify.
- Many techniques create only a local approximation, and never create an approximation designed to show the whole space of examples.
- When the objective function is very complex, it can always be described by a collection of local approximations.

Advantages of IBL:

- The most important advantage of delay (time classification of the example) is that instead of estimating the target function for the sample input spaces, these methods can estimate them locally and separately for each new event.
- Quite simple implementation.
- The present information in the set of examples is never lost.

Apart from the advantages, the IBL methods has disadvantages. The cost of classifying a new example can be high. It is due to the fact that most calculations are carried out during the classification of the example, and not during the collection of teaching examples. Therefore, various techniques for efficient indexing of teaching examples are used here. This helps in faster calculations of the result for the new examples during the query. IBL algorithms do not maintain a set of abstractions of model created from the instances. The k-NN algorithms have large space requirement [17].

Aha in [2] discussed how the storage requirement can be reduced significantly, with only a minor loss in learning rate and accuracy.

We distinguish three algorithms IB1, IB2 and IB3 that describe effect of Instance-Based Learning. IB1 is the simplest algorithm. IB2 improves IB1 so that the storage requirement is lowered. IB3 improves IB2 further to handle noisy instances. In this paper, we use IB3 algorithm which is described below.

IB3 Algorithm pseudocode [1]:
CD ← Ø
For each x∈ training set do

1. For each y ∈ CD do
Sim[y] ← Similarity(x,y)

2. y_{max} ← some y x∈ CD with maximal Sim[y]
3. If class(x) = class(y_{max})
   Then classification ← correct
   Else
   (a) classification ← incorrect
   (b) CD ← CD U {x}
4. For each y in CD do
   If Sim[y] > Sim[y_{max}]
   Then
   (a) Update y classification record
   (b) If y record is significantly poor

   Then CD ← C − {y}
where similarity function for two instances x and y is defined as

$$Similarty(x, y) = \sqrt{\sum_{i=1}^{n} f(x_i, y_i)}, \qquad (1)$$

where i is the attribute index. For numeric attributes, Euclidean distance is often used as a function f. For non-numeric attributes the value of function f is 1 if they are equal and 0 otherwise.

IB3 employs a "wait and see" evidence gathering method to determine which saved instances are expected to perform well during classification. IB3 maintains a classification record with each saved instance. A classification record saves an instance's classification performance on subsequently presented training instances. The record suggests how it will perform in the future. IB3 uses a significance test to determine which instances are well classified and which ones are believed to be noisy. The well-classified ones are used to classify subsequently presented instances. The noisy ones are discarded from the concept description.

Imperfect Knowledge Used to Acquire Knowledge
The methods of inference in the logic of predicates are subject to the use of a correct inference system - reliable when it comes to the truth of the conclusions drawn. However, they are based on the necessary assumption that the statements contained in the initial knowledge base and considered true are really true. Provided they are true, statements derived as a result of inference are also true. This assumption can often not be met in relation to the knowledge that a person has - and yet the initial content of the knowledge base must come from it.

Knowledge from a human being can be imperfect. Applying to this knowledge methods that assume perfect knowledge. We are also exposed to conclusions that do not have to be true and about which logical value we cannot say anything. Therefore, it is advisable to equip the application systems on the basis of imperfect knowledge with special mechanisms for its processing. Thanks to the mechanism it will be possible to characterize the type and degree of imperfection of knowledge derived from a person, as well as new knowledge derived from it by the applicant system. By penetrating deeper into the nature of the imperfection of human knowledge, at least three basic types knowledge imperfection can indicate:

- uncertainty: the accuracy of some statements is not certain
- incomplete: some true statements are not known, but it is not possible to assume their falsehood,
- inaccuracy: belonging to certain relations, corresponding to predicates appearing in the statements, is not known exactly.

3   Stages of Acquiring Knowledge from an Expert

Verification of results in the application process based on the IBL data acquisition methodologies may turn out to be very difficult because the process of obtaining knowledge from an expert, especially an expert in the field of dermatology, is a very difficult matter. According to the methodology of acquiring knowledge from an expert and, in principle, knowledge acquisition stages, it may be difficult to obtain knowledge from an expert. Sometimes there are no experts of a given problem.

The articulation of knowledge is connected with the record of the knowledge (know-how) possessed by the expert. Acquiring knowledge from an expert usually consists of several stages:

1. Observation of an expert at the workplace.
2. Discussion of the problem.
3. Describing the problem.
4. Analyzing the problem.
5. Improving the system.
6. System testing.
7. Legalization of the system.

At each of these stages we can meet with various problems that can make this knowledge uncertain, incomplete or inaccurate.

Difficulties in obtaining knowledge from an expert can be associated with many factors e.g. lack of an expert of a given problem, unsatisfactory expert knowledge or expert's unwillingness to cooperate, Sometimes expert may have a problems with articulation of his knowledge and experience.

We discuss the problem of acquiring knowledge from an expert in the field of dermatology. The problem concerns the correct verification of the condition of the skin lesion. The basis for dermatological diagnosis is the ability to assess and differentiate normal skin lesions. Sometimes it is very difficult. In the case of primary skin eruptions, which are usually a direct result of the development of skin disease in some cases, even in the early stages of the disease, we cannot detect these efflorescence, because they can occur for a short time. There are many factors that can lead to misdiagnosis, which often leads to very long and expensive clinical treatment. As it can be seen, correct diagnosis of skin lesions can be difficult even for an expert dermatologist. Therefore, obtaining reliable knowledge that can be used to design an automatic system that supports the diagnosis of lesions is very hard hence an attempt was made to verify the results in the inference process based on the IBL data acquisition methodologies [16].

4 Dermatological Asymmetry Measures (DAS) and Dermatological Asymmetry Measure of Shape (DASMShape)

Dermatologists in their assessment of the lesions use the ABCD, 3PCLD and 7CPL that were introduced basing on their experts experience. In both scales one of the factor feature is asymmetry of the shape, hue and structures distributions. The asymmetry of the lesion has discrete values: 0 for symmetric in 2 perpendicular axes; 1 for symmetric in 1 axis and 0 for having no symmetry ones. PH2 is one of the reference dermoscopic datasets with the data validated by the dermatology experts. The images we used are stored in the accessible PH2 [18] database with their reference data. It is used in the presented research as testing set. Only 167 images out of 200 images has lesions borders limited to the image area and only these ones are used in the research.

In the paper [21], the new approach to the asymmetry in the assessment of the lesion asymmetry was proposed, i.e. more precise definition of the value of symmetry in continuous form. The dermatological asymmetry measure (DASM) was defined as a function of dermatological asymmetry measures in shape (DASMShape) [21,22], in hue distribution (DASMHue) [19] and in structure distribution (DASMStruct). A more precise definition of the value of symmetry in continuous form. All three values have continuous values from the range of $<0, 2>$ to be consistent with the values used by dermatologists and defined above. In this work we present only the method of determining the value of the shape asymmetry i.e. DASMShape.

In the papers [20,21], we defined DASMShape as function depending on the number of the symmetry axes. The starting point of the method of deriving and calculating DASMShape value is binary mask achieved after segmentation of the image. The DASMShape is a function of a shape symmetry vector $W = [n(t_1), n(t_2), \ldots, n(t_k)]$, where k $\geq$ 2 and $n(t_i)$ is equal into the number of different symmetry axes achieved for 5 different similarity thresholds $\{0.9, 0.93, 0.94, 0.95, 0.97\}$ for a given binary mask (see Table 1).

Two DASMShape functions were proposed, rational:

$$DASMShape(W) = \frac{2}{f(W)},\qquad(2)$$

where inner function f: $f \colon R^k \to <1, \infty)$, and exponential:

$$DASMShape(W) = 2\exp(-f(W)),\qquad(3)$$

where inner function $f: R^k \rightarrow R^+ \cup \{0\}$. Both DASMShape functions have standardized values $(0, 2>$ for a chosen inner functions, where 0 means complete symmetry in 2 axes and 2 complete asymmetry.

After calculating the DASMShape value (using either Eq. 2 or 3) we could map that value into dermatological DAS values set $\{0, 1, 2\}$ using a set of two crisp shape thresholds (ST): ST $= \{lst, ust\}$, where

$$ST(DASMShape(W)) = \begin{array}{l} 0 \Leftrightarrow DASMShape(W) < lst \\ 1 \Leftrightarrow lst \le DASMShape(W) < ust \\ 2 \Leftrightarrow DASMShape(W) > ust \end{array} \qquad (4)$$

The values of lst and ust in Eq. 4 depend on the DASMShape function type and are derived after optimization of results. The objective of the optimization is to achieve minimum value in the number of misclassified images parallely with the number of underestimated cases. The results for some chosen cases are shown in Table 2. As it can be seen there are differences in the final values of DASMShape v1 and v2.

Table 1. The examples of calculating number of symmetry line w for selected images from PH2 [18] dataset and their corresponding DASMShape measures and their mapping $M(a)$ into $\{0, 1, 2\}$ DAS values

Image ID from PH2	Coefficient values $n(t_i)$ for the shape symmetry vector W					DAS (PH2)	DASMShape values for f(W) and coefficients a after mapping			
	$n(0.9)$	$n(.93)$	$n(.94)$	$n(.95)$	$n(.97)$		a_y	$M(a_y)$	a_z	$M(a_z)$
IMD003	10	4	3	0	0	0	0.37275	0	0.25491	0
IMD035	1	0	0	0	0	2	1.98010	2	1.99202	2
IMD002	8	3	3	1	0	1	0.50316	0	0.22623	0
IMD075	1	1	1	0	0	2	1.86479	2	1.66943	2
IMD155	2	2	2	2	1	0	0.09192	0	0.15523	0
IMD339	12	12	7	6	0	2	0.00000	0	0.00000	0
IMD211	2	1	1	1	0	1	1.48164	2	1.18193	1
IMD405	2	1	1	1	0	2	1.48164	2	1.18193	1
IMD406	13	6	5	2	0	2	0.02969	0	0.00290	0

For the IBL reason we prepared also the training set consisting of 19 values corresponding to the symmetry of the known regular figures and their number of symmetry axes. They consist of 5 attributes and a class values equal 0, 1 or 2. The maximum number of the symmetry axes is also connected with the clustering the closest symmetry lines into one if they differ by $10°$ angle. It means that for a wheel we can have maximum of 16 axes. Then for a wheel, the resulting attribute list $\{16, 16, 16, 16, 16\}$ and class value equal 0. A pentagram shape has resulting attribute list $\{5, 5, 5, 5, 5\}$ and class value equal 1.

5   Research and Discussion of the Results

Figure 1 presents the general approach to IBL using IB3 with and without cross validation for the data from PH2 and achieved by DASMShape algorithm. We conduct several subsequent tests using prepared training set and testing it on asymmetry values from PH2 dataset, the values from DASMShape defined in [21] for two the best optimized measures. The final test was done also using DAS as a training set and mentioned above DASMShape values [4].

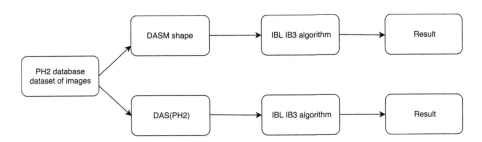

Fig. 1. DAS and DASMShape data processing and comparing with IB3 algorithm

Table 2. IBL classification results on DAS and DASMShape values, with true positive (TP) and false positive (FP) statistics

Type of DAS/DASMShape	Number of correctly classified instances	Number incorrectly classified instances	Class. ratio	TP rate for a class			FP rate for a class		
				0	1	2	0	1	2
DAS (PH2)	73	94	43.7	40.2	41.9	58.6	21.7	39	20.3
DAS (PH2) cv	97	70	58.1	66.4	35.5	51.7	30	24.3	13.8
DASMShape1	112	55	67.1	61.6	72	73.2	3.7	33.8	3.6
DASMShape1 cv	138	29	82.6	97.7	80	60.7	6.2	16.9	0
DASMShape2	108	59	64.8	53.3	63.3	87.2	10.4	34.3	3.3
DASMShape2 cv	141	26	84.4	92.2	80	72.3	7.8	14.6	0
DASMShape (DAS)	110	57	65.9	98.8	20	35.7	58	6.3	0.9
DASMShape2 (DAS)	108	59	64.7	100	20	25.5	54.5	5.8	7.5

In the Table 2, we present the results of IB3 testing the DAS and DASMShape for 167 cases from PH2 for which the asymmetry can be derived, the ones where the whole lesion is seen in the image. Table 2 shows that the classification ratio is around 58% even using cross validation. The proposed in [21] DASMShape measure and discussed in [20] gives better classification results. The results for two DASMShape measures are around 82–84%. Even using DAS as a training set and both DASMShape gives the classification rate around 66%.

In our research, we use PH2 as reference data sets. Calculation of the asymmetry is one of the most important factors improving the diagnosis of skin lesions i.e. according to 3 point check list. In the paper, we focus on estimating using IBL whether the proposed shape asymmetry measure DASMShape is defined properly. This is the starting point for estimating the asymmetry of changes, because the other factors of the asymmetry (hue and structures) depends on initial shape symmetry [15].

In the paper, we have combined the methods of acquiring knowledge especially Instance-Based Learning with dermatological asymmetry measure of shape. We have shown that Instance-Based Learning can be easily used in pattern matching [3,5,10].

6   Conclusions

In the paper, we have analyzed and presented the approach of dermatologists to melanocytic and non-melanocytic skin lesions. The asymmetry of skin lesions is described as a function of shape, color and structure. The paper discusses the symmetry of shape in the segmentation of skin lesions and defines a new dermatological measure.

The results of the method/algorithm using the PH2 data set and IBL methodology are presented. Mild and malignant skin changes are often very difficult to diagnose because there are many factors that can lead to misdiagnosis, which often leads to very long and costly clinical treatment. It is often difficult to recognize and diagnose melanocytic and non-melanocyte skin lesions. Different sets of factors must be considered, which leads to a correct diagnosis. Proper dermatological diagnosis is based on clinical symptoms, experience of doctors and proper differentiation of skin lesions.

In the paper, we have conducted our research concerning asymmetry of shape using Instance-Based Learning with dermatological asymmetry measure of shape using the reference PH2 data. Table 2 shows that the classification ratio is around 58% even using cross validation. The proposed in [21] DASMShape measure and discussed in [20] is better classified by 24–26%. The difference can be derived partially from the DAS definition. It gives the final asymmetry value. In order to get the full prediction we need to use full DASM values as defined in [15,19]. Nonetheless, the data verification using IBL algorithms is very important in order to design and develop reliable dermatological diagnosis supporting systems.

References

1. Aha, D.W.: Tolerating noisy, irrelevant and novel attributes in instance-based learning algorithms. Int. J. Man Mach. Stud. **36**, 267–287 (1992)
2. Aha, D.W., Kibler, D., Albert, M.K.: Instance-based learning algorithms. Mach. Learn. **6**, 37–66 (1991)

3. Batchelor, B.G.: Pattern Recognition: Ideas in Practice. Plenum Press, New York (1978)
4. Biberman, Y.: A context similarity measure. In: Bergadano, F., De Raedt, L. (eds.) ECML 1994. LNCS, vol. 784, pp. 49–63. Springer, Heidelberg (1994). https://doi.org/10.1007/3-540-57868-4_50
5. Bishop, C.: Pattern Recognition and Machine Learning. Information Science and Statistics. Springer, New York (2006). ISSN 1613-9011
6. Brent, M.: Instance-based Learning: nearest neighbour with generalisation. Department of Computer Science, University of Waikato (1995)
7. Brodley, C.E.: Addressing the selective superiority problem: automatic algorithm/model class selection. In: Proceedings of the Tenth International Machine Learning Conference, Amherst, MA, pp. 17–24 (1993)
8. Broomhead, D.S., Lowe, D.: Multi-variable functional interpolation and adaptive networks. Complex Syst. **2**, 321–355 (1988)
9. Cameron-Jones, R.M.: Instance selection by encoding length heuristic with random mutation hill climbing. In: Proceedings of the Eighth Australian Joint Conference on Artificial Intelligence, pp. 99–106 (1995)
10. Carpenter, G.A., Grossberg, S.A.: Massively parallel architecture for a self-organizing neural pattern recognition machine. Comput. Vis. Graph. Image Process. **37**, 54–115 (1987)
11. Chang, C.-L.: Finding prototypes for nearest neighbor classifiers. IEEE Trans. Comput. **23**(11), 1179–1184 (1974)
12. Gagliardi, F.: Instance-based classifiers applied to medical databases: diagnosis and knowledge extraction. J. Artif. Intell. Med. **52**(3), 123–139 (2011)
13. Gagliardi, F.: Instance-based classifiers to discover the gradient of typicality in data. In: Pirrone, R., Sorbello, F. (eds.) AI*IA 2011. LNCS (LNAI), vol. 6934, pp. 457–462. Springer, Heidelberg (2011). https://doi.org/10.1007/978-3-642-23954-0_47
14. Hastie, T., Tibshirani, R., Friedman, J.: Prototype methods and nearest-neighbors. In: Hastie, T., Tibshirani, R., Friedman, J. (eds.) The Elements of Statistical Learning. Data Mining, Inference, and Prediction. SSS, pp. 459–483. Springer, New York (2009). https://doi.org/10.1007/978-0-387-84858-7_13
15. Jaworek-Korjakowska, J., Tadeusiewicz, R.: Assessment of asymmetry in dermoscopic colour images of pigmented skin lesions. In: Proceedings of the IASTED International Conference on Biomedical Engineering, BioMed 2013, pp. 368–375 (2013)
16. Jebara, T.: Machine Learning Discriminative and Generative. Springer, New York (2004). https://doi.org/10.1007/978-1-4419-9011-2
17. Kanwal, N., Bostanci, E.: Comparative Study of Instance Based Learning and Back Propagation for Classification Problems. GRIN Verlag, USA (2016)
18. Mendonca, T., Ferreira, P.M., Marques, J.S., Marcal, A.R.S., Rozeira, J.: PH2-a dermoscopic image database for research and benchmarking. In: 35th Annual International Conference of the IEEE Engineering in Medicine and Biology Society (EMBC), Osaka, pp. 5437–5440 (2013)
19. Milczarski, P.: Skin lesion symmetry of hue distribution. In: Proceedings of the IEEE 9th International Conference on Intelligent Data Acquisition and Advanced Computing Systems: Technology and Applications, IDAACS 2017, pp. 1006–1013 (2017)

20. Milczarski, P., Stawska, Z., Maślanka, P.: Skin lesions dermatological shape asymmetry measures. In: Proceedings of the IEEE 9th International Conference on Intelligent Data Acquisition and Advanced Computing Systems: Technology and Applications, IDAACS 2017, pp. 1056–1062 (2017)
21. Milczarski, P., Stawska, Z., Was, L., Wiak, S., Kot, M.: New dermatological asymmetry measure of skin lesions. Int. J. Neural Netw. Adv. Appl. **4**, 32–38 (2017)
22. Was, L., Milczarski, P., Stawska, Z., Wyczechowski, M., Kot, M., Wiak, S., Wozniacka, A., Pietrzak, L.: Analysis of dermatoses using segmentation and color hue in reference to skin lesions. In: Rutkowski, L., Korytkowski, M., Scherer, R., Tadeusiewicz, R., Zadeh, L.A., Zurada, J.M. (eds.) ICAISC 2017. LNCS (LNAI), vol. 10245, pp. 677–689. Springer, Cham (2017). https://doi.org/10.1007/978-3-319-59063-9_61
23. Wilson, D.R., Martinez, T.R.: Reduction techniques for instance-based learning algorithms. Mach. Learn. **37**, 257–286 (2000)

A Fuzzy Measure for Recognition of Handwritten Letter Strokes

Michał Wróbel[1]([⊠]), Katarzyna Nieszporek[1], Janusz T. Starczewski[1,2], and Andrzej Cader[3,4]

[1] Institute of Computational Intelligence, Czestochowa University of Technology, Czestochowa, Poland
michal.wrobel@iisi.pcz.pl
[2] Radom Academy of Economics, Radom, Poland
[3] Information Technology Institute, University of Social Sciences, Łódź, Poland
[4] Clark University, Worcester, MA 01610, USA

Abstract. In this paper, we propose and compare a few methods of representing a stroke as a vector of numbers. For each method, we describe, how to calculate the fuzzy measure of two strokes similarity. Vectors are determined on the basis of polynomials calculated by a stroke approximation.

Keywords: Handwriting recognition · ICR · Stroke · Fuzzy Similarity

1 Introduction

Handwriting recognition (or HWR) is the ability of a computer to read handwriting as an actual text. Handwriting recognition is different than printed text recognition; however both of them can be combined with natural language understanding, e.g. [5]. Main issues of HWR are described in [7]. In the basic handwriting style, called a cursive script, the letters are connected by ligatures, so it is difficult to separate them. Moreover, even extracting a line of a text may be difficult, because letters from different lines of a text may be touching or even overlapping (techniques for text line extraction are still under development, one of such approaches is presented in [9]).

Generally, two kinds of handwriting recognition are distinguished: on-line and off-line. As the difference is explained in [8], in the on-line case a text is read from a device (for example from a tablet) during writing, hence the order in which the individual lines are created is actually known (some approaches to tracking moving objects can be helpful [4]). Off-line recognition is based on images gathered for example from a scan of a document.

One of the approaches used in handwriting recognition is stroke extraction. The stroke is a fragment of a letter (or generally of a picture) which may be created by a single move of a hand. In [1] there is presented an approach based on extracting only vertical and horizontal fragments of strokes. Another approach

© Springer International Publishing AG, part of Springer Nature 2018
L. Rutkowski et al. (Eds.): ICAISC 2018, LNAI 10841, pp. 761–770, 2018.
https://doi.org/10.1007/978-3-319-91253-0_70

relies on extracting all of the strokes. In [6] there is presented an algorithm for extraction of strokes. A simple example of a complete algorithm for recognizing handwritten texts using stroke extraction is presented in [11].

In addition to extraction of strokes, there is an issue of classifying the strokes received. We are in need of a method for comparing a given stroke with a pattern. The similarity of the two strokes is not a binary value in general, therefore a fuzzy logic approach may be especially useful. An original approach to model uncertainties is presented in [2]. As mentioned in [3,10], a convenient form of the object for classification is its representation in the form of a vector of features.

Therefore, we search for a method for representing a stroke and comparing it with another stroke. This method should satisfy the following properties:

- stroke should be represented by a vector of data, as short as possible;
- this vector should be as simple as possible to create and process;
- there should be a good method to compare two strokes – strokes, that people intuitive treat as similar must have better fuzzy similarity measure than different strokes.

In this paper, we present a few novel approaches to represent and compare strokes.

2   Defining the Problem

We have got a stroke given as an ordered set of points:

$$S = (P_1, P_2, \ldots, P_n) \tag{1}$$

Each point is just a couple of coordinates $P_i = (x_i, y_i)$. We want to express a stroke as a vector of several numbers. This vector should be short, easy to calculate and clearly describing a shape of the stroke.

We also have to construct a good procedure for comparing vectors of two strokes. We consider a result of this procedure in the form of a fuzzy measure of stroke identity, with the constraint that it should equal 1 in cases of identical strokes and it should tend to 0 whenever strokes differ strongly.

3   Polynomial Representation of the Stroke

We may express a stroke as a function of a parameter $t \in [0, 1]$. Let's define t_i as a distance between the beginning of the stroke and the point p_i divided by length of the stroke. We may express it in the following manner:

$$t_i = \frac{\sum\limits_{j=1}^{i-1} \sqrt{(x_j - x_{j+1})^2 + (y_j - y_{j+1})^2}}{\sum\limits_{j=1}^{n-1} \sqrt{(x_j - x_{j+1})^2 + (y_j - y_{j+1})^2}} \tag{2}$$

This approach was mentioned in [6]. Value t_1 is always 0 and value t_n is always 1. Now we may consider a stroke as a couple of functions $x = f_1(t)$ and $y = f_2(t)$. We know values of these functions in n points. Therefore, we are able to approximate these functions.

We may notice that functions $x = f_1(t)$ and $y = f_2(t)$ have convenient forms to approximate them by polynomials of a degree not greater than 3. Examples of approximation are shown in Fig. 1. We can see that the approximation is quite good – the shape of each stroke is visible.

$$x = a_3 t^3 + a_2 t^2 + a_1 t + a_0$$
$$y = b_3 t^3 + b_2 t^2 + b_1 t + b_0 \tag{3}$$

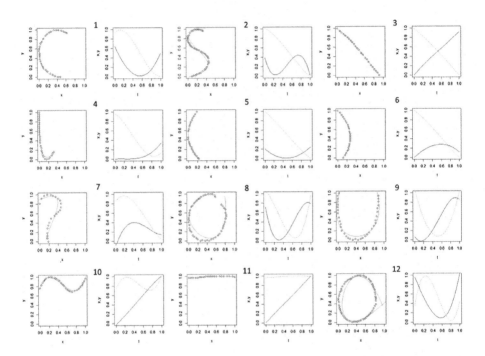

Fig. 1. Examples of 12 strokes and their approximations. For each stroke there are shown: set of points and approximated stroke (left graph) and functions $x(t), y(t)$ (right graph)

Values a_0 and b_0 represent only localization of the stroke, not a shape, so we may not consider them. So, we have a vector of six numbers.

$$\boldsymbol{v} = (a_3, a_2, a_1, b_3, b_2, b_1) \tag{4}$$

We assume stroke length is greater then 0 (it cannot be a single point), so at least one element of vector $\boldsymbol{v}$ is different than 0. So, vector length is always greater than 0 and the vector may be normalized. After normalization vector $\hat{\boldsymbol{v}} = \frac{\boldsymbol{v}}{\|\boldsymbol{v}\|}$ describes only the shape of the stroke, not their size or position.

3.1   Methods of Comparison

Our goal is to compare two strokes described by normalized vectors $\hat{u}$ and $\hat{v}$. Both vectors are unit, consequently, the distance between them belongs to $[0, 2]$. Therefore, we can express similarity between vectors by $1 - \frac{1}{2}\|\hat{v} - \hat{u}\|$, remembering that the same stroke may be represented by two different vectors, since both endpoints can be treated as the beginning of the stroke. Accordingly, the two following functions represent the shape of the same stroke:

$$
\begin{aligned}
x &= a_3 t^3 + a_2 t^2 + a_1 t + a_0 \\
x' &= a_3(1-t)^3 + a_2(1-t)^2 + a_1(1-t) + a_0
\end{aligned}
\tag{5}
$$

$$
\begin{aligned}
x' &= a_3(1-t)^3 + a_2(1-t)^2 + a_1(1-t) + a_0 \\
&= a_3(1 - 3t + 3t^2 - t^3) + a_2(1 - 2t^2 + t^2) + a_1(1-t) + a_0 \\
&= -a_3 t^3 + (3a_3 + a_2)t^2 + (-3a_3 + 2a_2 - a_1)t + (a_3 + a_2 + a_1 + a_0)
\end{aligned}
\tag{6}
$$

If we have vector v, we can calculate its alternative form v'.

$$
\begin{aligned}
v &= (a_3, a_2, a_1, b_3, b_2, b_1) \\
v' &= (a'_3, a'_2, a'_1, b'_3, b'_2, b'_1)
\end{aligned}
\tag{7}
$$

where

$$
\begin{array}{ll}
a'_3 = -a_3 & b'_3 = -b_3 \\
a'_2 = 3a_3 + a_2 & b'_2 = 3b_3 + b_2 \\
a'_1 = -3a_3 + 2a_2 - a_1 & b'_1 = -3b_3 + 2b_2 - b_1
\end{array}
\tag{8}
$$

Vector v' also should be normalized ($\hat{v}' = \frac{v'}{\|v'\|}$) even if v was a unit vector. Finally, this proposition of fuzzy measure is:

$$
F(\hat{u}, \hat{v}) = 1 - \frac{1}{2} min(\|\hat{v} - \hat{u}\|, \|\hat{v}' - \hat{u}\|)
\tag{9}
$$

3.2   Window Normalization

Instead of vector normalization, we may use window normalization. To do this, we have to move and scale the stroke to fit it to a square window $[0, 1]^2$. We have an ordered set of points presented in form 1. Now, we intend to transform it to a set $S_w = (P_{w1}, P_{w2}, \ldots, P_{wn})$, where:

$$
\begin{aligned}
s &= max(x_{max} - x_{min}, y_{max} - y_{min}) \\
P_{wi} &= \left(\frac{x_i - x_{min}}{s}, \frac{y_i - y_{min}}{s}\right)
\end{aligned}
\tag{10}
$$

After that, we are able to approximate the polynomials in similar forms to 3 and create a vector like in 4. It is important, that in this method, we should not normalize vector v.

If we want to compare two vectors u_w and v_w, we should calculate the distance between them. We also have to consider the alternative vector v' created as in 8. It is hard to provide the biggest possible distance between vectors, hence we will use the hyperbolic tangent $tanh$ to scale the distance to the range $[0, 1]$. Consequently, the fuzzy measure used in this method is:

$$F(\boldsymbol{u}_w, \boldsymbol{v}_w) = 1 - tanh(min(\|\boldsymbol{v}_w - \boldsymbol{u}_w\|, \|\boldsymbol{v}'_w - \boldsymbol{u}_w\|)) \tag{11}$$

3.3   Component Functions

The polynomial used in our method is a sum of component functions a_3t^3, a_2t^2, and a_1t. We may also express this polynomial as a sum of other functions, easier to imagine by an expert.

$$x = Aa(t) + Bb(t) + Cc(t) \tag{12}$$

The proposition of functions a, b, c is presented in form 13. Graphs of these functions are presented in Fig. 2.

$$\begin{aligned} a(t) &= t \\ b(t) &= 4(t - \tfrac{1}{2}) \\ c(t) &= \tfrac{32}{3}t(t - \tfrac{1}{2})(t - 1) \end{aligned} \tag{13}$$

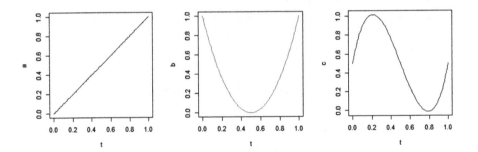

Fig. 2. Component functions a, b, and c

While observing coefficients A, B, C it is quite easy to imagine the shape of the function. We calculate these coefficients from polynomial $a_3t^3 + a_2t^2 + a_1t$ as follows:

$$\begin{aligned} A &= a_3 + a_2 + a_1 \\ B &= \tfrac{3}{8}a_3 + \tfrac{1}{4}a_2 \\ C &= \tfrac{3}{32}a_3 \end{aligned} \tag{14}$$

We have to recast this way both polynomials: $x(t)$ and $y(t)$. Thus, vector $\boldsymbol{v}$ in this method is:

$$\boldsymbol{v} = (A_x, B_x, C_x, A_y, B_y, C_y) \tag{15}$$

where we still keep in mind an alternative vector v'. At $t = \frac{1}{2}$, the function b has got its axis of symmetry and functions a and c have got their points of symmetry. Consequently, the vector v' is just:

$$v' = (-A_x, B_x, -C_x, -A_y, B_y, -C_y) \tag{16}$$

In this method, we also can calculate the fuzzy measure in two ways: with vector normalization, like in 9, or with window normalization, like in previous subsection.

4   Tests

In the previous section, we have presented four methods for representing and comparing strokes:

- **Method 1** – polynomial approximation, calculating alternative vector (Eq. 8), vector normalization, calculating the fuzzy measure (form 9);
- **Method 2** – window normalization (Eq. 10), polynomial approximation, calculating alternative vector (Eq. 8), calculating the fuzzy measure (Eq. 11);
- **Method 3** – polynomial approximation, calculating coefficients A, B, C (Eq. 13) calculating alternative vector (Eq. 16), vector normalization, calculating the fuzzy measure (Eq. 9);
- **Method 4** – window normalization (form 10), polynomial approximation, calculating coefficients A, B, C (Eq. 13) calculating alternative vector (form 16), calculating the fuzzy measure (Eq. 11);

In this section, we present the comparison between these methods and our primary method described in [11] (called here **Method 5**).

4.1   Quality Measure

Each method was tested by the following way:

- A few groups of strokes were loaded. In each group, there were a few similar strokes.
- Each couple of strokes was compared. If both strokes belonged to the same group, the result was stored in the first list, if strokes belonged to different groups, the result was stored in the second list.
- We expect from the perfect method that each value in the first list should be greater than each value in the second list. As variable n, we stored the number of values in the first list. Next, we merged both lists, sorted in descending order and got an n-th element. It was our threshold.
- We counted the number of values in the first list smaller than the threshold and the number of values in the second list greater than or equal to the threshold. Sum of these two numbers was the number of incorrectly classified couples of strokes.

- We divided the value calculated in the previous point by the sum of both lists lengths. Finally, we considered it as our measure of a method quality.

4.2 Results

All the methods were tested on groups of strokes presented in the Fig. 3. Results of the testing methods are presented in the Table 1.

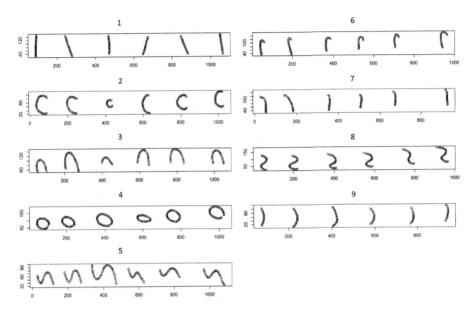

Fig. 3. Groups of strokes for testing

Table 1. Results of the experiments

Method	Quality	Threshold
1	0.950	0.779
2	0.904	0.081
3	0.918	0.884
4	0.906	0.742
5	0.915	0.902

In the Tables 2, 3, 4, 5 and 6, there are presented average results of comparing strokes from different groups.

Table 2. Results obtained by method 1

Group	1	2	3	4	5	6	7	8	9
1	0,81								
2	0,34	0,90							
3	0,44	0,62	0,85						
4	0,38	0,68	0,50	0,76					
5	0,39	0,73	0,32	0,61	0,93				
6	0,38	0,66	0,43	0,48	0,66	0,84			
7	0,58	0,41	0,52	0,52	0,38	0,37	0,78		
8	0,35	0,28	0,66	0,55	0,39	0,29	0,63	0,92	
9	0,50	0,27	0,57	0,33	0,35	0,46	0,48	0,55	0,79

Table 3. Results obtained by method 2

Group	1	2	3	4	5	6	7	8	9
1	0,63								
2	0,00	0,07							
3	0,00	0,00	0,05						
4	0,00	0,00	0,00	0,02					
5	0,00	0,00	0,00	0,00	0,02				
6	0,01	0,00	0,00	0,00	0,00	0,24			
7	0,23	0,00	0,00	0,00	0,00	0,00	0,46		
8	0,00	0,00	0,00	0,00	0,00	0,00	0,00	0,06	
9	0,14	0,00	0,00	0,00	0,00	0,01	0,14	0,00	0,35

Table 4. Results obtained by method 3

Group	1	2	3	4	5	6	7	8	9
1	0,86								
2	0,66	0,86							
3	0,35	0,34	0,88						
4	0,31	0,48	0,48	0,75					
5	0,48	0,38	0,60	0,38	0,90				
6	0,73	0,72	0,40	0,43	0,36	0,86			
7	0,88	0,63	0,36	0,31	0,51	0,70	0,90		
8	0,73	0,59	0,38	0,34	0,55	0,57	0,78	0,90	
9	0,86	0,60	0,31	0,28	0,43	0,72	0,88	0,70	0,96

Table 5. Results obtained by method 4

Group	1	2	3	4	5	6	7	8	9
1	0,73								
2	0,38	0,70							
3	0,09	0,08	0,74						
4	0,07	0,14	0,13	0,48					
5	0,13	0,06	0,08	0,05	0,59				
6	0,52	0,46	0,12	0,13	0,07	0,74			
7	0,75	0,34	0,09	0,07	0,14	0,47	0,79		
8	0,45	0,26	0,08	0,07	0,13	0,26	0,52	0,75	
9	0,73	0,32	0,08	0,06	0,11	0,51	0,76	0,40	0,92

Table 6. Results obtained by method 5

Group	1	2	3	4	5	6	7	8	9
1	0,84								
2	0,71	0,91							
3	0,42	0,73	0,91						
4	0,47	0,68	0,71	0,92					
5	0,43	0,58	0,78	0,54	0,91				
6	0,71	0,82	0,60	0,59	0,50	0,89			
7	0,86	0,73	0,46	0,47	0,49	0,71	0,88		
8	0,73	0,80	0,62	0,56	0,66	0,76	0,78	0,90	
9	0,86	0,78	0,47	0,50	0,44	0,79	0,88	0,73	0,97

5   Conclusions

As a main conclusion from our tests, the best results have been provided by
Method 1. We have identified that Methods 1–4 have essential problems with
cycles (see strokes from group 4). This situation has been occurred, because a
cycle recognized by a human has no endpoint of a stroke. However, the algorithm
distinguishes strokes according to their endpoints. This produces an unexpected
ambivalence and it makes the comparative process difficult and thus inefficient
in practical use. Nevertheless, the comparison of cycles looks promisingly in the
case of Method 5.

References

1. Álvarez, D., Fernández, R., Sánchez, L.: Stroke based handwritten character recognition. In: Corchado, E., Snášel, V., Abraham, A., Woźniak, M., Graña, M., Cho, S.-B. (eds.) HAIS 2012. LNCS (LNAI), vol. 7208, pp. 343–351. Springer, Heidelberg (2012). https://doi.org/10.1007/978-3-642-28942-2_31
2. Beg, I., Rashid, T.: Modelling uncertainties in multi-criteria decision making using distance measure and topsis for hesitant fuzzy sets. J. Artif. Intell. Soft Comput. Res. **7**(2), 103–109 (2017)
3. Bologna, G., Hayashi, Y.: Characterization of symbolic rules embedded in deep DIMLP networks: a challenge to transparency of deep learning. J. Artif. Intell. Soft Comput. Res. **7**(4), 265–286 (2017)
4. Chang, O., Constante, P., Gordon, A., Singana, M.: A novel deep neural network that uses space-time features for tracking and recognizing a moving object. J. Artif. Intell. Soft Comput. Res. **7**(2), 125–136 (2017)
5. Ke, Y., Hagiwara, M.: An English neural network that learns texts, finds hidden knowledge, and answers questions. J. Artif. Intell. Soft Comput. Res. **7**(4), 229–242 (2017)
6. L'Homer, E.: Extraction of strokes in handwritten characters. Pattern Recogn. **33**, 1147–1160 (2000)
7. Ostrowski, D.J., Cheung, P.Y.K.: A fuzzy logic approach to handwriting recognition. In: Patyra, M.J., Mlynek, D.M. (eds.) Fuzzy Logic, pp. 299–314. Vieweg+Teubner Verlag, Wiesbaden (1996). https://doi.org/10.1007/978-3-322-88955-3_10
8. Plamondon, R., Srihari, S.N.: Online and off-line handwriting recognition: a comprehensive survey. IEEE Trans. Pattern Anal. Mach. Intell. **22**(1), 63–84 (2000)
9. Ptak, R., Żygadło, B., Unold, O.: Projection-based text line segmentation with a variable threshold. Int. J. Appl. Math. Comput. Sci. **27**(1), 195–206 (2017)
10. Riid, A., Preden, J.S.: Design of fuzzy rule-based classifiers through granulation and consolidation. J. Artif. Intell. Soft Comput. Res. **7**(2), 137–147 (2017)
11. Wróbel, M., Starczewski, J.T., Napoli, C.: Handwriting recognition with extraction of letter fragments. In: Rutkowski, L., Korytkowski, M., Scherer, R., Tadeusiewicz, R., Zadeh, L.A., Zurada, J.M. (eds.) ICAISC 2017. LNCS (LNAI), vol. 10246, pp. 183–192. Springer, Cham (2017). https://doi.org/10.1007/978-3-319-59060-8_18

Author Index

Abdallah, Fatima   II-489
Abel, Mara   II-288
Al-Jawadi, Radhwan   I-279
Alomar, Miquel L.   I-226
Álvarez, Óscar   II-107
Aquino, Nelson Marcelo Romero   I-376
Argasiński, Jan K.   II-619
Atroszko, Jakub   II-265
Ávila, Bráulio Coelho   II-240

Baczyński, Michał   II-642
Barbosa, Carlos Eduardo M.   I-290
Bartczuk, Łukasz   II-504, II-516
Basurra, Shadi   II-489
Bauer, Joanna   II-179
Behr, Carina M.   I-302
Benjamin, Josephine B. M.   I-690
Berruezo, Pedro   II-642
Bielecka, Marzena   II-117
Bielecki, Andrzej   I-3, II-579
Bielecki, Włodzimierz   II-207
Bilski, Jarosław   I-15
Binek, Wojciech   II-168
Bobek, Szymon   II-168, II-276
Böhme, Hans-Joachim   I-657
Bujok, Petr   I-313
Burduk, Robert   I-585

Cader, Andrzej   I-761, II-354, II-504
Calderón, Juan M.   II-700, II-740
Canals, Vincent   I-226
Cao, Jinde   I-25
Capizzi, Giacomo   II-711
Carbonera, Joel Luís   II-288
Cardona, Gustavo A.   II-700, II-740
Castro-Gutierrez, Eveling   II-25
Cavalca, Diego L.   I-323, I-536
Cernea, Ana   II-107
Chaari, Lotfi   II-14
Chakuma, Bayanda   I-333
Cherri, Luiz Henrique   I-548
Chhipa, Rohan Hemansu   I-345
Chiverton, John   I-148

Chmielewski, Leszek J.   II-81
Cierniak, Robert   II-127
Clark, Patrick G.   II-301
Cpałka, Krzysztof   I-449, II-250
Créput, Jean-Charles   II-3
Cui, Beibei   II-3

d'Allonnes, Adrien Revault   II-579
da Silva, Lucas C.   I-201
Damaševičius, Robertas   I-58, I-615
Datko, Szymon   II-445
de Souza Alves, Demison Rolins   II-603
Dobosz, Piotr   II-127
Dordal, Osmar Betazzi   II-240
dos Santos Ferreira, Fabio   II-603
Dou, Zhumei   II-463
Dowdall, Shane   II-191
Drwiła, Zuzanna   II-677
Du, Yang   II-463
Duda, Piotr   II-311
Dzienisik, Patrycja   II-677
Dziwiński, Piotr   II-504, II-516

Ekpenyong, Moses   I-35
Ekseth, Ole Kristian   II-321
Espinosa-Curiel, Ismael Edrein   I-363
Estacio Cerquin, Laura Jovani   II-25

Fajardo-Delgado, Daniel   I-363
Fakhfakh, Mohamed   II-14
Fakhfakh, Nizar   II-14
Fernandes, Ricardo A. S.   I-323, I-536
Fernández-Brillet, Celia   II-107
Fernández-Martínez, Juan Luis   II-107
Fernández-Muñiz, Zulima   II-107
Filho, Pedro P. Rebouças   I-201
Filippov, Aleksey   II-799
Filutowicz, Zbigniew   I-130, II-127
Fonał, Krzysztof   II-333
Fränti, Pasi   I-593, I-604, I-680, II-343
Frasser, Christiam F.   I-226
Fujimoto, Yuki   I-47

Gaber, Mohamed Medhat   II-489
Gabryel, Marcin   I-615
Galkowski, Tomasz   II-354
Gallegos Guillen, Joel Oswaldo   II-25
Gao, Cheng   II-301
Garani, Shayan Srinivasa   I-136
Gavrilova, Marina   I-474
Geweniger, Tina   I-736
Ghosh, Indrajit   I-623
Gierdziewicz, Maciej   I-3
Gil, David   II-265
Goetzen, Piotr   II-376
Górski, Jarosław   II-81
Grabska-Gradzińska, Iwona   II-619
Grobler, Jacomine   I-302, I-572
Grycuk, Rafał   II-36
Grzanek, Konrad   I-15, I-449, II-73, II-250
Grzymala-Busse, Jerzy W.   II-301
Guskov, Gleb   II-799
Gutoski, Matheus   I-376

Hattori, Leandro Takeshi   I-376
Hatwagner, Miklós F.   II-630
Hatwágner, Miklós   I-386
Hegyháti, Máté   I-386
Helbig, Mardé   I-333, I-345, I-462
Helbin, Piotr   II-642
Hellbach, Sven   I-657
Herai, Roberto Hiroshi   II-240
Holloway, Eric   I-395
Horzyk, Adrian   I-76, I-170
Hsieh, June-Nan   I-690
Hvasshovd, Svein-Olav   II-321

Isern, Eugeni   I-226
Ismail, Ali   II-652
Ivanovas, Adomas   I-58

Jaimes, Luis G.   II-700, II-740
Jamroz, Dariusz   II-364
Jankowski, Maciej   I-633
Jankowski, Norbert   I-70
Janowski, Maciej   I-76
Jaworski, Maciej   II-376
Jelonkiewicz, Jerzy   I-130
Jobczyk, Krystian   II-532, II-544, II-665
Jurczuk, Krzysztof   I-514

Kacprzak, Magdalena   II-557
Kadavy, Tomas   I-405, I-486, I-524, I-561
Kaden, Marika   I-724
Kalita, Piotr   I-3
Kania, Kacper   I-97
Karczmarek, Paweł   I-645, II-137, II-148,
    II-570
Kazikova, Anezka   I-417
Kiersztyn, Adam   I-645, II-137, II-148,
    II-570
Klimek, Radosław   II-387, II-677
Kloczkowski, Andrzej   II-107
Kluza, Krzysztof   II-453, II-689
Kóczy, László T.   II-630
Kojecký, Lumír   I-427
Kolbusz, Janusz   I-108, I-190
Korbicz, Józef   II-157, II-592
Kordos, Mirosław   I-438
Korkosz, Mariusz   II-117
Korytkowski, Marcin   II-734
Kot, Marek   I-750
Kowal, Marek   II-157
Kowalczyk, Bartosz   I-15
Krawczak, Maciej   II-398
Kredens, Kelvin Vieira   II-240
Kretowski, Marek   I-514
Kruk, Michał   II-81
Krzyżak, Adam   I-118
Kurek, Jarosław   II-81
Kusy, Maciej   I-267
Kutt, Krzysztof   II-168

Lage, G. G.   I-536
Łapa, Krystian   I-438, I-449
Laskowska, Magdalena   I-130
Laskowski, Łukasz   I-130
Lazzaretti, André Eugênio   I-376
Lenart, Marcin   II-579
León, José   II-700
Lesot, Marie-Jeanne   II-579
Ligęza, Antoni   II-532, II-544, II-665, II-689,
    II-788
Lischke, Falko   I-657
Liu, Ying   I-148
Lo Sciuto, Grazia   II-711
Lopes, Heitor Silvério   I-376
Luzar, Marcel   II-592
Lysenko, Oleksandr   I-108, I-190

Machireddy, Amrutha I-136
Maciejewski, Henryk II-445
Maree, Armand I-462
Maret, Yann I-474
Mariescu-Istodor, Radu I-593
Markowitch, Olivier II-423
Markowska-Kaczmar, Urszula I-97
Marks, Robert I-395
Martins, Juliano Vieira II-240
Maskeliūnas, Rytis I-58
Maslanka, Pawel I-750
Massanet, Sebastia II-642
Masum, Shamsul I-148
Mazurkiewicz, Jacek II-179
Mi, Chuanmin II-408
Mierzwiak, Rafał II-408
Migasiewicz, Agnieszka II-179
Milczarski, Piotr I-750, II-48, II-191
Misiak, Piotr II-168, II-276
Modrzejewski, Mateusz II-723
Mollinetti, Marco Antonio Florenzano II-603
Mora, Higinio II-265
Moshkin, Vadim II-799
Mosion, Michal II-179
Mroczek, Teresa II-301
Murillo, Juan C. II-740

Najgebauer, Patryk II-36, II-376
Nakano, Ryohei I-214
Nakaya, Yuta I-160
Nalepa, Grzegorz J. II-168
Napoli, Christian II-711
Neruda, Roman I-257
Neumann, Thomas I-657
Nielek, Radosław II-752
Niemiec, Rafal II-301
Niemyska, Wanda II-642
Nieszporek, Katarzyna I-761, II-73
Niskanen, Vesa A. II-630
Nowak, Jakub II-734
Nowicki, Robert II-734

Obando, Javier Delgado II-25
Oberson, Daniel I-474
Obuchowicz, Rafał II-117
Ochoa-Ornelas, Raquel I-363
Ohira, Toru I-47

Orłowski, Arkadiusz II-81
Osana, Yuko I-160
Ostreika, Armantas I-58
Ősz, Olivér I-386

Pałkowski, Marek II-207
Papiez, Piotr I-170
Partyka, Marian I-118
Pedrycz, Witold I-645, II-137, II-148, II-570
Peng, Peng II-408
Pereira, Rodrigo Lisboa II-603
Petrisor, Teodora II-579
Piasecki, Maciej II-777
Piech, Henryk I-130
Pluhacek, Michal I-405, I-417, I-486, I-524, I-561
Pluta, Piotr II-127
Podbielska, Halina II-179
Połap, Dawid I-58, II-711
Porebski, Sebastian II-217
Postawka, Aleksandra II-229
Pozorska, Jolanta I-183
Przybył, Andrzej I-449
Przybyszewski, Krzysztof I-615, II-433, II-516
Pytel, Krzysztof I-495

Quesada, Wilson O. II-740

Rafajłowicz, Ewaryst I-668
Rafajłowicz, Wojciech I-505
Rakus-Andersson, Elisabeth II-92
Reska, Daniel I-514
Ribeiro, Manassés I-376
Riekert, Marius I-462
Roca, Miquel I-226
Rodriguez, Jonathan I. II-740
Rodriguez, Pablo Ospina II-700
Rokita, Przemysław II-723
Rokui, Jun I-88
Romanowski, Jakub II-752
Rosselló, Josep L. I-226
Rozycki, Pawel I-108, I-190
Rubach, Pawel I-247
Ruiz-Aguilera, Daniel II-642
Rutkowska, Danuta II-92
Rutkowski, Leszek I-25
Rutkowski, Tomasz II-752

Sadowski, Tomasz   II-62
Sampaio, Fausto   I-201
Sánchez, María Guadalupe   I-363
Saralajew, Sascha   I-724
Satoh, Seiya   I-214
Sawicka, Anna   II-557
Scalabrin, Edson Emilio   II-240
Scherer, Magdalena   I-183
Scherer, Rafał   II-36, II-734
Senkerik, Roman   I-405, I-417, I-486, I-524, I-561
Serra Neto, Mario Tasso Ribeiro   II-603
Shikler, Rafi   II-711
Siedlecki, Krzysztof   II-207
Sieranoja, Sami   I-680, II-343
Silva Jr., Elias T.   I-201
Sinaga, Kristina P.   I-690
Sisiaridis, Dimitrios   II-423
Siwocha, Agnieszka   II-36, II-734
Skibinsky-Gitlin, Erik S.   I-226
Skubalska-Rafajłowicz, Ewa   I-702
Śliwiński, Przemysław   II-229
Ślusarczyk, Grażyna   II-652
Śmietańska, Katarzyna   II-81
Smith, James S.   I-235
Souza, Daniel Leal   II-603
Spavieri, Guilherme   I-323, I-536
Starczewski, Artur   II-433
Starczewski, Janusz T.   I-761, II-73
Staszewski, Paweł   II-752
Stawska, Zofia   I-750, II-191
Straszecka, Ewa   II-217
Strug, Barbara   I-714, II-652
Strug, Joanna   I-714
Studniarski, Marcin   I-279
Suchenia, Anna   II-689
Świderski, Bartosz   II-81
Szkatuła, Grażyna   II-398
Szupiluk, Ryszard   I-247
Szuster, Marcin   II-763
Szymański, Julian   II-265

Teixeira, Otavio Noura   II-603
Tessaro Lunardi, Willian   I-548

Umoren, Imeh   I-35

Vasconcelos, Germano C.   I-290
Vastag, Gyula   II-630
Vidal, Vicente   I-363
Vidnerová, Petra   I-257
Viktorin, Adam   I-405, I-417, I-486, I-524, I-561
Viljoen, Hulda   I-572
Villmann, Andrea   I-724
Villmann, Thomas   I-657, I-724, I-736
Voos, Holger   I-548

Waga, Karol   I-593
Walkowiak, Tomasz   II-445, II-777
Was, Lukasz   I-750
Wiaderek, Krzysztof   II-92
Wiak, Slawomir   I-750
Wieczorek, Grzegorz   II-81
Wilamowski, Bogdan M.   I-108, I-190, I-235
Wiśniewski, Piotr   II-544, II-689, II-788
Woldan, Piotr   II-752
Woźniak, Marcin   I-58, II-711
Wróbel, Michał   I-761, II-73
Wydrzyński, Marcin   I-438
Wyrobek, Joanna   II-453, II-689

Yang, Jiawei   I-604
Yang, Miin-Shen   I-690
Yanguas-Rojas, David   II-740
Yarushkina, Nadezhda   II-799
Ye, Zhili   II-463

Zajdel, Roman   I-267
Zalasiński, Marcin   II-250
Zbrzezny, Andrzej   II-557
Zdunek, Rafał   II-62, II-333
Żejmo, Michał   II-157
Zelinka, Ivan   I-427, I-524
Zeng, Wanling   II-463
Zhadkovska, Khrystyna   II-570
Zhang, Dingqian   II-463
Zieliński, Sławomir K.   II-475

Printed in the United States
By Bookmasters